AutoCAD
and Its Applications

BASICS

by

Terence M. Shumaker
Faculty Emeritus
Former Chairperson
Drafting Technology
Autodesk Premier Training Center
Clackamas Community College, Oregon City, Oregon

David A. Madsen
President, Madsen Designs Inc.
Faculty Emeritus, Former Department Chairperson Drafting Technology
Autodesk Premier Training Center
Clackamas Community College, Oregon City, Oregon
Director Emeritus, American Design Drafting Association

David P. Madsen
Vice President, Madsen Designs Inc.
Computer-Aided Design and Drafting Consultant and Educator
Autodesk Developer Network Member
American Design Drafting Association Member

2010

Publisher
The Goodheart-Willcox Company, Inc.
Tinley Park, Illinois
www.g-w.com

Library of Congress Catalog Card Number 2009004709

ISBN 978-1-60525-161-5

1 2 3 4 5 6 7 8 9 – 10 – 14 13 12 11 10 09

The Goodheart-Willcox Company, Inc. Brand Disclaimer: Brand names, company names, and illustrations for products and services included in this text are provided for educational purposes only and do not represent or imply endorsement or recommendation by the author or the publisher.

The Goodheart-Willcox Company, Inc. Safety Notice: The reader is expressly advised to carefully read, understand, and apply all safety precautions and warnings described in this book or that might also be indicated in undertaking the activities and exercises described herein to minimize risk of personal injury or injury to others. Common sense and good judgment should also be exercised and applied to help avoid all potential hazards. The reader should always refer to the appropriate manufacturer's technical information, directions, and recommendations; then proceed with care to follow specific equipment operating instructions. The reader should understand these notices and cautions are not exhaustive.

The publisher makes no warranty or representation whatsoever, either expressed or implied, including but not limited to equipment, procedures, and applications described or referred to herein, their quality, performance, merchantability, or fitness for a particular purpose. The publisher assumes no responsibility for any changes, errors, or omissions in this book. The publisher specifically disclaims any liability whatsoever, including any direct, indirect, incidental, consequential, special, or exemplary damages resulting, in whole or in part, from the reader's use or reliance upon the information, instructions, procedures, warnings, cautions, applications, or other matter contained in this book. The publisher assumes no responsibility for the activities of the reader.

Cover Source: Dag Sundberg/GettyImages

Library of Congress Cataloging-in-Publication Data

Shumaker, Terence M.
    AutoCAD and Its Applications: BASICS 2010 / by Terence M.
Shumaker, David A. Madsen, David P. Madsen. – 17th ed.
        p. cm.

    Includes bibliographical references and index.
    ISBN 978-1-60525-161-5
    1. Computer graphics. 2. AutoCAD. I. Madsen, David A.
II. Madsen, David P.

T385.S461466 2010
    620'.00420285536--dc22   2009004709

Introduction

AutoCAD and Its Applications—Basics is a textbook providing complete instruction in mastering fundamental AutoCAD® 2010 tools and drawing techniques. Typical applications of AutoCAD are presented with basic drafting and design concepts. The topics are covered in an easy-to-understand sequence and progress in a way that allows you to become comfortable with the tools as your knowledge builds from one chapter to the next. In addition, *AutoCAD and Its Applications—Basics* offers the following features:

- Step-by-step use of AutoCAD tools.
- In-depth explanations of how and why tools function as they do.
- Extensive use of font changes to specify certain meanings.
- Examples and descriptions of industry practices and standards.
- Screen captures of AutoCAD features and functions.
- Professional tips explaining how to use AutoCAD effectively and efficiently.
- More than 250 exercises to reinforce the chapter topics. These exercises also build on previously learned material.
- Chapter tests for review of tools and key AutoCAD concepts.
- A large selection of drafting problems supplementing each chapter. Problems are presented as industrial drawings, engineering sketches, or architectural, civil, electrical, or other related industry drawings.

With *AutoCAD and Its Applications—Basics*, you learn AutoCAD tools and become acquainted with information in other areas:

- Preliminary planning and sketches.
- Drawing geometric shapes and constructions.
- Parametric drawing techniques.
- Special editing operations that increase productivity.
- Placing text and tables according to accepted industry practices.
- Making multiview drawings (orthographic projection).
- Dimensioning techniques and practices, based on accepted standards.
- Drawing section views and designing graphic patterns.
- Creating shapes and symbols.
- Creating and managing symbol libraries.
- Plotting and printing drawings.

Learning Objectives identify key items you will learn in the chapter.

Command Entry Graphics show ribbon, Application Menu, and keyboard entry options. Tool options are also shown where applicable.

Professional Tips increase your productivity in using AutoCAD tools and techniques.

Illustrations, including AutoCAD "screen shots," line art illustrations, and illustrated tables, make learning easy.

CHAPTER 18

Linear and Angular Dimensioning

Learning Objectives

After completing this chapter, you will be able to do the following:
- ✓ Add linear dimensions to a drawing.
- ✓ Add angular dimensions to a drawing.
- ✓ Draw datum and chain dimensions.
- ✓ Add dimensions for multiple items using the **QDIM** tool.

A drawing often requires a variety of dimensions to describe the size and shape of features and objects. Linear and angular dimensions are two of the most common. This chapter covers the process of adding linear and angular dimensions to a drawing using several dimensioning tools. You will also learn how to add a break symbol to a dimension line and use the **QDIM** tool.

Placing Linear Dimensions

Linear dimensions usually measure straight distances, such as distances between horizontal, vertical, or slanted surfaces. The **DIMLINEAR** tool allows you to place linear dimensions.

Dimension tools reference the current dimension style and the points or objects you select to create a single dimension object. When you use the **DIMLINEAR** tool, for example, you create a dimension object that includes all related dimension style characteristics, dimension and extension lines, arrowheads, and a dimension value associated with the distance between selected points.

Once you access the **DIMLINEAR** tool, pick a point to locate origin of the first extension line, and then pick a point to locate the origin of the second extension line. See Figure 18-1. Use object snap modes and other drawing aids to pick the exact points where extension lines begin. Once you establish the extension line origins, you can select from several options that appear at the Specify dimension line location or [Mtext/Text/Angle/Horizontal/Vertical/Rotated] prompt. To apply the default option and create a linear dimension, move the dimension line to the desired location and pick. See Figure 18-2.

Ribbon
Home
> Annotation
Annotate
> Dimensions
Linear
DIMLINEAR

Type
DIMLINEAR
DLI

475

Controlling Draw Order

Drawings often include overlapping objects. The overlap is difficult to see when all objects use a thin lineweight and when lineweight display is off. Controlling display order is better illustrated with an object that has width, such as the donuts shown in Figure 6-8. In this example, the donuts were originally drawn after the other objects. You can change the drawing order of the donuts, and all other objects, to place the items above or below selected objects and to the front or back of all objects. You can also set draw order by picking an object to select it, right-clicking, and choosing **Draw Order**. Figure 6-9 describes the options for changing draw order.

Ribbon
Home
> Modify
Bring to Front
DRAWORDER

Type
DRAWORDER
DR

NOTE

You may need to use the **DRAWORDER** tool on several objects until the objects display correctly. Objects move to the front of the drawing when they are modified.

PROFESSIONAL TIP

Use **DRAWORDER** to help display and select objects that are hidden by other objects.

Figure 6-8.
The order of objects can be changed to place any object under or above other objects.

Figure 6-9.
Four options are available for rearranging the order of objects in a drawing.

Option	Function
Bring to Front	Places the selected objects at the front of the drawing.
Send to Back	Places the selected objects at the back of the drawing.
Bring Above Objects	Moves the selected objects above the reference object.
Send Under Objects	Moves the selected objects below the reference object.

183

Chapter 6   View Tools

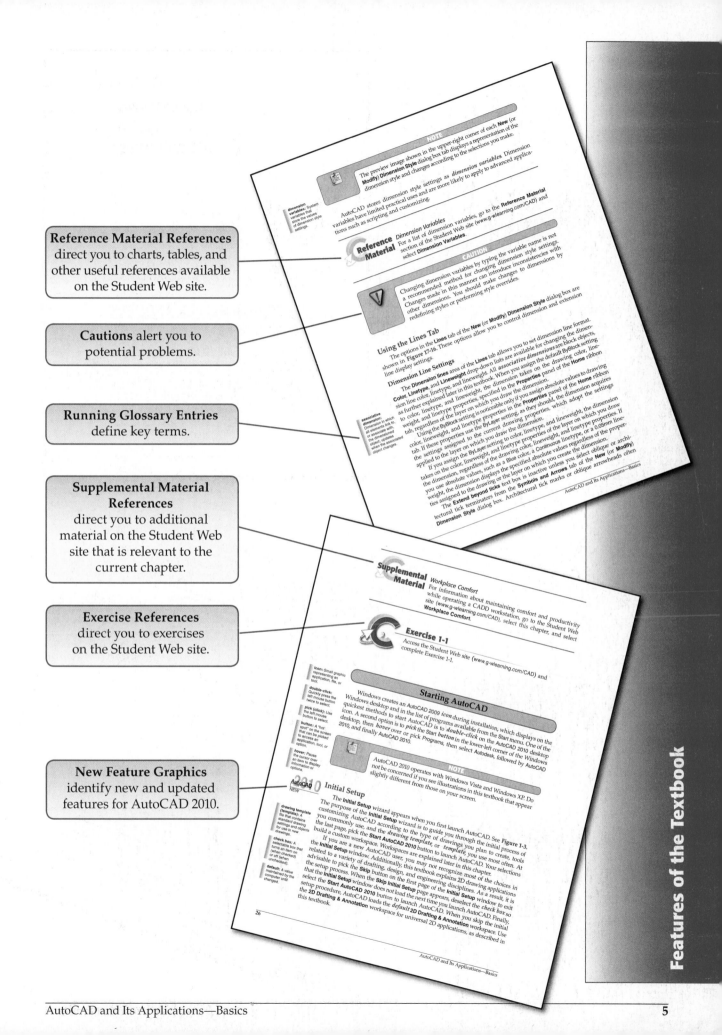

Reference Material References direct you to charts, tables, and other useful references available on the Student Web site.

Cautions alert you to potential problems.

Running Glossary Entries define key terms.

Supplemental Material References direct you to additional material on the Student Web site that is relevant to the current chapter.

Exercise References direct you to exercises on the Student Web site.

New Feature Graphics identify new and updated features for AutoCAD 2010.

Notes explain important aspects of a topic.

Express Tool References direct you to Express Tool material on the Student Web site.

Template Development References direct you to Template Development material on the Student Web site.

Chapter Tests reinforce the knowledge gained by reading the chapter and completing the exercises.

Figure 10-15.
The **Existing** option of the **SCALETEXT** tool scales text objects using their individual justification settings.

BL Justification	BL Justification
MC Justification	MC Justification
TR Justification	TR Justification
Original Text	Text Scaled Using Existing Base Point Option

points using the **Existing** option. Notice how the text scales in relation to its own justification setting.

After you specify the justification to use as the base point, AutoCAD prompts for the scaling type. The default **Specify new model height** option allows you to type a new value for the text height of non-annotative objects. If the selected text is annotative, the value you enter is ignored. Use the **Paper height** option to type a new paper text height value for the text height of annotative objects. If the selected text is non-annotative, the value you enter is ignored.

The **Match object** option allows you to pick an existing text object. The height of the selected text object adopts the text height from the text object you pick. Use the **Scale factor** option to scale text objects that have different heights relative to their current heights. For example, using a scale factor of 2 scales all of the selected text objects to twice their current size.

> **NOTE**
> You should only use the **SCALETEXT** tool to scale non-annotative text.

JUSTIFYTEXT

Ribbon
Annotate
> Text
Justify

Type
JUSTIFYTEXT

Changing Text Justification

Use the **JUSTIFYTEXT** tool to change the justification point without moving the text. Pick the text for which you want to change justification, and enter the new justification option.

Exercise 10-6
Access the Student Web site (www.g-wlearning.com/CAD) and complete Exercise 10-6.

Express Tools
Chapter 10
The **Express Tools** ribbon tab includes additional tools for improved functionality and productivity during the drawing processes. The following Express Tools represent the most useful text express tools. For information about these tools, go to the Student Web site (www.g-wlearning.com/CAD), select this chapter, and select **Using Text Express Tools**.

Text Fit	Arc-Aligned Text
Text Mask	Enclose Text with Object
Unmask Text	Change Text Case
Convert Text to Mtext	

Template Development
Chapter 7
For detailed instructions on setting object snaps and polar tracking to save time and increase efficiency, go to the Student Web site (www.g-wlearning.com/CAD), select this chapter, and select **Template Development**.

Chapter Test

Answer the following questions. Write your answers on a separate sheet of paper or go to the Student Web site (www.g-wlearning.com/CAD) and complete the electronic chapter test.

1. Define the term *object snap*.
2. What is an AutoSnap tooltip?
3. Name the following AutoSnap markers:

A. _____   E. _____   I. _____
B. _____   F. _____   J. _____
C. _____   G. _____   K. _____
D. _____   H. _____   L. _____

4. How do you set running object snaps?
5. Define the term *running object snap*.
6. How do you access the **Drafting Settings** dialog box to change object snap settings?
7. If you are using running object snaps and want to make several point specifications without the aid of object snap, but want to continue the same running object snaps after making the desired point selections, what is the easiest way to turn off the running object snaps temporarily?
8. If you are using running object snaps and you want to make a single point selection without the object snap override.
9. Describe the effects of the running object snaps, what do you do?
10. How do you activate the **Object Snap** shortcut menu?
11. Where are the four quadrant points on a circle?
12. What is the situation when the tooltip reads Extended Intersection?
13. What does it mean when the tooltip reads Deferred Perpendicular?

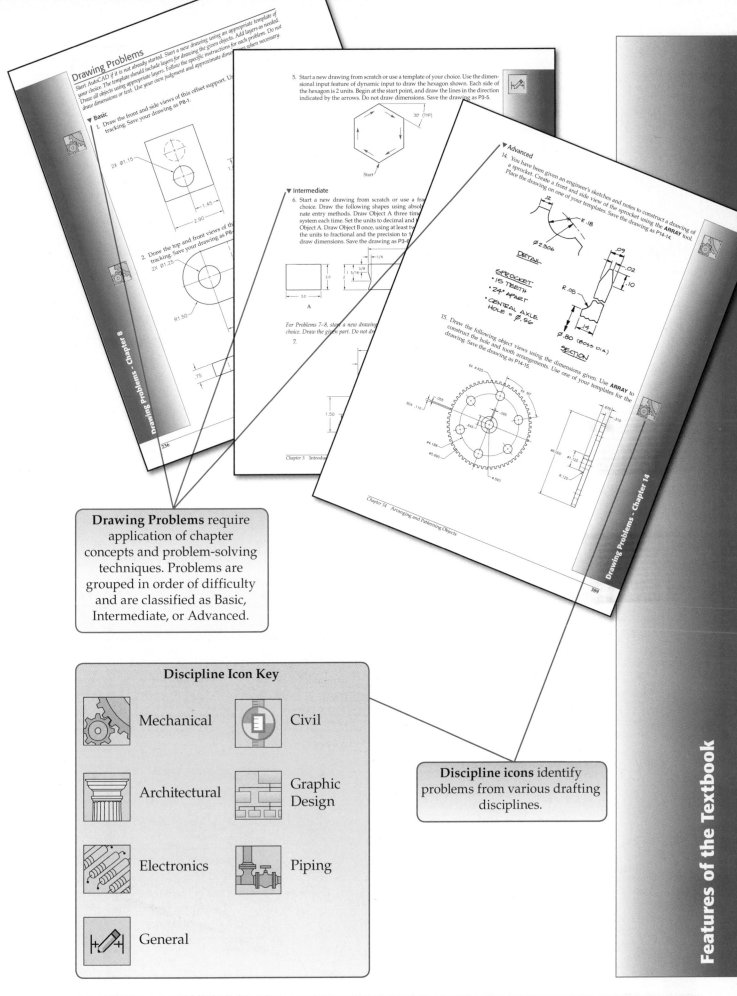

Drawing Problems require application of chapter concepts and problem-solving techniques. Problems are grouped in order of difficulty and are classified as Basic, Intermediate, or Advanced.

Discipline Icon Key

Mechanical

Civil

Architectural

Graphic Design

Electronics

Piping

General

Discipline icons identify problems from various drafting disciplines.

Features of the Textbook

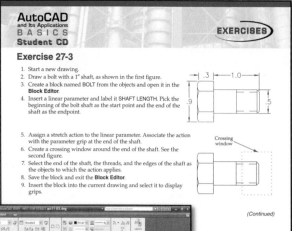

Exercise 27-3

1. Start a new drawing.
2. Draw a bolt with a 1″ shaft, as shown in the first figure.
3. Create a block named BOLT from the objects and open it in the **Block Editor**.
4. Insert a linear parameter and label it SHAFT LENGTH. Pick the beginning of the bolt shaft as the start point and the end of the shaft as the endpoint.
5. Assign a stretch action to the linear parameter. Associate the action with the parameter grip at the end of the shaft.
6. Create a crossing window around the end of the shaft. See the second figure.
7. Select the end of the shaft, the threads, and the edges of the shaft as the objects to which the action applies.
8. Save the block and exit the **Block Editor**.
9. Insert the block into the current drawing and select it to display grips.

(Continued)

Exercises. Chapter exercises are provided on the Student Web site, allowing you to switch between the exercise directions and AutoCAD on-screen.

AutoCAD Software. Pick this button to access a Web site from which you can download the AutoCAD Electrical software at no cost for use with this book.

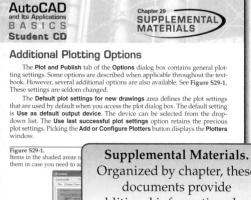

Additional Plotting Options

The **Plot and Publish** tab of the **Options** dialog box contains general plotting settings. Some options are described when applicable throughout the textbook. However, several additional options are also available. See **Figure S29-1**. These settings are seldom changed.

The **Default plot settings for new drawings** area defines the plot settings that are used by default when you access the plot dialog box. The default setting is **Use as default output device**. The device can be selected from the dropdown list. The **Use last successful plot settings** option retains the previous plot settings. Picking the **Add or Configure Plotters** button displays the **Plotters** window.

Figure S29-1.
Items in the shaded areas r...
them in case you need to ad...

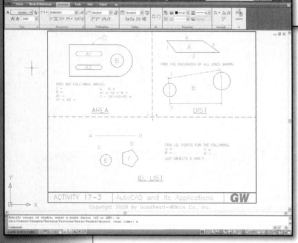

Supplemental Materials. Organized by chapter, these documents provide additional information about topics discussed in the textbook.

Copyright by Goodheart-Willcox Co., Inc.   Additional Plotting Options, page 1

Drafting Symbols

Symbols provide a "common language" for drafters all over the world. However, symbols can be meaningful only if they are created according to the relevant standards or conventions. This document describes and illustrates common dimensioning, GD&T, architectural, piping, and electrical symbols.

Standard Dimensioning Symbols

The size of dimensioning symbols varies with text size, but it should be consistent with the height of the text. In the following illustration, h = text height.

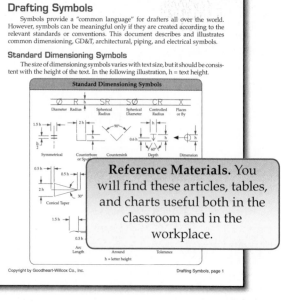

Reference Materials. You will find these articles, tables, and charts useful both in the classroom and in the workplace.

Copyright by Goodheart-Willcox Co., Inc.   Drafting Symbols, page 1

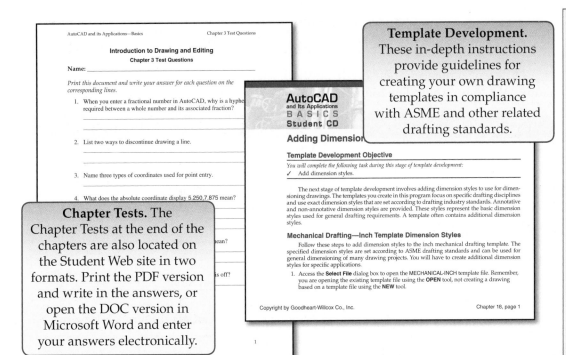

Chapter Tests. The Chapter Tests at the end of the chapters are also located on the Student Web site in two formats. Print the PDF version and write in the answers, or open the DOC version in Microsoft Word and enter your answers electronically.

Template Development. These in-depth instructions provide guidelines for creating your own drawing templates in compliance with ASME and other related drafting standards.

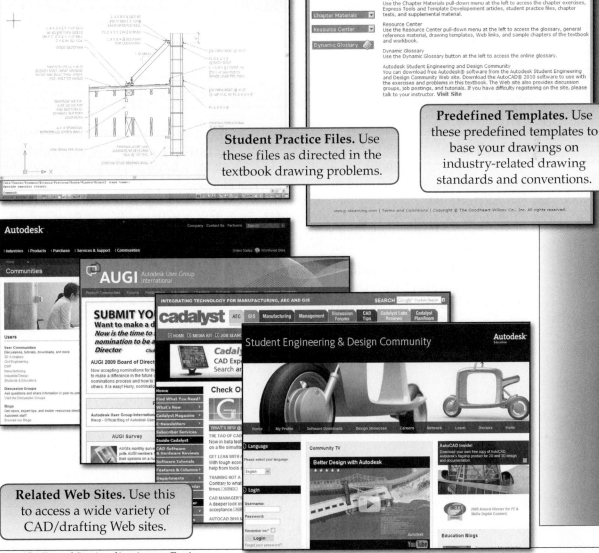

Student Practice Files. Use these files as directed in the textbook drawing problems.

Predefined Templates. Use these predefined templates to base your drawings on industry-related drawing standards and conventions.

Related Web Sites. Use this to access a wide variety of CAD/drafting Web sites.

Fonts Used in This Textbook

Different typefaces are used throughout this textbook to define terms and identify AutoCAD commands. The following typeface conventions are used in this textbook:

Text Element	Example
AutoCAD tools	**LINE** tool
AutoCAD menu browser menus	**Draw > Arc > 3 Points**
AutoCAD system variables	**LTSCALE** system variable
AutoCAD toolbars and buttons	**Quick Access** toolbar, **Undo** button
AutoCAD dialog boxes	**Insert Table** dialog box
Keyboard entry (in text)	Type LINE
Keyboard keys	[Ctrl]+[1] key combination
File names, folders, and paths	C:\Program Files\AutoCAD 2010\mydrawing.dwg
Microsoft Windows features	Start menu, Programs folder
Prompt sequence	Command:
Keyboard input at prompt sequence	Command: **L** *or* **LINE**↵
Comment at a prompt sequence	Specify first point: (*pick a point or press* [Enter])

Other Text References

For additional information, standards from organizations such as ANSI (American National Standards Institute) and ASME (American Society of Mechanical Engineers) are referenced throughout the textbook. Use these standards to create drawings that follow industry, national, and international practices.

Also for your convenience, other Goodheart-Willcox textbooks are referenced. Referenced textbooks include *AutoCAD and Its Applications—Advanced* and *Geometric Dimensioning and Tolerancing*. These textbooks can be ordered directly from Goodheart-Willcox.

AutoCAD and Its Applications—Basics covers basic AutoCAD applications. For a textbook covering the advanced AutoCAD applications, please refer to *AutoCAD and Its Applications—Advanced*.

Contents in Brief

About the Authors

Terence M. Shumaker is Faculty Emeritus, the former Chairperson of the Drafting Technology Department, and former Director of the Autodesk Premier Training Center at Clackamas Community College. Terence taught at the community college level for over 25 years. He has professional experience in surveying, civil drafting, industrial piping, and technical illustration. He is the author of Goodheart-Willcox's *Process Pipe Drafting* and coauthor of the *AutoCAD and Its Applications* series.

David A. Madsen is the president of Madsen Designs Inc. (www.madsendesigns.com). David is Faculty Emeritus, the former Chairperson of Drafting Technology and the Autodesk Premier Training Center at Clackamas Community College and former member of the American Design and Drafting Association (ADDA) Board of Directors. David was honored by the ADDA with Director Emeritus status at the annual conference in 2005. David was an instructor and a department chair at Clackamas Community College for nearly 30 years. In addition to community college experience, David was a Drafting Technology instructor at Centennial High School in Gresham, Oregon. David also has extensive experience in mechanical drafting, architectural design and drafting, and construction practices. He is the author of Goodheart-Willcox's *Geometric Dimensioning and Tolerancing* and coauthor of the *AutoCAD and Its Applications* series (Release 10 through 2008 editions), *Architectural Drafting Using AutoCAD, AutoCAD Architecture and Its Applications, Architectural Desktop and its Applications, Architectural AutoCAD*, and *AutoCAD Essentials*.

David P. Madsen holds a Master of Science degree in Educational Policy, Foundations, and Administrative Studies with a specialization in Postsecondary, Adult, and Continuing Education; a Bachelor of Science degree in Technology Education; and an Associate of Science degree in General Studies and Drafting Technology. Dave has been involved in providing Drafting and Computer-Aided Design and Drafting instruction to adult learners since 1999. Dave has extensive and varied experience in the drafting, design, and engineering fields. He has worked in the drafting industry for over ten years and has created everything from mechanical and electronic to architectural and civil drawings.

Acknowledgments

Technical Assistance and Contribution of Materials

Margo Bilson of Willamette Industries, Inc.
Fitzgerald, Hagan, & Hackathorn
Bruce L. Wilcox, Johnson and Wales University School of Technology

Contribution of Photographs or Other Technical Information

Arthur Baker
Autodesk
CADalyst magazine
CADENCE magazine
Chris Lindner
EPCM Services, Ltd.
Harris Group, Inc.

International Source for Ergonomics
Jim Webster
Kunz Associates
Myonetics, Inc.
Norwest Engineering
Schuchart & Associates, Inc.
Willamette Industries, Inc.

Trademarks

Autodesk, the Autodesk logo, 3ds max, Autodesk VIZ, AutoCAD, DesignCenter, AutoCAD Learning Assistance, AutoSnap, and AutoTrack are either registered trademarks or trademarks of Autodesk, Inc., in the U.S.A. and/or other countries.

Microsoft, Windows, and Windows NT are registered trademarks of Microsoft Corporation in the United States and/or other countries.

Contents

Basic Drawing and Printing

Creating Text and Tables

Editing Drawings

Additional AutoCAD Applications

Using Layouts

Student Web Site Content

Using the Web Site

Express Tools

Exercises

Template Development

Chapter Tests

Reference Materials

Supplemental Materials

Student Practice Files

Predefined Templates

Download Student AutoCAD

Related Web Links

CHAPTER 1

Introduction to AutoCAD

Learning Objectives

After completing this chapter, you will be able to do the following:

✓ Define computer-aided design and drafting.
✓ Describe typical AutoCAD applications.
✓ Explain the value of planning your work and system management.
✓ Describe the purpose and importance of drawing standards.
✓ Demonstrate how to start and exit AutoCAD.
✓ Describe the AutoCAD interface.
✓ Use a variety of methods to select AutoCAD tools.
✓ Use the features found in the **AutoCAD Help** window.

Computer-aided design and drafting (CADD) is the process of using a computer with software to design and produce models and drawings according to specific industry and company standards. The terms *computer-aided design (CAD)* and *computer-aided drafting (CAD)* refer to specific aspects of the CADD process. This chapter introduces the AutoCAD CADD system. In this chapter, you will learn how to begin working with AutoCAD and how to control the AutoCAD environment.

computer-aided design and drafting (CADD): The process of using a computer with software to design and produce models and drawings.

AutoCAD Applications

AutoCAD *tools* and *options* are available for drawing objects of any size or shape. Use AutoCAD to prepare two-dimensional (2D) drawings, three-dimensional (3D) models, and animations. AutoCAD is a universal CADD program that applies to any drafting, design, or engineering discipline. For example, you can use AutoCAD to design and document mechanical parts and assemblies, architectural buildings, civil and structural engineering projects, electronics, and technical illustration. Using the AutoCAD software and this textbook, you will learn how to construct, lay out, dimension, and annotate 2D drawings. *AutoCAD and Its Applications—Advanced* provides detailed instruction on 3D modeling and 3D rendering.

tool (command): An instruction issued to the computer to complete a specific task. For example, the **LINE** tool is used to draw lines.

option: A choice associated with a tool, or an alternative function of a tool.

2D Drawings

2D drawings display object length and width, or width and height, in a flat (two-dimensional) form. A 2D drawing typically includes dimensions and annotations that fully describe features on the drawing. This practice results in a document used to manufacture or construct a product. 2D drawings are common in all drafting, design, and engineering fields. **Figure 1-1** shows an example of an architectural building floor plan created using AutoCAD.

3D Models

2D drawings are useful for documenting engineering and design requirements. However, 3D models are often more appropriate for design, visualization, analysis, and testing. AutoCAD provides tools and options for developing *wireframe*, *surface*, and *solid* models. 3D models are virtual representations of actual products. Add color, lighting, and texture to display a model in a realistic format. See **Figure 1-2A.** Use view tools to rotate and adjust a model to view it from any direction. See **Figure 1-2B.** Apply animation to a model, such as a *walkthrough* of a model home, to show product design or function.

wireframe model: A 3D model consisting of lines and curves connecting at the corners of an object to form edges; contains no surface properties or solid mass.

surface model: A 3D model consisting of volumeless surfaces, such as planes and curved faces that represent the exterior of an object.

solid model: A 3D model defined by object surfaces and volume; includes physical properties, such as mass and density, that can be analyzed.

walkthrough: A computer simulation that follows a path through or around a 3D model.

Reference Material *Glossaries*

For detailed glossaries of CADD, AutoCAD, and computer terms, go to the **Reference Material** section of the Student Web site (www.g-wlearning.com/CAD) and select **Glossary of Computer Terms** or **Glossary of CADD Terms**.

Figure 1-1
AutoCAD provides tools and options for accurately creating 2D drawings such as the architectural floor plan shown here.

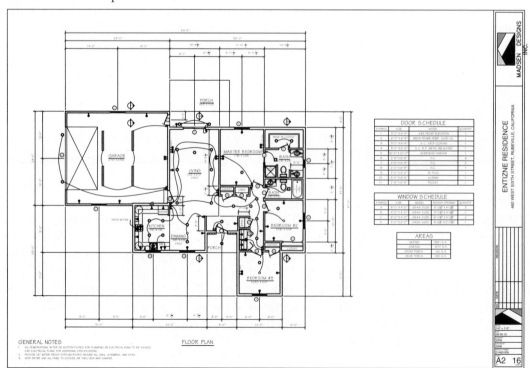

Figure 1-2.
A—A 3D wireframe model (left) with realistic colors and textures added (right). B—A model can be rotated, zoomed in and out, and viewed from any location in 3D space.

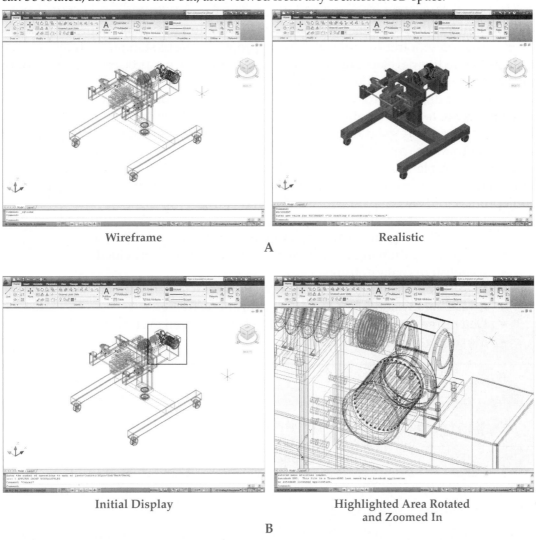

Wireframe   Realistic

A

Initial Display   Highlighted Area Rotated
                     and Zoomed In

B

Before You Begin

CADD offers greater speed, power, accuracy, and flexibility than traditional manual, or board, drafting. However, designing and drafting effectively with a computer requires a skilled CADD operator. Proficient AutoCAD users possess detailed knowledge of AutoCAD software operation. Therefore, it is important to be familiar with AutoCAD tools and to know how they work and when they are best suited for a specific task. There is no substitute for knowing AutoCAD tools. You must also have experience with design and drafting systems and conventions, and apply these rules when using AutoCAD.

As you begin your CADD training, develop effective methods for managing your work. First, plan your drawing sessions thoroughly to organize your thoughts. Second, learn and use industry and classroom, or office, standards. Third, save your work often. If you follow these three procedures, you will find it easier to use CADD tools and methods, and your experience with AutoCAD will be more productive and enjoyable.

Planning Your Work

A drawing plan involves thinking about the entire process or project in which you are involved. A plan determines how you approach a project. It focuses on the drawings you intend to create, the information you want to present, the types of symbols needed to represent information, and the appropriate use of standards. Take as much time as needed to develop drawing and project goals. Then proceed with the confidence of knowing where you are heading.

Plan your drawing projects carefully. You may want your applications to happen immediately or to be automatic, but if you hurry and do little or no planning, you may become frustrated. A good drawing plan can save time. During your early stages of AutoCAD training, consider creating a planning sheet, especially for your first few assignments. A planning sheet should document the tools you use, the selections you make, and the dimensions needed. You may also prepare a sketch as part of the planning process. A drawing plan and sketch can help you:

- Determine the drawing layout.
- Set the drawing area by laying out views and required free space.
- Confirm the drawing units, based on the dimensions provided.
- Establish drawing settings.
- Preset drawing variables, such as layers, text styles, and dimension styles.
- Establish how and when to perform various activities.
- Determine the best use of AutoCAD, resulting in an even workload.
- Maximize your use of equipment.

Reference Material *Planning Sheet*

For a sample project planning sheet, go to the **Reference Material** section of the Student Web site (www.g-wlearning.com/CAD) and select **Planning Sheet**.

Using Drawing Standards

standards:
Guidelines containing operating procedures, drawing techniques, and record keeping methods.

Most industries, schools, and companies have established *standards*. It is important that standards exist and are understood and used by all CADD personnel. Drawing standards can include:

- Methods of file storage (location and name)
- File naming conventions
- File backup methods and times
- Drawing templates with predefined settings
- Layout characteristics
- Borders and title blocks
- Drawing symbols
- Dimensioning styles and techniques
- Text styles
- Table styles
- Layer settings
- Plot styles

The standards you follow may vary in content. The most important aspect of standards is that people use them. When you follow drawing standards, drawings are consistent, you become more productive, and the classroom or office functions more efficiently.

The mechanical drafting standards used in this textbook are based on the American Society of Mechanical Engineers (ASME) and American National Standards Institute (ANSI) ASME Y series. Other drafting standards used in this textbook are based on appropriate discipline-specific standards, including those stated in the United States National CAD Standard.

Reference Material — *Drawing Standards*

For more information about drawing standards, go to the **Reference Material** section of the Student Web site (www. g-wlearning.com/CAD) and select **Drawing Standards**.

NOTE

You may consider other drafting standards when planning a drawing session and preparing drawings. *DIN* refers to the German standard *Deutsches Institut Für Normung*, established by the German Institute for Standardization. *Gb* refers to *Guo Biao* (Chinese) standards, *ISO* is the International Organization for Standardization, and *JIS* is the Japanese Industry Standard.

DIN: Deutsches Institut Für Normung.

Gb: Guo Biao (Chinese) standard.

ISO: International Organization for Standardization.

JIS: Japanese Industry Standard.

Saving Your Work

Drawings are lost due to software error, hardware malfunction, power failure, or accidents. Prepare for such an event by saving your work frequently. Develop the habit of saving your work at least every ten to fifteen minutes. The automatic save option, described in Chapter 2, can be set to save drawings automatically at predetermined intervals. However, you should also save your work manually at frequent intervals.

Working Procedures Checklist

As you begin learning AutoCAD, you will realize that several skills are required to become proficient. The following checklist provides you with some hints to help you become comfortable with AutoCAD. These hints also allow you to work quickly and efficiently.

- ✓ Carefully plan your work.
- ✓ Frequently check object and drawing settings, such as layers, styles, and properties, to see which object characteristics and drawing options are in effect.
- ✓ Read the prompts, tooltips, and *alerts* displayed by AutoCAD.
- ✓ Consistently check for the correct options, instructions, or keyboard entry of data.
- ✓ *Right-click* to access shortcut menus and review available options.
- ✓ Think ahead and know your next move.
- ✓ Learn new tools and options that can increase your speed and efficiency.
- ✓ Save your work every ten to fifteen minutes.
- ✓ Learn to use available resources, such as this textbook, to help solve problems and answer questions. You should also become familiar with the AutoCAD help system.

alert: A pop-up that indicates a potential problem or required action.

right-click: Use the right mouse button to select.

Supplemental Material

Workplace Comfort

For information about maintaining comfort and productivity while operating a CADD workstation, go to the Student Web site (www.g-wlearning.com/CAD), select this chapter, and select **Workplace Comfort**.

Exercise 1-1

Access the Student Web site (www.g-wlearning.com/CAD) and complete Exercise 1-1.

---

icon: Small graphic representing an application, file, or tool.

double-click: Quickly press the left mouse button twice to select.

pick (click): Use the left mouse button to select.

button: A "hot spot" on the screen that can be picked to access an application, tool, or option.

hover: Pause the cursor over an item to display information or options.

NEW

drawing template (template): A file that contains standard drawing settings and objects for use in new drawings.

check box: A selectable box that turns an item on (when checked) or off (when unchecked).

default: A value maintained by the computer until changed.

Starting AutoCAD

Windows creates an AutoCAD 2009 *icon* during installation, which displays on the Windows desktop and in the list of programs available from the Start menu. One of the quickest methods to start AutoCAD is to *double-click* on the AutoCAD 2010 desktop icon. A second option is to *pick* the Start *button* in the lower-left corner of the Windows desktop, then *hover* over or pick Programs, then select Autodesk, followed by AutoCAD 2010, and finally AutoCAD 2010.

NOTE

AutoCAD 2010 operates with Windows Vista and Windows XP. Do not be concerned if you see illustrations in this textbook that appear slightly different from those on your screen.

Initial Setup

The **Initial Setup** wizard appears when you first launch AutoCAD. See **Figure 1-3**. The purpose of the **Initial Setup** wizard is to guide you through the initial process of customizing AutoCAD according to the type of drawings you plan to create, tools you commonly use, and the *drawing template*, or *template*, you use most often. At the last page, pick the **Start AutoCAD 2010** button to launch AutoCAD. Your selections build a custom workspace. Workspaces are explained later in this chapter.

If you are a new AutoCAD user, you may not recognize most of the choices in the **Initial Setup** window. Additionally, this textbook explains 2D drawing applications related to a variety of drafting, design, and engineering disciplines. As a result, it is advisable to pick the **Skip** button on the first page of the **Initial Setup** window to exit the setup process. When the **Skip Initial Setup** page appears, deselect the *check box* so that the **Initial Setup** window does not load the next time you launch AutoCAD. Finally, select the **Start AutoCAD 2010** button to launch AutoCAD. When you skip the initial setup procedure, AutoCAD loads the *default* **2D Drafting & Annotation** workspace. Use the **2D Drafting & Annotation** workspace for universal 2D applications, as described in this textbook.

Figure 1-3.
Use the **Initial
Setup** window to
begin the process
of customizing
AutoCAD
according to your
discipline, interface,
and template
preferences.

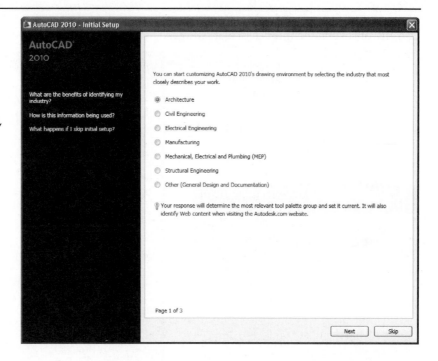

NOTE

To access the **Initial Setup** wizard, pick the **Initial Setup...** button on
the **User Preferences** tab of the **Options** dialog box. To display the
Options dialog box, pick the **Options** button at the bottom of the
Application Menu. The **Options** dialog box is described later in this
chapter.

Exiting AutoCAD

Use the **EXIT** tool to end an AutoCAD session. To exit, pick the program **Close**
button, located in the upper-right corner of the AutoCAD window; double-click the
Application Menu button, found in the upper-left corner of the AutoCAD window;
select the **Exit AutoCAD** button in the **Application Menu**; or with a file open, type EXIT
or QUIT and press [Enter]. See **Figure 1-4.**

NOTE

If you attempt to exit before saving your work, AutoCAD prompts
you to save or discard changes.

Exercise 1-2

Access the Student Web site (www.g-wlearning.com/CAD) and
complete Exercise 1-2.

Figure 1-4.
Use any of several techniques to exit AutoCAD when you finish a drawing session.

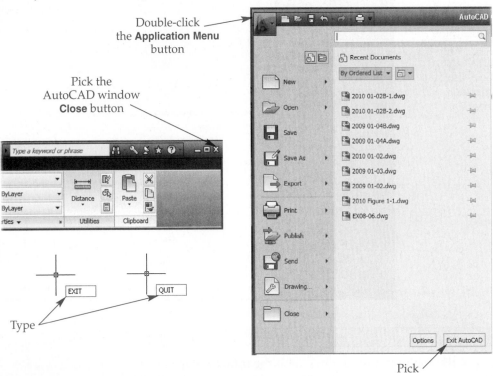

Double-click
the **Application Menu**
button

Pick the
AutoCAD window
Close button

Type

Pick

The AutoCAD Interface

interface: Items that allow users to input data to and receive outputs from a computer system.

graphical user interface (GUI): On-screen features that allow users to interact with a software program.

Interface items include devices to input data, such as the keyboard and mouse, and devices to receive computer outputs, such as the monitor. AutoCAD uses a Windows-style *graphical user interface (GUI)* with an **Application Menu**, ribbon, dialog boxes, and AutoCAD-specific items. See **Figure 1-5.** You will explore the unique AutoCAD interface in this chapter and throughout this textbook. Learn the format, appearance, and proper use of interface items to help quickly master AutoCAD.

workspace: Preset work environment containing specific interface items.

drawing window (graphics window): The largest area in the AutoCAD window, where drawing and modeling occurs.

NOTE

As you learn AutoCAD, you may want to customize the graphical user interface according to common tasks and specific applications. Customize AutoCAD manually or begin customization using the **Initial Setup** window. *AutoCAD and Its Applications—Advanced* explains customizing the user interface.

Workspaces

system variable: A command that configures AutoCAD to accomplish a specific task or exhibit a certain behavior. The value of each variable is saved with the drawing, so the next time the drawing is opened, the value remains the same.

The **2D Drafting & Annotation** *workspace,* shown in **Figure 1-5,** is active by default when you skip the initial setup procedure. The **2D Drafting & Annotation** workspace displays interface features above and below a large *drawing window* (also called the *graphics window*) and contains tools and options most often used for 2D drawing. To activate an alternative workspace, pick the **Workspace Switching** button on the status bar and select a different workspace. See **Figure 1-6.** You can also use the WSCURRENT *system variable.*

Figure 1-5.
The default AutoCAD window with the **2D Drafting & Annotation** workspace active.

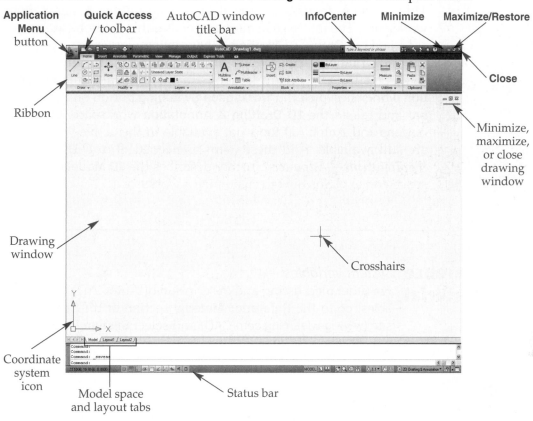

Figure 1-6.
Use the **Workspace Switching** button on the status bar to change to a different workspace, create a new workspace, or customize the user interface.

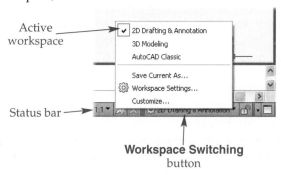

The **3D Modeling** workspace provides tools and options primarily used for 3D modeling applications. The **AutoCAD Classic** workspace displays the traditional AutoCAD menu bar, toolbars, and the **Tool Palettes** window with tools and options used for both 2D and 3D designs. The **Initial Setup Workspace**, if present, includes the default settings specified in the **Initial Setup** window. This workspace is available even if you skip the initial setup process. A new **Initial Setup Workspace** forms each time you use the **Initial Setup** function.

NEW

This textbook focuses on the default **2D Drafting & Annotation** workspace. The interface items explained in this textbook are those associated with the **2D Drafting & Annotation** workspace, except in specific situations that require additional items. To return interface items to their default locations in the **2D Drafting & Annotation** workspace, pick the **Workspace Switching** button on the status bar and reload the **2D Drafting & Annotation** workspace. Interface options and AutoCAD tools not available in the active workspace are still available. Add these items as needed. *AutoCAD and Its Applications—Advanced* further describes the **3D Modeling** workspace and explains how to customize a workspace.

Reference Material

System Variables

For a detailed listing and description of AutoCAD system variables, go to the **Reference Material** section of the Student Web site (www.g-wlearning.com/CAD) and select **System Variables**.

Exercise 1-3

Access the Student Web site (www.g-wlearning.com/CAD) and complete Exercise 1-3.

Crosshairs and Cursor

The AutoCAD crosshairs is the primary means of pointing to objects or locations within a drawing. The crosshairs changes to the familiar Windows cursor when moved outside of the drawing area or over an interface item, such as the status bar.

Control crosshair length using the *text box* or *slider* found in the **Crosshair size** area on the **Display** tab of the **Options** dialog box. Longer crosshairs can help to reference alignment between objects.

text box: A box in which you type a name, number, or single line of information.

slider: A movable bar that increases or decreases a value when you slide the bar.

tooltip: A pop-up that provides information about the item over which you are hovering.

Tooltips

A *tooltip* displays when you hover over most interface items. See **Figure 1-7**. The content presented in a tooltip varies depending on the item. Many tooltips expand as you continue to hover over a tool. The initial tooltip might only display the tool name, a brief description of the tool, and the command name. As you continue to hover, an explanation on how to use the tool and other information may appear.

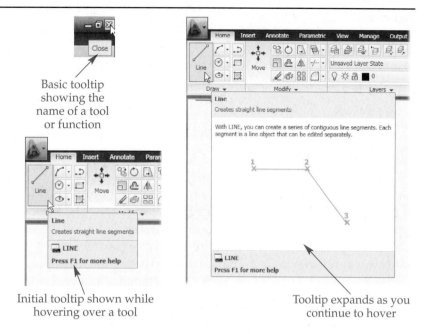

Figure 1-7.
Examples of tooltips displayed as you hover the cursor over an item.

Basic tooltip showing the name of a tool or function

Initial tooltip shown while hovering over a tool

Tooltip expands as you continue to hover

Controlling Windows

Control the AutoCAD and drawing windows using the same methods used to control other windows within the Windows operating system. To minimize, maximize, or close the AutoCAD window or individual drawing windows, pick the appropriate icon in the upper-right corner. You can also adjust the AutoCAD window by right-clicking on the title bar and choosing from the standard window control menu. Window sizing operations are also the same as those for other windows within the Windows operating system.

Floating and Docking

Several interface items, including the AutoCAD and drawing windows, can *float* or be *docked*. Floating features appear within a border. Some items, such as the drawing window, have a title bar at the top or side. You can move and resize floating windows in the same manner as other windows. However, drawing windows will only move and resize within the AutoCAD window. Different options are available depending on the particular interface item and the float or docked status of the item. Typically, the close and minimize or maximize options are available. Some floating items, such as sticky panels, include *grab bars*.

Locking

To prevent certain interface items from moving accidentally, lock the features in either a floating or a docked state. To access locking options, pick the **Toolbar/Window Positions** button on the status bar or select the **View** tab on the ribbon and then the **Window Locking** *flyout* from the **Windows** panel. **Figure 1-8** displays the **Lock Location** menu options.

Select an option to lock the interface items that reside in that group as floating or docked. To unlock a group, select the option again. To quickly lock or unlock all interface items, select **Locked** or **Unlocked** from the **All** cascading menu. Move a locked feature without unlocking it by holding down the [Ctrl] key while moving the feature.

float: Describes interface items that can be freely resized or moved about the screen.

docked: Describes interface items that are locked into position on an edge of the AutoCAD window (top, bottom, left, or right).

grab bars: Two thin bars at the top or left edge of a docked or floating feature; used to move the feature.

flyout: Set of related buttons that appears when you pick the arrow next to certain tool buttons.

Figure 1-8.
Some or all interface
items can be locked
in position.

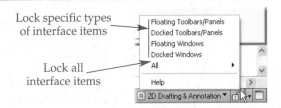

Lock specific types
of interface items

Lock all
interface items

Shortcut Menus

shortcut menus:
Context-sensitive
menus available
by right-clicking on
interface items or
drawing objects.
Menu content
varies based on
the location of the
cursor and the
current conditions,
such as whether
a tool is active or
whether an object is
selected.

**context-sensitive
menu options:**
Options specific to
the tool currently
in use.

cascading menu: A
menu that contains
options related to the
chosen menu item.

AutoCAD uses *shortcut menus*, also known as *cursor menus*, *right-click menus*, or *pop-up menus*, to simplify and accelerate tool and option access. When you right-click in the drawing area with no tool active, the first item displayed in the shortcut menu is typically an option to repeat the previous tool or operation. If you right-click while a tool is active, the shortcut menu contains *context-sensitive menu options*. See **Figure 1-9**. Some menu options have a small arrow to the right of the option name. Hover over the option to display a *cascading menu*. The **Recent Input** cascading menu shows a list of recently used tools, options, or values, depending on the specific shortcut menu. Pick from the list to reuse a function or value.

Exercise 1-4

Access the Student Web site (www.g-wlearning.com/CAD) and complete Exercise 1-4.

NEW

Application Menu

The **Application Menu** provides access to application- and file-related tools and settings through a system of menus and menu options. The **Application Menu** displays when you pick the **Application Menu** button, located in the upper-left corner of the AutoCAD window. See **Figure 1-10**.

Figure 1-9.
Shortcut menus
provide instant
access to tools and
options related to
the current drawing
or editing operation.

Pick to view and select
the most recent tools

Cascading menu
of recent tools

Tool-specific
options

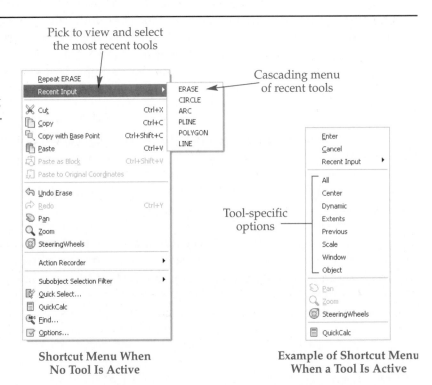

Shortcut Menu When
No Tool Is Active

Example of Shortcut Menu
When a Tool Is Active

Figure 1-10.
Use the **Application Menu** to access common application and file management tools and settings, search for commands, and view open and recently used documents.

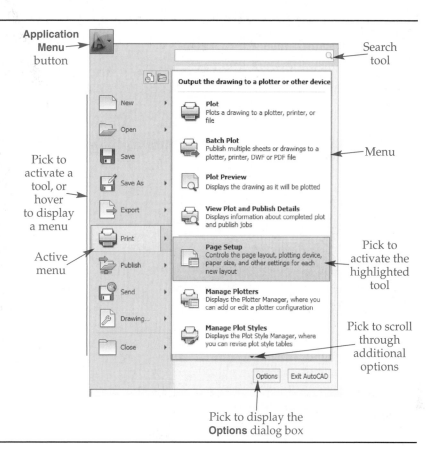

Items on the left side of the **Application Menu** function as buttons to activate common application tools, and except for the **Save** button, also display menus. For example, press the **New** button to begin a new file using the **QNEW** tool. To display a menu, hover the cursor over the menu name, or pick the arrow on the right side of the button. Long menus include small arrows at the top and bottom for scrolling through selections. Some options have a small arrow to the right of the item name that, when selected or hovered over, expands to provide a submenu. Pick the desired option to activate the tool.

Accessing Tools and Options

A tool or option accessible from the **Application Menu** appears as a graphic in the margin of this textbook. This graphic represents the process of picking the **Application Menu** button, then selecting a menu button, or hovering over a menu and picking a menu option or a submenu option. The example shown in this margin illustrates accessing the **PAGESETUP** tool from the **Application Menu**, as shown in **Figure 1-10.**

Searching for Commands

The **Application Menu** contains a search tool used to locate and access any AutoCAD command listed in the Customize User Interface (CUI) file. Type the name of the command you want to access in the **Search** text box. Commands that match the letters you enter appear as you type. Typing additional letters narrows the search, with the best-matched command listed first. **Figure 1-11** shows using the **Search** text box to locate the **SAVE** tool for saving a file. Pick a command from the list to start the command.

NOTE

The **Recent Documents** and **Open Documents** features of the **Application Menu** provide access to recently and currently open files. Chapter 2 describes these functions.

Figure 1-11.
Use the **Application Menu** to search for a command. Pick the command from the list to activate.

Type the name of a command here

Pick to start the command

Exercise 1-5

Access the Student Web site (www.g-wlearning.com/CAD) and complete Exercise 1-5.

Quick Access Toolbar

toolbars: Interface items that contain tool buttons or drop-down lists.

tool buttons: Interface items used to start tools.

Toolbars contain *tool buttons*. Each tool button includes an icon that represents an AutoCAD tool or option. As you move the cursor over a tool button, the button highlights and may display a border. Use the tooltip to become familiar with the tool icons. Select a tool button to activate the associated tool.

The default **Quick Access** toolbar is located on the title bar in the upper-left corner of the AutoCAD window, to the right of the **Application Menu** button. See **Figure 1-12.** The **Quick Access** toolbar provides fast, convenient access to some of the most commonly used tools. Activating most tools from the **Quick Access** toolbar requires only a single pick. Most other interface items require two or more picks to activate a tool.

Figure 1-12.
Use the **Quick Access** toolbar to access commonly used tools. Pick a tool button to activate the corresponding tool.

Pick to display a flyout

Pick to display options for customizing the **Quick Access** toolbar

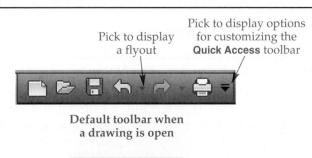

Default toolbar when a drawing is open

Default toolbar when no drawing is open

AutoCAD and Its Applications—Basics

When a drawing is open and the default **2D Drafting & Annotation** workspace is active, the toolbar contains **New**, **Open**, **Save**, **Undo**, **Redo**, and **Plot** buttons. When a drawing is not open, the **New**, **Open**, and **Sheet Set Manager** tool buttons display. The **Quick Access** toolbar is fully customizable by adding, removing, and relocating tool buttons. To make basic adjustments, pick the **Customize Quick Access Toolbar** flyout on the right side of the toolbar. *AutoCAD and Its Applications—Advanced* further explains customizing the user interface.

A tool or option accessible from the **Quick Access** toolbar appears as a graphic in the margin of this textbook. This graphic represents the process of picking a **Quick Access** toolbar button. The example shown in this margin illustrates accessing the **REDO** tool from the **Quick Access** toolbar to redo a previously undone operation.

NOTE

Several toolbars appear in the **AutoCAD Classic** workspace. These toolbars are usually application- or task-specific. The **Application Menu**, **Quick Access** toolbar, and ribbon replace classic toolbars in all other workspaces. Refer to *AutoCAD and Its Applications—Advanced* for information on customizing a workspace to display classic toolbars.

Exercise 1-6

Access the Student Web site (www.g-wlearning.com/CAD) and complete Exercise 1-6.

Ribbon

The ribbon, shown in **Figure 1-13**, is the primary means of accessing tools and options. The ribbon provides a convenient location from which to select tools and options that traditionally would require access by extensive typing, multiple tool-bars, or several menus. The ribbon allows you to spend less time looking for tools and options, while reducing clutter in the AutoCAD window and increasing valuable drawing window space.

The ribbon appears by default in all workspaces except the **AutoCAD Classic** work-space. Use the *tabs* along the top of the ribbon to access individual collections of related *ribbon panels*, or *panels*. Each panel houses groups of similar tools. For example, the

tab: A small stub at the top or side of a page, window, dialog box, or palette, allowing access to other portions of the item.

ribbon panels (panels): Palette divisions that group tools.

Figure 1-13.
The ribbon is the most often used palette and is docked at the top of the drawing window. Palettes provide access to tools, options, properties, and settings.

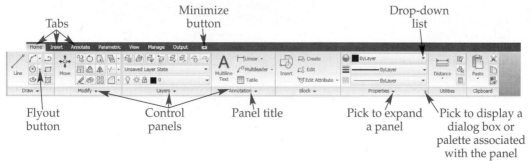

Annotate tab includes several panels, each with specific tools for creating, modifying, and formatting annotations, such as text. The tabs and panels shown when the **2D Drafting & Annotation** workspace is active provide access to 2D drawing tools.

A tool or option accessible from the ribbon appears in a graphic located in the margin of this textbook, like the example shown in this margin. The graphic identifies the tab and panel where the tool is located. You may need to expand the panel or pick a flyout to locate the tool. This example shows how to access the **LINE** tool using the ribbon.

Ribbon Panels

drop-down list: A list of options that appears when you pick a button that contains a down arrow.

The large tool button in a panel signifies the most often used panel tool. In addition to tool buttons, panels can contain flyouts, *drop-down lists*, and other items. Some panels have a triangle, or arrow, next to the panel name. If you see this arrow, pick the bottom, or title, of the panel to display additional, related tools and functions. See **Figure 1-14.** To show the expanded list on-screen at all times, select the push pin button.

> **NOTE**
>
> When you pick an option from a ribbon flyout, the option becomes the new default and appears in the ribbon. This makes it easier to select the same option the next time you use the tool.

Some panels include a small arrow in the lower-right corner or the panel. Pick this arrow to access a dialog box or palette closely associated with the panel function. For example, pick the arrow in the lower-right corner of the **Home** tab, **Properties** panel, as shown in **Figure 1-14,** to display the **Properties** palette. The **Properties** palette is one of the most often used tools for adjusting object properties.

Adjustments

Right-click on a portion of the ribbon that is not occupied by a panel to access a shortcut menu with a variety of ribbon display options. **Figure 1-15** provides a brief description of basic ribbon shortcut menu functions. Minimize options can also be activated by repeatedly pressing the **Minimize** button to the right of the tabs. You can also control the display of tabs and panels by right-clicking on a panel and selecting from the appropriate cascading submenu.

By default, the ribbon is docked horizontally below the AutoCAD window title bar. Use the **Undock** shortcut menu option described in **Figure 1-15** to change the ribbon to a floating state, as shown in **Figure 1-16.** Right-click on the title bar or pick the **Properties** button to select from a list of undocked ribbon control options. The **Auto-hide** option allows the ribbon to minimize when the cursor is away from the ribbon, conserving significant drawing space.

Figure 1-14.
An expanded panel provides additional, related tools and functions. In this case, the **Draw** panel has been expanded.

Pick to pin the expanded list to the screen

Pick to display the **Properties** palette

Figure 1-15.
Right-click options for displaying and organizing ribbon elements.

Selection	Result
Minimize > Minimize to Tabs	Shows only tabs. Pick a tab to show all panels in the tab.
Minimize > Minimize to Panel Titles	Displays tabs and panel titles. Pick a panel title to display the panel.
Minimize > Show Full Ribbon	Shows the default full ribbon.
Show Tabs	Allows you to choose which tabs to display.
Show Panels	Allows you to select which panels to display.
Show Panel Titles	Uncheck to hide panel titles.
Undock	Changes the ribbon to a floating state.
Close	Closes the ribbon. Use the **RIBBON** tool to redisplay the ribbon

Figure 1-16.
Floating palettes remain on-screen as you work in the drawing area.

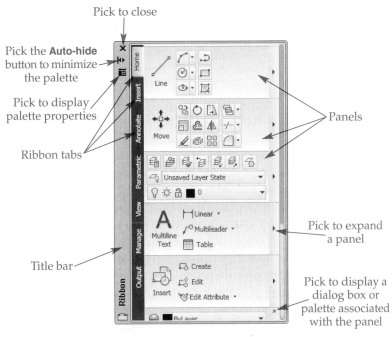

Pick to close

Pick the **Auto-hide** button to minimize the palette

Pick to display palette properties

Ribbon tabs

Panels

Pick to expand a panel

Title bar

Pick to display a dialog box or palette associated with the panel

---

PROFESSIONAL TIP

Resize the floating ribbon using the resizing arrows that appear when you move the cursor over the ribbon edge. Then pick the **Auto-hide** button to take full advantage of the ribbon while displaying the largest possible drawing area.

To reposition a ribbon tab, hold down the left mouse button on a tab, and drag the tab right or left if the ribbon is in a horizontal orientation, or up or down if the ribbon is in a vertical orientation. Release the button when the tab is in the desired location. To reposition a panel within a tab, hold down the left mouse button on a panel title and drag the panel to the desired location.

Figure 1-17.
A sticky panel created by dragging the **Draw** panel from the **Home** tab and dropping it into the drawing window.

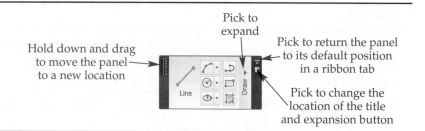

Hold down and drag to move the panel to a new location

Pick to expand

Pick to return the panel to its default position in a ribbon tab

Pick to change the location of the title and expansion button

You can also drag a panel from a tab and drop it in the drawing window to create a *sticky panel*. See **Figure 1-17**. A sticky panel is much like a toolbar and conveniently remains on-screen when you select a different ribbon tab. Hover over a sticky panel to reveal grab bars for moving the panel and buttons for returning the panel to its appropriate ribbon tab and adjusting the title orientation. You can also drag and drop a sticky panel back into the appropriate tab.

sticky panel: A ribbon panel moved out of a tab and made to float in the drawing window.

> **NOTE**
>
> The **Application Menu**, **Quick Access** toolbar, and ribbon replace the traditional menu bar in workspaces other than the **AutoCAD Classic** workspace. To display the menu bar, pick the **Customize Quick Access Toolbar** flyout on the right side of the **Quick Access** toolbar and choose **Show Menu Bar**.

Palettes

palette (modeless dialog box): Special type of window containing tool buttons and other features found in dialog boxes. Palettes can remain open while other tools are in use.

list box: A boxed area that contains a list of items or options from which to select.

scroll bar: A bar tipped with buttons used to scroll through a list of options or information.

Palettes, also known as *modeless dialog boxes*, control many AutoCAD functions. Palettes can look like extensive toolbars or more like dialog boxes, depending on the function and floating or docked state. You can consider the ribbon a palette used to access tools and options. Palettes can contain tool buttons, flyouts, drop-down lists, and many other features, such as *list boxes*, and *scroll bars*. Unlike a dialog box, you do not need to close a palette in order to use other tools and work on the drawing. Like the ribbon, panels divide some palettes into groups of tools. Large palettes are divided into separate pages or windows, which are commonly accessed using tabs.

To display a palette, pick a palette button from the **Palettes** panel in the **View** ribbon tab. You can also display most palettes using palette-specific access techniques. For example, to access the **Properties** palette, pick the arrow in the lower-left corner of the **Properties** panel in the **Home** ribbon tab; double-click on most objects in the drawing window; select an object, right-click and then select **Properties**; or type **PROPERTIES**.

When you display a palette for the first time, it is often in a floating state, although some palettes are dockable. Most of the palette control features available on the floating ribbon, such as docking and sizing, apply to other palettes as well. Deselect the **Allow Docking** palette property or menu option if you do not want to have the ability to dock a palette. The **Properties** button or shortcut menu on some palettes include other functions, such as the **Transparency...** option, which makes the palette transparent, allowing drawing geometry behind the palette to be viewed. See **Figure 1-18**.

> **NOTE**
>
> Palettes play a major role in the operation of AutoCAD. Specific palettes are described when applicable throughout this textbook or in *AutoCAD and Its Applications—Advanced*.

Figure 1-18.
To make a palette transparent, pick the **Properties** button or right-click in the title bar and
select **Transparency....**

Transparent tool palette

Properties button

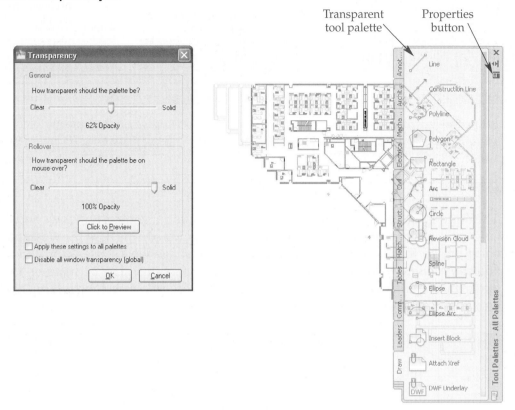

Exercise 1-7

Access the Student Web site (www.g-wlearning.com/CAD) and
complete Exercise 1-7.

Status Bars

AutoCAD provides two types of status bars. The application status bar applies to
all open drawings. A drawing status bar, when activated, appears above the *command
line* and is specific to each drawing. Status bars are the quickest and most effective
way to manage certain drawing settings.

command line:
Area at the bottom
of the screen where
commands (tool
names) and options
may be typed.

Application Status Bar

The application status bar is located along the bottom of the AutoCAD window. See
Figure 1-19. The application status bar is divided into areas that display and control a
variety of drawing aids and tools. The coordinate display field, located on the left side of the
application status bar, shows the XYZ coordinates of the crosshairs, identifying its location
in drawing space. *Status toggle buttons* are located next to the coordinate display field.

**status toggle
buttons:** Buttons
that toggle drawing
aids and tools on
and off.

> **NOTE**
>
> Status toggle buttons appear as icons by default. To change the
> display from icons to names, right-click on any button and deselect
> **Use Icons**. This option applies only to the status toggle buttons.

Figure 1-19.
Picking buttons on the application status bar is the quickest and most effective way to manage certain drawing settings.

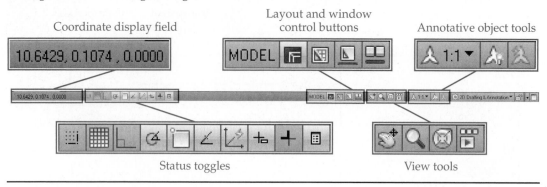

The buttons on the right side of the application status bar control windows and the drawing environment, activate tools, and adjust annotation scaling. Use the **Workspace Switching** button found in this area to change and manage workspaces. You can use the **Toolbar/Window Positions** button to lock interface items. The remaining tools and settings available on the application status bar are described when applicable throughout this textbook.

NOTE

Right-click on the application status bar, away from the coordinate display field or a button, to access a shortcut menu with options for modifying the application status bar display. Uncheck an item on the list to hide the item from the status bar. *AutoCAD and Its Applications—Advanced* further describes how to customize the status bar.

Drawing Status Bar

A **Drawing Status Bar** option is available from the application status bar shortcut menu and the **Status Bar** flyout in the **Windows** panel of the **View** ribbon tab. When this option is selected, a separate drawing status bar appears in the drawing window. The **Annotation Scale**, **Annotation Visibility**, and **AutoScale** tools move from the application status bar to the drawing status bar. See **Figure 1-20**. The settings are unique to each open file.

Figure 1-20.
The drawing status bar, when displayed, is specific to the current drawing. Each open drawing has its own drawing status bar.

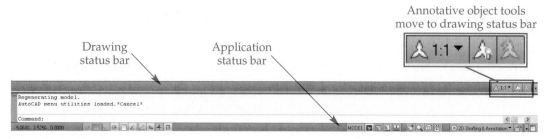

Right-click on the coordinate display field or a button in the application or drawing status bar to view a shortcut menu specific to the item. Picking options from a status bar shortcut menu is often the most efficient method of controlling drawing settings.

Exercise 1-8

Access the Student Web site (www.g-wlearning.com/CAD), and complete Exercise 1-8.

Dialog Boxes

You will see many *dialog boxes* during a drawing session, including those used to create, save, and open files. Dialog boxes contain many of the same features found in other interface items, including icons, text, buttons, and flyouts. **Figure 1-21** shows the dialog box that appears when you pick **Insert** from the **Block** panel of the **Insert** ribbon tab. This dialog box displays many common dialog box elements.

dialog box: A window-like part of the user interface that contains various kinds of information and settings.

A dialog box appears when you pick any menu selection or button displaying an ellipsis (…).

Use the cursor to set variables and select items in a dialog box. In addition, many dialog boxes include images, *preview boxes*, or other methods to help you to select appropriate options. When you pick a button in a dialog box that includes an ellipsis (…), another dialog box appears. You must make a selection from the second dialog box before returning to the original dialog box. A button with an arrow icon requires you to select in the drawing area.

Figure 1-21.
A dialog box appears when you pick an item that is followed by an ellipsis. The dialog box shown here displays when you select the **INSERT** tool.

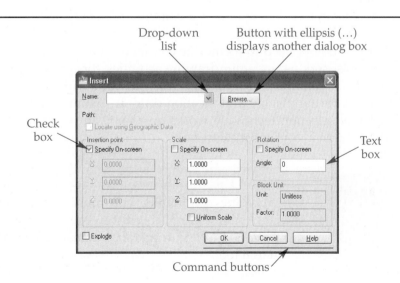

Exercise 1-9

Access the Student Web site (www.g-wlearning.com/CAD) and complete Exercise 1-9.

System Options

OPTIONS

Application Menu
Options
Options

Type
OPTIONS
OP

AutoCAD system options are contained in the **Options** dialog box. System options apply to the entire program and are not specific to a file. Many system options help configure the work environment. This textbook references the **Options** dialog box when applicable.

> **NOTE**
>
> The **Options** dialog box can also be accessed by right-clicking when no tool is active and selecting **Options...**.

Selecting Tools

dynamic input:
Area near the crosshairs where commands may be typed and context-oriented information is provided.

command aliases:
Abbreviated command names entered at the keyboard.

Type
LINE
L

Tools are available by direct selection from the ribbon, shortcut menus, the **Application Menu**, the **Quick Access** toolbar, palettes, and the status bar. An alternative is to type the command that activates a tool using *dynamic input* or the command line. To activate a tool by typing, type the single-word command name or *command alias* and press [Enter] or the space bar, or right-click. You can use uppercase, lowercase, or a combination of uppercase and lowercase letters. You can only issue one command at a time.

You can activate all tools and options by typing commands. All tool names and aliases, along with other access techniques available in the **2D Drafting & Annotation** workspace, appear in a graphic in the margin of this textbook. The example displayed in the margin shows the command name (**LINE**) and alias (**L**) you can use to access the **LINE** tool.

Accessing tools using one or a combination of the methods explained earlier in this chapter offers advantages over typing commands at the keyboard. A major benefit is that you do not need to memorize command names or aliases. Another advantage is that tools, options, and your drawing activities appear on-screen as you work, using visual icons, tooltips, and prompts. As you work with AutoCAD, you will become familiar with the display and location of tools. As an AutoCAD drafter, you decide which tool selection technique works best for you. A combination of tool selection methods often proves most effective.

> **PROFESSIONAL TIP**
>
>
>
> When typing commands, you must exit the current tool before issuing a new tool. In contrast, when you use the ribbon or other input methods, the current tool automatically cancels when you pick a different tool.

Even though you may not access tools by typing command names, you must still enter certain values by typing. For example, you may have to enter the diameter of a circle using the keyboard.

Reference Material *Command Aliases*
For a detailed list of command aliases, go to the **Reference Material** section of the Student Web site (www.g-wlearning.com/CAD) and select **Command Aliases**.

Dynamic Input

Dynamic input allows you to keep your focus at the point where you are drawing. When dynamic input is on, a temporary area for tool input and information appears in the drawing window, below and to the right of the crosshairs by default. See **Figure 1-22.** When a tool is in progress, regardless of how you access the tool, the next action needed to proceed appears, along with additional tool options, an input area, and additional information.

Depending on the tool in progress, different information and options are available in the dynamic input area. For example, in **Figure 1-23**, the **RECTANGLE** tool has been issued. The first part of the dynamic input area is the tooltip, which reads Specify first corner point or. In this case, to draw a rectangle, you need to pick in the drawing area, enter *coordinates* to specify the first corner of the rectangle, or access available options as suggested by the "or" portion of the prompt.

coordinates: Numerical values used to locate a point in the drawing area.

Figure 1-22.
Use dynamic input to type or select tools and values from a temporary input area next to the crosshairs.

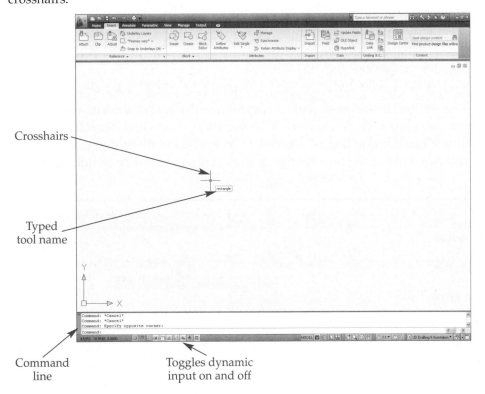

Crosshairs

Typed tool name

Command line

Toggles dynamic input on and off

Figure 1-23.
The dynamic input fields after the **RECTANGLE** tool has been started.

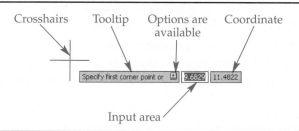

Crosshairs   Tooltip   Options are available   Coordinate

Specify first corner point or   9.6829   11.4822

Input area

Pressing the down arrow key displays the options available for the current tool. See **Figure 1-24.** Select an option using the cursor, or press the down arrow again to cycle through the available options, as indicated by a bullet next to the option. To select a bulleted option, press [Enter]. You can also select an option by right-clicking and selecting an option from the shortcut menu. The information displayed in the dynamic input area changes while you work with a tool, depending on the actions you choose. **Figure 1-25** shows the dynamic input display when the **LINE** tool is active.

> **NOTE**
>
> Dynamic input can be toggled on and off by picking the **Dynamic Input** button on the status bar or pressing the [F12] key. You can issue commands without dynamic input on.

Command Line

The command line, shown in **Figure 1-22**, provides the same function as dynamic input, but allows you to enter tools and context-specific information in a traditional window format. By default, the command line is docked at the bottom of the AutoCAD window, above the status bar. It acts like a palette and displays the Command: prompt and reflects any commands you issue. The command line also displays prompts that supply information or that request input.

When you issue a command, AutoCAD either performs the specified operation or displays prompts for additional information needed. The commands that activate AutoCAD tools have a standard format, structured as follows:

```
Command: COMMANDNAME↵
Current settings: Setting1 Setting2 Setting3
Instructional text [Option1/oPtion2/opTion3/...] <default option or value>:
```

Settings or options associated with a command display as shown. The prompt indicates what you should do to continue the operation. The square brackets contain available options. Each option has an alias, or unique combination of uppercase characters, that you can enter at the prompt rather than typing the entire option name. If a

Figure 1-24.
Press the down arrow key to expose tool options. Pick an option with the cursor, or use the up and down arrow keys to position the highlight at the desired option and press [Enter] to select.

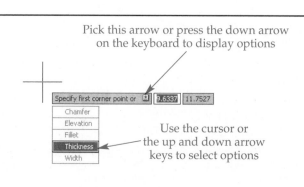

Pick this arrow or press the down arrow on the keyboard to display options

Specify first corner point or   9.6337   11.7527

Chamfer
Elevation
Fillet
Thickness
Width

Use the cursor or the up and down arrow keys to select options

Figure 1-25. Dynamic input fields change while a tool is in use. In this example, crosshair coordinates appear first. After you select the first endpoint, the distance and angle of the crosshairs relative to the first endpoint display.

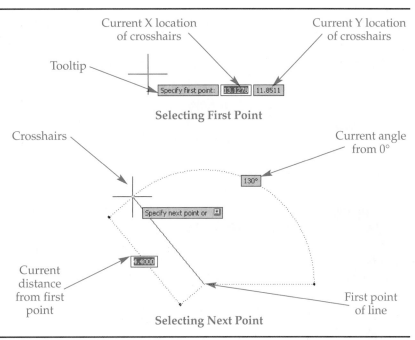

Current X location of crosshairs

Current Y location of crosshairs

Tooltip

Specify first point: 13.1278 11.8511

Selecting First Point

Crosshairs

Current angle from 0°

130°

Specify next point or

Current distance from first point

4.4000

First point of line

Selecting Next Point

default option is displayed in the angle brackets (<>), press [Enter] to accept the option rather than typing the value again.

Each default AutoCAD workspace includes the command line. The command line can float or be docked, resized, and locked. The floating command line contains the **Auto-hide** and **Properties** buttons found on palettes. Depending on your working preference, use the command line at the same time as dynamic input, or disable the command line if you use only dynamic input. To hide the command line, pick the **Close** button on the command line title bar, right-click on the command line and pick **Close**, type COMMANDLINEHIDE, or press [Ctrl]+[9].

PROFESSIONAL TIP

While learning AutoCAD, pay close attention to the prompts displayed in the dynamic input area and at the command line.

Keyboard Keys

Many keys on the keyboard, known as *shortcut keys* or *keyboard shortcuts*, allow you to perform AutoCAD functions quickly. Become familiar with these keys to improve your AutoCAD performance. Whenever it is necessary to cancel a tool or dialog box, press the *escape key* [Esc] in the upper-left corner of the keyboard. Some tool sequences require that you press [Esc] twice to cancel the operation.

Use the up and down arrow keys to select previously used tools. When no tool is active, press the up arrow to display the previously used tool. If dynamic input is active, previously used tools appear near the crosshairs by default. To display previously used tools at the command line, you must pick the command line before pressing the up arrow, or turn off dynamic input. If you continue to press the up arrow, AutoCAD continues to backtrack through the tools you have used. Press [Enter] to activate a displayed tool.

Function keys provide instant access to tools. They can also be programmed to perform a series of commands. Control and shift key combinations require that you press and hold the [Ctrl] or [Shift] key and then press a second character. You can

shortcut key (keyboard shortcut): Single key or key combination used to quickly issue a command or select an option.

escape key: Keyboard key used to cancel a tool or exit a dialog box.

function keys: The keys labeled [F1] through [F12] along the top of the keyboard.

activate several tools using [Ctrl] key combinations. A tooltip typically indicates if a key combination is available.

Reference Material *Shortcut Keys*
For a complete list of keyboard shortcuts, go to the **Reference Material** section of the Student Web site (www.g-wlearning.com/CAD) and select **Shortcut Keys**.

Exercise 1-10

Access the Student Web site (**www.g-wlearning.com/CAD**) and complete Exercise 1-10.

Getting Help

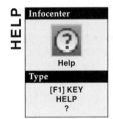

HELP

Infocenter

Help

Type
[F1] KEY
HELP
?

If you need help with a specific tool, option, or AutoCAD feature, use this textbook as a guide, or use the help system contained in the **AutoCAD Help** window. The graphic shown in the margin identifies several ways to access the **AutoCAD Help** window. You can also access the **AutoCAD Help** window from the **InfoCenter**, described later in this chapter, or by selecting **Help** from a shortcut menu.

The **AutoCAD Help** window consists of two frames. See **Figure 1-26.** The left frame has three tabs for locating help topics. The right frame displays the selected help topics. The **Contents** tab in the left frame displays a list of book icons and topic names. The book icons represent the organizational structure of books of topics within the AutoCAD documentation. Topics contain the actual help information; the icon used to represent a topic is a sheet of paper with a question mark. To open a book or a help topic, double-click on its name or icon.

> **NOTE**
>
> If you are unfamiliar with how to use a Windows help system, spend time now exploring all the topics under **AutoCAD Help** in the **Contents** tab of the **AutoCAD Help** window.

Although the **Contents** tab of the **AutoCAD Help** window is useful for displaying all the topics in an expanded table of contents manner, it may not be very useful when you are searching for a specific item. In this case, you should refer to the help file index. This is the function of the **Index** tab. Use the **Search** tab to search the help documents for specific words or phrases.

In addition to the two frames, six buttons reside at the top of the **AutoCAD Help** window. The **Hide/Show** button controls the visibility of the left frame. Use the **Back** button to view the previously displayed help topic. Use the **Forward** button to go forward to help pages you viewed before pressing the **Back** button. The **Home** button takes you to the AutoCAD Help page. The **Print** button prints the currently displayed help topic. The **Options** button presents a menu with a variety of items used to control other aspects of the **AutoCAD Help** window.

Figure 1-26.
Get help using the **AutoCAD Help** window or the **InfoCenter.**

InfoCenter

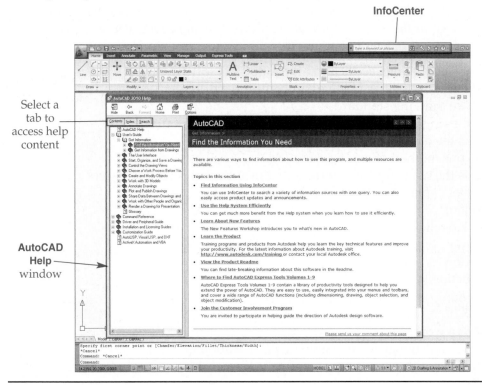

Select a tab to access help content

AutoCAD Help window

If you press the [F1] key while you are in the process of using a tool, help information associated with the active tool displays. This *context-oriented help* saves valuable time, since you do not need to scan through the help contents or perform a search to find the information.

context-oriented help: Help information for the active tool.

Using the InfoCenter

The **InfoCenter**, located on the right side of the title bar as shown in **Figure 1-26**, allows you to search for help topics without first displaying the **AutoCAD Help** window. It also provides buttons for access to the **Subscription Center**, **Communication Center**, and the **Favorites** list. Type a question in the text box to search for topics. Then select the appropriate topic from the list to display it in the **AutoCAD Help** window. To add a topic to the **Favorites** list, pick the star next to the topic. Pick the **Subscription Center** button to access information associated with your license and subscription eligibility, options, and services. Pick the **Communication Center** button to access content on a variety of help topics. Pick the **Favorites** button to access any help topics you have stored.

Exercise 1-11

Access the Student Web site (www.g-wlearning.com/CAD) and complete Exercise 1-11.

Chapter Test

Answer the following questions. Write your answers on a separate sheet of paper or go to the Student Web site (www.g-wlearning.com/CAD), and complete the electronic chapter test.

1. Describe at least one application for AutoCAD software.
2. Briefly explain what is involved in planning a drawing.
3. What are drawing standards?
4. Why should you save your work every ten to fifteen minutes?
5. What is the quickest method for starting AutoCAD?
6. Name one method of exiting AutoCAD.
7. What is the name for the interface that includes on-screen features?
8. Define or explain the following terms:
   A. Default
   B. Pick or click
   C. Hover
   D. Button
   E. Function key
   F. Option
   G. Tool
9. What is a workspace?
10. How do you change from one workspace to another?
11. What is the difference between a docked interface item and a floating interface item?
12. How do you select the locking options to lock the interface items in either their floating and docked state?
13. What is a flyout?
14. How do you access a shortcut menu?
15. What does it mean when a shortcut menu is described as context-sensitive?
16. Explain the basic function of the **Application Menu**.
17. Describe the **Application Menu** search tool and briefly explain how to use it.
18. Briefly describe an advantage of using the ribbon.
19. What is the function of tabs in the ribbon?
20. What is another name for a palette?
21. Describe the function of the application status bar.
22. What is the meaning of the ... (ellipsis) in a menu option or button?
23. What are the two primary methods for accessing AutoCAD tools? List interface items associated with each.
24. Briefly describe the function of dynamic input.
25. Briefly explain the function of the [Esc] key.
26. How do you access previously used tools when dynamic input is on?
27. Name the function keys that execute the following tasks. (Refer to the Shortcut Keys document in the **Reference Material** section on the Student Web Site.)
   A. Snap mode (toggle)
   B. Grid mode (toggle)
   C. Ortho mode (toggle)
28. Describe two ways to access the **AutoCAD Help** window.
29. Describe the purpose of the book icons in the **Contents** tab of the **AutoCAD Help** window.
30. What is context-oriented help, and how is it accessed?

Problems

Start AutoCAD if it is not already started. Follow the specific instructions for each problem.

▼ Basic

1. Perform the following tasks:
   A. Open the **AutoCAD Help** window.
   B. In the **Contents** tab, expand the **User's Guide** book.
   C. Expand the **Get Information** book.
   D. Expand the **Find the Information You Need** book.
   E. Pick and read each topic in the right pane.
   F. Close the **AutoCAD Help** window, and then close AutoCAD.

2. Launch AutoCAD using the Start button on the Windows task bar.
   A. Move the cursor over the buttons in the status bar and read the tooltip for each.
   B. Slowly move the cursor over each of the ribbon panels and read the tooltips.
   C. Pick the **Application Menu** to display it. Hover over the **File** menu, then use the right arrow key to move to the **File** options. Then use the down arrow key to move through all the menu options.
   D. Press the [Esc] key to dismiss the menu.
   E. Close AutoCAD.

▼ Intermediate

3. Interview your drafting instructor or supervisor and try to determine what type of drawing standards exist at your school or company. Write them down and keep them with you as you learn AutoCAD. Make notes as you progress through this textbook on how you use these standards. Also, note how the standards could be changed to match the capabilities of AutoCAD.

4. Research your drawing department standards. If you do not have a copy of the standards, acquire one. If AutoCAD standards have been created, make notes as to how you can use these in your projects. If no standards exist in your department or company, make notes about how you can help develop standards. Write a report on why your school or company should create CAD standards and how they would be used. Describe who should be responsible for specific tasks. Recommend procedures, techniques, and forms, if necessary. Develop this report as you progress through your AutoCAD instruction and as you read this textbook.

5. Develop a drawing planning sheet for use in your school or company. List items you think are important for planning a CAD drawing. Make changes to this sheet as you learn more about AutoCAD.

6. Create a freehand sketch of the default AutoCAD window with the **2D Drafting & Annotation** workspace active. Label each of the screen areas. To the side of the sketch, write a short description of the function of each screen area.

7. Create a freehand sketch showing three examples of tooltips displayed as you hover the cursor over an item. To the side of the sketch, write a short description of each example's function.

8. Using the **Application Menu** search tool, type the letter C and review the information provided in the **Application Menu**. Then add the letter L. How does the information change? Continue typing O, S, and E to complete the **CLOSE** command. Write a short paragraph explaining how you might use this search tool to find a command if you are unsure how the command is spelled or where it is located.

Drawing Problems - Chapter 1

9. Research and write a report of 250 words or less covering the U.S. National CAD Standard.

10. Research and write a report of 250 words or less covering workplace ethics, especially as related to CAD applications and CAD-related software.

11. Research and write a report of 150 words or less covering an ergonomically designed CAD workstation. Include a sketch of what you might consider a high-quality design for a workstation and label its characteristics.

Drawings and Templates

Learning Objectives

After completing this chapter, you will be able to do the following:

- ✓ Start a new drawing.
- ✓ Save your work.
- ✓ Close files.
- ✓ Open saved files.
- ✓ Work with multiple open documents.
- ✓ Create drawing templates.
- ✓ Determine and specify drawing units and limits.

In this chapter, you will learn how to start new drawings, save drawings, open existing drawings, and prepare drawing templates. This chapter also explains some of the basic drawing aids used in AutoCAD. You will find the drawing settings described in this chapter very useful as you begin working with drawing and drawing template files.

Starting a New Drawing

There are two primary AutoCAD file types: *drawing files*, which have a .dwg extension, and *drawing template files*, also known as *templates*, which have a .dwt extension. New drawings typically begin from templates that include standard drawing settings and objects. All template settings and contents are included in a new drawing. To help avoid confusion as you learn AutoCAD, remember that a new drawing file references a drawing template file, but the drawing file is where you draw.

Depending on your initial setup options or any customization you have done, when you launch AutoCAD, a new drawing file appears. The drawing references a default template. If you skip the initial setup process as advised in this textbook, the drawing you initially see references the acad.dwt template provided by AutoCAD. The drawing is appropriate for initial drawing applications and uses basic decimal U.S. Customary (inch) unit settings. You are ready to create a drawing, save, and close the file. To start another new drawing, you can use a template or start from scratch.

drawing files:
Files that contain the actual drawing geometry and information.

drawing template files (templates):
Files referenced to develop new drawings; templates contain standard drawing settings and objects.

Choosing a Template

The **QNEW** tool is the primary means of starting a new drawing. By default, the **Select template** dialog box appears when you access the **QNEW** tool. See **Figure 2-1.** The **Select template** dialog box lists the templates found in the specified drawing template folder. The default template folder shown in **Figure 2-1** includes a variety of templates supplied with AutoCAD.

NOTE

All file navigation dialog boxes, including the **Select template** dialog box and those used to save, close, and open files, support auto-complete. Type in the **File Name** text box to view a list of files matching the characters you enter.

The tutorial templates for manufacturing and architecture include a border and title block. All other templates are blank, but include drawing settings specific to the requirements of a certain industry or drawing. For general 2D drawing applications, use the acad.dwt template, which includes basic U.S. Customary (inch) unit settings, or the acadiso.dwt template, which includes basic metric unit settings according to ISO standards. To use a template to begin a new drawing, double-click on the file name, right-click on the file and pick **Select**, or select the file and pick the **Open** button.

NOTE

You can also access the **QNEW** tool from the **Quick View Drawings** tool described later in this chapter.

PROFESSIONAL TIP

Use the **Options** dialog box to change the default drawing template folder displayed in the **Select template** dialog box. In the **Files** tab, expand Template Settings, and then expand Drawing Template File Location. Pick the **Browse...** button to select a folder.

Figure 2-1.
Use the **Select template** dialog box to begin a new drawing.

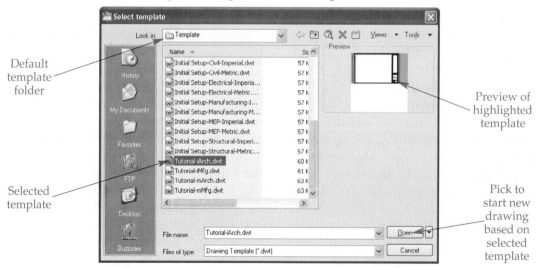

Default template folder

Selected template

Preview of highlighted template

Pick to start new drawing based on selected template

Starting from Scratch

For a new AutoCAD user, an effective way to begin a new drawing is to start "from scratch" using a blank drawing file without a border, title block, modified layouts, or customized drawing settings. Start from scratch when you are just beginning to learn AutoCAD, when you plan to create your own template, or when the start or end of a project is unknown. To start a drawing from scratch, pick the flyout next to the **Open** button in the **Select template** dialog box. See **Figure 2-2**. Then, to begin a drawing using basic inch unit settings, pick **Open with no Template-Imperial**, or to begin a drawing using basic metric unit settings, pick **Open with no Template-Metric**.

Setting a Quick Start Template

AutoCAD provides a quick start feature to begin a drawing using the **QNEW** tool and a specific template, skipping the **Select template** dialog box. Set this function in the **Options** dialog box. Pick the **Files** tab, expand the Template Settings option, and then expand the Default Template File Name for QNEW function. See **Figure 2-3**. None displays by default, which causes the **Select template** dialog box to appear. Pick the **Browse...** button to select a specific template to launch each time you use the **QNEW** tool.

PROFESSIONAL TIP

Use the **NEW** tool to override the quick start template and display the **Select template** dialog box. This allows you to use the **QNEW** tool to begin a drawing with the most frequently used template, but access other templates when necessary.

Type
NEW

Saving Your Work

You should save your drawing or template immediately after you begin work. Then, save every 10 to 15 minutes while working. Saving every 10 to 15 minutes results in less lost work if a software error, hardware malfunction, or power failure occurs. Several AutoCAD tools allow you to save your work. In addition, any tool or option ending the AutoCAD session provides an alert asking if you want to save changes to the drawing or template. This gives you a final option to save or not save changes.

Naming Drawings

Set up a system that allows you to determine the content of a drawing by the drawing name. Drawing names often identify a product by name and number—for example, VICE-101, FLOORPLAN, or 6DT1009. Use a standard drawing naming system that contains a clear and concise reference to the project, part number, process, sheet number, and revision level, depending on the product.

Figure 2-2.
Pick the arrow next to the **Open** button and select an **Open with no Template** option to start a drawing from scratch.

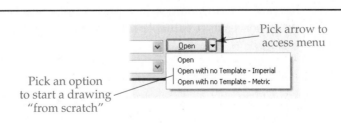

Pick an option to start a drawing "from scratch"

Pick arrow to access menu

Figure 2-3.
Specifying a template for the **QNEW** tool.

Files
tab

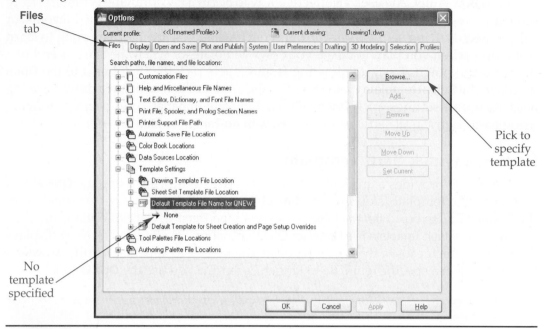

Pick to
specify
template

No
template
specified

Although it is possible to give a drawing file an extended name, such as Details for Top Half of Compressor Housing for ACME, Inc., Part Number 4011A, Revision Level C, a standardized naming system that uses a shorter name, such as ACME.4011A.C, provides all of the necessary information. If you need to add information for easier recognition, add the content to the base name. Using the previous example, you might name the file ACME 4011A.C Compressor Housing.Top.Casting Details.

Some rules and restrictions apply to naming drawings. You can use most alphabetic and numeric characters and spaces, as well as most punctuation symbols. Characters that cannot be used include the quotation mark ("), asterisk (*), question mark (?), forward slash (/), and backward slash (\). You can use a maximum of 256 characters. You do not have to include the file extension, such as .dwg or .dwt, with the file name. File names are not case sensitive. For example, you can name a drawing PROBLEM 2-1, but Windows interprets Problem 2-1 as the same file name.

NOTE

The U.S. National CAD standard includes a comprehensive file naming structure for architectural and construction-related drawings. The system can be adapted for other disciplines.

PROFESSIONAL TIP

Every school or company should have a drawing naming system. Record drawing names in a part numbering or drawing name log. A log serves as a valuable reference if you forget what the drawings contain.

Using the QSAVE Tool

The **QSAVE** tool is the most frequently used tool for saving a drawing. If the current file has not yet been named, the **QSAVE** tool displays the **Save Drawing As** dialog box. See **Figure 2-4.** The **Save Drawing As** dialog box is a standard file selection dialog box. To save a file, first choose the type of file to save from the **Files of type:** drop-down list. Select the AutoCAD 2010 Drawing (*.dwg) option for most applications. Use the AutoCAD Drawing Template (*.dwt) to save a template file. Choose the appropriate older AutoCAD version from the list to convert an AutoCAD 2010 file to an older AutoCAD format. This allows someone using an older version of AutoCAD to view your file. You also have the option of saving to a *drawing exchange file (DXF)* or *drawing standards file (DWS)* format.

Next, select the folder in which to store the file from the **Save in:** drop-down list. To move upward from the current folder, pick the **Up one level** button. To create a new folder in the current location, pick the **Create New Folder** button and type the folder name.

The name Drawing1 appears in the **File name:** text box, if the file is the first file started since the launch of AutoCAD. Change the name to the desired file name. You do not need to include the extension. Once you specify the correct location and file name, pick the **Save** button to save the file. You can also press the [Enter] key to activate the **Save** button. If you already saved the current file, accessing the **QSAVE** tool updates, or resaves, the file based on the current file state. In this situation, **QSAVE** issues no prompts and displays no dialog box.

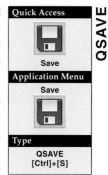

drawing exchange file (DXF): A file format often used by other CADD systems.

drawing standards file (DWS): A file used to check the standards of another file using AutoCAD standards-checking tools.

Using the SAVEAS Tool

Use the **SAVEAS** tool to save a copy of a file using a different name, or to save a file in an alternative format, such as a previous AutoCAD release format. You can also use the **SAVEAS** tool when you *open* a drawing template file to use as a basis for another drawing. This leaves the template unchanged and ready to use for starting other drawings.

The **SAVEAS** tool always displays the **Save Drawing As** dialog box. The name and location of a previously saved file appears. Confirm that the **Files of type:** drop-down list displays the desired file type and that the **Save in:** drop-down list displays the correct drive and folder. Type the new file name in the **File name:** text box and pick the **Save** button.

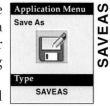

Figure 2-4.
The **Save Drawing As** dialog box.

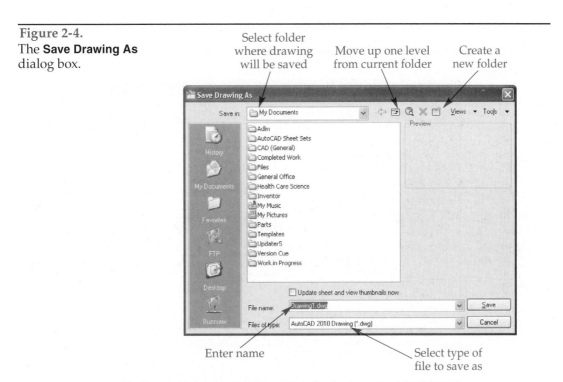

Automatic Saves

automatic save: A save procedure that occurs at specified intervals without your input.

AutoCAD provides an *automatic save* tool that automatically creates a temporary backup file during a work session. Settings in the **File Safety Precautions** area of the **Open and Save** tab in the **Options** dialog box control automatic saves. See **Figure 2-5.** Automatic save is on by default. Type the number of minutes between saves in the **Minutes between saves** text box. The default setting automatically saves every 10 minutes. By default, AutoCAD names automatically saved files *FileName_n_n_nnnn*.sv$ in the C:\Documents and Settings*user*\Local Settings\Temp folder.

The automatic save timer starts as soon as you make a change to the file and resets when you save the file. The file automatically saves when you start the first tool after reaching the automatic save time. Keep this in mind if you let the computer remain idle, as an automatic save does not execute until you return and use a tool. Be sure to save your file manually if you plan to be away from your computer for an extended period.

The automatic save feature is intended for use in case AutoCAD shuts down unexpectedly. Therefore, when you close a file, the automatic save file associated with that file automatically deletes. If AutoCAD shuts down unexpectedly, the automatic save file is available for use. By default, the next time you open AutoCAD after a system failure, the **Drawing Recovery Manager** displays, containing a node for every file to display all

Application Menu

Drawing
> Open the
  Drawing
  Recovery
  Manager

Type

DRAWING
RECOVERY

Figure 2-5.
Use the **Open and Save** tab in the **Options** dialog box to set up the automatic save feature and backup files.

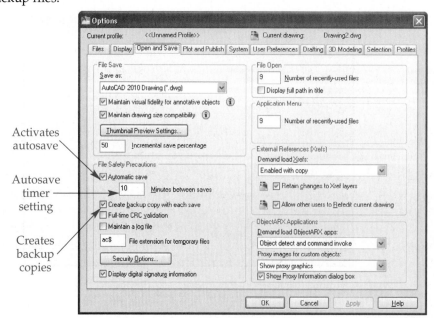

of the available versions of the file: the original file, the recovered file saved at the time of the system failure, the automatic save file, and the .bak file. Pick each version in the **Drawing Recovery Manager** to view it, determine which version you want to save, and then save that file. You can save the recovered file over the original file name.

NOTE

The **Automatic Save File Location** listing in the **Files** tab of the **Options** dialog box determines the folder where the automatic save files are stored.

Using Backup Files

By default, an AutoCAD backup file, with the extension .bak, automatically saves in the same folder where the drawing or template is located. When you save a drawing or template, the drawing or template file updates, and the old drawing or template file overwrites the backup file. Therefore, the backup file is always one save behind the drawing or template file.

The backup feature is on by default and is controlled using the **Create backup copy with each save** check box in the **Open and Save** tab of the **Options** dialog box. Refer again to **Figure 2-5.** If AutoCAD shuts down unexpectedly, you may be able to recover a file from the backup version using the **Drawing Recovery Manager.** You can also rename a backup and try to open it as a drawing or template. Use Windows Explorer to rename the file, changing the file extension from .bak to .dwg to restore a drawing, or from .bak to .dwt to restore a template.

Windows Explorer
For more information about using Windows Explorer, go to the Student Web site (www.g-wlearning.com/CAD), select this chapter, and select **Windows Explorer.**

Recovering a Damaged File
For information about recovering a damaged file, go to the Student Web site (www.g-wlearning.com/CAD) , select this chapter, and select **Recovering a Damaged File.**

Closing Files

Use the **CLOSE** tool to exit the current file without ending the AutoCAD session. One of the quickest methods of closing a file is to pick the **Close** button from the drawing window title bar. If you close a file before saving your work, AutoCAD prompts you to save or discard changes. Pick the **Yes** button to save the file. Pick the **No** button to discard any changes made to the file since the previous save. Pick the **Cancel** button if you decide not to close the drawing and want to return to the drawing area.

As you will learn, AutoCAD allows you to have multiple files open at the same time. Use the **CLOSEALL** tool to close all open files. You are prompted to save each file in which you made changes.

> **NOTE**
>
> You can also close files using the **Quick View Drawings** tool, described later in this chapter.

Exercise 2-1

Access the Student Web site (www.g-wlearning.com/CAD) and complete Exercise 2-1.

Opening Saved Files

Once you save and close a file, you will eventually want to reopen the file. To open a file, use the **OPEN** tool, select recently opened files from the **Application Menu**, or open a file from Windows Explorer.

Using the OPEN Tool

When you access the **OPEN** tool, the **Select File** dialog box appears, containing a list of folders and files. See **Figure 2-6.** Double-click on a file folder to open it, and then double-click on the desired file to open the file.

Figure 2-6.
Use the **Select File** dialog box to open an AutoCAD file. In this illustration, the AutoCAD 2010\ Sample folder is open. The Architectural – Annotation Scaling and Multileaders drawing is selected and appears in the File name: text box and in the preview area.

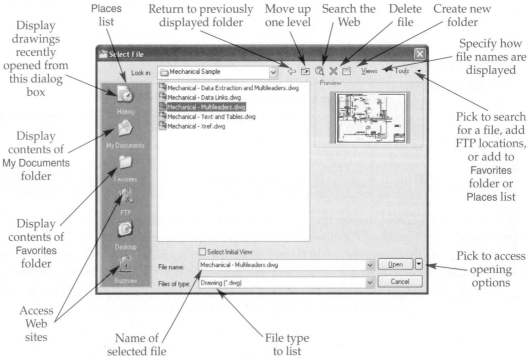

Places list — Return to previously displayed folder — Move up one level — Search the Web — Delete file — Create new folder

Display drawings recently opened from this dialog box

Specify how file names are displayed

Display contents of My Documents folder

Pick to search for a file, add FTP locations, or add to Favorites folder or Places list

Display contents of Favorites folder

Pick to access opening options

Access Web sites

Name of selected file

File type to list

An image of the selected file appears in the **Preview** area. This provides an easy way for you to view the contents of a file without loading it into AutoCAD. After picking a file name to highlight, you can quickly highlight other files and scan through previews using the keyboard arrow keys. Use the up and down arrow keys to move vertically between files, and use the left and right arrow keys to move horizontally. The **Select File** dialog box includes a list on the left side that provides instant access to certain folders. See **Figure 2-7**.

NOTE

You can also access the **Open** dialog box by picking the **Open...** button in the **Quick View Drawings** toolbar, described later in this chapter.

Figure 2-7.
Additional features in the **Select File** dialog box.

Button	Description
 History	Lists drawing files recently opened from the **Select File** dialog box.
 My Documents	Displays the files and folders contained in the My Documents folder for the current user.
 Favorites	Displays files and folders located in the Favorites folder for the current user. To add the folder displayed in the **Look in:** box to the favorites list, select **Add to Favorites** from the **Tools** flyout.
 FTP	Displays available FTP (file transfer protocol) sites. To add or modify the listed FTP sites, select **Add/Modify FTP Locations** from the **Tools** flyout.
 Desktop	Lists the files, folders, and drives located on the computer desktop.
 Buzzsaw	Displays projects on the Buzzsaw Web site. Buzzsaw.com is designed for the building industry. After setting up a project hosting account, users can access drawings from a given construction project on the Web site. This allows the various companies involved in the project to have instant access to the drawing files.

Exercise 2-2

Access the Student Web site (www.g-wlearning.com/CAD) and complete Exercise 2-2.

Finding Files

Pick **Find...** from the **Tools** flyout at the top of the **Select File** dialog box to search for files. The **Find** dialog box, shown in **Figure 2-8,** appears. If you know the file name, type it in the **Named:** text box. If you do not know the name, you can use wildcard characters, such as *, to narrow the search.

Choose the type of file to search for from the **Type:** drop-down list. You can search for DWG, DWS, DXF, or DWT files using the **Find** dialog box. Use Windows Explorer or Windows Search to search for other file types.

To complete a search more quickly, avoid searching the entire hard drive. If you know the folder in which the file is located, specify the folder in the **Look in:** text box. Pick the **Browse** button to select a folder from the **Browse for Folder** dialog box. Check the **Include subfolders** check box to search the subfolders within the selected folder. The **Date Modified** tab provides options to search for files modified within a certain time. This option is useful if you want to list all drawings modified within a specific week or month.

NOTE

Right-click on a folder or file to display a shortcut menu of options. Pick somewhere off the menu to close the menu.

PROFESSIONAL TIP

Certain file management capabilities are available in file dialog boxes, similar to those in Windows Explorer. For example, to rename an existing file or folder, pick it once and pause for a moment, then pick the name again. This places the name in a text box for editing. Type the new name and press [Enter].

Figure 2-8.
Use the **Find** dialog box to locate drawing files.

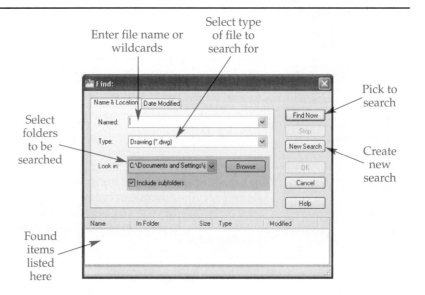

Exercise 2-3

Access the Student Web site (www.g-wlearning.com/CAD) and complete Exercise 2-3.

Opening Recent Drawings

Pick the **Recent Documents** button in the **Application Menu** to display a list of recently opened documents. See **Figure 2-9.** This provides convenient access to files that may be related to your current project. By default, the name and location of up to 9 recent files appears.

Use the **Ordered List** drop-down list to organize recent files. Select **By Ordered List**, **By Access Date**, **By Size**, or **By Type** according to how you want files arranged. Select from the display options flyout to specify how recent files appear. Choose the appropriate option to display files as icons, small images, medium images, or large images.

When you hover over a file in the list of recent documents, a tooltip with file information and a preview of the file appears. Pick a file from the list to open the file. Pick the pushpin icon to the right of the file to keep the file on the recent documents list. Unpinned files are eventually removed from the recent documents list as you open other files.

Figure 2-9.
Use the **Recent Documents** menu on the **Application Menu** to locate and open a recent file quickly.

Pick to display open files

Pick to select how to list recent files

Pick to display recent files

Recent Documents

By Ordered List

Pick to select a display option

New

Open

Save

Save As

Export

Print

Publish

Send

Drawing...

Close

Structural - Imperial.dwg

Mechanical - Imperial.dwg

Annotation - Metric.dwg

Mechanical - Data Extraction and...

Mechanical - Multileaders.dwg

Plot Screening and Fill Patterns.dwg

db_samp.dwg

Blocks and Tables - Imperial.dwg

Architectural - Annotation Scaling and...

Recently opened files

Pick to pin the file to the list

Options   Exit AutoCAD

You can specify the number of previous drawings displayed in the **Recent Documents** list by accessing the **Open and Save** tab of the **Options** dialog box. The number in the **Number of recently-used files** text box in the **Application Menu** area controls this function.

NOTE

If you try to open a deleted or moved file, AutoCAD displays the message Cannot find the specified drawing file. Please verify that the file exists. AutoCAD then opens the **Select File** dialog box.

Using Windows Explorer

Double-click on an AutoCAD file from Windows Explorer to open the file. If AutoCAD is not already running, it starts and the file opens. You can also drag-and-drop a file onto the AutoCAD **Command** line. If AutoCAD is not running, you can drag-and-drop the file onto the AutoCAD 2010 icon on your desktop. AutoCAD then starts and opens the drawing file.

Opening Old Files

In AutoCAD 2010, you can open AutoCAD files created in AutoCAD Release 12 or later. When you open and work on a file from a previous release, AutoCAD automatically updates the file to the AutoCAD 2010 file format when you save. A file saved in AutoCAD 2010 displays in the **Preview** image tile in the **Select File** dialog box.

NOTE

In order to view an older release file opened in AutoCAD 2010 in its original format, you must use the **SAVEAS** tool to save it back to the appropriate file format.

Opening as Read-Only or Partial Open

You can open files in various modes by selecting the appropriate option from the **Open** flyout in the **Select File** dialog box. When you open a file as *read-only*, you cannot save changes to the original file. However, you can make changes to the file and then use the **SAVEAS** tool to save the modified file using a different name. This ensures that the original file remains unchanged.

When opening a large drawing you may choose to issue a *partial open*. This allows you to open a portion of a drawing by selecting specific views and layers to open. Views and layers are described in later chapters. You can also partially open a drawing in the read-only mode.

read-only:
Describes a drawing file opened for viewing only. You can make changes to the drawing, but you cannot save changes without using the SAVEAS tool.

partial open:
Opening a portion of a file by specifying only the views and layers you need to see.

Working with Multiple Documents

Most drafting projects are composed of a number of files; each file presents or organizes a different aspect of the project. For example, an architectural drafting project might include a site plan, a floor plan, electrical and plumbing plans, and assorted detail drawings. Another example is a mechanical assembly composed of several unique parts. The required drawings in this example might include an overall assembly drawing, plus individual detail drawings of each component.

Drawings and other files associated with a project are usually closely related to each other. By opening two or more files at the same time, you can easily reference information contained in existing files while working in a new file. AutoCAD allows you to copy all or part of the contents from one file directly into another using a drag-and-drop or similar operation.

Controlling Windows

Each file you start or open in AutoCAD appears in its own drawing window. The file name displays on the drawing window title bar if the drawing window is floating, or on the AutoCAD window title bar if you minimize the drawing window. The drawing windows and the AutoCAD window have the same relationship that program windows have with the Windows desktop. When you maximize a drawing window, it fills the available area in the AutoCAD window. Minimizing a drawing window displays it as a reduced-size title bar along the bottom of the drawing area. Pick the title bar of a minimized or floating drawing window to activate the file. You cannot move drawing windows outside the AutoCAD window. **Figure 2-10** illustrates drawing windows in a floating state and minimized.

Using the Quick View Drawings Tool

The **Quick View Drawings** tool is one of many AutoCAD tools that you can use to switch between open drawings. The visual display of this tool allows you to see and control open drawing files without actually changing drawing windows. The quickest way to access the **Quick View Drawings** tool is to pick the **Quick View Drawings** button on the status bar.

Figure 2-10.
Display drawing windows as floating windows when more than one file is open to move quickly from drawing to drawing. Minimized drawing windows display as reduced-size title bars. Pick the title bar to display a window control menu.

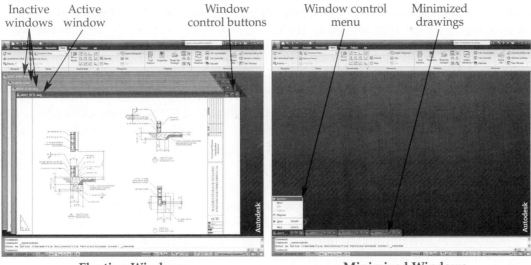

| Inactive windows | Active window | Window control buttons | Window control menu | Minimized drawings |

Floating Windows   **Minimized Windows**

Figure 2-11.
The **Quick View Drawings** tool offers an effective visual method for changing between open drawings.

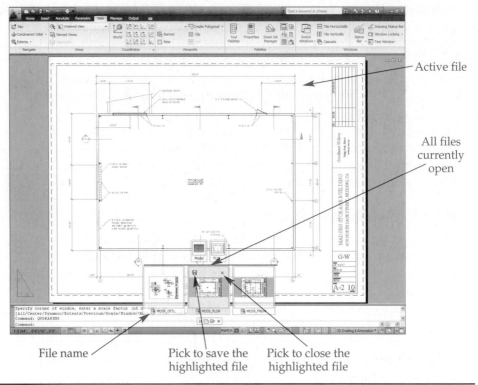

Active file

All files currently open

File name

Pick to save the highlighted file

Pick to close the highlighted file

The **Quick View Drawings** tool appears in the lower center of the AutoCAD window. See **Figure 2-11.** A thumbnail image and file name appear for each open file. Files are arranged in the order in which they were opened, with the file that was opened first on the left side of the row. The current file highlights when you initially access the **Quick View Drawings** tool. Pick the thumbnail of a different file to make the drawing window current, or move the cursor over a thumbnail to show additional options for controlling the drawing window.

The **Quick View Drawings** tool includes a small toolbar below the file thumbnail images. See **Figure 2-12.** By default, the **Quick View Drawings** tool hides when you pick a thumbnail to switch files. To keep the tool on-screen after you select a thumbnail, pick the **Pin Quick View Drawings** button on the left side of the toolbar. The **New...** button activates the **QNEW** tool and the **Open...** button activates the **OPEN** tool. The **New...** and **Open...** buttons are especially useful for starting a new drawing or opening an existing drawing that relates to the current project. Select the **Close Quick View Drawings** button to close the **Quick View Drawings** tool.

> **NOTE**
>
> If you pin the **Quick View Drawings** tool to the screen, close the tool and then access the tool again, the **Quick View Drawings** tool will still be in the pinned state.

You can also quickly display a specific drawing layout without first activating the drawing window. See **Figure 2-13.** Layouts are used to prepare a drawing for plotting and are fully described later in this textbook.

Right-click on a thumbnail image to access a shortcut menu of options for controlling open files. The **Windows** cascading menu provides options for arranging all open

Figure 2-12.
The **Quick View Drawings** toolbar provides access to basic functions directly from the **Quick View Drawings** feature.

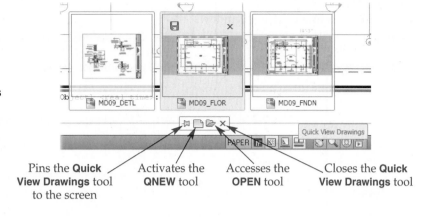

Pins the **Quick View Drawings** tool to the screen

Activates the **QNEW** tool

Accesses the **OPEN** tool

Closes the **Quick View Drawings** tool

Figure 2-13.
A—The initial display when you hover over a file thumbnail. B—Hover over the model space and layout thumbnails to enlarge.

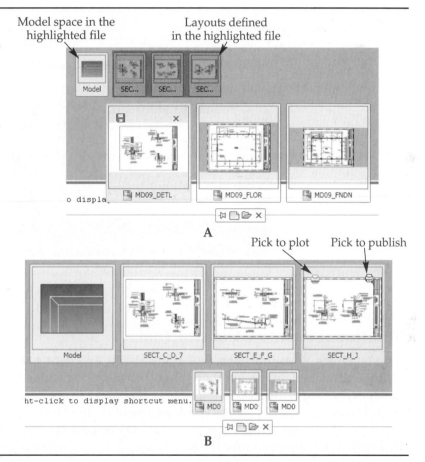

Model space in the highlighted file

Layouts defined in the highlighted file

A

Pick to plot   Pick to publish

B

files. Choose the **Arrange Icons** option to arrange minimized drawings neatly along the bottom of the drawing window area. Select the **Tile Vertically** option to tile unminimized drawing windows in a vertical arrangement, with the active window placed on the left. Pick the **Tile Horizontally** option to tile unminimized drawing windows in a horizontal arrangement, with the active drawing window placed on top. Pick the **Cascade** option to arrange the drawing windows in a cascading style. The effects of tiling the drawing windows vary depending on the number of open windows and whether they tile horizontally or vertically. See **Figure 2-14.**

The shortcut menu that appears when you right-click on a file's thumbnail image includes additional options. Pick **Copy File as a Link** to copy the entire file to the Clipboard for pasting into a drawing or document. Select **Close All** to close all open documents. Pick **Close other files** to close all files except the active file. Choose **Save All** to save all open documents. Select **Close** to close the active file.

Figure 2-14.
Tiled and cascading drawing windows.

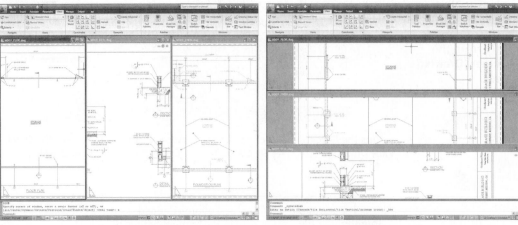

Vertical Tiling   Horizontal Tiling

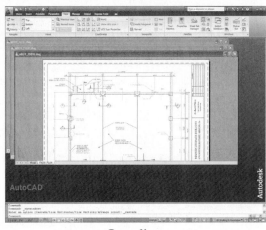

Cascading

Additional Window Control Tools

The **Quick View Drawings** tool provides an excellent method for working with multiple drawings. However, there are other ways to change the active drawing and manage drawing windows. In the ribbon, the **View** tab contains a **Windows** panel with several buttons. Pick the **Switch Windows** flyout to display a list of all open files. Select a file from the flyout to make it active. The **Tile Horizontally**, **Tile Vertically**, **Cascade**, and **Arrange Icons** buttons control the arrangement of open drawings. These are the same features found in the **Quick View Drawings** tool shortcut menu. The **Windows** panel also houses the status bar and window locking functions described in Chapter 1.

Pick the **Open Documents** button in the **Application Menu** to display currently open AutoCAD files. Files list alphanumerically in the order in which they were opened. You can display open files as icons, small images, medium images, or large images by picking the appropriate option from the display options flyout. To activate a different drawing window, pick the file from the list.

PROFESSIONAL TIP

Another technique for switching between open drawings is to press [Ctrl]+[F6]. This is a very effective way to cycle through open drawings quickly.

NOTE

Typically, you can change the active drawing window as desired. However, you cannot activate a different window while a dialog box is open, or in certain other cases. You must either complete the operation or cancel the dialog box before switching is possible.

Exercise 2-4

Access the Student Web site (www.g-wlearning.com/CAD) and complete Exercise 2-4.

Creating and Using Templates

Drawing templates allow you to use an existing drawing as a starting point for a new drawing. Templates are incredible productivity boosters, and help ensure that everyone in a department, class, school, or company uses the same drawing standards. When you use a well-developed template, drawing settings are set automatically or are available each time you begin a new drawing. Templates usually include the following:

✓ Units and angle values
✓ Drawing limits
✓ Grid, snap, and other drawing aid settings
✓ Standard layouts with a border and title block
✓ Text styles
✓ Table styles
✓ Dimension styles
✓ Layer definitions and linetypes
✓ Plot styles
✓ Commonly used symbols and blocks
✓ General notes and other annotations

Reference Material   *Drawing Sheets*

For tables describing *sheet* characteristics, including *sheet size*, drawing scale, and drawing limits, go to the Reference Material section of the Student Web site (www.g-wlearning.com/CAD) and select **Drawing Sheets**. Additional information regarding sheet parameters and selection is described later in this textbook.

sheet: The paper used to lay out and plot drawings.

sheet size: Size of the paper used to lay out and plot drawings.

Creating Templates

If none of the drawing templates supplied with AutoCAD meet your needs, you can create and save your own custom templates. AutoCAD allows you to save *any* drawing as a template. Develop a template whenever several drawing applications require the same setup procedure. The template then allows you to apply the same setup to any number of future drawings. Using templates offers several advantages, including increased productivity by decreasing setup requirements. You can modify and then save an existing AutoCAD template as a new custom template, or you can construct a template from scratch. As you learn more about working with AutoCAD, you will find many settings to include in your templates.

When you have everything needed in the template, the template is ready for a final save and use. Use the **SAVEAS** tool and **Save Drawing As** dialog box to save a template. To save the drawing as a template, pick AutoCAD Drawing Template (*.dwt) from the **Files of type:** drop-down list. By default, the file list box shows the drawing templates currently found in the Template folder. See **Figure 2-15.**

You can store custom templates in any appropriate location, but if you place them in the Template folder, they automatically appear in the **Select template** dialog box by default. After specifying the name and location for the new template file, pick the **Save** button to save the template. The **Template Options** dialog box now appears. See **Figure 2-16.** Type a description of the template file in the **Description** area. A brief description usually works best. In the **Measurement** drop-down list, specify English or Metric units, and then pick the **OK** button.

The template name should relate to the template, such as Mechanical Template for mechanical part and assembly drawings. The template might be named for the drawing application, such as Architectural floor plans, or it might be as simple as Template 1. Write the template name and contents in a reference manual. This provides future reference for you and other users.

> **NOTE**
>
> Pick **AutoCAD Drawing Template** from the **Save As** menu of the **Application Menu** to preset the template file type in the **Save Drawing As** dialog box.

Figure 2-15.
Saving a template in the AutoCAD Template folder.

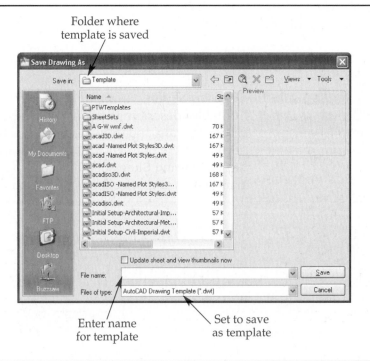

Folder where template is saved

Enter name for template

Set to save as template

Figure 2-16.
Type a description of the new template in the **Template Options** dialog box.

Enter description for template

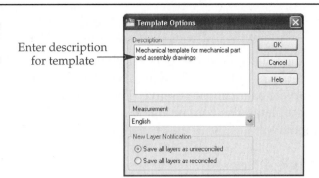

PROFESSIONAL TIP

As you refine your setup procedure, you can open and revise template files. Use the **SAVEAS** tool to save a new template from an existing template.

Template Development

Template Development
Chapter 2

The Template Development section of the Student Web site contains several predefined templates that you can use to create drawings in accordance with correct mechanical, architectural, and civil drafting standards. **Figure 2-17** describes each of the available templates. The mechanical drafting templates are based on the American Society of Mechanical Engineers (ASME) and American National Standards Institute (ANSI) ASME Y series of drafting standards. The architectural and civil drafting templates are based on appropriate architectural and civil drafting standards, including standards specified in the United States National CAD Standard.

In addition to the complete, ready-to-use templates in the Template Development section on the Student Web site, the Template Development sections at the end of several chapters refer you to the Student Web site for important template creation topics and procedures. Use the Template Development feature of this textbook to learn gradually how to prepare templates in accordance with correct mechanical, architectural, and civil drafting standards.

Figure 2-17.
The predefined drawing templates available on the Student Web site.

Template File	Discipline	Units	Layout Sheet Sizes
MECHANICAL-INCH.dwt	Mechanical	Decimal inches	A-size, B-size, C-size, D-size
MECHANICAL-METRIC.dwt	Mechanical	Metric	A4-size, A3-size, A2-size, A1-size
ARCHITECTURAL-US.dwt	Architectural	Feet and inches	Architectural C-size, Architectural D-size
ARCHITECTURAL-METRIC.dwt	Architectural	Metric	Architectural A2-size, Architectural A1-size
CIVIL-US.dwt	Civil	Decimal inches	C-size, D-size
CIVIL-METRIC.dwt	Civil	Metric	A2-size, A1-size

Basic Drawing Settings

Drawing settings determine the general characteristics of a drawing. You can change drawing settings within a drawing, but it is best to adhere to the settings defined in the template as much as possible. The most basic drawing settings include units and limits. Templates also contain many other settings, as explained when applicable in this textbook.

Working in Model Space

model: Any drawing composed of various objects, such as lines, circles, and text, and usually created at full size. However, this term is usually reserved for 3D drawings.

model space: The environment in AutoCAD where drawings and designs are created.

paper (layout) space: The environment in AutoCAD where layouts are created for plotting and display purposes.

layout: An arrangement in paper space of items drawn in model space.

Each drawing or template file includes two environments in which you can work. *Model space* is where you design and draft the *model* of a product. In mechanical drafting, for example, use model space to draw part and assembly views. In architectural drafting, use model space to draw building plans, elevations, and sections.

Once you complete the drawing or model in model space, you switch to *paper (layout) space*, where you prepare a *layout*. A layout represents the sheet of paper used to organize and scale, or lay out, and plot or export a drawing or model. Layouts typically include a border, title block, and general notes. A single drawing can have multiple layouts.

This textbook fully explains paper space when appropriate. For now, all of your drawing and drawing setup should take place in model space. Model space is active by default when you start a drawing from scratch and when you use many of the available templates.

A variety of on-screen characteristics indicate whether you are working in model space or paper space. You know you are in model space when you see the model space coordinate system icon, the active **Model** tab, the **MODEL** button on the status bar, and no representation of a sheet. See **Figure 2-18.** You know you are in a layout when you see the paper space coordinate system icon, an active layout tab, the **PAPER** button on

Figure 2-18.
Model space is the environment in which drawings and designs are created.

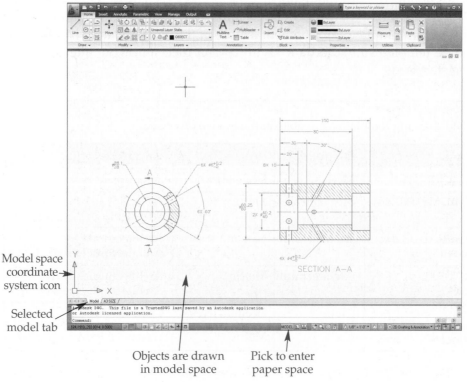

Model space coordinate system icon

Selected model tab

Objects are drawn in model space

Pick to enter paper space

Figure 2-19.
Paper space is the environment in which drawings and designs are laid out on paper for plotting.

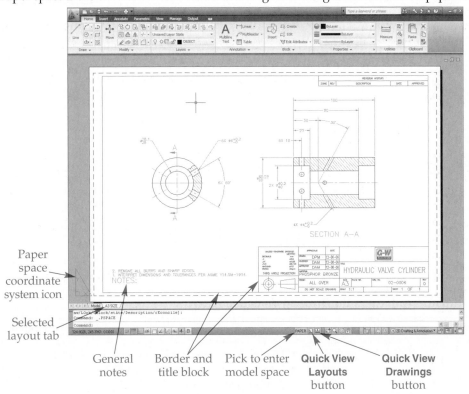

Paper space coordinate system icon

Selected layout tab

General notes

Border and title block

Pick to enter model space

Quick View Layouts button

Quick View Drawings button

the status bar, and a representation of a sheet. See **Figure 2-19.** If you find that you are not in model space, pick the **Model** tab, the **PAPER** button on the status bar, or use **Quick View Layouts** or **Quick View Drawings**. **Quick View Layouts** is fully described later in this textbook.

> ## CAUTION
>
> This textbook assumes that all drawing takes place in model space, until it is appropriate to use a layout. For now, if you find you are not in model space, activate model space in the current drawing and make model space active in all of your custom templates.

Setting Drawing Units

Drawing units define the linear and angular measurements used while drawing and the precision to which these measurements display. In AutoCAD, 1 unit could be 1″, 1 mm, 1 m, or 1 mile. Most AutoCAD users generally think of 1 unit to be either 1″ or 1 mm. Use the **Drawing Units** dialog box to set linear and angular units. See **Figure 2-20.** Specify linear unit characteristics in the **Length** area. Use the **Type:** drop-down list to set the linear units format and use the **Precision:** drop-down list to specify the precision of linear units. **Figure 2-21** describes linear unit formats.

Set the angular unit format and precision in the **Type:** and **Precision:** drop-down lists in the **Angle** area of the **Drawing Units** dialog box. Select the **Clockwise** check box to change the direction for angular measurements to clockwise from the default setting of counterclockwise.

Pick the **Direction...** button to access the **Direction Control** dialog box. See **Figure 2-22.** Pick the **East**, **North**, **West**, or **South** *radio button* to set the compass

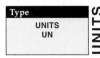

Type
UNITS
UN

UNITS

drawing units: The standard units for linear and angular measurements and the precision of the measurements.

radio button: A selection that activates a single item in a group of options.

Figure 2-20.
Use the **Drawing Units** dialog box to set linear and angular unit values.

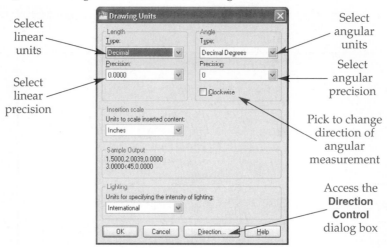

Select linear units

Select linear precision

Select angular units

Select angular precision

Pick to change direction of angular measurement

Access the **Direction Control** dialog box

Figure 2-21.
Linear unit formats available in the **Drawing Units** dialog box.

Type	Typical Applications	Characteristics	Example
Decimal	Mechanical—inch or metric, architectural, structural, civil—metric	• Decimal inches or millimeters. • Conforms to the ASME Y14.5M dimensioning and tolerancing standard. • Four decimal place default precision.	14.1655
Engineering	Civil—feet and inch	• Feet and decimal inches. • Four decimal place default precision.	1'-2.1655"
Architectural	Architectural, structural—feet and inch	• Feet, inches, and fractional inches. • 1/16" default precision.	1'-2 3/16"
Fractional	Mechanical—fractional	• Fractional parts of any common unit of measure. • 1/16" default precision.	14 3/16
Scientific	Chemical engineering, astronomy	• E+01 means the base number is multiplied by 10 to the first power. • Used when very large or small values are required. • Four decimal place default precision.	1.4166E+01

orientation. The **Other** radio button activates the **Angle:** text box and the **Pick an angle** button. Enter an angle for zero direction in the **Angle:** text box. The **Pick an angle** button allows you to pick two points on the screen to establish the angle zero direction.

CAUTION

Use the default direction of 0° East at all times, unless you have a specific need to change the compass direction angle, such as when measuring direction using azimuths (0° North). This textbook uses the 0° East direction, and this default should be set in order to complete most exercises and problems correctly.

AutoCAD and Its Applications—Basics

Figure 2-22.
The **Direction Control** dialog box.

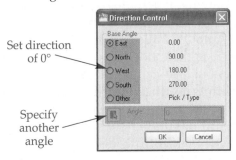

Set direction of 0°

Specify another angle

Figure 2-23.
Angular unit formats available in the **Drawing Units** dialog box.

Type	Applications and Characteristics	Example
Decimal Degrees	• Mechanical, architectural, and structural drafting applications. • Degrees and decimal parts of a degree. • Initial default setting.	45°
Deg/Min/Sec	• Civil and sometimes mechanical, architectural, and structural drafting applications. • Degrees, minutes, and seconds. • 1 degree = 60 minutes; 1 minute = 60 seconds.	45°0′0″
Grads	• Grad is the abbreviation for *gradient*. • One-quarter of a circle has 100 grads; a full circle has 400 grads.	50.000g
Radians	• A radian is an angular unit of measurement in which 2π radians = 360° and π radians = 180°. Pi (π) is approximately equal to 3.1416. • A 90° angle has $\pi/2$ radians and an arc length of $\pi/2$. • Changing the precision displays the radian value rounded to the specified decimal place.	0.785r
Surveyor	• Civil drafting applications. • Degrees, minutes, and seconds. • Uses bearings. A bearing is the direction of a line with respect to one of the quadrants of a compass. Bearings are measured clockwise or counterclockwise (depending on the quadrant), beginning from either north or south. • An angle measuring 55°45′22″ from north toward west is expressed as N55°45′22″W. • Set precision to degrees, degrees/minutes, degrees/minutes/seconds, or decimal display accuracy of the seconds part of the measurement.	N45°E

Figure 2-23 describes angular unit formats. After selecting the linear and angular units and precision, pick the **OK** button to exit the **Drawing Units** dialog box.

Exercise 2-5

Access the Student Web site (www.g-wlearning.com/CAD) and complete Exercise 2-5.

Adjusting Drawing Limits

You prepare an AutoCAD drawing at actual size, or full scale, regardless of the type of drawing, the units used, or the size of the final layout on paper. Use model space to draw full-scale objects. AutoCAD allows you to specify the size of a virtual model space drawing area, known as the model space drawing limits, or *limits*. You typically set limits in a template, but you can change them at any time during the drawing process. The concept of limits is somewhat misleading, because the AutoCAD drawing area is infinite in size. For example, if you set limits to 17″ × 11″, you can still create objects that extend past the 17″ × 11″ area, such as a line that is 1200′ long. As a result, you can choose not to consider limits while developing a template or creating a drawing. Conversely, as you learn AutoCAD, you may decide that setting appropriate drawing limits is helpful, especially when you are drawing large objects. Regardless of whether you choose to acknowledge limits, you should be familiar with the concept, and recognize that some AutoCAD tools, such as the **ZOOM** and **PLOT** tools discussed in later chapters, provide options for using limits.

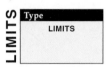

LIMITS

Type	
	LIMITS

Use the **LIMITS** tool to set model space drawing limits. The first prompt asks you to specify the coordinates for the lower-left corner of the drawing limits. For now, when setting limits, the lower-left corner is 0,0. Press [Enter] to accept the default 0,0 value or enter a new value and press [Enter]. The next prompt asks you to specify the coordinates for the upper-right corner of the virtual drawing area. For example, type 17,11 and then press [Enter] to set limits of 17″ x 11″. The first value is the horizontal measurement, and the second value is the vertical measurement of the limits. A comma separates the values.

In general, you should set limits larger than the objects you plan to draw. You can determine limits accurately by identifying the drawing scale, converting the scale to a scale factor, and then multiplying the scale factor by the size of sheet on which you plan to plot the drawing. For now, calculate the approximate total length and width of all the objects you plan to draw, adding extra space for dimensions and notes. For example, if you are drawing a 48′ × 24′ building floor plan, allow 10′ on each side for dimensions and notes to make a total virtual drawing area of 68′ × 44′.

> **NOTE**
>
> The **LIMITS** tool provides a limits-checking feature that, when turned on, restricts your ability to draw outside of the drawing limits. Turn on limits checking by entering or selecting the **ON** option of the **LIMITS** tool. Turn off limits checking using the **OFF** option.

Exercise 2-6

Access the Student Web site (www.g-wlearning.com/CAD) and complete Exercise 2-6.

Template Development

Chapter 2

For detailed instructions to begin the development of drawing templates, go to the Student Web site (www.g-wlearning.com/CAD), select this chapter, and select **Template Development**.

Chapter Test

Answer the following questions. Write your answers on a separate sheet of paper or go to the Student Web site (www.g-wlearning.com/CAD) and complete the electronic chapter test.

1. What is a drawing template?
2. What is the name of the dialog box that opens by default when you pick the **New** button on the **Quick Access** toolbar?
3. Briefly explain how to start a drawing from scratch.
4. How often should work be saved?
5. Explain the benefits of using a standard system for naming drawing files.
6. Name the tool that allows you to save your work quickly without displaying a dialog box.
7. What tool allows you to save a drawing file in an older AutoCAD format?
8. How do you set AutoCAD to save your work automatically at designated intervals?
9. Identify the tool you would use if you wanted to exit a drawing file, but remain in the AutoCAD session.
10. What is the quickest way to close an AutoCAD drawing file?
11. How can you close all open drawing windows at the same time?
12. Which **Application Menu** function allows you to select the name of a recently opened drawing file and open it?
13. How can you set the number of recently opened files listed in the **Application Menu**?
14. What does the term *read-only* mean?
15. Describe the advantages of using the **Quick View Drawings** tool to work with multiple open drawings.
16. How do you keep the **Quick View Drawings** tool on-screen after you pick a thumbnail image?
17. How do you quickly cycle through all the currently open drawings in sequence?
18. What is sheet size?
19. How can you convert a drawing file into a drawing template?
20. Name three settings you can specify in the **Drawing Units** dialog box.

Drawing Problems

Start AutoCAD if it is not already started. Follow the specific instructions for each problem.

▼ Basic

1. Start a new drawing using the **acad-Named Plot Styles** template supplied by AutoCAD. Save the new drawing as a file named P2-1.dwg.

2. Start a new drawing using the **acadiso** template supplied by AutoCAD. Save the new drawing as a file named P2-2.dwg.

3. Start a new drawing using the **Tutorial-iArch** template supplied by AutoCAD. Save the new drawing as a file named P2-3.dwg.

▼ Intermediate

*Problems 4 through 7 can be done if the **AutoCAD 2010\Sample** file folder is loaded. All drawings listed are found in that folder.*

4. Locate and preview the Lineweights drawing. Open the drawing and describe it in your own words.

5. Locate and preview the TrueType drawing. Open the drawing and describe it in your own words.

6. Locate and preview the Tablet drawing. Open the drawing and describe it in your own words.

7. Locate and preview the 3D House drawing. Open the drawing and describe it in your own words.

▼ Advanced

The following problems can be saved as templates for future use.

8. Create a template with decimal units with 0.0 precision, decimal degrees with 0.0 precision, default angle measure and orientation, and a limits setting of 0,0 × 17,11. Save the template as P2-8.dwt. Enter an appropriate description for the template.

9. Create a template with metric units with 0.0 precision, decimal degrees with 0.0 precision, default angle measure and orientation, and a limits of 0,0 × 22,17. Save the template as P2-9.dwt. Enter an appropriate description for the template.

10. Create a template with architectural units with 0'-0" precision, decimal degrees with 0 precision, and a limits of 0,0 × 1632,1056. Save the template as P2-10.dwt. Enter an appropriate description for the template.

Introduction to Drawing and Editing

Learning Objectives

After completing this chapter, you will be able to do the following:

✓ Use appropriate values when responding to prompts.
✓ Describe the Cartesian coordinate system.
✓ Determine and specify drawing snap and grid.
✓ Draw given objects using the **LINE** tool.
✓ Describe and use several point entry methods.
✓ Demonstrate an ability to use dynamic input and the command line.
✓ Use direct distance entry with polar tracking and **Ortho** mode.
✓ Revise objects using the **ERASE** tool.
✓ Create selection sets using various selection options.
✓ Use the **UNDO**, **U**, **REDO**, and **OOPS** tools appropriately.

This chapter introduces a variety of fundamental drawing and editing concepts and processes, including point entry and object selection methods. You will learn to pick points and draw objects using the **LINE** tool. You will learn to make selections and changes using the **ERASE** tool. You can learn much about AutoCAD using the basic tools and operations described in this chapter.

Responding to Prompts

Most drawing and editing tools require that you respond to a prompt. A prompt "asks" you to perform a specific task. For example, when you are drawing a line, a prompt asks you to specify line endpoints. When you erase an object, a prompt asks you to select objects to erase. Many prompts provide options that you can select instead of responding to the immediate request. For example, after picking the first line endpoint, you are promoted to select the next point, or you can choose the **Undo** option to remove the previous selection instead of picking another point.

Responding with Numbers

Many tools require you to enter specific numeric data, such as the endpoint location of a line, or the radius of a circle. Some prompts require you to enter a whole number. Other entries require whole numbers that are positive or negative. AutoCAD understands that a number is positive without the plus sign (+) in front of the value. However, you must add the minus sign (–) in front of a negative number.

Much of the data you enter may not be whole numbers. In these cases, you can use any real number expressed as decimal, as a fraction, or in scientific notation using positive or negative values. Examples of acceptable real number entries include:

    4.250
    –6.375
    1/2
    1-3/4
    2.5E+4 *(25,000)*
    2.5E–4 *(0.00025)*

For fractions, the numerator and denominator must be whole numbers greater than zero. For example, 1/2, 3/4, and 2/3 are all acceptable fraction entries. Fractional numbers greater than one must have a hyphen between the whole number and the fraction. For example, type 2-3/4 for two and three-quarters. The hyphen (-) separator is needed because a space acts just like pressing [Enter] and automatically ends the input. The numerator can be larger than the denominator, as in 3/2, but *only* if you do not include a whole number with the fraction. For example, 1-3/2 is not a valid input for a fraction.

When you enter coordinates or measurements, the values used depend on the units of measurement. For decimal or fractional length units, AutoCAD interprets an entry of 2.500 or 2-1/2 to be 2.500 or 2-1/2 *units*, not a specific unit of measure. You are responsible for applying the appropriate units. Most AutoCAD users generally think of 1 unit to be either 1" or 1 mm.

When using decimal or fractional units, you do not need to add a suffix after the numeral. Inch marks (") are unnecessary and time-consuming to add, and adding mm for millimeters, for example, is not recognized.

AutoCAD assumes inch measurements when you use architectural and engineering length units. In this case, any value greater than 1" expresses in inches, feet, or feet and inches. The values can be whole numbers, decimals, or fractions. For measurements in feet, the foot symbol (') must follow the number, as in 24'. If a value is in feet and inches, there is no space and it is unnecessary to add a hyphen (-) between the feet and inch value. For example, 24'6 is the proper input for the value 24'-6". If the inch part of the value contains a fraction, the inch and fractional part of an inch are separated by a hyphen, such as 24'6-1/2. Never mix feet with inch values greater than one foot. For example, 24'18 is an invalid entry. In this case, you should type 25'6.

NOTE

AutoCAD accepts the inch (") and foot (') symbols only when you specify architectural or engineering drawing units.

Ending and Canceling Tools

Some AutoCAD tools, such as the **LINE** tool, remain active until stopped. For example, you can continue to pick points to create new line segments until you end the **LINE** tool. You can usually end a tool by pressing [Enter] or the space bar, or by right-clicking and selecting **Enter**. Press [Esc] to cancel an active tool or abort data entry. It may be necessary to press [Esc] twice to cancel certain tools completely.

If you press the wrong key or misspell a word while using a tool or answering a prompt, use [Backspace] to correct the error. This works only if you notice your mistake *before* you accept the value. If you enter an inappropriate value or option, AutoCAD usually responds with an error message. Access the **AutoCAD Text Window** to view lengthy error messages, or review entries. Return to the graphics screen using the same method you used to access the text window, or pick any visible portion of the graphics screen.

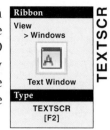
TEXTSCR

Ribbon
View
> Windows
Text Window
Type
TEXTSCR
[F2]

PROFESSIONAL TIP

You can cancel the active tool and access a new tool at the same time by picking a ribbon button or an **Application Menu** option.

Introduction to Drawing

You create most geometric shapes using drawing tools, such as the **LINE** tool. The most common method of locating and sizing objects while using drawing tools is through *point entry*. Specifying the two endpoints of a line segment is an example of point entry. To enter points, you use the *Cartesian (rectangular) coordinate system*. XYZ coordinate values describe the locations of points. These values, called *rectangular coordinates*, locate any point in 3D space.

In 2D drafting, the *origin* divides the coordinate system into four quadrants on the XY plane. The most basic point entry relates to the origin, where X = 0 and Y = 0, or 0,0. See **Figure 3-1.** The origin is usually at the lower-left corner of the drawing. This setup places all points in the upper-right quadrant of the XY plane, where both X and Y coordinate values are positive. See **Figure 3-2.** To locate a point in 3D space, a third dimension rises up from the surface of the XY plane and is given a Z value.

To describe a coordinate location, the X value is listed first, the Y value is second, and the Z value is third. A comma separates each value. For example, the coordinate location of 3,1,6 represents a point that is three units from the origin in the X direction, one unit from the origin in the Y direction, and six units from the origin in the Z direction. *AutoCAD and Its Applications—Advanced* describes how to use the Z axis to construct 3D models.

point entry: Identifying a point location in the AutoCAD coordinate system.

Cartesian (rectangular) coordinate system: A system in which points are located in space according to distances from three intersecting axes.

rectangular coordinates: A set of numerical values that identify the location of a point on the X, Y, and Z axes of the Cartesian coordinate system.

origin: The intersection point of the X, Y, and Z axes.

Figure 3-1.
The 2D Cartesian coordinate system consists of X and Y axes. The origin is located at the intersection of the axes.

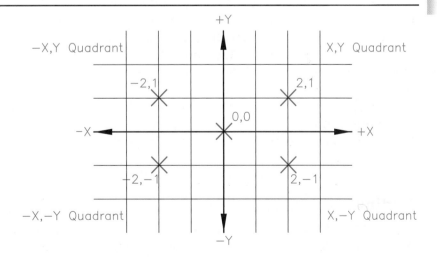

Figure 3-2.
By default, the upper-right quadrant of the Cartesian coordinate system fills the screen.

Increasing Y coordinate value

Increasing X coordinate value

Default origin

Exercise 3-1

Access the Student Web site (www.g-wlearning.com/CAD) and complete Exercise 3-1.

Introduction to Drawing Aids

AutoCAD includes many drawing aids that increase accuracy and productivity. The coordinate display, grid, and grid snap modes are examples of basic drawing aids. You will learn other useful drawing aids in later chapters. As an AutoCAD user, you will learn to apply whichever tools and options work best and quickest for a specific drawing task.

Using Grid Mode

GRID

Type
GRID
[Ctrl]+[G]
F7

DSETTINGS

Type
DSETTINGS
DS
SE

grid: A pattern of dots that appears on-screen to aid in the drawing process.

Turn on **Grid** mode to display a *grid* on the screen. See **Figure 3-3.** The grid is for reference only, for use as a visual aid to drawing layout. The quickest way to toggle **Grid** mode on and off is to pick the **Grid Display** button on the status bar.

Use the options on the **Snap and Grid** tab of the **Drafting Settings** dialog box to adjust the spacing between grid dots. See **Figure 3-4.** A quick way to access the **Drafting Settings** dialog box is to right-click on any of the status bar toggle buttons and select **Settings...**. Use the **Grid On** check box to turn the grid on or off. Grid spacing is set in the **Grid spacing** area. Type the desired values in the **Grid X spacing:** and **Grid Y spacing:** text boxes. Factors to consider when setting grid spacing are described later in this chapter.

The options in the **Grid behavior** area allow you to set how the grid appears on-screen. When the **Adaptive Grid** check box is selected, and the grid spacing is too dense, AutoCAD adjusts the display automatically so the grid can be shown on-screen. The **Display grid beyond Limits** option determines whether the grid appears only within the drawing limits. The **Allow subdivision below grid spacing** and **Follow Dynamic UCS** options apply to 3D applications.

Figure 3-3.
Dots represent the grid spacing when **Grid** mode is active.

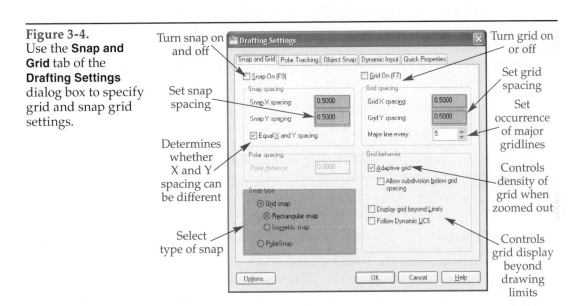

Grid pattern

Pick to toggle **Grid Display**

Figure 3-4.
Use the **Snap and Grid** tab of the **Drafting Settings** dialog box to specify grid and snap grid settings.

Turn snap on and off

Set snap spacing

Determines whether X and Y spacing can be different

Select type of snap

Turn grid on or off

Set grid spacing

Set occurrence of major gridlines

Controls density of grid when zoomed out

Controls grid display beyond drawing limits

Using Snap Mode

Turn on **Snap** mode to activate the *snap grid*, also known as *snap resolution* or *snap*. One of the quickest ways to toggle snap on and off is to pick the **Snap Mode** button on the status bar. By default, snap is off, and when you move the mouse, the crosshairs moves freely on the screen. Turn snap on to move the crosshairs in specific increments. Using the snap grid is different from using the on-screen grid. Snap controls the movement of the crosshairs, while the grid is only a visual guide. The grid and snap settings can, however, be used together.

The **Snap and Grid** tab of the **Drafting Settings** dialog box includes options for setting snaps. Refer again to **Figure 3-4.** Use the **Snap On** check box to turn snaps on or off. Snap increment is set in the **Snap spacing** area. Type the desired values in the **Snap X spacing:** and **Snap Y spacing:** text boxes.

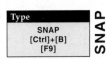

Type

SNAP
[Ctrl]+[B]
[F9]

SNAP

snap grid (snap resolution, snap): Invisible grid that allows the crosshairs to move only in exact increments.

Use the **Snap type** area to control how snaps function. The default snap type is **Grid snap** with the **Rectangular snap** style. This is the snap method previously described. Select the **Grid snap** type and **Isometric snap** style to aid in creating isometric drawings, explained in Chapter 24. The **PolarSnap** type allows you to snap to precise distances along alignment paths when you use polar tracking. Chapter 7 covers polar tracking.

Grid and Snap Considerations

You should consider several factors when setting and using **Grid** and **Snap** modes. For decimal units, set the grid and snap values to standard decimal increments such as .0625, .125, or .5 for an inch drawing, or 1, 10, or 20 for a metric drawing. For architectural units, use standard increments such as 1, 6, and 12 (for inches) or 1, 2, 4, 5, and 10 (for feet). A very large drawing might have a grid spacing of 12 (one foot), or 120 (ten feet), while a small drawing may use a spacing of .125 or less.

The **Grid** and **Snap** modes can be set at different values to complement each other. For example, the grid may be set at .5, and the snap may be set at .25. With these settings, each mode plays a separate role in assisting drawing layout. If the smallest dimension is .125, for example, then an appropriate snap value is .125 with a grid spacing of .25. Often the most effective use of grid and snap is to set equal X and Y spacing. However, if many horizontal features conform to one increment and most vertical features correspond to another, then you may choose to set different X and Y values.

You can change the snap and grid values at any time without changing the location of points or lines already drawn. You should do this when larger or smaller values would assist you with a certain part of the drawing. For example, suppose a few of the dimensions are multiples of .0625, but the rest of the dimensions are multiples of .250. Change the snap spacing from .250 to .0625 when laying out the smaller dimensions.

Exercise 3-2

Access the Student Web site (www.g-wlearning.com/CAD) and complete Exercise 3-2.

Drawing Lines

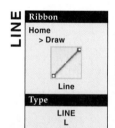

LINE

Ribbon
Home
> Draw

Line

Type
LINE
L

To draw a line, access the **LINE** tool and select a start point, which is the first line endpoint. As you move the mouse, a *rubberband line* appears connecting the first point and the crosshairs. Continue selecting additional points to connect a series of lines. Then press [Enter] or the space bar, or right-click and select **Enter** to end the **LINE** tool.

Undoing the Previously Drawn Line

If you make an error while still using the **LINE** tool, right-click and select **Undo**, pick the **Undo** dynamic input option, or type U and press [Enter]. This removes the previously drawn line and allows you to continue from the previous endpoint. You can use the **Undo** option repeatedly to delete line segments until the entire line is gone. See **Figure 3-5**.

Using the Close Option

To aid in drawing a *polygon* using the **LINE** tool, after you draw two or more line segments, use the **Close** option to connect the endpoint of the last line segment to the start point of the first line segment. To use this option, right-click and select **Close**, pick the **Close** dynamic input option, or type C or CLOSE and press [Enter]. In **Figure 3-6**, the last line is drawn using the **Close** option.

rubberband line: A stretch line that extends from the crosshairs with certain drawing tools to show where an object will be drawn.

polygon: Closed plane figure with at least three sides. Triangles and rectangles are examples of polygons.

Figure 3-5.
*Using the **Undo** option of the active **LINE** tool. Dashed lines are used here to represent the undone lines.*

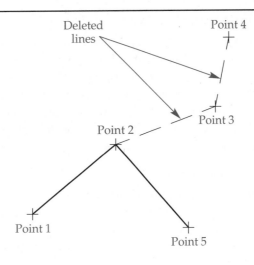

Figure 3-6.
*Using the **Close** option to complete a box.*

Exercise 3-3

Access the Student Web site (www.g-wlearning.com/CAD) and complete Exercise 3-3.

Point Entry Methods

The most basic method of point entry is to pick a point using the crosshairs. Picking a random point in space is typically not accurate. One method of creating an accurate drawing is to use drawing aids. For example, with **Snap** mode on, you can select a specific point corresponding to the snap increment. Another option is to use a point entry method that locates an exact point on the rectangular coordinate system. This typically requires specialized coordinate input.

Exercise 3-4

Access the Student Web site (www.g-wlearning.com/CAD) and complete Exercise 3-4.

Using Absolute Coordinates

Points located using *absolute coordinates* are measured from the origin (0,0). For example, a point located two units horizontally (X = 2) and two units vertically (Y = 2) from the origin is an absolute coordinate of 2,2. A comma separates the values. Drawing a line starting at 2,2 and ending at 4,4 creates the line shown in **Figure 3-7.** The first point you pick when drawing a line is often positioned using absolute coordinates. Remember, when using the absolute coordinate system, you locate each point from 0,0. If you enter negative X and Y values, the selection occurs outside of the upper-right XY plane quadrant.

Using Relative Coordinates

When using *relative coordinates*, you may want to think of the previous point as the "temporary origin." Use the @ symbol to enter relative coordinates. For example, if the first point of a line is located at 2,2 and the second point is positioned using a relative @2,2 coordinate entry, the second point is located 4,4 from the origin. Refer again to **Figure 3-7.**

Using Polar Coordinates

When using *polar coordinate* entry, you specify the length of the line followed by the angle at which the line is drawn. A less than (<) symbol separates the distance and angle. **Figure 3-8** shows the default angular values used for polar coordinate entry. By default, 0° is to the right, or east, and angles measure counterclockwise. When preceded by the @ symbol, a polar coordinate point locates relative to the previous point. If the @ symbol is not included, the coordinate locates relative to the origin. For example, to draw a line 2 units long at a 45° angle, starting 2 units from 0,0 at a 45° angle, type 2<45 for the first point, and @2<45 for the second point. See **Figure 3-9.**

Figure 3-7.
Locating points using absolute and relative coordinates.

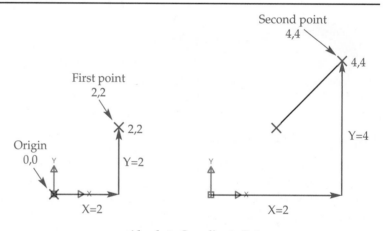

Absolute Coordinate Entry

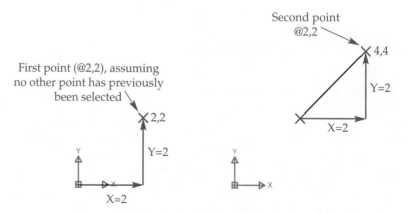

Relative Coordinate Entry

AutoCAD and Its Applications—Basics

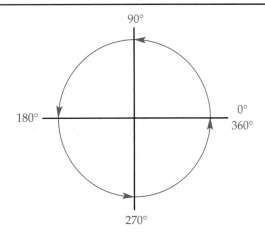

Figure 3-8.
Angles used when entering polar coordinates.

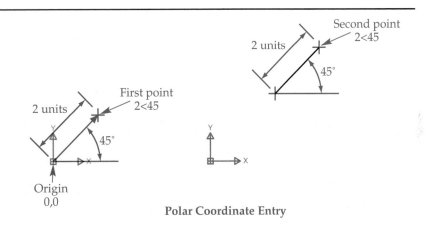

Figure 3-9.
Locating points using polar coordinates.

Polar Coordinate Entry

The Coordinate Display

The area on the left side of the status bar shows the coordinate display field. The drawing units setting determines the format and precision of the display. Pick the coordinate display in the status bar to toggle the display on and off. When coordinate display is on, the coordinates constantly change as the crosshairs moves. When coordinate display is off, the coordinates are "grayed out," but still update to identify the location of the last point selected.

When a tool such as **LINE** is active, you can choose the coordinate display mode. When the **Relative** mode is set and a tool is active, the coordinates of the crosshairs position display as polar coordinates relative to the previously picked point. The coordinates update each time you pick a new point. When you select the **Absolute** mode and a tool is active, the coordinate of the crosshairs location is set relative to the origin. The **Geographic** mode is available if you specify the geographic drawing location.

Using Dynamic Input

Dynamic input is on by default and is one of the most effective tools for entering coordinates. Dynamic input provides the same function as the command line, but it allows you to keep your focus at the point where you are drawing. Also, additional coordinate entry techniques are available with dynamic input. Common point entry methods function differently when dynamic input is active, depending on settings.

When you start the **LINE** tool, dynamic input prompts you to specify the first point. The X coordinate input field is active, and the Y coordinate input field appears. See **Figure 3-10.** The X and Y coordinates of the first point can be typed using *pointer input* with absolute, relative, or polar coordinate entry.

pointer input: The process of entering points using dynamic input.

Figure 3-10.
After you start the **LINE** tool, dynamic input displays these items. When you type the @ symbol to use relative coordinates, the symbol displays in the tooltip. When you use polar coordinates, the less than symbol (<) displays in the tooltip before the angle.

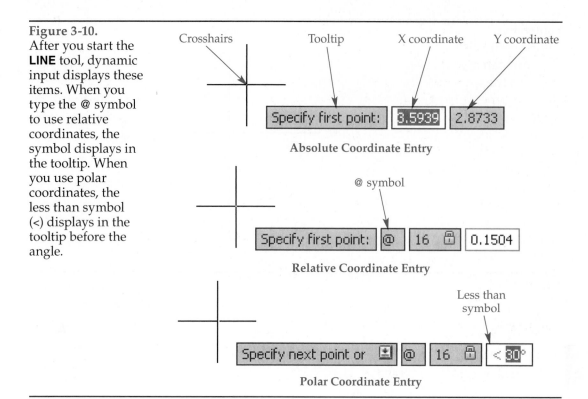

Crosshairs　　Tooltip　　X coordinate　　Y coordinate

Specify first point: 3.5939 2.8733

Absolute Coordinate Entry

@ symbol

Specify first point: @ 16 🔒 0.1504

Relative Coordinate Entry

Less than symbol

Specify next point or ⊞ @ 16 🔒 < 30°

Polar Coordinate Entry

Absolute coordinate entry is the default when you select the first point. Type the X value, and then press [Tab] or enter a comma to lock in the X value and move to the Y coordinate input field. Now type the Y value and right-click or press [Enter] to select the point. Polar coordinate entry is the other likely option for specifying the first point. Type the less than symbol (<) after entering the length of the line in the X coordinate input field. Then enter the angle of the line in the Y coordinate input field. Dynamic input fields automatically change to anticipate the next entry.

Once you enter the start point of the line, dynamic input provides a *dimensional input* feature that allows you to enter the length of a line and the angle at which the line is drawn, similar to polar coordinate entry. To use dimensional input, you must first establish the start point. Then, by default, distance and angle input fields appear. See **Figure 3-11.** Enter the length of the line in the active distance input field and press [Tab] to lock in the distance and move to the angle input field. Type the angle of the line and right-click or press [Enter] to select the point. The angular values used for

dimensional input: A method of entering points that is similar to polar coordinate entry, but uses dynamic input.

Figure 3-11.
Use dimensional input to define the length and angle of a line.

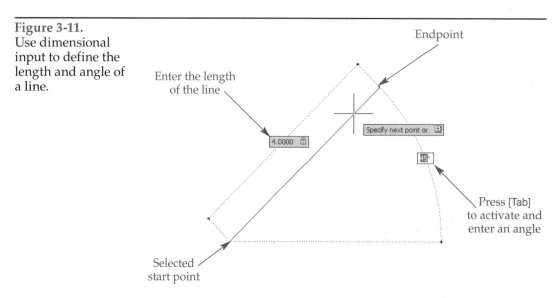

Endpoint

Enter the length of the line

Specify next point or ⊞

4.0000 🔒

15°

Press [Tab] to activate and enter an angle

Selected start point

dimensional input are the same as those for polar coordinate entry. Refer again to Figure 3-8.

You can also use pointer input to pick additional points once you select the start point of the line. Relative coordinates are active by default, which means you do not need to type @ before entering the X and Y values. To select the second point using a relative coordinate entry, type the X value in the distance input field, then type a comma to lock in the X value and move to the Y coordinate input field. Type the Y value and right-click or press [Enter] to select the point.

Dimensional input is on by default, but you can temporarily turn it off by typing the # symbol before entering values. This makes dynamic input default to polar format, which means you can enter the length of the line in the active X coordinate input field, and then press [Tab] to lock in the length and move to the Y coordinate input field. Enter the angle of the line and right-click or press [Enter] to select the point. In order to use an absolute coordinate entry with the default settings, type the # symbol, enter the X coordinate in the active field, type a comma, type the Y value, and right-click or [Enter] to select the point.

PROFESSIONAL TIP

Use [Tab] to cycle through dynamic input fields. You can make changes to values before accepting the coordinates.

Supplemental Material *Dynamic Input Settings*
For more information about adjusting dynamic input options, go to the Student Web site (www.g-wlearning.com/CAD), select this chapter, and select **Dynamic Input Settings**.

Using the Command Line

Dynamic input is a very effective tool for locating points because of its on-screen display and ease of use. You can also use the command line for point entry, but you must closely adhere to point entry methods. Neither dimensional input nor quick input settings are available with the command line. You can use dynamic input at the same time as the command line, or you can disable dynamic input to use only the command line for tool input and information. Another option is to hide the command line to free additional drawing space and focus on using dynamic input.

Absolute, relative, and polar point entry methods accomplish the same tasks whether they are entered using dynamic input or typed at the command line. The following content provides examples of point entry using the command line. You can apply the same examples to dynamic input. Even if you choose not to use the command line, review these examples to help better understand point entry techniques. You must disable dynamic input in order for these exact command sequences to work properly.

Figure 3-12.
Drawing a shape using the **LINE** tool and absolute coordinates.

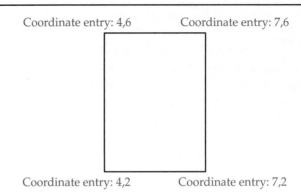

Coordinate entry: 4,6   Coordinate entry: 7,6

Coordinate entry: 4,2   Coordinate entry: 7,2

Absolute Coordinate Entry

Follow these commands and absolute coordinate entries at the command line as you refer to **Figure 3-12.**

> Command: **L** *or* **LINE**↵
> Specify first point: **4,2**↵
> Specify next point or [Undo]: **7,2**↵
> Specify next point or [Undo]: **7,6**↵
> Specify next point or [Close/Undo]: **4,6**↵
> Specify next point or [Close/Undo]: **4,2**↵
> Specify next point or [Close/Undo]: ↵
> Command:

Exercise 3-5

Access the Student Web site (www.g-wlearning.com/CAD) and complete Exercise 3-5.

Relative Coordinate Entry

Follow these commands and relative coordinate entry methods as you refer to **Figure 3-13.**

> Command: **L** *or* **LINE**↵
> Specify first point: **2,2**↵
> Specify next point or [Undo]: **@6,0**↵
> Specify next point or [Undo]: **@2,2**↵
> Specify next point or [Close/Undo]: **@0,3**↵
> Specify next point or [Close/Undo]: **@-2,2**↵
> Specify next point or [Close/Undo]: **@-6,0**↵
> Specify next point or [Close/Undo]: **@0,-7**↵
> Specify next point or [Close/Undo]: ↵
> Command:

Exercise 3-6

Access the Student Web site (www.g-wlearning.com/CAD) and complete Exercise 3-6.

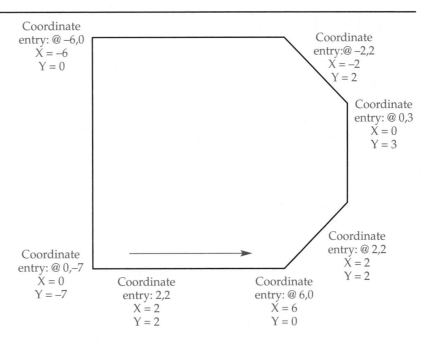

Figure 3-13.
Drawing a shape using the **LINE** tool and relative coordinates. Notice that negative (–) values are used and the coordinates are entered counterclockwise from the first point, in this case, 2,2.

Coordinate entry: @ –6,0
X = –6
Y = 0

Coordinate entry:@ –2,2
X = –2
Y = 2

Coordinate entry: @ 0,3
X = 0
Y = 3

Coordinate entry: @ 2,2
X = 2
Y = 2

Coordinate entry: @ 0,–7
X = 0
Y = –7

Coordinate entry: 2,2
X = 2
Y = 2

Coordinate entry: @ 6,0
X = 6
Y = 0

Polar Coordinate Entry

Follow these commands and polar coordinate entry methods as you refer to Figure 3-14.

```
Command: L or LINE↵
Specify first point: 2,6↵
Specify next point or [Undo]: @2.5<0↵
Specify next point or [Undo]: @3<135↵
Specify next point or [Close/Undo]: 2,6↵
Specify next point or [Close/Undo]: ↵
Command: ↵
LINE Specify first point: 6,6↵
Specify next point or [Undo]: @4<0↵
Specify next point or [Undo]: @2<90↵
Specify next point or [Close/Undo]: @4<180↵
Specify next point or [Close/Undo]: @2<270↵
Specify next point or [Close/Undo]: ↵
Command:
```

Exercise 3-7

Access the Student Web site (www.g-wlearning.com/CAD) and complete Exercise 3-7.

Figure 3-14.
Using polar coordinates to draw a shape.

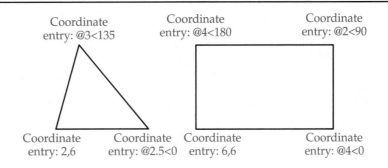

Coordinate entry: @3<135

Coordinate entry: @4<180

Coordinate entry: @2<90

Coordinate entry: 2,6

Coordinate entry: @2.5<0

Coordinate entry: 6,6

Coordinate entry: @4<0

Using Direct Distance Entry

To draw a line using *direct distance entry*, drag the crosshairs in any direction from the first point of the line. Then type a numerical value indicating the distance from that point. Direct distance entry is a very quick way to draw lines at a specific length. However, direct distance entry itself is not very useful unless you incorporate other drawing tools. Polar tracking and **Ortho** mode can be used to draw lines at accurate angles using direct distance entry.

Introduction to Polar Tracking

Polar tracking is on by default and causes the drawing crosshairs to "snap" to predefined angle increments. Pick the **Polar Tracking** button on the status bar or press [F10] to toggle polar tracking on and off. When polar tracking is on, as you move the crosshairs toward a polar tracking angle, AutoCAD displays an alignment path and tooltip. The default polar angle increments are 0°, 90°, 180°, or 270°. Polar tracking is an AutoTrack mode. Chapter 7 fully explains AutoTrack.

To use polar tracking in combination with direct distance entry, first access the **LINE** tool and specify a start point. Then move the crosshairs in alignment with a polar tracking angle. Type the length of the line and press [Enter] or right-click and select **Enter**. See **Figure 3-15.** Chapter 7 fully explains polar tracking.

Drawing in Ortho Mode

Ortho mode constrains points selected while drawing and editing to be only horizontal or vertical from the previous point entry. See **Figure 3-16.** Pick the **Ortho Mode** button on the status bar to toggle **Ortho** mode on and off. If **Ortho** mode is turned off, you can temporarily turn it on when drawing an object by holding down [Shift].

To use **Ortho** in combination with direct distance entry, access the **LINE** tool and specify a start point. Move the crosshairs to display a horizontal or vertical rubberband in the direction you want to draw. Then type the length of the line and press [Enter] or right-click and select **Enter**. See **Figure 3-17.**

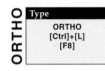

ORTHO	Type
	ORTHO
	[Ctrl]+[L]
	[F8]

Figure 3-15.
Using polar tracking to draw lines at predefined angle increments.

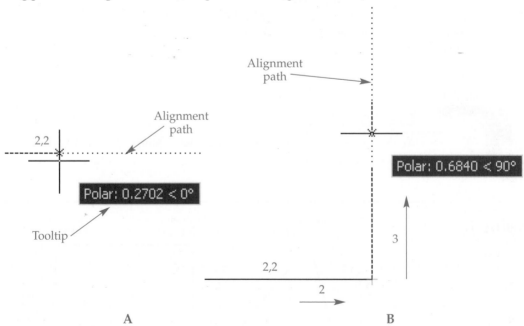

Figure 3-16.
You can only draw horizontal and vertical lines when **Ortho** mode is on. Notice the location of the crosshairs when **Ortho** mode is on and off.

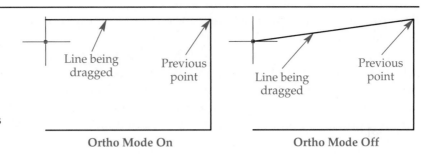

Ortho Mode On   Ortho Mode Off

Figure 3-17.
Using direct distance entry to draw lines a designated distance from a current point. With **Ortho** mode on, move the cursor in the desired direction and type the distance.

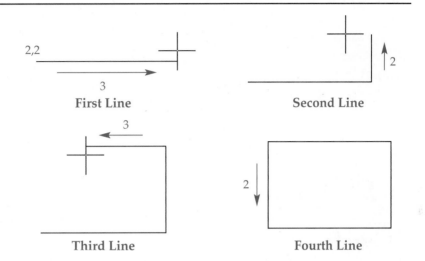

First Line   Second Line

Third Line   Fourth Line

Practice using different point entry techniques and decide which method works best for certain situations. Keep in mind that you can mix methods to help enhance your drawing speed. For example, absolute coordinates may work best to locate an initial point. Polar coordinates may work better to locate features in a circular pattern or in an angular relationship. Practice with direct distance entry using polar tracking and **Ortho** mode to see the advantages and disadvantages of each.

Exercise 3-8

Access the Student Web site (www.g-wlearning.com/CAD) and complete Exercise 3-8.

Using Previously Picked Points

Selecting previously picked points is a common need while drawing, especially if, for example, you draw a line, exit the **LINE** tool, and then decide to go back and connect a new line to the end of a previously selected point. The quickest way to reselect the last point entered is to first activate a tool, such as **LINE**. When you see the Specify first point: prompt, right-click or press [Enter] or the space bar. This action automatically connects the first endpoint of the new line segment to the endpoint of the previous line. You can also use the **Continue** function for drawing other objects such as arcs and polylines, as described in later chapters.

You can access the coordinates of many previously selected points, not just the last selected point. When using dynamic input, press the up arrow key at a point selection prompt to display the coordinates of the last picked point. You can continue to press the up arrow key to cycle through other previously picked points. As you scroll through previous point coordinates, a symbol appears at the location of each point. When the point symbol appears at the coordinates you want to pick, press [Enter] or right-click and choose **Enter**. Previous coordinates also display at the command line when a tool is active and you press the up arrow key.

PROFESSIONAL TIP

Another option to access the last selected point at a point selection prompt is to type @ and press [Enter]. This option is useful when you are connecting other shapes, such as circles, to the endpoint of a line.

Exercise 3-9

Access the Student Web site (www.g-wlearning.com/CAD) and complete Exercise 3-9.

Introduction to Editing

editing: Procedure used to modify an existing object.

selection set: A group of one or more drawing objects, typically defined to perform an editing operation.

Many *editing* tools exist to help increase productivity. This chapter introduces the basic editing tools **ERASE**, **OOPS**, **UNDO**, **U**, and **REDO**. To edit a drawing, you usually select one or more objects to create a *selection set*. The edit affects all objects in the selection set. The following information introduces several selection set methods using the **ERASE** tool. Keep in mind, however, that these techniques apply to most editing tools whenever the Select objects: prompt appears.

Erasing Objects

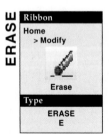

pick box: Small box that replaces the crosshairs when objects are to be selected.

Use the **ERASE** tool to remove unwanted drawing content. When you access the **ERASE** tool, the Select objects: prompt appears and an object selection target, or *pick box*, replaces the screen crosshairs. Move the pick box over the item you want to erase and pick that item. The object highlights and the Select objects: prompt redisplays, allowing you to select additional objects to erase. When you finish selecting objects, erase the selected objects by right-clicking, pressing [Enter], or pressing the space bar. See **Figure 3-18**.

Figure 3-18.
Using the **ERASE** tool to erase a single object.

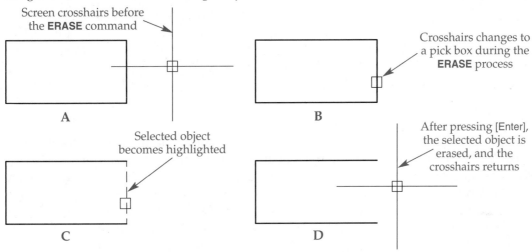

Screen crosshairs before
the **ERASE** command

A

Crosshairs changes to
a pick box during the
ERASE process

B

Selected object
becomes highlighted

C

After pressing [Enter],
the selected object is
erased, and the
crosshairs returns

D

NOTE

By default, when you move the crosshairs or pick box over an object and pause for a moment, the object changes to a thicker lineweight and becomes dashed. When you move the crosshairs or pick box off the object, the object display returns to normal. This allows you to preview the object before you select it. When many objects are in a small area, this feature helps you select the correct object the first time and often eliminates the need to cycle through stacked objects.

Exercise 3-10

Access the Student Web site (www.g-wlearning.com/CAD) and complete Exercise 3-10.

Window and Crossing Selection

Use window and crossing selection to select multiple objects, reducing the need to pick individual objects with the pick box. Window selection allows you to draw a box, or "window," around an object or group of objects to select for editing. Everything entirely within the window selects at the same time. Portions of objects that project outside the window remain unselected.

Crossing selection is similar to window selection, but with crossing selection, objects contained within the box *and crossing the box* are selected. The crossing selection box displays a dotted outline with a light green background to distinguish it from the window selection box, which displays a solid outline and light blue background.

The quickest and most effective way to use window or crossing selection is through a feature known as *automatic windowing*, or *implied windowing*, which is on by default. To apply automatic window selection, use the pick box to select a point clearly above or below and to the left of the objects to erase. A selection box replaces the pick box, and the Specify opposite corner: prompt appears. Move the corner of the selection box to the right and up or down so the box completely covers the objects to erase. Pick to locate the second corner. See **Figure 3-19**. All objects that lie completely within the box highlight. Right-click or press [Enter] or the space bar to complete the **ERASE** tool.

automatic windowing (implied windowing): Selection method that allows you to select multiple objects at one time without entering a selection option.

Figure 3-19.
Using automatic windowing to select all objects completely inside a window selection box.

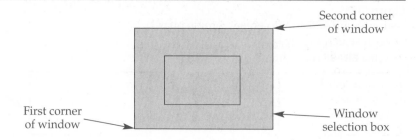

To apply automatic crossing selection, use the pick box to select a point clearly above or below and to the right of the object(s) to erase. A selection box replaces the pick box, and the Specify opposite corner: prompt appears. Move the corner of the selection box to the left and up or down, across the objects you want to select. Remember, the crossing box does not need to enclose the entire object to erase it, as does the window box. **Figure 3-20** shows how to use crossing selection to erase three of the four lines of a rectangle that was drawn using the **LINE** tool. Right-click or press [Enter] or the space bar to complete the **ERASE** tool.

> **NOTE**
>
> You can also type W or WINDOW at the Select objects: prompt to use manual window selection, or type C or CROSSING to use manual crossing selection. When you use manual window or crossing selection, the selection box remains in the window or crossing format whether the first pick is left or right of objects, and whether you move the cursor to the left or right.

Exercise 3-11

Access the Student Web site (www.g-wlearning.com/CAD) and complete Exercise 3-11.

Window Polygon and Crossing Polygon Selection

The window and crossing selection methods use a rectangular selection box, which may not allow you to select needed objects easily. An alternative is to form a window or crossing selection polygon. To use window polygon selection, type WPOLYGON or WP at the Select objects: prompt. Then pick points to draw a polygon

Figure 3-20.
Using crossing selection to select all objects inside and touching the crossing selection box.

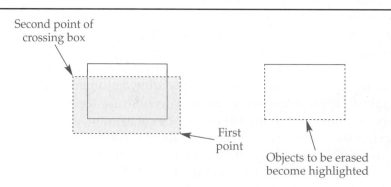

Figure 3-21.
A—Using window polygon selection to erase objects.
B—Using crossing polygon selection to erase objects. Everything enclosed within and crossing the polygon selects.

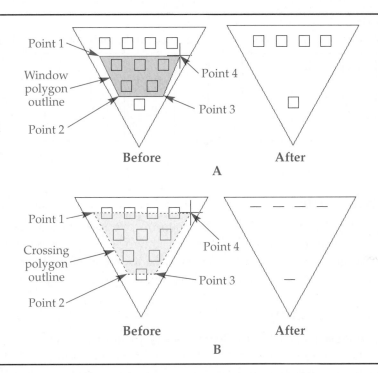

enclosing the objects you want to select. As you pick corners, the polygon drags into place. See **Figure 3-21A**.

Crossing polygon selection is similar to window polygon selection, but with crossing polygon selection, you create a crossing selection. To form a crossing selection polygon, type **CPOLYGON** or **CP** at the Select objects: prompt. Then pick points to draw a polygon. See **Figure 3-21B**.

PROFESSIONAL TIP

In window or crossing polygon selection, AutoCAD does not allow you to select a point that causes the lines of the selection polygon to intersect each other. Pick locations that do not result in an intersection. Use the **Undo** option if you need to go back and relocate a previous pick point.

Fence Selection

Fence selection is another method used to select several objects at the same time. When using fence selection, you place a fence, or connected lines, through the objects you want to select. Only the objects passing through the fence are included in the selection set. To use fence selection, type **FENCE** or **F** at the Select objects: prompt. Then pick points to draw a fence through the objects to select. The fence can be straight or staggered, as shown in **Figure 3-22**.

Exercise 3-12

Access the Student Web site (www.g-wlearning.com/CAD) and complete Exercise 3-12.

Figure 3-22.
Using fence selection to erase objects. The fence can be either straight or staggered.

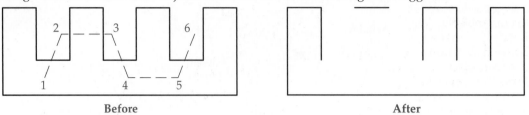

Before   After

Last Selection

Type LAST or L at the Select objects: prompt to select the last object drawn. The last selection feature selects only the last item drawn. You must repeatedly access a tool, such as **ERASE**, and use last selection every time to select individual items in reverse order. This is extremely slow compared to selecting the objects using other methods.

Previous Selection

Type Previous or P at the Select objects: prompt to reselect all the objects selected during the previous selection set. You usually use previous selection when you need to carry out more than one sequential editing operation on a specific group of objects. In this case, use previous selection to reselect the objects you just edited.

> **NOTE**
>
> Previous selection does not reselect erased objects.

Selecting All Objects

Use the **Select All** tool to quickly select every object in the drawing. Everything in the drawing that is not on a frozen layer is selected, even objects that are outside of the current drawing window display. Chapter 5 describes layers.

Cycling through Stacked Objects

One way to deal with *stacked objects* is to let AutoCAD *cycle* through the overlapping objects. Cycling works best when several objects cross at the same place or are very close together. To cycle through stacked objects, first access a tool, such as **ERASE**. Next, with the Select objects: prompt shown, move the pick box over the intersection of the stacked objects. Hold down [Shift] and repeatedly press the space bar to cycle through the stacked objects. When the desired object highlights, release [Shift] and pick (left-click) to select the item. See **Figure 3-23.**

stacked objects: Objects that overlap in the drawing. When you pick with the mouse, the topmost object selects by default.

cycle: Repeatedly select a series of stacked objects until the desired object highlights.

Changing the Selection Set

The quickest way to remove one or more objects from the current selection set is to hold down [Shift] and reselect the objects. This is possible only for individual picks and automatic windows. For automatic windowing, hold down [Shift] and pick the first corner. Then release [Shift] and pick the second corner. Select objects as usual to add them back to the selection set.

Another option for removing objects from a selection set is to type REMOVE or R at the Select objects: prompt. This enters the **Remove** option and changes the Select objects: prompt to Remove objects:, allowing you to pick the objects to remove from the selection set. To switch back to the selection mode, type ADD or A at the Remove objects:

Figure 3-23.
Cycling through a series of stacked circles until the desired object highlights.

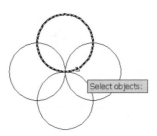

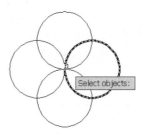

Original group of objects
with top item highlighted

Hold down shift key and press the
space bar to cycle through objects

prompt. This enters the **Add** option and restores the Select objects: prompt, allowing you to select additional objects.

PROFESSIONAL TIP

Removing items from a selection set is especially effective when used in combination with the **Select All** selection option. This allows you to keep very specific objects while erasing everything else.

Exercise 3-13

Access the Student Web site (www.g-wlearning.com/CAD) and complete Exercise 3-13.

Supplemental Material *Selection Display Options*
For detailed information about adjusting selection display options, go to the Student Web site (www.g-wlearning.com/CAD), select this chapter, and select **Selection Display Options**.

Using the UNDO Tool

The **UNDO** tool offers several options that allow you to undo a single operation or a number of operations at once. The **UNDO** tool is different from the **Undo** option of certain tools, such as the **LINE** tool. This quickest way to use the **UNDO** tool is to pick the **Undo** button on the **Quick Access** toolbar. Select the button as many times as needed to undo multiple operations. An alternative is to pick the flyout and select all of the tools to undo from the list.

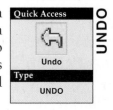

 **Supplemental Material**

UNDO *Options*

For detailed information about the options available when you access the **UNDO** tool from a source other than the **Quick Access** toolbar, go to the Student Web site (www.g-wlearning.com/CAD), select this chapter, and select **UNDO Options**.

Using the U Tool

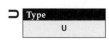

The **U** tool undoes the effect of the previously entered tool. You can reissue the **U** tool to continue undoing tool actions, but you can only undo one tool at a time. The actions are undone in the order in which they were used.

 NOTE

You can also activate the **U** tool by right-clicking in the drawing window and selecting **Undo** *current*.

Redoing the Undone

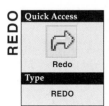

Use the **REDO** tool to reverse the action of the **UNDO** and **U** tools. The **REDO** tool works only *immediately* after undoing something. The **REDO** tool does not bring back line segments removed using the **Undo** option of the **LINE** tool. The quickest way to use the **REDO** tool is to pick the **Redo** button on the **Quick Access** toolbar. Select the button as many times as needed to redo multiple undone operations. An alternative is to pick the flyout and select one or more undone operations from the list to redo.

 **Exercise 3-14**

Access the Student Web site (www.g-wlearning.com/CAD) and complete Exercise 3-14.

Using the OOPS Tool

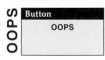

The **OOPS** tool brings back the last object you *erased*. Unlike the **UNDO** and **U** tools, **OOPS** only returns the objects erased in the most recent procedure. It has no effect on other modifications. If you erase several objects in the same tool sequence, all of the objects return to the screen.

Template Development
Chapter 3

For detailed instructions on setting grid and snap values for specific drawing templates, go to the Student Web site (www. g-wlearning.com/CAD), select this chapter, and select **Template Development**.

Chapter Test

1. When you enter a fractional number in AutoCAD, why is a hyphen required between a whole number and its associated fraction?
2. Briefly describe the Cartesian coordinate system.
3. Name the tool used to place a pattern of dots on the screen.
4. Name two ways to access the **Drafting Settings** dialog box.
5. How do you activate the **Snap** mode?
6. How do you set a grid spacing of .25?
7. List two ways to discontinue drawing a line.
8. Give the tools and entries to draw a line from Point A to Point B to Point C and back to Point A. Return to the Command: prompt.
   A. Command: _____
   B. Specify first point: _____
   C. Specify next point or [Undo]: _____
   D. Specify next point or [Undo]: _____
   E. Specify next point or [Close/Undo]: _____
9. Name three types of coordinates used for point entry.
10. What does the absolute coordinate display 5.250,7.875 mean?
11. What does the polar coordinate display @2.750<90 mean?
12. How can you turn on the coordinate display field if it is off?
13. What two general methods of point entry are available when dynamic input is active?
14. Explain, in general terms, how direct distance entry works.
15. What are the default angle increments for polar tracking?
16. How can you turn on the **Ortho** mode?
17. Explain how you can continue drawing another line segment from a previously drawn line.
18. When you access the **ERASE** tool, what replaces the screen crosshairs?
19. How does the appearance of window and crossing selection boxes differ?
20. List five ways to select an object to erase.
21. Define *stacked objects*.
22. What is the difference between pressing the **Undo** button on the **Quick Access** toolbar and entering the **UNDO** tool?
23. How many tool sequences can you undo at one time with the **U** tool?
24. Name the tool used to bring back an object that was previously removed using the **UNDO** tool.
25. Name the tool used to bring back the last object(s) erased before starting another tool.

Drawing Problems

Start AutoCAD if it is not already started. Follow the specific instructions for each problem. Do not draw dimensions or text.

▼ Basic

1. Start a new drawing from scratch or use a template of your choice. Use the **LINE** tool and draw the following objects on the left side of the screen. Accurately draw the specified objects with grid and snap turned off.
   - Right triangle.
   - Isosceles triangle.
   - Rectangle.
   - Square.
   Save the drawing as P3-1.

2. Start a new drawing from scratch or use a template of your choice. Draw the same objects specified in Problem 1 on the right side of the screen. This time, make sure the snap grid is turned on. Observe the difference between having snap on for this problem and off for the previous problem. Save the drawing as P3-2.

3. Start a new drawing from scratch or use a decimal unit template of your choice. Draw an object by connecting the following point coordinates. Use dynamic input to enter the coordinates. Save your drawing as P3-3.

Point	Coordinates	Point	Coordinates
1	2,2	8	@-1.5,0
2	@1.5,0	9	@0,1.25
3	@.75<90	10	@-1.25,1.25
4	@1.5<0	11	@2<180
5	@0,-.75	12	@-1.25,-1.25
6	@3,0	13	@2.25<270
7	@1<90		

4. Start a new drawing from scratch or use a template of your choice. Use direct distance entry and polar tracking to draw the outline shown. Each grid square is one unit. Do not draw the grid lines. Save the drawing as P3-4.

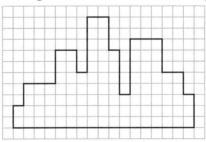

5. Start a new drawing from scratch or use a template of your choice. Use the dimensional input feature of dynamic input to draw the hexagon shown. Each side of the hexagon is 2 units. Begin at the start point, and draw the lines in the direction indicated by the arrows. Do not draw dimensions. Save the drawing as **P3-5**.

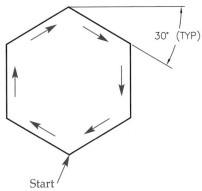

▼ Intermediate

6. Start a new drawing from scratch or use a fractional unit template of your choice. Draw the following shapes using absolute, relative, and polar coordinate entry methods. Draw Object A three times, using a different point entry system each time. Set the units to decimal and the precision to 0.0 when drawing Object A. Draw Object B once, using at least two methods of coordinate entry. Set the units to fractional and the precision to 1/16 when drawing Object B. Do not draw dimensions. Save the drawing as **P3-6**.

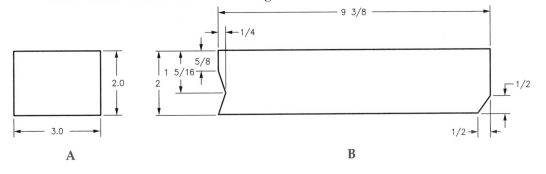

For Problems 7–8, start a new drawing from scratch or use a decimal unit template of your choice. Draw the given part. Do not draw dimensions. Save the drawings as **P3-7** *and* **P3-8**.

7.

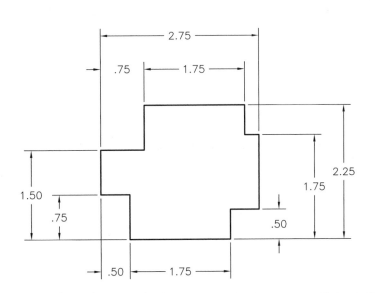

8.

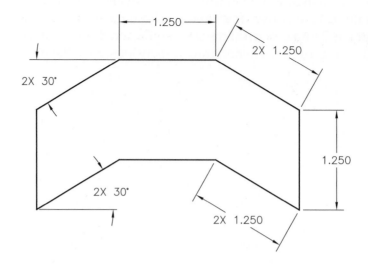

9. Start a new drawing from scratch or use a decimal unit template of your choice. Draw the objects shown in A and B. Begin at the start point and then discontinue the **LINE** tool at the point shown. Complete each object using the **Continue** option. Do not draw dimensions. Save the drawing as P3-9.

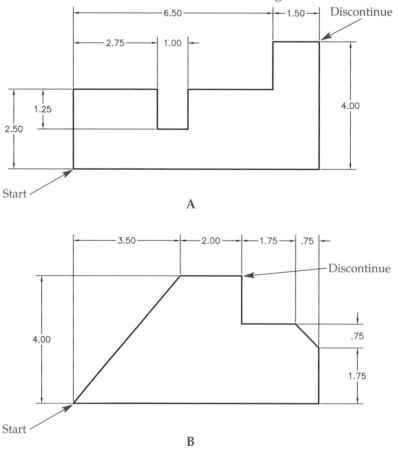

10. Sketch the X and Y axes on a sheet of paper. Label the origin, the positive values for X = 1 through X = 10, and the positive values for Y = 1 through Y = 10. Then sketch the object described by the following coordinate points:

2,2
8,2
8,7
7,7
7,3
6,3
6,6
4,6
4,3
3,3
3,7
2,7
2,2

▼ Advanced

11. Sketch the X and Y axes of the Cartesian coordinate system as you did for the previous problem. Then sketch an object outline of your choice within the axes. List, in order, the rectangular coordinates of the points a drafter would need to specify in AutoCAD to recreate the object in your sketch.

For Problems 12–14, start a new drawing from scratch or use a decimal unit template of your choice. Draw the given part. Do not draw dimensions. Save the drawings as P3-12, P3-13, *and* P3-14.

12.

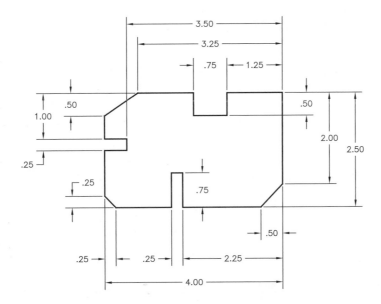

13.

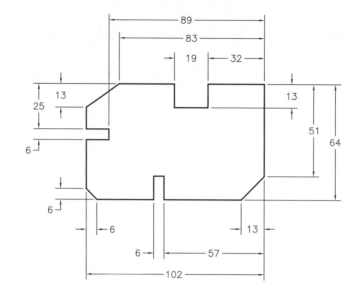

14.

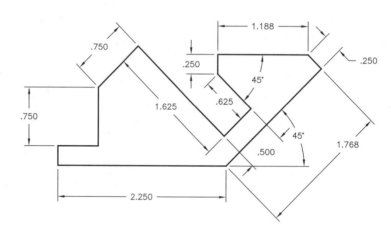

Basic Object Tools

Learning Objectives

After completing this chapter, you will be able to do the following:

✓ Draw circles using the **CIRCLE** tool options.
✓ Draw arcs using the **ARC** tool options.
✓ Use the **ELLIPSE** tool to draw ellipses and elliptical arcs.
✓ Use the **PLINE** tool to draw polylines.
✓ Draw polygons using the **POLYGON** tool.
✓ Draw rectangles using the **RECTANGLE** tool options.
✓ Draw donuts and filled circles using the **DONUT** tool.
✓ Draw true spline curves using the **SPLINE** tool.

This chapter describes how to use a variety of tools to draw basic shapes. This chapter presents the ribbon as the primary means of accessing basic object tool options. When you select an object tool option from the ribbon, specific prompts associated with the option appear. When you issue a tool using dynamic input or the command line, you must enter specific options when prompted.

Using Basic Object Tools

When you draw basic shapes, AutoCAD prompts you to select coordinates, just as when you draw a line. For example, the Specify center point: prompt appears when you draw a circle using the **Center, Radius** or **Center, Diameter** method. Your response to this prompt is similar to your response to the Select first point: prompt that appears when you use the **LINE** tool. Use any appropriate point entry method to pick a location.

Some basic object tools give you the option of entering specific values to define the size and shape of the object. For example, the Specify radius of circle: prompt appears after you locate the center point of a circle using the **Center, Radius** option of the **CIRCLE** tool. Usually, the most effective response to this type of prompt is to type a value, such as the circle radius in this example. An alternative is to use an appropriate point entry technique. Refer to Chapter 3 to review proper numerical responses and point entry methods.

Figure 4-1.
Dragging a circle to a desired size. The circle attached to the crosshairs stretches like a rubberband until you pick a point to define the radius.

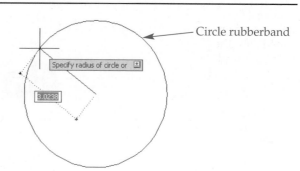

Circle rubberband

Specify radius of circle or

3.0238

Similar to the **LINE** tool, many object tools include a rubberband shape that connects the first point selection to the crosshairs. The image aids in sizing and locating the object. For example, when you draw a circle using the **Center, Radius** option, a circle image appears on-screen after you pick the center point. The image gets larger or smaller as you move the pointer. When you pick the radius, the actual circle object replaces the rubberband image. See **Figure 4-1**.

Drawing Circles

CIRCLE

Ribbon
Home
> Draw

Circle

Type
CIRCLE
C

The **CIRCLE** tool provides several methods for drawing *circles*. Choose the appropriate option based on how you want to locate the circle, and the information you know or want to use to construct the circle. The ribbon is an effective way to access circle tool options. See **Figure 4-2**.

Drawing a Circle by Radius

Use the **Center, Radius** option of the **CIRCLE** tool to specify the center point and the radius of a circle. After selecting the **Center, Radius** option, define the center point. Then type a value for the radius and press [Enter] or the space bar, or right-click and pick **Enter**. You can also define the radius using point entry. See **Figure 4-3**.

circle: A closed curve with a constant radius around a center point; size is usually dimensioned according to the diameter.

Figure 4-2.
The **Circle** options can be accessed from the **Draw** panel of the **Home** ribbon tab.

Pick the **Circle** flyout to display the **Circle** options

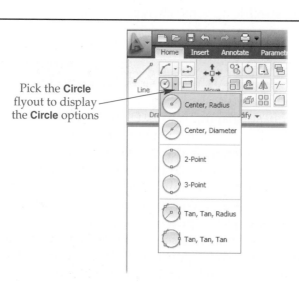

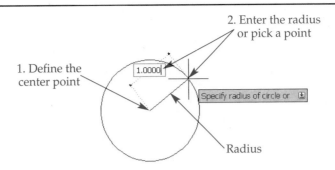

Figure 4-3.
Drawing a circle by specifying the center point and radius.

1. Define the center point

2. Enter the radius or pick a point

1.0000

Specify radius of circle or

Radius

NOTE

The radius value you enter is stored as the new default radius setting, allowing you to quickly draw another circle with the same radius.

Drawing a Circle by Diameter

You can also draw a circle by specifying the center point and the diameter. The **Center, Diameter** option is convenient because most circular holes, shafts, and features are sized according to diameter. After selecting the **Center, Diameter** option, define the center point. Then type a value for the diameter and press [Enter] or the space bar, or right-click and pick **Enter**. You can also define the diameter using point entry. In the **Center, Diameter** option, the crosshairs measures the diameter, but the rubberband circle passes midway between the center and the crosshairs. See **Figure 4-4.**

NOTE

If you use the **Center, Radius** option to draw a circle after using the **Diameter** option, AutoCAD changes the default to a radius measurement based on the previous diameter.

Drawing a Two-Point Circle

A two-point circle is drawn by picking two points on opposite sides of the circle to define its diameter. The **2-Point** option is useful if the diameter of the circle is known, but the center is difficult to find. One example of this is locating a circle between two lines. The process of drawing a two-point circle is very similar to drawing a line. After selecting the **2 Points** option, enter or select a point for the first endpoint of the circle's diameter. Then enter or select the second endpoint of the circle's diameter. See **Figure 4-5.**

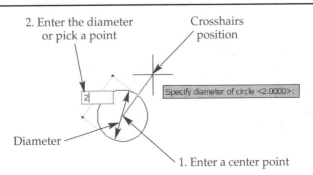

Figure 4-4.
When you use the **Center, Diameter** option, AutoCAD calculates the circle's position as you move the crosshairs.

2. Enter the diameter or pick a point

Crosshairs position

2

Specify diameter of circle <2.0000>:

Diameter

1. Enter a center point

Figure 4-5.
Drawing a circle by selecting two points.

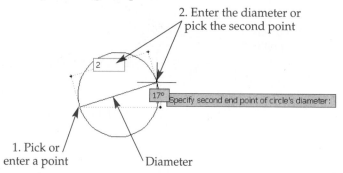

2. Enter the diameter or pick the second point

17°   Specify second end point of circle's diameter:

1. Pick or enter a point

Diameter

Drawing a Three-Point Circle

The **3-Point** option is the best method to use if you know three points on the circumference of a circle. After selecting the **3-Point** option, enter or select three points in any order to define the circumference of the circle. See **Figure 4-6.**

Drawing a Circle Tangent to Two Objects

tangent: A line, circle, or arc that meets another circle or arc at only one point.

point of tangency: The point shared by tangent objects.

object snap: A tool that snaps to exact points, such as endpoints or midpoints, when you pick a point near these locations.

The **Tan, Tan, Radius** option creates a circle of a specified radius *tangent* to two objects. You can draw a circle tangent to given lines, circles, or arcs. The circle automatically positions at the *point of tangency*. The **Tan, Tan, Radius** option uses an *object snap* known as **Deferred Tangent** to assist you in picking a point exactly tangent to other objects. Object snaps are covered in detail in Chapter 7. After accessing the **Tan, Tan, Radius** option, hover the crosshairs over the first line, arc, or circle to which the new circle will be tangent. Pick when the deferred tangent symbol appears. Repeat the process to select the second line, arc, or circle to which the new circle will be tangent. Then enter or select the radius of the circle. See **Figure 4-7.**

> **NOTE**
>
> If the radius you enter while using the **Tan, Tan, Radius** option is too small, AutoCAD displays the message Circle does not exist.

Drawing a Circle Tangent to Three Objects

The **Tan, Tan, Tan** option allows you to draw a circle tangent to three existing objects. This option creates a three-point circle using the three points of tangency. Like the **Tan, Tan, Radius** option, the **Tan, Tan, Tan** option uses the **Deferred Tangent** object snap to assist you in picking a point exactly tangent to other objects. After accessing the **Tan, Tan, Tan** option, pick three lines, arcs, or circles to which the new circle will be

Figure 4-6.
Drawing a circle by picking three points that lie on the circle.

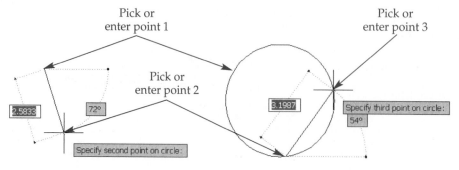

Pick or enter point 1

Pick or enter point 3

Pick or enter point 2

2.5833   72°

3.1987   Specify third point on circle: 54°

Specify second point on circle:

Figure 4-7.
Two examples of
drawing circles
tangent to two given
objects using the
Tan, Tan, Radius
option.

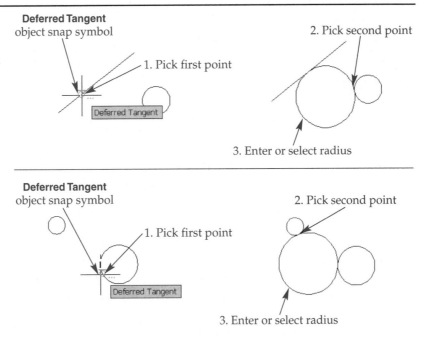

tangent. You must pick each item when the deferred tangent symbol appears, but the
order in which you pick the items is not critical. See **Figure 4-8.**

NOTE

Unlike the **Tan, Tan, Radius** option, the **Tan, Tan, Tan** option does
not automatically recover when you pick a point where no tangent
exists. In such a case, you must manually reactivate the **Tangent**
object snap to make additional picks. Chapter 7 describes the **Tangent**
object snap. For now, type TAN and press [Enter] at the point selection prompt to return the pick box so you can pick again.

Figure 4-8.
Two examples of
drawing circles
tangent to three
given objects.

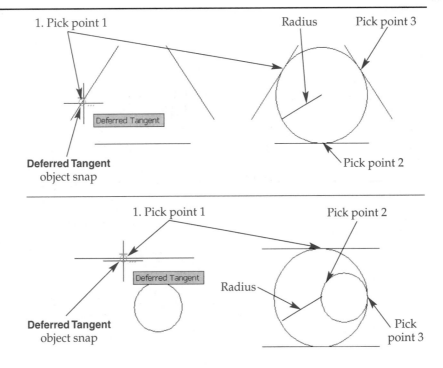

Exercise 4-1

Access the Student Web site (www.g-wlearning.com/CAD) and complete Exercise 4-1.

Drawing Arcs

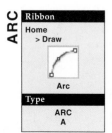

Ribbon

Home
> Draw

Arc

Type

ARC
A

---

arc: Any portion of a circle; usually dimensioned according to the radius.

included angle: The angle formed between the center, start point, and endpoint of an arc.

chord length: The linear distance between two points on a circle or arc.

The **ARC** tool offers a number of different options for drawing *arcs*. Select the appropriate option based on how you want to locate the arc, and the information you know or want to use to construct the arc. The ribbon is an effective way to access arc tool options. See **Figure 4-9.**

The various **ARC** tool options allow you to create an arc for any situation requiring an arc. **Figure 4-10** describes most methods. The step-by-step examples shown in **Figure 4-10** depict a single common application of each option. Arc placement is set according to your selections and the values you enter. Some arc options prompt for the *included angle*, and others prompt for the *chord length*. Locating points in a clockwise or counterclockwise pattern affects the result when using most arc options. The values you specify, including the use of positive or negative numbers, also affects the result.

NOTE

The **3-Point** option is default when you enter the **ARC** tool at the keyboard.

Figure 4-9.
Selecting an **Arc** option from the **Draw** panel of the **Home** ribbon tab.

Pick to display
Arc options

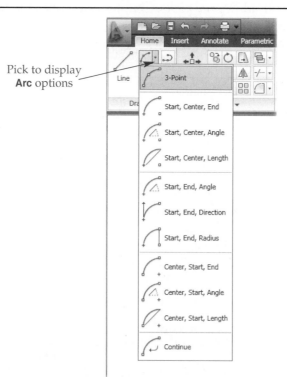

AutoCAD and Its Applications—Basics

Figure 4-10.
Select the appropriate arc creation method based on how you want to locate the arc and the information you know or want to use to construct the arc.

Option	Direction	Steps
3-Point	Clockwise or counterclockwise	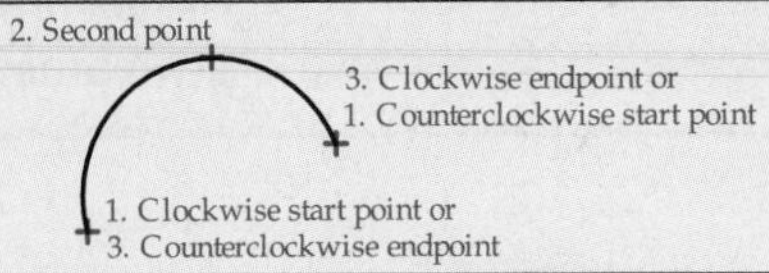 2. Second point 3. Clockwise endpoint or 1. Counterclockwise start point 1. Clockwise start point or 3. Counterclockwise endpoint
Start, Center, End	Counterclockwise	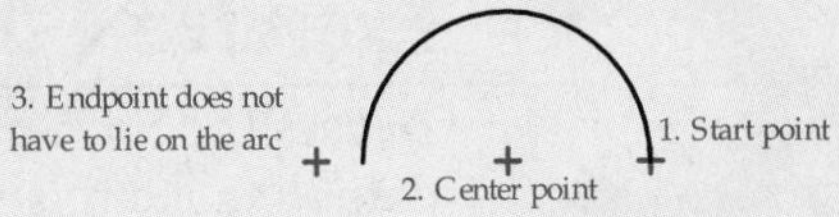 3. Endpoint does not have to lie on the arc 1. Start point 2. Center point
Start, Center, Angle	Positive angle = Clockwise Negative angle = Counterclockwise	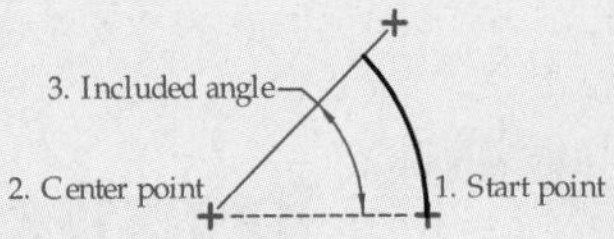 3. Included angle 2. Center point 1. Start point
Start, Center, Length	Counterclockwise	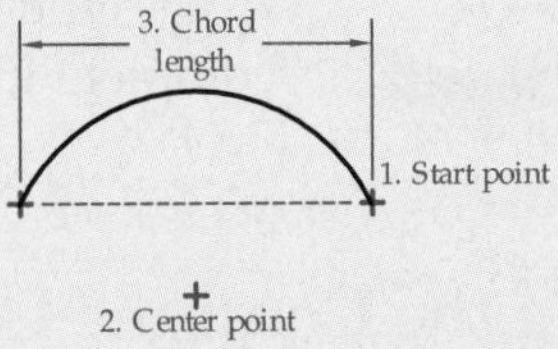 3. Chord length 1. Start point 2. Center point
Start, End, Angle	Positive angle = Clockwise Negative angle = Counterclockwise	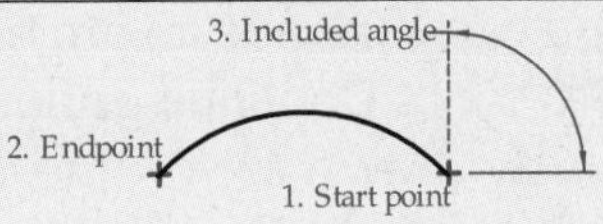 3. Included angle 2. Endpoint 1. Start point
Start, End, Direction	Tangent to specified direction	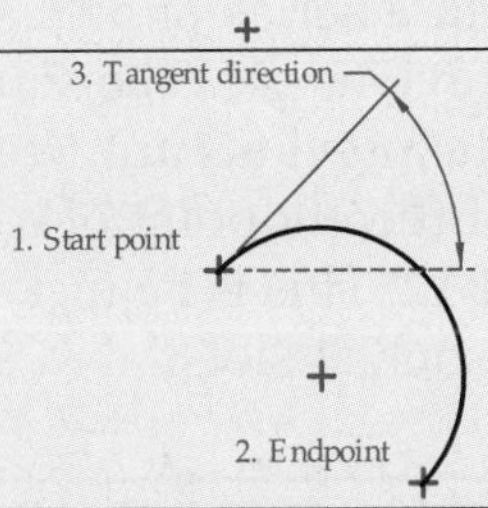 3. Tangent direction 1. Start point 2. Endpoint
Start, End, Radius	Counterclockwise	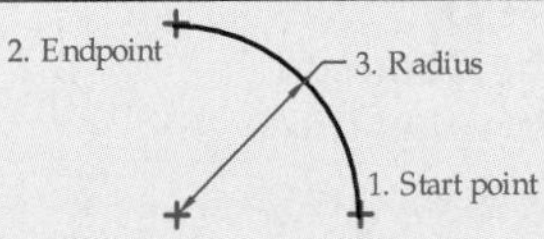 2. Endpoint 3. Radius 1. Start point
Center, Start, End	Counterclockwise	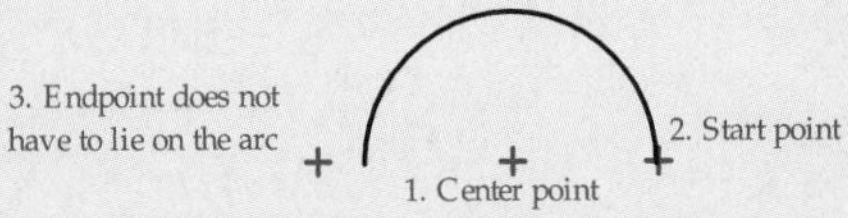 3. Endpoint does not have to lie on the arc 2. Start point 1. Center point
Center, Start, Angle	Positive angle = Clockwise Negative angle = Counterclockwise	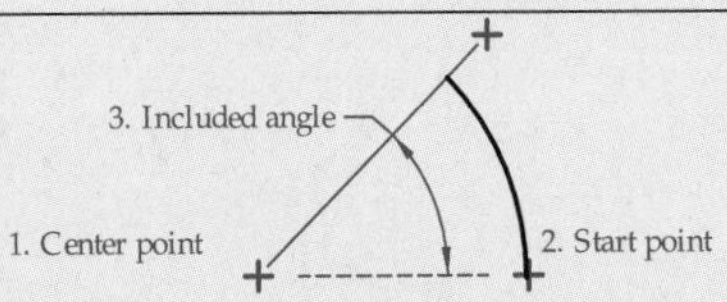 3. Included angle 1. Center point 2. Start point
Center, Start, Length	Counterclockwise	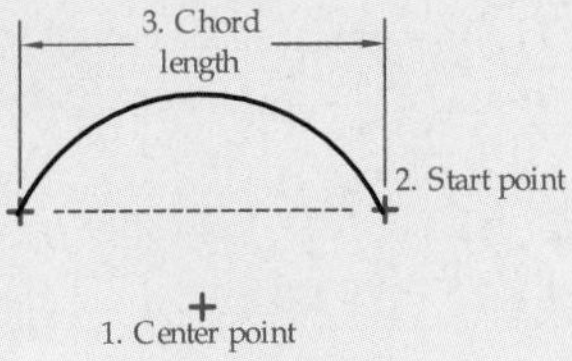 3. Chord length 2. Start point 1. Center point

Reference Material

Chord Length Table

For a chord length table and other reference tables, go to the **Reference Material** section of the Student Web site (www. g-wlearning.com/CAD) and select **Standard Tables**.

Exercise 4-2

Access the Student Web site (www.g-wlearning.com/CAD) and complete Exercise 4-2.

Exercise 4-3

Access the Student Web site (www.g-wlearning.com/CAD) and complete Exercise 4-3.

Using the Continue Option

You can continue an arc from the endpoint of a previously drawn arc or line using the **ARC** tool's **Continue** option. The arc automatically attaches to the endpoint of the previously drawn arc or line, and the Specify endpoint of arc: prompt appears. Pick the endpoint to create the arc.

When a series of arcs are drawn using the **Continue** option, each arc is tangent to the previous arc. The start point and direction occur from the endpoint and direction of the previous arc. See **Figure 4-11.** When you use the **Continue** option to begin an arc at the endpoint of a previously drawn line, the arc is tangent to the line. See **Figure 4-12.** This is a quick way to draw an arc tangent to a line for a variety of applications, such as slots.

Figure 4–11.
Using the **Continue** option to draw three tangent arcs.

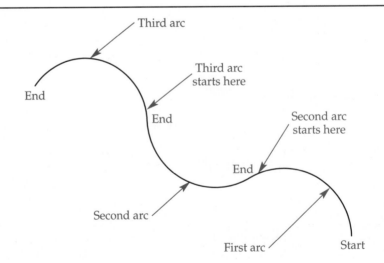

Figure 4–12.
An arc continuing from the previous line. Point 2 is the start of the arc, and Point 3 is the end of the arc. The arc and line are tangent at Point 2.

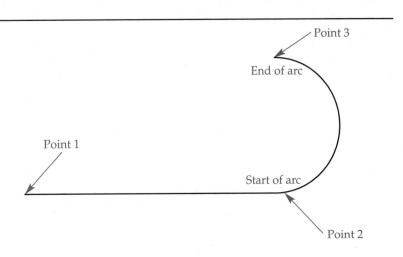

Point 3

End of arc

Point 1

Start of arc

Point 2

NOTE

You can also access the **Continue** option by beginning the **ARC** tool and then pressing [Enter] or the space bar, or by right-clicking and selecting **Enter** when prompted to specify the start point of the arc.

Exercise 4-4

Access the Student Web site (www.g-wlearning.com/CAD) and complete Exercise 4-4.

Drawing Ellipses

When you view a circle at an angle, it appears as an *ellipse* and contains both a *major axis* and a *minor axis*. For example, a 30° ellipse is a circle rotated 30° from the line of sight. **Figure 4-13** shows the parts of an ellipse.

The **ELLIPSE** tool provides several methods for drawing elliptical shapes. Choose the appropriate option based on how you want to locate the ellipse, the information you know or want to use to construct the ellipse, and whether the ellipse is whole or partial.

Using the Center Option

Use the **Center** option to construct an ellipse by specifying the center point and one endpoint for each of the two axes. Select the center point of the ellipse and then select the endpoint of one of the axes. Specify the endpoint of the other axis to complete the ellipse. See **Figure 4-14**.

ellipse: An oval shape that contains two centers of equal radius.

major axis: The longer of the two axes in an ellipse.

minor axis: The shorter of the two axes in an ellipse.

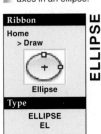

Figure 4–13.
The parts of an ellipse.

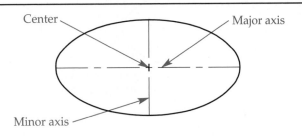

Center

Major axis

Minor axis

Figure 4–14.
Drawing an ellipse by picking the center and an endpoint for each axis. The order in which you enter or pick axis endpoints is not critical. The distance from each endpoint to the center point determines the major and minor axes.

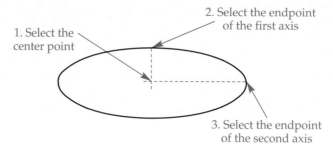

1. Select the center point

2. Select the endpoint of the first axis

3. Select the endpoint of the second axis

Using the Axis, End Option

The **Axis, End** option establishes the first axis and one endpoint of the second axis. Select the endpoint of one of the axes and then select the other endpoint of the same axis. Enter a distance from the midpoint of the first axis to the end of the second axis to complete the ellipse. The first axis can be the major or minor axis, depending on what you enter for the second axis. See **Figure 4-15.**

Using the Rotation Option

Use the **Rotation** option to create an ellipse by specifying the angle at which a circle is rotated from the line of sight. Begin by constructing an ellipse as usual, but be sure to select the major axis with the first axis endpoint. Then, when the Specify distance to other axis or: prompt appears, select the **Rotation** option instead of picking the second axis endpoint. Finally, enter the angle at which the circle is rotated from the line of sight to produce the ellipse.

For example, if you respond by entering 30 for a 30° rotation, an ellipse forms based on a circle rotated 30° from the line of sight. A 0 response draws an ellipse with the minor axis equal to the major axis, which is a circle. AutoCAD rejects any rotation angle between 89.99994° and 90.00006° or between 269.99994° and 270.00006°. **Figure 4-16** shows the relationship among several ellipses having the same major axis length, but different rotation angles.

> **NOTE**
>
> The **Rotation** option works with both the **Center** option and the **Axis, End** option.

Figure 4–15.
Constructing the same ellipse by choosing different axis endpoints.

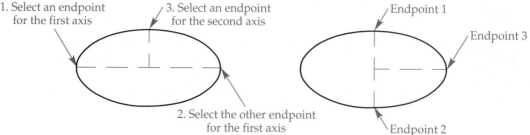

1. Select an endpoint for the first axis

3. Select an endpoint for the second axis

2. Select the other endpoint for the first axis

Endpoint 1

Endpoint 3

Endpoint 2

Figure 4–16.
Ellipse rotation angles.

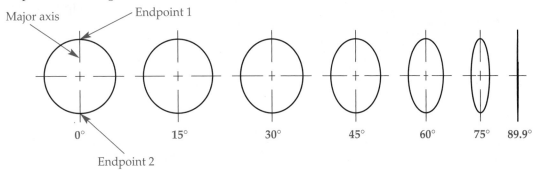

Major axis — Endpoint 1

0° 15° 30° 45° 60° 75° 89.9°

Endpoint 2

Exercise 4-5

Access the Student Web site (www.g-wlearning.com/CAD) and complete Exercise 4-5.

Drawing Elliptical Arcs

Ribbon
Home
> Draw
Elliptical Arc

Use the **Arc** option of the **ELLIPSE** tool to draw elliptical arcs. Creating an elliptical arc is just like drawing an ellipse, but with two additional steps that define the start point and endpoint of the elliptical arc. Several options are available for defining the size and shape of an elliptical arc.

The default elliptical arc uses axis endpoints to define the ellipse, and then start and end angles to produce the elliptical arc. After selecting the **Elliptical Arc** option, specify the endpoint of one of the ellipse axes. Then select the other endpoint of the same axis. Enter a distance from the midpoint of the first axis to the end of the second axis to form an ellipse. Finally, select the start and end angles for the elliptical arc.

The start and end angles are the angular relationships between the ellipse's center and the arc's endpoints. The angle of the first axis establishes the angle of the elliptical arc. For example, a 0° start angle begins the arc at the first endpoint of the first axis. A 45° start angle begins the arc 45° counterclockwise from the first endpoint of the first axis. End angles are also counterclockwise from the start point. **Figure 4-17** shows an elliptical arc drawn using a 0° start angle and a 90° end angle and displays sample arcs with different start and end angles.

Additional elliptical arc options are also available. **Figure 4-18** briefly describes each option. Use the **Center** option when appropriate instead of the default axis endpoint method. The **Parameter**, **Included angle**, and **Rotation** options are available during the process of creating axis endpoint or center elliptical arcs.

Exercise 4-6

Access the Student Web site (www.g-wlearning.com/CAD) and complete Exercise 4-6.

Figure 4–17.
Drawing elliptical arcs. Note the three examples at the bottom created using three different angles.

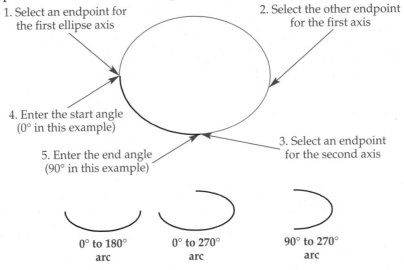

1. Select an endpoint for the first ellipse axis
2. Select the other endpoint for the first axis
4. Enter the start angle (0° in this example)
5. Enter the end angle (90° in this example)
3. Select an endpoint for the second axis

0° to 180° arc 0° to 270° arc 90° to 270° arc

Figure 4–18.
Additional options available for drawing elliptical arcs.

Option	Application	Process
Center	Lets you establish the center of the elliptical arc. **Rotation**, **Parameter**, and **Included angle** options are available.	1. Select the ellipse center point. 2. Select the endpoint of one of the ellipse axes. 3. Pick the endpoint of the other axis to form the ellipse. 4. Enter the start angle for the elliptical arc. 5. Select the end angle.
Parameter	Use instead of picking the start angle of the elliptical arc. AutoCAD uses a different means of vector calculation to create the elliptical arc.	1. Specify the start parameter point. 2. Specify the end parameter point.
Included angle	Establishes an included angle beginning at the start angle.	1. Specify the included angle.
Rotation	Allows you to rotate the elliptical arc about the first axis by specifying a rotation angle. **Parameter** and **Included angle** options are available.	1. Specify the rotation around the major axis. 2. Specify the start angle for the elliptical arc. 3. Specify the end angle.

Drawing Polylines

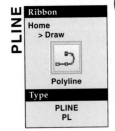

PLINE

Ribbon
Home
> Draw

Polyline

Type
PLINE
PL

polyline: A series of lines and arcs that constitute a single object.

Use the **PLINE** tool to draw *polylines*. When you use the default polyline settings, the process of drawing polyline segments is identical to the process of creating line segments using the **LINE** tool. Access the **PLINE** tool and use point entry to locate polyline endpoints. When you finish specifying points, press [Enter], the space bar, or [Esc], or right-click and select **Enter**. The difference between a polyline and a line is that once it is drawn, all segments of a continuous polyline act as a single object. The **PLINE** tool also provides more flexibility than the **LINE** tool, allowing you to draw a single object composed of straight lines and arcs of varying thickness.

AutoCAD and Its Applications—Basics

Undo and **Close** options are available for drawing polylines, and they function the same as those for drawing lines using the **LINE** tool. Use the **Undo** option to remove the last segment of a polyline without leaving the **PLINE** tool. This removes the previously drawn segment and allows you to continue from the previous endpoint. You can use the **Undo** option repeatedly to delete polyline segments until the entire object is gone. Use the **Close** option to connect the endpoint of the last polyline segment to the start point of the first polyline segment.

Setting the Polyline Width

The default polyline settings create a polyline with a constant width of 0. A polyline drawn using a constant width of 0 is similar to a standard line and accepts the lineweight applied to the layer on which the polyline is drawn. Chapter 5 explains layers. Adjust the polyline width to create thick and tapered polyline objects.

To change the width of a polyline segment, access the **PLINE** tool, select the first point, and then choose the **Width** option. AutoCAD prompts you to specify the starting width of the line, followed by the ending width of the line. Enter the same starting and ending width value to draw a polyline with constant width. The rubberband line from the first point reflects the width settings. **Figure 4-19A** shows a 4″ polyline with starting and ending widths of .25″. Notice that the starting and ending points of the line are located at the center of the line segment's width.

To create a tapered line segment, enter different values for the starting and ending widths. In the example shown in **Figure 4-19B**, starting width is .25 units, and the ending width is .5 units. One special use of a tapered polyline is the creation of arrowheads. To draw an arrowhead, use the **Width** option and specify 0 as the starting width, and then use an appropriate ending width.

NOTE

Using a starting or ending width value other than 0 overrides the lineweight applied to the layer on which the polyline is drawn.

Using the Halfwidth Option

The **Halfwidth** option allows you to specify the width of the polyline from the center to one side, as opposed to the total width of the polyline defined using the **Width** option. Access the **PLINE** tool, pick the first polyline endpoint, and then choose the **Halfwidth** option. Specify starting and ending values at the appropriate prompts.

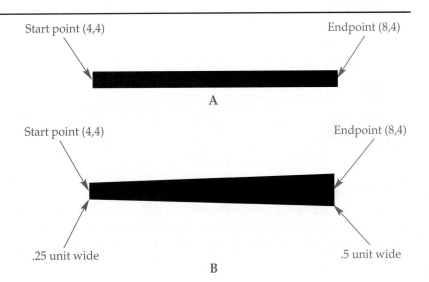

Figure 4-19.
A—A thick polyline drawn using the **Width** option of the **PLINE** tool.
B—Using the **Width** option of the **PLINE** tool to draw a polyline with a tapered width.

Start point (4,4) Endpoint (8,4)

A

Start point (4,4) Endpoint (8,4)

.25 unit wide .5 unit wide

B

Figure 4-20.
Specifying the width of a polyline with the **Halfwidth** option. A starting value of .25 produces a polyline width of .5 unit, and an ending value of .5 produces a polyline width of 1 unit.

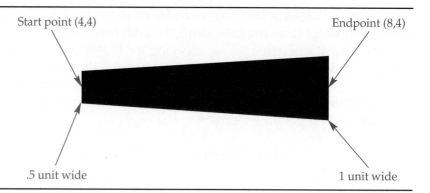

Start point (4,4)   Endpoint (8,4)

.5 unit wide   1 unit wide

Notice that the polyline in **Figure 4-20** is twice as wide as the polyline in **Figure 4-19B**, even though the same values are entered.

> **NOTE**
>
> All polyline objects with width, including polylines, polygons drawn using the **POLYGON** tool, rectangles drawn using the **RECTANGLE** tool, and donuts can appear filled or empty. The **Apply solid fill** setting in the **Display performance** area of the **Display** tab in the **Options** dialog box controls the appearance. Polyline objects are filled by default. You can also type FILL or FILLMODE and use the **On** or **Off** option. The fill display for previously drawn polyline objects updates when the drawing regenerates. You can regenerate the drawing manually by typing REGEN.

Using the Length Option

The **Length** option allows you to draw a polyline parallel to the previous polyline or line. After drawing a polyline, reissue the **PLINE** tool and pick a start point. Choose the **Length** option and enter the desired length. The second polyline draws parallel to the previous polyline using specified length.

Exercise 4-7

Access the Student Web site (www.g-wlearning.com/CAD) and complete Exercise 4-7.

Exercise 4-8

Access the Student Web site (www.g-wlearning.com/CAD) and complete Exercise 4-8.

Drawing Polyline Arcs

Use the **Arc** option of the **PLINE** tool to draw polyline arcs. Polyline arcs can include width, set according to the **Width** or **Halfwidth** option, and can continue from or to polyline segments drawn during the same operation to form a single object. Polyline

Figure 4-21.
An example of a polyline arc with different starting and ending widths.

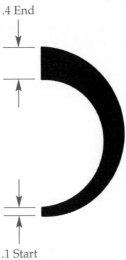

.4 End

.1 Start

arc width can range from 0 to the radius of the arc. You can set different starting and ending arc widths. See **Figure 4-21.** You can enter the **Width** or **Halfwidth** and **Arc** options in either order. Use the **Line** option to return the **PLINE** tool back to straight-line segment mode.

In addition to the **Close, Undo, Width,** and **Halfwidth** options, the **PLINE Arc** option includes functions for controlling the size and location of polyline arcs. Many of the **PLINE Arc** options allow you to create polyline arcs using the same methods available for drawing arcs using the **ARC** tool. Select the appropriate option and follow the prompts to create the polyline arc. Review the **ARC** tool options to help recognize the function of similar **PLINE Arc** options. **Figure 4-22** provides a brief description of each additional **PLINE Arc** option.

Figure 4–22.
Additional options available for drawing polyline arcs.

Option	Application	Options for Completion
Angle	Specify the polyline arc size according to an included angle.	1. Specify an endpoint. 2. Use the **Center** option to select the center point. 3. Use the **Radius** option to enter the radius.
Center	Specify the location of the polyline arc center point, instead of allowing AutoCAD to calculate the location automatically.	1. Specify an endpoint. 2. Use the **Angle** option to specify the included angle. 3. Use the **Length** option to specify the chord length.
Direction	Alter the polyline arc bearing, or tangent direction, instead of allowing the polyline arc to form tangent to the last object drawn.	1. Specify an endpoint.
Radius	Specify the polyline arc radius.	1. Specify an endpoint. 2. Use the **Angle** option to specify the included angle.
Second point	Draw a three-point polyline arc.	1. Pick the second point, followed by the endpoint.

Exercise 4-9

Access the Student Web site (www.g-wlearning.com/CAD) and complete Exercise 4-9.

Supplemental Material

Multilines

For information about drawing and editing multiline objects, go to the Student Web site (www.g-wlearning.com/CAD), select this chapter, and select **Multilines**.

Drawing Regular Polygons

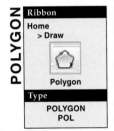
You can use the **POLYGON** tool to draw any *regular polygon* with up to 1024 sides. Polygons drawn using the **POLYGON** tool are single polyline objects. The first prompt asks for the number of sides. For example, to draw an octagon, which is a regular polygon with eight sides, enter 8. Next, decide how to describe the size and location of the polygon. The default setting involves choosing the center and radius of an imaginary circle. To use this method, after entering the number of polygon sides, specify a location for the polygon center point. A prompt then asks if you want to form an *inscribed polygon* or a *circumscribed polygon*. Select the appropriate option and specify the radius to create the polygon. See **Figure 4-23.**

regular polygon:
A closed geometric figure with three or more equal sides and equal angles.

inscribed polygon:
A polygon that is drawn inside an imaginary circle so that its corners touch the circle.

circumscribed polygon: A polygon that is drawn outside of an imaginary circle so that the sides of the polygon are tangent to the circle.

> **NOTE**
>
> The number of polygon sides you enter, the **Inscribed in circle** or **Circumscribed about circle** option you select, and the radius you enter are stored as the new default settings, allowing you to quickly draw another polygon with the same characteristics.

Figure 4–23. Polygons can be inscribed in a circle (left) or circumscribed around a circle (right).

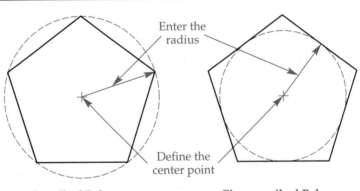

Inscribed Polygon   Circumscribed Polygon

PROFESSIONAL TIP

Regular polygons, such as the *hexagons* commonly drawn to represent bolt heads and nuts on mechanical drawings, are normally dimensioned across the flats. Use the **Circumscribed about circle** option to draw a polygon dimensioned across the flats. The radius you enter is equal to one-half the distance across the flats. Use the **Inscribed in circle** option to dimension a polygon across the corners, and to confine the polygon within a circular area.

Using the Edge Option

Use the **Edge** option to construct a polygon if you do not know the center point location or radius of the imaginary circle, but you do know the size and location of a polygon edge. After you access the **POLYGON** tool and enter the number of sides, choose the **Edge** option at the Specify center of polygon or [Edge]: prompt. Specify a point for the first endpoint of one of the polygon's sides. Then select the second endpoint of the polygon side. See **Figure 4-24.**

Exercise 4-10

Access the Student Web site (www.g-wlearning.com/CAD) and complete Exercise 4-10.

Figure 4–24.
Use the **Edge** option of the **POLYGON** tool to construct a polygon according to the location and size of an edge.

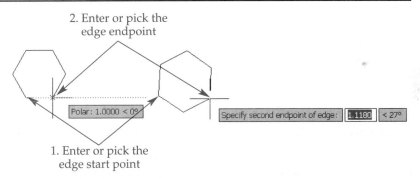

2. Enter or pick the edge endpoint

Polar: 1.0000 < 0°

Specify second endpoint of edge: 1.1180 < 27°

1. Enter or pick the edge start point

Drawing Rectangles

Use the **RECTANGLE** tool to draw rectangles easily. Rectangles drawn using the **RECTANGLE** tool are single polyline objects. To draw a rectangle using default settings, enter or pick one corner and then the opposite diagonal corner. See **Figure 4-25.** By default, the **RECTANGLE** tool draws a rectangle at a 0° angle with sharp corners.

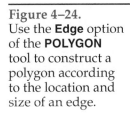

RECTANGLE

Drawing Chamfered Rectangles

Use the **Chamfer** option to include *chamfered* corners during rectangle construction. See **Figure 4-26A.** When prompted, enter the first chamfer distance, followed by the second chamfer distance. Entering a value of 0 at the first or second chamfer distance prompt creates a rectangle with sharp corners. After setting the chamfer distances, you can either draw the rectangle or select another option. However, using the **Fillet**

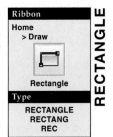

Figure 4–25.
Using the **RECTANGLE** tool.

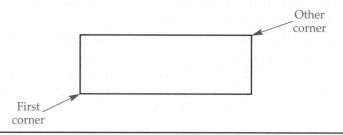

option overrides the **Chamfer** option. The rectangle you draw must be large enough to accommodate the specified chamfer distances. Otherwise, you will see sharp corners. New rectangles continue to produce chamfers until you reset the chamfer distances to 0 or use the **Fillet** option to create rounded corners.

Drawing Rounded Rectangles

fillet: A rounded interior corner.

round: A rounded exterior corner.

Use the **Fillet** option to include rounded corners during rectangle construction. See **Figure 4-26B.** AutoCAD uses the term *fillet* to describe both *fillets* and *rounds*. The **Fillet** option requires you to specify the round radius. When prompted, enter the round radius. After setting the round radius, you can either draw the rectangle or select another option. However, using the **Chamfer** option overrides the **Fillet** option. The rectangle you draw must be large enough to accommodate the specified round radius. Otherwise, you will see sharp corners. New rectangles continue to produce rounds until you reset the round radius to 0 or use the **Chamfer** option to create chamfered corners.

> **NOTE**
>
> This chapter introduces adding chamfers and rounds while creating rectangles. Chapter 12 covers adding chamfers and rounds using the **FILLET** tool in more detail.

Drawing Rectangles with Line Width

The **Width** option allows you to adjust rectangle line width, or "boldness." This should not to be confused with lineweight, described in Chapter 5. After you choose the **Width** option, a prompt asks you to enter the line width. For example, to create a rectangle with lines that are .5 wide, enter .5. After setting the rectangle width, you

Figure 4–26.
You can add chamfers or rounds to rectangles during construction.

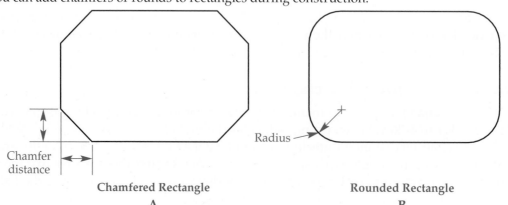

Chamfer distance

Radius

Chamfered Rectangle
A

Rounded Rectangle
B

AutoCAD and Its Applications—Basics

can either draw the rectangle or set additional options. All new rectangles are drawn using the specified width. Reset the **Width** option to 0 to create new rectangles using a standard "0-width" line.

Specifying Rectangle Area

Use the **Area** option to draw a rectangle when you know the area of the rectangle and the length of one of its sides. This option is available after you pick the first corner point. Enter the **Area** option, and then specify the total area for the rectangle using a value that corresponds to the current units. For example, enter 45 to draw a rectangle with an area of 45 in^2. Next, choose the **Length** option if you know the length of a side, or the X value, or choose the **Width** option if you know the width of a side, or Y value. When prompted, enter the length or width to complete the rectangle. AutoCAD calculates the unspecified dimension automatically and draws the rectangle.

Specifying Rectangle Dimensions *Part of HomeWork p. 133*

An alternative to using point entry to specify the opposite corner of a rectangle is to enter the length and width of a rectangle using the **Dimensions** option. The option is available after you pick the first corner of the rectangle. Enter the **Dimensions** option and specify the length of a side, which is the X value. Next, enter the width of a side, which is the Y value. After you specify the length and width, AutoCAD asks for the other corner point. If you want to change the dimensions, select the **Dimensions** option again. If the dimensions are correct, specify the other corner point to complete the rectangle. The second corner point determines which of four possible rectangles you draw. See **Figure 4-27.**

Drawing a Rotated Rectangle

Use the **Rotation** option, available after you pick the first corner of the rectangle, to draw a rectangle at an angle other than 0°. A prompt asks you to specify the rotation angle. Enter or select an angle to rotate the rectangle. Then pick the opposite corner of the rectangle. An alternative is to choose the **Pick points** option when the Specify rotation angle or [Pick points]: prompt appears. If you select the **Pick points** option, the prompt asks you to select two points to define the angle. A new rotation value becomes the default angle for using the **Rotation** option.

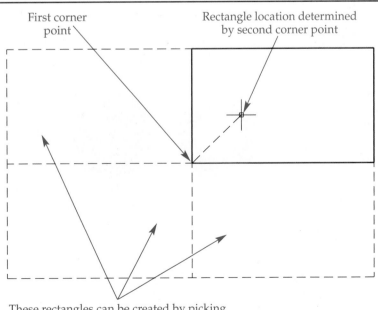

Figure 4–27.
The orientation of the rectangle relative to the first corner point is determined by the second corner point.

First corner point

Rectangle location determined by second corner point

These rectangles can be created by picking a different second corner point

NOTE

The **Elevation** and **Thickness** options of the **RECTANGLE** tool are appropriate for 3D applications and are explained in *AutoCAD and Its Applications—Advanced*.

PROFESSIONAL TIP

You can use a combination of rectangle settings to draw a single rectangle. For example, you can enter a width value, chamfer distances, and length and width dimensions to create a single rectangle.

Exercise 4-11

Access the Student Web site (www.g-wlearning.com/CAD) and complete Exercise 4-11.

Drawing Donuts and Filled Circles

DONUT

Ribbon
Home
> Draw
Donut
Type
DONUT
DO

The **DONUT** tool allows you to draw a thick or filled circle. See **Figure 4-28**. A donut is a single polyline object. After activating the **DONUT** tool, enter the inside diameter and then the outside diameter of the donut. Enter a value of 0 for the inside diameter to create a completely filled donut, or solid circle.

The center point of the donut attaches to the crosshairs, and the Specify center of donut or <exit>: prompt appears. Pick a location to place the donut. The **DONUT** tool remains active until you right-click or press [Enter], the space bar, or [Esc]. This allows you to place multiple donuts of the same size using a single instance of the **DONUT** tool.

Exercise 4-12

Access the Student Web site (www.g-wlearning.com/CAD) and complete Exercise 4-12.

Figure 4–28.
The appearance of a donut depends on its inside and outside diameters and the current **FILL** mode.

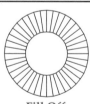  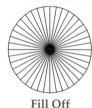

Fill On   Fill On   Fill Off   Fill Off
I.D. = 0   I.D. = 0

Drawing True Splines

The **SPLINE** tool is used to create a special type of curve called a *nonuniform rational B-spline (NURBS) curve*, or *spline*. To create a spline, access the **SPLINE** tool and specify control points using any standard point entry method. For example, the spline shown in **Figure 4-29** uses absolute coordinate values of 2,2 for the first point, 4,4 for the second point, and 6,2 for the last point.

After you specify all spline control points, press [Enter] or the space bar, or right-click and select **Enter**, to end the point entry process. To complete the spline, enter values at the Specify start tangent: and Specify end tangent: prompts. Specifying the start and end tangents changes the direction in which the spline curve begins and ends. Right-click or press [Enter] or the space bar at the tangent prompts to accept the default direction, as calculated by AutoCAD.

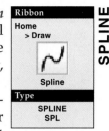

Ribbon
Home
> Draw

Spline

Type
SPLINE
SPL

nonuniform rational B-spline (NURBS) curve: A true (mathematically correct) spline.

NOTE

If you specify only two points for a spline curve and accept the AutoCAD default start and end tangents, an object that looks like a line is created, but the object is a spline.

Drawing Closed Splines

You can use the **Close** option of the **SPLINE** tool to draw a closed spline by connecting the last point to the first point. See **Figure 4-30**. After closing a spline, you are prompted to specify the tangent direction. Press [Enter] or the space bar or right-click and select **Enter** to accept the default calculated by AutoCAD.

Specifying the Spline Tangents

The previous **SPLINE** tool examples use the default AutoCAD tangent directions. You can set tangent directions by entering values at the prompts that appear after you pick spline points. The tangency is based on the tangent direction of the selected point. The results of specifying vertical and horizontal tangent directions are shown in **Figure 4-31**.

Exercise 4-13

Access the Student Web site (www.g-wlearning.com/CAD) and complete Exercise 4-13.

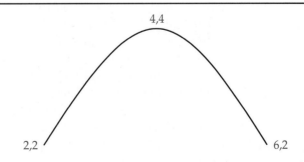

Figure 4-29.
A spline drawn with the **SPLINE** tool. AutoCAD's default start and end tangents were used for this spline.

Figure 4-30.
Using the **Close** option of the **SPLINE** tool with AutoCAD default tangents to draw a closed spline. Compare this spline to the object shown in Figure 4-29.

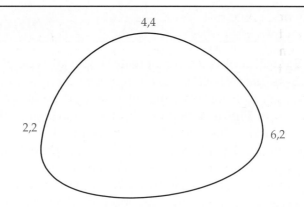

Figure 4-31.
These splines were drawn through the same points, but they have different start and end tangent directions. The arrows indicate the tangent directions.

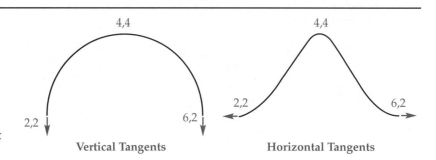

Vertical Tangents   Horizontal Tangents

Chapter Test

Answer the following questions. Write your answers on a separate sheet of paper or go to the Student Web site (www.g-wlearning.com/CAD) and complete the electronic chapter test.

1. Describe the rubberband display shown when you draw a **Center, Radius** circle. What is the purpose of the rubberband?
2. When you use the **CIRCLE** tool, what are the options for responding to the prompt Specify radius of circle?
3. Explain how to create a circle with a diameter of 2.5 units.
4. What option of the **CIRCLE** tool creates a circle of a specific radius that is tangent to two existing objects?
5. Define the term *point of tangency*.
6. Explain how to draw a circle tangent to three objects.
7. Briefly explain how to create a three-point arc.
8. Explain the procedure to draw an arc beginning with the center point and having a 60° included angle.
9. Define the term *included angle* as it applies to an arc.
10. What is the default option if you enter the **ARC** tool at the keyboard?
11. List three input options that you can use to draw an arc tangent to the endpoint of a previously drawn arc.
12. Name the two axes found on an ellipse.
13. Briefly describe the procedure to draw an ellipse using the **Axis, End** option.
14. What is the **ELLIPSE** rotation angle that causes you to draw a circle?
15. Identify two ways to access the **Arc** option for drawing elliptical arcs.
16. How do you draw a filled arrow using the **PLINE** tool?
17. Which **PLINE** tool option allows you to specify the width from the center to one side?
18. Explain how to turn the **FILL** mode off.
19. Briefly describe how to create a polyline parallel to a previously drawn polyline or line.

20. Explain how to draw a hexagon measuring 4″ (102 mm) across the flats.
21. Given the distance across the flats of a hexagon, would you use the **Inscribed** or **Circumscribed** option to draw the hexagon?
22. Name the ribbon tab and panel that contains the **RECTANGLE** tool.
23. Name at least three tools you could use to create a rectangle.
24. Name the tool option used to draw rectangles with rounded corners.
25. Name the tool option designed for drawing rectangles with a specific line thickness.
26. Give the easiest keyboard shortcut for the following tools:
    A. **CIRCLE**
    B. **ARC**
    C. **ELLIPSE**
    D. **POLYGON**
    E. **RECTANGLE**
    F. **DONUT**
27. Describe a method for drawing a solid circle.
28. Explain how to draw two donuts with an inside diameter of 6.25 and an outside diameter of 9.50.
29. Name the tool that can be used to create a true spline.
30. How do you accept the AutoCAD defaults for the start and end tangents of a spline?

Drawing Problems

Start AutoCAD if it is not already started. Start a new drawing from scratch or use an appropriate template of your choice. Follow the specific instructions for each problem. Do not draw centerlines. Do not draw dimensions or text. Use your own judgment and approximate dimensions when necessary.

▼ Basic

1. Use the **LINE** tool and the **CIRCLE** tool options to draw the objects below. Save the drawing as P4-1.

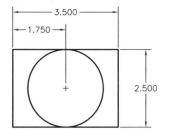

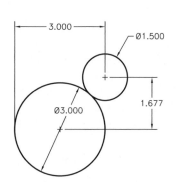

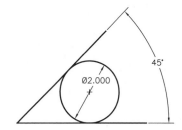

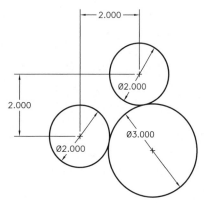

2. Use the **CIRCLE** and **ARC** tool options to draw the object below. Save the drawing as P4-2.

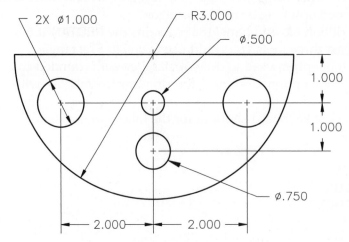

3. Draw the spacer below. Save the drawing as P4-3.

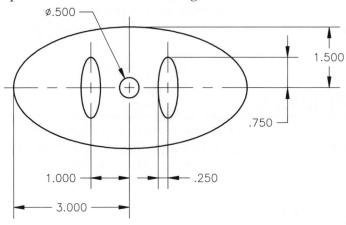

4. Draw the following object. Save the drawing as P4-4.

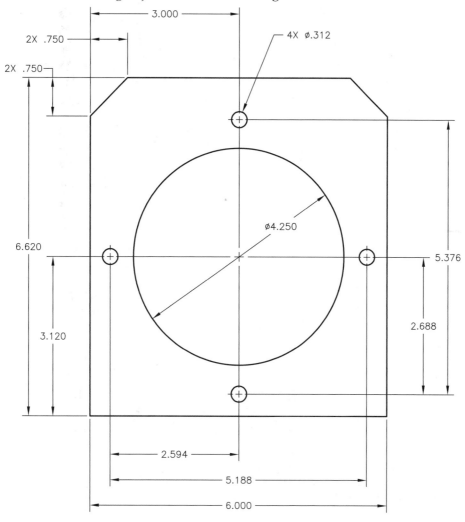

5. Draw the following object. Save the drawing as P4-5.

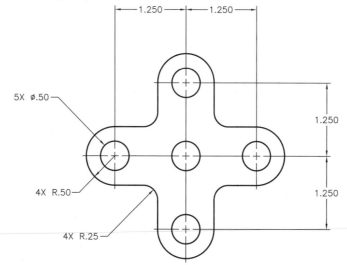

6. Draw the following object. Save the drawing as P4-6.

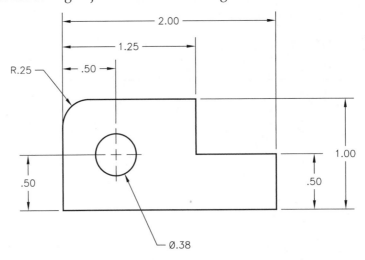

7. Draw the pipe spacer shown. Save the drawing as P4-7.

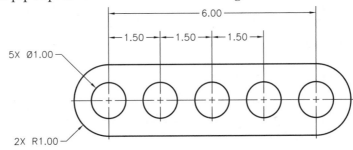

8. Use the **PLINE** tool to draw the following object with a .032 line width. Save the drawing as P4-8.

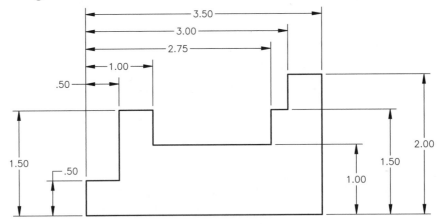

Drawing Problems - Chapter 4

9. Use the **PLINE** tool to draw the following object with a .032 line width. Save the drawing as P4-9.

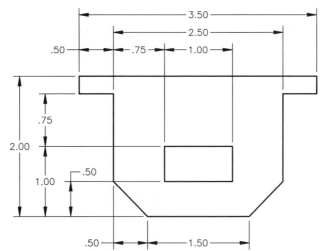

10. Use the **PLINE** tool to draw the following object with a .032 line width.
    A. Deactivate solid fills and use the **REGEN** tool, and reactivate solid fills and reissue the **REGEN** tool.
    B. Observe the difference with solid fills enabled.
    C. Save the drawing as P4-10.

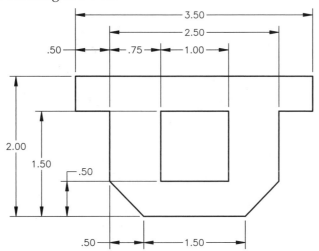

11. Use the **PLINE** tool to draw the filled rectangle shown below. Save the drawing as P4-11.

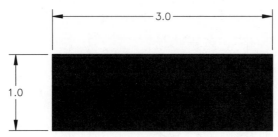

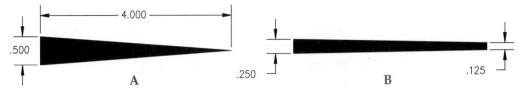

A

B

13. Draw the object shown below. Set decimal units, .25 grid spacing, .0625 snap spacing, and limits of 11,8.5. Save the drawing as P4-13.

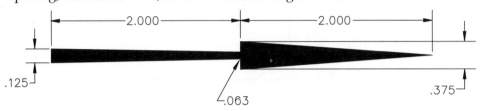

▼ Intermediate

14. Draw the following object. Save the drawing as P4-14.

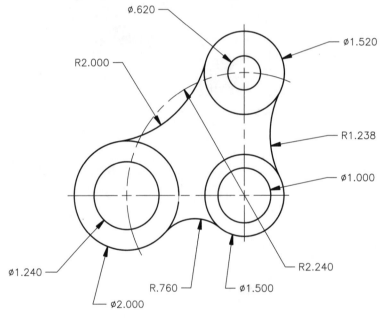

(Art courtesy of Bruce L. Wilcox)

15. Draw the pressure cylinder shown below. Use the **Arc** option of the **ELLIPSE** tool to draw the cylinder ends. Save the drawing as P4-15.

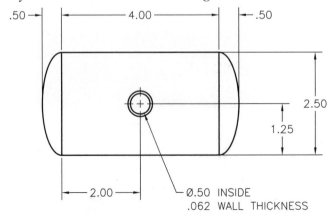

Ø.50 INSIDE
.062 WALL THICKNESS

16. Draw the gasket shown below. Save the drawing as P4-16.

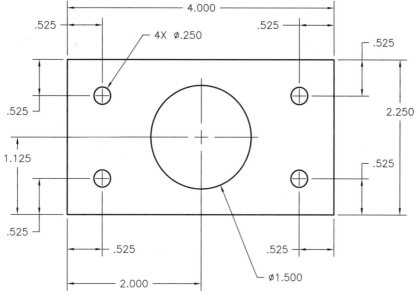

17. Draw the single polyline shown below. Use the **Arc**, **Width**, and **Close** options of the **PLINE** tool to complete the shape. Set the polyline width to 0, except at the points indicated. Save the drawing as P4-17.

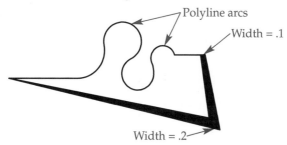

18. Draw the two curved arrows shown below using the **Arc** and **Width** options of the **PLINE** tool. The arrowheads should have a starting width of 1.4 and an ending width of 0. The body of each arrow should have a beginning width of .8 and an ending width of .4. Save the drawing as P4-18.

19. Draw the following object. Save the drawing as P4-19.

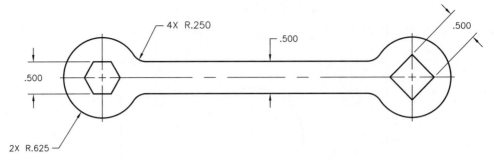

20. Draw the following object. Save the drawing as P4-20.

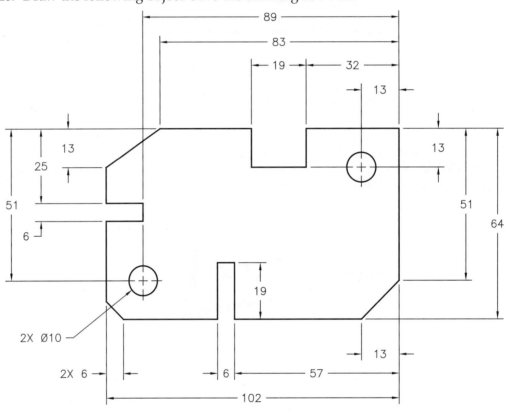

21. Draw the pipe fitting shown. Save the drawing as P4-21.

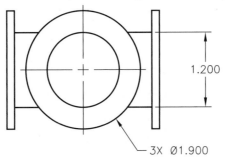

1.200

3X Ø1.900

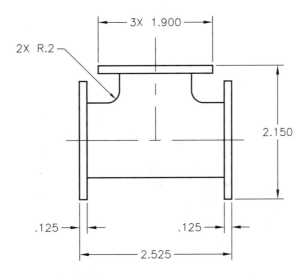

3X 1.900

2X R.2

2.150

.125   .125

2.525

22. Draw the ellipse template shown. Save the drawing as P4-22.

ELLIPSE TABLE		
KEY	MAJOR DIA	MINOR DIA
A	.9951	.5745
B	1.0717	.6187
C	1.1482	.6629
D	1.2247	.7071
E	1.3013	.7513
F	1.3778	.7955
G	1.4544	.8397
H	1.5309	.8839
I	1.6075	.9281
J	1.6840	.9723
K	1.7606	1.0165
L	1.8371	1.0607
M	1.9902	1.1490
N	2.1433	1.2374
O	2.2964	1.3258
P	2.4495	1.4142

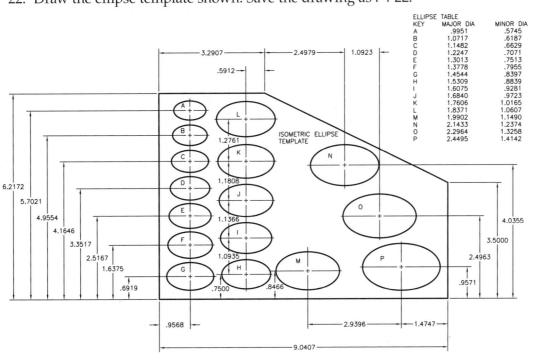

ISOMETRIC ELLIPSE TEMPLATE

3.2907   2.4979   1.0923

.5912

1.2761

1.1808

1.1366

1.0935

.7500   .8466

.6919

.9568

2.9396   1.4747

9.0407

6.2172

5.7021

4.9554

4.1646

3.3517

2.5167

1.6375

4.0355

3.5000

2.4963

.9571

23. Draw the gasket shown. Save the drawing as P4-23.

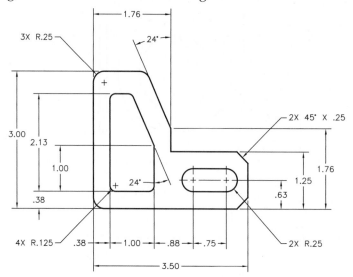

24. Use the **SPLINE** tool to draw the curve for the cam displacement diagram below. Use the following guidelines and the given drawing to complete this problem:
    A. The total rise equals 2.000.
    B. The total displacement can be any length.
    C. Divide the total displacement into 30° increments.
    D. Draw a half circle divided into 6 equal parts on one end.
    E. Draw a horizontal line from each division of the half circle to the other end of the diagram.
    F. Draw the displacement curve with the **SPLINE** tool by picking points where the horizontal and vertical lines cross.
    G. Label the displacement increments along the horizontal scale as shown. Save the drawing as P4-24.

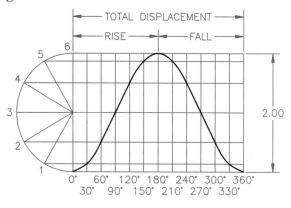

25. Draw the object shown below. Save the drawing as P4-25.

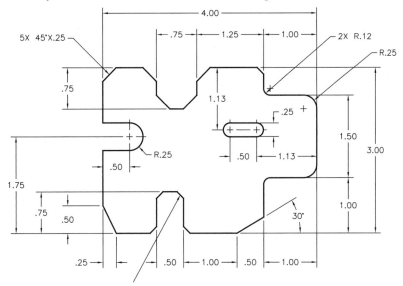

26. You have just been given the sketch of a new sports car design (shown below). You are asked to create a drawing from the sketch. Use the **LINE** tool and selected shape tools to draw the car. Do not be concerned with size and scale. Consider the tools and techniques used to draw the car, and try to minimize the number of objects. Save your drawing as P4-26.

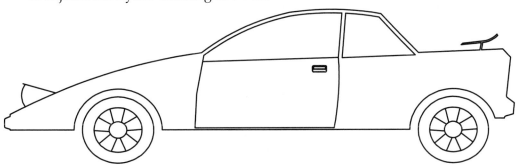

27. You have just been given the sketch of an innovative new truck design (shown below). You are asked to create a drawing from the sketch. Use the **LINE** tool and selected shape tools to draw a truck resembling the sketch. Do not be concerned with size and scale. Save your drawing as P4-27.

28. Draw the following object. Save the drawing as P4-28.

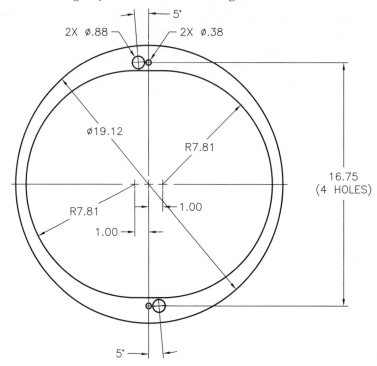

29. Draw this elevation using the **ARC**, **ELLIPSE**, **RECTANGLE**, and **DONUT** tools. Do not be concerned with size and scale. Save the drawing as P4-29.

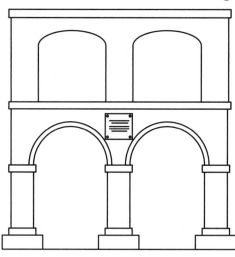

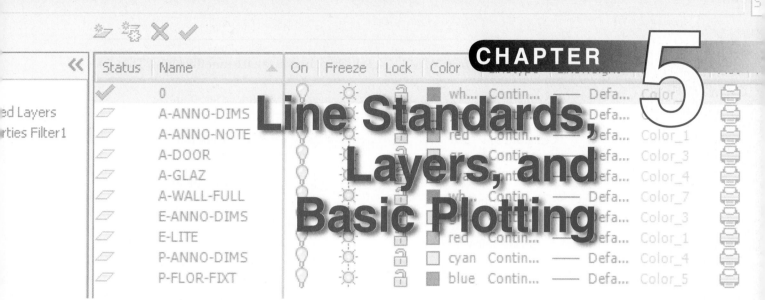

Line Standards, Layers, and Basic Plotting

Learning Objectives

After completing this chapter, you will be able to do the following:
✓ Describe basic line conventions.
✓ Create and manage layers.
✓ Draw objects on separate layers.
✓ Use **DesignCenter** to copy layers and linetypes between drawings.
✓ Print and plot your drawings.

AutoCAD uses a layer system to organize linetypes and other object characteristics to conform to accepted standards and conventions. You can also use layer display options to help create different drawing sheets, views, and displays from a single drawing. This chapter introduces line conventions and the AutoCAD layer system. It also introduces printing and plotting so that you can begin printing and plotting your drawings. Printing and plotting are described in detail later in this textbook.

Line Standards

Drafting is a graphic language that uses lines, symbols, and words to describe products to be manufactured. *Line conventions* are designed to enhance the readability of drawings. This section introduces the line standards that you will apply later in this chapter when you begin loading linetypes and defining layers, as well as throughout your drafting career.

Drawings include a variety of linetypes to classify drawing content. The ASME drafting standards recommend two line thicknesses to establish contrasting lines in a drawing. Lines are thick or thin. Thick lines are twice as thick as thin lines, with recommended widths of 0.6 mm and 0.3 mm, respectively. **Figure 5-1** shows recommended line width and type as defined in ASME Y14.2M, *Line Conventions and Lettering*. **Figure 5-2** describes the most often used linetypes. **Figure 5-3** shows an example of a drawing with several common linetypes. This textbook further explains and illustrates line standards when applicable.

> **line conventions:**
> Standards related to line thickness and type.

Figure 5-1.
Line conventions adapted from ASME Y14.2M. Thick lines use a 0.6 mm thickness. Thin lines use a 0.3 mm line thickness.

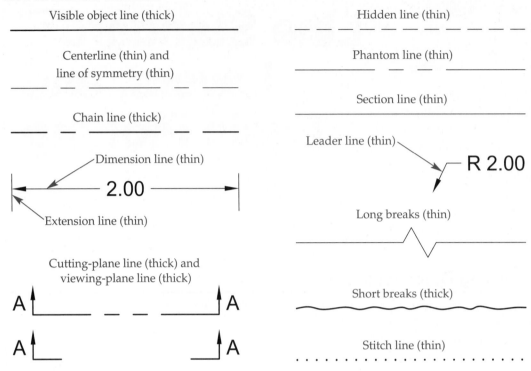

Figure 5-2.
Descriptions of common lines and line standards. Line characteristics and spacing are measured at full scale. Specifications vary according to drawing size.

Type	Purpose	Standards
Object lines (visible, outline)	Show the contour or outline of objects.	Thick, solid (Continuous).
Hidden lines	Represent features that are hidden in the current view.	Thin, dashed (HIDDEN or DASHED). Dashes are .125″ (3 mm) long and are spaced .06″ (1.5 mm) apart.
Centerlines	Locate the centers of circles and arcs, and show the axis of cylindrical or symmetrical shapes.	Thin (CENTER). Extend .125″ to .25″ (3 mm to 6 mm) past objects. Centerlines consist of one .125″ dash alternating with one .75″ to 1.5″ (19 mm to 38 mm) dash. A .06″ (1.5 mm) space separates the dashes. Small centerline dashes should cross only at the center of a circle.
Extension lines	Show the extent of a dimension.	Thin, solid (Continuous). Begin .06″ (1.5 mm) from an object and extend .125″ (3 mm) beyond the last dimension line. Can cross object lines, hidden lines, and centerlines, but should not cross dimension lines. Centerlines become extension lines when used to show the extent of a dimension.
Dimension lines	Show the distance being measured.	Thin, solid (Continuous). Broken near the center for placement of the dimension numeral in mechanical drafting. Unbroken in architectural and structural drawings, with dimension placed on top of the dimension line. Arrows terminate the ends of dimension lines, except in architectural drafting, where slashes (ticks) or dots are often used.

(Continued)

Figure 5-2.
(Continued)

Type	Purpose	Standards
Leader lines	Connect a specific note to a feature on a drawing.	Thin, solid (Continuous). Often terminate with an arrowhead at the feature. May be curved on architectural drawings. Straight leader lines often have a small shoulder at the note.
Cutting-plane lines	Identify the location and viewing direction of a section view.	Thick (PHANTOM or DASHED). Can be drawn in one of two ways.
Viewing-plane lines	Identify the location of a view	Thick (PHANTOM or DASHED). Can be drawn in one of two ways.
Section lines	In a section view, show where material has been cut away.	Thin, usually drawn in a pattern. Different linetypes can be used to indicate specific or different material.
Break lines	Show where a portion of an object has been removed for clarity or convenience.	Thin or thick depending on the symbol, solid (Continuous). Break representation is based on the object or material being broken.
Phantom lines	Identify repetitive details, show alternate positions of moving parts, and locate adjacent positions of related parts.	Thin, PHANTOM. Two .125″ (3 mm) dashes, alternating with one .75″ to 1.5″ (19 mm to 38 mm) dash. Spaces between dashes are .06″ (1.5 mm).
Chain lines	Indicate special features or unique treatment for a surface.	Thick (CENTER).

Figure 5-3.
An example of a mechanical drawing with several common types of lines.

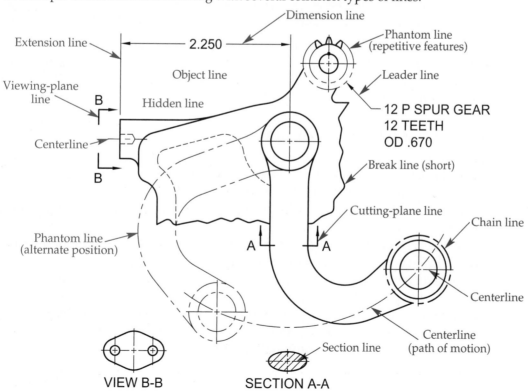

VIEW B-B   SECTION A-A

Figure 5-4.
A few of the most common U.S. National CAD Standard line thicknesses.

Thickness	Application
Thin, 0.25 mm	Dimensioning features, phantom lines, hidden lines, centerlines, long break lines, schedule grid lines, and background objects.
Medium, 0.35 mm	Object lines, text for dimension values, notes and schedules, terminator marks, door and window elevations, and schedule grid accent lines.
Wide, 0.5 mm	Major object lines at elevation edges, cutting-plane lines, short break lines, title text, minor title underlines, and border lines.
Extra wide, 0.7 mm	Major title underlines, schedule outlines, large titles, special emphasis object lines, elevation and section grade lines, property lines, sheet borders, and schedule borders.

The U.S. National CAD Standard (NCS) recommends a specific line thickness and characteristics for architectural and similar drawings. Thickness options range from extra fine, with a 0.13 mm thickness, to 4X wide, with a 2 mm thickness. You can use the range of NCS-recommended lineweights to provide accents to your drawings as needed. A common practice is to select a few of the lineweights that correlate best to specific applications. See **Figure 5-4.**

NOTE

Many AutoCAD tools simplify and automate the process of using and applying correct lines standards. You will learn applications and techniques for drawing specific types of lines in this book.

Introduction to Layers

overlay system:
A system of separating drawing components by layer.

layers:
Components of AutoCAD's overlay system that allow users to separate objects into logical groups for formatting and display purposes.

In AutoCAD, an *overlay system* separates different drawing components, referred to as *layers*. You can display all the layers together, or "overlaid," to reflect the entire design drawing. Display or hide individual layers as needed to show specific details or design components.

Increasing Productivity with Layers

Using layers increases productivity in several ways:
- ✓ Each layer can be assigned a different color, linetype, and lineweight to correspond to line conventions and to help improve clarity.
- ✓ Changes can be made to a layer promptly, affecting all objects drawn on the layer.
- ✓ Selected layers can be turned off or frozen to decrease the amount of information displayed on the screen or to speed screen regeneration.
- ✓ Each layer can be plotted in a different color, linetype, or lineweight, or it can be set not to plot at all.
- ✓ Specific information can be grouped on separate layers. For example, a floor plan can be drawn on specific floor plan layers, the electrical plan on electrical layers, and the plumbing plan on plumbing layers.

✓ Several plot sheets can be created from the same drawing file by controlling layer visibility to separate or combine drawing information. For example, a floor plan and electrical plan can be reproduced together and sent to an electrical contractor for a bid. The floor plan and plumbing plan can be reproduced together and sent to a plumbing contractor.

Layers Used in Different Drafting Fields

Typically, the type of drawing you create determines the function of each layer. In mechanical drafting, you usually assign a specific layer to each different type of line or object. For example, draw object lines on an Object layer that is black in color, uses a solid (Continuous) linetype, and is 0.6 mm wide. Draw hidden lines on a green Hidden layer that uses a 0.3 mm hidden (HIDDEN or DASHED) linetype.

Architectural and civil drawings may have hundreds of layers, each used to produce a specific item. For example, draw full-height floor plan walls on a black A-WALL-FULL layer that uses a 0.5 mm solid (Continuous) linetype. Add plumbing fixtures to a floor plan on a blue P-FLOR-FIXT layer that uses a 0.35 mm solid (Continuous) linetype.

You can create layers for any type of drawing, including detail parts, assemblies, floor plans, foundation plans, partition layouts, plumbing systems, electrical systems, structural systems, roof drainage systems, reflected ceiling systems, HVAC systems, site plans, profiles, topographic maps, and details. Interior designers may use floor plan, interior partition, and furniture layers. In electronics drafting, you can draw each level of a multilevel circuit board on its own layer.

Creating and Using Layers

The **LAYER** tool opens the **Layer Properties Manager** palette, which is used to create and delete layers and control layer properties. See **Figure 5-5.** Two panes divide the **Layer Properties Manager**. The list view pane on the right side uses a column format to list layers and provide layer property controls. Properties in each column appear as an

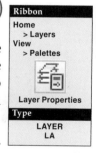

Ribbon

Home
> Layers
View
> Palettes

Layer Properties

Type

LAYER
LA

LAYER

Figure 5-5.
The **Layer Properties Manager**. Layer 0 is the AutoCAD default layer.

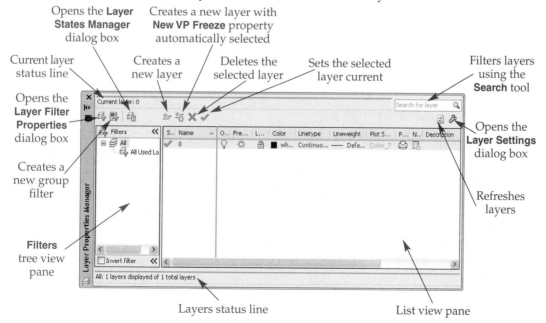

Opens the **Layer States Manager** dialog box

Creates a new layer with **New VP Freeze** property automatically selected

Current layer status line

Creates a new layer

Deletes the selected layer

Sets the selected layer current

Filters layers using the **Search** tool

Opens the **Layer Filter Properties** dialog box

Opens the **Layer Settings** dialog box

Creates a new group filter

Refreshes layers

Filters tree view pane

Layers status line

List view pane

Figure 5-6.
Pick the icons in the **Layer Properties Manager** to change layer settings.

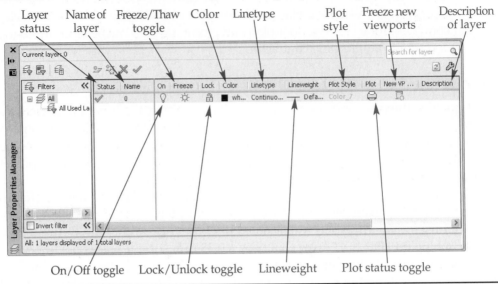

icon or as an icon and a name. See **Figure 5-6.** Pick a property to change layer settings. The tree view pane on the left side of the palette displays filters that you can use to limit the number of layers displayed in the list view pane.

Only one layer is required in an AutoCAD drawing. This default layer is named 0 and cannot be deleted, renamed, or purged from the drawing. The 0 layer is primarily reserved for drawing blocks, as described later in this textbook. Draw each object on a layer specific to the object. For example, draw object lines on an Object layer, draw floor plan walls on an A-WALL layer, and draw construction lines on a Construction or A-ANNO-NPLT layer.

Defining New Layers

The **Name** column in the list view shows the names of all the layers in the drawing. Add layers to a drawing to meet the needs of the current drawing project. To add a new layer, select an existing layer that contains properties similar to those you want to assign to the new layer. If this is the first new layer in a default template, only the 0 layer is available to reference. Then pick the **New Layer** button, right-click in the list view and select **New Layer**, or press [Enter] or [Alt] + [N]. A new layer appears, using a default name. See **Figure 5-7.** The layer name is highlighted when the listing appears, allowing you to type a new name. Pick away from the layer in the list or press [Enter] to accept the layer.

Figure 5-7.
A new layer is named Layer*n* by default.

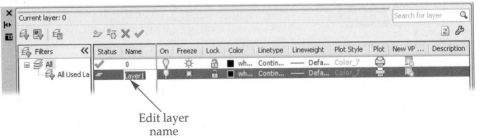

Layer Names

Layers should be given names to reflect drawing content. Layer names can have up to 255 characters and can include letters, numbers, and certain other characters, including spaces. Some examples of typical mechanical, architectural, and civil drafting layer names include:

Mechanical	Architectural	Civil
Object	A-WALL-FULL	C-BLDG
Hidden	A-GLAZ	C-WATR
Center	A-DOOR	C-TOPO
Dimension	E-LITE	C-PROP
Construction	P-FLOR-FIXT	C-NGAS
Section	S-FNDN	C-SSWR
Border	M-HVAC	C-ELEV

Layers names are usually set according to specific industry or company standards. However, for very simple drawings, layers might be named by linetype and color. For example, the layer name Continuous-White can have a continuous linetype drawn in white. The layer usage and color number, such as Object-7, may be used to indicate an object line with color 7. Another option is to assign the linetype a numerical value. For example, object lines can be 1, hidden lines can be 2, and centerlines can be 3. If you use this method, keep a written record of your numbering system for reference.

More complex layer names may be appropriate for some applications. The name might include the drawing number, color code, and layer content. The layer name Dwg100-2-Dimen, for example, could refer to drawing DWG100, color 2, for use when adding dimensions. The American Institute of Architects (AIA) *CAD Layer Guidelines*, associated with the NCS, specifies a layer naming system for architectural and related drawings. The system uses a highly detailed layer naming process that gives each layer a discipline designator and major group, and if necessary, one or two minor groups and a status field. The AIA system allows complete identification of drawing content.

Layer names automatically arrange alphanumerically as you create new layers. See **Figure 5-8.** Pick any column heading in the list view to sort layer names in ascending or descending order according to that column. The **Layer Properties Manager** is a palette, so new layers and changes made to existing layers automatically save and apply to the drawing. There is no need to "apply" changes or close the palette to see the effects of the changes in the drawing.

PROFESSIONAL TIP

If you need to create multiple layers, accelerate the process by pressing the comma key after typing each layer name to create another new layer.

Renaming Layers

To change an existing layer name using the **Layer Properties Manager**, pick the name in the **Name** column once to highlight it, pause for a moment, and then pick it again. Type the new name and press [Enter] or pick outside of the text box. You can also rename a layer by picking the name once to highlight it and then pressing [F2], or by right-clicking and selecting **Rename Layer**. You cannot rename layer 0 and layers associated with an external reference.

Figure 5-8.
Layer names are automatically placed in alphanumeric order when you create new layers or change layer names.

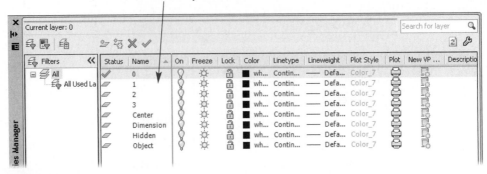

Layer names sort
automatically

Exercise 5-1

Access the Student Web site (www.g-wlearning.com/CAD) and complete Exercise 5-1.

Selecting Multiple Layers

Select multiple layers to speed the process of deleting or applying the same properties to several layers. To select multiple layers in the list view, use the same techniques you use to select files. You can highlight a single name by picking it, while picking another name deselects the previous name and highlights the new selection. You can use [Shift] to select two layers and all layers between the selections in the listing. Holding [Ctrl] while picking layer names highlights or deselects each selected name without affecting any other selections. You can also use a window selection to select all the layers that contact the window. The following additional selection options are available when you right-click in the list view:

- **Select All.** Selects all layers.
- **Clear All.** Deselects all layers.
- **Select All but Current.** Selects all layers except the current layer.
- **Invert Selection.** Deselects all selected layers and selects all deselected layers.

Exercise 5-2

Access the Student Web site (www.g-wlearning.com/CAD) and complete Exercise 5-2.

Layer Status

The icon in the **Status** column describes the status, or existing use of a layer. A green check mark indicates the *current layer*. The status line at the top of the **Layer Properties Manager** also identifies the current layer.

A white sheet of paper, or **Not In Use** icon, in the **Status** column indicates that the layer is not being used in any way by the drawing, the layer is not current, and no objects have been drawn on the layer. A blue sheet of paper, or **In Use** icon, in the

current layer:
The active layer. Whatever you draw is placed on the current layer.

Status column means that objects have been drawn with the layer, but the layer is not current. The **In Use** icon can also mean that the layer cannot be deleted or purged from the drawing, even if no objects are drawn on the layer.

Current

NOTE

If layers in use are not indicated in the **Layer Properties Manager**, pick the **Settings** button in the upper-right corner to display the **Layer Settings** dialog box and select the **Indicate layers in use** check box.

Not in Use

In Use

Setting the Current Layer

To set a different layer current using the **Layer Properties Manager**, double-click the layer name, pick the layer name in the layer list and select the **Set Current** button, or right-click on the layer and choose **Set Current**. You can also make a different layer current without using the **Layer Properties Manager** by selecting the layer you want to use from the **Layer Control** drop-down list of the **Home** ribbon tab. See **Figure 5-9.** This is often the most effective way to activate and manage layers while drawing. Pick the drop-down arrow and select a layer from the list to set current. Use the vertical scroll bar to move up and down through a long list.

NOTE

You can use the **Layer Properties Manager** or the **Layer Control** drop-down list to change the current layer or layer properties while a tool is active. For example, draw one line segment using the current layer, and then without exiting the **LINE** tool, make a different layer current to draw the next line segment on that layer.

Figure 5-9.
The **Layer Control** drop-down list allows you to change the current layer and change the properties of layers.

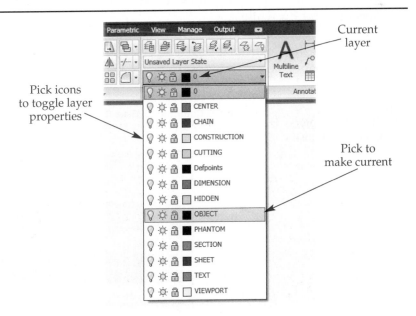

To assign existing objects to a different layer, select the objects and then choose a different layer using the **Layer Properties Manager** or the **Layer Control** drop-down list.

Exercise 5-3

Access the Student Web site (www.g-wlearning.com/CAD) and complete Exercise 5-3.

Setting Layer Color

You can assign a unique color to each layer to help differentiate drawing items on-screen. Layer colors can also affect the appearance of drawings plotted in color and can control object properties such as lineweight. Plotting using colors is not typical, but assigning color to layers is still very important for drawing clarity, organization, workability, and format. Layer colors should highlight the important features on the drawing and not cause eyestrain.

The **Color** column of the list view in the **Layer Properties Manager** indicates the color applied to each layer. Pick the color swatch to change the color of an existing layer using the **Select Color** dialog box. See **Figure 5-10.** This dialog box includes an **Index Color** tab, a **True Color** tab, and a **Color Books** tab from which you can select a color. Each tab uses a different method of obtaining colors.

The default **Index Color** tab includes 255 color swatches from which you can choose. This tab is commonly referred to as the AutoCAD Color Index (ACI) because colors are coded by name and number. The first seven colors in the ACI include both a numerical index number and a name: 1 = red, 2 = yellow, 3 = green, 4 = cyan, 5 = blue, 6 = magenta, and 7 = white.

To select a color, pick the appropriate color swatch or type the color name or ACI number in the **Color:** text box. The color white (number 7) shows up black with the default cream-colored drawing window background. The "white" name comes

Figure 5-10.
Use the **Select Color** dialog box to choose a layer color.

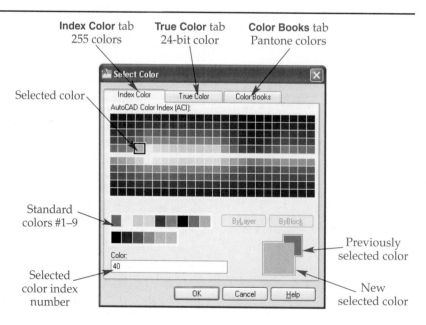

from the concept of using a black drawing window background, which was once the AutoCAD default display. If you change the drawing window background to a dark color such as black, color 7 shows up white on the screen.

As you move the cursor around the color swatches, the **Index color:** note updates to show you the number of the color over which the cursor is hovering. Beside the **Index color:** note is the **Red, Green, Blue:** (RGB) note. This indicates the RGB numbers used to mix the highlighted color. When you pick a color, the **Index color:** note appears in the **Color:** text box. A preview of the newly selected color and a sample of the previously assigned color appear in the lower right of the dialog box. An easy way to explore the ACI numbering system is to pick a color swatch and see what number appears in the **Color:** text box.

After selecting a color, pick the **OK** button to assign the specified color to the selected layer. The color appears as the color swatch for the layer. All objects drawn on this layer appear in the selected color, or ByLayer, by default.

NOTE

Your graphics card and monitor affect color display characteristics and sometimes the number of available colors. Most color systems usually support at least 256 colors.

PROFESSIONAL TIP

Use the color swatch in the **Layer Control** drop-down list to change the color assigned to a layer without accessing the **Layer Properties Manager**.

Type	
	COLOR

COLOR

CAUTION

The **COLOR** tool provides access to the **Select Color** dialog box, which you can use to set an *absolute value* for color. If the absolute color value is set to red, for example, all objects appear red regardless of the color assigned to the layers on which objects are drawn. However, use layers and the **Layer Properties Manager** and set color as ByLayer to control object color for most applications.

absolute value: In property settings, a value set directly instead of being referenced by layer or a block. The current layer settings are ignored when an absolute value is set.

Exercise 5-4

Access the Student Web site (www.g-wlearning.com/CAD) and complete Exercise 5-4.

Setting Layer Linetype

Different line thicknesses and linetypes enhance the readability of drawings. You can apply standard line conventions to objects by assigning a linetype and thickness to each layer. AutoCAD provides standard linetypes that you can use to match ASME, NCS, or other applicable standards. You can also create your own custom linetypes. To achieve different line thicknesses, it is necessary to assign lineweights to layers.

Figure 5-11.
The **Select Linetype** dialog box allows you to load linetypes for use in the current drawing.

List of loaded linetypes

Pick to load additional linetypes

The **Linetype** column of the list view in the **Layer Properties Manager** indicates the linetype applied to each layer. To change the linetype of an existing layer, pick the current linetype. This displays the **Select Linetype** dialog box, shown in **Figure 5-11.** Initially, the Continuous linetype is the only linetype listed in the **Loaded linetypes** list box. Use the Continuous linetype to draw solid lines with no breaks. AutoCAD maintains linetypes in external linetype definition files. Before you can apply a linetype other than Continuous to a layer, you must load the linetype into the **Select Linetype** dialog box.

Adding Linetypes

To add linetypes, pick the **Load...** button to display the **Load or Reload Linetypes** dialog box. See **Figure 5-12.** The acad.lin or acadiso.lin file is active, depending on the template you use to begin a new drawing. The acad.lin and acadiso.lin files are identical except that in the acadiso.lin file, the non-ISO linetypes are scaled up 25.4 times. The scale factor of 25.4 converts inches to millimeters for use in metric drawings. To switch to a different linetype definition file, pick the **File...** button in the **Load or Reload Linetypes** dialog box. Then use the **Select Linetype File** dialog box to select the desired file.

The **Available Linetypes** list displays the name and a description, which includes an image, of each linetype available from the specified linetype definition file. Use the scroll bars to view all available linetypes, and use the image in the **Description** column to aid in selecting the appropriate linetypes to load. Choose a single linetype, or select multiple linetypes using standard selection practices or the shortcut menu. Pick the **OK** button to return to the **Select Linetype** dialog box, where the linetypes you selected now appear. See **Figure 5-13.** In the **Select Linetype** dialog box, pick the desired linetype, and then pick **OK**. The HIDDEN linetype selected in **Figure 5-13** is now the linetype assigned to the layer named Hidden, as shown in **Figure 5-14.**

Figure 5-12.
The **Load or Reload Linetypes** dialog box displays linetypes available for loading.

Select file where linetype definitions are stored

Select linetypes to load into drawing

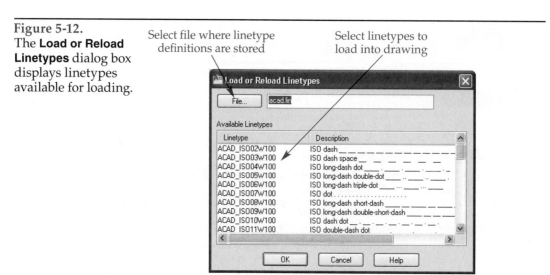

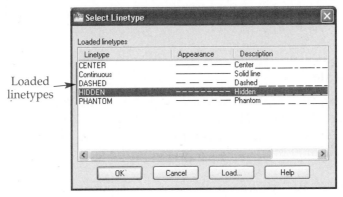

Loaded
linetypes

Figure 5-14.
Objects drawn on the Hidden layer now display a HIDDEN linetype.

Linetype changed
to HIDDEN

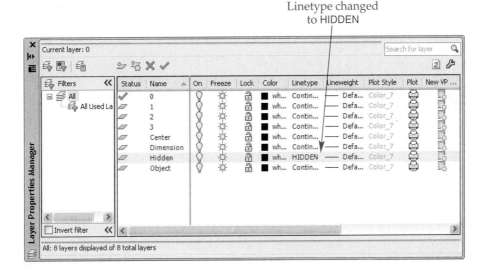

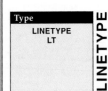

CAUTION

The current linetype can be an absolute linetype value. The **LINETYPE** tool provides access to the **Linetype Manager**, which you can use to control a variety of linetype characteristics, but you should set linetype as ByLayer for most applications. Changing linetype to a value other than ByLayer overrides layer linetype. Therefore, if the absolute linetype value is set to HIDDEN, for example, all objects are drawn using a HIDDEN linetype regardless of the linetype assigned to the layers on which objects are drawn. The **Linetype Manager** includes other options that are unnecessary for typical applications, or are more appropriately set using other techniques. Use layers and the **Layer Properties Manager** to control object linetype for most applications.

Exercise 5-5

Access the Student Web site (www.g-wlearning.com/CAD) and complete Exercise 5-5.

Setting Linetype Scale

linetype scale: The lengths of dashes and spaces in linetypes.

You can change *linetype scale* to increase or decrease the lengths of dashes and spaces in linetypes in order to make your drawing more closely match standard drafting practices. Changing the *global linetype scale* is the preferred method for adjusting linetype scale, though it is possible to change the linetype scale of individual objects.

global linetype scale: A linetype scale applied to every linetype in the current drawing.

You can use the **LTSCALE** system variable to make a global change to the linetype scale. The default global linetype scale factor is 1.0000. Any line with dashes initially assumes this factor. To change the linetype scale for the entire drawing, type LTSCALE and enter a new value. The drawing regenerates and the global linetype scale changes for all lines on the drawing. A value less than 1.0 makes the dashes and spaces smaller, and a value greater than 1.0 makes the dashes and spaces larger. See Figure 5-15. Experiment with different linetype scales until you achieve the desired results.

> **CAUTION**
>
> Be careful when changing linetype scales to avoid making your drawing look odd and not in accordance with drafting standards.

Setting Layer Lineweight

lineweight: The assigned width of lines for display and plotting.

Assign *lineweight* to a layer to manage the weight, or thickness, of objects. You can control the display of line thickness to match ASME, NCS, or other applicable standards. The **Lineweight** column of the list view in the **Layer Properties Manager** indicates the lineweight applied to each layer. To change the lineweight of an existing layer, pick the current lineweight to display the **Lineweight** dialog box. See Figure 5-16. The **Lineweight** dialog box displays fixed lineweights available in AutoCAD. Scroll through the **Lineweights:** list and select the lineweight you want to assign to the layer. Pick the **OK** button to apply the lineweight and return to the **Layer Properties Manager**.

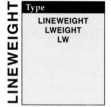

The **LINEWEIGHT** tool provides access to the **Lineweight Settings** dialog box, shown in Figure 5-17. The **Units for Listing** area allows you to set the lineweight thickness to **Millimeters (mm)** or **Inches (in)**. The units apply only to values in the **Lineweight** and **Lineweight Settings** dialog boxes, allowing you to select lineweights based on a known unit of measurement.

Check the **Display Lineweight** box to turn lineweight on. Lineweight appears on-screen when lineweight display is turned on. Use the **Adjust Display Scale** slider to

Figure 5-15.
The CENTER linetype at different linetype scales.

Scale Factor	Line
0.5	⎯ ⎯ ⎯ ⎯ ⎯ ⎯ ⎯ ⎯ ⎯ ⎯
1.0	⎯⎯ ⎯ ⎯⎯ ⎯ ⎯⎯ ⎯ ⎯⎯
1.5	⎯⎯⎯ ⎯ ⎯⎯⎯ ⎯ ⎯⎯⎯

Figure 5-16.
Use the **Lineweight** dialog box to assign a lineweight to a layer.

Select lineweight from list

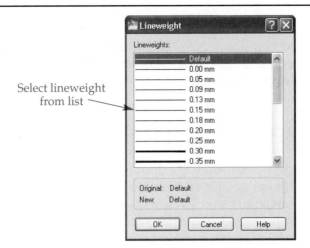

Figure 5-17.
The **Lineweight Settings** dialog box.

Select lineweight

Select units

Lineweight display

Default lineweight setting

Display scale

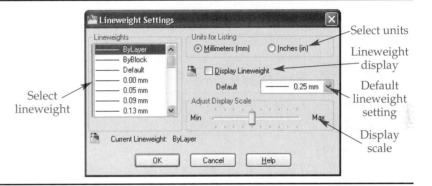

adjust the lineweight display scale to improve the appearance of different lineweights when lineweight display is on. When lineweight display is off, all objects display a 0, or one pixel, thickness regardless of the lineweight assigned to the object. You can also toggle screen lineweights by picking the **Show/Hide Lineweight** button on the status bar.

The value used when you assign the Default lineweight to a layer is set in the **Default** drop-down list. The Default lineweight is an application setting and applies to any drawing you open. The Default lineweight is not template-specific and remains set until you change the value. Do not assign layers the Default lineweight value if you anticipate using a different *default* lineweight for different drawing applications. Change each layer's lineweight individually. This rule maintains flexibility and consistency between drawings.

CAUTION

You can use the **Lineweights** area of the **Lineweight Settings** dialog box to set an absolute lineweight value. However, you should set lineweight as ByLayer for most applications. Changing lineweight to a value other than ByLayer overrides layer lineweight. Therefore, if the absolute linetype value is set to 0.30 mm, for example, all objects are drawn using a 0.30 mm weight regardless of the lineweight assigned to the layers on which objects are drawn. Use layers and the **Layer Properties Manager** to control object lineweight for most applications.

PROFESSIONAL TIP

Layers are meant to simplify the drafting process. They separate different details of the drawing and can reduce the complexity of drawing display. Set color, linetype, and lineweight using layers, and do not override these properties for individual objects. Also, once you establish layers, avoid resetting and mixing color, linetype, and lineweight properties. Doing so can mislead you and others who may try to find certain details.

Exercise 5-6

Access the Student Web site (www.g-wlearning.com/CAD) and complete Exercise 5-6.

Plot   No Plot

Layer Plotting Properties

The **Plot Style** column of the **Layer Properties Manager** lists the plot style assigned to each layer. By default, the plot style setting is disabled. Plot styles are described later in this textbook.

The **Plot** column displays icons to show whether the layer plots. Select the default printer icon to turn off plotting for a particular layer. The **No Plot** icon appears when the layer is not available for plotting. The layer is still displayed and selectable, but it does not plot.

Adding a Layer Description

The **Description** column provides an area to type a short description for each layer. To add or change a description, pick the description once to highlight it, pause for a moment, and then pick it again. Type an appropriate description, and press [Enter] or pick outside of the **Description** text box. You can also define the layer description by right-clicking and selecting **Change Description**.

On   Off

Turning Layers On and Off

The **On** column shows whether a layer is on or off. The yellow light bulb, or **On** icon, means the layer is on. Objects on a layer that is turned on display on-screen and can be selected and plotted. If you pick the icon, the light bulb "turns off" (becomes gray), turning the layer off. Objects on a layer that is turned off do not display on-screen and are not plotted. Objects on an layer that is turned off can still be edited using advanced selection techniques, and they regenerate when a drawing regeneration occurs.

Freezing and Thawing Layers

Freeze   Thaw

The **Freeze** column shows whether a layer is thawed or frozen. Objects on a frozen layer do not display, plot, or regenerate when the drawing regenerates. You cannot edit objects on a frozen layer. Freeze layers to ensure that you do not accidentally modify objects they contain, and to increase system performance. The snowflake, or **Freeze**, icon displays when a layer is frozen. When a layer is thawed, objects on the layer appear on-screen, and they can be selected and regenerated. The sun, or **Thaw**, icon appears for thawed layers. Pick the **Freeze** or **Thaw** icon to toggle thawing and freezing.

Icons in the **New VP Freeze** column control freezing or thawing of layers when you create a new viewport. Additional layer functions also apply to layouts and viewports. Layouts and viewports are described later in this textbook.

Locking and Unlocking Layers

Lock   Unlock

The unlocked and locked padlock symbols (**Unlock** and **Lock** icons) located in the **Lock** column of the **Layer Properties Manager** control layer locking and unlocking. Objects on a locked layer remain visible, and you can use a locked layer to draw new objects. However, you cannot edit existing objects on a locked layer. Lock a layer whenever you want to see objects on-screen, but eliminate the possibility of selecting those objects.

Layers are unlocked by default. Pick an **Unlock** icon to lock the layer. When you rest the crosshairs over an object on a locked layer, the **Lock** icon appears next to the cursor. To lock all layers except specific layers, select the layers you want to remain unlocked, and then right-click on the selection and pick **Isolate selected layers**.

Figure 5-18.
Locked layers fade by default. You can increase or decrease fading and enable or disable locked layer fading as needed.

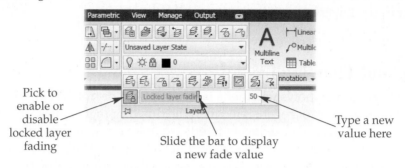

Pick to enable or disable locked layer fading

Slide the bar to display a new fade value

Type a new value here

Locked Layer Fading

By default, all locked layers fade, allowing unlocked layers to stand out on-screen. The quickest way to control locked layer fading is to use the options available in the expanded **Layers** panel of the **Home** ribbon tab. See **Figure 5-18.** Pick the **Locked layer fading** button to allow or disable locked layer fading. When you allow locked layer fading, use the **Locked Layer Fading** slider to increase or decrease fading, or type a new fading percentage. You can use a fade value between 0 and 90. A fade value of 0 fades the display of unisolated layers the least, while a fade value of 90 significantly fades unisolated layers. The default fade value is 50%. **Figure 5-19** shows the effect of locked layer fading on a drawing.

Deleting Layers

To delete a layer using the **Layer Properties Manager**, select the layer and pick the **Delete Layer** button, or right-click on the layer and choose **Delete Layer**. You cannot delete or purge the 0 layer, the current layer, layers containing objects, or layers associated with an external reference.

Figure 5-19.
All of the layers in this drawing are locked except A-WALL-FULL, which contains the walls. A—Locked layer fading is disabled. B—Locked layer fading is on and set to a fade value of 75.

Locked Layer Fading Disabled

Locked Layer Fading Enabled
Fade Value: 75

A

B

Adjusting Property Columns

To resize a column in the **Layer Properties Manager**, move the cursor over the column edge to display the resize icon and drag the column to the desired width. You can maximize the width of an individual column to show the longest value in the column, or in the case of columns that list properties as icons, to display the full column heading. To maximize the width of a single property column, right-click on the column heading and select **Maximize column**. To maximize the width of all columns, right-click on any property column heading and select **Maximize all columns**.

You can optimize the width of columns to show the longest value in the column list for properties displayed as text, while reducing the width of columns that list properties as icons. To optimize the width of a single property column, right-click on the column heading and select **Optimize column**. To optimize the width of all columns, right-click on any property column heading and select **Optimize all columns**.

By default, a vertical bar appears on the right side of the **Name** column. Any column left of the bar is "frozen". Frozen columns remain in position when you move the scroll bar near the bottom of the **Layer Properties Manager**. Scroll columns to the right of the vertical bar using the horizontal scroll bar. To turn off the freeze function, right-click on any property column heading and select **Unfreeze column**. To freeze columns, right-click on a property column heading and select **Freeze column** to turn on the freeze function for every column left of the selected column.

You can hide columns in the **Layer Properties Manager** by right-clicking on any property column heading and deselecting the property column name from the menu. Another option is to right-click on any property column heading and select **Customize…** to display the **Customize Layer Columns** dialog box. You can use this dialog box to hide property columns by the associated check boxes. To move a column left or right in the **Layer Properties Manager**, pick a column name and select the **Move Up** or **Move Down** button. Reset the display of all property columns to default settings by right-clicking on any property column heading and selecting **Restore all columns to defaults**.

Supplemental Material

Additional Layer Tools
For information about additional layer tools, go to the Student Web site (www.g-wlearning.com/CAD), select this chapter, and select **Additional Layer Tools**.

Filtering Layers

A single drawing often includes a very large number of layers. Displaying all layers at the same time in the list view pane can make it more difficult to work with the layers. The filter tree view pane in the **Layer Properties Manager** manages *layer filters* that you can be apply to reduce the number of layers that appear in the list view. See **Figure 5-20**.

Select the **All** node of the filter tree view to display all layers in the drawing. Layer filters are listed in alphabetical order inside the **All** node. Pick the **All Used Layers** filter to hide all the layers that have no objects on them. When you insert external references, an Xref filter node appears, allowing you to filter the display of layers associated with external references. External references are described later in this textbook. You can create other filters as needed.

layer filters: Filters that screen out, or filter, layers you do not want to display in the list view pane of the **Layer Properties Manager**.

Figure 5-20.
Create and restore layer filters using the filter tree view of the **Layer Properties Manager**.

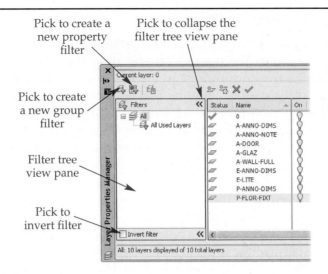

Pick to create a new property filter

Pick to collapse the filter tree view pane

Pick to create a new group filter

Filter tree view pane

Pick to invert filter

NOTE

The filter tree view of the **Layer Properties Manager** can be collapsed by picking the **Collapse Layer filter tree** button. To display all filters and layers in the list view, right-click in the layer list area and select **Show Filters in Layer List**.

Creating a Property Filter

property filter: A filter that screens layers according to a specific layer property.

An example of using a *property filter* is filtering all layers that are turned on, or have a name beginning with the letter *A*, or both. The default **All Used Layers** filter is a property filter that filters layers according to layer status. To create a property filter, pick the **New Property Filter** button to display the **Layer Filter Properties** dialog box. See **Figure 5-21.** Enter a name for the new filter in the **Filter name:** text box. Pick the appropriate **Filter definition** area field box to define properties to filter.

The Status, On, Freeze, Lock, Plot, and New VP Freeze fields display a flyout. Select an option from the flyout to filter according to the selection. For example, pick the **On** icon from the On field to display only layers that are turned on. In the Name field, type a layer name or a partial layer name using the * wildcard character to filter according to layer name. For example, to see all layers that start with an *A*, type a*. You can filter using the Color, Linetype, Lineweight, or Plot Style fields by typing an appropriate value or selecting the ellipses (…) button to select from the corresponding dialog box.

After you define a property filter, another row appears in the **Filter definition:** area. This allows you to create a more advanced filter. **Figure 5-21** shows a filter named Floor Plan, in which two rows are used to filter out all the layers except the layers beginning with the letter A and the P-FLOR-FIXT layer. To save the filter, pick the **OK** button. The new filter now displays in the filter tree view area.

Creating a Group Filter

group filter: A filter created by adding layers to the filter definition.

An example of using a *group filter* is dragging and dropping all layers used to draw an electrical plan into a group filter. All of the layers are selectable through the group filter regardless of their individual properties. To create a group filter, select the **New Group Filter** button. A new group filter appears in the filter tree view. Select the **All** node at the top of the filter tree area to display all the layers in the drawing. Then, to add a layer to the group filter, drag a layer from the list and drop it onto the group filter name.

Figure 5-21.
Use the **Layer Filter Properties** dialog box to create and edit a property filter. Use multiple rows in the **Filter definition:** area to add layers to the filter set.

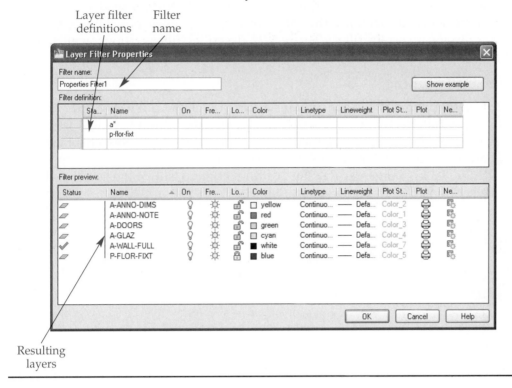

You can also add layers to a group filter by selecting the group filter, right-clicking, and choosing **Add** from the **Select Layers** cascading submenu. This feature allows you to select objects on the layers you want to add to the group filter. After selecting the objects, right-click or press [Enter] to add the layers to the group filter.

Pick **Replace** from the **Select Layers** cascading submenu to select objects on the layers to replace all other layers in the group filter. After selecting the objects, right-click or press [Enter] to add the layers to the group filter. To remove a layer from a group filter, right-click on the layer in the layer list area of the **Layer Properties Manager** and choose **Remove From Group Filter**.

Activating a Layer Filter

Select the filter from the filter tree view of the **Layer Properties Manager** to activate the filter. When a layer filter is active, only those layers associated with the filter appear in the layer list area of the **Layer Properties Manager** and the **Layer Control** drop-down list in the **Layers** panel of the **Home** ribbon tab. To view all the layers again, pick the **All** node at the top of the filter tree area of the **Layer Properties Manager**.

> **NOTE**
>
> The lower **Layer Properties Manager** status bar provides a description of the active layer filter settings.

Inverting Layer Filters

To invert, or reverse, a layer filter using the **Layer Properties Manager**, pick the **Invert filter** check box located in the lower-left corner or right-click in the layer list area and select **Invert Layer Filter**. For example, selecting the **All Used Layers** filter shows only the layers that have objects on them. You can invert the **All Used Layers** filter to show all unused layers without creating an additional filter.

Additional Layer Filter Options

Other options associated with filters are accessible from a shortcut menu that appears when you right-click on a filter in the **Layer Properties Manager.** Most of the options in the shortcut menu are the same for the filter types, but some options are available only for a certain filter. **Figure 5-22** describes filter options available and applied only to the layers associated with the filter.

Filtering by Searching

The **Layer Properties Manager** contains a search tool you can use to filter layers in the list view without actually creating a filter. To use the search feature, type a layer name or a partial layer name using the * wildcard character in the **Search for layer** text box. For example, to see all the layers that start with an *A*, type a*. As you type, layers that match the letters you enter display. Adding letters narrows the search, with the most relevant or best-matched layers listed first.

Exercise 5-7

Access the Student Web site (www.g-wlearning.com/CAD) and complete Exercise 5-7.

Layer States

Once you save a *layer state*, you can readjust layer settings to meet your needs, with the option to restore a previously saved layer state at any time. For example, a basic architectural drawing might use the layers shown in **Figure 5-20**. Three different drawings can be plotted using this drawing file: a floor plan, a plumbing plan, and an electrical plan. The following chart shows the layer settings for each of the three drawings:

layer state: A saved setting, or state, of layer properties for all layers in the drawing.

Layer	Description	Floor Plan	Plumbing Plan	Electrical Plan
0		Off	Off	Off
A-ANNO-DIMS	Floor Plan Dimensions	On	Frozen	Frozen
A-ANNO-NOTE	Floor Plan Notes	On	Frozen	Frozen
A-DOOR	Doors	On	Frozen	Locked
A-GLAZ	Windows	On	Frozen	Locked
A-WALL-FULL	Full Height Walls	On	Locked	Locked
E-ANNO-DIMS	Electrical Plan Dimensions	Frozen	Frozen	On
E-LITE	Electrical Plan Lights	Frozen	Frozen	On
P-ANNO-DIMS	Plumbing Plan Dimensions	Frozen	On	Frozen
P-FLOR-FIXT	Plumbing Plan Fixtures	Locked	On	Locked

You can save each of the three groups of settings as an individual layer state. You can then restore a layer state to return the layer settings for a specific drawing. This is easier than changing the settings for each layer individually.

Option	Function
Visibility	Changes the **On/Off** and **Thawed/Frozen** states of the unfiltered layers.
Lock	Locks or unlocks the unfiltered layers.
Viewport	Freezes or thaws the unfiltered layers in the current layout viewport.
Isolate Group	Freezes all layers except those associated with the filter and the current layer.
New Properties Filter	Opens the **Layer Filter Properties** dialog box.
New Group Filter	Creates a new group filter.
Convert to Group Filter	Converts a property filter to a group filter.
Rename	Renames the selected filter.
Delete	Deletes the selected filter.
Properties	Edits a property filter.
Select Layers	Provides options to add layers or replace layers in an existing group filter.

Use the **Layer States Manager**, shown in **Figure 5-23,** to create a new layer state. Pick the **New...** button to display the **New Layer State to Save** dialog box. See **Figure 5-24.** Type a name for the layer state in the **New layer state name:** text box and enter a description. Pick the **OK** button to save the new layer state. Once you create a layer state, you can adjust layer properties as needed. **Figure 5-25** describes the areas, options, and buttons available in the **Layer States Manager**.

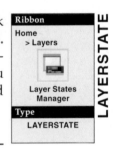

Figure 5-23.
The **Layer States Manager** allows you to save, restore, and manage layer settings.

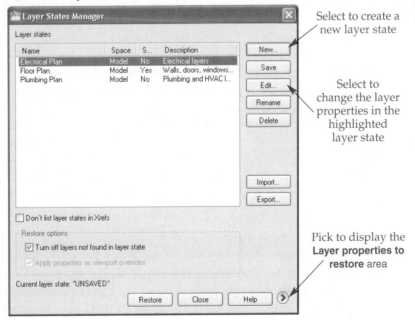

Figure 5-24.
Creating a new layer state.

Enter the layer state name →

Enter a description for the layer state →

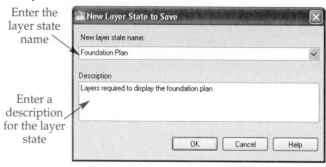

Figure 5-25.
Layer state options available in the **Layer States Manager**.

Item	Feature
Layer states	Displays saved layer states. The **Name** column provides the name of the layer state. The **Space** column indicates whether the layer state was saved in model space or paper space. The **Same as DWG** column indicates whether the layer state is the same as the current layer properties. The **Description** column lists the layer state description added when the layer state was saved.
Save	Pick to resave and override the selected layer state with the current layer properties.
Edit	Opens the **Edit Layer State** dialog box, where you can adjust the properties of each layer state without exiting the **Layer States Manager**.
Rename	Activates a text box that allows you to rename the current layer state.
Delete	Deletes the selected layer state.
Import	Opens the **Import layer state** dialog box, used to import an LAS file containing an existing layer state into the **Layer States Manager**.
Export	Opens the **Export layer state** dialog box, used to save a layer state as an LAS file. The file can be imported into other drawings, allowing you to share layer states between drawings containing identical layers.
Don't list layer states in Xrefs	Hides layer states associated with external reference drawings. External references are described in Chapter 30.
Restore options	Check the **Turn off layers not found in layer state** check box to turn off new layers or layers removed from a layer state when the layer state is restored. Check **Apply properties as viewport overrides** to apply layer viewport overrides when you are adjusting layer states within a layout.
Layer properties to restore	Check the layer properties that you want to restore when the layer state is restored. Pick the **Select All** button to pick all properties. Pick the **Clear All** button to deselect all properties.

NOTE

You can also access the **Layer States Manager** by picking the **Layer States Manager** button from the **Layer Properties Manager** or right-clicking in the layer list view of the **Layer Properties Manager** and selecting the **Restore Layer State** option. To save a layer state outside the **Layer States Manager**, pick the **New Layer State...** option from the **Layer States** drop-down list in the **Layers** panel on the **Home** ribbon tab.

AutoCAD and Its Applications—Basics

After you create a layer state, you can restore layer properties to the settings saved in the layer state at any time. To activate a layer state using the ribbon, select the layer state from the **Layer States** drop-down list in the **Layers** panel on the **Home** ribbon tab. You can also restore a layer state using the **Layer States Manager** by selecting the layer state from the list and picking the **Restore** button.

 *Layer Settings*

For information about options available in the **Layer Settings** dialog box, go to the Student Web site (www.g-wlearning.com/CAD), select this chapter, and select **Layer Settings**.

Reusing Drawing Content

In nearly every drafting discipline, individual drawings created as part of a given project are likely to share a number of common elements. All the drawings within a specific drafting project generally have the same set of standards. *Drawing content*, such as layer names and properties, text size and font used for annotation, dimensioning methods and appearances, drafting symbols, drawing layouts, and even drawing details, is often duplicated in many different drawings. One of the most fundamental advantages of CADD systems is the ease with which you can share content between drawings. Once you define a commonly used drawing feature, you can reuse the item as needed in any number of drawing applications.

drawing content: All of the objects, settings, and other components that make up a drawing.

Drawing templates represent one way to reuse drawing content. Creating your own customized drawing template files provides an effective way to start each new drawing using standard settings. Drawing templates, however, provide only a starting point. During the course of a drawing project, you may need to add previously created content to the current drawing. Some drawing projects require you to revise an existing drawing rather than start a completely new drawing. For other projects, you may need to duplicate the standards used in a drawing a client has supplied.

Introduction to DesignCenter

AutoCAD provides a powerful drawing content manager called **DesignCenter**. The **DesignCenter** palette allows you to reuse drawing content defined in previous drawings using a drag-and-drop operation. **Figure 5-26** displays the main features of **DesignCenter**. **DesignCenter** can be used to manage several types of drawing content, including layers, linetypes, blocks, dimension styles, layouts, table styles, text styles, and externally referenced drawings. **DesignCenter** allows you to load content directly from any accessible drawing, without opening the drawing in AutoCAD.

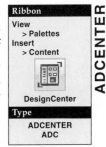

Ribbon
View
> Palettes
Insert
> Content

DesignCenter

Type
ADCENTER
ADC

ADCENTER

Figure 5-26.
Use **DesignCenter** to copy content from one drawing to another.

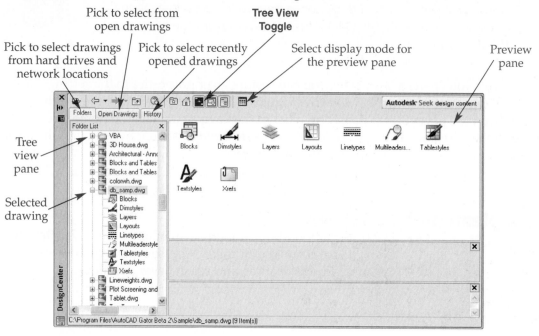

Copying Layers and Linetypes

To copy content using **DesignCenter**, first use the tree view pane to locate the drawing that includes the content you want to reuse. If the tree view is not already visible, toggle it on by picking the **Tree View Toggle** button in the **DesignCenter** toolbar. The three tabs on the **DesignCenter** toolbar control the tree view display. Select the **Folders** tab to display the folders and files found on the hard drive and network. Pick the **Open Drawings** tab to list only drawings that are currently open. Select the **History** tab to list recently opened drawings.

Pick the plus sign (+) next to a drawing icon to view the content categories for the drawing. Each category of drawing content includes a representative icon. Pick the **Layers** icon to load the preview pane with the layer content in the selected drawing. See **Figure 5-27**.

To use drag-and-drop to import layers into the current drawing, move the cursor over the desired icon in the preview pane in **DesignCenter**. Press and hold down the

Figure 5-27.
Displaying the layers found in a drawing using **DesignCenter**.

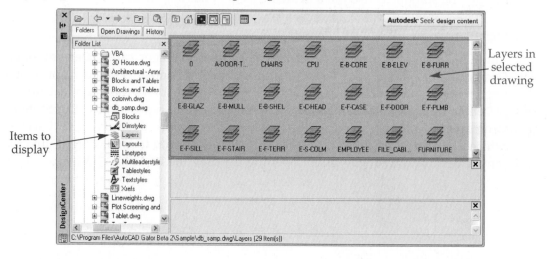

Figure 5-28.
To copy layers shown in **DesignCenter** into the current drawing, select the layers to be copied and then drag and drop them into the drawing area of the current drawing.

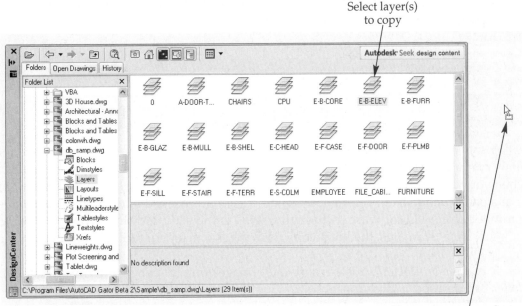

Select layer(s) to copy

Cursor appearance during drag-and-drop operation

pick button, and then drag the cursor to the open drawing. See **Figure 5-28.** Release the pick button to add the selected content to your current drawing file. You can also import layers into the current drawing by selecting the desired icons in the preview pane, right-clicking, and picking **Add Layer(s)**.

Use **DesignCenter** to copy linetypes from one file to another using the same procedure used to copy layers. In the tree view, select the drawing containing the linetypes you want to copy. Select the **Linetypes** icon to display the linetypes in the preview palette. Select the linetypes to copy, and then use drag-and-drop or the shortcut menu to add the linetypes to the current drawing.

NOTE

You cannot import drawing content if the content uses the same name as existing content. For example, if the name of a layer you try to reuse already exists in the destination drawing, the layer is ignored. The existing settings for the layer are preserved, and a message displays at the command line indicating that duplicate settings were ignored.

 **Supplemental Material**

DesignCenter
For more information about the tools and features available for using **DesignCenter**, go to the Student Web site (www.g-wlearning. com/CAD), select this chapter, and select **DesignCenter**.

Exercise 5-8

Access the Student Web site (www.g-wlearning.com/CAD) and complete Exercise 5-8.

Introduction to Printing and Plotting

soft copy: The electronic data file of a drawing.

hard copy: A physical drawing produced by a printer or plotter.

A *soft copy* appears on the computer monitor, making it inconvenient to use for many manufacturing and construction purposes. The soft-copy drawing is unavailable when you turn off the computer. A *hard copy* of a drawing is useful on the shop floor or at a construction site. A hard-copy drawing can be checked and redlined without a computer or CADD software. Although CADD is the standard throughout the world for generating drawings, the hard-copy drawing is still a vital tool for communicating the design.

A printer or plotter transfers soft-copy images onto paper. The terms *printer* and *plotter* can be used interchangeably, although *plotter* typically refers to a large-format printer. Desktop printers generally print 8 1/2″ × 11″ and sometimes 11″ × 17″ drawings. These are the printers common to computer workstations. Desktop printers print small drawings and reduced-size test prints. Large-format printers print larger drawings, such as C-size and D-size drawings. The most common types of both desktop and large-format printers are inkjet and laser printers. Pen plotters, which "draw" with actual ink pens, are still in use, but are less common.

Plotting in Model Space

You typically plot final drawings using a layout in paper space. A layout represents the sheet of paper used to organize and scale, or lay out, and plot or export a drawing or model. However, you can plot from model space as well as from a layout. The following content describes plotting from model space only.

Plotting from model space is common when a layout is unnecessary, to view how model space objects will appear on paper, and to make quick hard copies, such as when submitting basic assignments to your instructor or supervisor. The information in this chapter gives you only the basics, so you can make your first plot. This textbook fully explains creating and plotting layouts and additional printing and plotting information when appropriate.

Making a Plot

This section describes one of the many methods for creating a plot from model space. Refer to **Figure 5-29** as you read the following plotting procedure.

1. Access the **Plot** dialog box. If the column on the far right of the dialog box shown in **Figure 5-29** is not displayed, pick the **More Options** button (>) in the lower-right corner.
2. Check the plot device and paper size specifications in the **Printer/plotter** and **Paper size** areas.

PLOT

Quick Access

Plot

Ribbon

Output
> Plot

Plot

Application Menu

Print
Print
> Plot

Type

PLOT
[Ctrl]+[P]

Figure 5-29.
The **Plot** dialog box.

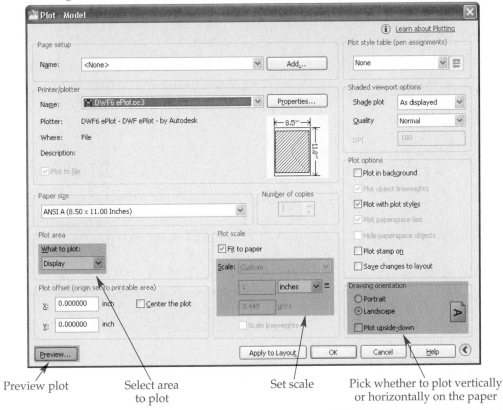

Preview plot   Select area   Set scale   Pick whether to plot vertically
                    to plot   or horizontally on the paper

3. Select what to plot in the **Plot area** section. The **Limits** option displays when you plot from model space. Select this option to plot everything inside the defined drawing limits. Pick the **Extents** option to plot the furthest extents of objects in the drawing. Select the **Display** option to plot the current screen display, exactly as it is shown. When you select the **Window** option, the **Page Setup** dialog box disappears temporarily so you can pick two opposite corners to define a window around the area to plot. Once you define the window, a **Window...** button appears in the **Plot area** section. Pick the button to redefine the opposite corners of a window around the portion of the drawing to plot.

4. Select an option in the **Drawing orientation** area. Choose **Portrait** to orient the drawing vertically (*portrait*) or **Landscape** to orient the drawing horizontally (*landscape*). The **Plot upside-down** option rotates the paper 180°.

5. Set the scale in the **Plot scale** area. Scale is measured as a ratio of either inches or millimeters to drawing units. Select a predefined scale from the **Scale:** drop-down list or enter values into the custom fields. Choose the **Fit to paper** check box to let AutoCAD automatically increase or decrease the plot area to fill the paper.

6. If desired, use the **Plot offset (origin set to printable area)** area to set additional left and bottom margins around the plot or to center the plot.

7. Pick the **Preview...** button to display the sheet as it will look when it is plotted. See **Figure 5-30**. The cursor appears as a magnifying glass with + and − symbols. Hold the left mouse button and move the cursor to increase or decrease the displayed image to view more or less detail. Press [Esc] to exit the preview.

8. Pick the **OK** button in the **Plot** dialog box to send the data to the plotting device.

portrait: A vertical paper orientation.

landscape: A horizontal paper orientation.

Figure 5-30.
A preview of the plot shows exactly how the drawing will appear on the paper.

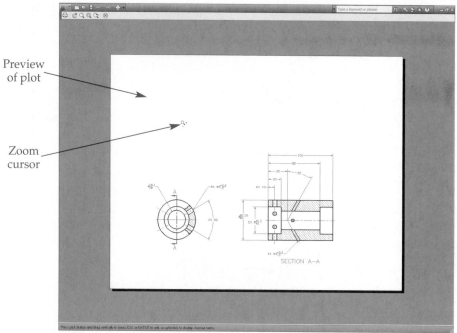

Preview
of plot

Zoom
cursor

Exercise 5-9

Access the Student Web site (www.g-wlearning.com/CAD) and complete Exercise 5-9.

Template Development
Chapter 5

For detailed instructions on adding layers to each of your drawing templates, go to the Student Web site (www.g-wlearning.com/CAD), select this chapter, and select **Template Development**.

Chapter Test

Answer the following questions. Write your answers on a separate sheet of paper or go to the Student Web site (www.g-wlearning.com/CAD), and complete the electronic chapter test.

1. Identify the following linetypes:

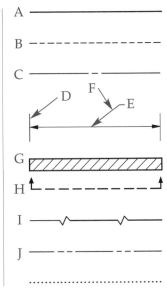

2. Identify two ways to access the **Layer Properties Manager**.
3. How can you tell if a layer is off, thawed, or unlocked by looking at the **Layer Properties Manager**?
4. Should you draw on layer 0? Explain.
5. How can several new layer names be entered consecutively without using the **New Layer** button in the **Layer Properties Manager**?
6. How do you make another layer current in the **Layer Properties Manager**?
7. How do you make another layer current using the ribbon?
8. How can you display the **Select Color** dialog box from the **Layer Properties Manager**?
9. List the seven standard color names and numbers.
10. How do you change a layer's linetype in the **Layer Properties Manager**?
11. What is the default linetype in AutoCAD?
12. What condition must exist before a linetype can be used in a layer?
13. Describe the basic procedure to change a layer's linetype to HIDDEN.
14. What is the function of the linetype scale?
15. Explain the effects of using a global linetype scale.
16. Why do you have to be careful when changing linetype scales?
17. What is the state of a layer *not* displayed on the screen and *not* calculated by the computer when the drawing is regenerated?
18. Explain the purpose of locking a layer.
19. Identify the following layer status icons:

A.   D.

B.   E.

C.   F.

20. Identify at least three layers that cannot be deleted from a drawing.
21. Describe the purpose of layer filters.
22. Name the two basic types of filters.
23. Which button in the **Layer Properties Manager** allows you to save layer settings so they can be restored later?
24. In the tree view area of **DesignCenter**, how do you view the content categories of one of the listed open drawings?
25. How do you display all the available layers in a drawing using the **DesignCenter** preview pane?
26. Briefly explain how drag-and-drop works.
27. Define *hard copy* and *soft copy*.
28. Identify four ways to access the **Plot** dialog box.
29. Describe the difference between the **Display** and **Window** options in the **Plot area** section of the **Plot** dialog box.
30. Explain how to examine what a plot will look like before you actually print the drawing. What is the major advantage of doing a plot preview?

Drawing Problems

Start AutoCAD if it is not already started. Start a new drawing using an appropriate template of your choice. The template should include layers for drawing the given objects. Add layers as needed. Draw all objects using appropriate layers. Follow the specific instructions for each problem. Do not draw dimensions or text. Use your own judgment and approximate dimensions when necessary.

▼ Basic

1. Draw the hex head bolt pattern shown below. Save the drawing as P5-1.

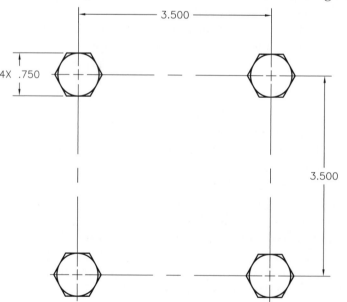

2. Create a 1/2″ hex nut with 3/4″ across the flats and a .422″ root diameter as shown. Save the drawing as P5-2.

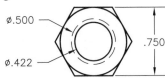

3. Draw the part shown below. Save the drawing a̶

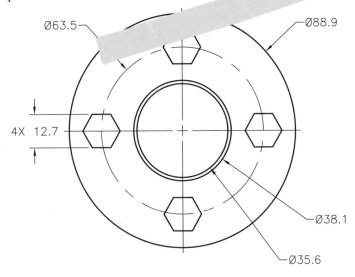

4. Draw the part shown below. Save the drawing as P5-4.

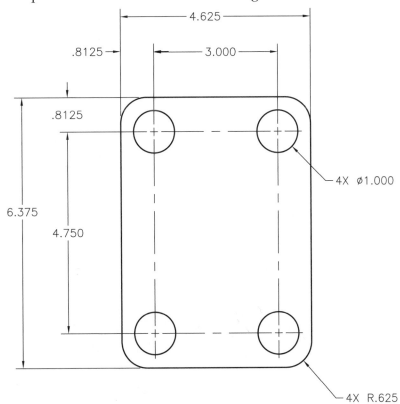

▼ Intermediate

5. Open P4-3, create or import a new layer for centerlines, and draw the centerlines. Save the drawing as P5-5.

6. Open P4-7, create or import a new layer for centerlines, and draw the centerlines. Save the drawing as P5-6.

▼ Advanced

7. Open P4-19, create or import a new layer for centerlines, and insert the centerlines. Change the global linetype scale to achieve an effect similar to the centerlines shown in Chapter 4. Save the drawing as P5-7.

8. Draw the plot plan shown below. Use the linetypes shown, which include Continuous, HIDDEN, PHANTOM, CENTER, FENCELINE2, and GAS_LINE. Make your drawing proportional to the example. Save the drawing as P5-8.

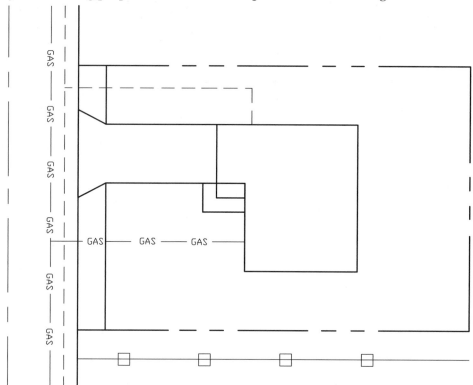

9. Draw the line chart shown below. Use the linetypes shown, which include Continuous, HIDDEN, PHANTOM, CENTER, FENCELINE1, and FENCELINE2. Make your drawing proportional to the given example. Save the drawing as P5-9.

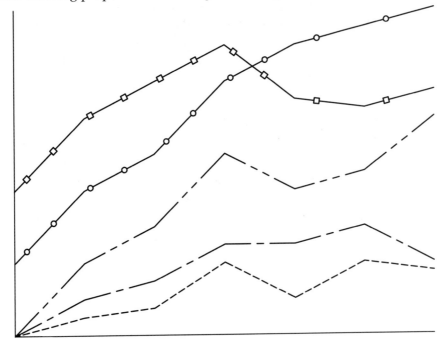

10. Create the controller integrated circuit diagram. Use a ruler or scale to keep the proportion as close as possible. Save the drawing as P5-10.

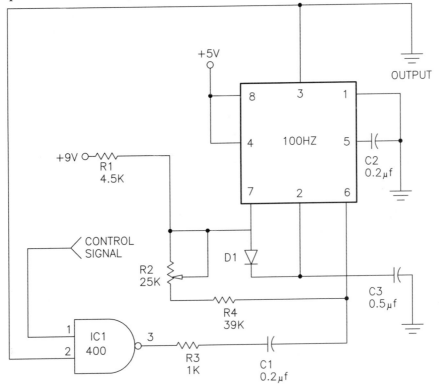

11. Draw a drift boat similar to the one shown below. Estimate dimensions. Save the drawing as P5-11.

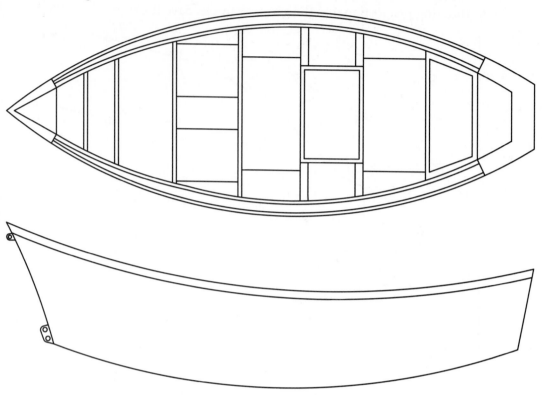

12. Draw the fishing boat shown. Save the drawing as P5-12.

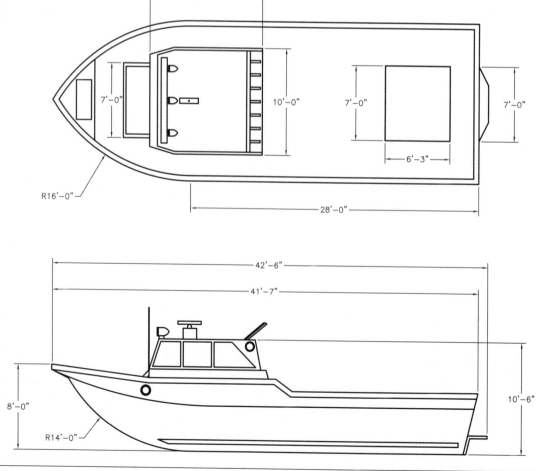

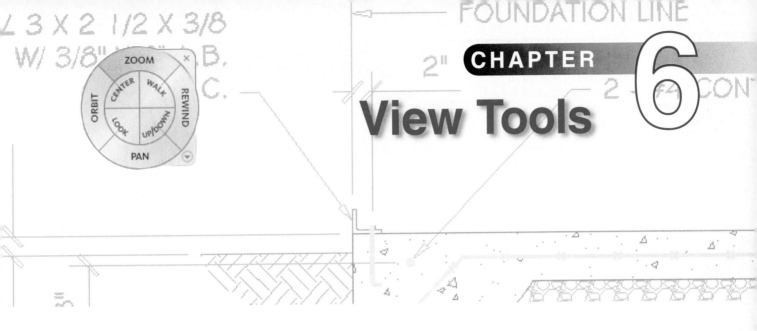

View Tools

Learning Objectives

After completing this chapter, you will be able to do the following:
- ✓ Increase and decrease the displayed size of objects.
- ✓ Adjust the display window to view other portions of a drawing.
- ✓ Use SteeringWheels for 2D applications.
- ✓ Use transparent display tools and control display order.
- ✓ Create named views that can be recalled instantly.
- ✓ Create multiple viewports in the drawing window.
- ✓ Explain the difference between redrawing and regenerating the display.
- ✓ Use the **Clean Screen** tool.

View tools allow you to observe and work more efficiently with a specific portion of a drawing. As you create drawings that are more complex and draw large and small objects, you will realize the importance of adjusting the drawing display. This chapter describes a variety of view tools that you will use frequently during the drawing process.

zooming: Making objects appear bigger (zoom in) or smaller (zoom out) on the screen without affecting their actual sizes.

Zooming

The **ZOOM** tool provides several methods for *zooming*. Choose the appropriate zoom option based on the portion of the drawing you want to display and whether you want to *zoom in* or *zoom out*. This chapter focuses on the ribbon as the primary means of accessing **ZOOM** tool options. See **Figure 6-1.** When you select a zoom option from the ribbon, all prompts are specific to the selected option. When you use dynamic input or the command line to access zoom tools, you need to enter specific options when prompted.

zoom in: Change the display area to show a smaller part of the drawing at a higher magnification.

zoom out: Change the display area to show a larger part of the drawing at a lower magnification.

Ribbon
View
> Navigate

Zoom
Type
ZOOM
Z

ZOOM

NOTE

You can also activate zoom tools from various shortcut menus, or by picking the **Zoom** button on the status bar.

Figure 6-1.
ZOOM tool options
found in the **Zoom**
flyout button on the
Navigate panel of the
View ribbon tab.

Zoom flyout

Navigate panel

Realtime Zooming

realtime zoom:
A zoom that can
be viewed as it is
performed.

When you access the **Realtime** zoom option, known as *realtime zooming*, the zoom cursor appears as a magnifying glass with a plus and minus. Press and hold the left mouse button and move the cursor up to zoom in and down to zoom out. When you achieve the appropriate display, release the mouse button. Repeat the process to make further adjustments. Right-click while the zoom cursor is active to display a shortcut menu with several view options. This is a quick way to access alternative zoom options and related view options. **Figure 6-2** briefly describes each view option. This chapter further explains most of these options when applicable. When finished zooming, press [Esc], [Enter], or the space bar, or right-click and pick **Exit**.

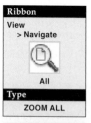

Ribbon

View
> Navigate

All

Type

ZOOM ALL

PROFESSIONAL TIP

AutoCAD supports most mice that have a scroll wheel between the two mouse buttons. Roll the wheel forward (away from you) to zoom in. Roll the wheel backward (toward you) to zoom out. This function also pans to the location of the crosshairs while zooming.

Ribbon

View
> Navigate

Extents

Type

ZOOM EXTENTS

Additional Zoom Options

Depending on your drawing task, you may choose to use one or more of the additional zoom options instead of the **Realtime** option or the navigation wheels described later in this chapter. The **All** option zooms to the edges of the drawing limits. If objects are drawn beyond the limits, the **All** option zooms to the edges of your geometry. Always use this option after you change the drawing limits.

Ribbon

View
> Navigate

Object

Type

ZOOM OBJECT

The **Extents** option zooms to the extents, or edges, of objects in a drawing. If you have a mouse with a scroll wheel, double-click the wheel to zoom to the drawing extents. The **Object** option allows you to select an object or set of objects. The selection is zoomed and centered to fill the display area.

Figure 6-2.
Options on the shortcut menu that are displayed when you right-click while the zoom cursor is active.

Selection	Cursor	Option	Function
Pan		Pan Realtime	Adjust the placement of the drawing on the screen.
Zoom		Zoom Realtime	Toggle between **Pan** and **Zoom Realtime** to adjust the view.
3D Orbit		3D Orbit	Move around a 3D object. 3D navigation tools are described in *AutoCAD and Its Applications—Advanced*.
Zoom Window		Zoom Window	Unlike the typical zoom window, described later in this chapter, this option requires you to press and hold the pick button while dragging the window box to the opposite corner, then release the pick button.
Zoom Original	(No cursor displayed)	(No tool name)	Restores the previous display before any realtime zooming or panning occurred; useful if the modified display is not appropriate.
Zoom Extents	(No cursor displayed)	Zoom Extents	Zoom to the extents of the drawing geometry.

The **Window** option allows you to pick opposite corners of a box. Objects in the box enlarge to fill the display. The **Window** option is the default if you pick a point on the screen after entering the **ZOOM** tool at the keyboard.

The **Scale** option allows you to zoom in or out according to a specific magnification scale factor. The **nX** option scales the display relative to the current display. The **nXP** option scales a drawing in model space relative to paper space, as described later in this textbook.

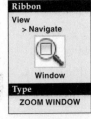

Ribbon
View
> Navigate

Window

Type
ZOOM WINDOW

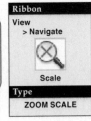

Ribbon
View
> Navigate

Scale

Type
ZOOM SCALE

> **NOTE**
>
>
>
> This textbook does not describe the **Previous**, **In**, **Out**, **Center**, and **Dynamic** options of the **ZOOM** tool. These options provide functions that you can achieve more easily using other view tools.

Exercise 6-1

Access the Student Web site (www.g-wlearning.com/CAD) and complete Exercise 6-1.

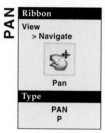

Ribbon
View
> Navigate

Pan

Type
PAN
P

Panning

The **PAN** tool allows you to *pan*. Panning is usually required while drawing, especially while you are creating large objects, and while you are zoomed in on a part of the drawing to view fine detail. You often use the **PAN** and **ZOOM** tools together to change the display.

panning: Changing the drawing display so that different portions of the drawing are visible on-screen.

NOTE

You can also activate the **PAN** tool from various shortcut menus, or by picking the **Pan** button on the status bar.

realtime panning: A panning operation in which you can see the drawing move on the screen as you pan.

When you first access the **PAN** tool, you are using the **Realtime** option, known as *realtime panning*. The **Realtime** option is the quickest and easiest method of adjusting the on-screen display. After starting the tool, press and hold the left mouse button and move the pan cursor in the direction to pan. A right-click displays the same shortcut menu available for realtime zooming. To exit realtime panning, press [Esc], [Enter], or the space bar, or right-click and pick **Exit**.

PROFESSIONAL TIP

If you have a mouse with a scroll wheel, press and hold the wheel button and move the mouse to perform a realtime pan.

NOTE

An alternative realtime panning method involves using drawing window scroll bars. To display the scroll bars, select the **Display scroll bars in drawing window** check box in the **Window Elements** area of the **Display** tab in the **Options** dialog box. Realtime panning is more efficient than using the scroll bars.

NOTE

By default, when you use the **U** tool after multiple zooming and panning operations, zooms and pans are grouped together, allowing you to return to the original view. To make each zoom and pan operation count individually, deselect the **Combine zoom and pan commands** check box in the **Undo/Redo** area of the **User Preferences** tab in the **Options** dialog box.

Exercise 6-2

Access the Student Web site (www.g-wlearning.com/CAD) and complete Exercise 6-2.

Introduction to SteeringWheels

AutoCAD SteeringWheels provide an alternative means of accessing and using certain view tools. Individual SteeringWheels are known as *navigation wheels*. Some navigation wheels and many of the tools available from navigation wheels are more appropriate for preparing 3D models. The **ZOOM**, **CENTER**, **PAN**, and **REWIND** tools are effective for 2D drafting applications. All other **SteeringWheels** tools and options are covered in *AutoCAD and Its Applications—Advanced*.

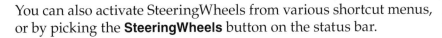

NOTE

You can also activate SteeringWheels from various shortcut menus, or by picking the **SteeringWheels** button on the status bar.

When you access SteeringWheels in model space, the **Full Navigation Wheel** appears by default, and the UCS icon changes to a 3D display. In layout space, the **2D Navigation Wheel** appears. See **Figure 6-3.** Navigation wheels display next to the cursor and are divided into *wedges*. Each wedge houses a navigation tool, similar to a tool button. Hover over a wedge to highlight the wedge. You can pick certain wedges to activate a tool. Other wedges require that you hold down the left mouse button to use the tool.

wedges: The parts of a wheel that contain navigation tools.

A navigation wheel remains on-screen until closed. This allows you to use multiple navigation tools. To close a navigation wheel, pick the **Close** button in the upper-right corner of the wheel, press [Esc], [Enter], or the space bar, or right-click and pick **Close Wheel**.

Zooming with the Navigation Wheel

The **ZOOM** navigation tool offers realtime zooming. Press and hold the left mouse button on the **ZOOM** wedge to display the pivot point icon and zoom navigation cursor. See **Figure 6-4.** The pivot point is the location where you press the **ZOOM** wedge. Move the zoom navigation cursor up to zoom in and down to zoom out. The pivot point icon also zooms in or out as a visual aid to zooming. When you achieve the appropriate display, release the left mouse button.

Using the Center Navigation Tool

The **CENTER** navigation tool centers the display screen at a picked point, without zooming. Press and hold the left mouse button on the **CENTER** wedge. The pivot point

Figure 6-3.
The **Full Navigation Wheel** appears by default when you access **SteeringWheels** in model space. The **2D Navigation Wheel** appears in a layout.

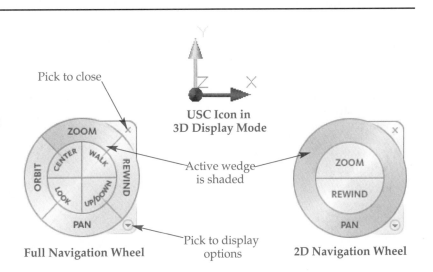

Pick to close

USC Icon in
3D Display Mode

Active wedge
is shaded

Pick to display
options

Full Navigation Wheel

2D Navigation Wheel

Figure 6-4.
To use the **ZOOM** navigation tool, move the cursor up to zoom in and down to zoom out.

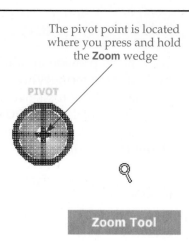

The pivot point is located where you press and hold the **Zoom** wedge

PIVOT

Zoom Tool

icon appears when you move the cursor over an object. Release the mouse button to pan so the location of the pivot point relocates to the center of the drawing window when you release the mouse button. See **Figure 6-5.**

Panning with the Navigation Wheel

The **PAN** navigation tool uses realtime panning to adjust the on-screen display. Press and hold the left mouse button on the **PAN** wedge to display the pan navigation cursor. Move the pan navigation cursor in the direction to pan. Release the left mouse button when you achieve the desired display.

Rewinding

The **REWIND** navigation tool allows you to observe the effects view tools have made on the drawing display, and return to a previous display. For example, if you use the navigation wheel to zoom in, then pan, then zoom out, you can rewind through each action and return to the original display, the zoomed-in view, the panned display, and then back to the current zoomed-out view. By default, you can rewind through view actions created using most view tools, not just those accessed from a navigation wheel.

Pick the **REWIND** tool once to return to the previous display. Thumbnail images appear in frames as the previous view restores. The orange-framed thumbnail surrounded by brackets indicates the restored display and its location in the sequence of events. See **Figure 6-6.** Repeatedly pick the **REWIND** button to cycle back through prior views. Another option is to press and hold the left mouse button on the **REWIND** wedge to display the framed view thumbnails. Then, while still holding the left mouse

Figure 6-5.
To use the **CENTER** navigation tool, move the cursor over an object at the point you want to center in the drawing area and release the left mouse button.

Pivot icon appears only when you move the cursor over an object

This point pans to the exact center of the drawing window

PIVOT

Center Tool

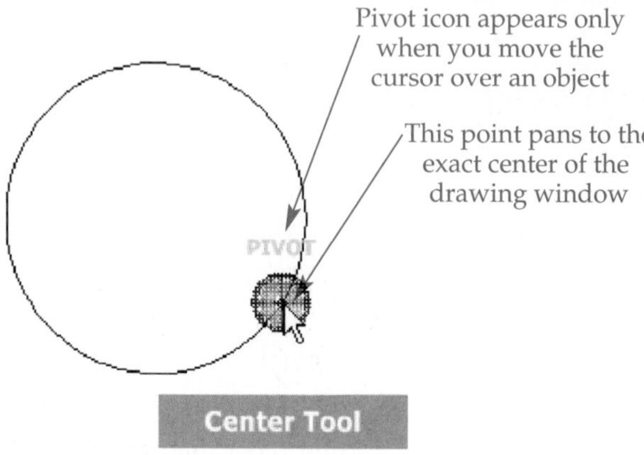

Figure 6-6.
Use the **REWIND** navigation tool to step back through and restore previous display configurations.

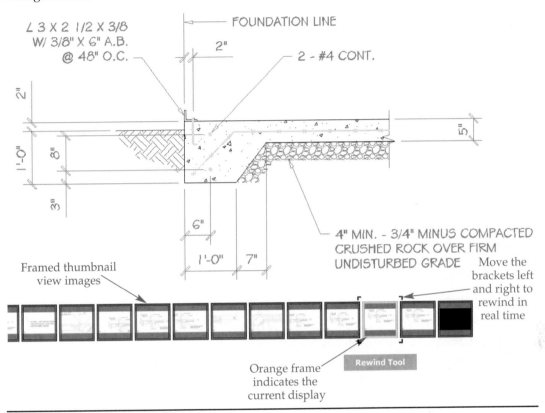

Framed thumbnail view images

Move the brackets left and right to rewind in real time

Rewind Tool

Orange frame indicates the current display

button, move the brackets left over the thumbnails to cycle through earlier views, and right to return to later views. Release the left mouse button when you achieve the desired display.

NOTE

By default, if you access and use a view tool, such as **ZOOM**, from a source other than the navigation wheel, such as the ribbon, a rewind icon appears in place of the thumbnail. A thumbnail displays as you move the brackets over the rewind icon.

Exercise 6-3

Access the Student Web site (www.g-wlearning.com/CAD) and complete Exercise 6-3.

SteeringWheel Options

A shortcut menu of options displays when you right-click while using a navigation wheel or when you pick the options button in the lower-right corner of a wheel. The options vary depending on the current work environment. Most of the options provide access to options and navigation wheels specifically for 3D modeling. The following options apply to 2D drafting applications:

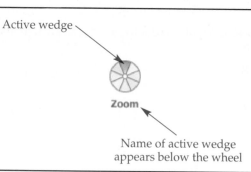

Figure 6-7.
The **Full Navigation Wheel** in mini mode.

Active wedge

Zoom

Name of active wedge appears below the wheel

- **Mini Full Navigation Wheel.** Displays the **Full Navigation Wheel** in mini format. See **Figure 6-7.**
- **Full Navigation Wheel.** Displays the **Full Navigation Wheel** in the default big format.
- **Fit to Window.** Zooms and pans to show all objects centered in the drawing window.
- **SteeringWheel Settings....** Displays the **SteeringWheel Settings** dialog box.
- **Close Wheel.** Closes the navigation wheel.

Supplemental Material

SteeringWheels Settings

For information about options found in the **SteeringWheel Settings** dialog box that are specific to 2D drafting, go to the Student Web site (www.g-wlearning.com/CAD), select this chapter, and select **SteeringWheels Settings**.

Using Transparent Display Tools

transparently:
When referring to tool access, a tool describes using while another tool is in progress.

Selecting a new tool usually cancels the tool in progress and then starts the new tool. However, you can use some tools *transparently*, temporarily interrupting the active tool. After the transparent tool has completed, the interrupted tool resumes. Therefore, it is not necessary to cancel the initial tool. You can use many display tools transparently, including **PAN**, **ZOOM**, and **SteeringWheels**.

An example of when transparent tools are useful is drawing a line when one end of the line is somewhere off the screen. One option is to cancel the **LINE** tool, zoom out to see more of the drawing, and select **LINE** again. A more efficient method is to use **PAN** or **ZOOM** transparently with the **LINE** tool. To do so, begin the **LINE** tool and pick the first point. At the Specify next point: prompt, access the **PAN** or **ZOOM** tool. You can use any access method, though right-clicking and selecting **Pan** or **Zoom**, or using the wheel mouse, is often quickest. Once you display the correct view, pick the second line endpoint. You can also activate tools transparently by typing an apostrophe (') before the tool name. For example, to enter the **ZOOM** tool transparently, type 'Z or 'ZOOM.

PROFESSIONAL TIP

You can use tools such as **Grid**, **Snap**, and **Ortho** transparently, but it is quicker to activate these modes with the appropriate button on the status bar or using a function key.

Controlling Draw Order

Drawings often include overlapping objects. The overlap is difficult to see when all objects use a thin lineweight and when lineweight display is off. Controlling display order is better illustrated with an object that has width, such as the donuts shown in **Figure 6-8.** In this example, the donuts were originally drawn after the other objects. You can change the drawing order of the donuts, and all other objects, to place the items above or below selected objects and to the front or back of all objects.

Use the **DRAWORDER** tool to change the order of objects in a drawing. You can also set draw order by picking an object to select it, right-clicking, and choosing **Draw Order. Figure 6-9** describes the options for changing draw order.

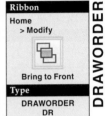

Ribbon
Home
> Modify

Bring to Front

Type
DRAWORDER
DR

DRAWORDER

Figure 6-8.
The order of objects can be changed to place any object under or above other objects.

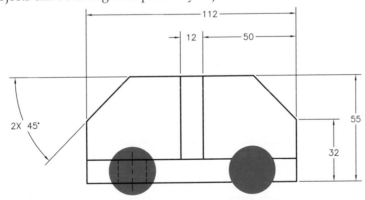

Figure 6-9.
Four options are available for rearranging the order of objects in a drawing.

Option	Function
Bring to Front	Places the selected objects at the front of the drawing.
Send to Back	Places the selected objects at the back of the drawing.
Bring Above Objects	Moves the selected objects above the reference object.
Send Under Objects	Moves the selected objects below the reference object.

Exercise 6-4

Access the Student Web site (www.g-wlearning.com/CAD) and complete Exercise 6-4.

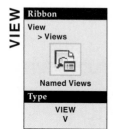
Creating Named Views

Use the **View Manager** to create and name specific views of the drawing. A view can be a portion of the drawing, such as the upper-left quadrant or a separate detail, or it can represent an enlarged area. When further drawing or editing operations are required, you can quickly and easily recall named views.

The left side of the **View Manager** contains a list of view types, or nodes. See **Figure 6-10.** You can expand each node, except **Current**, to reveal saved views. The **Current** node displays the properties of the current view. The **Model Views** node contains a list of saved model views. The **Layout Views** node contains a list of saved layout views. The **Preset Views** node lists all preset orthogonal and isometric views. Pick one of the view nodes to display information about the view type.

The right side of the **View Manager** contains buttons to control or modify the selected view or view type. These actions are also available in a shortcut menu when you right-click on the view or view type. Pick one of the view names to display information related to the current view in the middle area of the dialog box. Refer again to **Figure 6-10.** The first section, **General**, contains details such as the name of the view, layer settings saved with the view, and other specific view type settings. The **General** section is not visible when you select the **Current** node. The **Animation** section sets the properties for animated drawings and slide shows. The settings in the **View** section include camera position, target position, and perspective status. The **Clipping** section controls front plane and back plane location and the clipping status. Some of these items are covered in more detail in this chapter; others are reserved for later chapters or *AutoCAD and Its Applications—Advanced*, where the information is more relevant.

Figure 6-10.
The view nodes of the **View Manager** dialog box help organize saved and preset drawing views. Select a named view to see its properties and a preview image.

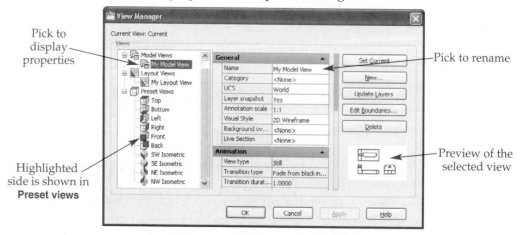

The lower-right corner of the **View Manager** shows a preview image of the selected view. This image is visible only when you select one of the named model or layout views.

New Views

To save the current display as a view, pick the **New...** button or right-click on a view node and select **New...** to access the **New View/Shot Properties** dialog box. See **Figure 6-11.** Type the view name in the **View name:** text box. If the named view is associated with a category in the **Sheet Set Manager**, you can select the category from the **View category** drop-down list. The **Sheet Set Manager** is described later in this textbook.

The **New View/Shot Properties** dialog box provides many options that are applicable to 3D modeling animations. To create a basic 2D view, select **Still** from the **View type** drop-down list, and focus on the settings in the **View Properties** tab. The **Current display** radio button is the default. Click **OK** to add the view name to the list. AutoCAD creates a view from the current display.

To use a window to define the view, pick the **Define window** radio button in the **New View** dialog box, and then pick the **Define view window** button. Pick two points to define a window. After you select the second corner, the **New View/Shot Properties** dialog box reappears. When you pick the **OK** button, the **View Manager** updates to reflect the new view.

Figure 6-11.
Use the **New View/Shot Properties** dialog box to save the current display as a view or define a window to create a view.

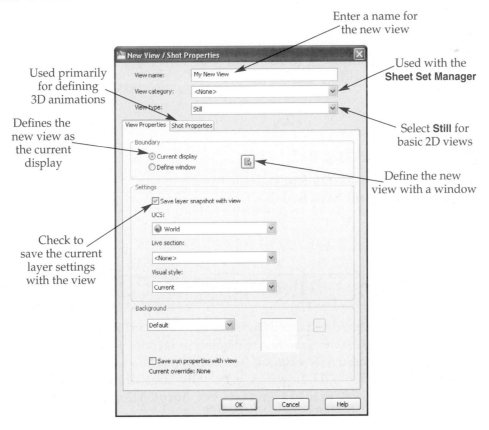

Select the **Save layer snapshot with view** check box to save the current layer settings when you save a new view. Saved layer settings are recalled each time the view is set current.

Activating Views

To display one of the listed views from inside the **View Manager**, pick the view name from the list in the **Views** area of the **View Manager** and pick the **Set Current** button. The name of the current view appears in the **Current View:** label above the **Views** area. Pick the **OK** button to display the selected view. To display a view without accessing the **View Manager**, select the name of the view from the list in the **Views** panel of the **View** ribbon tab.

NOTE

The **Preset Views** node allows you to choose one of the ten preset orthogonal or isometric views. Orthogonal and isometric views are covered in greater depth in *AutoCAD and Its Applications—Advanced*.

PROFESSIONAL TIP

Part of your project planning should include view names. A consistent naming system guarantees that all users know the view names without having to list them. The views can be set as part of the template drawings.

Exercise 6-5

Access the Student Web site (www.g-wlearning.com/CAD) and complete Exercise 6-5.

Tiled Viewports

tiled viewports:
Viewports created in model space.

viewport: The window or frame within which a drawing is visible.

floating viewports:
Viewports created in paper space.

You can divide the model space drawing window into *tiled viewports*. Another type of *viewport*, *floating viewports*, are created in layouts. Floating viewports and layouts are describe later in this textbook. You can use tiled viewports in model space for 2D and 3D applications. 2D drawings, especially large drawings with significant detail, lend themselves well to tiled viewports. See *AutoCAD and Its Applications— Advanced* for examples of tiled viewports in 3D.

By default, the drawing window contains one viewport. Additional viewports divide the drawing window into separate tiles that butt against one another like floor tile. Tiled viewports cannot overlap. Multiple viewports contain different views of the same drawing, displayed at the same time. Only one viewport can be active at any given time. The active viewport has a bold outline around its edges. See **Figure 6-12.**

Creating Tiled Viewports

The **Viewports** dialog box provides one method for creating tiled viewports. **Figure 6-13** shows the **New Viewports** tab of the **Viewports** dialog box. The **Standard viewports:** list contains many preset viewport configurations. The configuration name identifies the number of viewports and the arrangement or location of the largest viewport. Select a configuration to see a preview of the tiled viewports in the **Preview** area on the right side of the **New Viewports** tab. Select *Active Model Configuration* to preview the current configuration. Pick the **OK** button to divide the drawing window into the selected viewport configuration.

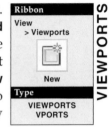

> **NOTE**
>
> You can activate the same preset viewport configurations available from the **New Viewports** tab of the **Viewport** dialog box using the **Set Viewports** drop-down list in the **Viewports** panel on the **View** ribbon tab. Prompts guide you through the process of creating the correct configuration. The arrangement you choose applies to the active viewport only.

The additional options in the **Viewports** dialog box are useful when two or more viewports already exist. The **Apply to:** drop-down list allows you to specify whether the viewport configuration applies to the entire drawing window or to the active viewport only. Select **Display** to apply the configuration to the entire drawing area.

Figure 6-12.
An example of three tiled viewports in model space. All of the viewports contain the same objects, but the display in each viewport can be unique.

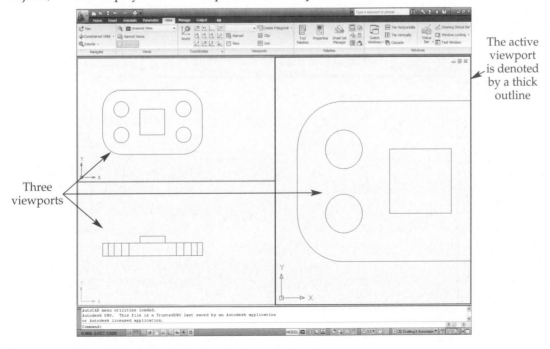

Three viewports

The active viewport is denoted by a thick outline

Figure 6-13.
Specify the number and arrangement of tiled viewports in the **New Viewports** tab of the **Viewports** dialog box.

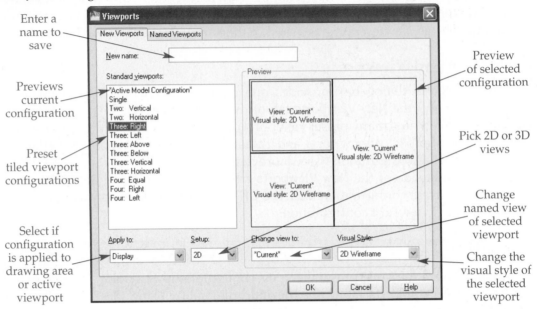

Enter a name to save

Previews current configuration

Preset tiled viewport configurations

Select if configuration is applied to drawing area or active viewport

Preview of selected configuration

Pick 2D or 3D views

Change named view of selected viewport

Change the visual style of the selected viewport

Select **Current Viewport** to apply the new configuration in the active viewport only. See **Figure 6-14**.

The default setting in the **Setup:** drop-down list is **2D**. When you select the **2D** option, all viewports show the top view of the drawing. If you choose the **3D** option, the different viewports display various 3D views of the drawing. At least one viewport is set up with an isometric view. The other viewports have different views, such as a top view or side view. The viewport configuration displays in the **Preview** image. To

Figure 6-14.
You can subdivide a viewport by choosing **Current Viewport** in the **Apply to:** drop-down list. Here, the top-left viewport is subdivided using the **Two: Vertical** preset configuration.

Configuration applied to active viewport

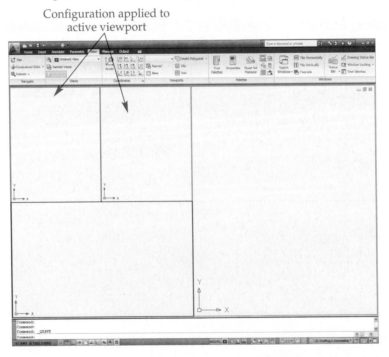

AutoCAD and Its Applications—Basics

change a view in a viewport, pick the viewport in the **Preview** image and then select the new viewpoint from the **Change view to:** drop-down list.

If none of the preset configurations is appropriate, you can create and save a unique viewport configuration. After you create the custom viewport configuration, enter a descriptive name in the **New name:** text box. When you pick the **OK** button, the new named viewport configuration records and displays in the **Named Viewports** tab the next time you access the **Viewports** dialog box. See **Figure 6-15.** Select a different named viewport configuration and pick **OK** to apply it to the drawing area. Named viewport configurations can apply to the active viewport only.

NOTE

Apply the **1 Viewport** option to return the current viewport configuration to the default single viewport.

Working in Tiled Viewports

After you select the viewport configuration and return to the drawing area, move the pointing device around and notice that only the active viewport contains crosshairs. The cursor is an arrow in the other viewports. To make an inactive viewport active, move the cursor into the inactive viewport and pick.

As you draw in one viewport, the image appears in all viewports. Try drawing lines and other shapes and notice how the viewports are affected. Use a display tool, such as **ZOOM**, in the active viewport and notice the results. Only the active viewport reflects the use of the **ZOOM** tool.

Joining Tiled Viewports

Use the **Join Viewports** tool to join two viewports. When you select the **Join** tool, AutoCAD prompts you to select the dominant viewport. Select the viewport that has the view you want to display in the joined viewport, or press [Enter] to select the active viewport. Then select the viewport to join with the dominant viewport. AutoCAD "glues" the two viewports together and retains the dominant view.

Ribbon
View
> Viewports
Join Viewports
Type
VIEWPORTS
VPORTS

Figure 6-15.
The **Named Viewports** tab displays custom viewports.

List of named viewport configurations

Preview of selected configuration

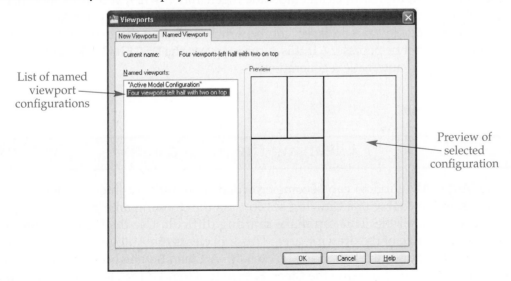

The two viewports you join cannot create an L-shape viewport. The adjoining edges of the viewports must be the same size in order to join them.

Exercise 6-6

Access the Student Web site (www.g-wlearning.com/CAD) and complete Exercise 6-6.

Redrawing and Regenerating the Screen

redrawing:
Refreshing the display of objects on the screen without recalculating the vectors.

regenerating:
Recalculating all objects based on the current zoom magnification and redisplaying them.

Redrawing refreshes the display of objects. *Regenerating* recalculates all object coordinates and displays them based on the current zoom magnification. For example, if curved objects appear as straight segments when you zoom in, you can regenerate the display to smooth the curves. Each viewport is a separate virtual screen. As a result, if you are using two or more viewports, you must decide whether to redraw or regenerate a single viewport or all the viewports.

Tool	Function
REDRAW	Redraws the display of the current viewport only.
REGEN	Regenerates the display in the current viewport only.
REDRAWALL	Redraws the display of the entire drawing.
REGENALL	Regenerates the display of the entire drawing.

PROFESSIONAL TIP

AutoCAD does an automatic regeneration when you use a tool that changes certain aspects of objects. This regeneration can take considerable time on large, complex drawings, and the regeneration may not be necessary. If this is the case, use the **REGENAUTO** tool to turn off automatic regenerations.

Type
[CTRL]+[0]

Cleaning the Screen

The AutoCAD window can become crowded with multiple interface items, such as palettes, in the course of a drawing session. As the drawing area gets smaller, less of the drawing is visible. This can make drafting difficult. Use the **Clean Screen** tool to maximize the size of the drawing area. This tool clears the AutoCAD window of all toolbars, palettes, and title bars. See **Figure 6-16.** A **Clean Screen** button is also available in the status bar. Accessing the **Clean Screen** tool toggles the clean screen display on and off.

Figure 6-16.
Using the **Clean Screen** tool. A—Initial display with the **Properties** palette and **Layer Properties Manager** displayed. B—Display after using the **Clean Screen** tool.

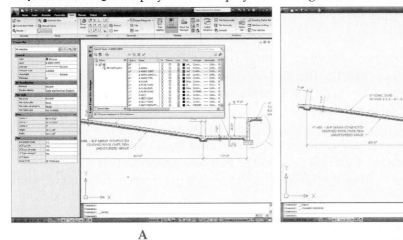

A   B

PROFESSIONAL TIP

The **Clean Screen** tool can be helpful when you have multiple drawings displayed. Only the active drawing appears when you use the **Clean Screen** tool. This allows you to work more efficiently within one of the drawings.

Supplemental Material

View Transitions and Resolution

For information about view transitions and view resolution, go to the Student Web site (www.g-wlearning.com/CAD), select this chapter, and select **View Transitions and Resolution**.

Template Development
Chapter 6

For detailed instructions on zooming to drawing limits in each template, go to the Student Web site (www.g-wlearning.com/CAD), select this chapter, and select **Template Development**.

Chapter Test

Answer the following questions. Write your answers on a separate sheet of paper or go to the Student Web site (www.g-wlearning.com/CAD) and complete the electronic chapter test.

1. During the drawing process, when should you use **ZOOM**?
2. Briefly explain how to use the **Realtime** zoom option.
3. What is the difference between the **Extents** and **All** zoom options?
4. What is the purpose of the **PAN** tool?
5. What is the difference between zooming and panning?
6. Which **SteeringWheels** navigation tools can be used in 2D drafting applications?
7. Explain how to use the **CENTER** tool on the **Full Navigation Wheel**.
8. What feature of the **Full Navigation Wheel** allows you to return to previous display settings?
9. How can you display a miniature version of the **Full Navigation Wheel**?
10. How is a transparent display command entered at the keyboard?
11. Name at least three display tools that can be used transparently.
12. Which tool changes the order in which objects are displayed in a drawing?
13. How can you obtain a list of existing views?
14. How do you create a named view of the current screen display?
15. How do you display an existing view?
16. What type of viewport is created in model space?
17. How can you specify whether a new viewport configuration applies to the entire drawing window or the active viewport?
18. Explain the procedures and conditions that need to exist for joining viewports.
19. What is the difference between the **REDRAW** and **REGEN** tools?
20. Which tool regenerates all of the viewports?

Drawing Problems

Start AutoCAD if it is not already started. Follow the specific instructions for each problem.

▼ Basic

1. Perform the following tasks:
   A. Draw a circle.
   B. Use realtime zooming to zoom in and out on the circle.
   C. Use realtime panning to pan the screen display.

2. Create a freehand sketch of the full-size **Full Navigation Wheel**. Label each of the wedges.

▼ Intermediate

3. Create a freehand sketch of the clean-screen AutoCAD window. Label each of the screen areas. To the side of the sketch, write a short description of each screen area's function.

4. Open the drawing named 3D House.dwg found in the AutoCAD 2010\Sample folder.
   Perform the following display functions on the drawing:
   A. Zoom to the drawing extents.
   B. Create a view named Rendering.
   C. Replace the view with the Top view.
   D. Create a view named Plan using **Define Window** in the **New View** dialog box.
   E. Use realtime pan and realtime zoom to create a display of the dining room in the top-right area of the Plan view.
   F. Create a view of this display named Dining Room.
   G. Display the view named Rendering.
   H. Save the drawing as P6-4.

▼ Advanced

5. Use the **Select Template** dialog box to load the Tutorial-iMfg.dwt template or another mechanical template you have access to that contains a border and title block. Do the following:
   A. Zoom into the title block area. Create and save a view named Title.
   B. Zoom to the extents of the drawing and create and save a view named All.
   C. Determine the areas of the drawing that will contain notes, parts list, and revisions. Zoom into these areas and create views with appropriate names, such as Notes, Partlist, and Revisions.
   D. Divide the drawing area into commonly used multiview sections. Save the views with descriptive names such as Top, Front, Rightside, and Leftside.
   E. Restore the view named All.
   F. Save the drawing as P6-5.

---

6. Use the **Select Template** dialog box to load the Tutorial-iArch.dwt template or another architectural drafting template you have access to that contains a border and title block. Do the following:
   A. Zoom into the title block area. Create and save a view named Title.
   B. Zoom to the extents of the drawing and create and save a view named All.
   C. Determine the area of the drawing that will contain notes, schedules, or revisions. Zoom into these areas and create views with appropriate names such as Notes, Schedules, and Revisions.
   D. Restore the view named All.
   E. Save the drawing as P6-6.

AutoCAD and Its Applications—Basics

CHAPTER 7

Object Snaps and AutoTrack

Learning Objectives

After completing this chapter, you will be able to do the following:

✓ Set running object snap modes for continuous use.
✓ Use object snap overrides for single point selections.
✓ Select appropriate object snaps for various drawing tasks.
✓ Use AutoSnap™ features to speed up point specifications.
✓ Use AutoTrack™ to locate points relative to other points in a drawing.

This chapter explains how to use the powerful object snap and AutoTrack tools to draw accurate geometry. Object snap and AutoTrack tools are some of the most useful and efficient drawing aids in AutoCAD. Like coordinate entry, object snaps and AutoTrack allow you to produce accurate geometric constructions, but they do not constrain, or apply relationships between, objects. Chapter 22 explains how to use parametric tools to constrain objects.

Object Snap

Object snap increases your drafting performance and accuracy through the concept of *snapping*. Object snap *modes* identify the object snap point. The AutoSnap feature is on by default and displays snap mode information while you draw. AutoSnap uses visual signals that appear as *markers* displayed at the snap point. **Figure 7-1** shows two examples of visual AutoSnap cues. After a brief pause, a tooltip appears, indicating the object snap mode. **Figure 7-2** describes each object snap mode. Refer to **Figure 7-2** constantly as you learn to use object snaps.

object snap: A tool that locates exact points, such as endpoints or midpoints, on or in relation to existing objects.

snapping: Picking a point near the intended position to have the crosshairs "snap" exactly to the specific point.

markers: Visual cues to confirm object snap points.

NOTE

If you cannot see an AutoSnap marker because of the size of the current screen display, you can still confirm the point before picking by reading the tooltip, which indicates if a point acquires beyond the visible area.

Figure 7-1.
AutoSnap displays markers and related tooltips for object snap modes.

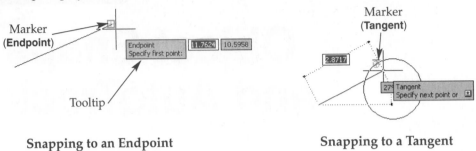

Snapping to an Endpoint   Snapping to a Tangent

Figure 7-2.
The object snap modes.

Mode	Marker	Description
Endpoint	□	Locates the nearest endpoint of a line, arc, polyline, elliptical arc, spline, ellipse, ray, solid, or multiline.
Midpoint	△	Finds the middle point of any object having two endpoints, such as a line, polyline, arc, elliptical arc, polyline arc, spline, xline, or multiline.
Center	○	Finds the center point of radial objects, including circles, arcs, ellipses, elliptical arcs, and radial solids.
Node	⊗	Locates a point object drawn with the **POINT**, **DIVIDE**, or **MEASURE** tool, or a dimension definition point.
Quadrant	◇	Locates the closest of the four quadrant points on circles, arcs, elliptical arcs, ellipses, and radial solids. (Some of these objects may not have all four quadrants.)
Intersection	✕	Locates the closest intersection of two objects.
Extension	+	Finds a point along the imaginary extension of an existing line, polyline, arc, polyline arc, elliptical arc, spline, ray, xline, solid, or multiline.
Insertion	⌐⌐	Locates the insertion point of text objects and blocks.
Perpendicular	⌐	Finds a point that is perpendicular to an object from the previously picked point.
Tangent	○	Finds points of tangency between radial and linear objects.
Nearest	⊠	Locates the point on an object that is closest to the crosshairs.
Apparent Intersection	⊠	Locates the intersection between two objects that appear to intersect on-screen in the current view, but may not actually intersect in 3D space. Creating and editing 3D objects is described in *AutoCAD and Its Applications—Advanced*.
Parallel	∥	Finds any point along an imaginary line parallel to an existing line or polyline.
None		Temporarily turns running object snap off during the current selection.

You can use object snaps for many drawing and editing applications. Practice with the different object snap modes to find the ones that work best for specific situations. With practice, object snap use becomes second nature, greatly increasing your productivity and accuracy.

Running Object Snaps

Running object snaps are on by default and are often the quickest and most effective way to use object snap. The **Endpoint**, **Center**, **Intersection**, and **Extension** running object snap modes are active by default. The quickest way to activate or deactivate running object snap modes is to right-click on the **Object Snap** or **Object Snap Tracking** button on the status bar and select the running object snaps to turn on or off.

You can also set running object snap modes using the **Object Snap** tab in the **Drafting Settings** dialog box. See **Figure 7-3.** To display the **Drafting Settings** dialog box, right-click on any of the status bar toggle buttons and selecting **Settings...**. Pick individual check boxes and use the **Select All** and **Clear All** buttons to help choose only those object snaps you want to run.

To use running object snaps, move the crosshairs near the location on an existing object where the object snap is to occur. When you see the appropriate marker, and if necessary, the tooltip, pick to locate the point at the exact position on the existing object. For example, with the **Endpoint** running object snap on, move the crosshairs toward the end of a curve to display the endpoint marker. Pick to locate the point at the exact endpoint of the existing object. See **Figure 7-4.**

Toggle running object snaps off and on by picking the **Object Snap** button on the status bar, pressing [F3], or using the **Object Snap On (F3)** check box on the **Object Snap** tab of the **Drafting Settings** dialog box. Turn off running object snaps to locate points without the aid, or to avoid possible confusion of object snap modes. The selected running object snap modes restore when you turn running object snaps back on.

running object snaps: Object snap modes that run in the background during all drawing and editing procedures.

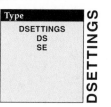

Type
DSETTINGS
DS
SE

DSETTINGS

Figure 7-3.
Running object snap modes can also be set in the **Drafting Settings** dialog box.

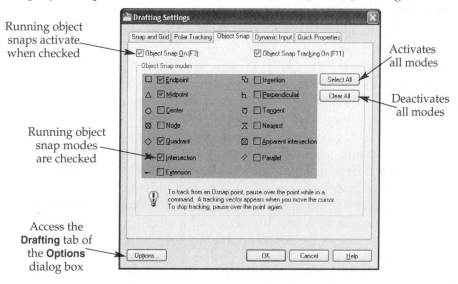

Figure 7-4.
Using the **Endpoint** object snap. When using running object snaps, be sure the correct snap marker and tooltip appear before you pick.

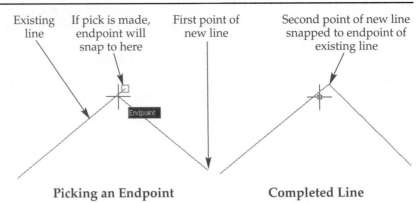

Existing line

If pick is made, endpoint will snap to here

First point of new line

Second point of new line snapped to endpoint of existing line

Picking an Endpoint   Completed Line

PROFESSIONAL TIP

Activate only the running object snap modes that you use most often. Too many running object snaps can make it difficult to snap to the appropriate location, especially on detailed drawings with several objects near each other. Use object snap overrides to access object snap modes that you use less often.

NOTE

By default, a keyboard point entry overrides running object snaps. Use the **Priority for Coordinate Data Entry** area on the **User Preferences** tab of the **Options** dialog box to adjust the default setting.

Object Snap Overrides

object snap override: A method of isolating a specific object snap mode while a tool is in use. The selected object snap temporarily overrides the running object snap modes.

Use an *object snap override* to select a specific point when running object snaps conflict with each other, or to use an object snap that is not running. Running objects snaps return after you make the object snap override selection. Most object snap modes are available as running object snaps or object snap overrides, although some modes are available only as object snap overrides.

After you access a tool, the preferred technique for activating an object snap override is to use the **Object Snap** shortcut menu. To use this method, press and hold [Shift] and then right-click to display the shortcut menu. See **Figure 7-5.** Select an object snap mode and then move the crosshairs near the location on an existing object where the object snap is to occur. When you see the marker, pick to locate the point at the exact position on the existing object. You can use this technique for accessing the **Object Snap** shortcut menu regardless of the active tool or whether you are picking the first point or an additional point.

When you locate a start point using some tools, an alternative for selecting an object snap is to right-click without holding [Shift]. This option functions the same except that the **Object Snap** shortcut menu is available from the **Snap Overrides** cascading submenu. Examples include selecting the center of a circle or placing an additional point using other tools, such as the second point of a line.

NOTE

You can activate an object snap override by typing the first three letters of the name of the object snap. For example, enter END to activate the **Endpoint** object snap or CEN to activate the **Center** object snap.

Figure 7-5.
The **Object Snap**
shortcut menu
provides quick
access to object snap
overrides.

holding down
Ball going to
mbutton pan
tool

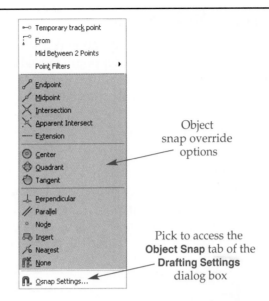

○—○ Temporary track point	
🗗 From	
Mid Between 2 Points	
Point Filters ▶	
✓ Endpoint	
✗ Midpoint	
✗ Intersection	
✗ Apparent Intersect	
---- Extension	
⊙ Center	
◈ Quadrant	
⟳ Tangent	
⊥ Perpendicular	
∥ Parallel	
° Node	
🖪 Insert	
⊅ Nearest	
🖾 None	
🔲 Osnap Settings...	

Object
snap override
options

Pick to access the
Object Snap tab of the
Drafting Settings
dialog box

PROFESSIONAL TIP

Remember that object snap modes are not tools, but are used with
tools. An error message appears if you activate an object snap mode
when no tool is active.

Endpoint Object Snap

The **Endpoint** object snap mode is available as a running object snap or object snap
override. To snap to an endpoint, move the crosshairs near the endpoint of a line, arc,
polyline, or spline. When the endpoint marker appears, pick to locate the point at the
exact endpoint. Refer again to **Figure 7-4**.

Midpoint Object Snap

The **Midpoint** object snap mode is available as a running object snap or object snap
override. To snap to a midpoint, move the crosshairs near the midpoint of a line, arc,
or polyline. When the midpoint marker appears, pick to locate the point at the exact
midpoint. See **Figure 7-6**.

Exercise 7-1

Access the Student Web site (www.g-wlearning.com/CAD) and
complete Exercise 7-1.

Figure 7-6.
Using the **Midpoint**
object snap.

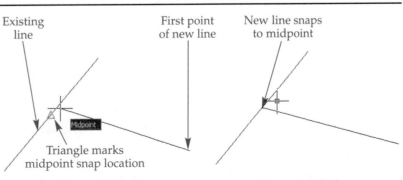

Existing line · First point of new line · New line snaps to midpoint · Midpoint · Triangle marks midpoint snap location

Picking a Midpoint · **Completed Line**

Center Object Snap

The **Center** object snap mode is available as a running object snap or object snap override. To snap to a center point, move the crosshairs near the *perimeter*, not the center point, of a circle, donut, ellipse, elliptical arc, polyline arc, or arc. The **Center** object snap mode will *not* locate the center of a large circle if the crosshairs is not near the perimeter of the circle. When the center marker appears, pick to locate the point at the exact center. See **Figure 7-7**.

Quadrant Object Snap

quadrant: Quarter section of a circle, donut, or ellipse.

The **Quadrant** object snap mode is available as a running object snap or object snap override. To snap to a *quadrant*, move the crosshairs near the appropriate 0°, 90°, 180°, or 270° point of a circle, donut, ellipse, elliptical arc, polyline arc, or arc. When you see the quadrant marker, pick to locate the point at the exact quadrant position. See **Figure 7-8**.

> **NOTE**
>
> Quadrant positions are unaffected by the current angle zero direction, but they always coincide with the angle of the X and Y axes. The quadrant points of circles, donuts, and arcs are at the right (0°), top (90°), left (180°), and bottom (270°), regardless of the rotation of the object. The quadrant points of ellipses and elliptical arcs, however, rotate with the objects.

Figure 7-7.
Using the **Center** object snap.

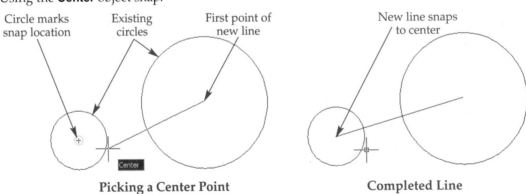

Figure 7-8.
Using the **Quadrant** object snap.

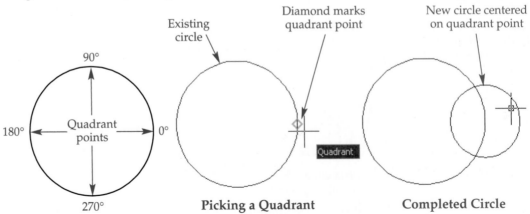

Exercise 7-2

Access the Student Web site (www.g-wlearning.com/CAD) and complete Exercise 7-2.

Intersection Object Snap

The **Intersection** object snap mode is available as a running object snap or object snap override. To snap to an intersection, move the crosshairs near the intersection of two or more objects. When you see the intersection marker, pick to locate the point at the exact intersection. See **Figure 7-9.**

Extension Object Snap

The **Extension** object snap mode is available as a running object snap or object snap override. The **Extension** object snap differs from most other object snaps because it uses *acquired points*, instead of direct point selection. To snap to an extension, move the crosshairs near the endpoint of a curve, but do not select. When an acquired point is found, a point symbol (+) marks the location. Move the crosshairs away from the acquired point to display an *extension path*. Pick a point along the extension path. **Figure 7-10** shows an example of using an **Extension** object snap twice to draw a line a specific distance away from two acquired points. In this example, type the .8 distance when you see the extension path. Dynamic input is not required to enter this value.

acquired point:
A point found by moving the crosshairs over a point on an existing object to reference for use when picking a new point.

extension path:
Dashed line or arc that extends from the acquired point to the current location of the crosshairs.

Exercise 7-3

Access the Student Web site (www.g-wlearning.com/CAD) and complete Exercise 7-3.

Figure 7-9.
Using the **Intersection** object snap.

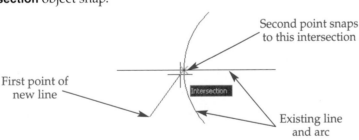

Second point snaps to this intersection

First point of new line

Intersection

Existing line and arc

Figure 7-10.
Using the **Extension** object snap to create a line .8 unit away from a rectangle.

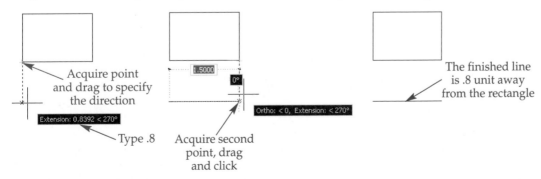

Acquire point and drag to specify the direction

Extension: 0.8392 < 270°

Type .8

1.5000

0°

Ortho: < 0, Extension: < 270°

Acquire second point, drag and click

The finished line is .8 unit away from the rectangle

Extended Intersection Object Snap

Even if objects do not intersect, you can snap to the location where the objects would intersect, if they were long enough. Make this selection using the **Extension** object snap or **Extended Intersection** object snap override. When using the **Extension** object snap mode, move the crosshairs near the endpoint of one curve to acquire the first point, and then move the crosshairs near the endpoint of another curve to acquire the second point. Now, move the crosshairs away from the acquired point, near the location of where the objects would intersect. When you see two extension paths and an intersection icon, pick to locate the point. See **Figure 7-11A**.

The **Extended Intersection** object snap override works by selecting objects one at a time using the **Intersection** object snap override. Once you activate the **Intersection** object snap override, move the cursor over one of the objects to display the intersection marker with an ellipsis (...), and pick the object. Then move the cursor over the other object to display the intersection marker at the extended intersection, and pick. See **Figure 7-11B**.

Figure 7-11.
Locating the center of a circle at the extended intersection of two objects. A—Using the **Extension** object snap. B— Using the **Extended Intersection** object snap.

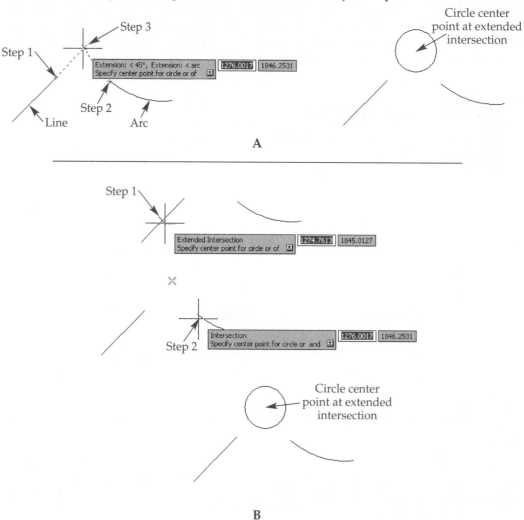

Exercise 7-4

Access the Student Web site (www.g-wlearning.com/CAD) and complete Exercise 7-4.

Perpendicular Object Snap

The **Perpendicular** object snap mode is available as a running object snap or object snap override. To snap to perpendicular, move the crosshairs near the point of perpendicularity on a line, arc, elliptical arc, ellipse, spline, xline, multiline, polyline, solid, trace, or circle, or the endpoint of a line, arc, polyline, or spline. When the perpendicular marker appears, pick to locate the point exactly perpendicular to the existing object. See **Figure 7-12**.

Figure 7-13 shows using the **Perpendicular** object snap mode to begin a line perpendicular to an existing object. The tooltip reads Deferred Perpendicular, and the perpendicular marker includes an ellipsis (...). The second endpoint determines the location of the line in a *deferred perpendicular* condition.

> **NOTE**
>
> Perpendicularity measures from the point of intersection. Therefore, it is possible to draw a line perpendicular to a circle or arc.

deferred perpendicular: A condition in which calculation of the perpendicular point is delayed until another point is picked.

Exercise 7-5

Access the Student Web site (www.g-wlearning.com/CAD) and complete Exercise 7-5.

Figure 7-12.
Drawing a line from a point perpendicular to an existing line.

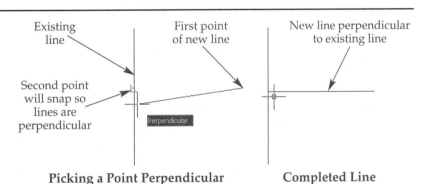

Figure 7-13.
Deferring the second point of a line to establish a perpendicular construction.

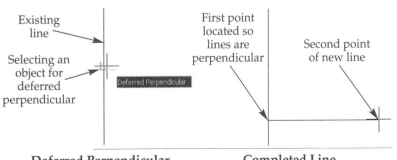

Tangent Object Snap

The **Tangent** object snap mode is available as a running object snap or object snap override. To snap to the point of tangency, move the crosshairs near an arc, circle, ellipse, elliptical arc, or spline. When the tangent marker appears, pick to locate the point at the exact point of tangency. See **Figure 7-14.**

When drawing an object tangent to two objects, you may need to pick multiple points to fix the point of tangency. For example, the point at which a line is tangent to a circle is found according to the locations of both ends of the line. Until both points are identified, the object snap specification is for *deferred tangency.* Once both endpoints are known, the tangency calculates, and the object is drawn in the correct location. See **Figure 7-15.**

deferred tangency: A condition in which calculation of the point of tangency is delayed until both points are picked.

Exercise 7-6

Access the Student Web site (www.g-wlearning.com/CAD) and complete Exercise 7-6.

Parallel Object Snap

The **Parallel** object snap mode is available as a running object snap or object snap override. To snap to a point parallel to a line or polyline, move the crosshairs near the existing object to display the parallel marker. Then, move the crosshairs away from and near parallel to the existing object. As you near a position parallel to the existing object, a *parallel alignment path* extends from the location of the crosshairs, and the parallel marker reappears to indicate acquired parallelism. Pick a point along the parallel alignment path. See **Figure 7-16.**

parallel alignment path: A dashed line, parallel to the existing line, that extends from the location of the crosshairs.

Exercise 7-7

Access the Student Web site (www.g-wlearning.com/CAD) and complete Exercise 7-7.

Figure 7-14.
Using the **Tangent** object snap.

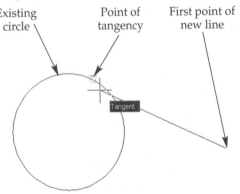

Existing circle

Point of tangency

First point of new line

Picking a Tangent Point

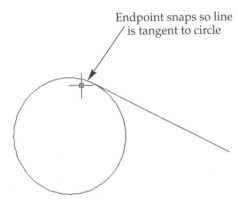

Endpoint snaps so line is tangent to circle

Completed Line

Figure 7-15.
Drawing a line tangent to two circles.

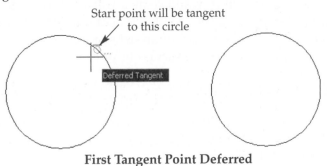

Start point will be tangent
to this circle

First Tangent Point Deferred

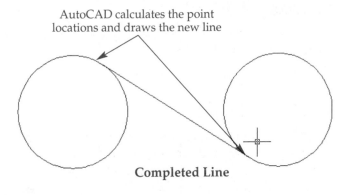

Endpoint will be
tangent to this circle

Picking Second Tangent Point

AutoCAD calculates the point
locations and draws the new line

Completed Line

Figure 7-16.
Using the **Parallel** object snap option to draw a line parallel to an existing line. A—Select the
first endpoint for the new line, select the **Parallel** object snap, and then move the crosshairs
near the existing line to acquire a point. B—Once the parallel point is acquired, move the
crosshairs near the location of the parallel line to display an extension path.

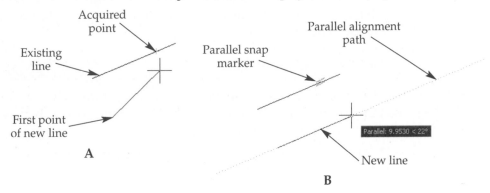

Acquired
point

Existing
line

Parallel snap
marker

Parallel alignment
path

First point
of new line

A

New line

B

Node Object Snap

The **Node** object snap mode is available as a running object snap or object snap override. To snap to a node, move the crosshairs near a point drawn using the **POINT**, **DIVIDE**, or **MEASURE** tool, or the origin of an extension line. When the node marker appears, pick to locate the point at the exact node, or point.

NOTE

In order for the **Node** object snap to find a point object, the point must be in a visible display mode. Chapter 8 explains point display mode controls.

Nearest Object Snap

The **Nearest** object snap mode is available as a running object snap or object snap override. Use the **Nearest** mode to specify a point that is directly on an object, but cannot be located with any of the other object snap modes, or when the location of the intersection is not critical. To snap to a nearest point, move the crosshairs near an existing object. When the nearest marker appears, pick to locate the point at a location on object closest to the crosshairs.

Exercise 7-8

Access the Student Web site (www.g-wlearning.com/CAD) and complete Exercise 7-8.

Temporary Track Point Snap

The **Temporary track point** snap mode is available only as an object snap override. It allows you to locate a point aligned with or relative to another point. For example, use the **Temporary track point** snap to place the center of a circle at the center of an existing rectangle. At the Specify center point for circle or [3P/2P/Ttr (tan tan radius)]: prompt, select the **Temporary tracking point** snap, and then use the **Midpoint** object snap to pick the midpoint of one of the vertical lines. This establishes the Y coordinate of the rectangle's center. See **Figure 7-17A.** When the Specify center point for circle or [3P/2P/Ttr (tan tan radius)]: prompt reappears, select the **Temporary tracking point** snap mode, and then use the **Midpoint** object snap mode to pick the midpoint of one of the horizontal lines. This establishes the X coordinate of the rectangle's center. See **Figure 7-17B.** Finally, pick to locate the center of the circle where the two tracking vectors intersect, and specify the circle radius. See **Figure 7-17C.**

NOTE

The direction in which you move the crosshairs from the temporary tracking point determines the X or Y alignment. Switch between horizontal or vertical tracking as needed.

Figure 7-17.
Using temporary tracking to locate the center of a rectangle. A—The midpoint of the left line is acquired. B—The midpoint of the bottom line is acquired. C—The center point of the circle is located at the intersection of the tracking vectors.

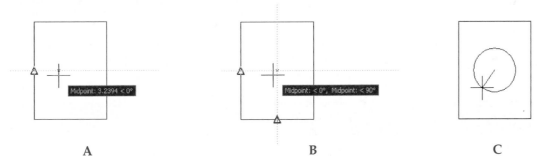

A   B   C

Snap From

The **From** snap mode is available only as an object snap override. It allows you to locate a point using coordinate entry from a specified reference base point. For example, use the **From** snap to place the center of a circle using a polar coordinate entry from the midpoint of an existing line. At the Specify center point for circle or [3P/2P/Ttr (tan tan radius)]: prompt, select the **From** snap mode, and then use the **Midpoint** object snap to pick the midpoint of the line. At the <Offset>: prompt, enter the polar coordinate @2<45 to establish the center of the circle 2 units and at a 45° angle from the midpoint of the line. Specify the radius of the circle to complete the operation. See **Figure 7-18**.

Mid Between 2 Points Snap

The **Mid Between 2 Points** snap feature is available only as an object snap override, and is very effective for locating a point exactly between two specified points. Use object snaps or coordinate point entry to pick reference points accurately. The example in **Figure 7-19** locates the center of a circle between two line endpoints.

Exercise 7-9

Access the Student Web site (www.g-wlearning.com/CAD) and complete Exercise 7-9.

Figure 7-18.
An example of using the **From** point selection mode to locate the center of a circle using a midpoint object snap and polar coordinate entry.

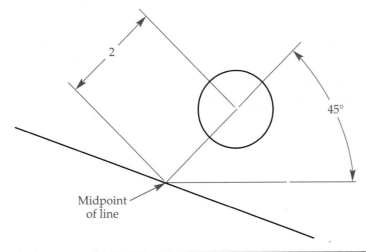

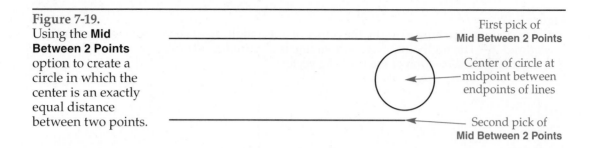

Figure 7-19.
Using the **Mid Between 2 Points** option to create a circle in which the center is an exactly equal distance between two points.

First pick of **Mid Between 2 Points**

Center of circle at midpoint between endpoints of lines

Second pick of **Mid Between 2 Points**

AutoTrack

AutoTrack offers an object snap tracking mode and a polar tracking mode. These tools are helpful for common drafting and design tasks, including basic geometric constructions. AutoTrack uses *alignment paths* and *tracking vectors* as drawing aids. You can use AutoTrack with any tool that requires a point selection.

alignment paths: Temporary lines and arcs that coincide with the position of existing objects.

tracking vectors: Temporary lines that display at specific angles, typically 0°, 90°, 180°, and 270°.

object snap tracking: Mode that provides horizontal and vertical alignment paths for locating points after a point is acquired with object snap.

Object Snap Tracking

Object snap tracking has two requirements: running object snaps must be active, and the crosshairs must pause over the intended selection long enough to acquire the point. Pick the **Object Snap Tracking** button on the status bar, press [F11], or use the **Object Snap Tracking On (F11)** check box in the **Object Snap** tab of the **Drafting Settings** dialog box to toggle object snap tracking on and off. Object snap tracking mode works with running object snaps. You must activate running object snaps and the appropriate running object snap modes in order for object snap tracking to function properly.

In **Figure 7-20**, object snap tracking is used with the **Perpendicular** and **Midpoint** running object snaps to draw a line 2 units long, perpendicular to the existing slanted

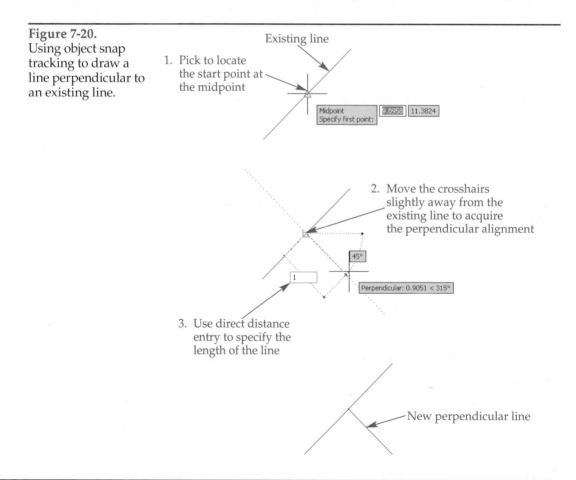

Figure 7-20.
Using object snap tracking to draw a line perpendicular to an existing line.

Existing line

1. Pick to locate the start point at the midpoint

Midpoint
Specify first point: 9.8958 11.3824

2. Move the crosshairs slightly away from the existing line to acquire the perpendicular alignment

45°

1

Perpendicular: 0.9051 < 315°

3. Use direct distance entry to specify the length of the line

New perpendicular line

line. Running object snaps, the **Perpendicular** and **Midpoint** running object snap modes, and object snap tracking are active before using the **LINE** tool to draw the new line. Select the midpoint of the existing line to locate the start point of the new line. Then move the crosshairs slightly away from the existing line to display the perpendicular marker and the alignment path. Use direct distance entry along the alignment path to complete the line.

In **Figure 7-21**, object snap tracking is used with the **Midpoint** running object snap to position a circle directly above the midpoint of a horizontal line and to the right of the midpoint of an angled line. With running object snaps, the **Midpoint** running object snap mode, and object snap tracking active, use the **Circle, Radius** tool to draw the circle. Pause the crosshairs near the midpoint of the horizontal line to acquire the first point, and then pause the crosshairs near the midpoint of the angled line to acquire the second point. Move the crosshairs to the position as shown in the second step of **Figure 7-21** until two tracking vectors appear. Pick to locate the center of the circle, and complete the operation by entering a radius.

PROFESSIONAL TIP

Use object snap tracking whenever possible to complete tasks that require you to reference locations on existing objects. Often the combination of running object snaps and object snap tracking is the quickest way to construct geometry.

Exercise 7-10

Access the Student Web site (www.g-wlearning.com/CAD) and complete Exercise 7-10.

Figure 7-21.
Using object snap tracking to position a circle in line with the midpoints of two lines.

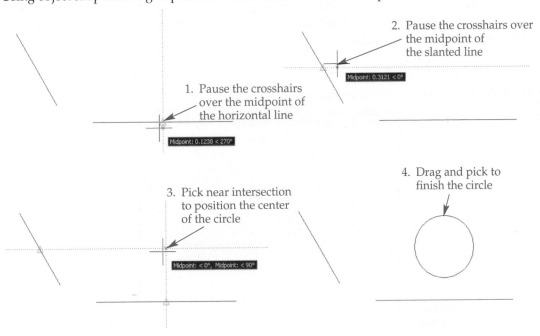

1. Pause the crosshairs over the midpoint of the horizontal line
   Midpoint: 0.1238 < 270°

2. Pause the crosshairs over the midpoint of the slanted line
   Midpoint: 0.3121 < 0°

3. Pick near intersection to position the center of the circle
   Midpoint: < 0°, Midpoint: < 90°

4. Drag and pick to finish the circle

Polar Tracking

Chapter 3 introduces *polar tracking* as an accurate method of using direct distance entry. Pick the **Polar Tracking** button on the status bar, press [F10], or use the **Polar Tracking On (F10)** check box in the **Polar Tracking** tab of the **Drafting Settings** dialog box to toggle polar tracking on and off. When polar tracking is on, the crosshairs snaps to preset incremental angles when you locate a point relative to another point. For example, when you draw a line, polar tracking is not active for the first point selection, but it is available for the second and additional point selections. Polar tracking vectors appear as dotted lines whenever the crosshairs aligns with any of these preset angles.

To set incremental angles, use the **Polar Tracking** tab in the **Drafting Settings** dialog box. See **Figure 7-22**. To display the **Drafting Settings** dialog box, right-click on any of the status bar toggle buttons and select **Settings...**. The **Polar Angle Settings** area sets polar angle increments. Use the **Increment angle** drop-down list to select the angle increments at which polar tracking vectors occur. A variety of preset angles is available. The default increment is 90, which provides angle increments every 90°. The 30° setting shown in **Figure 7-22** provides polar tracking in 30° increments.

NOTE

The preset angle increments available in the **Polar Angle Settings** area of the **Drafting Settings** dialog box are also available in the shortcut menu displayed when you right-click on the **Polar Tracking** button on the status bar.

To add specific polar tracking angles, pick the **New** button in the **Polar Angle Settings** area, and type a new angle value in the text box that appears in the **Additional angles** window. Repeat the process to add other angles. Additional angles work with the increment angle setting when you use polar tracking. Only the specific additional angles you enter are recognized, not each increment of the angle. Use the **Delete** button to remove angles from the list. Make the additional angles inactive by unchecking the **Additional angles** check box.

The **Object Snap Tracking Settings** area sets the angles available with object snap tracking. If you select **Track orthogonally only**, only horizontal and vertical alignment paths are active. If you select **Track using all polar angle settings**, alignment paths are active for all polar snap angles.

Figure 7-22.
The **Polar Tracking** tab of the **Drafting Settings** dialog box.

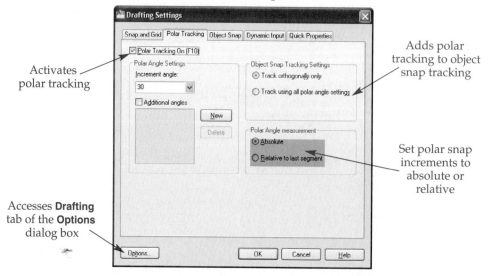

The **Polar Angle measurement** setting determines whether the polar snap increments are constant or relative to the previous segment. If you choose **Absolute**, polar snap angles measure from the base angle of 0° set for the drawing. If you pick **Relative to last segment**, each increment angle measures from a base angle established by the previously drawn segment.

Figure 7-23 shows how to draw a parallelogram using polar tracking and 30° angle increments. Access the **LINE** tool and select the first point. Then move the crosshairs to the right while the polar alignment path indicates <0°. Enter a direct distance value. Move the crosshairs to the 60° polar alignment path and enter a direct distance value. Move the crosshairs to the 180° polar alignment path and enter a value. Finally, use polar tracking with an angle of 240° and a specified line distance, or use the **Close** option to finish the parallelogram.

> **NOTE**
>
> You cannot use polar tracking and **Ortho** at the same time. AutoCAD automatically turns **Ortho** off when polar tracking is on, and it turns polar tracking off when **Ortho** is on.

Exercise 7-11

Access the Student Web site (www.g-wlearning.com/CAD) and complete Exercise 7-11.

Polar Tracking with Polar Snaps

You can also use polar tracking with polar snaps. For example, if you use polar tracking and polar snaps to draw the parallelogram in **Figure 7-23**, there is no need to type the length of the line, because you set the angle increment with polar tracking and a length increment with polar snaps. Establish angle and length increments using the **Snap and Grid** tab of the **Drafting Settings** dialog box. See **Figure 7-24**.

To activate polar snap, pick the **PolarSnap** radio button in the **Snap type & style** area of the dialog box. **PolarSnap** mode activates the **Polar spacing** area and deactivates the **Snap** area. Set the length of the polar snap increment in the **Polar distance:** text box. If the **Polar distance:** setting is 0, the polar snap distance is the orthogonal snap distance. **Figure 7-25** shows a parallelogram drawn with 30° angle increments and length increments of .75. The lengths of the parallelogram sides are 1.5 and .75.

Figure 7-23.
Using polar tracking with 30° angle increments to draw a parallelogram. A—After the first side is drawn, the alignment path and direct distance entry are used to create the second side. B—A horizontal alignment path is used for the third side. C—The parallelogram is completed with the Close option.

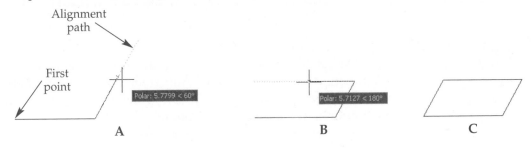

Figure 7-24.
The **Snap and Grid** tab of the **Drafting Settings** dialog box is used to set polar snap spacing.

Activates snap

Polar snap spacing

Select grid or polar snap

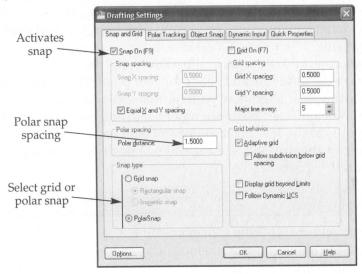

Figure 7-25.
Drawing a parallelogram with polar snap.

| A | B | C | D |

Using Polar Tracking Overrides

It takes time to set up polar tracking and polar snap options, but it is worth the effort if you draw several objects that can take advantage of this feature. Use polar tracking overrides to perform polar tracking when you need to define only one point. Polar tracking overrides work for the specified angle whether polar tracking is on or off. To activate a polar tracking override, type a less than symbol (<) followed by the desired angle when AutoCAD asks you to specify a point. For example, after you access the **LINE** tool and pick a first point, enter <30 to set a 30° override. Then move the crosshairs in the desired 30° direction and enter a distance, such as 1.5 to draw a 1.5-unit line.

Exercise 7-12

Access the Student Web site (www.g-wlearning.com/CAD) and complete Exercise 7-12.

Supplemental Material *AutoSnap and AutoTrack Options*

For information about options for controlling the appearance and function of AutoSnap and AutoTrack, go to the Student Web site (www.g-wlearning.com/CAD), select this chapter, and select **AutoSnap and AutoTrack Options**.

Template Development
Chapter 7

For detailed instructions on setting object snaps and polar tracking to save time and increase efficiency, go to the Student Web site (www.g-wlearning.com/CAD), select this chapter, and select **Template Development**.

Chapter Test

Answer the following questions. Write your answers on a separate sheet of paper or go to the Student Web site (www.g-wlearning.com/CAD) and complete the electronic chapter test.

1. Define the term *object snap*.
2. What is an AutoSnap tooltip?
3. Name the following AutoSnap markers:

A.   E.   I.

B.   F.   J.

C.   G.   K.

D.   H.   L.

4. How do you set running object snaps?
5. Define the term *running object snap*.
6. How do you access the **Drafting Settings** dialog box to change object snap settings?
7. If you are using running object snaps and want to make several point specifications without the aid of object snap, but want to continue the same running object snaps after making the desired point selections, what is the easiest way to turn off the running object snaps temporarily?
8. If you are using running object snaps and you want to make a single point selection without the effects of the running object snaps, what do you do?
9. Describe the object snap override.
10. How do you activate the **Object Snap** shortcut menu?
11. Where are the four quadrant points on a circle?
12. What is the situation when the tooltip reads Extended Intersection?
13. What does it mean when the tooltip reads Deferred Perpendicular?

14. Give the tool and entries needed to draw a line tangent to an existing circle and perpendicular to an existing line:
    A. Tool: _____
    B. Specify first point: _____
    C. to _____
    D. Specify next point or [Undo]: _____
    E. to _____
15. What is a deferred tangency?
16. What conditions must exist for the tooltip to read Tangent?
17. Which object snaps depend on acquired points to function?
18. What two display features does AutoTrack use to help you line up new objects with existing geometry?
19. What are the two requirements to use object snap tracking?
20. When are polar tracking vectors displayed as dotted lines?

Drawing Problems

Start AutoCAD if it is not already started. Start each new drawing using an appropriate template of your choice. Add layers as needed. Draw all objects on the appropriate layers. Follow the specific instructions for each problem. Do not draw dimensions or text. Use your own judgment and approximate dimensions when necessary.

▼ Basic

1. Draw the object below using object snap modes. Save the drawing as P7-1.

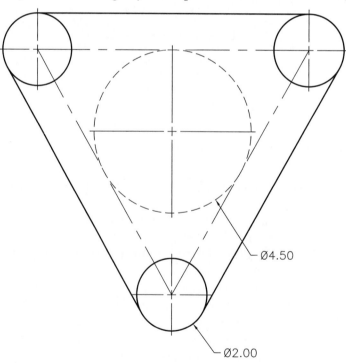

2. Draw the highlighted objects below, and then use the object snap modes indicated to draw the remaining objects. Save the drawing as P7-2.

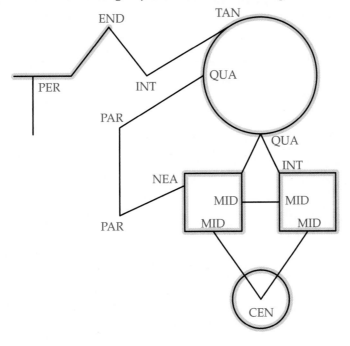

3. Draw the object below using the **Endpoint**, **Tangent**, **Perpendicular**, and **Quadrant** object snap modes. Save the drawing as P7-3.

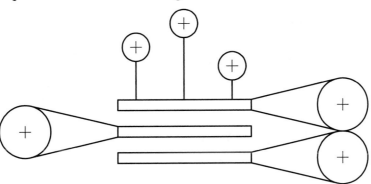

4. Draw the pipe separator shown below. Use object snaps and tracking to place the objects correctly. Save the drawing as P7-4.

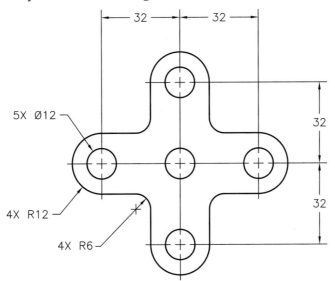

▼ Intermediate

5. Use the **Midpoint**, **Endpoint**, **Tangent**, **Perpendicular**, and **Quadrant** object snap modes to draw these electrical switch schematics. Save the drawing as P7-5.

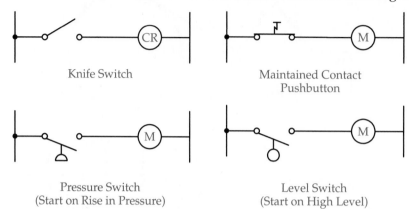

Knife Switch

Maintained Contact
Pushbutton

Pressure Switch
(Start on Rise in Pressure)

Level Switch
(Start on High Level)

6. Draw the elbow shown. Save the drawing as P7-6.

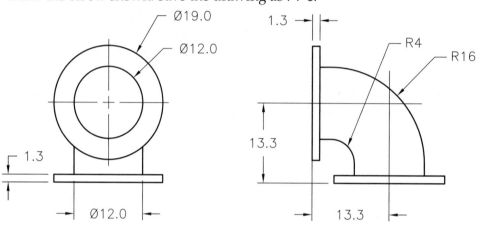

7. Draw the elbow shown. Save the drawing as P7-7.

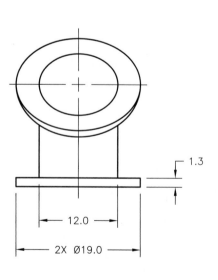

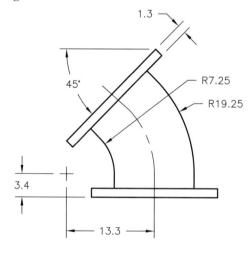

8. Draw the object shown. Save the drawing as **P7-8**.

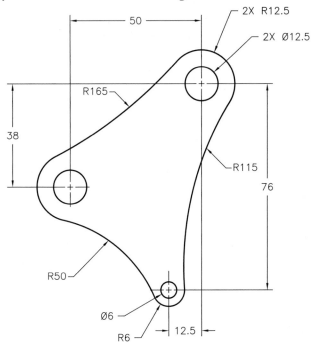

9. Draw the object shown. Save the drawing as **P7-9**.

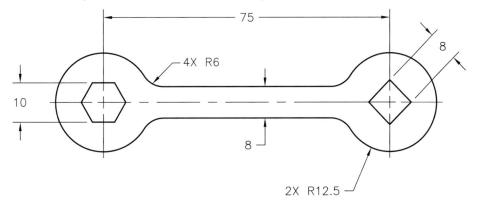

10. Draw the object shown. Save the drawing as P7-10.

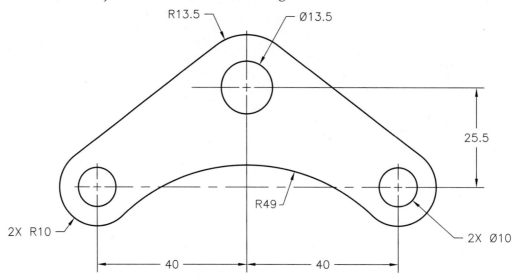

▼ Advanced

11. Use object snap modes to draw this elementary diagram. Save the drawing as P7-11.

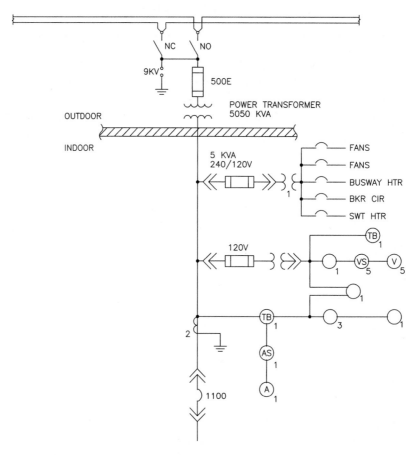

12. Draw the object shown. Save the drawing as **P7-12**.

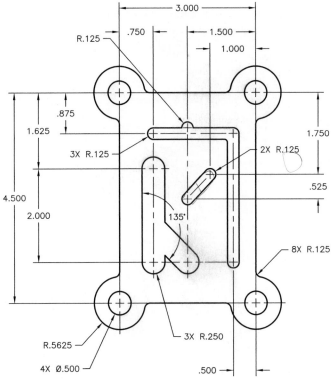
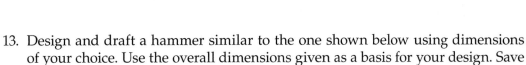

13. Design and draft a hammer similar to the one shown below using dimensions of your choice. Use the overall dimensions given as a basis for your design. Save the drawing as **P7-13**.

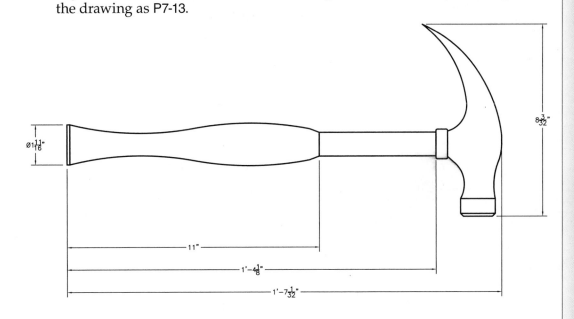

14. Draw the sailboat shown. Save the drawing as P7-14.

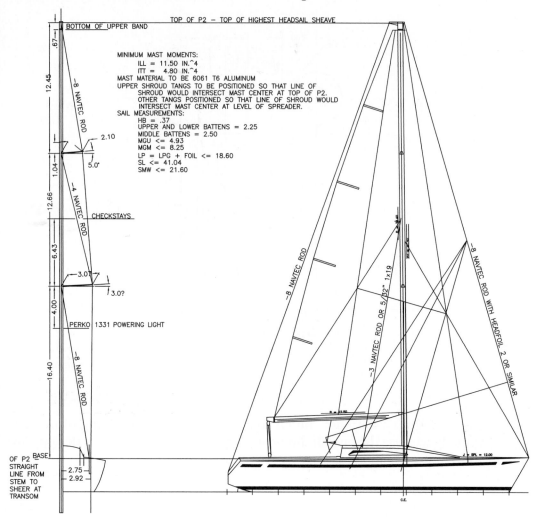

Construction Tools and Multiview Drawings

Learning Objectives

After completing this chapter, you will be able to do the following:

✓ Use the **OFFSET** tool to draw parallel objects.
✓ Place construction points.
✓ Mark points on objects at equal lengths using the **DIVIDE** tool.
✓ Mark points on objects at designated increments using the **MEASURE** tool.
✓ Create construction lines using the **XLINE** and **RAY** tools.
✓ Create orthographic multiview drawings.

This chapter explains how to create parallel offsets, divide objects, place point objects, and use construction lines. You can use these skills and the other geometric construction skills you have acquired to create multiview drawings. The tools described in this chapter allow you to produce accurate geometric constructions, but they do not apply relationships between objects. Chapter 22 explains how to use parametric tools to constrain objects.

Creating Parallel Offsets

The **OFFSET** tool is one of the most commonly used geometric construction tools. Offset lines and polylines for a variety of applications, such as constructing the thickness of architectural floor plan walls. Offset circles, arcs, and curves to form concentric objects. For example, offset a circle to form the wall thickness of a pipe.

Specifying the Offset Distance

Often the best way to use the **OFFSET** tool is to enter an offset value at the Specify offset distance or [Through/Erase/Layer] <*current*>: prompt. For example, to draw two concentric circles 1 unit apart, access the **OFFSET** tool, and specify an offset distance of 1. Pick the circle to offset, and then pick the side of the circle on which the offset occurs. See **Figure 8-1.** The **OFFSET** tool remains active, allowing you to pick another object to offset using the same offset distance. To exit the tool, press [Enter], [Esc], or the space bar, or choose the **Exit** option.

Ribbon
Home
> Modify

Offset

Type
OFFSET
O

Figure 8-1.
Drawing an offset circle using a designated distance.

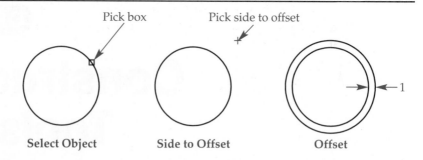

Pick box

Pick side to offset

1

Select Object   Side to Offset   Offset

NOTE

Select objects to offset individually. No other selection option, such as window or crossing selection, works to select objects to offset.

PROFESSIONAL TIP

When using most tools that prompt you to specify a value, such as distance or height, if you do not know the numerical value, an alternative is to pick two points. The distance between your selections sets the value. Typically, the two points are located on existing objects, and you recognize that your selections will result in the appropriate value. Use object snaps, AutoTrack, or coordinate entry to make accurate selections.

Using the Through Option

Another option to specify the offset distance is to pick a point through which the offset occurs. After you access the **OFFSET** tool, activate the **Through** option at the Specify offset distance or [Through/Erase/Layer] <*current*>: prompt instead of picking an object to offset. Then pick the object to offset, and pick the point through which the offset occurs. See **Figure 8-2.** The **OFFSET** tool remains active, allowing you to pick another object to offset using the **Through** option. Exit the **OFFSET** tool when you are finished.

Erasing the Original Object

Use the **Erase** option of the **OFFSET** tool to erase the original, or source, object during the offset. Initiate the **OFFSET** tool, activate the **Erase** option, and choose **Yes** at the Erase source object after offsetting? prompt to erase the source object. The **Yes** option

Figure 8-2.
Drawing an offset through a given point.

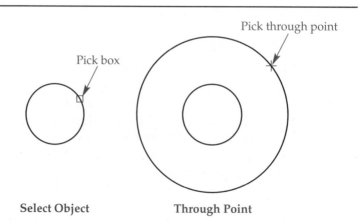

Pick through point

Pick box

Select Object   Through Point

remains set as default until you change the setting to **No**. Be sure to change the **Erase** setting back to **No** if you do not want the source offset object to erase the next time you use the **OFFSET** tool. Exit the **OFFSET** tool when you are finished.

Using the Layer Option

By default, offsets generate using the same properties, including layer, as the source object. Use the **Layer** option of the **OFFSET** tool to place the offset object on the current layer, regardless of the layer used to draw the source object. First, make the layer that you want to apply to the offset current. Then initiate the **OFFSET** tool, activate the **Layer** option, and choose **Current** at the Enter layer option for offset objects: prompt. The **Current** option remains set as default until you change the setting to **Source**. Be sure to change the **Layer** setting back to **Source** if you do not want the current layer applied to the offset the next time you use the **OFFSET** tool. Exit the **OFFSET** tool when you are finished.

Offsetting Multiple Times

After you select the object to offset, use the **Multiple** option to offset an object more than once, using the same distance between objects, without reselecting the object to offset. Initiate the **OFFSET** tool, specify the offset distance, and pick the source object. You can then select the **Multiple** option and begin picking to specify the offset direction. See **Figure 8-3.** Exit the **OFFSET** tool when finished.

NOTE

You can use the **Undo** option, when available, to undo the last offset without exiting the **OFFSET** tool.

Exercise 8-1

Access the Student Web site (www.g-wlearning.com/CAD) and complete Exercise 8-1.

Figure 8-3.
Use the **Multiple** option to create multiple offsets of the same distance, without picking the source object again.

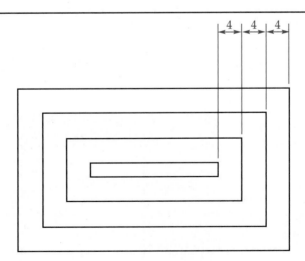

Drawing Points

POINT

Ribbon
Home
> Draw

Multiple Points

Type
POINT
PO

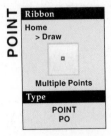

Point objects are useful for identifying specific locations on a drawing and for marking positions on objects. You can draw points anywhere on the screen using the **POINT** tool. Use any appropriate method to specify the location of a point object. To place a single point object and then exit the **POINT** tool, enter **POINT** or **PO** at the keyboard. To draw multiple points without exiting the **POINT** tool, access the **Multiple Points** function from the ribbon. Press [Esc] to exit the tool.

Setting Point Style

DDPTYPE

Ribbon
Home
> Utilities

Point Style

Type
DDPTYPE

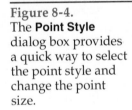

Point style and size are set using the **Point Style** dialog box. See **Figure 8-4**. By default, points appear as one-pixel dots and typically do not show up very well on-screen. Change the point style to make points more visible. The **Point Style** dialog box contains twenty different point styles. Pick the image of the desired style to make the point style current. All existing and new points change to the current style.

Setting Point Size

Set the point size by entering a value in the **Point Size:** text box of the **Point Style** dialog box. Pick the **Set Size Relative to Screen** button to change the point size in relation to different screen magnifications (zooming in or out). You may need to regenerate the display to view the relative sizes. Pick the **Set Size in Absolute Units** option button to make points appear the same size regardless of the screen magnification. See **Figure 8-5**.

Figure 8-4.
The **Point Style** dialog box provides a quick way to select the point style and change the point size.

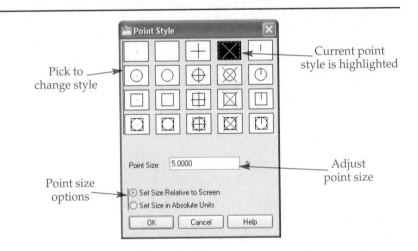

Pick to change style

Current point style is highlighted

Adjust point size

Point size options

Figure 8-5.
Points sized with the **Set Size Relative to Screen** setting change size as the drawing is zoomed. Points sized with the **Set Size in Absolute Units** setting remain a constant size.

Size Setting	Original Point Size	2X Zoom	.5 Zoom
Relative to Screen	⊠	⊠	⊠
Absolute Units	⊠	⊠	⊠

AutoCAD and Its Applications—Basics

Exercise 8-2

Access the Student Web site (www.g-wlearning.com/CAD) and complete Exercise 8-2.

Marking an Object at Specified Increments

You can use the **DIVIDE** tool to place point objects or blocks at equally spaced locations on a line, circle, arc, or polyline. The **DIVIDE** tool *does not* break an object into an equal number of segments. Access the **DIVIDE** tool and select the object to mark. Enter the number of segments to mark with points and exit the tool. The point style determines the style of the marks placed on the object. **Figure 8-6** shows an example of using the **DIVIDE** tool to place points at seven equal increments.

The **Block** option of the **DIVIDE** tool allows you to place a *block* at each increment. Select the **Block** option at the Enter the number of segments or [Block]: prompt to insert a block. AutoCAD asks if the block should align with the object. Blocks are described in detail later in this textbook.

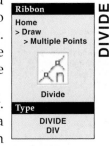

Ribbon
Home
> Draw
> Multiple Points

Divide

Type
DIVIDE
DIV

block: A previously drawn symbol or shape.

Marking an Object at Specified Distances

While the **DIVIDE** tool marks a line, circle, arc, or polyline according to a specified number of increments, the **MEASURE** tool places marks a specified distance apart. Access the **MEASURE** tool and select the object to mark. Measurement begins at the end closest to where you pick the object. Enter the distance between points to place points and exit the tool. All increments are equal to the specified segment length, except the last segment, which may be shorter. The point style determines the style of the marks placed on the object. The line shown in **Figure 8-7** was measured using a length of .75 unit. Use the **Block** option of the **MEASURE** tool to place a block at each measurement.

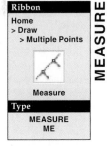

Ribbon
Home
> Draw
> Multiple Points

Measure

Type
MEASURE
ME

Exercise 8-3

Access the Student Web site (www.g-wlearning.com/CAD) and complete Exercise 8-3.

Figure 8-6.
Using the **DIVIDE** tool. Note that the default point style has been changed to ×.

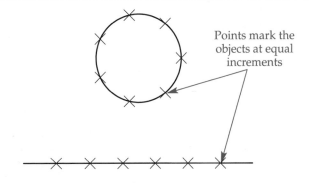

Points mark the objects at equal increments

Figure 8-7.
Using the **MEASURE** tool. Notice that the last segment may be shorter than the others, depending on the total length of the object.

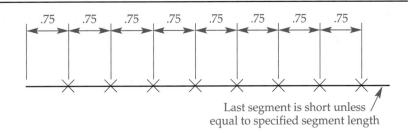

Last segment is short unless equal to specified segment length

construction lines: Lines commonly used to lay out a drawing.

The tracking vectors and alignment paths available with object snap and AutoTrack tools are examples of *construction lines* generated by AutoCAD. These tools are very efficient for creating objects because vector and alignment lines appear only when needed. Often, however, drawings require construction lines that stay visible while you continue to create geometry. You can draw construction geometry using any drawing tool, such as **LINE**, **ARC**, or **CIRCLE**. However, the **XLINE** and **RAY** tools are specifically designed for adding construction lines to help lay out a drawing. See **Figure 8-8.** Use the **XLINE** tool to draw an AutoCAD construction line, or *xline*. Use the **RAY** tool to draw a *ray*.

xline: A line in AutoCAD that is infinite in both directions and is used to help build accurate geometry.

ray: An AutoCAD line object that is infinite in one direction only; considered semi-infinite.

PROFESSIONAL TIP

Create construction geometry on a separate construction layer, named CONST, CONSTRUCTION, or A-ANNO-NPLT for example. Turn off or freeze the construction layer when unneeded, or you can easily recognize and erase objects drawn on the construction layer if necessary.

Using the XLINE Tool

XLINE

Ribbon
Home
> Draw

Construction Line

Type
XLINE
XL

Use the **XLINE** tool to draw infinitely long construction lines, often called *xlines*. To draw a basic xline, specify the location of two points through which the xline passes. After you pick the *root point*, you can select as many points as needed to create additional xlines. When you place multiple xlines from the same root point, the root point acts as an axis point, through which all additional xlines pass. See **Figure 8-9.** To exit the tool, right-click or press [Enter], [Esc], or the space bar.

root point: The first point specified to create a construction line.

Using the Hor and Ver Options

Xline options are available as alternatives to selecting two points to create the xline. The **Hor** option draws a horizontal xline through a single specified point. The **Ver** option draws a vertical xline through a specified point. After the first xline is drawn, the **XLINE** tool remains active, allowing you to create multiple xlines using the same option. Place as many horizontal or vertical xlines as needed, and then exit the **XLINE** tool.

Figure 8-8.
An example of a drawing laid out using construction lines.

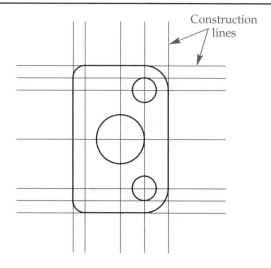

Construction lines

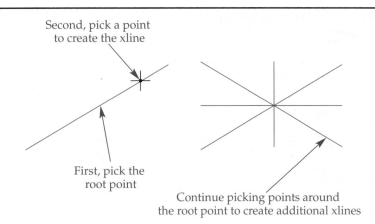

Figure 8-9.
Using the **XLINE** tool to draw an infinitely long construction line by picking two points through which the line passes.

Second, pick a point to create the xline

First, pick the root point

Continue picking points around the root point to create additional xlines

Using the Ang Option

Use the **Ang** option to draw an xline at a specified angle through a selected point. Access the **XLINE** tool and activate the **Ang** option. Enter an angle and then pick a point through which the xline passes. Another method is to select two points to define the angle. The **Reference** option of the **Ang** option allows you to reference the angle of an existing line object to use as the xline angle. This option is useful when you do not know the angle of the xline, but you know the angle between an existing object and the xline. See **Figure 8-10.**

Using the Bisect Option

The **Bisect** option draws an xline that bisects a specified angle, using the root point as the vertex. This is a convenient option for use in some geometric constructions. See **Figure 8-11.**

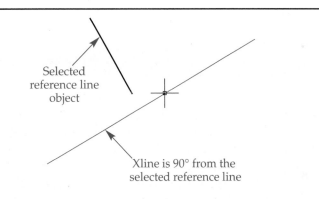

Figure 8-10.
Using **Reference** with the **Ang** option of the **XLINE** tool. A value of 90° is used in this example to make an xline perpendicular to the existing line.

Selected reference line object

Xline is 90° from the selected reference line

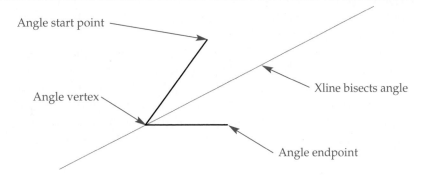

Figure 8-11.
Using the **Bisect** option of the **XLINE** tool.

Angle start point

Angle vertex

Xline bisects angle

Angle endpoint

Using the Offset Option

The **Offset** option draws an xline a specified distance from a selected line object. The **Offset** option of the **XLINE** tool functions much like the **OFFSET** tool. The difference is that the **Offset** option of the **XLINE** tool offsets an xline from the selected object, instead of the object itself. As with the **OFFSET** tool, specify an offset distance or use the **Through** option to pick a point through which to draw the construction line.

> **NOTE**
>
> Although xlines are infinite, they do not change the drawing extents, and they have no effect on zooming operations.

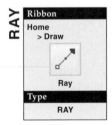

Exercise 8-4

Access the Student Web site (www.g-wlearning.com/CAD) and complete Exercise 8-4.

Using the RAY Tool

RAY

Ribbon
Home
> Draw
Ray
Type
RAY

The **RAY** tool allows you to specify the point of origin and a point through which the ray passes. In this respect, the **RAY** tool works much like the default option of the **XLINE** tool. The ray, however, is infinite only in the direction of the second pick point. Once you access the **RAY** tool, specify the root point of the ray. Then pick a second point to create the ray. Continue locating points to create additional rays from the same root point if necessary. To exit the tool, right-click or press [Enter], [Esc], or the space bar.

Multiview Drawings

Each drafting field has its own methods of presenting views of a product. Architectural drafting uses plan views, exterior elevations, and sections. In electronics drafting, a schematic diagram with electronic symbols shows circuit layout. In civil drafting, plan views and profiles show the topography of land. Mechanical drafting uses *multiview drawings*.

multiview drawings: Presentation of drawing views created through orthographic projection.

This textbook explains multiview drawings based on the ASME Y14.3M, *Multiview and Sectional View Drawings* standard. Multiview drawing views are created through *orthographic projection*. An imaginary *projection plane* is placed parallel to the object. Thus, the line of sight is perpendicular to the object. This results in 2D views of a 3D object. See **Figure 8-12.**

orthographic projection: Projecting object features onto an imaginary plane.

Six 2D orthographic views are possible: front, right side, left side, top, bottom, and rear. These views show all sides of an object, drawn in a standard arrangement for readability. The front view is the central, or most important, view. Other views occur around the front view and are usually directly aligned with or projected from it. See **Figure 8-13.** Notice in this figure that the horizontal and vertical edges in the front view align with the corresponding edges in the other views.

projection plane: The imaginary projection plane that is parallel to the object.

Figure 8-12.
Obtaining a
front view with
orthographic
projection.

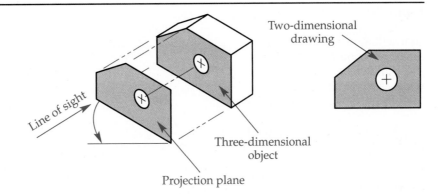

Figure 8-13.
Arrangement of the
six orthographic
views.

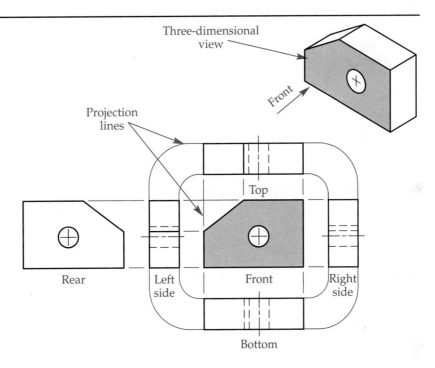

Selecting the Front View

The front view is usually the most descriptive view. Consider the following rules when selecting the front view:

- ✓ Most descriptive
- ✓ Most natural position
- ✓ Most stable position
- ✓ Provides the longest dimension
- ✓ Contains the least number of hidden features

Choosing Additional Views

Select additional views relative to the front view. Very few products require all six views. The required number of views depends on the complexity of the object. Use only enough views to completely describe the object. Drawing too many views is time-consuming and can clutter the drawing. The object shown in **Figure 8-14** needs only two views. These two views completely describe the width, height, depth, and features of the object. In some cases, a single view is enough to describe the object. You can often draw a thin part that has as uniform thickness, such as a gasket, with one view. **Figure 8-15** shows an example of a one-view drawing in which the part thickness is given as a note in the drawing or in the title block, eliminating the need for a second view.

Figure 8-14.
The views you choose to describe the object should show all height, width, and depth dimensions.

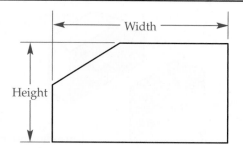
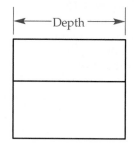

Figure 8-15.
A one-view drawing of a gasket. The thickness is uniform, so it can be given in a note.

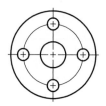

NOTE: THICKNESS 1.5mm

Auxiliary Views

Often you can completely describe an object using a combination of one or more of the six standard views. Sometimes, however, the orthographic views are not enough to properly identify some object surfaces. A *foreshortened* surface is typically described using an *auxiliary view*.

An auxiliary view is drawn by projecting lines perpendicular to a slanted surface. One projection line is often included on the drawing. It connects the auxiliary view to the view where the slanted surface appears as a line. The resulting auxiliary view shows the surface at its true size and shape. For most applications, a *partial auxiliary view* is enough. See **Figure 8-16.**

In some situations, there is not enough room on the drawing to project directly from the slanted surface. This requires that you locate the auxiliary view elsewhere. See **Figure 8-17.** A *viewing-plane line* is drawn next to the view where the slanted surface appears as a line. It terminates with bold arrowheads that point toward the slanted surface. A letter labels each end of the viewing-plane line. The letters relate the viewing-plane line with the proper auxiliary view. The auxiliary view includes a title, such as VIEW A-A, below the view to key the viewing plane with the view. When more

foreshortened: A surface at an angle to the line of sight. Foreshortened surfaces appear shorter than their true size and shape.

auxiliary view: View used to show a foreshortened surface at its true size and shape.

partial auxiliary view: An auxiliary view that shows only a single inclined surface of an object, rather than the entire object.

viewing-plane line: A thick dashed or phantom line identifying the viewing direction of a related view.

Figure 8-16.
Auxiliary views show the true size and shape of an inclined surface.

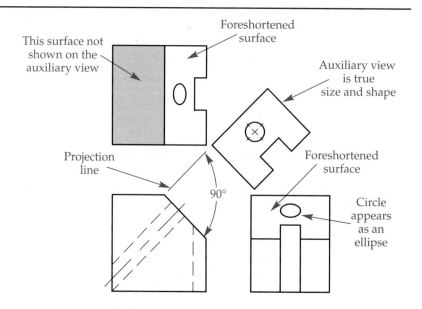

Figure 8-17.
When you add
a viewing-plane
line, you have the
flexibility to move
a view to a location
where there is
enough space to
draw the view.

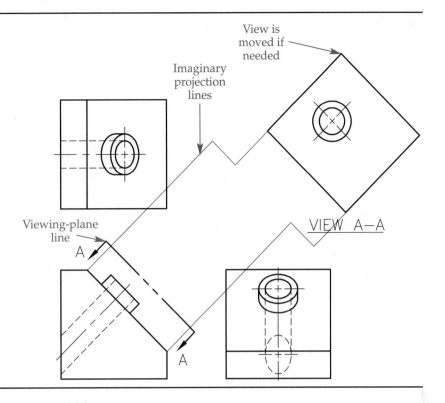

VIEW A—A

than one auxiliary view is drawn, labels continue with B-B through Z-Z, if necessary. The letters *I*, *O*, and *Q* are not used because they can be confused with numbers. An auxiliary view drawn away from the standard view retains the same angle as if it were projected directly.

NOTE

Use viewing plane lines whenever you cannot use direct projection to locate a secondary view, including one of the standard views.

Showing Hidden Features

Hidden features of an object are typically shown, even though they are not visible in the view at which you are looking. Visible edges appear as object lines. Hidden edges appear as hidden lines. Hidden lines are thin to provide contrast with thick object lines. See **Figure 8-18.**

Showing Symmetry and Circle Centers

Centerlines indicate the centerlines of symmetrical objects and the centers of circles. For example, in the circular view of a cylinder, centerlines cross to show the center of the cylinder. In the other view, a centerline identifies the axis. See **Figure 8-19.** The only place the small centerline dashes should cross is at the center of a circle, arc, ellipse, or other circular feature.

Figure 8-18.
Draw hidden features using hidden lines.

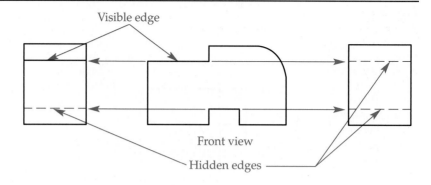

Visible edge

Front view

Hidden edges

Figure 8-19.
Use of centerlines in multiview drawings.

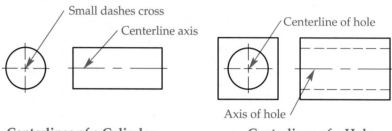

Small dashes cross

Centerline axis

Centerline of hole

Axis of hole

Centerlines of a Cylinder

Centerlines of a Hole

Constructing Multiview Drawings

You can construct multiview drawings using a variety of techniques, depending on the objects needed, personal working preference, and the information you know about the size and shape of items. Often you will use a combination of methods, including various coordinate point entries, object snaps, AutoTrack, and construction lines to produce drawings.

Drawing Orthographic Views

Figure 8-20 shows an example of how you can use object snap tracking to locate points for a new view by referencing points on existing views. In this example, a left-side view is constructed from an existing front view using object snap tracking and a running **Endpoint** object snap mode. Notice that the AutoTrack alignment path in **Figure 8-20A** provides a temporary construction line. Polar tracking vectors offer a similar type of temporary construction line. **Figure 8-20B** shows the completed front and left-side views.

Figure 8-21 shows an example of how you can use construction lines to form three views. In this example, vertical and horizontal xlines are offset to form an xline grid. Use the xline intersections to locate line and arc endpoints and the center point of the arc. Use the **Intersection** object snap mode to select the intersecting xlines. Notice that a single infinitely long xline can provide construction geometry for multiple views.

Exercise 8-5

Access the Student Web site (www.g-wlearning.com/CAD) and complete Exercise 8-5.

Figure 8-20.
An example of using object snap tracking to create an additional view. A—Referencing points from the front view to establish the first line of the left-side view. B—The completed front and left-side views.

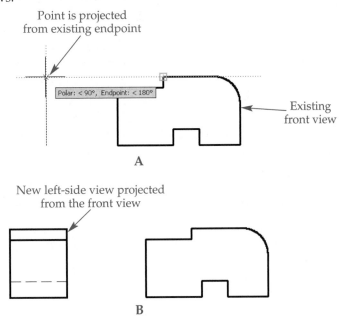

Figure 8-21.
Using a complete grid of construction lines to form a multiview drawing by "connecting the dots" at the construction line intersections. You can quickly draw the rectangular outlines of the right-side, left-side, and top views using the **RECTANGLE** tool.

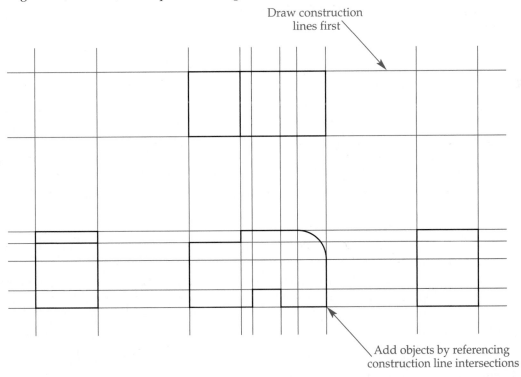

Drawing Auxiliary Views

Construct auxiliary views using the same tools and options you use to draw any of the six primary views. However, constructing auxiliary views presents unique challenges. Auxiliary view projection is 90° from an inclined surface. One of the most effective ways to draw a new auxiliary view, even without knowing or calculating the angle of the inclined surface, is to project construction lines from features on an existing view, perpendicular to the inclined surface.

Access the **XLINE** tool, and when the Specify a point or [Hor/Ver/Ang/Bisect/Offset]: prompt appears, use the **Perpendicular** object snap mode to select the inclined surface. A construction line perpendicular to the inclined surface attaches to the crosshairs. Use the appropriate object snap modes to select features on the existing view. You can then use object snaps or additional perpendicular construction lines to complete the auxiliary view. See **Figure 8-22.** You can also make a construction line perpendicular to a line object using the **Reference** option of the **XLINE** tool **Ang** option. Select the line object when prompted and enter an xline angle of 90.

NOTE

You can also use parametric tools, explained in Chapter 22, in addition or as an alternative to other techniques for constructing multi-view drawings.

Exercise 8-6

Access the Student Web site (www.g-wlearning.com/CAD) and complete Exercise 8-6.

Figure 8-22.
Using construction lines drawn perpendicular to the inclined surface on an existing view to construct an auxiliary view. The completed top and auxiliary views are for reference.

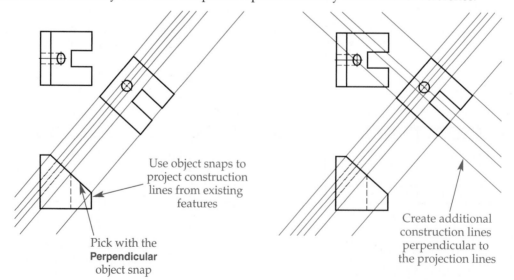

Use object snaps to project construction lines from existing features

Pick with the **Perpendicular** object snap

Create additional construction lines perpendicular to the projection lines

AutoCAD and Its Applications—Basics

Template Development
Chapter 8

For detailed instructions on choosing a more appropriate point style to use for construction purposes, go to the Student Web site (www.g-wlearning.com/CAD), select this chapter, and select **Template Development**.

Chapter Test

Answer the following questions. Write your answers on a separate sheet of paper or go to the Student Web site (www.g-wlearning.com/CAD) and complete the electronic chapter test.

1. List two ways to establish an offset distance using the **OFFSET** tool.
2. Which option of the **OFFSET** tool allows you to remove the source offset object?
3. How do you draw a single point, and how do you draw multiple points?
4. How do you access the **Point Style** dialog box?
5. If you use the **DIVIDE** tool and nothing appears to happen, what should you do?
6. How do you change the point size in the **Point Style** dialog box?
7. What tool can you use to place point objects that divide a line into 24 equal parts?
8. What is the difference between the **DIVIDE** and **MEASURE** tools?
9. Why is it a good idea to put construction lines on their own layer?
10. Name the tool that allows you to draw infinite construction lines.
11. Name the option that allows you to bisect an angle with a construction line.
12. What is the difference between the construction lines drawn with the tool identified in Question 10 and rays drawn with the **RAY** tool?
13. What ASME drafting standard applies to multiview drawings?
14. Provide at least four guidelines for selecting the front view of an orthographic multiview drawing.
15. How do you determine how many views of an object are necessary in a multiview drawing?
16. When can you describe a part with only one view?
17. When does a drawing require an auxiliary view, and what does an auxiliary view show?
18. What is the angle of projection from the slanted surface into the auxiliary view?
19. List two methods of aligning the views in a multiview drawing.
20. Describe an effective method of constructing an auxiliary view even if you do not know the angle of the inclined surface.

Drawing Problems

Start AutoCAD if it is not already started. Start a new drawing using an appropriate template of your choice. The template should include layers for drawing the given objects. Add layers as needed. Draw all objects using appropriate layers. Follow the specific instructions for each problem. Do not draw dimensions or text. Use your own judgment and approximate dimensions when necessary.

▼ Basic

1. Draw the front and side views of this offset support. Use object snap modes and tracking. Save your drawing as P8-1.

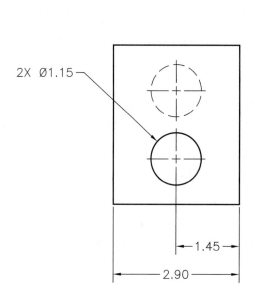
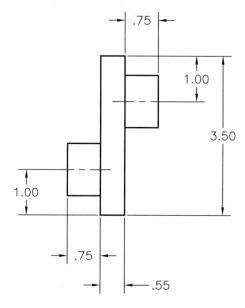

2. Draw the top and front views of this hitch bracket. Use object snap modes and tracking. Save your drawing as P8-2.

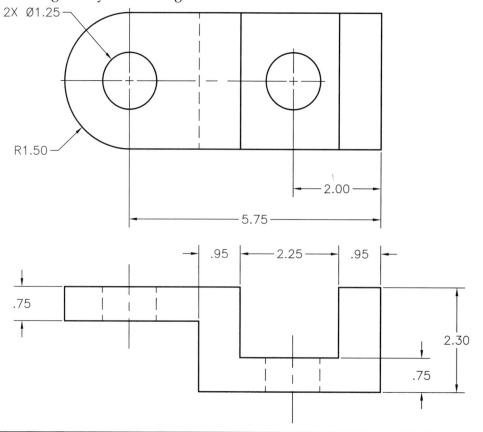

3. Draw this spring using the **OFFSET** tool for material thickness. Save the drawing as P8-3.

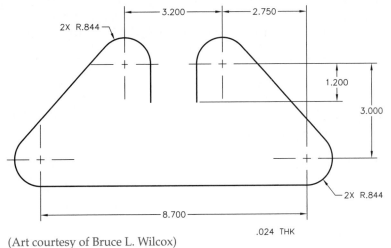

(Art courtesy of Bruce L. Wilcox)

4. Draw this sheet metal chassis. Use object snap tracking and polar tracking to your advantage. Save the drawing as P8-4.

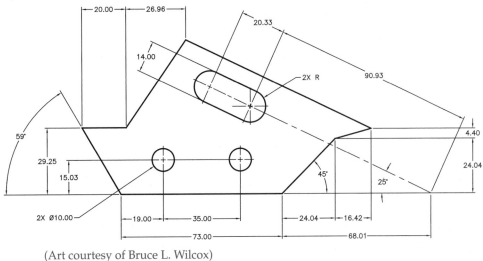

(Art courtesy of Bruce L. Wilcox)

5. Use the **OFFSET** tool to draw the elevation of the desk shown. Center 1″ × 4″ rectangular drawer handles 2″ below the top of each drawer. The top of the legs begin 1″ from the edge of the bottom of the desk. Save the drawing as **P8-5**.

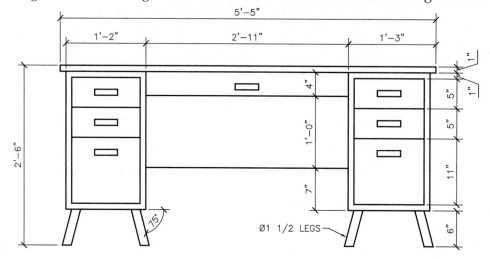

6. Draw this aluminum spacer. Use object snap modes and tracking. Save the drawing as **P8-6**.

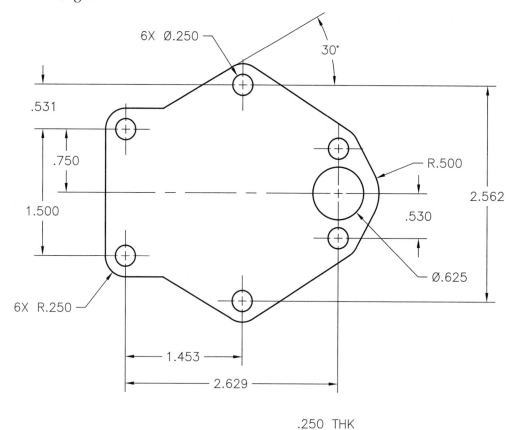

7. Draw this gasket. Save the drawing as **P8-7**.

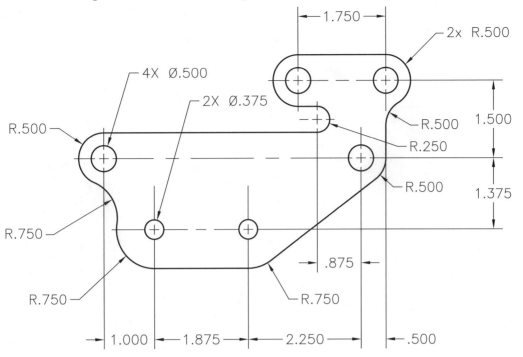

8. Draw this cup. Save the drawing as **P8-8**.

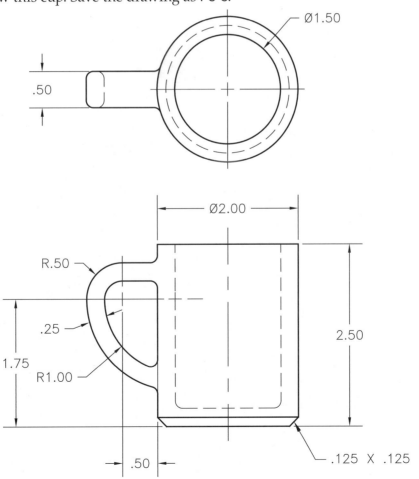

FILLETS AND ROUNDS R.10

9. Draw this bushing. Save the drawing as P8-9.

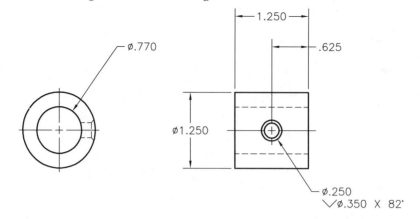

10. Draw this wrench. Save the drawing as P8-10.

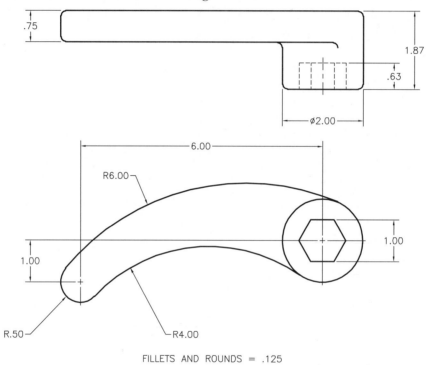

FILLETS AND ROUNDS = .125

11. Draw this support. Save the drawing as **P8-11**.

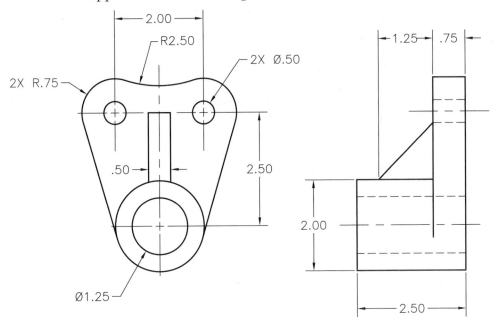

▼ Advanced

In Problems 12 through 16, draw the views needed to completely describe the objects. Use object snap modes, AutoTrack modes, xlines, rays, and offsets as needed. Save the drawings as P8-(problem number).

12.

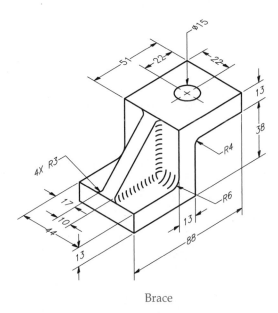

Brace

13.

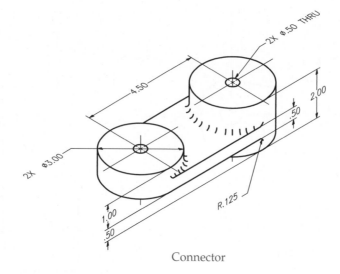

Connector

14.

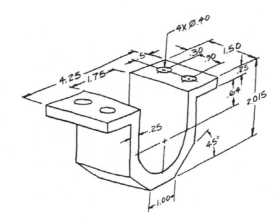

Journal Bracket (Engineer's Rough Sketch)

15.

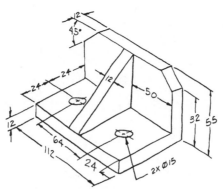

Angle Bracket (Engineer's Rough Sketch)
(Metric)

16.

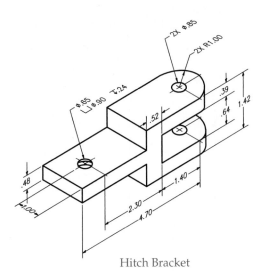

Hitch Bracket

17. Draw the views of this pillow block, including the auxiliary view. Save your drawing as P8-17.

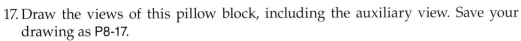

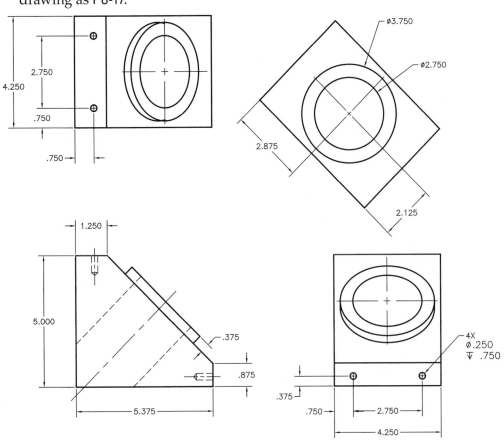

18. Draw the folded and flat pattern views of the sheet metal bracket shown. The part is made from 18-gauge steel. Save the drawing as P8-18.

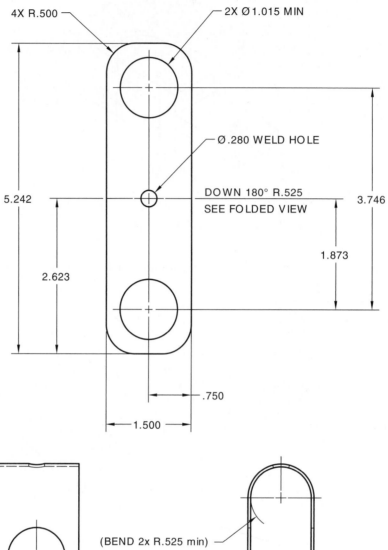

4X R.500

2X Ø1.015 MIN

Ø.280 WELD HOLE

DOWN 180° R.525
SEE FOLDED VIEW

5.242

3.746

2.623

1.873

.750

1.500

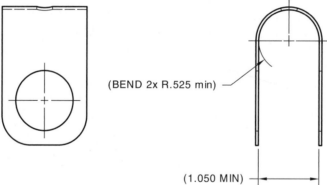

(BEND 2x R.525 min)

(1.050 MIN)

CHAPTER

9

Text Styles and Multiline Text

Learning Objectives

After completing this chapter, you will be able to do the following:

✓ Describe and use proper text standards.
✓ Calculate drawing scale and text height.
✓ Develop and use text styles.
✓ Use the **MTEXT** tool to create multiline text objects.

Letters, numbers, words, and notes, known as *annotation*, describe drawing information that is not shown using objects and symbols. CADD programs significantly reduce the tedious nature of adding *text* to a drawing. When drawn correctly, AutoCAD text is consistent and easy to read. This chapter introduces text standards and composition, and explains how to use the **MTEXT** tool to prepare a single text object that may consist of multiple lines of text, such as paragraphs or a list of general notes. Chapter 10 describes how to create single-line text using the **TEXT** tool, as well as other tools for working with text.

annotation: Letters, numbers, words, and notes used to describe information on a drawing.

text: Lettering on a CADD drawing.

Text Standards and Composition

Industry and company standards dictate how text appears on a drawing. **Figure 9-1** lists minimum letter heights based on the ASME Y14.2M, *Line Conventions and Lettering* standard. The U.S. National CAD Standard (NCS) and many companies, especially those who produce architectural and civil drawings, depart slightly from the ASME standard. The NCS specifies a minimum text height of 3/32″ (2.4 mm). Most text is drawn 1/8″ (3 mm) high, with titles and other unique text 1/4″ (3 mm) high. Regardless of the text height, all text should be consistent and easy to read.

Vertical text is standard on engineering drawings. Inclined text is used by some companies, especially those in structural drafting, and for specific drawing requirements, such as water features labeled on maps. **Figure 9-2** shows examples of vertical and inclined text. The recommended slant for inclined text is 68° from horizontal, though some drafters find 75° more appropriate. Text on a drawing is normally uppercase. Lowercase letters are sometimes used, most often on civil plans or maps. Typically, a drawing displays the same text format throughout. In some cases, a combination of text formats is used, such as text in architectural title blocks or on maps.

Figure 9-1.
Minimum letter heights based on the ASME Y14.2M, *Line Conventions and Lettering* standard.

Application	Height INCH	Height METRIC (mm)
Most text (dimension values, notes)	.12	3
Drawing title, drawing size, CAGE code, drawing number, revision letter	.24* .12**	6* 3**
Section and view letters	.24	6
Zone letters and numerals in borders	.24	6
Drawing block headings	.10	2.5

*D, E, F, H, J, K, A0, and A1 size sheets

**A, B, C, G, A2, A3, and A4 size sheets

Figure 9-2.
Vertical and inclined text.

ABC.. abc.. 123..
ABC.. abc.. 123..

Numbers in dimensions and notes are the same height as standard text. AutoCAD provides several methods for stacking text in fractions. When dimensions contain fractions, the fraction bar is usually placed horizontally between the numerator and denominator. However, many notes placed on drawings have fractions displayed with a diagonal fraction bar (/). In this case, use a dash or space between the whole number and the fraction. **Figure 9-3** shows examples of text for numbers and fractions in different unit formats.

AutoCAD text tools provide great control over text *composition*. You can lay out text horizontally, as is typical when adding notes, or draw text at any angle according to specific requirements. AutoCAD automatically spaces letters and lines of text. This helps maintain the identity of individual notes.

composition: The spacing, layout, and appearance of text.

> **NOTE**
>
> Several drawing variables can affect text height and format. Refer to appropriate industry, company, or school standards when adding text to your drawings.

Figure 9-3.
Examples of fractional text for different unit formats.

Decimal Inch	Fractional Inch	Millimeter
2.750 .25	$2\frac{3}{4}$ 2–3/4 2 3/4	2.5 3 0.7

Text presentation is important. Consider the following tips when adding text:

- Plan your drawing using rough sketches to allow room for text and notes.
- Arrange text to avoid crowding.
- Place related notes in groups to make the drawing easy to read.
- Place all general notes in a common location. Locate notes in the lower-left corner or above the title block when using ASME standards. Place notes in the upper-left corner when using military standards.
- Always use the spell checker.

Drawing Scale and Text Height

Ideally, you should determine drawing scale, scale factors, and text heights before you begin drawing. Incorporate these settings into your drawing template files, and make changes when necessary. The drawing scale factor is important because it determines how text appears on-screen and plots. To help understand the concept of drawing scale, look at the portion of a floor plan shown in **Figure 9-4.** You should draw everything in model space at full scale. This means that the bathtub, for example, is actually drawn 5′ long. However, at this scale, text size becomes an issue, because full scale text that is 1/8″ high is extremely small compared to the other full-scale objects. See **Figure 9-4A.** As a result, you must adjust the height of the text according to the drawing scale. See **Figure 9-4B.** You can calculate the scale factor manually and apply it to the text height, or you can allow AutoCAD to calculate it by using annotative text.

Figure 9-4.
An example of a portion of a floor plan drawn at full scale in model space. Text drawn at full scale, as shown in A, is very small compared to the large objects. You must scale text, as shown in B, in order to display and plot text correctly.

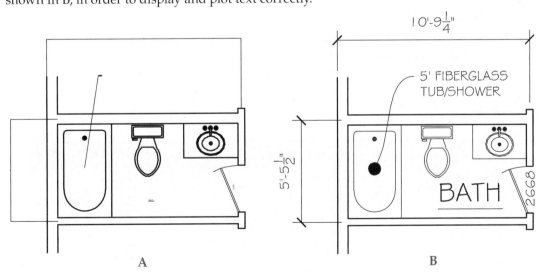

A

B

Scaling Text Manually

text height: The specified height of text, which may be different from the plotting size for text scaled manually.

scale factor: The reciprocal of the drawing scale.

paper text height: The plotted text height.

To adjust *text height* manually according to a specific drawing scale, you must calculate the drawing *scale factor*. **Figure 9-5** provides examples of calculating scale factor. Once you determine the scale factor, you then multiply the scale factor by the desired *paper text height* to get the model space text height.

For example, if you are working on a civil engineering drawing with a 1″ = 60′ scale, text drawn 1/8″ high is almost invisible. Remember, the drawing you are working on is 720 times larger than it is when plotted at the proper scale. Therefore, you must multiply the 720 scale factor by the desired text height. Text height multiplied by scale factor equals the model space scaled text height. So, in this example, multiply: 1/8″ (.125″) × 720 = 90″. The proper height of 1/8″ text in model space is 90″.

Annotative Text

annotative text: Text that is scaled by AutoCAD according to the specified annotation scale.

annotation scale: The drawing scale AutoCAD uses to calculate the height of annotative text.

AutoCAD scales *annotative text* according to the *annotation scale* you select, which eliminates the need for you to calculate the scale factor. Once you choose an annotation scale, AutoCAD determines the scale factor and applies it automatically to annotative text and all other annotative objects. For example, if you manually scale 1/8″ text for a drawing with a 1/4″ = 1′-0″ scale, or a scale factor of 48, you must draw the text using a text height of 6″ (1/8″ × 48 = 6″) in model space. When placing annotative text, using this example, you set an annotation scale of 1/4″ = 1′-0″. Then you draw the text using a paper text height of 1/8″ in model space. The 1/8″ high text scales to 6″ automatically because of the preset 1/4″ = 1′-0″ annotation scale.

Annotative text offers several advantages over manually scaled text, including the ability to control text scale based on scale, not scale factor. Annotative text is especially effective when the drawing scale changes or when a single sheet includes objects viewed at different scales.

PROFESSIONAL TIP

If you anticipate preparing scaled drawings, you should use annotative text and other annotative objects instead of traditional manual scaling. However, scale factor does influence non-annotative items and is still an important value to identify and use throughout the drawing process.

Setting the Annotation Scale

You should usually set annotation scale before you begin typing text so that the text height scales automatically. However, this is not always possible. It may be necessary to adjust the annotation scale throughout the drawing process, especially if you

Figure 9-5.
Examples of calculating scale factor.

Example	Scale	Conversion	Calculation	Scale Factor
Mechanical	1:2	None	2 ÷ 2 = 2	2
Civil	1″ = 60′	1″ = 720″ (60′ contains 720″)	720 ÷ 1 = 720	720
Architectural	1/4″ = 1′-0″	1/4″ (.25″) = 12″ (1′ contains 12″)	12 ÷ .25 = 48	48
Metric to Inch	1:1	1″ - 25.4 mm	25.4 ÷ 1 = 25.4	25.4
Metric to Inch	1:2	1″ = 25.4 mm × 2 (50.8)	50.8 ÷ 1 = 50.8	50.8

prepare multiple drawings with different scales on one sheet. This chapter approaches annotation scaling in model space only, using the process of selecting the appropriate annotation scale before typing text. To draw text using another scale, pick the new annotation scale and then type the text.

When you access a text tool and an annotative text style is current, the **Select Annotation Scale** dialog box appears. This is a very convenient way to set annotation scale before typing. Text styles are described later in this chapter. You can also select the annotation scale from the **Annotation Scale** flyout on the status bar. See **Figure 9-6**. The annotation scale is typically the same as the drawing scale.

Editing Annotation Scales

If a certain scale is not available, or if you want to change existing scales, pick the **Annotation Scale** flyout on the status bar and choose **Custom...** to access the **Edit Scale List** dialog box. From this dialog box, you can move the highlighted scale up or down in the list by picking the **Move Up** or **Move Down** button. To remove the highlighted scale from the list, pick the **Delete** button.

Ribbon
Annotate
> Annotation
Scaling
Scale List
Type
SCALELISTEDIT

SCALELISTEDIT

Figure 9-6.
Pick the **Annotation Scale** flyout on the status bar to activate an annotation scale.

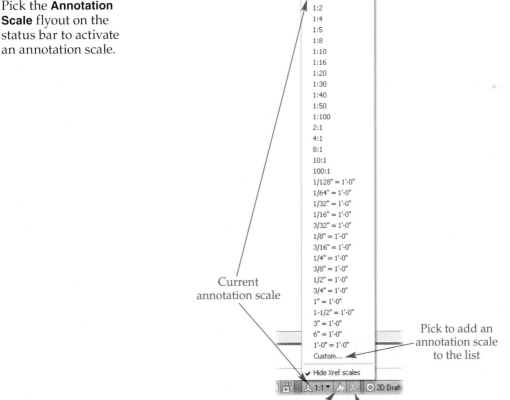

Current annotation scale

Pick to add an annotation scale to the list

Show or hide all annotative objects

Add annotative scales automatically to annotative objects when you change the annotation scale

Select the **Edit...** button to open the **Edit Scale** dialog box. Here you can change the name of the scale and adjust the scale by entering the paper and drawing units. For example, a scale of 1/4″ = 1′-0″ uses a paper units value of .25 or 1 and a drawing units value of 12 or 48.

To create a new annotation scale, pick the **Add...** button to display the **Add Scale** dialog box, which functions the same as the **Edit Scale** dialog box. Pick the **Reset** button to restore the default annotation scale. When the annotation scale is set current, you are ready to type annotative text that automatically displays at the correct text height according to the drawing scale.

Text Styles

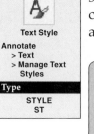

text style: A saved collection of settings for text height, width, oblique angle (slant), and other text effects.

Text styles control many text characteristics. You may have several text styles in a single drawing, depending on drawing requirements. Though you can adjust text format independently of a text style, you should create a text style for each unique application. For example, you may use a specific text style to draw most text objects, such as adding notes and dimensions, and a separate text style that uses different characteristics for adding text to a title block. You may also want to create an annotative and a non-annotative text style. Add text styles to drawing templates for repeated use.

Working with Text Styles

Create, modify, and delete text styles using the **Text Style** dialog box. See **Figure 9-7**. The **Styles** list box displays existing text styles. By default, Annotative and Standard text styles are available. The Annotative text style is preset to create annotative text, as indicated by the icon to the left of the style name. The Standard text style does not use the annotative function.

> **NOTE**
>
> Several panels on the ribbon contain small arrows in the lower-right corner that open appropriate dialog boxes. When a dialog box is accessible using one of these arrows, the arrow will be displayed in a margin graphic as shown next to this paragraph. In this case, picking the arrow opens the **Text Style** dialog box.

To make a text style current, double-click the style name; right-click the name and select **Current**; or pick the name and select the **Current** button. Below the **Styles** list box is a drop-down list that you can use to filter the number of text styles displayed in the **Text Style** dialog box. Pick the **All Styles** option to show all text styles in the file or pick the **Styles in use** option to show only the current style and styles used in the drawing.

Creating New Text Styles

To create a new text style, first select an existing text style from the **Styles** list box to use as a base for formatting the new text style. Then pick the **New...** button in the **Text Style** dialog box to open the **New Text Style** dialog box. See **Figure 9-8**. Notice that style1 appears in the **Style Name** text box. You can keep the default name, but you should replace it with a more descriptive name. For example, to create a text style for mechanical drawings that uses the Romans font and characters .12″ high, enter a name such as ROMANS-12. You could name a text style for architectural drawings that uses the Stylus BT font and characters .125″ high ARCHITECTURAL-125.

Figure 9-7.
Use the **Text Style** dialog box to create, rename, delete, and set the characteristics of a text style.

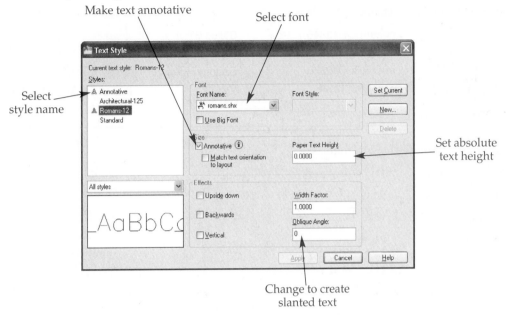

Make text annotative

Select font

Select style name

Set absolute text height

Change to create slanted text

Figure 9-8.
Enter a descriptive name for the new text style in the **New Text Style** dialog box.

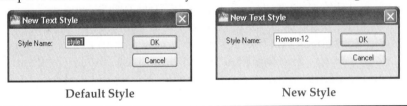

Default Style

New Style

Text style names can have up to 255 characters, including uppercase and lowercase letters, numbers, dashes (–), underlines (_), and dollar signs ($). After entering the text style name, pick the **OK** button. The new text style appears in the **Styles** list box of the **Text Style** dialog box, and you are ready to adjust text style characteristics.

PROFESSIONAL TIP

It is a good idea to record the names and details about the text styles you create and keep this information in a log for future reference.

Setting the Font Style

The **Font** area of the **Text Style** dialog box is where you select a *font* and the style of the selected font. Use the **Font Name** drop-down list to access available fonts. Most fonts, including the default Arial font, are TrueType fonts identified by the TrueType icon. TrueType fonts are *scalable fonts* and have an outline. By default, TrueType fonts appear and plot filled. Examples of TrueType fonts are shown in **Figure 9-9**.

Fonts linked to AutoCAD shape files have .shx file extensions and display the AutoCAD compass icon. The Romans (roman simplex) font closely duplicates the single-stroke lettering that has long been the standard for most drafting. **Figure 9-10** shows examples of SHX fonts.

font: A letter face design.

scalable fonts: Fonts that can be displayed or printed at any size while retaining proportional letter thickness.

Figure 9-9.
A few of the many TrueType fonts available.

Styles Suitable for Mechanical Drawings		Architectural Hand Lettered	
Arial	abcdABCD12345	Stylus BT	abcdABCD12345
Monospac821	abcdABCD12345	CitiBlueprint	abcdABCD12345
Swis721 BT	abcdABCD12345	CountryBlueprint	abcdABCD12345
Technic Bold	ABCDABCD12345	Vineta	**abcdABCD12345**

Figure 9-10.
Examples of AutoCAD SHX fonts.

	Fast Fonts				Simplex Fonts
	abcdABCD12345	Txt	Romans		abcdABCD12345
Monotxt	abcdABCD12345		Italic		*abcdABCD12345*

	Triplex Fonts				Complex Fonts
Romant	abcdABCD12345		Romanc		abcdABCD12345
	abcdABCD12345	Italict	Italicc		*abcdABCD12345*

The **Font Style** drop-down list is inactive unless the selected font includes options, such as bold or italic. The SHX fonts do not provide style options, but some of the TrueType fonts do. For example, the SansSerif font has Regular, Bold, BoldOblique, and Oblique options. Select a style or combination of styles to change the appearance of the font.

The **Use Big Font** check box enables when you select an SHX font. Pick the check box to activate *big fonts*. The **Big Font:** drop-down list, shown in **Figure 9-11,** is a supplement used to define many symbols not available in normal font files.

big fonts: Asian and other large-format fonts that have characters not present in other font files.

> ## CAUTION
>
> When a drawing contains a significant amount of text, especially TrueType font text, display changes may be slower and drawing regeneration time may increase.

Figure 9-11.
When you check the **Use Big Font** check box, the **Font Style:** drop-down list changes to display a list of available big fonts.

The **Big Font:** drop-down list displays the available nonstandard fonts

 **Reference Material** *AutoCAD Fonts*

For a sample of the many fonts available with AutoCAD, go to the **Reference Material** section on the Student Web site (www.g-wlearning.com/CAD) and select **AutoCAD Fonts**.

Text Style Height Options

The **Size** area of the **Text Style** dialog box contains options for defining text style height. Select the **Annotative** check box to set the text style as annotative and display the **Paper Text Height** text box. Deselect the **Annotative** check box to scale text manually and display the **Height** text box.

The default text height is 0. If you set a value other than 0, the text height is fixed for the text style and applies each time you use the text style. As a result, you can only use the specified text height to create single-line text, and you do not have the option of assigning a different text height to certain objects that include text, such as dimensions.

When you pick the **Annotative** check box, the **Match text orientation to layout** check box enables. Check this box to match the orientation of text in layout viewports with the layout orientation. Layouts are described later in this textbook.

PROFESSIONAL TIP

Use a text style text height of 0 to provide the greatest flexibility when drawing. A prompt asks you to specify a text height when you create single-line text, and you can specify a text height for dimension, multileader, and table styles. As an alternative, use a specific value to limit text height. In this case, the text style fully controls the height applied to single-line text, and dimension, multileader, and table styles. This restricts text objects to a certain height, which can increase productivity and accuracy, but requires that you create a text style for each unique height. Dimension, multileader, and table styles are explained later in this textbook.

Adjusting Text Style Effects

The **Effects** area of the **Text Style** dialog box includes options to set the text format. It contains options for printing text upside-down, backwards, and vertically. See **Figure 9-12**. The Vertical check box is inactive for all TrueType fonts. A check in this box makes

Figure 9-12.
Special effects for text styles can be set in the **Effects** area of the **Text Style** dialog box.

ꓕՍᑭꓴIᗡƎꓷOWИ
Upside-Down Text

ꓷᗡᴙAWꓘƆAᗺ
Backwards Text

V
E
R
T
I
C
A
L
Vertical Text

SHX font text vertical. Text on drawings is normally horizontal, but vertical text can be used for special effects and graphic designs. Vertical text usually works best with a 270° rotation angle.

Use the **Width Factor** text box to specify the text character width relative to its height. A width factor of 1 is default and is recommended for most drawing applications A width factor greater than 1 expands characters, and a factor less than 1 compresses characters. See **Figure 9-13.** You can set the width factor between 0.01 and 100.

The **Oblique Angle** text box allows you to set the angle at which text inclines. The 0 default draws characters vertically. A value greater than 0 slants characters to the right, and a negative value slants characters to the left. See **Figure 9-14.** Some fonts, such as the italic shape font, are already slanted.

PROFESSIONAL TIP

AutoCAD text slant measures from vertical. Use a 22° oblique angle to slant text according to the 68° horizontal incline standard.

NOTE

The **Preview** area of the **Text Style** dialog box displays an example of the selected font and font effects. This is a convenient way to see what the font looks like before using it in a new style.

Exercise 9-1

Access the Student Web site (www.g-wlearning.com/CAD) and complete Exercise 9-1.

Figure 9-13.
Examples of width factor settings for text.

Width Factor	Text
1	ABCDEFGHIJKLM
.5	ABCDEFGHIJKLMNOPQRSTUVWXY
1.5	ABCDEFGHI
2	ABCDEFG

Figure 9-14.
Examples of oblique angle settings for text.

Obliquing Angle	Text
0	ABCDEFGHIJKLM
15	*ABCDEFGHIJKLM*
–15	ABCDEFGHIJKLM

AutoCAD and Its Applications—Basics

Changing, Renaming, and Deleting Text Styles

If you make changes to a text style, such as selecting a different font, all existing text objects drawn using the modified text style update to reflect the changes. Use a different text style with unique characteristics when appropriate. To rename a text style using the **Text Style** dialog box, slowly double-click the name or right-click on the name and select **Rename**.

To delete a text style using the **Text Style** dialog box, right-click on the name and select **Delete**, or pick the style and select the **Delete** button. You cannot delete a text style that is assigned to text objects. To delete a style that is in use, assign a different style to the text objects that reference the style you want to delete. You cannot delete or rename the Standard style.

NOTE

You can also rename styles using the **Rename** dialog box. Select **Text styles** in the **Named Objects** list to rename the style.

Setting a Text Style Current

You can set a text style current using the **Text Style** dialog box by double-clicking the style in the **Styles** list box, right-clicking on the name and selecting **Set current**, or picking the style and selecting the **Set current** button. To set a text style current without opening the **Text Style** dialog box, use the **Text Style** list located in the expanded **Annotation** panel of the **Home** ribbon tab or the **Text** panel of the **Annotate** ribbon tab. See **Figure 9-15**.

PROFESSIONAL TIP

You can import text styles from existing drawings using **Design-Center**. See Chapter 5 for more information about using **Design-Center** to reuse drawing content.

Figure 9-15.
The fastest way to set a style current is to use the drop-down list on the ribbon.

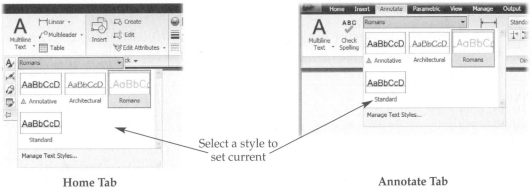

Select a style to set current

Home Tab   Annotate Tab

Creating Multiline Text

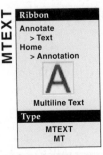

Ribbon

Annotate
> Text
Home
> Annotation

A

Multiline Text

Type

MTEXT
MT

text boundary: An imaginary box that sets the location and width for multiline text.

The **MTEXT** tool draws a single multiline text, or mtext, object that can include extensive paragraph formatting, lists, symbols, and columns. When you access the **MTEXT** tool, grayed-out letters appear next to the crosshairs to indicate the current text style and height settings. Pick the first corner of the *text boundary*. A prompt then asks you to specify the opposite corner or choose an available option. Use the options to preset specific mtext characteristics, including text height and style, text boundary justification, line spacing, rotation, width, and use of columns. You can control the same settings while typing, as explained in this chapter, or while editing mtext.

By default, mtext is set to use dynamic columns, which allow you to organize multiple text columns in a single mtext object. Although you can use dynamic columns to form a single paragraph, for typical text requirements without columns, disable columns. Before selecting the second corner of the text boundary, choose the **Columns** option and the **No columns** option. An arrow in the boundary shows the direction of text flow, where the boundary will expand as you type, if necessary. Pick the opposite corner of the text boundary to continue. See **Figure 9-16**. Columns are fully explained later in this chapter.

Using the Text Editor

text editor: The part of the multiline or single-line text system where text is typed.

Once you select the opposite corner of the text boundary, the **Text Editor** contextual ribbon tab and the *text editor* appear. **Figure 9-17** shows the display when you are not using columns. The controls and features for typing mtext are similar to those in software programs such as Microsoft® Word. The **Text Editor** ribbon tab provides tools for adjusting text typed into the text editor. You can access the shortcut menu shown in **Figure 9-18** by right-clicking anywhere outside of the ribbon. Many of the same options are available on the **Text Editor** ribbon tab.

> **NOTE**
>
> If you close the ribbon, the **Text Formatting** toolbar appears instead of the **Text Editor** ribbon tab. This textbook focuses on using the **Text Editor** ribbon tab to add mtext. The **Text Formatting** toolbar provides the same functions.

The text editor indicates the initial size of the area in which you type. When you are not using columns, long words and paragraphs extend past the text editor limits.

Figure 9-16.
The text boundary is the box within which you type. Consider the text boundary the extents of paragraph. Long words extend past the boundary in the direction the arrow indicates.

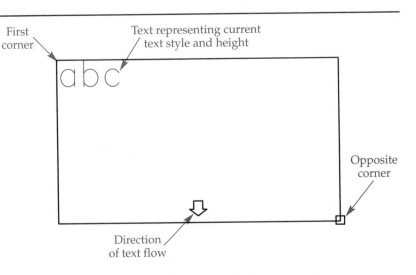

First corner

Text representing current text style and height

Opposite corner

Direction of text flow

Figure 9-17.
The **Text Editor** ribbon tab provides many options for creating mtext.

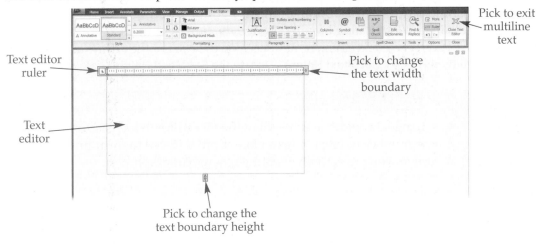

Text editor ruler

Text editor

Pick to change the text width boundary

Pick to exit multiline text

Pick to change the text boundary height

Figure 9-18.
Display the text editor shortcut menu by right-clicking anywhere except on the ribbon while the text editor is active.

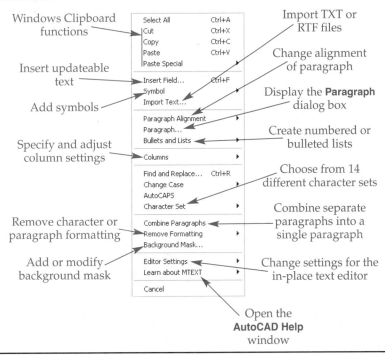

Windows Clipboard functions

Insert updateable text

Add symbols

Specify and adjust column settings

Remove character or paragraph formatting

Add or modify background mask

Import TXT or RTF files

Change alignment of paragraph

Display the **Paragraph** dialog box

Create numbered or bulleted lists

Choose from 14 different character sets

Combine separate paragraphs into a single paragraph

Change settings for the in-place text editor

Open the **AutoCAD Help** window

The text editor is transparent by default so that you can see how the text you type appears on-screen in relation to other objects. To make the text editor appear opaque, select the **Opaque Background** option available from the **Editor Settings** cascading submenu of the shortcut menu. The text editor includes a ruler where indent and tab stops and indent and tab markers are located.

> **NOTE**
>
> Pick the **Ruler** button in the **Options** panel of the **Text Editor** ribbon tab, or select the **Ruler** option from the **Editor Settings** cascading submenu of the shortcut menu to turn the ruler on or off. Use the **Undo** and **Redo** buttons, also found in the **Options** panel, to undo or redo text editor operations.

To change the width of the text editor, when you are not using columns, drag the arrows on the right-side end of the paragraph ruler. You can also right-click on the text editor ruler or the arrows at the bottom of the text editor and choose **Set Mtext Width...** to use the **Set Mtext Width** dialog box. To change the height of the text editor, drag the arrows at the bottom of the text editor, or right-click on the ruler or the arrows at the bottom of the text editor and select **Set Mtext Height...** to use the **Set Mtext Height** dialog box.

PROFESSIONAL TIP

Change the width and height of the text editor to increase or decrease the number of lines of text. Do not press [Enter] to form lines of text, unless you are actually creating a new paragraph.

The familiar Windows text editor cursor displays within the text editor at the current text height. Begin typing or editing as needed. The procedure for selecting and editing existing text is the same as in standard Windows text editors. To quickly select all text in the text editor, right-click inside the text editor and choose **Select All**. You can change the selected text highlight color by selecting the **Text Highlight Color...** option available from the **Editor Settings** cascading submenu of the shortcut menu.

When you finish typing text, exit the mtext system by picking the **Close Text Editor** button in the **Close** panel of the **Text Editor** ribbon tab or pick outside of the text editor. You can also press [Esc] or right-click and select **Cancel** to exit, but you are prompted to save changes. The easiest way to reopen the text editor to make changes to text content is to double-click on an mtext object.

NOTE

The text editor displays text horizontally, right-side up, and forward. Any special effects such as vertical, backwards, or upside-down take effect when you exit the text editor.

Reference Material *Shortcut Keys*

For a complete list of keyboard shortcuts for text editing, go to the **Reference Material** section of the Student Web site (www.g-wlearning.com/CAD) and select **Shortcut Keys**.

Exercise 9-2

Access the Student Web site (www.g-wlearning.com/CAD) and complete Exercise 9-2.

Stacking Text

When you enter a fraction in the **Text Editor**, the **AutoStack Properties** dialog box appears by default. See **Figure 9-19**. This dialog box allows you to activate AutoStacking, which causes the entered fraction to stack with a horizontal or diagonal fraction bar. You can also remove the leading space between a whole number and the fraction. This dialog box appears each time you enter a fraction. If you decide that you do not want this dialog box to pop up each time you create a fraction, pick the **Don't show this dialog again; always use these settings** check box.

Figure 9-19.
The **AutoStack**
Properties dialog
box.

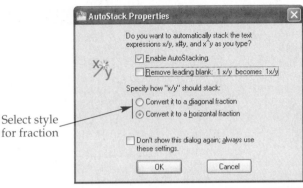

Select style
for fraction

Manual Stacking

You can use the **Stack** feature to manually stack selected text vertically or diagonally. To use this feature to draw a vertically stacked fraction, place a forward slash between the top and bottom characters. Then select the text and pick the **Stack** option from the shortcut menu. To form a *tolerance stack*, use the caret (^) character between characters. To use a diagonal fraction bar, type a number sign (#) between characters. See **Figure 9-20**. To unstack text, select stacked text and pick the **Unstack** option from the shortcut menu. The stacked characters are placed on a single line with the appropriate character (^, #, or /) between the characters.

tolerance stack:
Text that is stacked
vertically without a
fraction bar.

Stack Settings

The **Stack Properties** dialog box controls stack settings. To access the dialog box, select the stacked text and pick the **Stack Properties** option from the shortcut menu. **Figure 9-21** briefly describes the **Stack Properties** dialog box features.

Adding Symbols

You can insert a variety of common drafting symbols and other unique characters not found on a typical keyboard into the text editor. To insert a symbol, select the **Symbol** flyout on the **Insert** panel of the **Text Editor** ribbon tab, as shown in **Figure 9-22** or the **Symbol** cascading submenu available from the shortcut menu.

The **Symbol** menu allows the insertion of symbols at the text cursor location. The first two sections in the **Symbol** menu contain common symbols. The third section contains the **Non-breaking Space** option, which keeps two separate words together on one line. Pick a symbol to insert the symbol at the text cursor location and hide the **Symbol** menu. The **Other...** option opens the **Character Map** dialog box, shown in **Figure 9-23**. Use the following steps to insert a symbol from the **Character Map** dialog box:

1. Pick the font to display from the **Font:** drop-down list.
2. Pick the desired symbol and then pick the **Select** button. The symbol appears in the **Characters to copy:** box.
3. Pick the **Copy** button to copy the selected symbol or symbols to the Clipboard.

Figure 9-20.
Different types of
stack characters.
ASME standards
recommend that the
text height of stacked
fraction numerals
be the same as the
height of other
dimension numerals.

	Selected Text	Stacked Text
Vertical Fraction	1/2	$\frac{1}{2}$
Tolerance Stack	1^2	$\frac{1}{2}$
Diagonal Fraction	1#2	$\frac{1}{2}$

Figure 9-21.
The **Stack Properties** dialog box. Select 100% from the **Text size** drop-down list to conform to ASME standards.

Top and bottom numbers in stack

Set horizontal, diagonal, or tolerance style

Select bottom, center, or top alignment

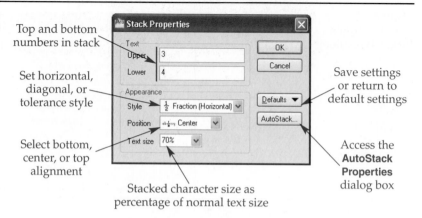

Save settings or return to default settings

Access the **AutoStack Properties** dialog box

Stacked character size as percentage of normal text size

Figure 9-22.
The **Symbol** menu options.

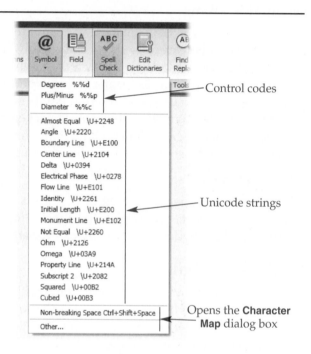

Control codes

Unicode strings

Opens the **Character Map** dialog box

Figure 9-23.
The **Character Map** dialog box.

Select font from drop-down list

Available symbols

Select to return to the **In-Place Text Editor**

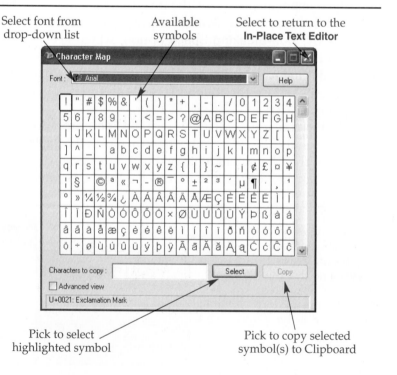

Pick to select highlighted symbol

Pick to copy selected symbol(s) to Clipboard

4. Close the **Character Map** dialog box.
5. In the text editor, place the cursor where you want the symbols to display.
6. Right-click and select **Paste** to paste the symbol at the cursor location.

Exercise 9-3

Access the Student Web site (www.g-wlearning.com/CAD) and complete Exercise 9-3.

Style Settings

The **Style** panel of the **Text Editor** ribbon tab includes options for changing the text style, annotative setting, and text height. Activate text style options before typing, or select existing text and make adjustments as necessary. A single mtext object can use a combination of text style settings. Remember, however, that making changes to some style settings overrides the settings specified in the text style, which is often not appropriate.

The text style flyout provides access to existing text styles. Use the scroll buttons to locate styles, or pick the expansion arrow to display a temporary window of styles. This allows you to use a text style other than the current text style while you are typing. Use the **Annotative** button to override the annotative setting of the current text style. Use the **Size** drop-down list to set the text height. If the current text style is annotative, or if you pick the **Annotative** button, the height you enter is the paper text height. If the current text style is not annotative, or if you deselect the **Annotative** button, the height you enter is the text height multiplied by the scale factor.

Character Formatting

The **Formatting** panel of the **Text Editor** ribbon tab includes options for adjusting character formatting. Activate character format options before typing, or select existing text and make adjustments as necessary. A single mtext object can use a combination of character formats. Remember, however, that making changes to character formatting overrides some of the settings specified in the text style and preset object properties, such as color. You should usually avoid this practice.

Pick the **Bold** button to make text bold. Select the **Italic** button to make text italic. The **Bold** and **Italic** settings work only with some TrueType fonts. Pick the **Underline** button to underline text, and select the **Overline** button to place a line over text. The **Font** drop-down list allows you to override the text font. The text color is set to ByLayer by default, but you can change the color by picking one of the colors in the **Color** drop-down list. Though you should usually define color as ByLayer, a single mtext object can use a combination of text colors.

Additional character formatting options are available from the expanded **Formatting** panel. The **Oblique Angle** text box overrides the angle at which text inclines. The value in the **Tracking** text box determines the amount of space between text characters. The default tracking value is 1, which results in normal spacing. Increase the value to add space between characters, or decrease the value to tighten the spacing between characters. You can enter any value between 0.75 and 4.0. See **Figure 9-24.** The value in the **Width Factor** text box overrides the text character width.

Figure 9-24.
The **Tracking** option for mtext determines the spacing between characters.

AutoCAD tracking
Normal Spacing

AutoCAD tracking
Tracking = 0.75

A u t o C A D   t r a c k i n g
Tracking = 2.0

Using a Background Mask

background mask: A mask that hides a portion of objects behind and around text so that the text is unobstructed.

Sometimes drawings require text to be placed over existing objects, such as graphic patterns, making the text hard to read. A *background mask* can solve this problem. Select the **Background Mask...** button to display the **Background Mask** dialog box. See **Figure 9-25.** To apply the mask settings to the current mtext object, check **Use background mask**. The **Border offset factor:** text box sets the amount of mask. This value, from 1 to 5, works with the text height value. If the border offset factor is set to 1, then the mask occurs directly within the boundary of the text. To offset the mask beyond the text boundary, use a value greater than 1. The formula is: border offset factor × text height = total masking distance from the bottom of the text. See **Figure 9-26.** The **Fill Color** area of the **Background Mask** dialog box allows you to apply color to the mask using the background color or a different color.

> **NOTE**
>
> The **Character Set** cascading submenu, available from the shortcut menu or the **More** flyout in the **Options** panel, displays a menu of code pages. A code page provides support for character sets used in different languages. Select a code page to apply it to the selected text.

Exercise 9-4

Access the Student Web site (www.g-wlearning.com/CAD) and complete Exercise 9-4.

Paragraph Formatting

justify: Align the margins or edges of text. For example, left-justified text is aligned along an imaginary left border.

The **Paragraph** panel of the **Text Editor** ribbon tab includes options for adjusting paragraph formatting. *Justify* the text boundary to control the arrangement and location of text within the text editor. You can also justify the text within the boundary independently of the text boundary justification. This provides great flexibility when

Figure 9-25.
The **Background Mask** dialog box controls text mask settings.

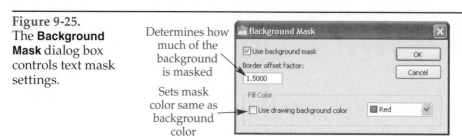

Determines how much of the background is masked

Sets mask color same as background color

Figure 9-26.
The border offset factor determines the size of the background mask. The text in the figure is 1/8" with different border offset factors.

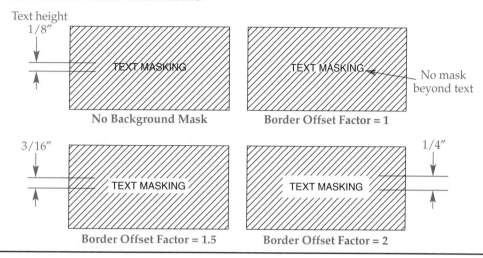

Text height
1/8"

TEXT MASKING

No Background Mask

TEXT MASKING

No mask beyond text

Border Offset Factor = 1

3/16"

TEXT MASKING

Border Offset Factor = 1.5

1/4"

TEXT MASKING

Border Offset Factor = 2

determining the location and arrangement of text. To justify the text boundary, select a justification option from the **Justification** flyout. Justification also determines the direction of text flow. **Figure 9-27** displays the options for justifying the mtext boundary vertically and horizontally.

Paragraph alignment occurs inside the text boundary. For example, when you apply a **Middle Center** text box justification, then set the paragraph alignment to **Left**, the text inside the boundary aligns to the left edge of the text boundary, while the text boundary remains positioned according to the **Middle Center** justification. See **Figure 9-28**.

paragraph alignment: The alignment of multiline text inside the text boundary.

Figure 9-27.
Options for justifying the mtext boundary.

✗ This symbol represents the insertion point

■ This symbol represents grips

– – This symbol represents the text boundary

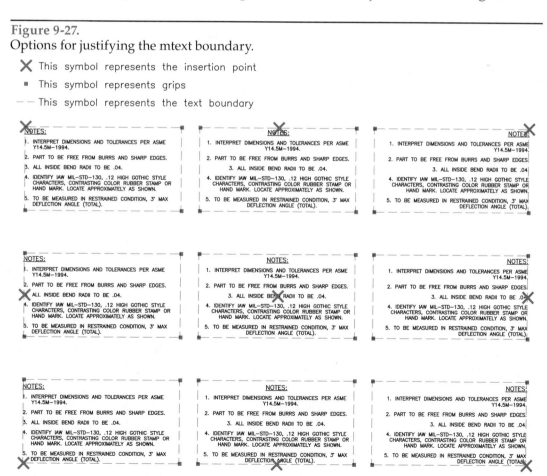

Figure 9-28.
You can adjust paragraph alignment independently of text boundary justification.

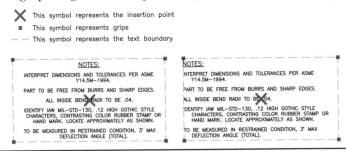

To adjust paragraph alignment, select one of the paragraph alignment buttons on the **Paragraph** panel, or pick one of the options from the **Paragraph Alignment** cascading submenu available from the shortcut menu. You can also control paragraph alignment using the **Paragraph** dialog box, shown in **Figure 9-29**. To display the **Paragraph** dialog box, pick the **Paragraph** button on the **Paragraph** panel or select the **Paragraph** option available from the shortcut menu. To set paragraph alignment in the **Paragraph** dialog box, pick the **Paragraph Alignment** check box, then choose the appropriate paragraph alignment radio button. You can choose from five paragraph alignment options, as shown in **Figure 9-30**.

Tabs, indents, paragraph spacing, and paragraph line spacing can also be set in the **Paragraph** dialog box. Use the **Tab** area to set custom tab stops. Pick the tab type radio button, enter a value for the tab in the text box, and pick the **Add** button to add it to the list and insert the tab on the ruler. **Figure 9-31** shows and briefly describes each tab option. Add as many custom tabs as necessary. You can also add custom tabs to the ruler by picking the tab button on the far left side of the ruler until the desired tab symbol appears. Then pick a location on the ruler to insert the tab.

The options in the **Left Indent** area set the indentation for the first line of a paragraph of text as well as the remaining portion of a paragraph. The **First line** indent is used each time you start a new paragraph. As text wraps to the next line, the **Hanging** indent is used. The options in the **Right Indent** area set the indentation for the right side of a paragraph. As text is typed, the right indent value, not the right edge of the text boundary, determines when the text wraps to the next line.

Figure 9-29.
The **Paragraph** dialog box.

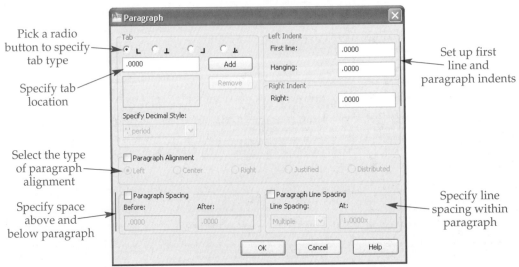

Figure 9-30.
Paragraph alignment options for mtext. In each of these examples, the text boundary justification is set to Top Left.

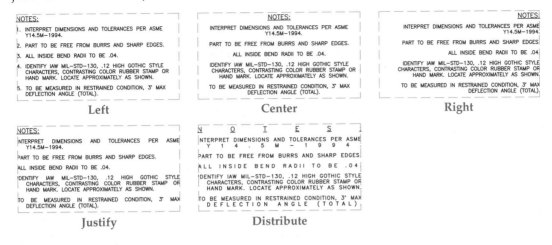

Left

Center

Right

Justify

Distribute

Figure 9-31.
Using custom tabs to position text in the text editor. When you press [Tab], the cursor moves to the tab position. The type of tab determines text behavior.

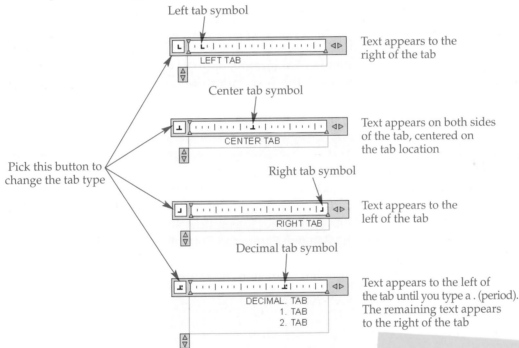

Left tab symbol

Text appears to the right of the tab

Center tab symbol

Text appears on both sides of the tab, centered on the tab location

Pick this button to change the tab type

Right tab symbol

Text appears to the left of the tab

Decimal tab symbol

Text appears to the left of the tab until you type a . (period). The remaining text appears to the right of the tab

The options in the **Paragraph Spacing** area define the amount of space before and after paragraphs. To set paragraph line spacing, pick the **Paragraph Spacing** check box. Then enter the spacing above a paragraph in the **Before** text box, and the spacing below a paragraph in the **After** text box. **Figure 9-32** shows examples of paragraph spacing settings.

The options in the **Paragraph Line Spacing** area adjust *line spacing*. Default line spacing for single lines of text is equal to 1.5625 times the text height. To adjust the line spacing, pick the **Paragraph Line Spacing** check box. Select the **Multiple** option from the **Line Spacing** drop-down list to enter a multiple of the text height in the **At** text box. For example, lines with a text height of .12" are spaced .1875" apart. To double-space lines, you could enter a value of 3.125x, making the space between lines of text .375".

line spacing: The vertical distance from the bottom of one line of text to the bottom of the next line.

Figure 9-32.
Examples of paragraph spacing. Each example uses a text height of .1875″ and a first line left indent of .5″.

Paragraph one typed with no paragpah spacing. Paragraph two typed with no paragraph spacing	Paragraph one typed with .25 before spacing and no after spacing.	Paragraph one typed with .125 before spacing and .5 after spacing.
	Paragraph two typed with .25 before spacing and no after spacing.	Paragraph two typed with .125 before spacing and .5 after spacing.
Before: 0″ After: 0″	Before: .25″ After: 0″	Before: .125″ After: .5″

To force the line spacing to be the same for all lines of text, select the **Exactly** option from the **Line Spacing** drop-down list and enter a value in the **At** text box. If you enter an exact line spacing that is less than the text height, lines of text stack on top of each other. To add spaces between lines automatically based on the height of the characters in the line, choose the **At Least** option from the **Line Spacing** drop-down list and enter a value in the **At** text box. The result is an equal spacing even between lines of different height text.

You can also set line spacing using the **Line Spacing** flyout button on the **Paragraph** panel of the **Text Editor** ribbon tab. Select one of the available multiple options, pick the **More...** button to display the **Paragraph** dialog box, or choose the **Clear Line Spacing** option to apply an automatic spacing similar to the **At Least** function.

NOTE

Combine multiple selected paragraphs to form a single paragraph using the **Combine Paragraphs** option available from the expanded **Paragraph** panel of the **Text Editor** ribbon tab or the shortcut menu.

PROFESSIONAL TIP

Quickly remove formatting from selected text using the **Remove Formatting** cascading submenu available from the shortcut menu or the **More** flyout in the **Options** panel. Pick the **Remove Character Formatting** option to remove character formatting, such as bold, italic, or underline. Select the **Remove Paragraph Formatting** option to remove paragraph formatting, including lists. Pick the **Remove All Formatting** option to remove all character and paragraph formatting.

Exercise 9-5

Access the Student Web site (www.g-wlearning.com/CAD) and complete Exercise 9-5.

Creating Lists

Drawings often use lists to organize information. Lists provide a way to arrange related items in a logical order and help make lines of text more readable. General notes are usually in list format.

You can create lists as you enter text or apply list formatting to existing text. Numbering or lettering automatically adjusts when you add or remove listed items. You must select the **Allow Bullets and Lists** option in order to create a list. This option

AutoCAD and Its Applications—Basics

is active by default. Unchecking this option converts any list items in the text object to plain text characters and disables the other options in the menu.

Lists can also contain sublevel items designated with double numbers, letters, or bullets. Default tab settings apply unless you adjust paragraph options. List tools are available from the **Numbering** flyout on the **Paragraph** panel of the **Text Editor** ribbon tab or the **Bullets and Lists** cascading submenu of the shortcut menu. These tools allow you to create numbered, bulleted, and alphabetical lists.

Choose an option from the **Lettered** cascading submenu to create an alphabetical list. Pick **Uppercase** to use uppercase lettering or choose **Lowercase** to use lowercase lettering. The **Uppercase** option is the default. Pick the **Numbered** option to form a numbered list. To create a bulleted list in the default style, select the **Bulleted** option. This places the default solid circle bullet symbol at the beginning of each line of text.

Another method of creating lists is to use the **Allow Auto-list** option, which is active by default. When using the **Allow Auto-list** option, AutoCAD detects characters that frequently start a list and automatically assigns the first list item. For example, if a line of text begins with a number or letter and a period, AutoCAD assumes that you are starting a list and formats any additional lines of text to continue the list.

To create a numbered or lettered auto-list, you must include punctuation, such as a period, parenthesis, or colon, and press [Tab] after the number or letter that begins the first item. When you press [Enter] to start a new line of text, the new line uses the same formatting as the previous line, and the next consecutive number or letter appears. To end the list, press [Enter] twice. **Figure 9-33** shows an example of a numbered list.

When creating a bulleted auto-list, you can use typical keyboard characters, such as a hyphen [-], tilde [~], bracket [>], or asterisk [*], at the beginning of a line. Another option is to insert a symbol at the beginning of a line. Then, to form the list, press [Tab] and type the line of text. When you press [Enter] to start a new line of text, the new line uses the same formatting bullet symbol as the previous line. To end the list, press [Enter] twice. See **Figure 9-34.**

Figure 9-33.
Framing notes arranged in a numbered list.

FRAMING NOTES:
1. ALL FRAMING NOTES TO BE DFL #2 OR BETTER.
2. ALL HEATED WALLS @ HEATED LIVING AREA TO BE 2 X 6 @ 16" OC. FRAME ALL EXTERIOR NON-BEARING WALLS W/2 X 6 STUDS @ 24" OC.
3. USE 2 X 6 NAILER AT THE BOTTOM OF ALL 2-2 X 12 OR 4 X HEADERS @ EXTERIOR WALLS, BACK HEADER W/2" RIGID INSULATION.
4. BLOCK ALL WALLS OVER 10'-0" HIGH AT MID HEIGHT.

Figure 9-34.
In addition to the regular bullet symbol, other keyboard characters can be used for items in bulleted lists.

•   An elevation of the beam with end views or sections
•   Complete locational dimensions for holes, plates, and angles
•   Length dimensions

Bulleted List with Bullet Symbols

~ Connection specifications
~ Cutouts
~ Miscellaneous notes for the fabricator

Bulleted List with Tilde Characters

Picking the **Use Tab Delimiter Only** option limits unwanted list formatting by instructing AutoCAD to recognize only tabs to start a list. If the **Use Tab Delimiter Only** option is unchecked, list formatting occurs when a space or tab follows the initial list item character.

You can convert multiple lines of text to a list by selecting all of the lines of text and then picking a list formatting option. AutoCAD detects where you type [Enter] to start a new line of text and lists the lines in sequence. When you create a list in this manner, a tab automatically occurs after the number, letter, or symbol preceding the text. Set tabs and indents to adjust spacing and appearance.

Additional options are available for creating lists. Use the **Off** option to remove any list characters or bulleting from selected text. Pick the **Restart** option to renumber or re-letter selected items in a new sequence. The numbering or lettering restarts from the beginning, using 1 or A, for example. Choose the **Continue** option to add selected items to a list that exists above the currently selected item. The number of the selected item continues from the previous list. Items below the selected item also renumber.

Exercise 9-6

Access the Student Web site (www.g-wlearning.com/CAD) and complete Exercise 9-6.

Forming Columns

Sometimes it is necessary to break up text into multiple sections, or columns. This is especially true when you add lengthy general notes or when you must group information together. See **Figure 9-35.** Mtext columns are created in the text editor as a single object. This eliminates the need to create multiple text objects to form separate columns of text. You can create columns as you enter text or apply column formatting to existing text.

This chapter previously suggests turning columns off before accessing the text editor. This approach is appropriate for typical text requirements without columns, especially as you learn to create mtext. By default, however, mtext is set to form

Figure 9-35.
An example of drawing notes created as a single mtext object and divided into three columns.

dynamic columns, with the associated **Manual height** option. Column tools are also available while the text editor is active, from the **Columns** flyout on the **Insert** panel of the **Text Editor** ribbon tab or from the **Columns** cascading submenu of the shortcut menu. You can create *dynamic columns* or *static columns*.

dynamic columns: Columns calculated automatically by AutoCAD according to the amount of text and the height and width of the columns.

static columns: Columns in which you divide the text into a specified number of columns.

Forming Dynamic Columns

To form dynamic columns, choose an option from the **Dynamic Columns** cascading submenu. Pick the **Auto height** option to produce columns of equal height. **Figure 9-36** shows methods for adjusting dynamic columns using **Auto height**. Increasing column width or height reduces the number of columns, while decreasing column width or height produces more columns. Pick the **Manual height** option to produce columns you can adjust individually for height to produce distinct groups of information. Pick and drag the arrows at the bottom of each column to adjust column height. See **Figure 9-37**.

Forming Static Columns

To form static columns, choose the number of columns from the **Static Columns** cascading submenu. The display of text in static columns depends on how much text is in the text editor and the height and width of the columns. However, the selected number of columns does not change even if text does not fill or extends past a column. **Figure 9-38** shows methods for adjusting static columns. Increasing column width or height rearranges the text in the specified number of columns, but the number of static columns does not change based on column width or height.

NOTE

To create more than six static columns, pick the **More...** option to access the **Column Settings** dialog box and enter the number of columns in the **Column Number** text box.

Figure 9-36.
Controlling columns using the dynamic column **Auto Height** option. Notice how column text flows automatically from one column to the next.

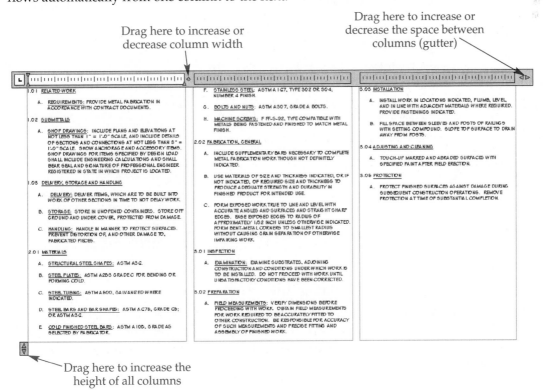

Drag here to increase or decrease column width

Drag here to increase or decrease the space between columns (gutter)

Drag here to increase the height of all columns

Figure 9-37.
Controlling the length of dynamic columns individually (manually).

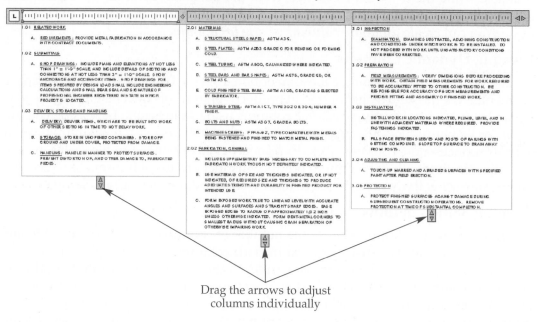

Drag the arrows to adjust columns individually

Figure 9-38.
Controlling static columns.

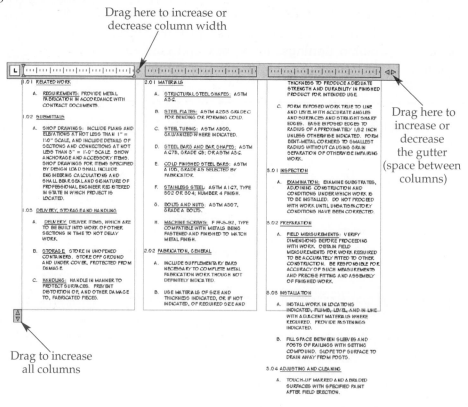

Drag here to increase or decrease column width

Drag here to increase or decrease the gutter (space between columns)

Drag to increase all columns

Using the Column Settings Dialog Box

You can use the **Column Settings...** dialog box as an alternative method for creating columns. To create dynamic columns, select the **Dynamic Columns** radio button, and then select either the **Auto height** or **Manual height** radio button. To create static columns, choose the **Static Columns** radio button and enter the number of static columns in the **Column Number** text box.

Additional controls become available depending on the selected column type radio buttons. Enter the number of static columns in the **Column Number** text box. The **Height** text box allows you to enter the height for all static or dynamic columns. The **Width** area allows you to set column width and the *gutter*. Enter the column width in the **Column** text box and the gutter width in the **Gutter** text box. The **Total** text box is available only with static columns. It allows you to enter the total width of the text editor, which is the sum of the width of all columns and the gutter spacing between columns. To eliminate columns, pick the **No Columns** radio button.

gutter: The space between columns of text.

NOTE

If you choose to remove columns using the **No Columns** option, any column breaks added using the **Insert Column Break** function remain set. Backspace to remove column breaks.

Controlling Column Breaks

You can identify the line of text at which a new column begins using the **Insert Column Break** option. To apply this technique, you must first form a dynamic or static column. Then place the cursor at a location in the text editor where you want a new column to start, such as the start of a paragraph. Pick the **Insert Column Break** option to form the break. The text shifts to the next column at the location of the break. Continue applying column breaks as needed to separate sections of information.

Exercise 9-7

Access the Student Web site (www.g-wlearning.com/CAD) and complete Exercise 9-7.

Importing Text

The **Import Text** option available from the expanded **Tools** panel of the **Text Editor** ribbon tab, or from the shortcut menu, allows you to import text from an existing text file directly into the text editor. The text file can be either a standard ASCII text file (TXT) or a rich text format (RTF) file. The imported text becomes a part of the current mtext object. The **Select File** dialog box appears when you access the **Import Text...** option. Select the text file to be imported and pick the **Open** button. The text inserts at the current cursor location.

PROFESSIONAL TIP

Importing text is useful if someone has already created specification notes in a program other than AutoCAD and you want to place the same notes in drawings.

Template Development
Chapter 9

For detailed instructions on adding text styles to each drawing template, go to the Student Web site (www.g-wlearning.com/CAD), select this chapter, and select **Template Development**.

Chapter Test

Answer the following questions. Write your answers on a separate sheet of paper or go to the Student Web site (www.g-wlearning.com/CAD) and complete the electronic chapter test.

1. Which ASME standard contains guidelines for lettering?
2. What is text composition?
3. Determine the AutoCAD text height for text to be plotted .188″ high using a half (1:2) scale. Show your calculations.
4. Determine the AutoCAD text height for text to be plotted .188″ high using a scale of 1/4″ = 1′-0″. Show your calculations.
5. Explain the function of annotative text and give an example.
6. What is the relationship between the drawing scale and the annotation scale for annotative text?
7. Define *text style*.
8. Describe how to create a text style that has the name ROMANS-12_15, uses the romans.shx font, has a fixed height of .12, a text width of 1.25, and an oblique angle of 15.
9. Define *font*.
10. What are "big fonts"?
11. When setting text height in the **Text Style** dialog box, what value do you enter so text height can be altered each time the **TEXT** tool is used?
12. How would you specify text to display vertically on the screen?
13. What does a width factor of .5 do to text when compared to the default width factor of 1?
14. Explain how to make a text style current quickly.
15. Name the tool that lets you create multiline text objects.
16. How does the width of the mtext boundary affect what you type?
17. What happens if the mtext you are entering exceeds or is not as long as the boundary length that you initially establish?
18. In the text editor, how do you open the text editor shortcut menu?
19. What happens when you enter a fraction in the mtext editor, and what does this allow you to do?
20. How can you draw stacked fractions manually when using the **MTEXT** tool?
21. What happens when you pick the **Other...** option in the **Symbol** cascading menu of the text editor shortcut menu?
22. What is the purpose of tracking?
23. What text feature allows you to hide parts of objects behind and around text?
24. What is the difference between text boundary justification and paragraph alignment?
25. Define *line spacing*.
26. Explain the function of the **Allow Auto-list** option.
27. Explain how to convert multiple lines of text into a numbered list.
28. Briefly describe the difference between dynamic columns and static columns.
29. How can you insert a column break in static columns?
30. In what two formats can text be imported into the text editor?

Drawing Problems

Start AutoCAD if it is not already started. Start a new drawing using an appropriate template of your choice. The template should include layers and text styles, when necessary, for drawing the given objects. Add layers and text styles as needed. Draw all objects using appropriate layers, and text styles, justification, and format. Follow the specific instructions for each problem. Use your own judgment and approximate dimensions when necessary.

▼ Basic

1. Use the **MTEXT** tool to type your name using a text style of your choosing and a text height of 1". Print or plot your name as a name tag. Save the drawing as P9-1.

2. Use the **MTEXT** tool to type the definition of the following terms using a text style with the Romand font and a .12 text height. Save the drawing as P9-2.
   - scale factor
   - annotative text
   - annotation scale
   - text height
   - paper text height

3. Use the **MTEXT** tool to type the following text using a text style with the Stylus BT font and a .125 text height. The heading text height is .25. Save the drawing as P9-3.

   ### KEY NOTES
   1. SLOPING SURFACE
   2. DIAGONAL SUPPORT STRUT
   3. VENT- PROVIDE NEW CANT FLASHING
   4. BRICK CHIMNEY- REMOVE TO BELOW
      DECK SURFACE

▼ Intermediate

4. Use the **MTEXT** tool to type the following text using a text style with the Romans font and a .12 text height. The heading text height is .24. Check your spelling. Save the drawing as P9-4.

   ### NOTES:

   1. INTERPRET ALL DIMENSIONS AND TOLERANCES PER ANSI Y14.4M−1994.
   2. REMOVE ALL BURRS AND SHARP EDGES.

   CASTING NOTES UNLESS OTHERWISE SPECIFIED:
   1. .31 WALL THICKNESS
   2. R.12 FILLETS
   3. R.06 CORNERS
   4. 1.5°−3.0° DRAFT
   5. TOLERANCES
      ± 1° ANGULAR
      ± .03 TWO−PLACE DIMENSIONS
   6. PROVIDE .12 THK MACHINING STOCK ON ALL MACHINED SURFACES.

5. Use the **MTEXT** tool to type the following text using a text style with the Stylus BT font and a .125 text height. The heading text height is .188. After typing the text exactly as shown, edit the text with the following changes:
   A. Change the \ in item 7 to 1/2.
   B. Change the [in item 8 to 1.
   C. Change the 1/2 in item 8 to 3/4.
   D. Change the ^ in item 10 to a degree symbol.
   E. Check your spelling after making the changes.
   F. Save as drawing P9-5.

COMMON FRAMING NOTES:
1. ALL FRAMING LUMBER TO BE DFL #2 OR BETTER.
2. ALL HEATED WALLS @ HEATED LIVING AREAS TO BE 2 X 6 @ 24" OC.
3. ALL EXTERIOR HEADERS TO BE 2-2 X 12 UNLESS NOTED, W/ 2" RIGID INSULATION BACKING UNLESS NOTED.
4. ALL SHEAR PANELS TO BE 1/2" CDX PLY W/8d @ 4" OC @ EDGE, HDRS, & BLOCKING AND 8d @ 8" OC @ FIELD UNLESS NOTED.
5. ALL METAL CONNECTORS TO BE SIMPSON CO. OR EQUAL.
6. ALL TRUSSES TO BE 24" OC. SUBMIT TRUSS CALCS TO BUILDING DEPT. PRIOR TO ERECTION.
7. PLYWOOD ROOF SHEATHING TO BE \ STD GRADE 32/16 PLY LAID PERP TO RAFTERS. NAIL W/8d @ 6" OC @ EDGES AND 12" OC @ FIELD.
8. PROVIDE [1/2" STD GRADE T&G PLY FLOOR SHEATHING LAID PERP TO FLOOR JOISTS. NAIL W/10d @ 6" OC @ EDGES AND BLOCKING AND 12" OC @ FIELD.
9. BLOCK ALL WALLS OVER 10'-0" HIGH AT MID.
10. LET-IN BRACES TO BE 1 X 4 DIAG BRACES @ 45^ FOR ALL INTERIOR LOAD-BEARING WALLS.

6. Draw the general caulking notes shown below. Save your drawing as P9-6.

CAULKING NOTES:

CAULKING REQUIREMENTS BASED ON 1992 OREGON RESIDENTIAL ENERGY CODE

1. SEAL THE EXTERIOR SHEATHING AT CORNERS, JOINTS, DOORS, WINDOWS, AND FOUNDATION SILL WITH SILICONE CAULK.
2. CAULK THE FOLLOWING OPENINGS W/ EXPANDED FOAM, BACKER RODS, OR SIMILAR:
   • ANY SPACE BETWEEN WINDOW AND DOOR FRAMES
   • BETWEEN ALL EXTERIOR WALL SOLE PLATES AND PLY SHEATHING
   • ON TOP OF RIM JOIST PRIOR TO PLYWOOD FLOOR APPLICATION
   • WALL SHEATHING TO TOP PLATE
   • JOINTS BETWEEN WALL AND FOUNDATION
   • JOINTS BETWEEN WALL AND ROOF
   • JOINTS BETWEEN WALL PANELS
   • AROUND OPENINGS

7. Draw the basic organizational chart shown below. Save your drawing as P9-7.

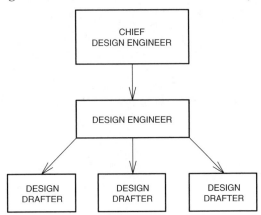

▼ Advanced

8. Draw the controller schematic shown below. Save your drawing as P9-8.

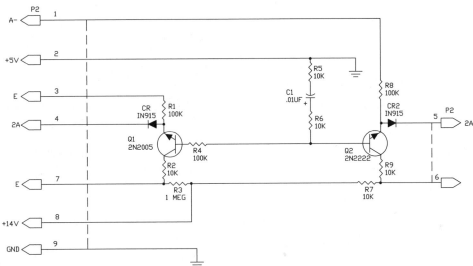

NOTES:

1. INTERPRET ELECTRICAL AND ELECTRONICS DIAGRAMS PER ANSI Y14.15.

2. UNLESS OTHERWISE SPECIFIED:

   RESISTANCE VALUES ARE IN OHMS.
   RESISTANCE TOLERANCE IS 5%.
   RESISTORS ARE 1/4 WATT.
   CAPACITANCE VALUES ARE IN MICROFARADS.
   CAPACITANCE TOLERANCE IS 10%.
   CAPACITOR VOLTAGE RATING IS 20V.
   INDUCTANCE VALUES ARE IN MICROHENRIES.

REFERENCE DESTINATIONS
LAST USED
R9
C1
CR2
Q2

9. Draw the electrical notes shown below. Save your drawing as P9-9.

ELECTRICAL NOTES:

1. ALL GARAGE AND EXTERIOR PLUGS AND LIGHT FIXTURES TO BE ON GFCI CIRCUIT.

2. ALL KITCHEN PLUGS AND LIGHT FIXTURES TO BE ON GFCI CIRCUIT.

3. PROVIDE A SEPARATE CIRCUIT FOR MICROWAVE OVEN.

4. PROVIDE A SEPARATE CIRCUIT FOR PERSONAL COMPUTER. VERIFY LOCATION WITH OWNER.

5. VERIFY ALL ELECTRICAL LOCATIONS W/ OWNER.

6. EXTERIOR SPOTLIGHTS TO BE ON PHOTOELECTRIC CELL W/ TIMER.

7. ALL RECESSED LIGHTS IN EXTERIOR CEILINGS TO BE INSULATION COVER RATED.

8. ELECTRICAL OUTLET PLATE GASKETS SHALL BE INSULATED ON RECEPTACLE, SWITCH, AND ANY OTHER BOXES IN EXTERIOR WALL.

9. PROVIDE THERMOSTATICALLY CONTROLLED FAN IN ATTIC WITH MANUAL OVERRIDE. VERIFY LOCATION WITH OWNER.

10. ALL FANS TO VENT TO OUTSIDE AIR. ALL FAN DUCTS TO HAVE AUTOMATIC DAMPERS.

11. HOT WATER TANKS TO BE INSULATED TO R-11 MINIMUM.

12. INSULATE ALL HOT WATER LINES TO R-4 MINIMUM. PROVIDE ALTERNATE BID TO INSULATE ALL PIPES FOR NOISE CONTROL.

13. PROVIDE 6 SQ. FT. OF VENT FOR COMBUSTION AIR TO OUTSIDE AIR FOR FIREPLACE CONNECTED DIRECTLY TO FIREBOX. PROVIDE FULLY CLOSABLE AIR INLET.

14. HEATING TO BE ELECTRIC HEAT PUMP. PROVIDE BID FOR SINGLE UNIT NEAR GARAGE OR FOR A UNIT EACH FLOOR (IN ATTIC).

15. INSULATE ALL HEATING DUCTS IN UNHEATED AREAS TO R-11. ALL HVAC DUCTS TO BE SEALED AT JOINTS AND CORNERS.

10. Draw the flowchart shown below. Save your drawing as P9-10.

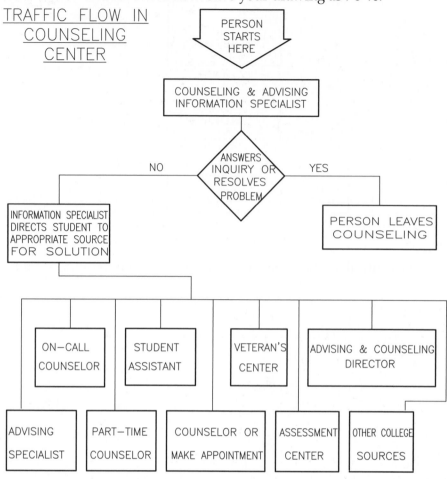

TRAFFIC FLOW IN
COUNSELING
CENTER

PERSON
STARTS
HERE

COUNSELING & ADVISING
INFORMATION SPECIALIST

ANSWERS
INQUIRY OR
RESOLVES
PROBLEM

NO   YES

INFORMATION SPECIALIST
DIRECTS STUDENT TO
APPROPRIATE SOURCE
FOR SOLUTION

PERSON LEAVES
COUNSELING

ON—CALL
COUNSELOR

STUDENT
ASSISTANT

VETERAN'S
CENTER

ADVISING & COUNSELING
DIRECTOR

ADVISING
SPECIALIST

PART—TIME
COUNSELOR

COUNSELOR OR
MAKE APPOINTMENT

ASSESSMENT
CENTER

OTHER COLLEGE
SOURCES

11. Draw the electrical legend shown below. Save your drawing as P9-11.

ELECTRICAL LEGEND:

	110 VOLT DUPLEX CONVENIENCE OUTLET
GFCI	110 VOLT GROUND FAULT CIRCUIT INTERRUPT DUPLEX OUTLET
GFCI WP	110 VOLT WATERPROOF GFCI DUPLEX OUTLET
	110 VOLT SPLIT WIRED OUTLET
	220 VOLT OUTLET
	JUNCTION BOX
TV	CABLE TELEVISION OUTLET
	CLOCK OUTLET
	DOOR BELL
$	SINGLE POLE SWITCH
$³	THREE-WAY SWITCH
O	CEILING-MOUNTED LIGHT
	WALL-MOUNTED LIGHT
	FLUORESCENT LIGHT
o	CIRCULAR RECESSED LIGHT
▢	SQUARE RECESSED LIGHT
	LIGHT, FAN COMBINATION
	LIGHT, FAN, HEAT COMBINATION
SD	CEILING-MOUNTED SMOKE DETECTOR
SD	WALL-MOUNTED SMOKE DETECTOR

Single-Line Text and Additional Text Tools

Learning Objectives

After completing this chapter, you will be able to do the following:

✓ Use the **TEXT** tool to create single-line text.
✓ Insert and use fields.
✓ Check your spelling.
✓ Edit existing text.
✓ Search for and replace text automatically.

This chapter describes how to use the **TEXT** tool to place single-line text objects. The **TEXT** tool is most useful for text items that require a single character, word, or line of text. You should typically use mtext to type paragraphs or if the text requires mixed fonts, sizes, colors, or other characteristics. This chapter also presents text editing functions and other valuable tools, including fields, spell checking, and methods for finding and replacing text.

Creating Single-Line Text

Access the **TEXT** tool to create a single-line text object. Pick a point to define the lower-left corner of the text, using default justification. Next, if the current text style uses a height of 0, enter the text height. If the current text style is annotative, enter the paper text height. If the current text style is not annotative, enter the text height multiplied by the scale factor. The next prompt asks for the text rotation angle. The default value is 0, which draws horizontal text. Other values rotate text in a counterclockwise direction. The text pivots about the start point, as shown in **Figure 10-1**.

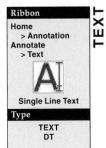

Ribbon

Home
> Annotation
Annotate
> Text

TEXT

Single Line Text

Type

TEXT
DT

NOTE

Changes you make to the default text angle orientation or direction affect the text rotation.

After you set the text height and rotation angle, a text editor and cursor equal in height to the text height appears on-screen at the start point. As you type, the text editor increases in size to display the characters. See **Figure 10-2.** Type additional lines

Figure 10-1.
Rotation angles for text. The plus sign indicates the start point.

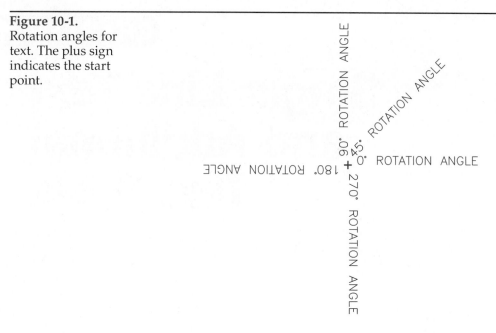

Figure 10-2.
Entering text with the **TEXT** tool.

Text cursor box

Text cursor

Start point

AutoCAD single—line text

Text editor

of text by pressing [Enter] at the end of each line. The text cursor automatically moves to a start point one line below the preceding line. Press [Enter] twice to exit the **TEXT** tool and keep what you have typed. You can cancel the tool at any time by pressing [Esc]. This action removes any incomplete lines of text.

While typing, you can right-click to display a shortcut menu of text options. These options function much like those for the **MTEXT** tool. Options for accessing help files and canceling the **TEXT** tool are also available from the shortcut menu.

NOTE

Set the text style you want to use current before accessing the **TEXT** tool. A **Style** option is available before you pick the text start point to set the text style, but it is difficult to use.

Text Justification

The **TEXT** tool offers a variety of justification options. Left justification is the default. To use a different justification option, choose the **Justify** option at the Specify start point of text [Justify/Style]: prompt before you pick the text start point.

The **Center** option allows you to select the center point for the baseline of the text. The **Middle** option allows you to center text horizontally and vertically at a given point. The **Right** option justifies text at the lower-right corner. You can also change the letter height and rotation when using these options. **Figure 10-3** compares the **Center**, **Middle**, and **Right** options. A number of text alignment options allow you to place text on a drawing in relation to the top, bottom, middle, left side, or right side of the text. **Figure 10-4** shows these alignment options.

Figure 10-3.
The **Center**, **Middle**, and **Right** text justification options.

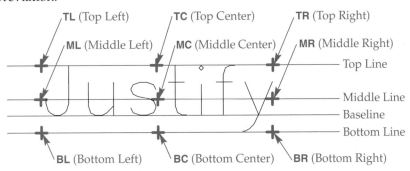

AUTOCAD CENTERED TEXT
Center Option

AUTOCAD MIDDLE TEXT
Middle Option

AUTOCAD RIGHT–JUSTIFIED TEXT
Right Option

Figure 10-4.
Using the **TL**, **TC**, **TR**, **ML**, **MC**, **MR**, **BL**, **BC**, and **BR** text alignment options. Notice the meaning of each abbreviation.

TL (Top Left)　　　**TC** (Top Center)　　　**TR** (Top Right)

ML (Middle Left)　　**MC** (Middle Center)　　**MR** (Middle Right)

Justify

Top Line

Middle Line
Baseline
Bottom Line

BL (Bottom Left)　　**BC** (Bottom Center)　　**BR** (Bottom Right)

When you use the **Align** justification option, AutoCAD automatically adjusts the text height to fit between the start point and endpoint. The height varies according to the distance between the points and the number of characters. The **Fit** option is similar to the **Align** option, except you can select the text height. AutoCAD adjusts character width to fit between the two given points, while keeping text height constant. **Figure 10-5** shows the effects of the **Align** and **Fit** options.

PROFESSIONAL TIP

The **Align** and **Fit** options of the **TEXT** tool are not recommended because the text height or width is inconsistent from one line of text to another. In addition, text height adjusted by the **Fit** option can cause one line of text to run into another.

Exercise 10-1

Access the Student Web site (www.g-wlearning.com/CAD) and complete Exercise 10-1.

Figure 10-5.
Examples of aligned and fit text. In aligned text, the text height is adjusted. In fit text, the text width is adjusted.

AUTOCAD ALIGNED TEXT
WHEN USING ALIGNED TEXT
THE TEXT HEIGHT
IS ADJUSTED SO THAT THE TEXT
FITS BETWEEN
TWO PICKED POINTS

Text height varies between lines

Align Option

AUTOCAD FIT TEXT
WHEN USING FIT TEXT
THE TEXT WIDTH
IS ADJUSTED SO THAT THE TEXT
HEIGHT REMAINS THE SAME
FOR EACH LINE

Text width varies between lines

Fit Option

Inserting Symbols

In order to insert a symbol with the **TEXT** tool, you must type a *control code sequence*. For example, to add the note ⌀2.75, type %%C2.75 in the text editor. In this example, %%C is the code used to add the diameter symbol. **Figure 10-6** shows many symbol codes and the symbols the codes create. Add a single percent sign normally. However, when a percent sign must precede another control code sequence, you can use %%% to force a single percent sign.

Figure 10-6.
Common control code sequences used to add symbols to single-line text.

Control Code or Unicode	Type of Symbol	Appearance
%%d	Degrees	°
%%p	Plus/Minus	±
%%c	Diameter	⌀
%%%	Percent	%
\U+2248	Almost equal	≈
\U+2220	Angle	∠
\U+E100	Boundary line	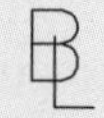 BL
\U+2104	Centerline	℄
\U+0394	Delta	△
\U+0278	Electrical phase	φ
\U+E101	Flow line	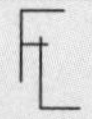 FL
\U+2261	Identity	≡
\U+E200	Initial length	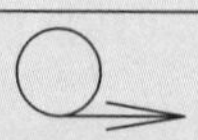 ⟲→
\U+E102	Monument line	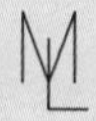 ML
\U+2260	Not equal	≠
\U+2126	Ohm	Ω
\U+03A9	Omega	Ω
\U+214A	Property line	PL
\U+2082	Subscript 2	$_2$
\U+00B2	Squared	^{2}
\U+00B3	Cubed	^{3}

Drawing Underscored or Overscored Text

Type %%O in front of the line of text to overscore text and %%U in front of the line of text to create underscored (underlined) text. To create the note UNDERSCORING TEXT, for example, type %%UUNDERSCORING TEXT. A line of text may require both underscoring and overscoring. To do this, use both control code sequences. For example, the control code sequence %%O%%ULINE OF TEXT produces LINE OF TEXT.

The %%O and %%U control codes are toggles that turn overscoring and underscoring on and off. Type %%U preceding a word or phrase to turn underscoring on. Type %%U after the desired word or phrase to turn underscoring off. Any text following the second %%U appears without underscoring. For example, enter DETAIL A HUB ASSEMBLY as %%UDETAIL A%%U HUB ASSEMBLY.

PROFESSIONAL TIP

Many drafters prefer to underline labels such as SECTION A-A or DETAIL B. Rather than draw line or polyline objects under the text, use the **Middle** or **Center** justification mode and underscoring. The view labels automatically underline and center under the views or details they identify.

Exercise 10-2

Access the Student Web site (www.g-wlearning.com/CAD) and complete Exercise 10-2.

Working with Fields

Fields display information related to a specific object, general drawing properties, or the current user or computer system. You can set field information to update automatically. This makes fields useful tools for displaying information that may change throughout the course of a project. For example, you could insert the **Date** field into a title block. The field updates automatically with the current date throughout the life of the drawing file.

field: A special type of text object that can display a specific property value, setting, or characteristic.

Inserting Fields

Use the **FIELD** tool and **Field** dialog box, shown in **Figure 10-7**, to add fields to mtext or text objects. To insert a field in an active mtext editor, pick the **Field** button from the **Insert** panel of the **Text Editor** ribbon tab, pick the **Insert Field** option available from the shortcut menu, or press [Ctrl]+[F]. To insert a field in an active single-line text editor, right-click and select **Insert Field...** or press [Ctrl]+[F].

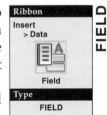

The **Field** dialog box includes many preset fields. Notice that fields are grouped into categories. When you select a category from the **Field category** drop-down list, only the fields within the category appear in the **Field names** list box. This makes it much easier to locate the desired field. Pick the field category, and then pick the field to insert from the **Field names** list box. You can also select from a list of formats to determine the display of the field. The **Format** list varies, depending on the selected field.

Figure 10-7.
Select fields using the **Field** dialog box.

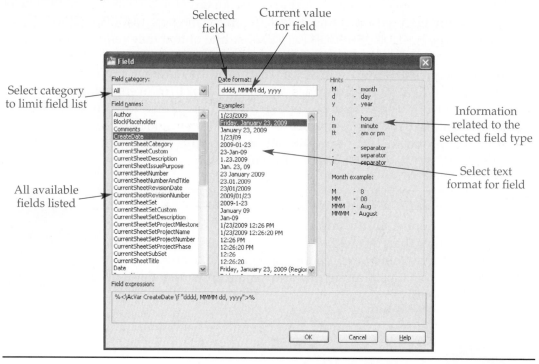

After you select the field and format, pick the **OK** button to insert the field. The field assumes the current text style. By default, field text displays a gray background. See **Figure 10-8.** This keeps you aware that the text is actually a field, and the value may change. You can deactivate the background in the **Fields** area of the **User Preferences** tab of the **Options** dialog box. See **Figure 10-9.**

Updating Fields

After you insert a field into a drawing, the displayed value may change. For example, a field indicating the current date changes every day. A field displaying the file name changes if the file name changes. A field displaying the value of an object property changes if modifications to the object cause the property to change.

Automatic or manual field *updating* is possible. Automatic updating is set using the **Field Update Settings** dialog box. To access this dialog box, pick the **Field Update Settings...** button in the **Fields** area of the **User Preferences** tab of the **Options** dialog box. Whenever a selected event (such as saving or regenerating) occurs, all associated fields automatically update.

Update fields manually using the **Update Fields** tool. After selecting the tool, pick the fields to update. You can use the **All** selection option to update all fields in a single operation. You can also update a field within the text editor by right-clicking on the field and selecting **Update Field**.

Updating:
AutoCAD's procedure for changing text in a field based on the field's current value.

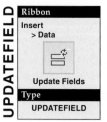

Figure 10-8.
A date and time field inserted into multiline text. The gray background identifies the text as a field.

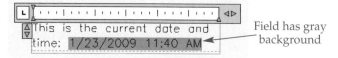

Figure 10-9.
Control the background display for fields in the **User Preferences** tab of the **Options** dialog box.

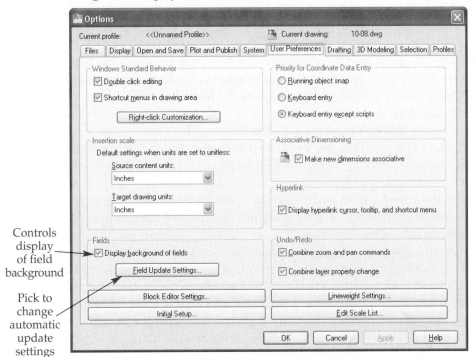

Controls
display
of field
background

Pick to
change
automatic
update
settings

Editing Fields

To edit a field, you must first select the text object containing the field for editing. You can do this quickly by double-clicking on the text object. Then double-click on the field to display the **Field** dialog box. You can also right-click in the field and pick **Edit Field...**. Use the **Field** dialog box to modify the field settings and pick **OK** to apply the changes.

You can also convert a field to standard text. When you convert a field, the currently displayed value becomes text, the association to the field is lost, and the value no longer updates. To convert a field to text, select the text for editing, right-click in the field, and pick **Convert Field To Text**.

NOTE

You can use fields with many AutoCAD tools, including inquiry tools, drawing properties, attributes, and sheet sets. Specific field applications are described where appropriate throughout this textbook.

Exercise 10-3

Access the Student Web site (www.g-wlearning.com/CAD) and complete Exercise 10-3.

Checking Your Spelling

The quickest way to check for correct spelling in text objects is to use the **Spell Check** tool available in a current text editor. This tool is active by default. To toggle the **Spell Check** tool on or off in mtext or single-line text editors, select the **Check Spelling** option from the **Editor Settings** cascading submenu available from the shortcut menu or select **Spell Check** from the **Spell Check** panel on the **Text Editor** ribbon tab. You can also turn the **Spell Check** tool on and off in an active mtext editor by deselecting the **Spell Check** button on the **Options** panel of the **Text Editor** ribbon tab.

A red dashed line appears under a word that may be spelled incorrectly. Right-click on the underlined word to display options for adjusting the spelling. See **Figure 10-10**. The first section at the top of the shortcut menu provides suggested replacements for the word. Pick the correct word to change the spelling in the text editor. If none of the initial suggestions are correct, you may be able to find the correct spelling from the **More Suggestions** cascading submenu.

If none of the spelling suggestions are appropriate, the word either is spelled correctly or is spelled so incorrectly that AutoCAD cannot recommend the right spelling. If the word is spelled correctly, pick the **Add to Dictionary** option to add the current word to the custom dictionary. You can add words with up to 63 characters. If you want to use the current spelling, recognized as incorrect by AutoCAD, but do not want to add the word to the dictionary, pick the **Ignore All** option. All words that match the currently found misspelled word in the active text editor are ignored and the underline is hidden. Common drafting words and abbreviations, such as the abbreviation for the word SCHEDULE (SCH) in **Figure 10-10** can be added to the dictionary or ignored.

Figure 10-10.
Quickly checking your spelling using the **Spell Check** tool. Spell checking in the multiline text editor is shown. The tool functions the same in the single-line text editor.

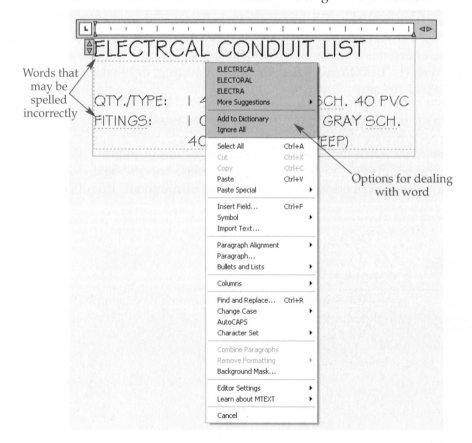

Using the SPELL Tool

An alternative method to check spelling is to use the **SPELL** tool to access the **Check Spelling** dialog box, shown in **Figure 10-11**. The **Check Spelling** dialog box checks spelling of all text objects, without activating a text editor. To check spelling, first identify the portion of the drawing you want to spell-check by selecting an option from the **Where to Check** drop-down list. Pick the **Entire drawing** option to check the spelling of all text objects in the drawing file, including model space and all layouts, or choose the **Current space/layout** to check spelling only of text objects in the active layout or in model space, if model space is active. You can also check the spelling of certain text objects by picking the **Select text objects** button, next to the **Where to Check** drop-down list, to enter the drawing window and select all the text objects for which you want to check the spelling. You do not need to choose the **Selected objects** option from the **Where to Check** drop-down list to check selected objects.

After you define what and where to check, pick the **Start** button to begin to check spelling. The first word that may be misspelled is highlighted in the drawing window and is active in the **Check Spelling** dialog box. **Figure 10-11** briefly describes the options found in the **Check Spelling** dialog box.

PROFESSIONAL TIP

Before you check spelling, you may want to adjust some of the spell-checking preferences provided in the **Check Spelling Settings** dialog box. Access this dialog box by picking the **Settings...** button in the **Check Spelling** dialog box, or select the **Check Spelling Settings...** option from the **Editor Settings** cascading submenu available from the text editor shortcut menu. If the mtext editor is already open, pick the small arrow in the lower-right corner of the **Spell Check** panel on the **Text Editor** ribbon tab. The settings apply to spelling checked using the **Spell Check** tool in an active text editor and using the **Check Spelling** dialog box.

Figure 10-11.
The **Check Spelling** dialog box.

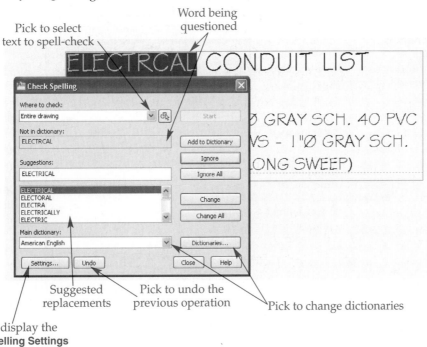

Pick to select text to spell-check

Word being questioned

Suggested replacements

Pick to undo the previous operation

Pick to change dictionaries

Pick to display the **Check Spelling Settings** dialog box

Changing Dictionaries

AutoCAD provides a variety of spelling dictionaries, including dictionaries for several non-English languages. Pick the **Dictionaries...** button of the **Check Spelling** dialog box or select the **Dictionaries...** option from the **Editor Settings** cascading submenu available from the text editor shortcut menu to access the **Dictionaries** dialog box. See **Figure 10-12.**

You can use the **Main dictionary** list to select one of the many language dictionaries to use as the current main dictionary. The main dictionary is protected; you cannot add definitions to it. You can use the **Custom dictionary** list to select the active custom dictionary. The default custom dictionary is sample.cus. Type a word in the **Content** text box that you want to either add or delete from the custom dictionary. For example, ASME Y14.5M is custom text used in engineering drafting. Pick the **Add** button to accept the custom words in the text box, or pick the **Delete** button to remove the words from the custom dictionary. Custom dictionary entries may be up to 63 characters in length.

You can create and manage a custom dictionary by picking the **Manage Custom Dictionaries...** option from the drop-down list to access the **Manage Custom Dictionary** dialog box. Pick the **New** button to create a new custom dictionary by entering a new file name with a .cus extension. You can add and delete words and combine dictionaries using any standard text editor. If you use a word processor such as Microsoft® Word, be sure to save the file as *text only*, with no special text formatting or printer codes. Add a custom dictionary by picking the **Add** button, and choose the **Remove** button to delete a custom dictionary from the list. You can also add existing custom dictionaries by picking the **Import...** button from the **Custom dictionary** area.

PROFESSIONAL TIP

You can create custom dictionaries for various disciplines. For example, you might add common abbreviations and brand names for mechanical drawings to a mech.cus file. A separate file named arch.cus might contain common architectural abbreviations and frequently used brand names.

Figure 10-12.
The **Dictionaries** dialog box.

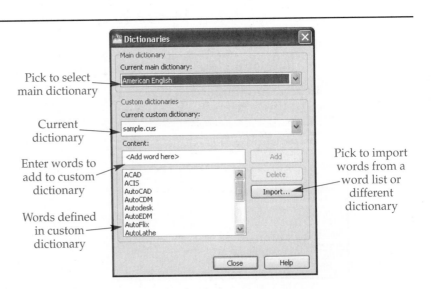

Pick to select main dictionary

Current dictionary

Enter words to add to custom dictionary

Words defined in custom dictionary

Pick to import words from a word list or different dictionary

Revising Text

The easiest way to reopen the text editor to make changes to text content is to double-click an mtext or text object. Another technique to re-enter the text editor is to pick the text object to modify and then right-click and select **Mtext Edit...** to revise mtext, or **Edit...** to modify single-line text.

NOTE

You can also type **MTEDIT** to edit mtext or **DDEDIT** to edit either single-line text or mtext.

Exercise 10-4

Access the Student Web site (www.g-wlearning.com/CAD) and complete Exercise 10-4.

Exercise 10-5

Access the Student Web site (www.g-wlearning.com/CAD) and complete Exercise 10-5.

Changing Case

If you forget to type text using uppercase letters, you can quickly set all text to uppercase by selecting the text and picking the **UPPERCASE** option from the **Change Case** cascading submenu available from the mtext or text editor shortcut menu. While editing mtext, you can also select the **Make Uppercase** button from the **Formatting** panel of the **Text Editor** ribbon tab. Select the **lowercase** option, also found in the **Change Case** cascading submenu, to change selected text to lowercase characters. While editing mtext, you can also select the **Make Lowercase** button from the **Formatting** panel of the **Text Editor** ribbon tab.

The **AutoCAPS** option available from the mtext editor shortcut menu or the expanded **Tools** panel of the **Text Editor** ribbon tab turns [Caps Lock] on when you open an mtext editor. [Caps Lock] turns off when you exit the text editor so that text in other programs is not all uppercase.

Cutting, Copying, and Pasting Text

Clipboard functions allow you to copy, cut, and paste text from any text-based application, such as Microsoft® Word, into a text editor. You can quickly access clipboard functions from the shortcut menu when an mtext or text editor is active. Pasted text retains its properties. You can also copy or cut and paste text from the text editor into other text-based applications.

AutoCAD provides three additional paste options for pasting text into the mtext text editor. These options are available from the **Paste Special** cascading submenu of the text editor shortcut menu. Pick the **Paste without Character Formatting** option to paste text without applying preset character formatting such as bold, italic, or underline. Select the **Paste without Paragraph Formatting** option to paste text without applying current paragraph formatting, including lists. Pick the **Paste without Any Formatting** option to paste text without applying any current character and paragraph formatting.

Finding and Replacing Text

AutoCAD provides tools for searching for a piece of text in your drawing and replacing it with an alternative piece of text. You can search for and replace text in an active mtext or single-line text editor, or without opening a text editor.

Using the Find and Replace Tool

You can activate the **Find and Replace** tool in the mtext and single-line text editors by selecting the **Find and Replace...** option available from the shortcut menu. Another way to turn on the tool in an active mtext editor is by selecting the **Find & Replace** button from the **Tools** panel of the **Text Editor** ribbon tab. The **Find and Replace** dialog box displays. See **Figure 10-13.**

Enter the text you are searching for in the **Find what:** text box. Enter the text to substitute in the **Replace with:** text box. Then pick the **Find Next** button to highlight the next instance of the search text. You can then pick the **Replace** or **Replace All** button to replace just the highlighted text or all words that match your search criteria. Check boxes control the characters and words recognized when finding and replacing text.

FIND

Ribbon
Annotate
> Text
Find Text
Type
FIND

Using the FIND Tool

Use the **FIND** tool to find text throughout the entire drawing and replace it with a different piece of text. Access the **FIND** tool when a text editor is not active to search the entire drawing. AutoCAD displays the **Find and Replace** dialog box when you access the **FIND** tool. See **Figure 10-14.** Another method for accessing the **Find and Replace** dialog box is to enter the text string you want to find in the **Find Text** text box in the **Text** panel of the **Annotation** ribbon tab, and press [Enter].

Figure 10-13.
Using the **Find and Replace** dialog box in an active text editor.

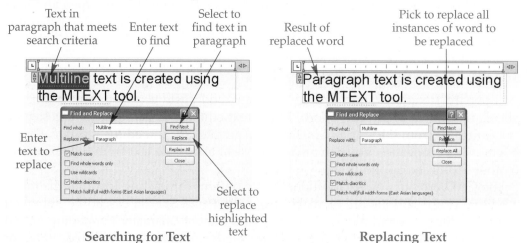

Searching for Text   Replacing Text

Figure 10-14.
Using the version of the **Find and Replace** dialog box that appears when you use the **Find** tool.

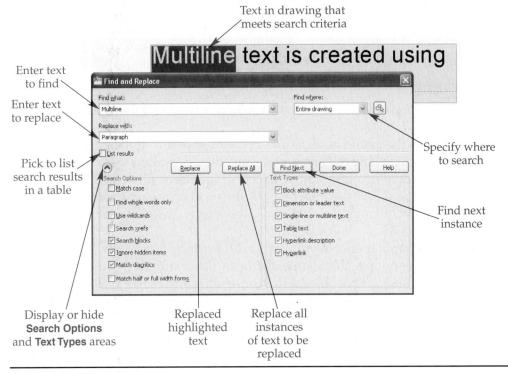

Text in drawing that meets search criteria

Enter text to find

Enter text to replace

Pick to list search results in a table

Display or hide **Search Options** and **Text Types** areas

Specify where to search

Find next instance

Replaced highlighted text

Replace all instances of text to be replaced

You can use the **Find and Replace** dialog box to locate and replace text in multiple text objects. As a result, to find and replace text, you must first identify the portion of the drawing you want to search by selecting an option from the **Find Where** drop-down list. These options are similar to those described for the **Find Where** drop-down list in the **Check Spelling** dialog box.

The **Find and Replace** dialog box is much like the dialog box of the same name that appears when you find and replace text within a text editor. However, this version allows you to display the search results in a table within the dialog box. It also provides more search options. Pick the **More Options** button to display several check boxes used to control the characters and words recognized when finding and replacing text.

NOTE

The find and replace strings are saved with the drawing file and can be reused.

Scaling Text

One option for changing the height of text objects is to use the **SCALETEXT** tool. This tool allows you to scale text objects in relation to their individual insertion points or in relation to a single base point. The **SCALETEXT** tool works with mtext and text objects. You can also select both types of text objects at the same time. After you access the **SCALETEXT** tool, select the text objects to scale. Then, at the Enter a base point option for scaling [Existing/Left/Center/Middle/Right/TL/TC/TR/ML/MC/MR/BL/BC/BR] <Existing>: prompt, specify the justification for the base point.

Figure 10-4 and **Figure 10-5** show all of the justification options except **Left** and **Existing.** The **Left** option scales text objects using their lower-left corner point as the base point. The **Existing** option scales text objects using their existing justification setting as the base point. **Figure 10-15** shows scaling text with different justification

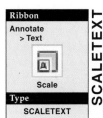

Ribbon
Annotate
> Text

Scale

Type
SCALETEXT

SCALETEXT

Figure 10-15.
The **Existing** option
of the **SCALETEXT**
tool scales text
objects using
their individual
justification settings.

Original Text

Text Scaled Using
Existing Base Point Option

points using the **Existing** option. Notice how the text scales in relation to its own justification setting.

After you specify the justification to use as the base point, AutoCAD prompts for the scaling type. The default **Specify new model height** option allows you to type a new value for the text height of non-annotative objects. If the selected text is annotative, the value you enter is ignored. Use the **Paper height** option to type a new paper text height value for the text height of annotative objects. If the selected text is non-annotative, the value you enter is ignored.

The **Match object** option allows you to pick an existing text object. The height of the selected text object adopts the text height from the text object you pick. Use the **Scale factor** option to scale text objects that have different heights relative to their current heights. For example, using a scale factor of 2 scales all of the selected text objects to twice their current size.

NOTE

You should only use the **SCALETEXT** tool to scale non-annotative text.

Changing Text Justification

Use the **JUSTIFYTEXT** tool to change the justification point without moving the text. Pick the text for which you want to change justification, and enter the new justification option.

Exercise 10-6

Access the Student Web site (www.g-wlearning.com/CAD) and complete Exercise 10-6.

Express Tools
Chapter 10

The **Express Tools** ribbon tab includes additional tools for improved functionality and productivity during the drawing processes. The following Express Tools represent the most useful text express tools. For information about these tools, go to the Student Web site (www.g-wlearning.com/CAD), select this chapter, and select **Using Text Express Tools**.

Text Fit	**Arc-Aligned Text**
Text Mask	**Enclose Text with Object**
Unmask Text	**Change Text Case**
Convert Text to Mtext	

Chapter Test

Answer the following questions. Write your answers on a separate sheet of paper or go to the Student Web site (www.g-wlearning.com/CAD) and complete the electronic chapter test.

1. List two ways to access the **TEXT** tool.
2. Write the control code sequence required to draw the following symbols when using the **TEXT** tool:
   A. 30°
   B. 1.375 ± .005
   C. ∅24
   D. <u>NOT FOR CONSTRUCTION</u>
3. Briefly explain the function and purpose of fields.
4. What is different about the on-screen display of fields compared to that of text?
5. How can you access the **Field Update Settings** dialog box?
6. Explain how to convert a field to text.
7. What is the quickest way to check your spelling within a current text editor?
8. How do you change the **Current word** if you do not think the word that is displayed in the **Suggestions:** text box of the **Check Spelling** dialog box is the correct word, but one of the words in the list of suggestions is the correct word?
9. Identify three ways to access the AutoCAD spell checker.
10. How do you change the main dictionary for use in the **Check Spelling** dialog box?
11. What appears if you double-click on multiline text?
12. What is the purpose of the **AutoCAPS** option available from the mtext editor shortcut menu?
13. Briefly describe how to find and replace text when an mtext or single-line text editor is open.
14. Name the tool that allows you to find a piece of text and replace it with an alternative piece of text in a single instance or for every instance in your drawing.
15. When using the **SCALETEXT** tool, which base point option would you select to keep the text object's current justification point?

Drawing Problems

Start AutoCAD if it is not already started. Start a new drawing using an appropriate template of your choice. The template should include layers and text styles, when necessary, for drawing the given objects. Add layers and text styles as needed. Draw all objects using appropriate layers, and text styles, justification, and format. Follow the specific instructions for each problem. Use your own judgment and approximate dimensions when necessary.

▼ Basic

1. Use the **TEXT** tool to type the following information. Change the text style to represent each of the four fonts named. Use a .25 unit text height and 0° rotation angle. Save the drawing as P10-1.

   TXT–AUTOCAD'S DEFAULT TEXT FONT, WHICH IS AVAILABLE FOR USE WHEN YOU BEGIN A DRAWING.
   ROMANS–SMOOTHER THAN TXT FONT AND CLOSELY DUPLICATES THE SINGLE-STROKE LETTERING THAT HAS BEEN THE STANDARD FOR DRAFTING.
   ROMANC–A MULTISTROKE DECORATIVE FONT THAT IS GOOD FOR USE IN DRAWING TITLES.
   ITALICC–AN ORNAMENTAL FONT SLANTED TO THE RIGHT AND HAVING THE SAME LETTER DESIGN AS THE COMPLEX FONT.

2. Use the list below to create text styles. Then use the **TEXT** tool to type the text, changing the text style to represent each style. Use a .25 unit text height. Save the drawing as P10-2.

TXT–EXPAND THE WIDTH BY THREE.
MONOTXT–SLANT TO THE LEFT –30°.
ROMANS–SLANT TO THE RIGHT 30°.
ROMAND–BACKWARDS.
ROMANC–VERTICAL.
ITALICC–UNDERSCORED AND OVERSCORED.
ROMANS–USE 16d NAILS @ 10" OC.
ROMANT–Ø32 (812.8).

3. Open P5-10 and add text to the circuit diagram. Use a text style with the Romans font. Save the drawing as P10-3.

▼ Intermediate

4. Create text styles with a .375 height and the following fonts: Arial, BankGothic Lt BT, CityBlueprint, Stylus BT, Swis721 BdOul BT, Vineta BT, and Wingdings. Use the **TEXT** tool to type the complete alphabet and numbers 1–10 for the text styles. Also, type all symbols available on the keyboard and the diameter, degree, and plus/minus symbols. Save the drawing as P10-4.

5. Create the window schedule shown below. Create the text using a text style with the Stylus BT font. Draw the hexagonal symbols in the SYM column. Save the drawing as P10-5.

WINDOW SCHEDULE				
SYM.	SIZE	MODEL	ROUGH OPEN	QTY.
A	12 x 60	JOB BUILT	VERIFY	2
B	96 x 60	W4N5 CSM.	8'-0 3/4" x 5'-0 7/8"	1
C	48 x 60	W2N5 CSM.	4'-0 3/4" x 5'-0 7/8"	2
D	48 x 36	W2N3 CSM.	4'-0 3/4" x 3'-6 1/2"	2
E	42 x 42	2N3 CSM.	3'- 6 1/2" x 3'-6 1/2"	2
F	72 x 48	G64 SLDG.	6'-0 1/2" x 4'-0 1/2"	1
G	60 x 42	G536 SLDG.	5'-0 1/2" x 3'-6 1/2"	4
H	48 x 42	G436 SLDG.	4'-0 1/2" x 3'-6 1/2"	1
J	48 x 24	A41 AWN.	4'-0 1/2" x 2'-0 7/8"	3

6. Create the door schedule shown below. Create the text using a text style with the Stylus BT font. Draw the circle symbols in the SYM column. Save the drawing as P10-6.

DOOR SCHEDULE			
SYM.	SIZE	TYPE	QTY.
1	36 x 80	S.C. R.P. METAL INSULATED	1
2	36 x 80	S.C. FLUSH METAL INSULATED	2
3	32 x 80	S.C. SELF CLOSING	2
4	32 x 80	HOLLOW CORE	5
5	30 x 80	HOLLOW CORE	5
6	30 x 80	POCKET SLDG.	2

7. Create the interior finish schedule shown below. Create the text using a text style with the Stylus BT font. Save the drawing as P10-7.

INTERIOR FINISH SCHEDULE												
ROOM	FLOOR					WALLS				CEILING		
	VINYL	CARPET	TILE	HARDWOOD	CONCRETE	PAINT	PAPER	TEXTURE	SPRAY	SMOOTH	BROCADE	PAINT
ENTRY					•							
FOYER			•			•			•			•
KITCHEN			•					•			•	•
DINING				•		•			•	•		•
FAMILY		•				•			•	•		•
LIVING		•				•		•		•		•
MSTR. BATH			•			•					•	
BATH #2			•			•				•	•	
MSTR. BED		•				•		•		•		•
BED #2		•				•				•		•
BED #3		•				•				•		•
UTILITY	•					•				•	•	•

8. Create the block diagram shown below. Create the text using a text style with the Romans font. Save the drawing as P10-8.

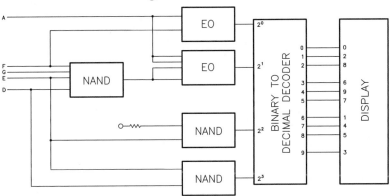

9. Create the block diagram shown below. Create the text using a text style with the Romans font. Use polylines to create the arrowheads. Save the drawing as P10-9.

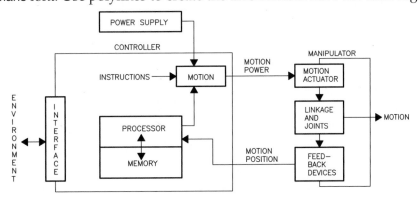

Chapter 10   Single-Line Text and Additional Text Tools

Drawing Problems – Chapter 10

10. Draw the AND/OR schematic shown below. Save your drawing as P10-10.

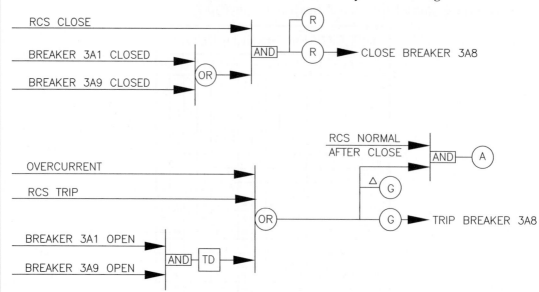

11. Draw the finish schedule shown below. Save your drawing as P10-11.

INTERIOR FINISH SCHEDULE										
ROOM	FLOOR				WALLS				CEIL		
	CARPET	VINYL	TILE	HARDWOOD	PAINT	PAPER	TEXTURE	SPRAY	SMOOTH	BROCADE	PAINT
FOYER			•		•		•			•	•
KITCHEN			•			•		•	•		•
DINING				•	•		•			•	•
FAMILY	•				•		•			•	•
LIVING	•				•		•			•	•
MASTER BED	•				•		•			•	•
MASTER BATH			•				•		•	•	•
BATH 2		•					•		•	•	
BED 2	•				•		•			•	•
BED 3	•				•		•			•	•
UTILITY		•					•		•	•	•

▼ Advanced

12. Create the title block shown below.

R -	CHANGE	DATE	ECN

HYSTER COMPANY

THIS PRINT CONTAINS CONFIDENTIAL INFORMATION WHICH IS THE PROPERTY OF HYSTER COMPANY. BY ACCEPTING THIS INFORMATION THE BORROWER AGREES THAT IT WILL NOT BE USED FOR ANY PURPOSE OTHER THAN THAT FOR WHICH IT IS LOANED.

SPECIFICATIONS

UNLESS OTHERWISE SPECIFIED DIMENSIONS ARE IN ~~INCHES~~ MILLIMETERS AND TOLERANCES FOR:

_____ PLACE DIMS± _____ : _____ PLACE DIMS± _____

ANGLES ± _____ : WHOLE DIMS± _____

DR.	SCALE	DATE
CK. MAT'L.	CK. DESIGN	REL. ON ECN

NAME

MODEL	DWG. FIRST USED	SIMILAR TO

DEPT.	PROJECT	LIST DIVISION

H   PART NO.   R

13. Start a new drawing using the C-size mechanical drawing template located on the Student Web site. Draw a small parts list connected to the title block, similar to the one shown below. Save the drawing as P10-13.

A. Enter PARTS LIST with a style containing a complex font.

B. Enter the other information using text and the **TEXT** tool.

3	HOLDING PINS	12
2	SIDE COVERS	3
1	MAIN HOUSING	1
KEY	DESCRIPTION	QTY

PARTS LIST

UNLESS OTHERWISE SPECIFIED
ALL DIMENSIONS IN

INCHES

AND TOLERANCES FOR:

1 PLACE DIMS: ±.1
2 PLACE DIMS: ±.01
3 PLACE DIMS: ±.005
ANGULAR: ±30'
FRACTIONAL: ±.1/32
FINISH: 125? in.

JANE'S DESIGN

DR: JANE	SCALE: FULL	DATE: XX–XX–XX	APPD:

MATERIAL: MILD STEEL

NAME: XXX–XXXX

FIRST USED ON: SIMILAR TO:

B   PART NO: 123–321   REV: 0

14. Create the architectural title block shown below. Use the same guidelines given for Problem 13. Save the drawing as P10-14.

15. Draw title blocks with borders for your electrical, piping, and general drawings. Use the same guidelines provided in Problem 13. The title block can be similar to the one displayed with Problem 12, but the area for mechanical drafting tolerances is not required. Research sample title blocks to come up with your design. Save the drawings as P10-15A, P10-15B, and P10-15C.

16. Draw the engineering change notice form shown below. Save your drawing as P10-16.

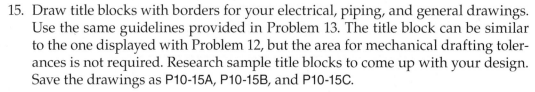

Engineering Change Notice

ECN NO.

Disposition of production stock:
A =Alter or rework U=Use in production
T=Transfer to service stock S=Scrap

Qty.	Drawing Size Part No.	R/N	Description	Change	Other Usage in Production	D/S
01						
02						
03						
04						
05						
06						
07						
08						
09						
10						
11						
12						
13						
14						
15						
16						
17						
18						

Reason:

Castings & forgings affected? ☐ Yes ☐ No	Design engineer:	Supervisor approval:	Release date:	Page

CHAPTER 11

Tables

Learning Objectives

After completing this chapter, you will be able to do the following:

✓ Create and modify table styles.
✓ Insert tables into a drawing.
✓ Edit tables.
✓ Insert formulas into table cells to perform calculations on numeric data.

This chapter describes how to create *tables*, which are common to a variety of drafting applications including bills of materials, door and window schedules, legends, and title block information. Review the tables and terminology shown in **Figure 11-1**. This will help you better understand tables and table information as you read this chapter.

> **table:** An arrangement of rows and columns that organize data to make it easier to read.

Table Styles

Table styles control a variety of table characteristics. You may have several table styles in a single drawing, depending on drawing requirements. Though you can adjust table format independently of a table style, you should create a table style for each unique application. For example, you can have one table style for creating door and window schedules, and another table style with different characteristics for adding

> **table style:** A saved collection of table settings, including direction, text appearance, and margin spacing.

Figure 11-1.
You can create tables in AutoCAD with the title and header rows at the top or the bottom.

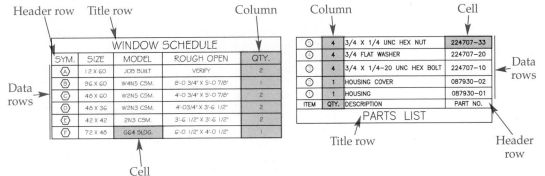

297

an interior finish schedule. In mechanical drafting, you might prepare a table style for parts lists and another table style for gear data tables. Add table styles to drawing templates for repeated use.

Working with Table Styles

TABLESTYLE

Ribbon
Home
> Annotation

Table Style
Annotate
> Tables

Table Style

Type
TABLESTYLE
TS

Create, modify, and delete table styles using the **Table Style** dialog box. See **Figure 11-2.** You can also open the **Table Style** dialog box from the **Insert Table** dialog box, described later in this chapter, by picking the **Launch the Table Style dialog** button.

The **Styles** list box displays existing table styles. The Standard table style is available and current by default. To make a table style current, double-click the style name; right-click the name and select **Set current**; or pick the name and select the **Set current** button. Below the **Styles** list box is a drop-down list that you can use to filter the number of table styles displayed in the **Table Style** dialog box. Pick the **All Styles** option to show all table styles in the file or pick the **Styles in use** option to show only the current style and styles used in the drawing.

Creating New Table Styles

To create a new table style, first select an existing table style from the **Styles** list box to use as a base for formatting the new table style. Then pick the **New...** button in the **Table Style** dialog box to open the **Create New Table Style** dialog box. See **Figure 11-3.** In the **New Style Name** text box, type a name for the new table style. You can base the new table on the formatting settings from a different table style by selecting the name of the table style from the **Start With** drop-down list.

The default new table style name is Copy of followed by the name of the selected existing style. You can keep the default name, but you should usually enter a more descriptive name, such as Parts List, Parts List No Heading, or Door Schedule. Table style names can have up to 255 characters, including uppercase or lowercase letters, numbers, dashes (–), underlines (_), and dollar signs ($). After entering the table style name, pick the **Continue** button to open the **New Table Style** dialog box and adjust table style settings. See **Figure 11-4.**

> **PROFESSIONAL TIP**
>
> It is a good idea to record the names and details about the table styles you create and keep this information in a log for future reference.

Figure 11-2.
The **Table Style** dialog box.

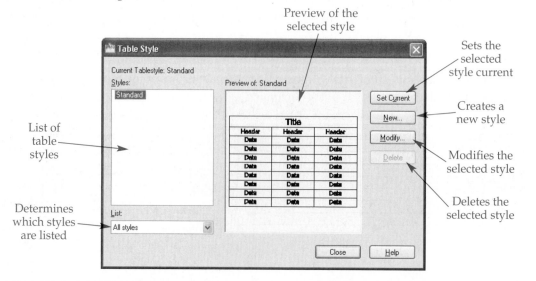

List of table styles

Determines which styles are listed

Preview of the selected style

Sets the selected style current

Creates a new style

Modifies the selected style

Deletes the selected style

Figure 11-3.
In the **Create New Table Style** dialog box, specify the name of the new table style and the existing style that will be copied as a basis for the new style.

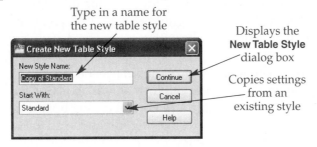

Type in a name for the new table style

Displays the **New Table Style** dialog box

Copies settings from an existing style

Figure 11-4.
The formatting properties for a new style are specified in the **New Table Style** dialog box. The **Data** cell style is shown in the **Cell styles** area in this figure.

Pick to create a starting table style

Pick to remove the starting table reference

Select a cell style

Pick to create a new cell style

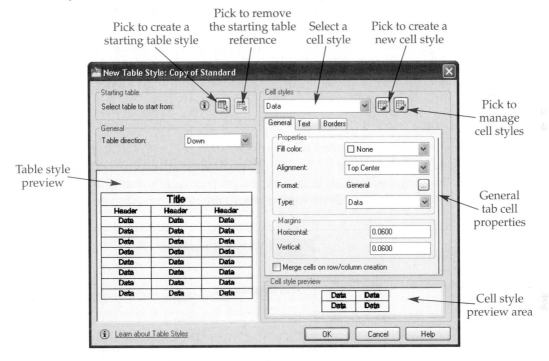

Pick to manage cell styles

Table style preview

General tab cell properties

Cell style preview area

Adjusting Table Direction

The **Table direction** setting in the **General** area of the **New Table Style** dialog box determines the placement of title and header rows and the order of data rows. Select **Down** from the drop-down list to place data rows below the title and header rows. Select **Up** to place data rows above the title and header rows. Refer again to **Figure 11-1**.

Cell Styles

Cell styles allow data cell rows, the column header row, and the title row to have their own formatting properties. Three default cell styles are available in the **Cell Styles** area of the **New Table Style** dialog box: **Data**, **Header**, and **Title**. Pick a cell style from the drop-down list to display the properties for the corresponding element. **Figure 11-4** shows the **Data** cell style selected. Cell formatting properties are set using the **General**, **Text**, and **Borders** tabs. The options in these tabs are the same for adjusting data, header, and title cell styles.

General Tab Settings

The **General** tab, shown in **Figure 11-4,** allows you to set general table characteristics. The **Fill color** drop-down provides options for adjusting the color used to fill cells. You can fill cells with color to highlight or organize table information. The default setting is None, which does not fill cells with a color. The drawing window color determines the on-screen table display. Pick a color from the drop-down list to fill the cells with the selected color. The **Alignment** drop-down list specifies text justification within the cell.

Format shows the current cell format, which is General by default. Pick the ellipsis (...) button to access the **Table Cell Format** dialog box, shown in **Figure 11-5.** The **Data Type** area lists options for formatting the selected table cell. Pick the appropriate format, such as **Text** or **Currency**, to view options for adjusting the format characteristics. Different options are available depending on the selected format.

Use the **Type** drop-down list to specify the cell data type. Pick **Data** to define a data cell type. Choose **Label** if the cell is a label type, such as a column heading or the table title. The **Margins** area provides a **Horizontal** and **Vertical** text box for controlling the horizontal and vertical space between cell content and borders. The default varies depending on the current units, but is .06 (1.5 mm) when decimal units are used.

Pick the **Merge cells on row/column creation** check box to merge the row of cells together to form a single cell. This check box is selected by default for the **Title** cell style. The title cell provides an example of when you may want to merge cells. The title applies to the entire table, or to each column.

Text Tab Settings

The **Text** tab, shown in **Figure 11-6,** allows you to set text characteristics for the selected cell style. The **Text style** drop-down list displays all of the text styles found in the current drawing. Select a style or pick the ellipsis (...) button to the right of the drop-down list to open the **Text Style** dialog box to create or modify a text style.

Figure 11-5.
Many different data types are available to format a table cell.

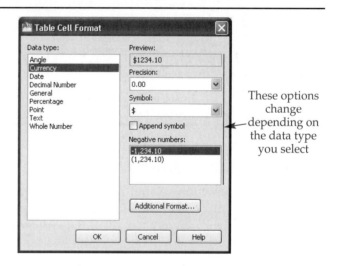

These options change depending on the data type you select

Figure 11-6.
The **Text** tab in the **New Table Style** dialog box allows you to set text properties.

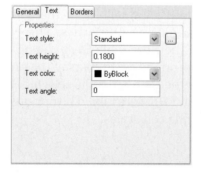

Use the **Text height** text box to specify the height of the text. The default setting for data row and column header cells varies depending on the current units, but is .18 (4.5 mm) when decimal units are used. If you assign a text height other than 0 to the text style, the **Text height** text box is grayed out. Use the **Text color** drop-down list to set the color of the text. The **Text angle** text box controls the rotation angle of text within the table cell. **Figure 11-7** shows an example of a 90° text angle applied to the **Header** cell style.

Borders Tab Settings

The **Borders** tab, shown in **Figure 11-8,** allows you to control the border display and characteristics for the selected cell style. Use the **Lineweight** drop-down list to assign a unique lineweight to cell borders. Use the **Linetype** drop-down list to assign a unique linetype to cell borders. As when creating layers, you must load linetypes if they are not currently loaded in the file in order to apply them to the cell border. Use the **Color** drop-down list to set the color of the cell borders.

Pick the **Double line** check box to add another line around the default single line border style. When checked, the **Spacing** edit box becomes available, allowing you to enter the distance between the double lines. The default spacing varies depending on the current units, but is .045 (1.125 mm) when decimal units are used.

The **Border** buttons control how the **Lineweight**, **Linetype**, **Color**, and **Double line** border properties apply to the cell borders. From right to left, the options are **All Borders**, **Outside Borders**, **Inside Borders**, **Bottom Border**, **Left Border**, **Top Border**, **Right Border**, and **No Borders**. Once you set the desired border properties, select or deselect these buttons according to how you want cell borders to display. **Figure 11-9** shows an example of each border style applied to the data cell borders.

> **NOTE**
>
> Lineweights appear on-screen only if you display lineweights. Pick the **Show/Hide Lineweight** button on the status bar to display lineweights.

Figure 11-7.
In this table, a 90° text angle has been applied to the header cell style.

ROOM SCHEDULE					
NUMBER	NAME	LENGTH	WIDTH	HEIGHT	AREA
1	BEDROOM 1	11'-0"	10'-0"	9'-0"	110 SQ. FT.
2	BEDROOM 2	10'-0"	11'-0"	9'-0"	110 SQ. FT.
3	MASTER BEDROOM	12'-0"	14'-0"	9'-0"	168 SQ. FT.
4	LIVING ROOM	12'-0"	16'-0"	9'-0"	192 SQ. FT.
5	DINING ROOM	11'-0"	12'-0"	9'-0"	132 SQ. FT.
6	KITCHEN	11'-0"	10'-0"	9'-0"	110 SQ. FT.

Figure 11-8.
The **Borders** tab in the **New Table Style** dialog box allows you to set cell border properties.

Figure 11-9.
Several border options are available for table cells. The settings shown are for data row borders.

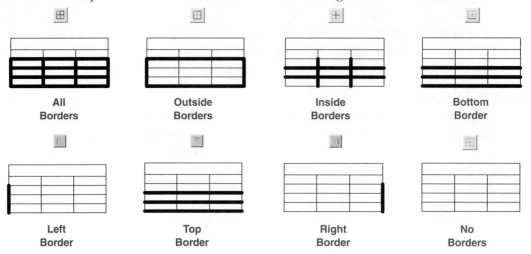

| All Borders | Outside Borders | Inside Borders | Bottom Border |

| Left Border | Top Border | Right Border | No Borders |

Creating Cell Styles

The default **Data**, **Header**, and **Title** cell styles are adequate for typical table applications. However, you can increase the flexibility and options for creating tables by developing additional cell styles. For example, you can create a cell style called Data Yellow that is the same as the **Data** cell style but fills cells with a yellow color. Then when you draw a table, you can choose from either the **Data** or the **Data Yellow** cell style, depending on the application.

To create a new cell style, first select an existing cell style from the **Cell Styles** area drop-down list to use as a base for formatting the new cell style. Then pick the **Create new cell style...** button from the **Cell Styles** area, or select **Create new cell style...** from the **Cell Styles** area drop-down list to display the **Create New Cell Style** dialog box. In the **New Style Name** text box, type a name for the new cell style. You can base the new cell style on the formatting settings of a different cell style by selecting the name of the cell style from the **Start With** drop-down list.

Use the **Manage Cell Styles** dialog box, shown in **Figure 11-10,** to create, rename, and delete cell styles. To access this dialog box, pick the **Manage Cell Style dialog...** button from the **Cell Styles** area, or select **Manage cell styles...** from the **Cell Styles** area drop-down list.

Figure 11-10.
Use the **Manage Cell Styles** dialog box to create new cell styles and to rename and delete existing cell styles.

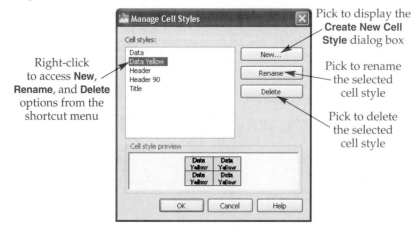

Right-click to access **New**, **Rename**, and **Delete** options from the shortcut menu

Pick to display the **Create New Cell Style** dialog box

Pick to rename the selected cell style

Pick to delete the selected cell style

Exercise 11-1

Access the Student Web site (www.g-wlearning.com/CAD) and complete Exercise 11-1.

Developing a Starting Table Style

One technique for creating a table is to use a starting table style to base a new table on an existing table. You can consider a starting table style a table template that includes preset table style characteristics and specific rows, columns, and data entries. Using a starting table style is much like copying a complete table and editing the table as needed. A starting table style can save time if you often prepare similar tables. For example, use a starting table style if you already created a complete door schedule, and want to add a very similar door schedule to a new drawing project that contains most of the same doors.

Before you can create a starting table style, an existing table must be available in the drawing. A starting table style references the characteristics of a selected table, including the number of columns and rows and the table direction. Other table style characteristics, such as text style, are set according to the selected base table style. As a result, it is usually most appropriate to create a new starting table style using a base table style that is the same as that used to draw the reference table. For example, if you create a door schedule using a table style named Door Schedule, you should base the new starting table style on the Door Schedule table style.

To create a starting table style, pick the **Select table to start from** button in the **Starting table** area. Then pick a border line of the table you want to reference to form the starting table style. The preview displays the selected table and the table style settings of the base table style. See **Figure 11-11.** Modify the table direction and cell style options using the **General** and **Cell Styles** areas. Pick the **Remove Table** button to remove the table reference from the table style. Pick the **Start from Table style** insertion option, described later in this chapter, to add a table to a drawing using a starting table style.

Changing, Renaming, and Deleting Table Styles

To change the characteristics of an existing table style, select the table style and pick the **Modify** button to access the **Modify Table Style** dialog box, which is the same as the **New Table Style** dialog box. If you change the characteristics of an existing table style, all tables added using that style redraw with the new values.

To rename a table style using the **Table Style** dialog box, slowly double-click the name or right-click on the name and select **Rename**. To delete a table style, right-click on the name and choose **Delete**, or pick the style and select the **Delete** button. You cannot delete a table style that is assigned to a table. To delete a style that is in use, assign a different style to the tables that reference the style you want to delete. You cannot delete or rename the Standard style.

Figure 11-11.
Creating a starting table style that references an existing table.

Pick to create a
starting table style

Pick to remove the
starting table reference

Select existing
table

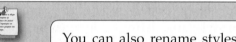

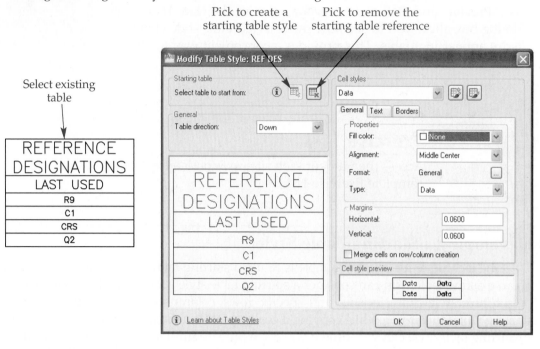

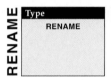

Setting a Table Style Current

You can set a table style current using the **Table Style** dialog box by double-clicking the style in the **Styles** list box, right-clicking the style and selecting **Set current**, or picking the style and selecting the **Set current** button. To set a table style current without opening the **Table Style** dialog box, use the **Table Style** flyout located in the expanded **Annotate** panel of the **Home** ribbon tab, or the **Tables** panel of the **Annotate** ribbon tab. See **Figure 11-12.**

Figure 11-12.
The fastest way to set a style current is to use one of the drop-down lists on the ribbon.

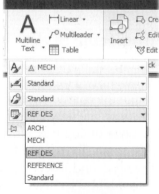

**Access from the
Home Ribbon Tab**

**Access from the
Annotate Ribbon Tab**

You can import table styles from existing drawings using **Design-Center**. See Chapter 5 for more information about using **Design-Center** to reuse drawing content.

Ribbon

Home
> Annotation
Annotate
> Tables

Table

Type
TABLE
TB

TABLE

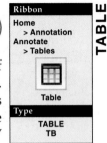

Inserting Tables

The **TABLE** tool allows you to insert an empty table by specifying the number of rows and columns. After you insert the table, you can type text into the table cells. You can also insert blocks and fields into table cells. The **TABLE** tool also provides other methods for inserting tables, such as beginning a table using a starting table style, forming a table from data in an existing Microsoft® Excel spreadsheet or CSV (comma-separated) file, and creating a table by referencing AutoCAD data. Accessing the **TABLE** tool opens the **Insert Table** dialog box, shown in **Figure 11-13**.

Placing an Empty Table

An empty table is the most basic type of table and requires you to specify all table characteristics and content. To place an empty table, first select a table style from the **Table Style** drop-down list, or pick the ellipsis (...) button to create or modify a style. The preview area shows a preview of a table with the current table style settings. The preview area does not adjust to the column and row settings, but it does show table style properties.

Next, pick the **Start from empty table** radio button in the **Insert options** area. The empty table insertion option is set in the **Insertion Behavior** area. Pick the **Specify insertion point** radio button to create a table using the values in the **Column & row settings** area, and then select a single point to place the table in the drawing.

Figure 11-13.
The **Insert Table** dialog box, shown with the **Start from empty table** insert option selected.

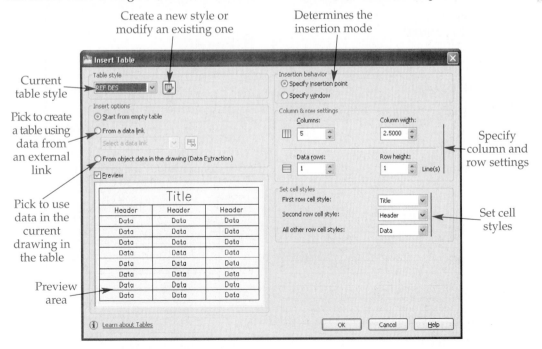

The number in the **Columns** text box determines the total number of table columns. The **Column width** value specifies the initial width of each column. You may want to enter a width larger than necessary in the **Column width** text box and then resize the columns later. The number in the **Data rows** text box determines the total number of data rows. The **Row height** value specifies the initial height of each row based on the number of lines typed and the table style margin settings. When you pick the **OK** button, AutoCAD prompts you to specify the table insertion point. **Figure 11-14A** shows a table created with three columns and five data rows using the **Insertion point** option.

Pick the **Specify window** radio button to create a table that fits within a rectangular area. The radio buttons that appear in the **Column & row settings** area control which column and row settings are active. To set a fixed number of columns, choose the **Columns** radio button. The selected table width determines the width of the specified number of columns. The alternative is to set a fixed column width by selecting the **Column width** radio button. The selected table width determines the total number of columns.

To set a fixed number of rows, choose the **Data rows** radio button. The selected table height determines the height of the specified number of data rows. The alternative is to set a fixed row height by selecting the **Row height** radio button. The selected table height determines the total number of rows.

When you pick the **OK** button, AutoCAD prompts you to select the upper-left and lower-right table corners. The fixed **Column & row settings** values are used and the other settings adjust to fit the window. **Figure 11-14B** shows a table created with three columns and five data rows using the **Specify window** option.

Figure 11-14.
Two ways to insert a table. A—Specifying a single insertion point. B—Windowing an area with two pick points.

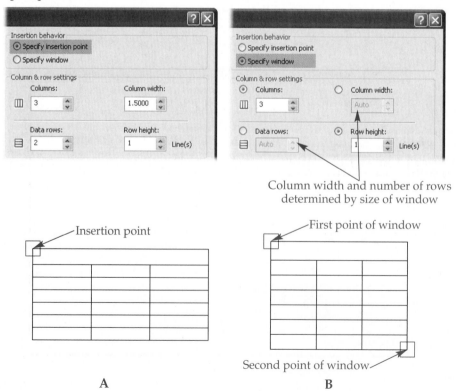

The number of data rows you set in the **Insert Table** dialog box does not include the title and header rows. If you specify 1 data row, for example, the table will have three rows because the top two rows are used for the table title and content headers, depending on cell style settings. The current table style text height and cell margin settings determine the default value for the row height. For example, enter a row height of 1 if you plan to have a single line of text in each cell.

NOTE

You can add, delete, and fully adjust rows and columns as needed. As a result, it is not critical that you enter the exact number and size of columns and rows before inserting the table.

Exercise 11-2

Access the Student Web site (www.g-wlearning.com/CAD) and complete Exercise 11-2.

Adding Cell Content

When you insert a table, the **Text Editor** ribbon tab appears by default, with the text editor cursor in the title cell ready for typing. See **Figure 11-15**. A dashed line around the border and a light gray background indicates the active cell. The *table indicator* identifies individual cells in the table. The identification system helps you to

table indicator: The grid of letters and numbers that appear around the table while the table is being edited to identify individual cells.

Figure 11-15.
The **Text Editor** ribbon tab is used to add and modify text table cell text. The active cell is indicated by a blinking cursor, a dashed border, and a light gray background.

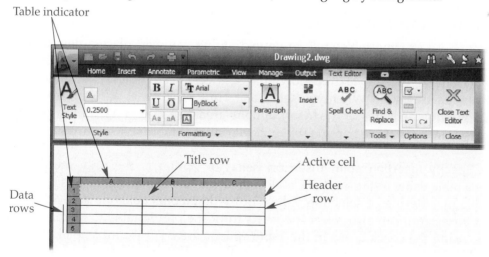

assign formulas to table cells for calculation purposes. Formulas are described later in this chapter. Before adding content to a cell, adjust the text settings in the **Text Editor** ribbon, if needed. Remember, however, that making changes to some text characteristics overrides the settings specified in the text style or table style, which is often not appropriate.

Hold [Alt] and press [Enter] to insert a return within the cell. When you finish entering text in the active cell, press [Tab] to move to the next cell. Hold [Shift] and press [Tab] to move the cursor backward (to the left or up) and make the previous cell active. Press [Enter] to make the cell directly below the current cell active, or exit the text editor if the cursor is at bottom cell. You can also use the arrow keys to navigate through table cells.

When you finish typing in the table, exit the text editor by picking the **Close Text Editor** button in the **Close** panel of the **Text Editor** ribbon tab, or by picking outside of the table. You can also press [Esc] or right-click and select **Cancel** to exit, but you are prompted to save changes. The easiest way to reopen the text editor to make changes to text in a cell is to double-click in the cell. **Figure 11-16** shows a completed table.

NOTE

The options and settings available in the **Text Editor** ribbon tab and shortcut menu function the same in table cells as they do when you edit mtext.

Exercise 11-3

Access the Student Web site (www.g-wlearning.com/CAD) and complete Exercise 11-3.

Using a Starting Table Style

If you add a starting table style to your drawing, you can place a new table by referencing the starting table style. To start from a table style, first select a starting table style from the **Table Style** drop-down list or select the ellipsis (...) button to create or modify a starting style. When you select a starting table style, the **Start from Table Style** radio button activates in the **Insert options** area. See **Figure 11-17**. The preview area shows a preview of the parent table with the current table style settings and table options.

The **Specify insertion point** insertion behavior option is the only method for inserting a table using a starting table style. However, you can add columns and rows to the table by entering or selecting values in the **Additional columns** and **Additional rows** text boxes. You can also select the items from the parent table to include in the new table using the check boxes in the **Table options** area. For example, pick the **Data cell text** check box to create a new table that contains all the text entries added to the data cells of the parent table.

Figure 11-16.
A completed parts list table.

1	PARTS LIST		
2	PART NUMBER	PART TYPE	QTY.
3	100−SCR−45	SCREW	18
4	202−BLT−32	BOLT	18
5	340−WSHR−06	WASHER	18

Figure 11-17.
You can use the **Insert Table** dialog box to create a new table using a starting table style.

Pick to create a new table using a starting table style

Start from Table Style becomes active

Preview of parent table

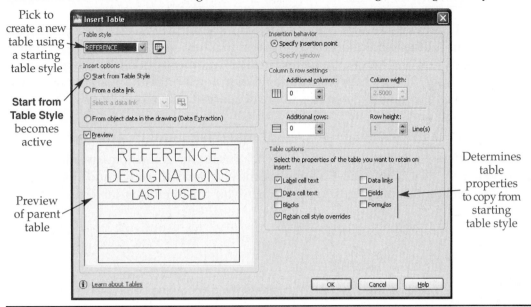

Determines table properties to copy from starting table style

Pick the **OK** button and specify the insertion point of the table. The **Text Editor** ribbon tab appears with the text editor cursor in the title cell ready for adding new content or editing existing values. Exit the text editor when you finish typing text. **Figure 11-18** shows a table created by referencing an existing starting table style, with two additional rows.

Exercise 11-4

Access the Student Web site (www.g-wlearning.com/CAD) and complete Exercise 11-4.

Figure 11-18.
You can create a new table quickly by referencing a starting table style.

Existing table style used to form a starting table style

REFERENCE DESIGNATIONS
LAST USED
R9
C1
CRS
Q2

In the new table, all table options are retained and two rows are added

REFERENCE DESIGNATIONS
LAST USED
R9
C1
CRS
Q2
T4
R6

AutoCAD provides several options for editing existing tables. One option is to re-enter the multiline text editor to edit the text in a table cell. For example, you may need to modify cell content or change text format. Another option is to make changes to the table layout. These changes include adding, removing, and resizing rows and columns, and wrapping table columns to break a large table into sections.

Editing Cell Content

To edit the text in a table cell, double-click inside the cell or pick inside the cell, right-click, and select **Edit Text**. This makes the selected cell active and displays the **Text Editor** ribbon tab. See **Figure 11-19**. When you finish editing table text, exit the text editor.

Exercise 11-5

Access the Student Web site (www.g-wlearning.com/CAD) and complete Exercise 11-5.

Picking Inside a Cell to Edit Table Layout

You can access several table layout settings by picking (single-clicking) *inside* a cell to make the cell active and display the **Table Cell** ribbon tab. The highlighted cell includes *grips*. The **Table Cell** ribbon tab contains options for adjusting table and individual cell layout. The same tools and options found in the **Table Cell** ribbon tab, as well as additional options, are available from a shortcut menu. See **Figure 11-20**. To display the shortcut menu, select a cell and then right-click anywhere in the drawing window. The first section of the shortcut menu contains Windows Clipboard functions. These options affect the entire cell contents. Select **Recent Input** to display a list of previous entries.

grips: Small boxes that appear at strategic points on an object, allowing you to edit the object directly.

Figure 11-19.
Double-click inside a cell to open the text editing function, allowing you to make changes to the text.

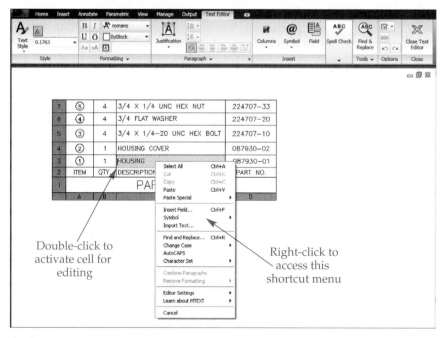

Figure 11-20.
Several table layout modification options are available when you pick inside a cell.

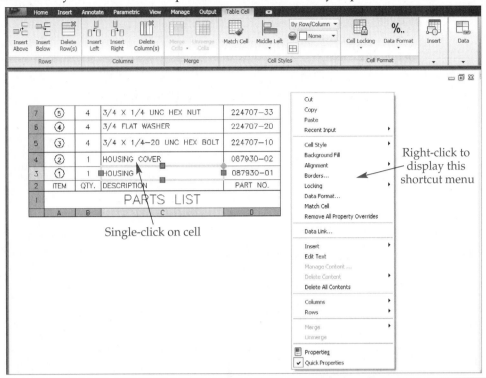

Single-click on cell

Right-click to display this shortcut menu

AutoCAD provides several ways to select multiple cells and apply changes to all the cells at once, as shown in **Figure 11-21.** One option is to pick in a cell and drag the window over the other cells. When you release the pick button, all of the cells touching the window become selected. You can also select multiple cells by picking a cell, holding down [Shift], and then picking another cell. This process selects the picked cells and all of the cells between. You can select entire columns or rows by picking the column or row number in the table indicator. To select the entire table, pick the corner of the table indicator.

> **NOTE**
>
> Make sure you pick completely inside of the cell. If you accidentally select one of the cell borders, the entire table becomes the selected object. Editing table layout by selecting a cell border is described later in this chapter.

Quickly Copying Cell Content

The most effective method for copying the content of one cell to multiple cells is to use the *auto-fill* function. To use auto-fill, pick inside the cell that contains the content you want to copy. The auto-fill grip is a diamond-shaped grip located in a corner of the cell. See **Figure 11-22.** Select the auto-fill grip, as shown in **Figure 11-23A,** and then right-click directly on the auto-fill grip and select an option to define how to adjust the cell content of auto-fill copies.

auto-fill: A table function that automatically fills selected cells based on the contents of a specified cell.

The **Fill Series** option fills cells with the content of the selected cell and applies format overrides. This option also automatically increases or decreases values of certain data types, such as dates, as the fill occurs. See **Figure 11-23B.** The **Fill Series Without Formatting** option fills cells with the content of the selected cell, but does not include format overrides.

Figure 11-21.
Selecting multiple cells in a table for editing. A—Using the pick-and-drag method.
B—Picking a range of cells using [Shift]. C—Selecting a row, column, or the entire table.

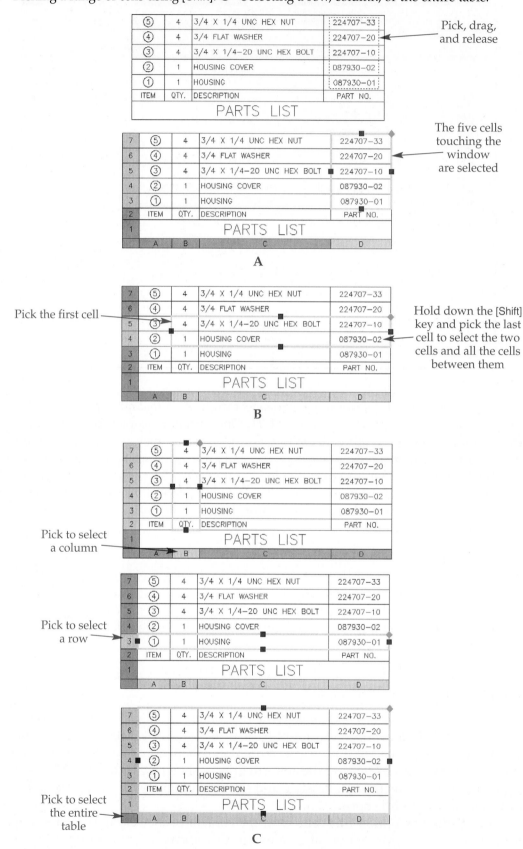

Pick, drag, and release

The five cells touching the window are selected

A

Pick the first cell

Hold down the [Shift] key and pick the last cell to select the two cells and all the cells between them

B

Pick to select a column

Pick to select a row

Pick to select the entire table

C

Figure 11-22.
Using the auto-fill function to copy cell content to multiple cells.

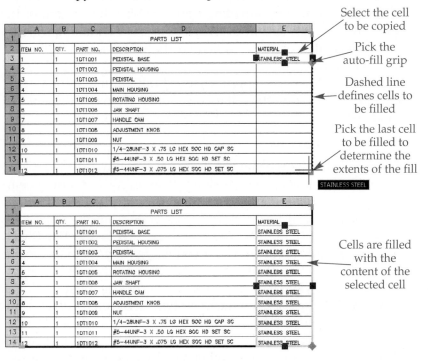

Select the cell to be copied

Pick the auto-fill grip

Dashed line defines cells to be filled

Pick the last cell to be filled to determine the extents of the fill

Cells are filled with the content of the selected cell

Figure 11-23.
Using the auto-fill options to control fill characteristics.

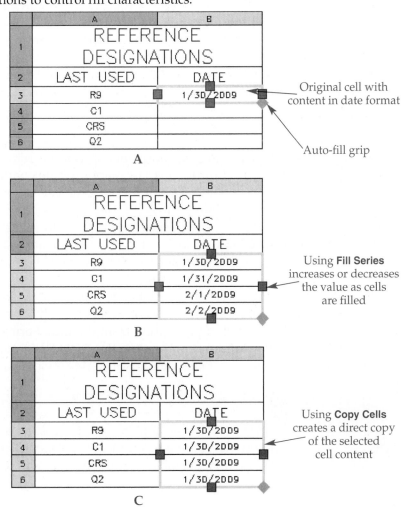

Original cell with content in date format

Auto-fill grip

Using **Fill Series** increases or decreases the value as cells are filled

Using **Copy Cells** creates a direct copy of the selected cell content

The **Copy Cells** option copies the content of the selected cell and applies format overrides, but creates a static cell copy that does not adjust data values. See **Figure 11-23C.** The **Copy Cells Without Formatting** option copies the content of the selected cell, but does not include any format overrides. The **Fill Formatting Only** option fills the cells only with format overrides applied to the selected cell, allowing you to enter cell content manually.

After you choose the appropriate fill option, move the cursor to the last cell to which you want to copy the cell content and pick inside the cell. All cells between the first and last cell fill with the content of the active cell.

Modifying Cell Styles

Most cell style properties are defined according to the table style used to create the table. However, you can override cell styles for individual cells or groups of cells as needed. Cell style options are available from the **Cell Styles** panel of the ribbon and from the shortcut menu. Select a cell style from the **Cell Style** drop-down list to override the cell style used for the active cell. **Create New Cell Style...** and **Manage Cell Style...** options are also available. If you made changes to the active cell, you can save the changes as a new cell style that you can use to adjust the display of other cells or apply to a table style. To save the cell properties as a new cell style, right-click, pick **Save as New Cell Style...** from the **Cell Style** cascading submenu, and enter a name for the style in the **Save as New Cell Style** dialog box.

Pick the **Background Fill** drop-down list to change the cell background color. Select a cell alignment option from the **Alignment** flyout to override the justification of content within the selected cell. Cell content is located in relation to the cell borders. Pick the **Cell Borders** button to open the **Cell Border Properties** dialog box. This dialog box allows you to override cell border display properties and contains the same options found in the **Border** tab of the **Table Style** dialog box.

Select the **Match Cell** button to copy format settings from one cell to another. First select the cell that has the settings you want to copy, and then pick the **Match Cell** button. AutoCAD prompts you to select a destination cell. Pick the cell to which you want to copy the settings. Select another cell or right-click to exit.

Adjusting Cell Format

Cell format options are available from the **Cell Format** panel of the ribbon and from the shortcut menu. The **Cell Locking** flyout lists options for locking cells. You can lock cells to protect them from unintended or inappropriate changes. The locked icon appears when you move the cursor over a locked cell. The **Unlocked** option unlocks the cell so that you can make changes to cell content and format. The **Content Locked** option locks only the content of the cell, allowing you to make changes to the cell format. The **Format Locked** option locks only the cell format, allowing you to make changes to the cell content. The **Content and Format Locked** option locks cell content and format so that you cannot make changes.

The table style you use to create the table defines the data format. However, you can override the format of individual cells or groups of cells as needed. Override the data format of the cells by picking an option from the **Data Format...** flyout, or choose the **Custom Table Cell Format...** option to open the **Table Cell Data Format** dialog box. This is the same dialog box available for adjusting table style data cell format.

PROFESSIONAL TIP

The current table style controls most cell style and format properties. If you plan to make significant changes to cell properties, it is better to modify the table style or create a new style.

Inserting Fields and Blocks

In addition to text, table cells can contain fields, blocks, and formulas. Formulas are described later in this chapter. Blocks are useful for creating a legend or for displaying images of parts in a parts list. The options for inserting a block in a table are briefly described here. Detailed information about blocks is provided later in this textbook. To insert fields, blocks, and formulas, use the tools available from the **Insert** panel of the **Table Cell** ribbon tab, or from the **Insert** cascading submenu of the shortcut menu.

Pick the **Field...** button to open the **Field** dialog box, which allows you to insert a field into a table cell. This is the same dialog box you use to insert fields into mtext and text objects. You can also insert fields into a cell using the mtext text editor.

Select the **Block...** button to insert a block into a table cell using the **Insert a Block in a Table Cell** dialog box. **Figure 11-24** describes the options available in this dialog box. A cell can contain text and blocks. Double-clicking a block in a cell opens the **Insert a Block in a Table Cell** dialog box. Double-clicking on text in a cell opens the **Text Formatting** function.

Adding and Resizing Columns and Rows

You can add, delete, and resize columns and rows after you create a table. The following options are available from the **Columns** panel on the ribbon or the **Columns** cascading submenu in the shortcut menu:

- **Insert Left.** Adds a new column to the left of the selected cell.
- **Insert Right.** Adds a new column to the right of the selected cell.
- **Delete.** Deletes the entire column (or set of columns) containing the selected cell(s).
- **Size Equally.** Automatically sizes the selected columns to the same width. This option is only available when you select cells belonging to multiple columns.

The following options are available from the **Rows** panel of the ribbon or the **Rows** cascading submenu in the shortcut menu:

Figure 11-24.
Options in the **Insert a Block in a Table Cell** dialog box.

Feature	Description
Name	Used to choose the block from a drop-down list of the blocks stored in the current drawing.
Browse	Displays the **Select Drawing File** dialog box, where a drawing file can be selected and inserted into the table cell as a block.
AutoFit	Scales the block automatically to fit inside the cell.
Scale	Sets the block insertion scale. For example, a value of 2 inserts the block at twice its original size. A value of .5 inserts the block at half its created size. The **Scale** option is not available if the **AutoFit** check box is checked.
Rotation angle	Rotates the block to the specified angle.
Overall cell alignment	Determines the justification of the block in the cell and overrides the current cell alignment setting.

- **Insert Above.** Adds a new row above the selected cell.
- **Insert Below.** Adds a new row below the selected cell.
- **Delete.** Deletes the row (or set of rows) containing the selected cell(s).
- **Size Equally.** Automatically sizes the selected rows to the same height. This option is only available when you select cells belonging to multiple rows.

NOTE

A quick way to insert a new row at the bottom of a table using a **Down** table direction is to position the cursor in the lower-right cell and press [Tab]. To insert a new row at the top of a table using an **Up** table direction, position the cursor in the upper-right cell and press [Tab].

PROFESSIONAL TIP

You can also adjust column and row size using grips. Grips are the boxes located in the middle of cell border lines. To resize a column or row, select a grip, move the mouse, and pick. Chapter 15 explains grips in detail.

Merging Cells

Merging allows you to combine adjacent cells. The default tile cell style is an example of a merged cell. Merge tools are available from the **Merge** panel of the ribbon and from the **Merge** cascading submenu in the shortcut menu. To merge cells, select the cells to merge, and then select the appropriate merge option from the **Merge cells** flyout. Select the **All** option to merge all cells into one cell. The **By Row** and **By Column** options allow you to merge cells in multiple rows or columns without removing the horizontal or vertical borders. Pick the **Unmerge Cells** button to separate merged cells back into individual cells.

NOTE

Selecting the **Delete All Contents** option deletes the contents in the selected cell. This is the same as picking a cell and pressing [Delete].

Exercise 11-6

Access the Student Web site (www.g-wlearning.com/CAD) and complete Exercise 11-6.

Picking a Cell Edge to Edit Table Layout

Additional methods for adjusting table layout are available when you pick the edge, or border, of a cell. This displays the table indicator grid, grips you can use to adjust row height and column width, and the table break function. After picking a cell border, right-click anywhere in the drawing window to display the shortcut menu shown in **Figure 11-25**.

Adjusting Table Style

Select a table style from the **Table Style** cascading submenu of the shortcut menu to apply a different table style to the selected table. Picking the **Set as Table in Current Table Style** option creates a starting table style based on the selected table and the current table style. This is an alternative technique for creating a starting table style without opening the **Table Style** dialog box. If the selected table was drawn using a starting table style, selecting the **Set as Table in Current Table Style** redefines the starting table. To save modifications made to the table as a new table style, pick the **Save as New Table Style...** option and enter a name for the style in the **Save as New Table Style** dialog box.

Resizing Columns and Rows

You can use the grips boxes that appear when you select a table cell border to adjust column and row size. Grips are the boxes and arrowheads located at the corners of columns and the table. To resize a column or row, select a grip box, move the cursor, and pick. You can use the arrowhead grips to increase or decrease row height and/or column width uniformly. Chapter 15 explains grips in detail.

The **Size Columns Equally** option sizes all columns to the same width. The total width of the table is divided evenly among the columns. The **Size Rows Equally** option sizes all rows to the same height. All rows increase in height to match the height of the tallest row in the table.

Using Table Breaks

Use the table break function to break a table into separate sections while maintaining a single table object. This is a common requirement when it is necessary to fit a long table in a specific area or on a certain size sheet. Pick a cell edge to access table breaking. The table breaking grip is located midway between the sides of the table at the top or the bottom of the table, depending on the table direction. See **Figure 11-26.**

To break a table, select the table breaking grip and then move the cursor into the table to display a preview of the table sections and a dashed line with crosshairs.

Figure 11-25.
Several additional table layout modification options are available when you pick any cell border.

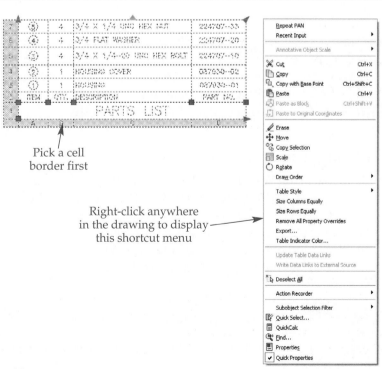

Pick a cell border first

Right-click anywhere in the drawing to display this shortcut menu

Figure 11-26.
The procedure for breaking, or wrapping, a table into sections.

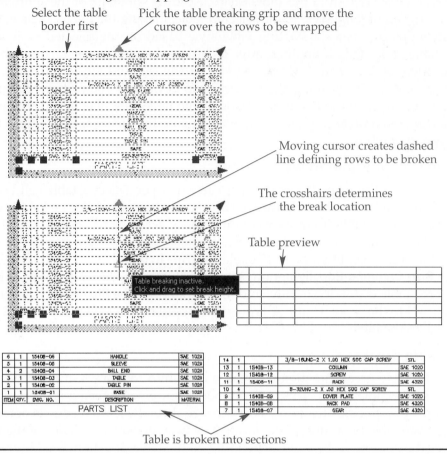

Select the table border first

Pick the table breaking grip and move the cursor over the rows to be wrapped

Moving cursor creates dashed line defining rows to be broken

The crosshairs determines the break location

Table preview

Table breaking inactive. Click and drag to set break height.

Table is broken into sections

The crosshairs determines the location of the break. The closer to the table title and headers the crosshairs appears, the more sections you create, as shown in the table preview. When the preview of the table looks correct, pick the location to form the table breaks.

> **NOTE**
>
> After you add table breaks, several options become available from the **Properties** palette for adjusting the table sections. Chapter 15 covers using the **Properties** palette.

Additional Table Layout Options

The following additional table options are available from the shortcut menu:
- **Remove All Property Overrides.** Restores the table to its original properties, defined according to the selected table style.
- **Export.** Exports the table as a CSV file.
- **Table Indicator Color....** Allows you to change the color of the table indicator shown when you pick inside a cell.

Exercise 11-7

Access the Student Web site (www.g-wlearning.com/CAD) and complete Exercise 11-7.

Calculating Values in Tables

Performing calculations on table data using *formulas* is a common requirement. For example, in a parts list, you can add the number of parts and show the total. In a door or window schedule, you can calculate the total number of doors or windows. A room schedule often includes square footage calculations for various areas. Formulas calculate operations based on numeric data in table cells. AutoCAD allows you to write formulas for sums, averages, counts, and other mathematical functions.

The table indicator grid that appears when you edit a table cell provides a numbering system for the cells. Letters identify columns, and numbers identify rows. Together, the column letter and row number describe cells in formulas. For example, A3 identifies the cell located in Column A, Row 3. **Figure 11-27** illustrates the naming system. Cell C6 is highlighted in the table shown.

Creating Formulas

When you enter a formula in a cell, it evaluates values from other cells and displays the resulting value. Formulas are field objects. As with other types of fields, the expression and the resulting value display a gray background. The value changes if values in the corresponding expression change. This enables you to update data in a table cell automatically when you update the data in other cells.

Typically, you reference table cells with existing numeric values when writing formulas. For example, you can add the values of all cells in a single column and display the total at the bottom of the column, in a new cell. You can define a formula that evaluates a range of continuous cells or cells that do not share a common border.

In a formula, you enter common symbols used for mathematical functions as operators in the expression: + for addition, – for subtraction, * for multiplication, / for division, and ^ for exponentiation. Use parentheses to enclose expressions for table cell formulas. To perform an operation correctly, you must enter the proper syntax in the table cell. The syntax uses the following conventions:

Entry	Description
=	The equal sign is placed at the beginning of a formula. This tells AutoCAD that you want to perform a calculation.
(	An open parenthesis after the equal sign begins the expression.
)	A closing parenthesis ends the expression.
(*expression*)	Write the expression by typing the cells to evaluate and the desired operator symbols.

Figure 11-27. Table cells are identified by column letter and row number. The table indicator grid provides a reference for identifying each cell individually.

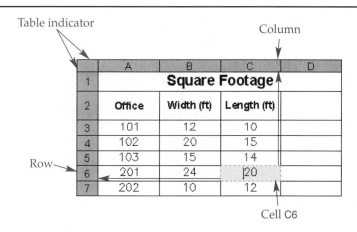

An example of a complete expression is =(C3+D4). This expression tells AutoCAD to add the value of cell C3 to the value of cell D4 and display the sum in the current cell. When identifying a cell in an expression, you must enter the letter before the number. For example, you cannot enter 3C to designate the cell C3. If you enter an incorrect expression or an expression evaluating cells without numeric data, AutoCAD displays the pound character (#) to indicate the error.

NOTE

Parentheses are unnecessary in some expressions, but other expressions will not calculate without them. It is good practice to use parentheses in all expressions.

You can use the text editor to input a formula. **Figure 11-28** shows an example of a multiplication formula. The expression =(B3*C3) is entered in cell D3. The resulting value appears after the text editor is closed.

You can also use grouped expressions in formulas by enclosing expression sets in parentheses. For example, the expression =(E1+F1)*E2 multiplies the sum of E1 and F1 by E2. Another example: =(E1+F1)*(E2+F2)/G6 multiplies the sum of E1 and F1 by the sum of E2 and F2 and divides the product by G6.

Creating Sum, Average, and Count Formulas

In addition to entering basic mathematical formulas in table cells manually, you can select from one of AutoCAD's formulas to calculate the sum, average, or count of a range of cells. Select a cell and choose an option from the **Formula** flyout on the **Table Cell** ribbon tab, or pick **Formula** from the **Insert** cascading submenu of the shortcut menu to display a cascading menu with formula options. See **Figure 11-29.**

The **Sum** option allows you to add the values of a range of cells by specifying a selection window on-screen. AutoCAD prompts you to pick the first corner of a window defining the cell range. The range you specify can include cells from several columns and rows. Pick inside the top or bottom cell that you want to include in the calculation. Then move the cursor and pick inside the lowest or highest cell, making sure that all of the desired cells are included in the window selection. See **Figure 11-30.** When you select the second point, the expression automatically appears in the cell. In the example

Figure 11-28.
Entering a multiplication formula. A—Type the expression in the table cell using the correct syntax. B—The resulting value displays after the expression is calculated.

	A	B	C	D
1	Square Footage			
2	Office	Width (ft)	Length (ft)	Sq Ft
3	101	12	10	=(B3*C3)
4	102	20	15	
5	103	15	14	
6	201	24	20	
7	202	10	12	

← Expression

A

Square Footage			
Office	Width (ft)	Length (ft)	Sq Ft
101	12	10	120
102	20	15	
103	15	14	
201	24	20	
202	10	12	

← Result

B

Figure 11-29.
The **Formula** flyout
on the **Table Cell**
ribbon tab allows
you to insert a
formula into a cell.

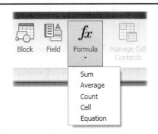

Figure 11-30.
Creating a sum
formula in a table
cell. A—Pick a cell
to hold the formula
and select a range of
cells for the formula
by windowing
around the cells. B—
After you pick the
second point of the
window, the formula
displays in the cell.

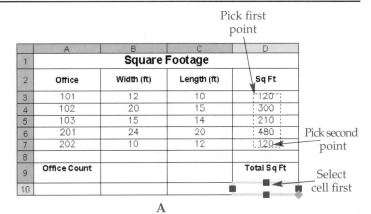

shown, the square footage for each office is first calculated in the Sq Ft column on the right in **Figure 11-30A.** The values are then selected for a sum formula that calculates the total square footage of all of the offices, as shown in **Figure 11-30B.**

Notice in **Figure 11-30B** that the resulting expression is =Sum(D3:D7) This formula specifies that the selected cell is equal to the sum of cells D3 through D7. The colon (:) indicates the range of cells for the calculation.

The **Average** option creates a formula that calculates the average value of the cells you select. The average is the sum of the selected cells divided by the number of cells selected. The **Count** option creates a formula that counts the number of selected cells. Only cells that contain a value are included in the count.

You can type sum, average, and count formulas directly into a cell without using the **Formula** flyout button or the **Insert** cascading menu. If you calculate a value over a range of cells, use the colon symbol to designate the range. You can also write an expression that evaluates individual cells instead of a range. The cells do not have to share a common border. To write an expression using nonadjacent cells, use a comma to separate the cell names. For example, to average cells D1, D3, and D6, type =Average(D1,D3,D6). This formula calculates the average of the cell values for cells D1, D3, and D6.

Figure 11-31.
Sum, average, and
count formulas
and their resulting
values.

	A	B	C	D
1		Square Footage		
2	Office	Width (ft)	Length (ft)	Sq Ft
3	101	12	10	120
4	102	20	15	300
5	103	15	14	210
6	201	24	20	480
7	202	10	12	120
8				
9	Office Count	Average Sq Ft Per Room		Total Sq Ft
10	5	246.000000		1230

=Count(A3:A7)   =Average(D3:D7)   =Sum(D3:D7)

You can include a range of cells and individual cells in the same expression. For example, to count cells A1 through B10 in addition to cells C4 and C6, enter =Count(A1: B10,C4,C6). **Figure 11-31** shows examples of sum, average, and count formulas.

NOTE

When using architectural units in a drawing, you can type the foot (') and inch (") symbols in table cells for use in values and formulas. When you use the foot symbol for a cell value, a formula in another cell automatically converts the resulting value to inches and feet.

Other Formula Options

The **Formula** flyout button and **Insert** cascading submenu contain additional options for writing table formulas. The **Cell** option allows you to select a table cell from a different table and insert its contents in the current cell. You can then use the cell value in a new formula. When you select the **Cell** option, AutoCAD prompts you to select the cell. The value of the selected cell appears in the current cell.

The **Equation** option allows you to enter an expression manually. Selecting this option places an equal sign (=) in the current cell. You can then type the expression.

NOTE

You can use the **Field** tool to insert and edit table cell formulas. Selecting **Formula** from the **Field names** list in the **Field** dialog box displays option buttons for creating sum, average, and count formulas. You can also select a cell value from a different table as a starting point. Select table cells on-screen to define the formula. You can use the **Formula** text box in the **Field** dialog box to add to or edit the formula. Unit format options are also available.

Exercise 11-8

Access the Student Web site (www.g-wlearning.com/CAD) and complete Exercise 11-8.

Supplemental Material *Linking a Table to Excel Data*
For information about using existing data entered in a Microsoft® Excel spreadsheet or a CSV file to create an AutoCAD table, go to the Student Web site (www.g-wlearning.com/CAD), select this chapter, and select **Linking a Table to Excel Data**.

Supplemental Material *Extracting Table Data*
For information about using existing AutoCAD text to create a table, go to the Student Web site (www.g-wlearning.com/CAD), select this chapter, and select **Extracting Table Data**.

Template Development For detailed instructions on adding table styles to each drawing template, go to the Student Web site (www.g-wlearning.com/CAD), select this chapter, and select **Template Development**.

Chapter 11

Chapter Test

Answer the following questions. Write your answers on a separate sheet of paper or go to the Student Web site (www.g-wlearning.com/CAD) and complete the electronic chapter test.

1. What is the purpose of creating a table style?
2. Briefly describe the procedure for creating a table style based on an existing table style.
3. What does the **Alignment** setting in the **New Table Style** dialog box do?
4. Which setting would you adjust in the **New Table Style** dialog box to increase the spacing between the text and the top of the cell?
5. How can creating a new table using a starting table style save time?
6. How can you make a table style current without opening the **Table Style** dialog box?
7. List two ways to open the **Insert Table** dialog box.
8. Describe the two ways to insert a table and explain how the methods differ.
9. By default, what two types of rows are at the top of a table when it is inserted?
10. What ribbon tab opens when you insert a table?
11. If you are finished typing in one cell and want to move to the next cell in the same row, what two keyboard keys can you use?
12. List two ways to make a cell active for editing.
13. Explain how to insert a block into a table cell.
14. How can you insert a new row quickly at the bottom of a table?
15. How are table cells identified in formulas?
16. Write the table cell formula that adds the value of C3 plus the value of D4.
17. What is the function of the colon symbol (:) in the formula =Sum(D3:D7)?
18. What is the difference between a sum formula and a count formula?
19. Write the table cell formula that averages the values of cells D1, D3, and D6.
20. Explain how to write a formula that calculates a function for cells that do not share common borders.

Drawing Problems

Follow these instructions to complete the drawing problems for this chapter:

- *Start AutoCAD if it is not already started.*
- *Start a new drawing using an appropriate template of your choice. The template should include layers, text styles, and table styles when necessary, for drawing the given tables.*
- *Add layers, text styles, and table styles as needed. Draw all objects using appropriate layers, text styles, table styles, justification, and format.*
- *Follow the specific instructions for each problem. Use your own judgment and approximate dimensions when necessary.*
- *Make the measurements for rows and columns approximately the same as in the given table.*

▼ Basic

1. Create the parts list shown below. Save the drawing as P11-1.

	1	CAPS	1/2–12UNC HEX NUT	210014–29
	1	CAPS	1/2 FLAT WASHER	320014–33
	2	CAPS	7/16 EXTERNAL SNAP RING	632043–43
	2	CAPS	1/4–20UNC WING NUT	255010–41
	2	CAPS	3/4 X 1/4–20UNC BOLT	803010–11
KEY	QTY	NAME	DESCRIPTION	PART NO.

PARTS LIST

2. Create the table shown below. Save the drawing as P11-2.

REFERENCE DESIGNATIONS	
LAST USED	DATE
R9	1/30/2009
C1	1/30/2009
CRS	1/30/2009
Q2	1/30/2009

▼ Intermediate

3. Create the door schedule shown below. Save the drawing as P11-3.

DOOR SCHEDULE			
SYM.	SIZE	TYPE	QTY.
①	36x80	S.C. RP. METAL INSULATED	1
②	36x80	S.C. FLUSH METAL INSULATED	2
③	32x80	S.C. SELF CLOSING	2
④	32x80	HOLLOW CORE	5
⑤	30x80	HOLLOW CORE	5
⑥	30x80	POCKET SLDG.	2

4. Create the window schedule shown below. Save the drawing as P11-4.

SYM.	SIZE	MODEL	ROUGH OPEN	QTY.
Ⓐ	12x60	JOB BUILT	VERIFY	2
Ⓑ	96x60	W4N5 CSM.	8'–0 3/4" x 5'–0 7/8"	1
Ⓒ	48x60	W2N5 CSM.	4'–0 3/4" x 5'–0 7/8"	2
Ⓓ	48x36	W2N3 CSM.	4'–0 3/4" x 3'–6 1/2"	2
Ⓔ	42x42	2N3 CSM.	3'–6 1/2" x 3'–6 1/2"	2
Ⓕ	72x48	G64 SLDG.	6'–0 1/2" x 4'–0 1/2"	1
Ⓖ	60x42	G536 SLDG.	5'–0 1/2" x 3'–6 1/2"	4
Ⓗ	48x42	G436 SLDG.	4'–0 1/2" x 3'–6 1/2"	1
Ⓙ	48x24	AA1 AWN.	4'–0 1/2" x 2'–0 7/8"	3

WINDOW SCHEDULE

5. Create the door schedule shown below. Save the drawing as P11-5.

SYMBOL	SIZE	MODEL	QUANTITY	SYMBOL	SIZE	MODEL	QUANTITY
1	3'-0" X 6'-8"	S.C. R.P. METAL INSULATED	1	11	4'-0" X 6'-8"	BI-FOLD	1
2	3'-0" X 6'-8"	S.C.-FLUSH-METAL INSULATED	2	12	2'-0" X 6'-0"	SHATTER PROOF	1
3	2'-8" X 6'-8"	S.C.-SELF CLOSING	2	13	6'-0" X 6'-8"	WOOD FRAME-TEMP. SLDG GL.	1
4	2'-8" X 6'-8"	H.C.	5	14	9'-0" X 7'-0"	OVERHEAD GARAGE	2
5	2'-6" X 6'-8"	H.C.	3				
6	2'-6" X 6'-8"	POCKET	2				
7	2'-4" X 6'-8"	POCKET	1				
9	5'-0" X 6'-0"	BI-PASS	2				
10	3'-0" X 6'-8"	BI-FOLD	1				

DOOR SCHEDULE   DOOR SCHEDULE

6. Open P11-5 and make the following changes:
   - Change the DOOR SCHEDULE so that SYMBOL items 11, 12, 13, and 14 continue directly below SYMBOL items 1 through 10, with only one DOOR SCHEDULE heading at the top.
   - Change the following abbreviations to full words:
   - S.C.R.P. – SOLID CORE RAISED PANEL
   - S.C. – SOLID CORE
   - H.C. – HOLLOW CORE

   Save the drawing as P11-6.

▼ Advanced

7. Create a parts list for a mechanical drawing with the content of your choice, or locate a drawing with a parts list and make a similar drawing. Save the drawing as P11-7.

8. Create a door and window schedule for an architectural drawing with the content of your choice, or locate a drawing with a door and window schedule and make a similar drawing. Save the drawing as P11-8.

9. Create a legend for a civil drawing with the content of your choice, or locate a drawing with a legend and make a similar drawing. Save the drawing as P11-9.

10. Create a parts list for a mechanical drawing with the content of your choice, or locate a drawing with a parts list and make a similar drawing. Save the drawing as P11-10.

Drawing Problems - Chapter 11

11. Open P11-1 and make the following changes:
    - Change PARTS LIST to PURCHASE PARTS LIST.
    - Change Part Number 803010-11 as follows: Key: 7, Name: HEX HD, Description: $\frac{1}{4}$20UNC-2 X $\frac{3}{4}$ BOLT.
    - Change Part Number 255010-41 as follows: Key: 11, Name: WING NUT, Description: $\frac{1}{4}$20UNC.
    - Change Part Number 632043-43 as follows: Key: 15, Name: SNAP RING, Description: Ø $\frac{7}{16}$ EXTERNAL.
    - Change Part Number 320014-33 as follows: Key: 19, Name: WASHER, Description: Ø $\frac{1}{2}$ FLAT.
    - Change Part Number 210014-29 as follows: Key: 21, Name: NUT, Description: $\frac{1}{2}$-12UNC-2 HEX.

    Save the drawing as P11-11.

12. Create a table of your own design and use at least six of the applications of calculating values in tables described in this chapter. Save the drawing as P11-12.

13. Access the Student Web site content for this chapter and review Supplement 11A, "Linking a Table to Excel Data." Use this information to create a table by linking to Microsoft® Excel data. To do this, create your own Excel spreadsheet or find an existing Excel spreadsheet containing the desired data. Link a table to the Excel data. Save the drawing as P11-13.

14. Access the Student Web site content for this chapter and review Supplement 11B, "Extracting Table Data." Use the description to create a table the same as or similar to the given examples and then extract the table data into an AutoCAD drawing. Save the drawing as P11-14.

CHAPTER 12

Basic Object Editing Tools

Learning Objectives

After completing this chapter, you will be able to do the following:

✓ Use the **FILLET** tool to draw fillets, rounds, and other rounded corners.
✓ Place chamfers and angled corners with the **CHAMFER** tool.
✓ Separate objects using the **BREAK** tool and combine objects using the **JOIN** tool.
✓ Use the **TRIM** and **EXTEND** tools to edit objects.
✓ Modify objects using the **STRETCH** and **LENGTHEN** tools.
✓ Change the size of objects using the **SCALE** tool.
✓ Use the **EXPLODE** tool.

This chapter explains methods for changing objects using basic editing tools. You will learn to use various editing tools to increase drawing efficiency. The editing tools described in this chapter include many options. As you work through this chapter, experiment with each option to see which is the most effective in different situations.

Using the FILLET Tool

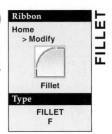

Drafting terminology refers to a rounded interior corner as a *fillet*, and a rounded exterior corner as a *round*. AutoCAD refers to all rounded corners as fillets. The **FILLET** tool draws a rounded corner between intersecting and nonintersecting lines, circles, arcs, and polylines. See **Figure 12-1.**

Setting Fillet Radius

After initiating the **FILLET** tool, use the **Radius** option to enter the fillet radius dimension. The fillet radius determines the size of a fillet and must be set before you select objects. Once you specify the radius, select the objects to fillet. The specified fillet radius is stored as the new default radius, allowing you to place additional fillets of the same size.

fillet: A rounded interior corner used to relieve stress or ease the contour of inside corners.

round: A rounded exterior corner used to remove sharp edges or ease the contour of exterior corners.

Figure 12-1.
Using the **FILLET** tool.

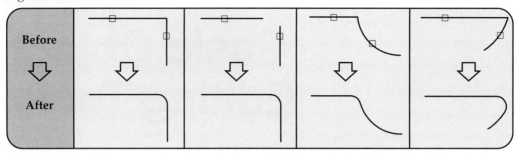

Exercise 12-1

Access the Student Web site (www.g-wlearning.com/CAD) and complete Exercise 12-1.

Forming Sharp Corners

Use a fillet radius of 0 to connect two objects at a sharp corner. You can also create a zero-radius fillet without setting the radius to 0 by holding down the [Shift] key when you pick the second object. This is a convenient way to connect objects at a corner, or to form a square corner if edges are perpendicular.

Filleting Parallel Lines

You can use the **FILLET** tool to draw a full radius between parallel lines. When you set the **Trim** option to **Trim**, a longer line trims to match the length of a shorter line. The radius of a fillet between parallel lines is always half the distance between the two lines, regardless of the radius setting. You can use this method to create a full radius, such as the end radii applied to a slot.

Rounding Polyline Corners

You can use the **Polyline** option to fillet all corners of a closed polyline. See **Figure 12-2.** Remember to set the appropriate radius before filleting. If you drew the polyline without using the **Close** option, the beginning corner does not fillet, as shown in **Figure 12-2.**

Fillet Trim Settings

The **Trim** option controls whether the **FILLET** tool trims object segments that extend beyond the fillet radius point of tangency. See **Figure 12-3.** Use the default **Trim** setting to trim objects. When you set the **Trim** option to **No Trim**, the fillet occurs, but filleted objects do not change.

Figure 12-2.
Using the **Polyline** option of the **FILLET** tool.

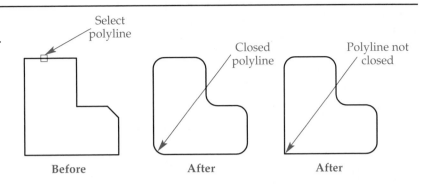

Select polyline

Closed polyline

Polyline not closed

Before   After   After

Figure 12-3.
Comparison of the **Trim** and **No trim** options of the **FILLET** tool.

Before Fillet	Fillet with Trim	Fillet with No Trim

PROFESSIONAL TIP

You can fillet objects even when the corners do not meet. If the **Trim** option is set to **Trim,** objects extend as required to generate the fillet and complete the corner. If the **Trim** option is set to **No Trim**, objects do not extend to complete the corner.

Making Multiple Fillets

Use the **Multiple** option to make several fillets without exiting the **FILLET** tool. The prompt for a first object repeats. To exist, press [Enter], the space bar, [Esc], or right-click and select **Enter**. When you use **Multiple** mode, use the **Undo** option to discard the previous fillet.

Exercise 12-2

Access the Student Web site (www.g-wlearning.com/CAD) and complete Exercise 12-2.

Ribbon

Home
> Modify

Chamfer

Type
CHAMFER
CHA

CHAMFER

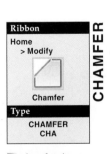

Using the CHAMFER Tool

Drafting terminology often refers to a *chamfer* as a small, angled surface used to relieve a sharp corner. The **CHAMFER** tool allows you to draw an angled corner between intersecting and nonintersecting lines, polylines, xlines, and rays. Chamfer size is determined based on the distance from the corner. A 45° chamfer is the same distance from the corner in each direction. See **Figure 12-4.** Typically, two distances or one distance and one angle identify the size of a chamfer. The defaults are zero units for the length and the angle. A value of .5 for both distances produces a 45° × .5 chamfer.

chamfer: In mechanical drafting, a small, angled surface used to relieve a sharp corner.

Figure 12-4.
Examples of chamfers.

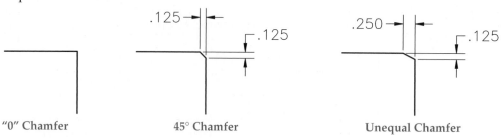

| "0" Chamfer | 45° Chamfer | Unequal Chamfer |

Setting Chamfer Distances

After initiating the **CHAMFER** tool, use the **Distance** option to enter the chamfer distances. Chamfer distances determine the size of a chamfer from a corner and must be set before you select objects. Once you specify the distances, select the objects to chamfer. **Figure 12-5** shows several chamfering operations. The specified chamfer distances are stored as the new default distances, allowing you to place additional chamfers of the same size.

Setting the Chamfer Angle

Instead of setting two chamfer distances, you can use the **Angle** option to set the chamfer distance for one edge and set an angle to determine the chamfer to the second edge. See **Figure 12-6.** After entering the distance and angle, select the two objects to chamfer. The specified distance and angle remain active until changed, allowing you to place additional chamfers of the same size.

Figure 12-5.
Using the **CHAMFER** tool.

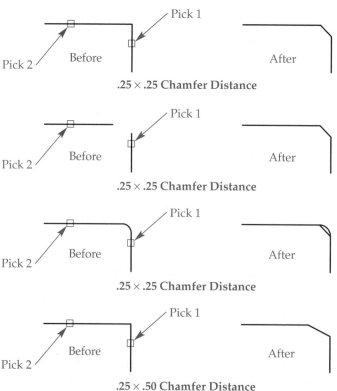

Figure 12-6.
Using the **Angle** option of the **CHAMFER** tool with the chamfer length set at .5 and the angle set at 45°.

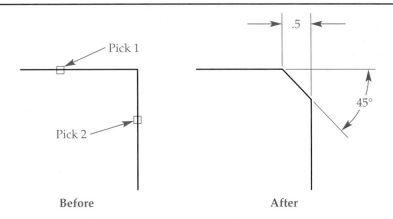

Before

After

Setting the Chamfer Method

AutoCAD maintains the specified chamfer distances, or a distance and an angle, until you change the values. You can set the values for each method without affecting the other. Use the **Method** option to toggle between drawing chamfers using the **Distance** and **Angle** options.

PROFESSIONAL TIP

You can use the **CHAMFER** tool to form sharp corners by specifying chamfer distances or an angle and distance of 0; or by holding [Shift] when you pick the second object.

NOTE

Only corners large enough to accept the specified chamfer size are eligible for chamfering. If the chamfer is too large, AutoCAD displays a message, such as Distance is too large *Invalid*.

Exercise 12-3

Access the Student Web site (www.g-wlearning.com/CAD) and complete Exercise 12-3.

Additional Chamfer Options

The **CHAMFER** tool includes the same **Polyline**, **Trim**, and **Multiple** options available with the **FILLET** tool, and similar rules apply when using these options with the **CHAMFER** tool. Use the **Polyline** option to chamfer all corners of a closed polyline, as shown in **Figure 12-7**. The **Trim** option controls whether the **CHAMFER** tool trims object segments that extend beyond the intersection, as shown in **Figure 12-8**. Use the **Multiple** option to make several chamfers without exiting the **CHAMFER** tool.

Exercise 12-4

Access the Student Web site (www.g-wlearning.com/CAD) and complete Exercise 12-4.

Figure 12-7.
Using the **Polyline** option of the **CHAMFER** tool. If you drew the polyline without using the **Close** option, the beginning corner does not chamfer.

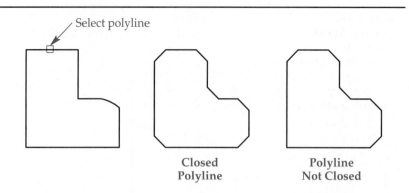

Select polyline

Closed Polyline

Polyline Not Closed

Figure 12-8.
Use the default **Trim** setting to trim objects. When you set the **Trim** option to **No Trim**, the chamfer occurs, but chamfered objects do not change.

Before Chamfer	Chamfer with Trim	Chamfer with No Trim

Using the BREAK Tool

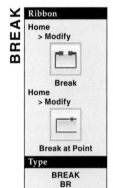

BREAK

Ribbon
Home
> Modify

Break

Home
> Modify

Break at Point

Type
BREAK
BR

You can use the **BREAK** tool to separate a single object into two objects. A break can remove a portion of an object or split the object at a single point, depending on the selected points. The **BREAK** tool requires you to select the object to break and the first and second break points. By default, the point you pick when you select the object to break also locates the first break point. To select a different first break point, use the **First point** option at the Specify second break point or [First point]: prompt. The portion of the object between the two points deletes. See **Figure 12-9.**

If you select the same point for the first and second break points, the **BREAK** tool splits the object into two pieces without removing a portion. You can accomplish this by entering @ at the Specify second break point or [First point]: prompt. The @ symbol repeats the coordinates of the previously selected point. You can also pick the **Break at Point** button on the expanded **Modify** panel of the **Home** ribbon tab. **Figure 12-10** shows the process of breaking without removing a portion of the object.

Always work in a counterclockwise direction when breaking arcs or circles. Otherwise, you may break the portion of the arc or circle you want to keep. If you want to break off the end of a line or an arc, pick the first point on the object. Pick the second point slightly beyond the end to be cut off. See **Figure 12-11.** When you pick a second point not on the object, AutoCAD selects a point on the object nearest the point you pick.

Figure 12-9.
Using the **BREAK** tool to break an object. You can use the first pick to select both the object and the first break point.

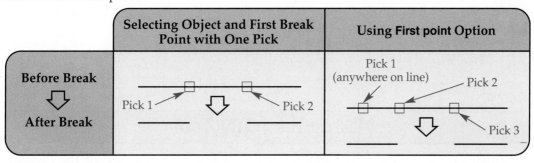

Figure 12-10.
Using the **BREAK** tool to break an object at a single point without removing any of the object. Select the same point as the first and second break points, or use the **Break at Point** button.

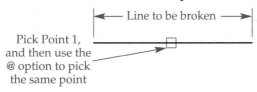

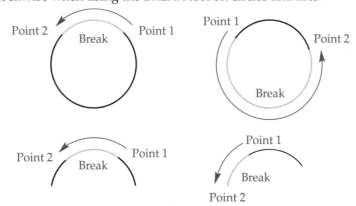

Figure 12-11.
Work counterclockwise when using the **BREAK** tool on circles and arcs.

PROFESSIONAL TIP

Use object snaps to pick a point accurately when using the **First point** option of the **BREAK** tool. However, in some cases it is necessary to turn running object snaps off if they conflict with points you are trying to pick.

Exercise 12-5

Access the Student Web site (www.g-wlearning.com/CAD) and complete Exercise 12-5.

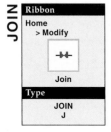
Using the JOIN Tool

Often multiple objects that should be one object form because of the drawing and editing process. These multiple objects make the drawing file size larger and the drawing more cumbersome. You can use the **JOIN** tool to join lines, polylines, splines, arcs, and elliptical arcs together to make one object. You can only join objects of the same type. For example, you can join a line to another line, but you cannot join a line to a polyline. In addition, joined objects must be in the same 2D plane.

Each object type has different rules for joining. Lines must be collinear, but they can touch, overlap, or have gaps between segments. See **Figure 12-12.** Polylines and splines must share a common endpoint and cannot have gaps between segments or overlap. **Figure 12-13** shows rules for joining polylines. The same rules apply for joining splines.

Arcs and elliptical arcs must share the same center point and circular path, but they can touch, overlap, or have gaps between segments. **Figure 12-14** shows joining two arcs separated by a gap. The same rules apply for joining elliptical arcs. Pick arcs or elliptical arcs in a clockwise direction to close the nearest clockwise gap, or pick in a counterclockwise direction to close the nearest counterclockwise gap. Depending on your selections, you may receive a prompt to convert arcs to a circle. If you do not want to form a circle, choose the **No** option and reselect the arcs in a counterclockwise direction.

Figure 12-12.
Lines must be collinear to join, but the lines can overlap, and there can be gaps between segments.

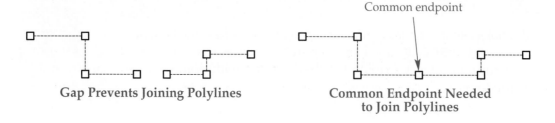

Two Collinear Lines

One Line after Joining

Figure 12-13.
You can join polylines only if they share an endpoint. The same rule applies to joining splines.

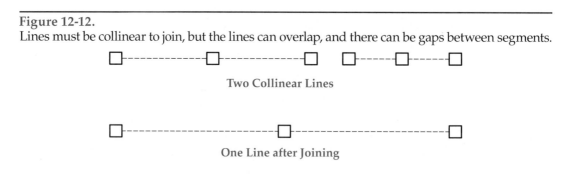

Common endpoint

Gap Prevents Joining Polylines

Common Endpoint Needed to Join Polylines

AutoCAD and Its Applications—Basics

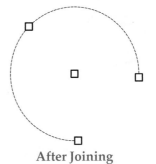

Figure 12-14.
Arcs can have a gap or be overlapping, as shown on the left, but they must share the same circular path. On the right, the two arcs are joined.

Before Joining   **After Joining**

NOTE

After selecting an arc or elliptical arc segment to join, you can use the **cLose** option to form a circle from the arc, or to form an ellipse from the elliptical arc. This option closes the two ends of the selected segment and does not join the segment with any other objects.

Trimming Objects

The **TRIM** tool cuts lines, polylines, circles, arcs, ellipses, splines, xlines, and rays that extend beyond a desired point of intersection. Once you access the **TRIM** tool, pick as many *cutting edges* as necessary and then right-click or press [Enter] or the space bar. Then pick the objects to trim to the cutting edges. To exit, right-click or press [Enter] or the space bar. See **Figure 12-15.**

The **TRIM** tool presents specific **Crossing** and **Fence** options that function the same as standard crossing and fence selection overrides, described in Chapter 3. **Figure 12-16** shows using the **Crossing** option to trim multiple objects. **Figure 12-17** shows an example of using the **Fence** option to trim multiple objects. However, automatic windowing with the crossing function is often the quickest and most effective method for trimming multiple objects. You can also use window or crossing polygons.

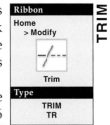

Ribbon
Home
> Modify

Trim

Type
TRIM
TR

cutting edge: An object such as a line, an arc, or text that defines the point (edge) at which the object you trim will be cut.

NOTE

To access the **EXTEND** tool while using the **TRIM** tool, after selecting the cutting edge(s), hold [Shift] and pick objects to extend to the cutting edge. The **EXTEND** tool is described later in this chapter.

Figure 12-15.
Using the **TRIM** tool. Note the cutting edges.

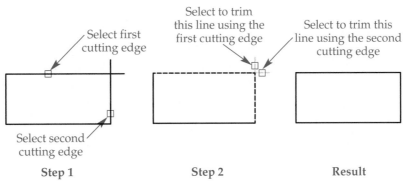

Select first cutting edge

Select to trim this line using the first cutting edge

Select to trim this line using the second cutting edge

Select second cutting edge

Step 1   Step 2   Result

Figure 12-16.
The only objects trimmed with the **Crossing** option are those that cross the edges of the crossing window. Automatic windowing accomplishes the same task.

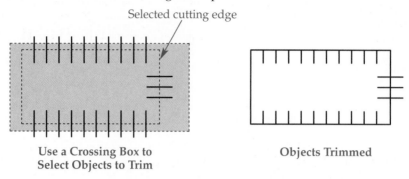

Selected cutting edge

Use a Crossing Box to Select Objects to Trim

Objects Trimmed

Figure 12-17.
The **Fence** option allows you to select around objects. In this case, the **RECTANGLE** tool was used to create the rectangle, so the cutting edge consists of the entire rectangle.

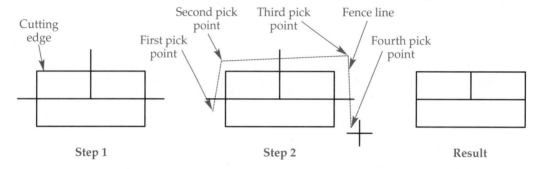

Cutting edge

Second pick point

Third pick point

Fence line

First pick point

Fourth pick point

Step 1   Step 2   Result

Trimming without Selecting a Cutting Edge

To trim objects to the nearest intersection without selecting a cutting edge, access the **TRIM** tool and, at the first Select objects or <select all>: prompt, right-click or press [Enter] or the space bar instead of picking a cutting edge. Then pick the objects to trim. You can continue selecting objects to trim without restarting the **TRIM** tool. To exit, right-click or press [Enter], the space bar, or [Esc].

Trimming to an Implied Intersection

implied intersection: The point at which objects would meet if they were extended.

Trimming to an *implied intersection* is possible using the **Edge** option of the **TRIM** tool. Access the **TRIM** tool, pick the cutting edges, and then select the **Extend** option. The **No extend** mode is active by default, and as a result, you cannot trim objects that do not intersect. Choose the **Extend** mode to recognize implied intersections, and then pick objects to trim. See **Figure 12-18**. This does not change the selected cutting edges.

> **NOTE**
>
> The **TRIM** tool includes additional options. Use the **eRase** option to erase objects selected to trim. Use the **Undo** option to restore previously trimmed objects without leaving the tool. You must activate the **Undo** option immediately after performing an unwanted trim. The **Project** option applies to trimming 3D objects. *AutoCAD and Its Applications—Advanced* explains drawing and editing 3D objects.

Figure 12-18.
Trimming to an
implied intersection
with the **Extend**
mode active.

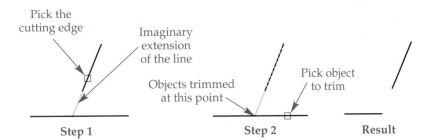

Pick the
cutting edge

Imaginary
extension
of the line

Pick object
to trim

Objects trimmed
at this point

| Step 1 | Step 2 | Result |

Extending Objects

The **EXTEND** tool allows you to extend lines, elliptical arcs, rays, open polylines, and arcs to meet another object. **EXTEND** does not work on closed polylines because an unconnected endpoint does not exist. Once you access the **EXTEND** tool, pick as many *boundary edges* as necessary, and then right-click or press [Enter] or the space bar. Then pick the objects to extend to the boundary edges. To exit, right-click or press [Enter] or the space bar. See **Figure 12-19**.

Like the **TRIM** tool, the **EXTEND** tool presents specific **Crossing** and **Fence** options. **Figure 12-20** shows using the **Crossing** option to extend multiple objects. **Figure 12-21** shows an example of using the **Fence** option to extend multiple objects. However, automatic windowing with the crossing function is often the quickest and most effective method for extending multiple objects. You can also use window or crossing polygons.

The **EXTEND** tool includes the same **Edge**, **Undo**, and **Project** options available for the **TRIM** tool, and similar rules apply when using these options with the **EXTEND** tool. **Figure 12-22** illustrates how to combine the **EXTEND** and **TRIM** tools, without selecting

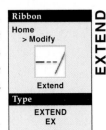

Ribbon

Home
> Modify

Extend

Type

EXTEND
EX

EXTEND

boundary edge:
The edge to which
objects such as
lines, arcs, and
polylines are
extended.

Figure 12-19.
Using the **EXTEND** tool. Note the boundary edges.

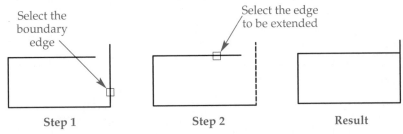

Select the
boundary
edge

Select the edge
to be extended

| Step 1 | Step 2 | Result |

Figure 12-20.
Selecting objects to extend using the **Crossing** option. Automatic windowing accomplishes the same task.

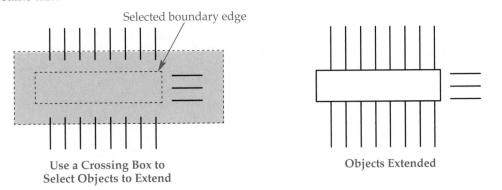

Selected boundary edge

**Use a Crossing Box to
Select Objects to Extend**

Objects Extended

Figure 12-21.
Extending multiple lines to a boundary edge using the **Fence** option.

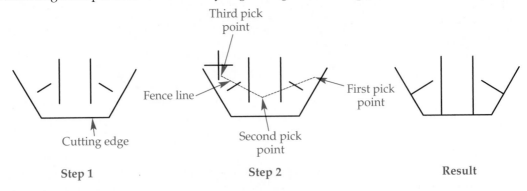

Third pick point

Fence line

First pick point

Cutting edge

Second pick point

Step 1   Step 2   Result

Figure 12-22.
To extend objects to the nearest intersection without selecting a boundary edge, right-click or press [Enter] or the space bar instead of picking a boundary edge. Then pick object(s) to extend. Hold down the [Shift] key to toggle between extending and trimming.

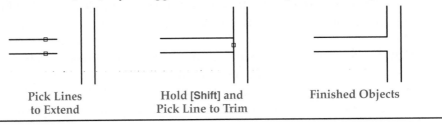

Pick Lines
to Extend

Hold [Shift] and
Pick Line to Trim

Finished Objects

a boundary edge, to insert a wall in a floor plan. You can apply this process to a variety of applications.

Use the **Extend** mode of the **Edge** option to extend to an implied intersection, as shown in **Figure 12-23**. Select the **Undo** option immediately after performing an unwanted extend to restore previous objects without leaving the tool. The **Project** option applies to extending 3D objects, as explained in *AutoCAD and Its Applications—Advanced.*

PROFESSIONAL TIP

Construction lines created using the **XLINE** and **RAY** tools are modified using standard editing tools. When you trim one infinite end of an xline, the object becomes a ray. When you trim both infinite ends of an xline, or the infinite end of a ray, the object becomes a line object. Therefore, in many cases, you can modify construction lines to become a portion of the actual drawing. This approach can save a significant amount of time for a variety of applications.

Figure 12-23.
Extending to an implied intersection with the **Extend** mode.

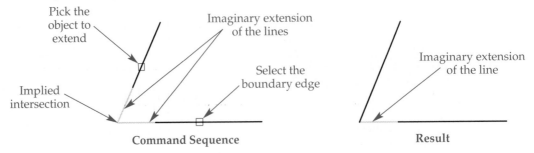

Pick the object to extend

Imaginary extension of the lines

Imaginary extension of the line

Implied intersection

Select the boundary edge

Command Sequence   Result

Exercise 12-6

Access the Student Web site (www.g-wlearning.com/CAD) and complete Exercise 12-6.

Stretching Objects

Ribbon

Home
> Modify

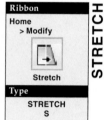

Stretch

Type

STRETCH
S

STRETCH

The **STRETCH** tool allows you to modify certain dimensions of an object while leaving other dimensions the same. In mechanical drafting, for example, you can stretch a screw body to create a longer or shorter screw. In architectural design, you can stretch room sizes to increase or decrease square footage.

Once you access the **STRETCH** tool, you must use a crossing box or polygon to select only the objects to be stretched. This is a very important requirement and is different from selection using other editing tools. See **Figure 12-24**. If you select using the pick box or a window, the **STRETCH** tool works like the **MOVE** tool, described in Chapter 14.

After selecting the objects to stretch, specify the *base point* from which the objects will stretch. Although the position of the base point is often not critical, you may want to select a point on an object, the corner of a view, or the center of a circle. As you move the crosshairs, the selection stretches or compresses. Pick a second point to complete the stretch.

base point: The initial reference point AutoCAD uses when stretching, moving, copying, and scaling objects.

PROFESSIONAL TIP

Use object snap modes to your best advantage while editing. For example, to stretch a rectangle to make it twice as long, use the **Endpoint** object snap to select the endpoint of a rectangle for the base point, and another **Endpoint** object snap to select the opposite endpoint of the rectangle.

Figure 12-24.
Using the **STRETCH** tool.

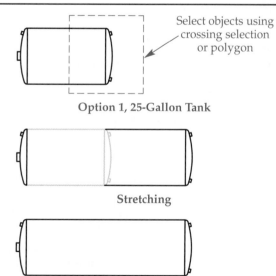

Select objects using crossing selection or polygon

Option 1, 25-Gallon Tank

Stretching

Option 2, 50-Gallon Tank

Using the Displacement Option

The **Displacement** option allows you to stretch objects relative to the origin, or 0,0,0 point. To stretch using a *displacement*, access the **STRETCH** tool and use a crossing box or polygon to select only the objects to stretch. Then select the **Displacement** option instead of defining the base point. At the Specify displacement <0,0,0>: prompt, enter an absolute coordinate to stretch the objects from the origin to the coordinate point. See **Figure 12-25.**

Using the First Point As Displacement

Another method for stretching an object is to use the first point as the displacement. This means the coordinates you use to select the base point automatically define the coordinates for the direction and distance for stretching the object. To apply this technique, access the **STRETCH** tool and use a crossing box or polygon to select only the objects to stretch. Then specify the base point, and instead of locating the second point, right-click or press [Enter] or the space bar to accept the <use first point as displacement> default. See **Figure 12-26.**

Figure 12-25.
Using the **Displacement** option of the **STRETCH** tool. A—An example of a 1 × 1 rectangle to stretch. B—Stretching the rectangle using a 1,0 displacement.

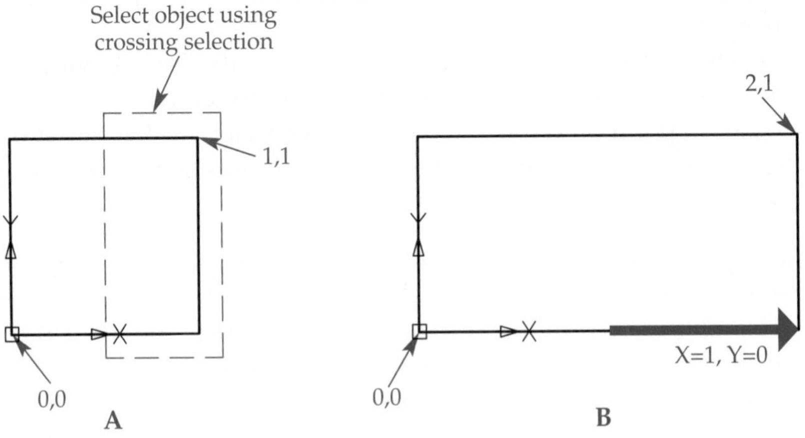

Figure 12-26.
A—An example of a 1 × 1 rectangle to stretch. B—Stretching using the selected base point (1,1) as the displacement.

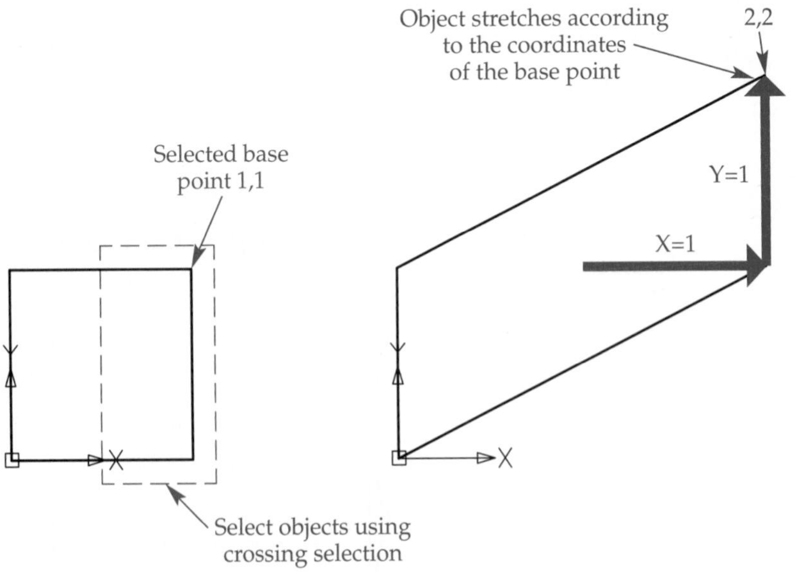

AutoCAD and Its Applications—Basics

Objects often do not line up in a convenient manner for using crossing box selection to pick objects to stretch. Consider using crossing polygon selection to make selection easier. If the stretch is not as expected, press [Esc] to cancel. The **STRETCH** tool and other editing tools work well with polar tracking or **Ortho** mode.

Exercise 12-7

Access the Student Web site (www.g-wlearning.com/CAD) and complete Exercise 12-7.

Using the LENGTHEN Tool

You can use the **LENGTHEN** tool to change the length of objects and the included angle of an arc. The **LENGTHEN** tool does not affect closed objects. For example, you can lengthen a line, a polyline, an arc, an elliptical arc, or a spline, but you cannot lengthen a closed ellipse, polygon, or circle. You can only lengthen one object at a time.

Once you access the **LENGTHEN** tool, select the object to change. AutoCAD gives you the current length if the object is linear or the included angle if the object is an arc. Choose one of the four options and follow the prompts. The **DElta** option allows you to specify a positive or negative change in length, measured from the endpoint of the selected object. The lengthening or shortening happens closest to the selection point and changes the length by the amount entered. See **Figure 12-27**. The **DElta** option has an **Angle** function that lets you change the included angle of an arc according to a specified angle. See **Figure 12-28**.

The **Percent** option allows you to change the length of an object or the angle of an arc by a specified percentage. The original length is 100 percent. Make the object shorter by specifying less than 100 percent or longer by specifying more than 100 percent. See **Figure 12-29**.

The **Total** option allows you to set the total length or angle of the object after the **LENGTHEN** operation. See **Figure 12-30**. The **DYnamic** option lets you drag the endpoint of the object to the desired length or angle using the crosshairs. See **Figure 12-31**. It is

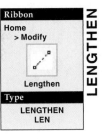

Ribbon

Home
> Modify

Lengthen

Type
LENGTHEN
LEN

LENGTHEN

Figure 12-27.
Using the **DElta** option of the **LENGTHEN** tool with values of .75 and –.75.

Select the object closest to the end you want lengthened or shortened

Original Object

.75

Lengthened by an Increment of .75

–.75

Shortened by an Increment of –.75

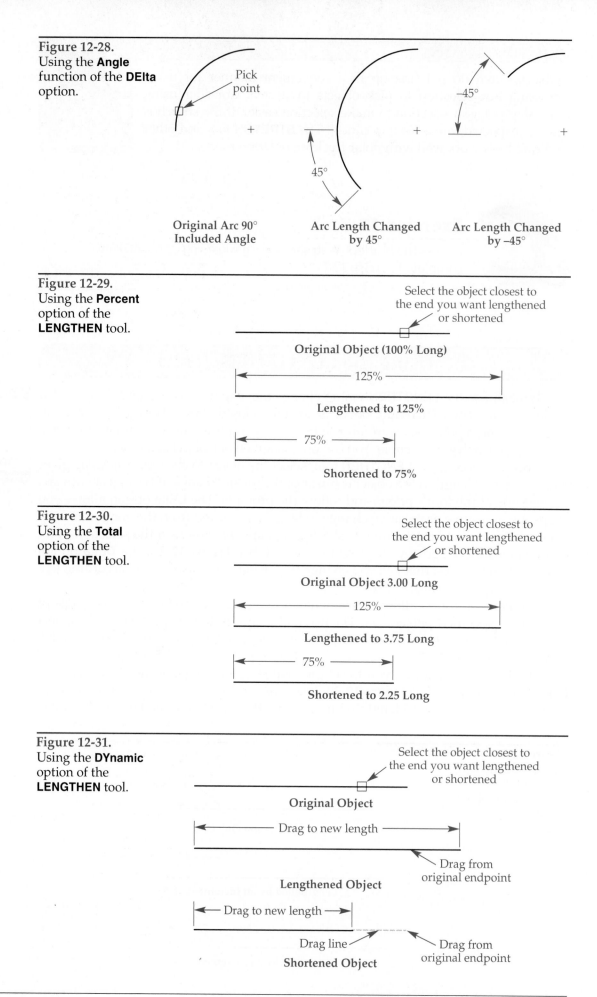

Figure 12-28.
Using the **Angle** function of the **DElta** option.

Pick point

−45°

45°

Original Arc 90° Included Angle

Arc Length Changed by 45°

Arc Length Changed by −45°

Figure 12-29.
Using the **Percent** option of the **LENGTHEN** tool.

Select the object closest to the end you want lengthened or shortened

Original Object (100% Long)

125%

Lengthened to 125%

75%

Shortened to 75%

Figure 12-30.
Using the **Total** option of the **LENGTHEN** tool.

Select the object closest to the end you want lengthened or shortened

Original Object 3.00 Long

125%

Lengthened to 3.75 Long

75%

Shortened to 2.25 Long

Figure 12-31.
Using the **DYnamic** option of the **LENGTHEN** tool.

Select the object closest to the end you want lengthened or shortened

Original Object

Drag to new length

Lengthened Object

Drag from original endpoint

Drag to new length

Drag line

Drag from original endpoint

Shortened Object

helpful to use dynamic input with polar tracking or **Ortho** mode, or have the grid and snap set to usable increments when using this option.

NOTE

You can only lengthen lines and arcs dynamically, and you can only decrease the length of a spline.

PROFESSIONAL TIP

You do not have to select the object before entering one of the **LENGTHEN** tool options, but doing so lets you know the current length and, if it is an arc, the angle of the object. This is especially helpful when you are using the **Total** option.

Exercise 12-8

Access the Student Web site (www.g-wlearning.com/CAD) and complete Exercise 12-8.

Using the SCALE Tool

The **SCALE** tool allows you to proportionately enlarge or reduce the size of objects. After you access the **SCALE** tool, pick a base point to define where the increase or decrease in size occurs. The selected objects move away from or toward the base point during scaling. The next step is to specify the scale factor. Enter a number to indicate the amount of enlargement or reduction. For example, to make the selection twice the current size, type 2 at the Specify scale factor or [Copy/Reference] <current>: prompt, as shown in **Figure 12-32. Figure 12-33** provides examples of scale factors.

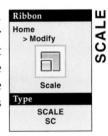

Ribbon
Home > Modify
Scale
Type
SCALE SC

SCALE

Using the Reference Option

You can use the **Reference** option instead of entering a scale factor by specifying a new size in relation to an existing dimension. For example, suppose you want to proportionately change the size of a part with an overall dimension of 2.50″ to an overall dimension of 3.00″. Access the **Reference** option and enter the current length, in this case 2.5, at the Specify reference length: prompt. Next, enter the length you want the object to be, in this example 3. See **Figure 12-34.**

Figure 12-32.
Using the **SCALE** tool. The base point does not move, but every other point in the object does.

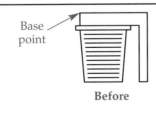
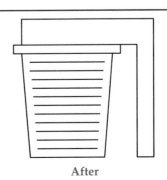

Base point

Before

After

Figure 12-33.
Scale factors and the resulting sizes.

Scale Factor	Resulting Size
10	10 times bigger
5	5 times bigger
2	2 times bigger
1	Equal to existing size
.75	3/4 of original size
.50	1/2 of original size
.25	1/4 of original size

Figure 12-34.
Using the **Reference** option of the **SCALE** tool.

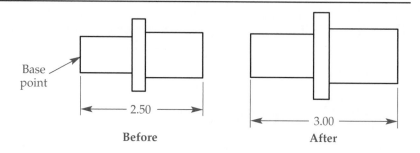

Base point

2.50

Before

3.00

After

PROFESSIONAL TIP

Specify the current and reference lengths using specific values; or choose points, often on existing objects. Picking points is especially effective when you do not know the exact current and reference lengths.

Copying While Scaling

The **Copy** option of the **SCALE** tool copies and scales the selected object, leaving the original object unchanged. The copy is moved to a location you specify.

NOTE

The **SCALE** tool changes all dimensions of an object proportionately. Use the **STRETCH** or **LENGTHEN** tool to change only the length, width, or height.

Exercise 12-9

Access the Student Web site (www.g-wlearning.com/CAD) and complete Exercise 12-9.

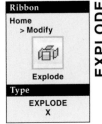
Exploding Objects

Ribbon
Home
> Modify

Explode

Type
EXPLODE
X

The **EXPLODE** tool allows you to change a single object that consists of multiple items into a series of individual objects. For example, you can explode a polyline object into individual line and arc segments that you can edit individually. Explode a multi-line text object to convert each line of text to a single-line text object. You can explode a variety of other objects, including multilines, regions, dimensions, leaders, and blocks. This textbook explains these objects when appropriate.

Access the **EXPLODE** tool, pick the object to explode, and right-click or press [Enter] or the space bar to cause the explosion. **Figure 12-35** shows an example of exploding a polyline object. In this example, the polyline becomes two collinear arcs with no polyline width or tangency information. Exploded polyline lines and arcs occur along the centerline of the original polyline.

Exercise 12-10

Access the Student Web site (www.g-wlearning.com/CAD) and complete Exercise 12-10.

Figure 12-35.
Exploding a polyline converts the object into individual lines and arcs and removes all polyline information.

Original Polyline   Exploded Polyline

Chapter Test

Answer the following questions. Write your answers on a separate sheet of paper or go to the Student Web site (www.g-wlearning.com/CAD) and complete the electronic chapter test.

1. How do you specify the size of a fillet?
2. Explain how to set the radius of a fillet to .50.
3. Which option of the **CHAMFER** tool would you use to specify a .125 × .125 chamfer?
4. What is the purpose of the **Method** option in the **CHAMFER** tool?
5. Describe the difference between the **Trim** and **No trim** options when using the **CHAMFER** and **FILLET** tools.
6. How can you split an object in two without removing a portion?
7. In what direction should you pick points to break a portion out of a circle or arc?
8. What tool can you use to combine two collinear lines into a single line object?
9. What two requirements must be met before two arcs can be joined?
10. Which tool performs the opposite function of the **EXTEND** tool?
11. Name the tool that trims an object to a cutting edge.
12. Name the tool associated with boundary edges.
13. Name the option in the **TRIM** and **EXTEND** tools that allows you to trim or extend to an implied intersection.

14. Which panel of the ribbon contains the **TRIM**, **EXTEND**, and **STRETCH** tools?
15. List two locations drafters normally choose as the base point when using the **STRETCH** tool.
16. Define the term *displacement*, as it relates to the **STRETCH** tool.
17. Identify the **LENGTHEN** tool option that corresponds to each of the following descriptions:
    A. Allows a positive or negative change in length from the endpoint.
    B. Changes a length or an arc angle by a percentage of the total.
    C. Sets the total length or angle to the value specified.
    D. Drags the endpoint of the object to the desired length or angle.
18. What tool would you use to reduce the size of an entire drawing by one-half?
19. Write the command aliases for the following tools:
    A. **CHAMFER**
    B. **FILLET**
    C. **BREAK**
    D. **TRIM**
    E. **EXTEND**
    F. **SCALE**
    G. **LENGTHEN**
20. Which tool removes all width characteristics and tangency information from a polyline?

Drawing Problems

Follow these instructions to complete the drawing problems for this chapter:

- *Start AutoCAD if it is not already started.*

- *Start a new drawing using an appropriate template of your choice. The template should include layers and text styles when necessary for drawing the given objects.*

- *Add layers and text styles as needed. Draw all objects using appropriate layers and text styles, justification, and format.*

- *Follow the specific instructions for each problem. Do not draw dimensions. Use your own judgment and approximate dimensions when necessary.*

▼ **Basic**

1. Draw Object A using the **LINE** and **ARC** tools. Make sure the corners overrun and the arc is centered on the lines, but does not touch the lines. Use the **TRIM**, **EXTEND**, and **STRETCH** tools to make Object B. Save the drawing as P12-1.

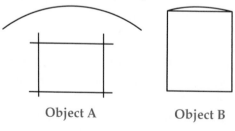

Object A   Object B

2. Open drawing P12-1 for further editing (Object A). Using the **STRETCH** tool, change the shape to create Object B. Save the drawing as P12-2.

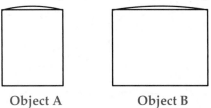

Object A   Object B

3. Refer to **Figure 12-24** in this chapter. Draw the object shown in Option 1. Stretch the object to twice its length, as shown in Option 2. Stretch the object again to one and a half times its length. *Hint:* use endpoint and midpoint object snap modes to stretch accurately. Save the drawing as P12-3.

4. Draw the object shown. Use the **CHAMFER** tool to create the inclined surface. Save the drawing as P12-4.

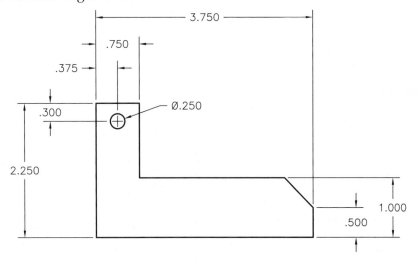

▼ Intermediate

5. Draw the object shown using the **ELLIPSE** and **LINE** tools. Use the **BREAK** or **TRIM** tool when drawing and editing the lower ellipse. Save the drawing as P12-5.

6. Open drawing P12-4 for further editing. Shorten the height of the object using the **STRETCH** tool, as shown below. Next, add to the object as indicated. Save the drawing as P12-6.

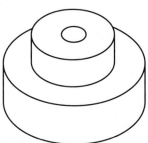

7. Use the **TRIM** and **OFFSET** tools to assist you in drawing this object. Do not draw centerlines or dimensions. Save the completed drawing as P12-7.

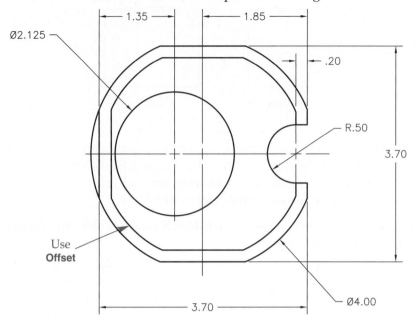

8. Draw the following plate. Use the **FILLET** tool where appropriate. Save the drawing as P12-8.

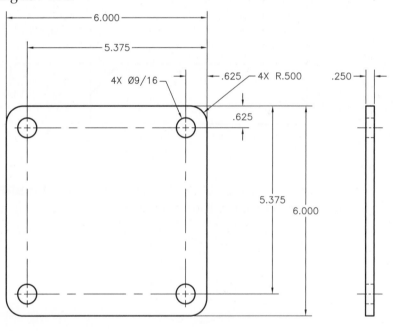

9. Draw the following toilet. Do not include dimensions. Use dimensions of your choice for objects that are not fully dimensioned. Save the drawing as P12-9.

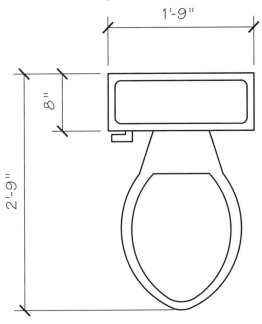

▼ Advanced

10. Draw the following object. Add rounds using the **FILLET** tool and chamfers using the **CHAMFER** tool. Use the trim mode setting to your advantage. Save the drawing as P12-10.

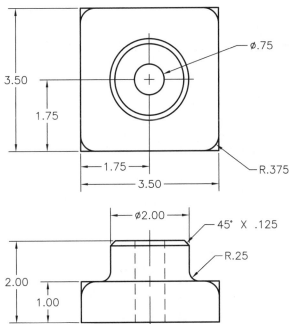

11. Draw the following bracket. Use the **FILLET** tool where appropriate. Save the drawing as P12-11.

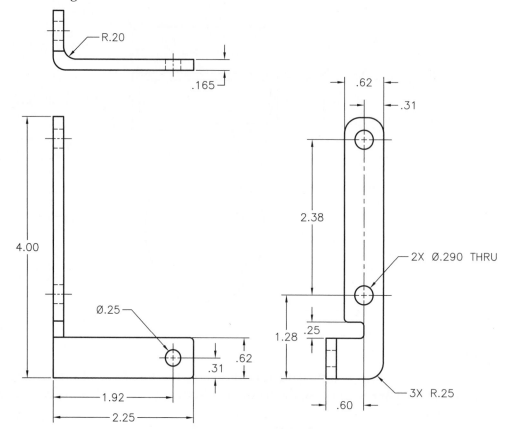

12. Draw the beam wrap detail shown. Save the drawing as P12-12.

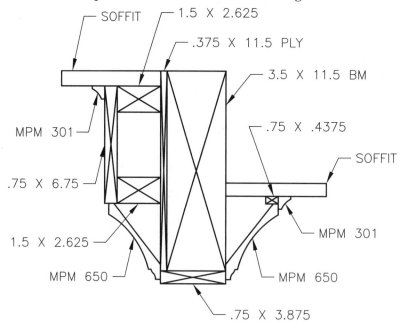

CHAPTER 13

Polyline and Spline Editing Tools

Learning Objectives

After completing this chapter, you will be able to do the following:

✓ Edit polylines with the **PEDIT** tool.
✓ Create polyline boundaries.
✓ Edit splines with the **SPLINEDIT** tool.
✓ Convert polylines and splines.

You can modify polyline and spline objects using standard editing tools such as **ERASE**, **STRETCH**, and **SCALE**. In addition to normal editing practices, you can modify polylines using the **PEDIT** tool and adjust splines using the **SPLINEDIT** tool. This chapter also describes how to create polyline boundaries and explores additional options for converting polylines and splines.

Using the PEDIT Tool

Access the **PEDIT** tool and select the polyline to edit, or activate the **Multiple** option to edit multiple polylines. To select a wide polyline, pick the edge of a polyline segment rather than the center. Choose from the list of options to activate the appropriate editing function.

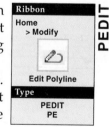

Ribbon

Home
> Modify

Edit Polyline

Type

PEDIT
PE

PEDIT

You can also use the **PEDIT** tool to convert a line, arc, or spline into a polyline. Access the **PEDIT** tool and select the object to convert. A prompt asks you if you want to turn the object into a polyline. Select the **Yes** option to make the conversion. Once the object is converted, the **PEDIT** tool continues normally.

> **NOTE**
>
>
>
> You can also access the **PEDIT** tool by selecting a polyline, right-clicking, and choosing **Polyline Edit**.

Opening and Closing Polylines

The **Open** and **Close** options allow you to close an open polyline or open a closed polyline, as shown in **Figure 13-1**. The **Open** option is unavailable if you closed the polyline by drawing the final segment manually. Instead, the **Close** option appears. The **Open** option is also available if you used the **Close** option of the **PLINE** tool. If you select an open polyline, the **Close** option appears instead of the **Open** option. Enter the **Close** option to close the polyline.

Joining Polylines

Use the **Join** option to create a single polyline object from connected but ungrouped polylines or from a polyline connected to lines or arcs. The **Join** option works only if objects connect appropriately. Segments cannot cross and there cannot be any spaces or breaks between the segments. See **Figure 13-2**. You can include the original polyline in the selection set, but it is not necessary. See **Figure 13-3**. AutoCAD automatically converts selected lines and arcs to polylines.

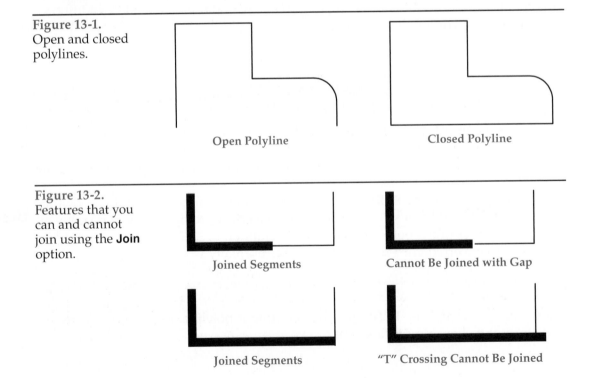

Figure 13-1.
Open and closed polylines.

Open Polyline   Closed Polyline

Figure 13-2.
Features that you can and cannot join using the **Join** option.

Joined Segments   Cannot Be Joined with Gap

Joined Segments   "T" Crossing Cannot Be Joined

Figure 13-3.
Joining a polyline to other connected lines and arcs.

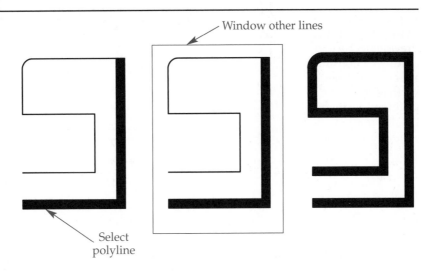

Window other lines

Select polyline

Changing Polyline Width

The **Width** option allows you to assign a new width to a polyline or donut. The width of the original polyline can be constant, or it can vary, but *all* segments change to the constant width you specify. See **Figure 13-4.**

Exercise 13-1

Access the Student Web site (www.g-wlearning.com/CAD) and complete Exercise 13-1.

Editing a Polyline Vertex or Point of Tangency

The **Edit vertex** option allows you to edit a *polyline vertex* and a *point of tangency.* When you enter the **Edit vertex** option, an "X" marker appears on-screen at the first polyline vertex or point of tangency. The **Edit vertex** option contains several functions that affect only the point identified by the "X" marker.

In **Figure 13-5,** the marker is moved clockwise through the points using the **Next** option and counterclockwise using the **Previous** option. If you edit the vertices of a polyline and nothing appears to happen, use the **Regen** option to regenerate the poly-line. Select the **eXit** option to return to the **PEDIT** prompt.

polyline vertex:
The point at which two straight polyline segments meet.

point of tangency:
The point at which a polyline arc meets another polyline arc or a straight polyline segment.

Making Breaks

You can use the **Break** function to break a polyline into separate polylines. Enter the **Edit vertex** option, move the "X" marker to the first vertex where you want to break the polyline, and activate the **Break** function. Move the "X" marker to the second vertex of the break, and enter the **Go** option to remove the portion of the polyline between

Figure 13-4.
Changing the width of a polyline.

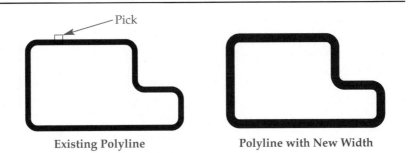

Pick

Existing Polyline

Polyline with New Width

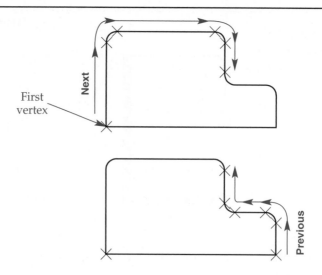

Figure 13-5.
Using the **Next** and **Previous** vertex editing options to specify polyline vertices. Note the different positions of the "X" marker.

First vertex

the two points. See **Figure 13-6.** To break the polyline at a point, without removing a segment, choose **Go** without moving to a second vertex.

Inserting a New Vertex

The **Insert** function allows you to add a new vertex to a polyline. Enter the **Edit vertex** option, move the "X" marker to the appropriate point near the desired new vertex, and activate the **Insert** function. Specify the location of the new vertex on or away from an existing polyline segment. See **Figure 13-7.**

Moving a Vertex

The **Move** function allows you to move a polyline vertex to a new location. Enter the **Edit vertex** option, move the "X" marker to the vertex you want to move, and enter the **Move** function. Then, specify the new vertex location on or away from an existing polyline segment. See **Figure 13-8.**

Straightening Polyline Segments or Arcs

The **Straighten** function allows you to straighten polyline segments or arcs between two points. Enter the **Edit vertex** option, move the "X" marker to one end of

Figure 13-6.
Using the **Break** vertex editing option to break a polyline and remove a portion.

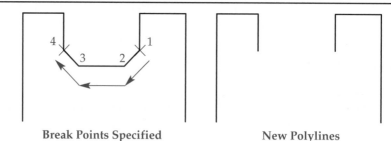

Break Points Specified

New Polylines

Figure 13-7.
Using the **Insert** vertex editing option to add a new vertex to a polyline.

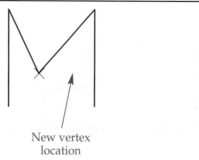
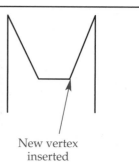

New vertex location

New vertex inserted

Figure 13-8.
Using the **Move**
vertex editing option
to move a polyline
vertex to a new
location.

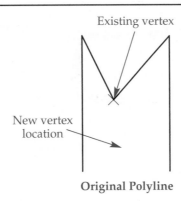
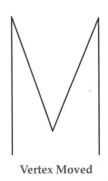

Existing vertex

New vertex
location

Original Polyline   Vertex Moved

the polyline segment you want to straighten, and enter the **Straighten** function. Move the "X" marker to the other end of the segment you want to straighten, and enter the **Go** option to straighten the polyline between the two points. See **Figure 13-9.** To straighten between two consecutive vertices, choose **Go** without moving to a second vertex. This provides a quick way to straighten an arc, as shown in **Figure 13-9.**

Changing Polyline Segment Widths

You can use the **Width** function to change the starting and ending widths of a polyline segment. Enter the **Edit vertex** option and move the "X" marker to the first vertex where you want to change the polyline width. Activate the **Width** function and specify the starting and ending width of the polyline segment. See **Figure 13-10.** If nothing appears to happen to the segment, press [Enter] and use the **Regen** option of the **PEDIT** tool to regenerate the polyline.

Exercise 13-2

Access the Student Web site (www.g-wlearning.com/CAD) and complete Exercise 13-2.

Fitting a Curve to a Polyline

In some situations, you may need to convert a polyline into a series of smooth curves. A graph, for example, may show a series of plotted points as a smooth curve instead of straight segments. You can accomplish this process, known as *curve fitting*, using the **Fit** option and the **Tangent** function of the **Edit vertex** option. The **Fit** option

curve fitting:
Converting a
polyline into a series
of smooth curves.

Figure 13-9.
The **Straighten**
vertex editing
option allows you to
straighten polyline
segments and arcs.

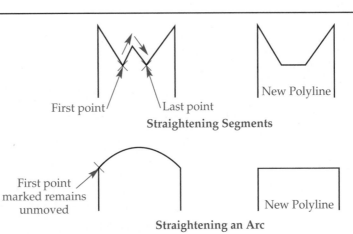

First point   Last point
Straightening Segments

New Polyline

First point
marked remains
unmoved

New Polyline

Straightening an Arc

Figure 13-10.
Changing the width of a polyline segment with the **Width** vertex editing option. Use the **Regen** option to display the change.

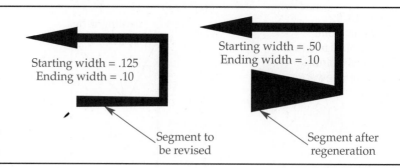

Starting width = .125
Ending width = .10

Starting width = .50
Ending width = .10

Segment to
be revised

Segment after
regeneration

fit curve: A curve that passes through all of its control points.

creates a *fit curve* by constructing pairs of arcs that pass through control points. You can specify the control points, or use the vertices of the polyline.

Prior to curve fitting, you have the option of assigning each vertex a tangent direction. AutoCAD then fits the curve based on the tangent directions you set. Specifying tangent directions is a way to edit vertices when the **Fit** option of the **PEDIT** tool does not produce the best results.

The **Tangent** function of the **Edit vertex** option allows you to edit tangent directions. After entering the **PEDIT** tool and the **Edit vertex** option, move the "X" marker to the first vertex to change. Enter the **Tangent** option and specify a tangent direction in degrees or pick a point in the expected direction. An arrow placed at the vertex then indicates the selected direction.

Continue by moving the marker to each vertex to change, entering the **Tangent** option for each vertex and selecting a tangent direction. Once all vertices to change include a specified tangent direction, enter the **Fit** option to create the curve.

You can also enter the **PEDIT** tool, select a polyline, and enter the **Fit** option without adjusting tangencies. **Figure 13-11** shows a polyline formed into a smooth curve using the **Fit** option. If the resulting curve does not look correct, enter the **Edit vertex** option and make changes as necessary.

Using the Spline Option

When you edit a polyline with the **Fit** option, the resulting curve passes through each polyline vertex. The **Spline** option also smoothes the corners of a straight-segment polyline. This option, however, creates a *spline curve* that passes through the first and last control points or vertices only. The curve pulls toward the other vertices, but does not necessarily pass through them. See **Figure 13-12**.

The **Spline** option creates a curve that approximates a true B-spline. You can choose a *cubic* and *quadratic* calculation to create the curve. Like a cubic curve, a quadratic curve passes through the first and last control points. The remainder of the

spline curve: A curve that passes through the first and last control points and is influenced by the other control points.

cubic curve: A very smooth curve created by the **PEDIT** **Spline** option with **SPLINETYPE** set at 6.

quadratic curve: A curve created by the **PEDIT Spline** option with **SPLINETYPE** set at 5.

Figure 13-11.
Using the **Fit** option of the **PEDIT** tool to turn a polyline into a smooth curve.

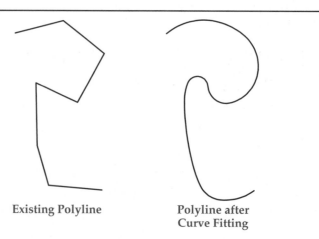

Existing Polyline

Polyline after
Curve Fitting

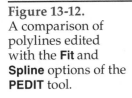

Figure 13-12.
A comparison of polylines edited with the **Fit** and **Spline** options of the **PEDIT** tool.

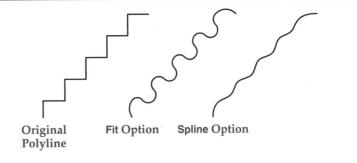

Original   Fit Option   Spline Option
Polyline

Figure 13-13.
The **SPLINETYPE** system variable controls whether the **Spline** option uses a quadratic or cubic curve.

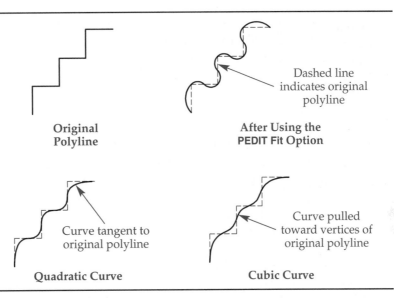

Original   After Using the
Polyline   PEDIT Fit Option

Dashed line indicates original polyline

Curve tangent to original polyline

Curve pulled toward vertices of original polyline

Quadratic Curve   Cubic Curve

curve is tangent to the polyline segments between the intermediate control points, as shown in **Figure 13-13.**

The **SPLINETYPE** system variable determines whether AutoCAD draws cubic or quadratic curves. The default setting is 6. At this setting, the **Spline** option of the **PEDIT** tool draws a cubic curve. Set the **SPLINETYPE** system variable to 5 to generate a quadratic curve. The only valid values for **SPLINETYPE** are 5 and 6.

You can set the number of line segments used to construct spline curves by entering a value in the **Segments in a polyline curve** text box in the **Display resolution** area of the **Display** tab of the **Options** dialog box. After changing the value, you must reissue the **Spline** option of the **PEDIT** tool and select the polyline to see the result. The default value is 8, which creates a smooth spline curve with moderate regeneration time. If you decrease the value, the resulting spline curve is less smooth. The resulting spline curve is smoother if you increase the value, but the regeneration time and drawing file size increase. See **Figure 13-14.**

NOTE

The **Fit** and **Spline** options of the **PEDIT** tool create approximations of a B-spline curve. Use the **SPLINE** tool to create a true B-spline curve.

Figure 13-14.
A comparison of curves drawn with different display resolution settings.

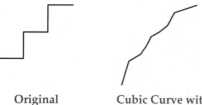
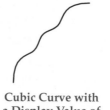

| Original Polyline | Cubic Curve with a Display Value of 2 Segments | Cubic Curve with a Display Value of 8 Segments (Default) | Cubic Curve with a Display Value of 20 Segments |

Exercise 13-3

Access the Student Web site (www.g-wlearning.com/CAD) and complete Exercise 13-3.

Straightening All Polyline Segments

The **Decurve** option returns a polyline edited with the **Fit** or **Spline** option to its original form. The information entered for tangent directions remains, however, for future reference. You can also use the **Decurve** option to straighten the curved segments of a polyline. See **Figure 13-15**.

Exercise 13-4

Access the Student Web site (www.g-wlearning.com/CAD) and complete Exercise 13-4.

Changing the Appearance of Polyline Linetypes

The **Ltype gen** (linetype generation) option determines how linetypes other than Continuous appear in relation to the vertices of a polyline. For example, if you use a Center linetype and disable the **Ltype gen** option, the polyline has a long dash at each vertex. When you activate the **Ltype gen** option, the polyline generates with a constant pattern in relation to the polyline as a whole. See **Figure 13-16**.

Figure 13-15.
The **Decurve** option allows you to straighten the curved segments of a polyline.

| Original Polyline | Polyline after Using the Decurve Option |

Figure 13-16.
A comparison
of polylines and
splined polylines
with the **Ltype gen**
option of the **PEDIT**
tool on and off.

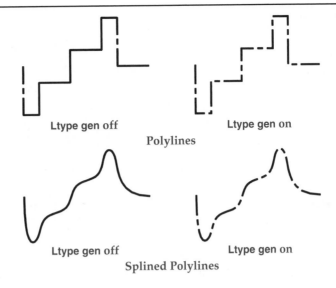

Ltype gen off   Ltype gen on

Polylines

Ltype gen off   Ltype gen on

Splined Polylines

Creating a Polyline Boundary

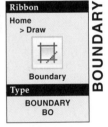

You can use the **BOUNDARY** tool to create a polyline boundary from line segments that form a closed area. The **Boundary Creation** dialog box appears when you access the **BOUNDARY** tool. See **Figure 13-17.**

Select the default **Polyline** option from the **Object type:** drop-down list to create a polyline around the specified area. Select **Region** from the **Object type:** drop-down list to create a *region* that you can use for area calculations, shading, extruding a solid model, and other purposes.

In the **Boundary set** drop-down list, the **Current viewport** setting is active. The **Current viewport** option defines the *boundary set* from everything visible in the current viewport, even if it is not in the current display. The **New** button allows you to define a different boundary set. The **Boundary Creation** dialog box closes and the Select objects: prompt appears allowing you to select the objects to use to create a boundary set. When you are finished, right-click or press [Enter] or the space bar. The **Boundary Creation** dialog box returns with **Existing set** active in the **Boundary set** drop-down list. This means the boundary set references the selected objects.

The **Island detection** setting specifies whether *islands* within the boundary apply as boundary objects. See **Figure 13-18.** Check **Island detection** to form separate boundaries from islands within a boundary.

When you select the **Pick Points** button, located in the upper-left corner, the **Boundary Creation** dialog box closes and the Pick internal point: prompt appears. If the

region: A closed 2D area that can have physical properties such as centroids and products of inertia.

boundary set: The part of the drawing AutoCAD evaluates to define a boundary.

island: A closed area inside a boundary.

Figure 13-17.
The **Boundary Creation** dialog box.

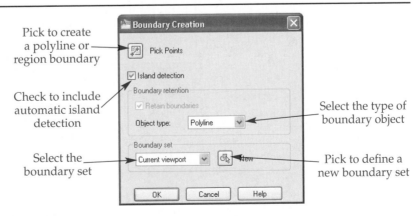

Pick to create a polyline or region boundary

Check to include automatic island detection

Select the boundary set

Select the type of boundary object

Pick to define a new boundary set

Figure 13-18.
When you define
a boundary set,
you can include or
exclude islands.

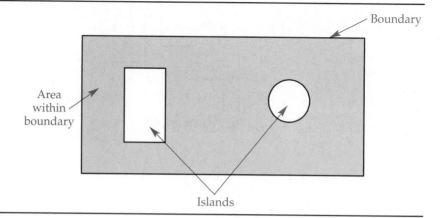

point you pick is inside a closed polygon, the boundary highlights, as shown in **Figure 13-19.** The **Boundary Definition Error** alert box appears if the point you pick is not within a closed polygon. Pick **OK** and try again.

Unlike an object created with the **Join** option of the **PEDIT** tool, a polyline boundary created with the **BOUNDARY** tool does not replace the original objects. The polyline traces over the defining objects with a polyline. The separate objects still exist underneath the newly created boundary. To avoid duplicate geometry, move the boundary to another location on the screen, erase the original objects, and then move the boundary back to its original position.

PROFESSIONAL TIP

You can simplify area calculations by using the **BOUNDARY** tool or by joining objects with the **Join** option of the **PEDIT** tool before issuing the **AREA** tool. To retain the original objects, explode the polyline after the area calculation if you used the **Join** option of the **PEDIT** tool. Erase the polyline boundary after the calculation if you used the **BOUNDARY** tool. Chapter 16 covers the **AREA** tool.

Figure 13-19.
When you select
a point inside a
closed polygon,
the boundary
highlights.

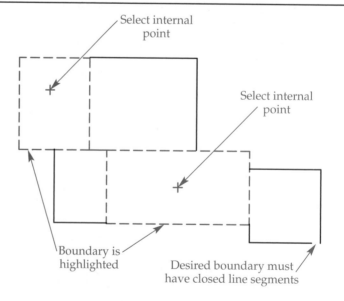

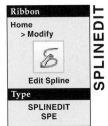
Using the SPLINEDIT Tool

Ribbon
Home
> Modify

Edit Spline
Type
SPLINEDIT
SPE

The **SPLINEDIT** tool allows you to edit spline objects. Access the **SPLINEDIT** tool and select the spline to edit. Grips identify spline control points, as shown in **Figure 13-20**. Choose from the list of options to activate the appropriate editing function.

NOTE

You can also access the **SPLINEDIT** tool by selecting a spline, right-clicking, and choosing **Spline**.

PROFESSIONAL TIP

Select the **Undo** option immediately after performing an unwanted edit to restore the previous spline without leaving the tool. Use the **Undo** option more than once to step back through each operation.

Editing Fit Data

The **Fit data** option allows you to edit spline *fit points*. The **Fit data** option has several functions. Use the **eXit** function to return to the **SPLINEDIT** option prompt. **Figure 13-21** provides examples of using fit data options.

fit points: Spline control points.

The **Add** option adds new fit points to a spline definition. You can locate a new fit point by picking a point or entering coordinates. Fit points appear as unselected grips. When you select a fit point, it highlights, along with the next spline fit point. You can then add a fit point between the two highlighted points. If you select the endpoint of the spline, only the endpoint highlights. If you select the start point, choose the **After** or **Before** option to insert the new fit point after or before the existing point. When you add a fit point, the spline curve updates according to the new point.

The **Add** option is in a running mode, which means you can continue to add points as needed. Press [Enter] or the space bar, or right-click and select **Enter** at a Specify new point <exit>: prompt to select other existing fit points and add points anywhere on the spline.

The **Delete** option deletes fit points as needed. However, at least two fit points must remain to define the spline. Like the **Add** option, the **Delete** option operates in a running mode, allowing as many deletions as needed. The spline curve updates to reflect changes made by deleting each point.

The **Move** option allows you to move fit points as necessary. The start point of the spline highlights. You can specify a different location by picking a new point. You can also specify other fit points to move. Pick the **Specify new location** option to move the

Figure 13-20.
The control points for a spline appear as grips.

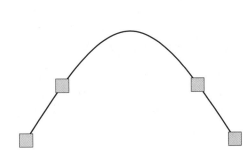

Figure 13-21.
Examples of using the **Fit data** options of the **SPLINEDIT** tool to edit a spline. Compare the original spline to each of the edited objects.

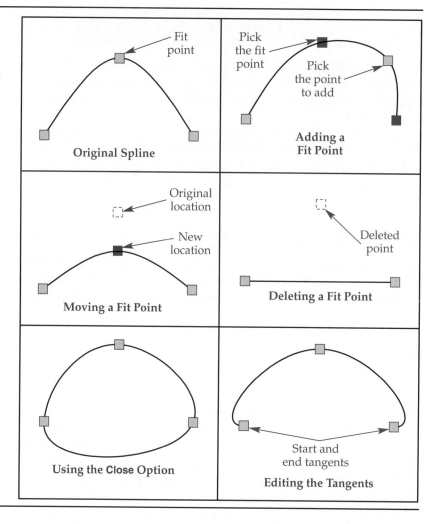

highlighted point to a new location. Select the **Next** option to highlight the next fit point, and then press [Enter]. Select the **Previous** option to highlight the previous fit point. Pick the **Select point** option to select a different fit point to move. Pick the **eXit** option to return to the **Fit data** option prompt.

The **Purge** option lets you remove fit point data from a spline. In very large drawings that contain many complex splines, purging fit point data reduces the file size by simplifying spline definitions. However, after you use this option, the resulting spline is not as easy to edit. After you purge a spline, the **Fit data** option is no longer available in the **SPLINEDIT** tool.

The **Tangents** option allows you to edit the start and end tangents for an open spline and the start tangent for a closed spline. The direction of the selected point determines the tangency. You can also use the **System default** option to set the tangency values to the AutoCAD defaults.

You can use the **toLerance** option to adjust fit tolerance values. The results are immediate, so you can adjust the fit tolerance as necessary to produce different results.

Opening or Closing a Spline

The **Open** and **Close** options are alternately displayed, depending on the status of the spline object. If the spline is open, the **Close** option displays. The **Open** option appears if the spline is closed. Use the available option to open or close the selected spline.

Moving a Vertex

The **Move vertex** option allows you to move the fit points of a spline. When you access this option, you can specify a new location for a selected fit point. The options displayed are identical to those used with the **Move** function of the **Fit data** option. You can pick a new location for the highlighted fit point, or you can enter another option. The following **Move vertex** options are available:

- **Specify new location.** Moves the currently highlighted point to a specified location.
- **Next.** Highlights the next fit point.
- **Previous.** Highlights the previous fit point.
- **Select point.** Picks a different fit point to move; provides an alternative to cycling through points with the **Next** or **Previous** functions.
- **eXit.** Returns to the **SPLINEDIT** prompt.

Exercise 13-5

Access the Student Web site (www.g-wlearning.com/CAD) and complete Exercise 13-5.

Smoothing or Reshaping a Spline Section

The **Refine** option allows you to fine-tune the spline curve and includes several functions. If necessary, use the **Add control point** option to add fit points to help smooth or reshape a section of the spline.

The **Elevate order** option causes more control points to appear on the curve for greater control and spline *order* refinement. For example, a cubic spline has an order of 4. In **Figure 13-22A**, the order of the spline is elevated from 4 to 6. You can use an order setting from 4 to 26, but you cannot adjust the value downward. For example, if the order is set to 24, the only remaining settings are 25 and 26.

order: In a spline, the degree of the spline polynomial + 1.

The **Weight** option changes the weight of a control point. When all control points have the same weight, they exert the same amount of pull on the spline. When you reduce a weight value for a control point, the spline does not pull as close to the point. When you increase a weight value, the control point exerts more pull on the spline.

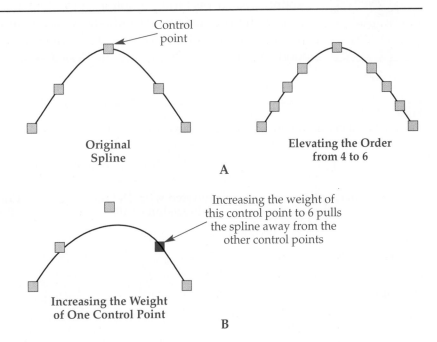

Figure 13-22.
A—The effects of elevating the order of a spline. B—Increasing the weight of an individual control point.

Control point

Original Spline

Elevating the Order from 4 to 6

A

Increasing the weight of this control point to 6 pulls the spline away from the other control points

Increasing the Weight of One Control Point

B

See **Figure 13-22B.** You can adjust the default setting of 1, but the weight setting must be positive. The control point selection functions of the **Weight** option are the same as those used with the **Move vertex** option of the **SPLINEDIT** tool. You can use the **Enter new weight** option to specify a new weight for the highlighted control point.

Exercise 13-6

Access the Student Web site (www.g-wlearning.com/CAD) and complete Exercise 13-6.

Converting Polylines and Splines

You can use the **Object** option of the **SPLINE** tool to convert a spline-fitted polyline object, created using the **PEDIT** tool, to a spline object. Once you access the **SPLINE** tool, activate the **Object** option instead of defining control points. Then pick a spline-fitted polyline object to convert the polyline to a spline.

The **Convert to Polyline** option of the **SPLINEDIT** tool allows you to convert a spline to a polyline. Select the spline to convert and enter a value at the Specify a precision <current>: prompt. The higher the specified precision, the more vertex points are added to the polyline, making the polyline smoother. See **Figure 13-23.**

NOTE

You can achieve different spline results by altering the specifications used with the **Fit tolerance** option. The setting specifies a tolerance within which the spline curve falls as it passes through the control points.

PROFESSIONAL TIP

A spline created by fitting a spline curve to a polyline is a linear approximation of a true spline and is not as accurate as a spline drawn using the **SPLINE** tool. An additional advantage of spline objects over smoothed polylines is that splines use less disk space.

Figure 13-23.
The effects of changing precision when converting a spline to a polyline.

| Original Spline | Converted with Default Precision of 10 | Converted with Precision of 1 |

Chapter Test

Answer the following questions. Write your answers on a separate sheet of paper or go to the Student Web site (www.g-wlearning.com/CAD) and complete the electronic chapter test.

1. Name the tool and option required to turn three connected lines into a single polyline.
2. When you enter the **Edit vertex** option of the **PEDIT** tool, where does AutoCAD place the "X" marker?
3. How do you move the "X" marker to edit a different polyline vertex?
4. Name the **Edit vertex** option of the **PEDIT** tool that relates to each definition below.
   A. Moves the "X" marker to the next position.
   B. Moves a polyline vertex to a new location.
   C. Breaks a polyline at a point or between two points.
   D. Generates the revised version of a polyline.
   E. Specifies a tangent direction.
   F. Adds a new polyline vertex.
   G. Returns to the **PEDIT** tool prompt.
5. Which **PEDIT** tool option and function allow you to change the starting and ending widths of a polyline?
6. Why might it appear that nothing happens when you change the starting and ending widths of a polyline?
7. Name the **PEDIT** tool option and function used for curve fitting.
8. Explain the difference between a fit curve and a spline curve.
9. Compare a quadratic curve, cubic curve, and fit curve.
10. Which **SPLINETYPE** system variable setting allows you to draw a quadratic curve?
11. Explain how you can adjust the way polyline linetypes are generated using the **PEDIT** tool.
12. Name the tool used to create a polyline boundary.
13. Name the tool that allows you to edit splines.
14. What is the purpose of the **Add** function of the **Fit data** option of the **SPLINEDIT** tool?
15. What is the minimum number of fit points for a spline?
16. Name the **SPLINEDIT** option that allows you to move the fit points in a spline.
17. Which function of the **SPLINEDIT Fit data** option allows you to reduce file size, but also makes the resulting spline harder to edit?
18. Identify the **Refine** option of the **SPLINEDIT** tool that lets you increase, but not decrease, the number of control points appearing on a spline curve.
19. Name the **Refine** function of the **SPLINEDIT** tool that controls the pull exerted by a control point on a spline.
20. Name the **SPLINE** tool option that allows you to turn a spline-fitted polyline into a true spline.

Drawing Problems

Follow these instructions to complete the drawing problems for this chapter:

- *Start AutoCAD if it is not already started.*
- *Start a new drawing using an appropriate template of your choice. The template should include layers and text styles necessary for drawing the given objects.*
- *Add layers and text styles as needed. Draw all objects using appropriate layers and text styles, justification, and format.*
- *Follow the specific instructions for each problem. Do not draw dimensions. Use your own judgment and approximate dimensions when necessary.*

▼ Basic

1. Use the **LINE** tool to draw two connected lines. Use the **PEDIT** tool to convert one of lines to a polyline, and then use the **Join** option to convert the line and polyline into a single polyline object. Use the **LINE** tool to draw a rectangle. Use the **PEDIT** tool to convert one of lines to a polyline, and then use the **Join** option to convert the three remaining lines and polyline into a single polyline object. Save the completed drawing as P13-1.

2. Use the **SPLINE** tool to draw a spline of your own design with at least four control points. Use the **Convert to Polyline** option of the **SPLINEDIT** tool to convert the spline to a polyline. Save the completed drawing as P13-2.

▼ Intermediate

3. Open drawing P4-17 and save it as P13-3. In the P13-3 drawing file, use the **PEDIT** tool to change the object into a rectangle. Use the **Decurve** and **Width** options and the **Straighten**, **Insert**, and **Move** vertex editing options of the **PEDIT** tool. Save the completed drawing as P13-3.

Original

After **PEDIT** Operations

4. Open drawing P4-18 and make the following changes. Save the drawing as P13-4.
   A. Combine the two polylines using the **Join** option of the **PEDIT** tool.
   B. Change the beginning width of the left arrow to 1.0 and the ending width to .2.
   C. Draw a polyline .062 wide, similar to Line A, as shown.

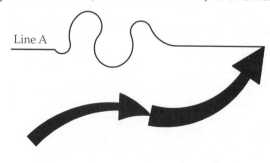

Line A

5. Draw four polylines .032 wide, using the following absolute coordinates for all four objects.

Point	Coordinates	Point	Coordinates	Point	Coordinates
1	1,1	5	3,3	9	5,5
2	2,1	6	4,3	10	6,5
3	2,2	7	4,4	11	6,6
4	3,2	8	5,4	12	7,6

Leave the first polyline as drawn. Use the **Fit** option of the **PEDIT** tool to smooth the second polyline. Use the **Spline** option of the **PEDIT** tool to turn the third polyline into a quadratic curve. Make the fourth polyline into a cubic curve. Use the **Decurve** option of the **PEDIT** tool to return one of the three edited polylines to its original form. Save the drawing as P13-5.

6. Use the **PLINE** tool to draw four copies of a patio plan similar to the one shown in Example A below. Draw the house walls 6″ wide. Leave the first plan as drawn. Use the **PEDIT** tool to create the designs shown. Use the **Fit** option for Example B, a quadratic spline for Example C, and a cubic spline for Example D. Change the **SPLINETYPE** system variable as required. Save the drawing as P13-6.

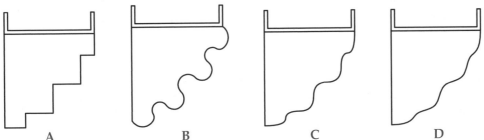

   A   B   C   D

7. Open drawing P13-6 and create four new patio designs. This time, use grips to edit the polylines and create designs similar to Examples A, B, C, and D below. Save the drawing as P13-7.

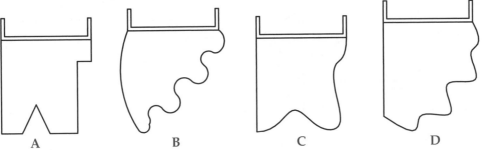

   A   B   C   D

8. Draw a spline similar to the original spline shown below seven times in a layout similar to the layout shown. Perform the **SPLINEDIT** operations identified under each of the seven splines. Save the drawing as P13-8.

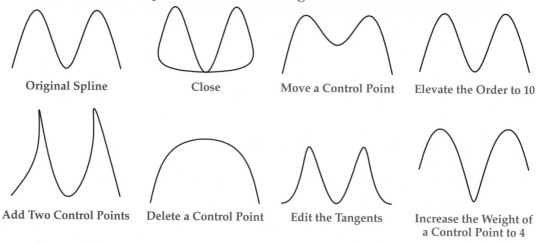

Original Spline   Close   Move a Control Point   Elevate the Order to 10

Add Two Control Points   Delete a Control Point   Edit the Tangents   Increase the Weight of a Control Point to 4

▼ Advanced

9. Draw the flow chart shown below. Use polylines to draw the connecting lines, arrows, and diamonds. Save the drawing as P13-9.

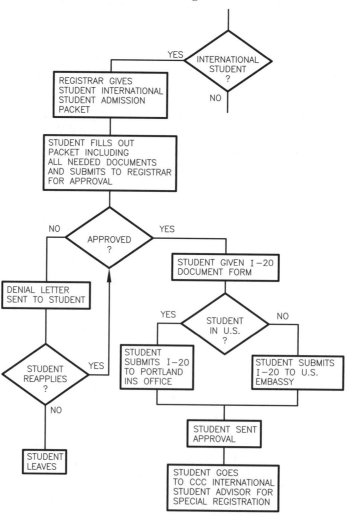

Drawing Problems - Chapter 13

10. Draw the flow chart shown below. Use polylines to draw the connecting lines, arrows, and diamonds. Save the drawing as P13-10.

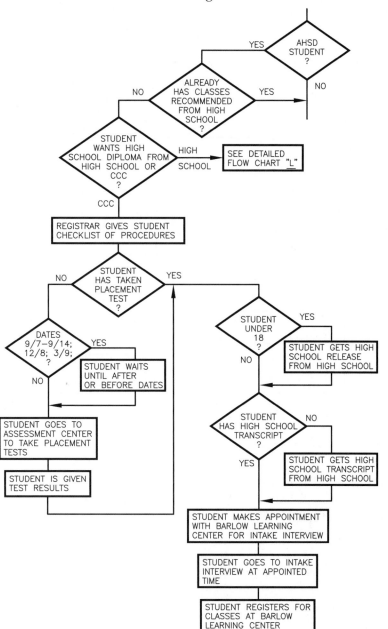

11. Draw the following roof plan. Save the drawing as P13-11.

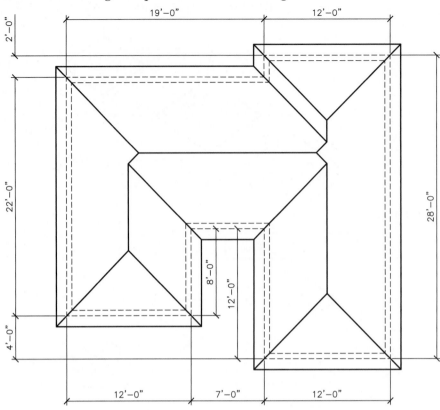

CHAPTER 14

Arranging and Patterning Objects

Learning Objectives

After completing this chapter, you will be able to do the following:

✓ Relocate objects using the **MOVE** tool.
✓ Change the angular positions of objects using the **ROTATE** tool.
✓ Use the **ALIGN** tool to simultaneously move and rotate objects.
✓ Make copies of objects using the **COPY** tool.
✓ Draw mirror images of objects using the **MIRROR** tool.
✓ Use the **REVERSE** tool.
✓ Create patterns of objects using the **ARRAY** tool.

This chapter explains methods for arranging and patterning objects using basic editing tools. You will learn to use various editing tools to increase drawing efficiency. The editing tools described in this chapter include many options. As you work through this chapter, experiment with each option to see which is the most effective in different situations.

Using the MOVE Tool

The **MOVE** tool provides an easy way for you to move a view or feature to a more appropriate location. The **MOVE** tool functions similar to the **STRETCH** tool, except that when objects move, they retain the same size and shape. Access the **MOVE** tool and select the objects to move. Proceed to the next prompt and specify the base point from which the objects will move. Though the position of the base point is often not critical, you may want to select a point on an object, the corner of a view, or the center of a circle. The selection moves as you move the crosshairs. Pick a second point to complete the move. See **Figure 14-1**.

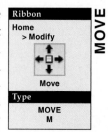

Ribbon
Home
> Modify
Move

Type
MOVE
M

MOVE

Using the Displacement Option

The **Displacement** option allows you to move objects relative to the origin, or 0,0,0 point. To move using a displacement, access the **MOVE** tool and select objects to move. Then select the **Displacement** option instead of defining the base point. At the Specify displacement <0,0,0>: prompt, enter an absolute coordinate to stretch the objects from the origin to the coordinate point. See **Figure 14-2**.

Figure 14-1.
Using the **MOVE** tool.

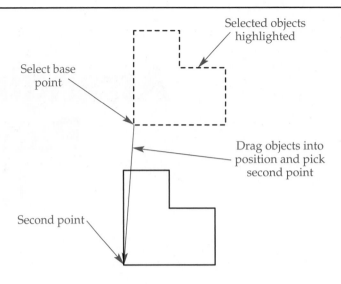

Selected objects highlighted

Select base point

Drag objects into position and pick second point

Second point

Figure 14-2.
Using the **Displacement** option of the **MOVE** tool to move objects. In this example, the origin is the base point and the absolute coordinate point 2,2 is the displacement.

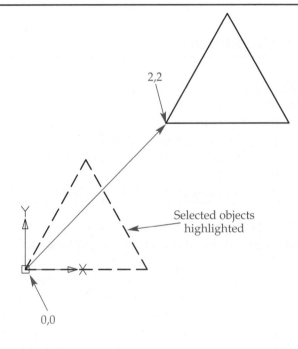

2,2

Selected objects highlighted

Y

X

0,0

Using the First Point As Displacement

Another method for moving an object is to use the first point as the displacement. This means the coordinates you use to select the base point automatically define the coordinates for the direction and distance for moving the object. To apply this technique, access the **MOVE** tool and select objects to move. Then specify the base point, and instead of locating the second point, right-click or press [Enter] or the space bar to accept the <use first point as displacement> default. See **Figure 14-3.**

PROFESSIONAL TIP

Use object snap modes to your best advantage while editing. For example, to move an object to the center of a circle, use the **Center** object snap mode to select the center of the circle.

Figure 14-3.
Moving a circle using the selected base point, 1,1 in this example, as the displacement.

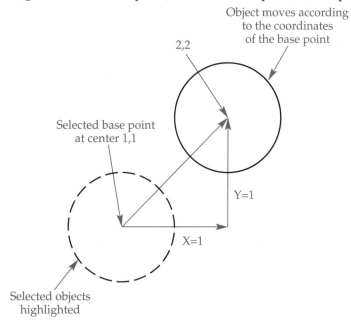

Object moves according
to the coordinates
of the base point

2,2

Selected base point
at center 1,1

Y=1

X=1

Selected objects
highlighted

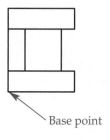

Exercise 14-1

Access the Student Web site (www.g-wlearning.com/CAD) and
complete Exercise 14-1.

Rotating Objects

Design changes often require you to rotate an object, feature, or view. For example,
you may have to rotate the furniture in an office layout for an interior design. Use the
ROTATE tool to revise the layout to obtain the final design. Access the **ROTATE** tool
and select the objects to rotate. Proceed to the next prompt and specify the base point,
or axis of rotation, around which the objects rotate. Next, enter a rotation angle at the
Specify rotation angle or [Copy/Reference] <current>: prompt. A negative rotation angle
revolves the object clockwise. A positive rotation angle revolves the object counterclockwise. See **Figure 14-4**.

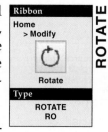

Ribbon
Home
> Modify

Rotate

Type
ROTATE
RO

ROTATE

Figure 14-4.
Rotation angles.

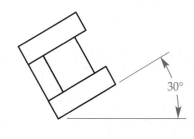

Base point

−30°

30°

−30° Rotation

30° Rotation

Using the Reference Option

You can use the **Reference** option instead of entering a rotation angle by specifying a new angle in relation to an existing angle. For example, suppose you want to rotate an object currently drawn at a 135° angle to a 180°. Access the **Reference** option and enter the angle at which the object is currently rotated, in this example 135. Next, enter the angle you want the object to rotate to, in this example 180. See **Figure 14-5A**.

If you do not know the angle at which the object is currently drawn or the angle at which you want to rotate the object, use the **Reference** option to rotate according to reference lines. To use this technique, access the **ROTATE** tool, select the objects to rotate, and pick the base point. Then select the **Reference** option and pick the two endpoints of a reference line that forms the existing angle. Finally, enter the new angle, as shown in **Figure 14-5B**, or select a point, such as a point on a correctly rotated object.

Creating a Copy While Rotating

The **Copy** option of the **ROTATE** tool copies and rotates the selected object. This option leaves the original object unchanged and rotates only the copy of the object.

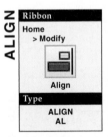

Exercise 14-2

Access the Student Web site (www.g-wlearning.com/CAD) and complete Exercise 14-2.

Using the ALIGN Tool

ALIGN

Ribbon
Home
> Modify

Align

Type
ALIGN
AL

source points: Points to define a reference line relative to the object's original position for an **ALIGN** operation.

destination points: Points to define the location of the reference line of the object's new location in an **ALIGN** operation.

Use the **ALIGN** tool to move and rotate an object with one operation. **ALIGN** is primarily a 3D tool, but it can be used for 2D drawings. After you access the **ALIGN** tool, select objects to align. Then, you must specify *source points* and *destination points*. Pick the first source point and then the first destination point. Next, pick the

---

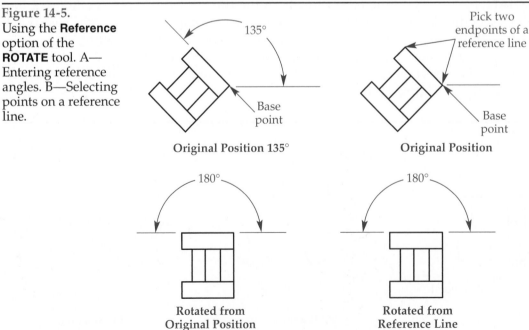

Figure 14-5.
Using the **Reference** option of the **ROTATE** tool. A—Entering reference angles. B—Selecting points on a reference line.

Pick two endpoints of a reference line

135°

Base point

Original Position 135°

180°

Rotated from Original Position

A

Base point

Original Position

180°

Rotated from Reference Line

B

Figure 14-6.
Using the **ALIGN** tool to move and rotate a kitchen cabinet layout against a wall.

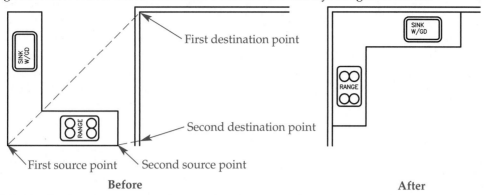

Before

After

second source point, followed by the second destination point. For 2D applications, you only need two source points and two destination points. Right-click or press [Enter] or the space bar when the prompt requests the third source and destination points. See **Figure 14-6**. The last prompt allows you to change the size of the selected object. Choose the **Yes** option to scale the object if the distance between the source points is different from the distance between the destination points. **Figure 14-7** illustrates using the **Scale** option of the **ALIGN** tool.

Exercise 14-3

Access the Student Web site (www.g-wlearning.com/CAD) and complete Exercise 14-3.

Using the COPY Tool

The **COPY** tool allows you to copy existing objects. The **COPY** tool functions similar to the **MOVE** tool, except that when you pick a second point, the original object remains in place and a copy is drawn. See **Figure 14-8**. Access the **COPY** tool, select objects to

Ribbon
Home
> Modify
Copy
Type
COPY
CO
CP

COPY

Figure 14-7.
The **Scale** option of the **ALIGN** tool allows you to change the size of an object during the alignment.

First source point   First destination point

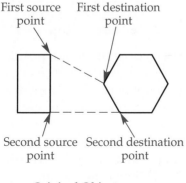

Second source point   Second destination point

Original Objects

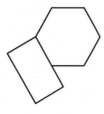

Rectangle Not Scaled

Rectangle Scaled

Figure 14-8.
Using the **COPY** tool.

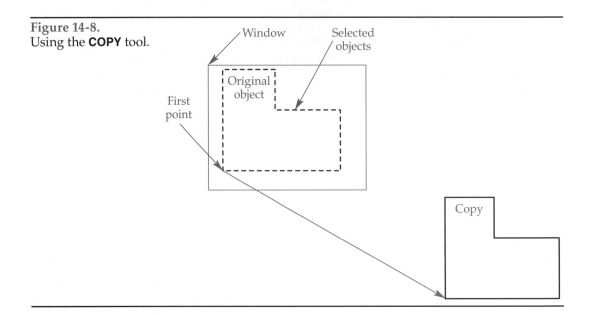

copy, specify a base point, and pick a location to place the first copy. By default, you can continue creating copies of the selected objects by specifying additional "second" points. Press [Enter] or the space bar or right-click and select **Enter** to exit.

As with the **MOVE** tool, you can specify a base point and a second point, specify a displacement using the **Displacement** option, or define the first point as displacement. These techniques function the same when making copies as when moving objects.

> **NOTE**
>
> By default, the **Multiple** copy mode is active, allowing you to create several copies of the same object using a single **COPY** operation. To make a single copy and exit the tool after placing the copy, use the **mOde** option and activate the **Single** function.

Exercise 14-4

Access the Student Web site (www.g-wlearning.com/CAD) and complete Exercise 14-4.

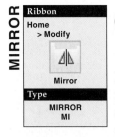

MIRROR

Ribbon
Home
> Modify

Mirror

Type
**MIRROR
MI**

Mirroring Objects

The **MIRROR** tool allows you to draw objects in a reflected, or mirrored, position. Mirroring is common to a variety of drafting applications. For example, in mechanical drafting you can mirror a part to form the opposite component of a symmetrical assembly. In architectural drafting, you can mirror an entire floor plan to create a duplex residence or to accommodate a different site orientation.

Once you access the **MIRROR** tool, select the objects to mirror. Then specify a **mirror line** at any angle by picking two points. After you pick the second mirror line point, you have the option to delete the original objects. The objects and any space between the objects and the mirror line are reflected. See **Figure 14-9**.

mirror line: The line of symmetry across which objects mirror.

Figure 14-9.
The **MIRROR** tool gives you the option to delete old objects. When you reflect an object about a mirror line, the space between the object and the mirror line also mirrors.

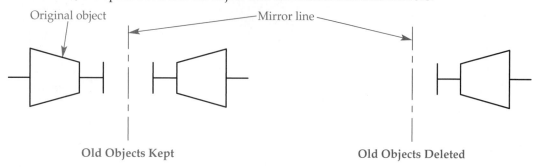

Original object

Mirror line

Old Objects Kept

Old Objects Deleted

NOTE

By default, the **MIRRTEXT** system variable is set to **0**, which prevents text from reversing during a mirror operation. Change the **MIRRTEXT** value to 1 to mirror text in relation to the original object. See **Figure 14-10.** Backward text is generally not acceptable, although it is used for reverse imaging.

Exercise 14-5

Access the Student Web site (www.g-wlearning.com/CAD) and complete Exercise 14-5.

Figure 14-10.
The **MIRRTEXT** system variable options.

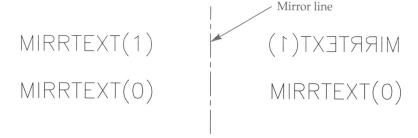

Mirror line

MIRRTEXT(1)

(1)TXƎTЯЯIM

MIRRTEXT(0)

MIRRTEXT(0)

Using the REVERSE Tool

AutoCAD 2010
NEW

Ribbon
Home
> Modify

Reverse

Type
REVERSE

REVERSE

You can use the **REVERSE** tool to reverse the calculation of points along lines, polylines, splines, and helixes. This makes the previous start point the new endpoint and the previous endpoint the new start point. You can also use the **rEverse** option of the **PEDIT** tool to reverse polylines, and the **rEverse** option of the **SPLINEDIT** tool to reverse splines.

As shown in **Figure 14-11,** reversing is most apparent when you apply a linetype that includes text or specific objects, or when you draw polylines with varying width. Reversing also affects the various vertex options of the **PEDIT** tool and control point options of the **SPLINEDIT** tool. Typically, it is improper to reverse text included with linetypes, although you may find specific applications where this is an appropriate requirement.

Figure 14-11.
Examples of objects before and after using the **REVERSE** tool.

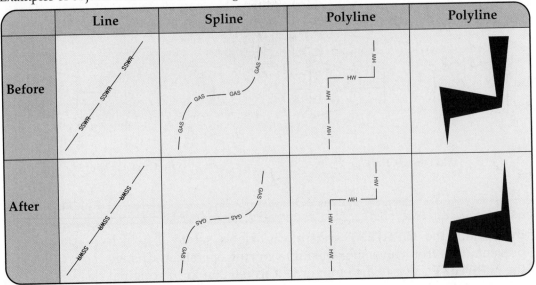

	Line	Spline	Polyline	Polyline
Before				
After				

Patterning with ARRAY

array: Multiple copies of an object arranged in a pattern.

Designs often require a rectangular or circular pattern objects. In interior planning, for example, you might draw a rectangular pattern, or *array*, of office desks. Suppose your design calls for four rows, each having four desks. You can create the pattern by drawing one desk and copying it fifteen times, but this operation is time-consuming. A quicker method is to create a rectangular pattern known as a *rectangular array*. A circular pattern is a *polar array*. **Figure 14-12** shows some basic arrays.

rectangular array: A pattern made up of columns and rows of objects.

Arranging Objects in a Rectangular Pattern

polar array: A circular pattern of objects.

To create arrays, use the **ARRAY** tool and the corresponding **Array** dialog box, shown in **Figure 14-13**. The **Rectangular Array** radio button is selected by default, allowing you to form a rectangular pattern of rows and columns. You can create an

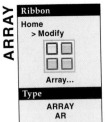

Figure 14-12.
Examples of arrays.

Rectangular Arrays

Polar Arrays

AutoCAD and Its Applications—Basics

Figure 14-13.
The **Array** dialog box options for a rectangular array.

Enter numbers of rows and columns

Select type of array

Pick to select objects

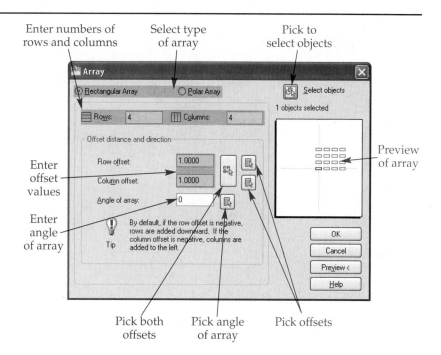

Enter offset values

Enter angle of array

Pick both offsets

Pick angle of array

Pick offsets

Preview of array

array that has a single row, a single column, or multiple rows and columns. Pick the **Select objects** button to return to the drawing area and select the objects to array. Right-click or press [Enter] or the space bar to return to the **Array** dialog box.

Next, specify the array characteristics. For example, suppose you want to create a rectangular pattern of a .5-unit square having four rows, four columns, and a .5 spacing between squares. Enter 4 in the **Rows:** and **Columns:** text boxes and 1.0000 in the **Row offset:** and **Column offset:** text boxes. See **Figure 14-14.** Notice that the distances do not refer to the space between the objects, but the distances between the same point on each object.

You can also enter row and column distances by picking points. One option is to use the **Pick Row Offset** and **Pick Column Offset** buttons in the **Array** dialog box to select each distance separately. The second option is to select the **Pick Both Offsets** button to specify both distances in one pick, as illustrated in **Figure 14-15. Figure 14-16** shows the four directions in which an array can grow. The direction is based on the use of positive and negative distance values for row and column offsets.

Figure 14-14.
The original object (in dashed lines) and the rectangular array. Note how the distances between rows and columns are determined.

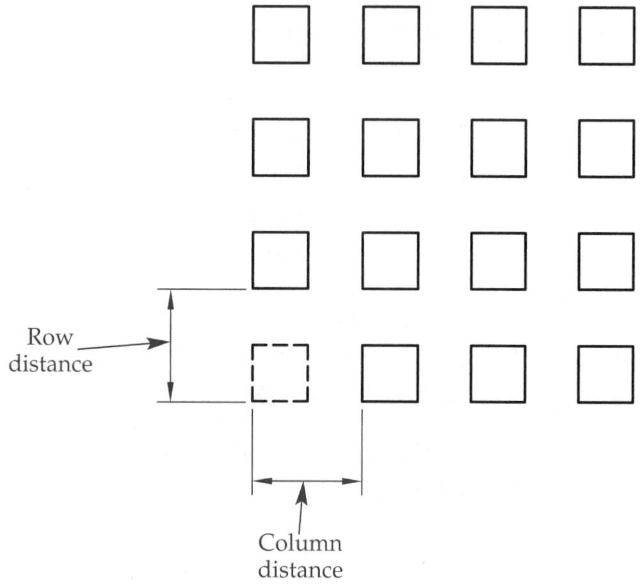

Row distance

Column distance

Figure 14-15.
The spacing of rows and columns in an array can be specified with a single point.

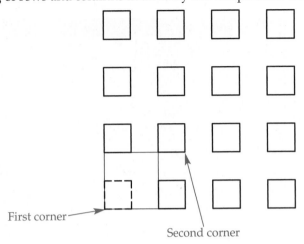

First corner

Second corner

Figure 14-16.
Positive and negative offset distances determine the direction in which an array will grow.

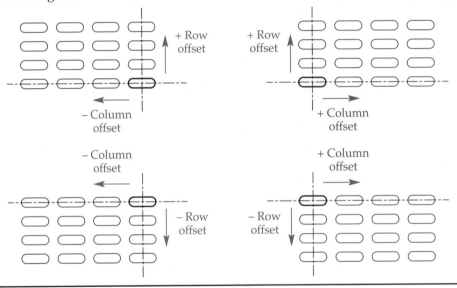

You can also create an angled rectangular array. Enter the angle in the **Angle of array:** text box or choose the **Pick Angle of Array** button to pick the angle on-screen. The column and row alignment rotate, not the objects. See **Figure 14-17.**

Arranging Objects around a Center Point

Pick the **Polar Array** radio button to create a circular pattern of objects around a center point. See **Figure 14-18.** Pick the **Select objects** button to return to the drawing area and select the objects to array. Right-click or press [Enter] or the space bar to return to the **Array** dialog box.

Next, specify the center point about which the objects in the array rotate. Enter the coordinates for the center point in the **X:** and **Y:** text boxes or select the **Pick Center Point** button to pick the center point on-screen. After locating the center point, you must specify the type of polar array to create using the **Method:** drop-down list. The selected method determines which settings in the dialog box are available. Three methods are available:

- Total number of items & Angle to fill
- Total number of items & Angle between items
- Angle to fill & Angle between items

AutoCAD and Its Applications—Basics

Figure 14-17.
Rectangular arrays can be set at an angle using the **Angle of array:** setting.

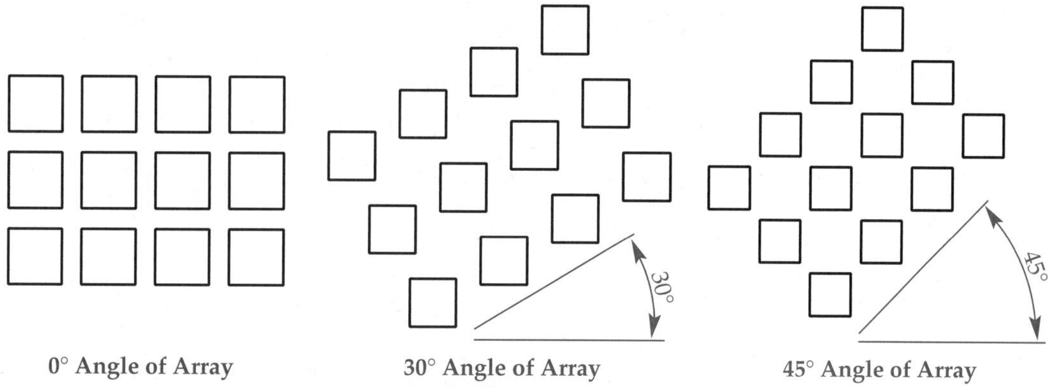

0° Angle of Array   30° Angle of Array   45° Angle of Array

Figure 14-18.
The **Array** dialog box options for a polar array.

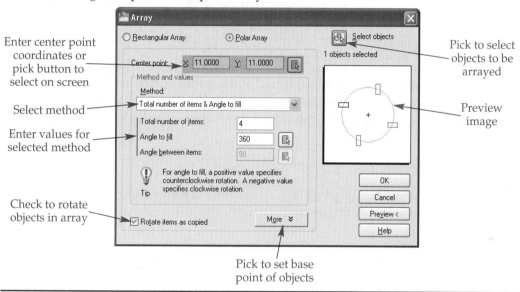

Enter center point coordinates or pick button to select on screen

Select method

Enter values for selected method

Check to rotate objects in array

Pick to select objects to be arrayed

Preview image

Pick to set base point of objects

The **Total number of items:** setting is the total number of objects in the array, including the original object. Use the Angle to fill & Angle between items method if you do not know the number of items to array. Enter a positive angle in the **Angle to fill:** text box to array the object in a counterclockwise direction, or enter a negative angle to array the object in a clockwise direction. Enter 360 to create a complete circular array. The **Angle between items:** setting specifies the angular distance between adjacent objects in the array. For example, to create a circular pattern of five items spaced 18° apart, enter 5 in the **Total number of items:** text box and 18 in the **Angle between items:** text box.

You can set objects to rotate as they array by checking **Rotate items as copied**. This keeps the same face of each object pointing toward the center point. If you do not rotate objects as they array, they remain in the same orientation as the original object. See **Figure 14-19.**

When you create a polar array, the base point of the object rotates and remains at a constant distance from the center point. The default base point varies for different types of objects, as shown in **Figure 14-20.**

Figure 14-19.
Rotating objects in a polar array. A—The square rotates as it arrays. B—The square does not rotate as it arrays.

A

B

Figure 14-20.
Default base points for objects.

Object Type	Default Base Point
Arc, circle, ellipse	Center
Rectangle, polygon	First corner
Line, polyline, donut	Start point
Block, text	Insertion point

If the default base point does not produce the desired array, you can choose a different base point for the selected object. Pick the **More** button to display the **Object base point** area. Deactivate the **Set to object's default** check box. Enter a new base point in the text boxes or pick the button to select a base point on-screen.

Exercise 14-6

Access the Student Web site (www.g-wlearning.com/CAD) and complete Exercise 14-6.

Chapter Test

Answer the following questions. Write your answers on a separate sheet of paper or go to the Student Web site (www.g-wlearning.com/CAD) and complete the electronic chapter test.

1. List two locations drafters normally choose as the base point when using the **MOVE** tool.
2. How would you go about rotating an object 45° clockwise?
3. Briefly describe the two methods of using the **Reference** option of the **ROTATE** tool.
4. Name the tool that can be used to move and rotate an object simultaneously.
5. How many points must you select to align an object in a 2D drawing?
6. Which ribbon tab and panel contains the **MOVE** and **COPY** tools?
7. Explain the difference between the **MOVE** and **COPY** tools.
8. Briefly explain how to make several copies of the same object.
9. Which tool allows you to draw a reverse image of an existing object?
10. What is the purpose of the **REVERSE** tool?
11. What is the difference between polar and rectangular arrays?
12. What four values should you know before you create a rectangular array?
13. Suppose an object is 1.5″ (38 mm) wide and you want to create a rectangular array with .75″ (19 mm) spacing between objects. What should you specify for the distance between columns?
14. How do you specify a clockwise polar array rotation?
15. What values should you know before you create a polar array?

Drawing Problems

Follow these instructions to complete the drawing problems for this chapter:

- *Start AutoCAD if it is not already started.*
- *Start a new drawing using an appropriate template of your choice. The template should include layers and text styles necessary for drawing the given objects.*
- *Add layers and text styles as needed. Draw all objects using appropriate layers and text styles, justification, and format.*
- *Follow the specific instructions for each problem. Do not draw dimensions. Use your own judgment and approximate dimensions when necessary.*

▼ Basic

1. Open P12-1. Rotate the object 90 degrees to the right and mirror the object to the left. Use the vertical base of the object as the mirror line. Your final drawing should look like the example below. Save the drawing as P14-1.

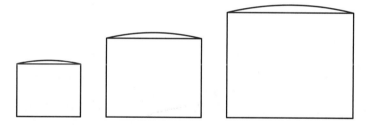

2. Open drawing P12-2. Make two copies of the object to the right of the original object. Scale the first copy 1.5 times size of the original object. Scale the second copy 2 times the size of the original object. Move the objects so they are approximately centered in your drawing area. Move the objects as needed to align the bases of all objects and provide an equal amount of space between the objects. Save the drawing as P14-2.

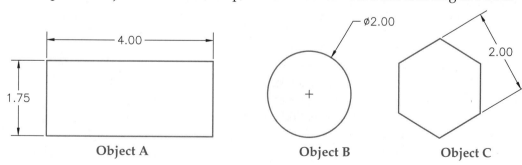

3. Draw Objects A, B, and C at the sizes shown below. Make a copy of Object A two units up. Make four copies of Object B three units up, center to center. Make three copies of Object C three units up, center to center. Save the drawing as P14-3.

Object A	Object B	Object C

Object A: 4.00 wide, 1.75 tall

Object B: ⌀2.00

Object C: 2.00

4. Open P12-4. Draw a mirror image as Object B. Then remove the original view and move the new view so that Point 2 is at the original Point 1 location. Save the drawing as P14-4.

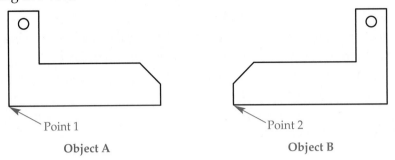

Object A   Object B

▼ Intermediate

5. Draw the object shown below. The object is symmetrical; therefore, draw only one-half. Mirror the other half into place. Use the **CHAMFER** and **FILLET** tools to your best advantage. All fillets and rounds are .125. Use the **JOIN** tool where necessary. Save the drawing as P14-5.

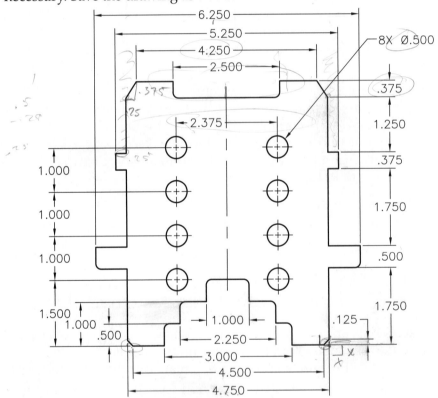

6. Draw the object shown below. Mirror the right half into place. Use the **CHAMFER** and **FILLET** tools to your best advantage. Save the drawing as P14-6.

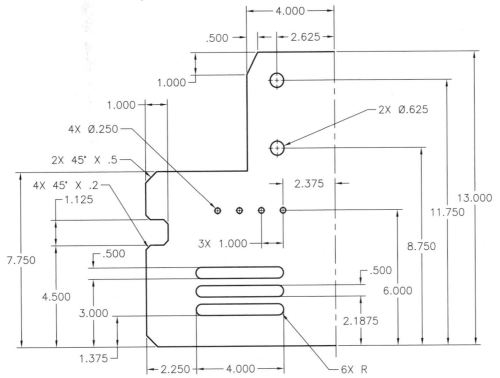

7. Redraw the objects shown below. Mirror the drawing, but make sure the text remains readable. Delete the original image during the mirroring process. Save the drawing as P14-7.

TRANSFER

LTS.HTRS.FANS

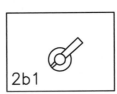

RESET

BYPASS

8. Draw this timer schematic. Save the drawing as P14-8.

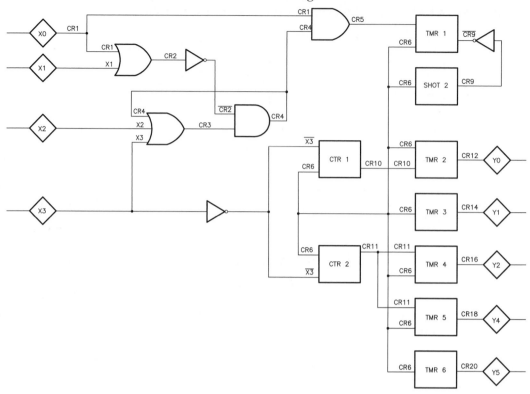

9. Use tracking and object snaps to draw the object shown below based on these instructions:
   A. Draw the outline of the object first, followed by the ten ⌀.500 holes (A).
   B. The holes labeled B are located vertically halfway between the centers of the holes labeled A. They have a diameter one-quarter the size of the holes labeled A.
   C. The holes labeled C are located vertically halfway between the holes labeled A and B. Their diameter is three-quarters of the diameter of the holes labeled B.
   D. The holes labeled D are located horizontally halfway between the centers of the holes labeled A. These holes have the same diameter as the holes labeled B.
   E. Draw the rectangles around the circles as shown.
   F. Do not draw dimensions, notes, or labels.
   G. Save the drawing as P14-9.

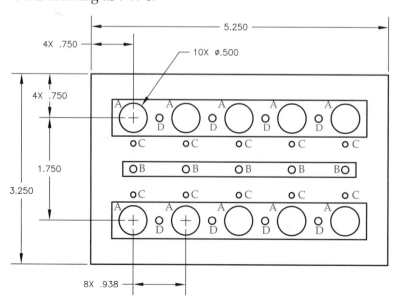

Drawing Problems - Chapter 14

10. Draw the portion of the gasket shown on the left. Use the **MIRROR** tool to complete the gasket as shown on the right. Save the drawing as P14-10.

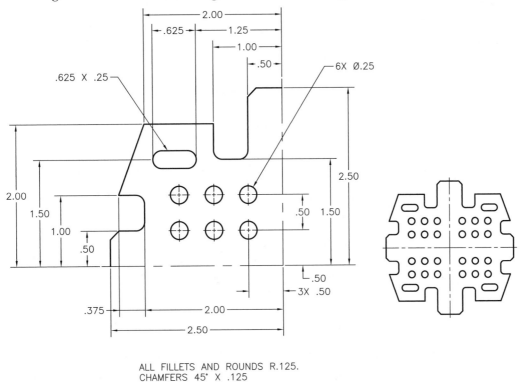

ALL FILLETS AND ROUNDS R.125.
CHAMFERS 45° X .125

11. Draw the padded bench. Use the **COPY** and **ARRAY** tools as needed. Save the drawing as P14-11.

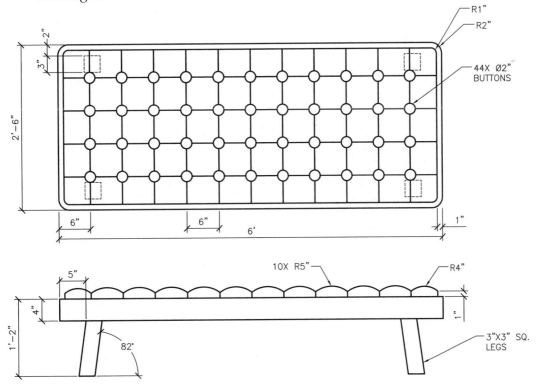

12. Draw the hand wheel shown below. Use the **ARRAY** tool to draw the spokes. Save the drawing as P14-12.

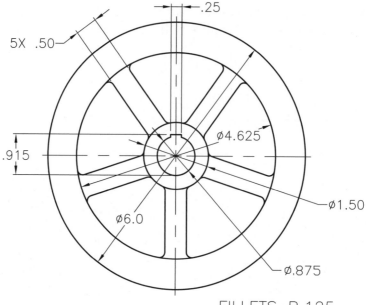

FILLETS R.125

13. Draw the control diagram. Draw one branch (including text) and use the **COPY** tool to your advantage. Use text editing tools as needed. Save the drawing as P14-13. (Design and drawing by EC Company, Portland, Oregon)

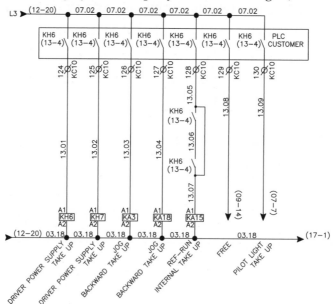

▼ Advanced

14. You have been given an engineer's sketches and notes to construct a drawing of a sprocket. Create a front and side view of the sprocket using the **ARRAY** tool. Place the drawing on one of your templates. Save the drawing as P14-14.

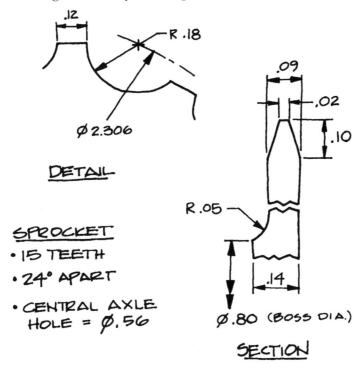

DETAIL

SPROCKET
- 15 TEETH
- 24° APART
- CENTRAL AXLE HOLE = Ø.56

SECTION

15. Draw the following object views using the dimensions given. Use **ARRAY** to construct the hole and tooth arrangements. Use one of your templates for the drawing. Save the drawing as P14-15.

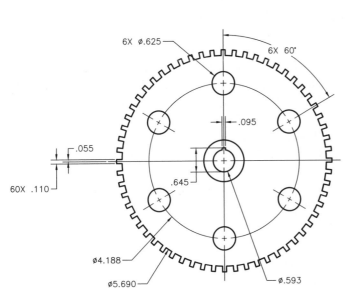

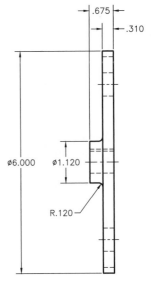

16. Draw this refrigeration system schematic. Save the drawing as P14-16.

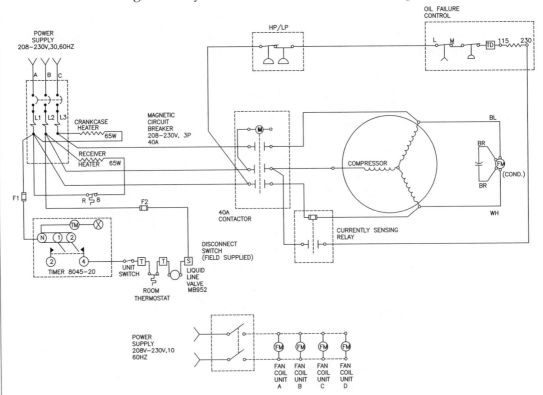

17. The following structural sketch shows a steel column arrangement on a concrete floor slab for a new building. The *I*-shaped symbols represent the steel columns. The columns are arranged in "bay lines" and "column lines." The column lines are numbered 1, 2, and 3. The bay lines are labeled *A* through *G*. The width of a bay is 24'-0". Line balloons, or tags, identify the bay and column lines. Draw the arrangement, using **ARRAY** for the steel column symbols and the tags. The following guidelines will help you:

A. Begin a new drawing using an architectural template.

B. Select architectural units and set up the drawing to print on a 36 × 24 sheet size. Determine the scale required for the floor plan to fit on this sheet size and specify the drawing limits accordingly.

C. Draw the steel column symbol to the dimensions given.

D. Set the grid spacing at 2'-0" (24").

E. Set the snap spacing at 12".

F. Draw all other objects.

G. Place text inside the balloon tags. Set the running object snap mode to **Center** and justify the text to **Middle**. Make the text height 6".

H. Save the drawing as P14-17.

AutoCAD and Its Applications—Basics

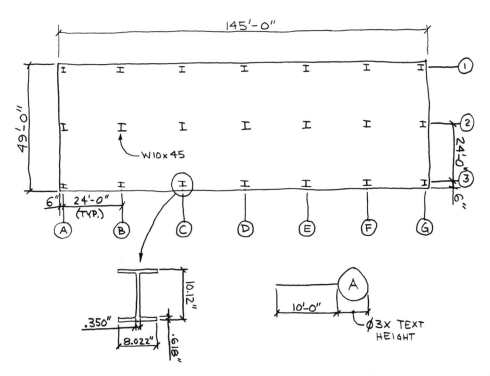

18. The sketch shown below is a proposed classroom layout of desks and chairs. One desk is shown with the layout of a chair, keyboard, monitor, and tower-mounted computer (drawn with dotted lines). All of the desk workstations should have the same configuration. The exact sizes and locations of the doors and windows are not important for this problem. Use the following guidelines to complete this problem:

A. Begin a new drawing.
B. Choose architectural units.
C. Set up the drawing to print on a C-size sheet, and be sure to create the drawing in model space.
D. Use the appropriate drawing and editing tools to complete this problem quickly and efficiently.
E. Draw the desk and computer hardware to the dimensions given.
F. Do not dimension the drawing.
G. Save the drawing as P14-18.

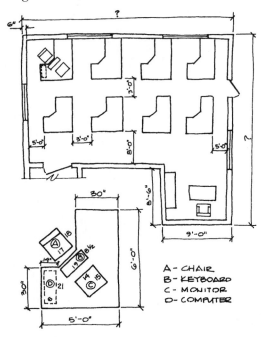

A - CHAIR
B - KEYBOARD
C - MONITOR
D - COMPUTER

Drawing Problems - Chapter 14

19. Draw the front elevation of this house. Create the features proportional to the given drawing. Use the **ARRAY** and **TRIM** tools to place the siding and porch rails evenly. Save the drawing as P14-19.

20. Use **SPLINE** and other tools, such as **ELLIPSE**, **MIRROR**, and **OFFSET**, to design an architectural door knocker similar to the one shown. Use an appropriate text tool and font to place your initials in the center. Save the drawing as P14-20.

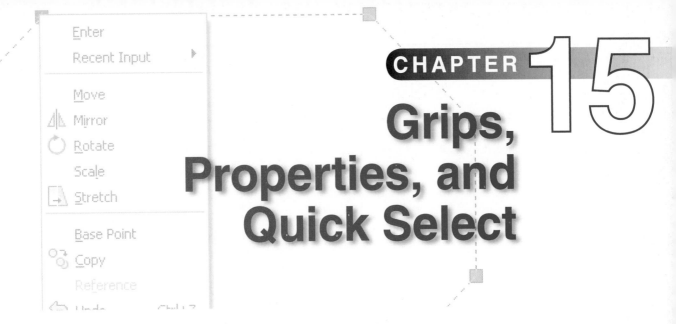

Grips, Properties, and Quick Select

Learning Objectives

After completing this chapter, you will be able to do the following:

✓ Use grips to stretch, move, rotate, scale, mirror, and copy objects.
✓ Adjust object properties using the **Quick Properties** panel and the **Properties** palette.
✓ Use the **MATCHPROP** tool to match object properties.
✓ Edit between drawings.
✓ Create selection sets using the **Quick Select** dialog box.

Typically, when using tools such as **ERASE**, **FILLET**, **MOVE**, and **COPY,** you access the tool first and then follow prompts that require you to select the object to edit. This chapter describes the process of editing objects and changing object properties by selecting an object first and then performing editing operations. This chapter also explains tools for selecting objects using selection set filters.

Using Grips

Grips appear on an object when you select the object while no tool is active. **Figure 15-1** shows the locations of grips on several different types of objects. When you select an object, the object highlights and grips appear in an *unselected grip* state. By default, unselected grips display as blue (Color 150) filled squares or arrows. Some objects include other, specialized grips, as described when applicable in this textbook. Move the crosshairs over an unselected grip to snap to the grip. Then pause to change the color of the grip to pink (Color 11). Hovering over an unselected grip and allowing it to change color helps you select the correct grip, especially when multiple grips are close together.

A *selected grip* appears as a red (Color 12) filled square or arrow by default. You can use a selected grip to perform several editing operations. If you select more than one object displaying unselected grips, what you do with the selected grips affects all of the selected objects. Objects having unselected and selected grips highlight and become part of the current selection set.

grips: Small boxes that appear at strategic points on an object when you select it, allowing you to edit the object.

unselected grips: Grips that have not yet been picked to perform an operation.

selected grip: A grip that has been picked to perform an operation.

Figure 15-1.
Grips appear at specific locations on objects.

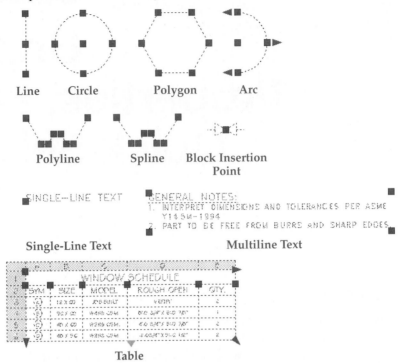

Line   Circle   Polygon   Arc

Polyline   Spline   Block Insertion
Point

Single-Line Text   Multiline Text

Table

To remove highlighted objects from a selection, hold down the [Shift] key and pick the objects to remove. [Shift] also allows you to add or remove selected grips. To make a second grip selected, hold [Shift] down while selecting both the first and second grip. To add more grips to the selected grip set, continue to hold [Shift] down and pick additional grips. With [Shift] held down, picking a selected (red) grip returns it to the unselected (blue) state. **Figure 15-2** shows an example of modifying two circles at the same time using selected grips.

Return objects to the selection set by picking them again. Remove all selected grips from the selection set by pressing [Esc] to cancel. Press [Esc] again to deselect all objects and remove grips from the selection set. You can also right-click and pick **Deselect All** to remove all grips.

Figure 15-2.
You can modify multiple objects simultaneously by pressing [Shift] to select additional grips.

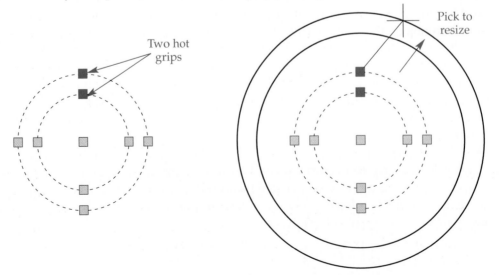

Pick to resize

Two hot grips

NOTE

Use the appropriate options in the **Grip Size** and **Grips** areas of the **Selection** tab in the **Options** dialog box to control grip size and color.

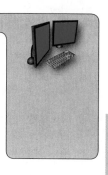

PROFESSIONAL TIP

You can perform many conventional editing operations when you pick an object to display unselected grips. For example, you can use the **ERASE** tool to remove selected objects that display unselected grips by first picking the objects and then activating the **ERASE** tool or pressing [Delete]. This technique is available when the **Noun/verb selection** check box is selected in the **Selection Modes** area of the **Selection** tab in the **Options** dialog box.

noun/verb selection: Performing tasks in AutoCAD by selecting the objects before entering a tool.

verb/noun selection: Performing tasks in AutoCAD by entering a tool before selecting objects.

Using Grip Tools

Grips provide access to the **STRETCH, MOVE, ROTATE, SCALE**, and **MIRROR** tools. In addition, the **Copy** option of the **MOVE** tool and sometimes, depending on the selected grip, the **STRETCH** tool imitate the **COPY** tool. Grip tools become available at the dynamic input cursor and the command line when you select a grip. Do not attempt to use conventional means of tool access, such as the ribbon. The first tool is **STRETCH**, as indicated by the ** STRETCH ** Specify stretch point or [Base point/Copy/Undo/eXit]: prompt. You can use the **STRETCH** tool at this prompt, or press [Enter] or the space bar or right-click and select **Enter** to cycle through additional tool options:

```
** STRETCH **
Specify stretch point or [Base point/Copy/Undo/eXit]: ↵
** MOVE **
Specify move point or [Base point/Copy/Undo/eXit]: ↵
** ROTATE **
Specify rotation angle or [Base point/Copy/Undo/Reference/eXit]: ↵
** SCALE **
Specify scale factor or [Base point/Copy/Undo/Reference/eXit]: ↵
** MIRROR **
Specify second point or [Base point/Copy/Undo/eXit]: ↵
** STRETCH **
Specify stretch point or [Base point/Copy/Undo/eXit]:
```

As an alternative to cycling through the tool options, you can right-click to access a grips shortcut menu, which is available only after you select a grip. The shortcut menu allows you to access the five grip editing tools without using the keyboard. A third option to activate a tool is to enter the first two characters of the desired tool. Type MO for **MOVE**, MI for **MIRROR**, RO for **ROTATE**, SC for **SCALE**, or ST for **STRETCH**.

Stretching Objects

The process of stretching using grips is similar to stretching using the traditional **STRETCH** tool. The main difference is that the selected grip acts as the stretch base point. Move the crosshairs to stretch the selected object. See **Figure 15-3**. If you pick the middle grip of a line or arc, the center grip of a circle, or the insertion point grip of a block, single-line text, multiline text, or table, the object moves instead of stretching.

Figure 15-4 shows two methods to stretch features of an object. Step 1 in **Figure 15-4A** stretches the first corner, and Step 2 stretches the second corner. You can combine the two operations by holding down [Shift] to select two grips, as shown in **Figure 15-4B**.

Figure 15-3.
Using the **STRETCH** grip tool. Note the selected grip in each case.

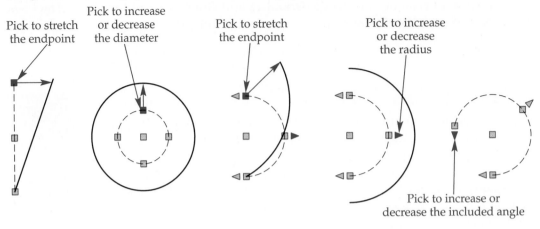

Figure 15-4.
Stretching an object. A—Select corners to stretch individually. B—Select several grips by holding down [Shift].

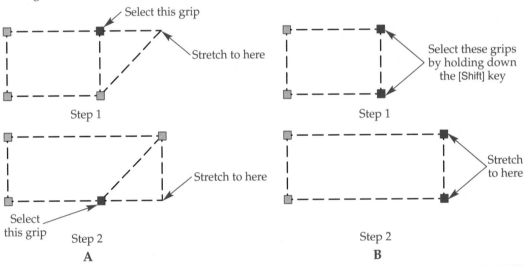

Use the **Base point** option to specify a base point instead of using the selected grip as the base point. Activate the **Undo** option to undo the previous operation. Choose the **eXit** option or press [Esc] to exit without completing the stretch. When you finish stretching, or exit the tool, the selected grip is gone, but unselected grips remain. Press [Esc] to remove the unselected grips.

Dynamic input is especially effective with the **STRETCH** grip tool. **Figure 15-5** shows an example of how you can use dimensional input to quickly modify the size of a circle or offset a circle by a specific distance using the **STRETCH** grip tool. In this example, enter the new radius of the circle in the distance input field, or press [Tab] to enter an offset in the other distance input field. This is just one example of how you can use dynamic input with grips. Most objects include similar operations.

> **PROFESSIONAL TIP**
>
> You can use grid snaps, coordinate entry techniques, polar tracking, object snaps, and object snap tracking with any of the grip editing tools to improve accuracy. Remember to use these functions as you edit your drawings.

Figure 15-5.
Using the dimensional input feature of dynamic input with the **STRETCH** grip tool.

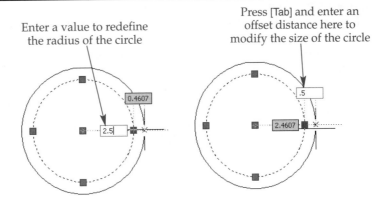

Enter a value to redefine the radius of the circle

Press [Tab] and enter an offset distance here to modify the size of the circle

0.4607

2.5

.5

2.4607

Exercise 15-1

Access the Student Web site (www.g-wlearning.com/CAD) and complete Exercise 15-1.

Moving Objects

To move an object with grips, select the object, pick a grip to use as the base point, and then activate the **MOVE** tool. Specify a new location for the base point to move the object. See **Figure 15-6**. The **Base point**, **Undo**, and **eXit** options are similar to those for the **STRETCH** tool.

Exercise 15-2

Access the Student Web site (www.g-wlearning.com/CAD) and complete Exercise 15-2.

Rotating Objects

To rotate an object using grips, select the object, pick a grip to use as the base point, and then activate the **ROTATE** tool. Specify a rotation angle from the base point to rotate the object. The **Base point**, **Undo**, and **eXit** options are similar to those for the **STRETCH** tool.

You can use the **Reference** option to specify a new angle in relation to an existing angle. The reference angle is the current angle of the object. If you know the value of

Figure 15-6.
When you specify the **MOVE** tool, the selected grip becomes the base point for the move.

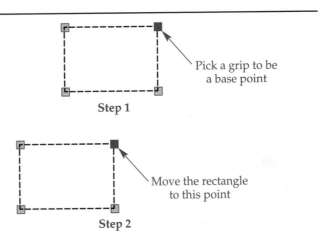

Pick a grip to be a base point

Step 1

Move the rectangle to this point

Step 2

Figure 15-7.
The rotation angle
and **Reference**
options of the
ROTATE grip tool.

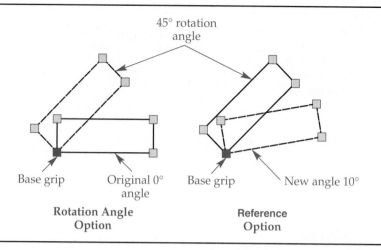

45° rotation
angle

Base grip   Original 0°   Base grip   New angle 10°
     angle

Rotation Angle   **Reference**
Option   **Option**

the current angle, enter the value at the prompt. Otherwise, pick two points on the reference line to identify the existing angle. Enter a value for the new angle or pick a point. **Figure 15-7** shows the **ROTATE** options.

Exercise 15-3

Access the Student Web site (www.g-wlearning.com/CAD) and complete Exercise 15-3.

Scaling Objects

To scale an object with grips, select the object, pick a grip to use as the base point, and then activate the **SCALE** tool. Enter a scale factor or pick a point to increase or decrease the size of the object. The **Base point**, **Undo**, and **eXit** options are similar to those for the **STRETCH** tool.

You can use the **Reference** option to specify a new size in relation to an existing size. The reference size is the current length, width, or height of the object. If you know the current size, enter the value at the prompt. Otherwise, pick two points on the reference line to identify the existing size. Enter a value for the new size or pick a point. **Figure 15-8** shows the **ROTATE** options.

Exercise 15-4

Access the Student Web site (www.g-wlearning.com/CAD) and complete Exercise 15-4.

Mirroring Objects

To mirror an object with grips, select the object, pick a grip to use as the first point of the mirror line, and activate the **MIRROR** tool. Then pick another grip or any point on-screen to locate the second point of the mirror line. See **Figure 15-9**. Unlike the standard **MIRROR** tool, the grips version of the **MIRROR** tool does not give you the immediate option to delete the old objects. Old objects delete automatically. To keep the original object while mirroring, use the **Copy** option of the **MIRROR** tool. The **Base point**, **Undo**, and **eXit** options are similar to those for the **STRETCH** tool.

Figure 15-8.
When using the **SCALE** tool with grips, you can enter a scale factor or use the **Reference** option.

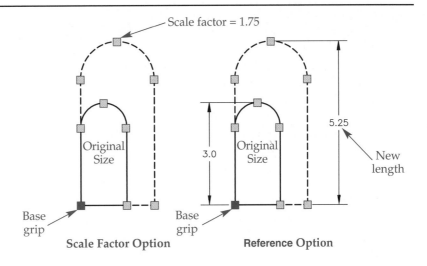

Scale factor = 1.75

Original Size

Base grip

Scale Factor Option

Original Size

Base grip

3.0

5.25

New length

Reference Option

Figure 15-9.
When you use grips to access the **MIRROR** tool, the selected grip becomes the first point of the mirror line, and the original object automatically deletes.

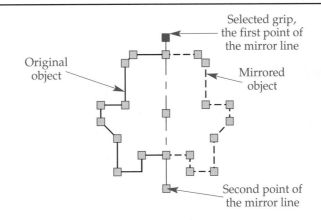

Selected grip, the first point of the mirror line

Original object

Mirrored object

Second point of the mirror line

Exercise 15-5

Access the Student Web site (www.g-wlearning.com/CAD) and complete Exercise 15-5.

Copying Objects

The **Copy** option is included in each of the grip editing tools. The effect of using the **Copy** option depends on the selected grip and tool. The original selected object remains unchanged, and the copy stretches when the **STRETCH** tool is active, rotates when the **ROTATE** tool is active, or scales when the **SCALE** tool is active. The **Copy** option of the **MOVE** tool is the true form of the **COPY** tool, allowing you to copy from any selected grip. The selected grip acts as the copy base point. Create as many copies of the selected object as needed, and then exit the tool.

Exercise 15-6

Access the Student Web site (www.g-wlearning.com/CAD) and complete Exercise 15-6.

Adjusting Object Properties

Every drawing object has specific properties. Some objects, such as lines, have few properties, while other objects, such as multiline text or tables, contain many properties. Properties include geometry characteristics, such as the location of the endpoint of a line in X,Y,Z space, the diameter of a circle, or the area of a rectangle. Layer is another property associated with all objects. The layer on which you draw an object defines other properties, including color, linetype, and lineweight. Most objects also include object-specific properties. For example, text objects have text properties, and tables have column, row, and cell properties.

You can edit object properties using modification tools or a variety of other methods, depending on the property. For example, you can adjust layer characteristics using layer tools. The multiline text editor allows you to adjust existing multiline text properties. Another technique to view and make changes to object properties is to use the **Quick Properties** panel or the **Properties** palette. These tools are especially effective for modifying a particular property or set of properties for multiple objects at once.

PROFESSIONAL TIP

You can view object, color, layer, and linetypes properties by hovering over an object. This is a quick way to reference basic object information. See **Figure 15-10**.

Using the Quick Properties Panel

The **Quick Properties** panel, shown in **Figure 15-11**, appears by default when you pick an object. The **Quick Properties** panel only floats, and by default, it appears above and to the right of the crosshairs. The drop-down list at the top of the **Quick Properties** panel indicates the type of object selected. Properties associated with the selected object display below the drop-down list in rows.

Figure 15-10.
Hover over an object to quickly view object, color, layer, and linetype properties.

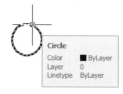

Figure 15-11.
You can use the **Quick Properties** panel to modify some object properties. A—The initial display of the **Quick Properties** panel for the **LINE** tool. B—The expanded list of properties.

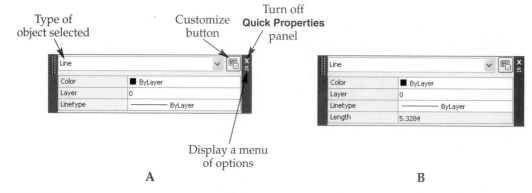

A   B

Figure 15-12.
The **Quick Properties** panel with three objects selected. You can edit the objects individually or select All (3) to edit them all together.

Total number of objects selected

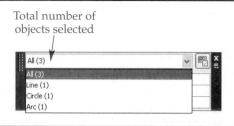

If you pick a circle, for example, rows of circle properties are listed. When you pick multiple objects, use the **Quick Properties** panel to modify all of the objects, or pick only one of the object types from the drop-down list to modify. See **Figure 15-12.** Select All (*n*) to change the properties of all selected objects. Only properties shared by all selected objects appear when you choose All (*n*). Select the appropriate object type to modify a single type of object.

The **Quick Properties** panel lists the most common properties associated with the selected objects. You should recognize most of the properties; they are the same values you use to create the objects. By default, three properties display, unless the selected objects contain fewer properties. If more properties are available, hover over a row or a **Quick Properties** panel side bar to expand the list.

To change a property, pick the property or its current value. The way you change a value depends on the property. A text box opens when you select certain properties, such as the **Radius** property of an arc or circle or the **Text height** property of a single-line or multiline text object. Enter a new value in the text box to change the property. Most text boxes display a calculator icon on the right side that opens the **QuickCalc** tool for calculating values. Chapter 16 covers using **QuickCalc**. Other properties, such as the **Layer** property, display a drop-down list of selections. A pick button is available for geometric properties, such as the **Center X** and **Center Y** properties of a circle. Select the pick button to specify a new coordinate location. When you choose an available ... (ellipsis) button, a dialog box related to the property opens.

Press [Esc] or pick the **Close** button in the upper-right corner of the panel to hide the **Quick Properties** panel. Closing the **Quick Properties** panel does not disable the tool. If you choose not to use the **Quick Properties** panel, a quick way to disable or enable the panel is to pick the **Quick Properties** button on the status bar.

Exercise 15-7

Access the Student Web site (www.g-wlearning.com/CAD) and complete Exercise 15-7.

Supplemental Material

Quick Properties Panel Options
For information about adjusting **Quick Properties** panel options, go to the Student Web site (www.g-wlearning.com/CAD), select this chapter, and select **Quick Properties Panel Options**.

Ribbon
View
> Palettes

Properties
Home
> Properties

Properties
Type
PROPERTIES
PROPS
CH
MO
[Ctrl]+[1]

Using the Properties Palette

The **Properties** palette, shown in **Figure 15-13,** provides the same function as the **Quick Properties** panel, but it allows you to view and adjust all properties related to the selected objects. You can dock, lock, and resize the **Properties** palette in the drawing area. You can access tools and continue to work while the **Properties** palette is displayed. To close the palette, pick the **X** in the top-left corner, select **Close** from the options menu, or press [Ctrl]+[1].

> **NOTE**
>
> If you have already selected an object, you can access the **Properties** palette by right-clicking and selecting **Properties**. You can also double-click many objects to select the object and open the **Properties** palette automatically.

Categories divide the **Properties** palette. When no object is selected, the **General**, **3D Visualization**, **Plot style**, **View**, and **Misc** categories list the current settings for the drawing. Underneath each category are rows of object properties. For example, in **Figure 15-13,** the Color row in the **General** category displays the current color. The color property in this example is ByLayer.

The upper-right portion of the **Properties** palette contains three buttons. Pick the **Quick Select** button to access the **Quick Select** dialog box, where you can create object selection sets, as described later in this chapter. Picking the **Select Objects** button deselects the currently selected objects and changes the crosshairs to a pick box, allowing you to select other objects. The third button toggles the value of the **PICKADD** system variable, which determines whether you need to hold down [Shift] when adding objects to a selection set.

In order to modify an object using the **Properties** palette, you must open the palette and select an object. For example, if your drawing includes a circle and a line and you want to modify the circle, either double-click the circle or pick the circle and then open the **Properties** palette. Working with properties in the **Properties** palette is very similar working with properties in the **Quick Properties** panel. When you select objects, the drop-down list at the top of the **Properties** palette indicates the types of

Figure 15-13.
You can use the **Properties** palette to modify object properties.

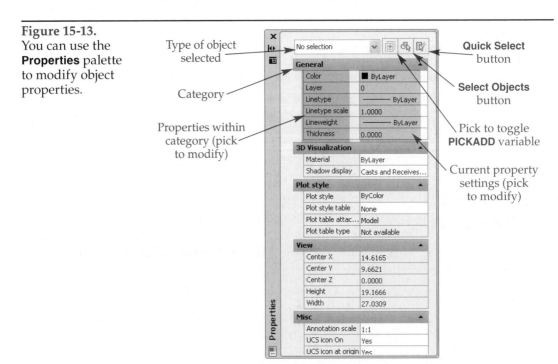

objects selected. The categories and property rows update to display properties associated with your selections.

If you pick a circle, for example, circle properties are listed. When you select multiple objects, use the **Properties** palette to modify all of the objects, or pick only one of the object types from the drop-down list to modify. See **Figure 15-14**. This drop-down list works exactly like the drop-down list in the **Quick Properties** panel.

You will recognize many of the properties listed in the **Properties** palette. However, the **Properties** palette lists all properties related to the object, including some you may not recognize. For example, the **3D Visualization** category and any properties related to the Z axis are for use in 3D applications. You should not adjust these properties or any other properties that you are not familiar with for basic drawing applications.

Change properties using the **Properties** palette the same way you change properties using the **Quick Properties** palette. Use the appropriate text box, drop-down list, or button to modify the value. After you make all the changes to the object, press [Esc] to clear the grips and remove the object from the **Properties** palette. The object now appears in the drawing window with the desired changes.

General Properties

All objects have a **General** category in the **Properties** palette. Refer again to **Figure 15-13**. The **General** category allows you to modify properties such as color, layer, linetype, linetype scale, plot style, lineweight, and thickness. The **Quick Properties** panel also lists certain general properties.

NOTE

You can change the layer of a selected object by choosing a layer from the **Layer Control** drop-down list in the **Layers** panel on the **Home** ribbon tab. You can change color, linetype, plot style, and lineweight by choosing from the appropriate drop-down list in the **Properties** panel on the **Home** ribbon tab.

CAUTION

Colors, linetypes, and lineweights should typically be set as ByLayer. Changing color, linetype, or lineweight to a value other than ByLayer overrides logical properties, making the property an *absolute value*. Therefore, if the color of an object is set to red, for example, it appears red regardless of the layer on which it is drawn.

For most applications, linetype scale should be set globally so the linetype scale of all objects is constant. Adjusting the linetype scale of individual objects can create nonstandard drawings and make it difficult to adjust linetype scale globally. For most applications, you should not override color, linetype, linetype scale, plot style, lineweight, or thickness.

absolute value: In property settings, a value set directly instead of being referenced by layer or a block. An absolute value ignores the current layer settings.

Figure 15-14.
The **Properties** palette with four objects selected. You can edit the objects individually or select All (4) to edit them all together.

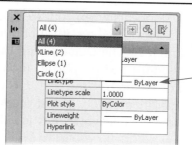

When All is selected only properties common to all selected objects appear

Figure 15-15.
The **Properties** palette with a Line object selected. Line objects have options in only three property categories.

Type of object selected

General properties

These values cannot be directly modified, but change if endpoints are modified

Start point and endpoint coordinates

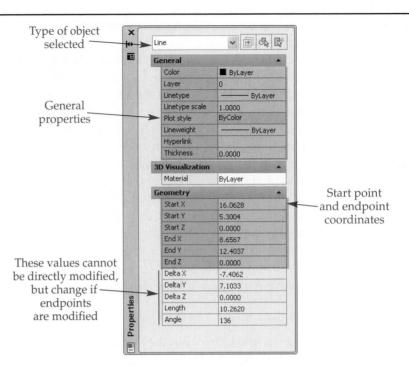

Geometry Properties

One of the most common categories in the **Properties** palette is **Geometry**. See **Figure 15-15.** Although most objects have a **Geometry** category, the properties within the category vary depending on the object. The **Quick Properties** panel also lists certain geometry properties. Typically, three properties allow you to change the absolute coordinates for the object by specifying the X, Y, and Z coordinates. When you select one of these properties, a pick button appears, allowing you to pick a point in the drawing for the new location. In addition to choosing a point with the pick button, you can change the value of the coordinate in a text box or use the calculator button to calculate a new location.

Figure 15-16 shows an example of the properties displayed when you select a circle. The **Geometry** category displays the current location of the center of the circle by showing three properties: **Center X**, **Center Y**, and **Center Z**. To choose a new center location for the circle, select the appropriate property. Pick a new point or type or calculate the coordinate values. You can also modify other circle size properties, such as the radius, diameter, circumference, and area.

Exercise 15-8

Access the Student Web site (www.g-wlearning.com/CAD) and complete Exercise 15-8.

Text Properties

The **Text** category appears when you select single-line or multiline text, and includes properties such as text style and justification. **Figure 15-17** shows text properties associated with multiline text. The **Quick Properties** panel also lists certain text-specific properties. You can adjust text properties using traditional techniques, such as reentering the text editor to modify the text. However, the **Properties** palette provides a convenient way to modify a variety of text properties without reentering the text editor. It is especially effective when you want to adjust a particular property for multiple

Figure 15-16.
The **Properties** palette with a Circle object selected for editing.

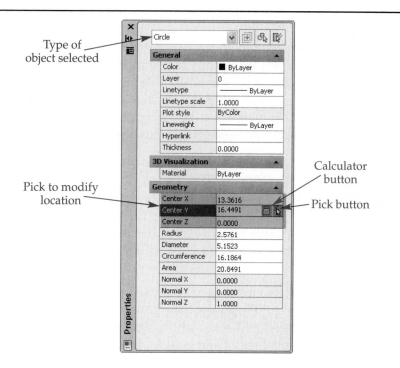

Type of object selected

Calculator button

Pick to modify location

Pick button

Figure 15–17.
The **Properties** palette shows the properties of the selected text. The properties of single-line text are slightly different from those of multiline text.

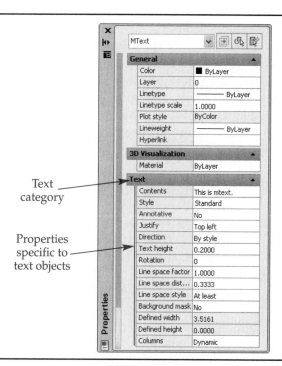

Text category

Properties specific to text objects

selected text objects. For example, you can change the annotative setting of all text in the drawing using the **Annotative** property row, or reset the height of multiple single-line or multiline text objects using the **Height** property row.

Exercise 15-9

Access the Student Web site (www.g-wlearning.com/CAD) and complete Exercise 15-9.

Table Properties

The **Properties** palette displays certain table properties depending on whether you select inside a cell or pick a cell edge to edit table layout. See **Figure 15-18**. The **Cell** and **Content** categories, which appear when you select inside a cell, allow you to adjust the selected cell properties. The **Table** and **Table Breaks** categories shown when you select a cell edge include common table settings and table break properties. The **Quick Properties** panel also lists certain table-specific properties.

> **PROFESSIONAL TIP**
>
> If the height of table rows is taller than desired, or if rows become unequal in height, enter a very small value in the **Table height** row of the **Table** category to return all rows to the smallest height possible based on the margin spacing between cell content and cell borders.

After you add table breaks, several useful options become available in the **Table Breaks** category of the **Properties** palette for adjusting table sections. The **Enabled** option toggles between the broken and unbroken table display. The **Yes** value displays when you create table breaks and enable breaking. Pick **No** to return the table to an unbroken display. The **Direction** option defines direction of broken table flow, or wrap. The default **Right** option wraps the table to the right. Select **Left** to wrap the table to the left or pick **Up** to wrap the table above.

The **Repeat top labels** option of the **Table Breaks** category repeats cells that use a **Label** cell type at the beginning of each section. Typically, the title cell and header cells use a **Label** cell type. Choose **Yes** to add the title and header cells to the wrapped table sections. The **Repeat bottom labels** option repeats cells that use a **Label** cell type at the end of each section. The **Manual positions** option allows you to move table sections independently while maintaining the table as a single object. When you select **No**, table sections move as a group.

Figure 15-18.
Table properties. A—The properties displayed when you select inside a cell to edit table layout. B—The properties displayed when you pick a cell edge to edit table layout.

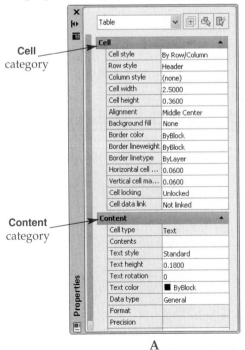

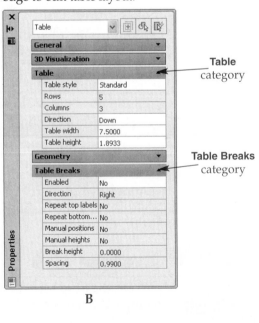

The **Manual height** option of the **Table Breaks** category adds a table-breaking grip to each section. This allows you to adjust the number of rows in each section independently and add additional breaks. When you select **No**, the table-breaking grip appears at the original section only and controls the number of breaks. The **Break height** text box allows you to define the height of each table section. The selected height determines the number of sections. The **Spacing** text box allows you to define the spacing between table sections. A value of 0 places the sections together.

Exercise 15-10

Access the Student Web site (www.g-wlearning.com/CAD) and complete Exercise 15-10.

Matching Properties

Ribbon

Home
> Clipboard

Match Properties

Type

MATCHPROP
PAINTER
MA

MATCHPROP

The **MATCHPROP** tool allows you to copy, or "paint," properties quickly from one object to other objects. You can match properties in the same drawing or between drawings. When you first access the **MATCHPROP** tool, AutoCAD prompts you for the *source object*. After you select the source object, AutoCAD displays the properties it will paint. The next prompt allows you to pick the *destination object*.

To change the paint properties, select the **Settings** option before picking the destination objects. The **Property Settings** dialog box appears, showing the properties to paint. See **Figure 15-19**. The **Basic Properties** area lists the general properties of the source object. The **Special Properties** area contains check boxes that allow you to paint over a variety of additional properties related to object styles. Properties replace in the destination object if the corresponding **Property Settings** dialog box check boxes are active. For example, if you want to paint only the layer property and text style of one text object to another text object, uncheck all boxes except the **Layer** and **Text** property check boxes.

source object:
When matching properties, the object with the properties you want to copy to other objects.

destination object:
When matching properties, the object that receives the properties of the source object.

Figure 15-19.
The **Property Settings** dialog box for the **MATCHPROP** tool. Select the properties to paint onto a new object.

Properties to be painted to other objects

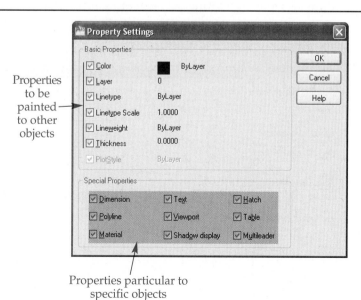

Properties particular to specific objects

Exercise 15-11

Access the Student Web site (www.g-wlearning.com/CAD) and complete Exercise 15-11.

Editing between Drawings

You can edit in more than one drawing at a time and edit between open drawings. For example, you can copy objects from one drawing to another. You can also refer to a drawing to obtain information, such as a distance, while working in a different drawing.

Figure 15-20 shows two drawings (the drawings created during Exercises 15-8 and 15-11) tiled horizontally. The Windows *copy and paste* function allows you to copy an object from one drawing to another. To use this feature, select the object you intend to copy. For example, if you want to copy the pentagon from drawing EX15-11 to drawing EX15-8, first select the pentagon. Then right-click to display the shortcut menu shown in **Figure 15-21.**

The shortcut menu has two options that allow you to copy to the Windows Clipboard. The **Copy** option copies selected objects from AutoCAD onto the Windows Clipboard to use in an AutoCAD drawing or another application. The **Copy with Base Point** option also copies the selected objects to the Clipboard, but it allows you to specify a base point to position the copied object for pasting. When you use this option, AutoCAD prompts you to select a base point. Select a logical base point, such as a corner or center point of the object.

After you select one of the copy options, make the second drawing active by picking the drawing or by pressing [Tab]+[Ctrl]. Right-click to display the shortcut menu shown in **Figure 15-22.** Notice that the copy options remain available, but three

copy and paste: A Windows function that allows an object to be copied from one location and pasted into another.

Figure 15-20.
Tile multiple drawings to make editing between drawings easier.

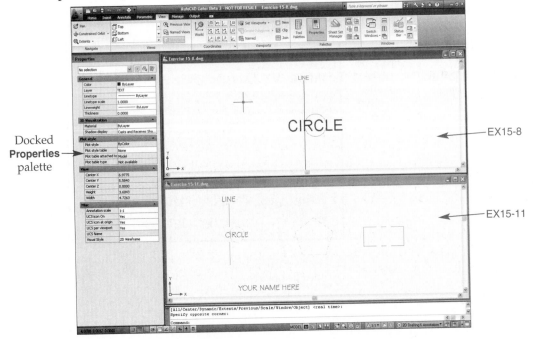

Docked **Properties** palette

EX15-8

EX15-11

AutoCAD and Its Applications—Basics

Figure 15-21.
Two copy options appear on the shortcut menu.

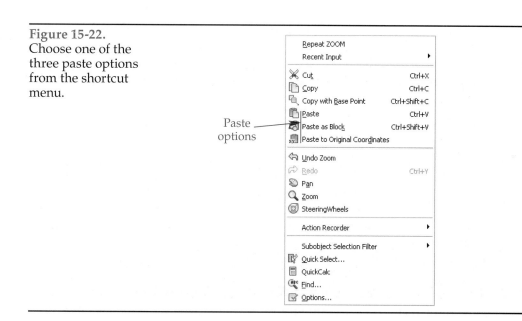

Copy options

	Repeat ZOOM	
	Recent Input	▶
⬭	Polyline Edit	
	Annotative Object Scale	▶
✂	Cut	Ctrl+X
🗐	Copy	Ctrl+C
🗐	Copy with Base Point	Ctrl+Shift+C
📋	Paste	Ctrl+V
🗐	Paste as Block	Ctrl+Shift+V
	Paste to Original Coordinates	
✎	Erase	
✛	Move	
	Copy Selection	
	Scale	
↻	Rotate	
	Draw Order	▶
	Deselect All	
	Action Recorder	▶
	Subobject Selection Filter	▶
	Quick Select...	
	QuickCalc	
	Find...	
	Properties	
✓	Quick Properties	

Figure 15-22.
Choose one of the three paste options from the shortcut menu.

Paste options

	Repeat ZOOM	
	Recent Input	▶
✂	Cut	Ctrl+X
🗐	Copy	Ctrl+C
🗐	Copy with Base Point	Ctrl+Shift+C
📋	Paste	Ctrl+V
🗐	Paste as Block	Ctrl+Shift+V
	Paste to Original Coordinates	
↩	Undo Zoom	
↪	Redo	Ctrl+Y
	Pan	
🔍	Zoom	
	SteeringWheels	
	Action Recorder	▶
	Subobject Selection Filter	▶
	Quick Select...	
	QuickCalc	
	Find...	
	Options...	

paste options are now available. These options are available only if there is something on the Clipboard. The **Paste** option pastes the information on the Clipboard into the current drawing. If you used the **Copy with Base Point** option, the objects to paste are attached to the crosshairs at the specified base point.

The **Paste as Block** option "joins" all objects on the Clipboard when they are pasted into the drawing. The pasted objects act like a block in that they are single objects grouped together to form one object. Blocks are covered later in this textbook. Use the **EXPLODE** tool to break up the block so that the objects act individually. The **Paste to Original Coordinates** option pastes the objects from the Clipboard to the same coordinates at which they were located in the original drawing.

NOTE

You can also copy and paste between documents using the Windows-standard [Ctrl]+[C] and [Ctrl]+[V] keyboard shortcuts, or use the buttons available from the **Clipboard** panel on the **Home** ribbon tab.

PROFESSIONAL TIP

You may find it more convenient to use the **MATCHPROP** tool to match properties between drawings. To use the **MATCHPROP** tool between drawings, select the source object from one drawing and the destination object from another.

Exercise 15-12

Access the Student Web site (www.g-wlearning.com/CAD) and complete Exercise 15-12.

QSELECT

Ribbon
Home
> Utilities

Quick Select

Type
QSELECT

Using Quick Select

When creating complex drawings, you often need to perform the same editing operation on many objects. For example, suppose you designed a complex metal part with more than 40 holes to accept 1/8″ screws. A design change occurs, and you are notified that 3/16″ screws are to be used instead. Therefore, the hole size must also change. You could select and modify each circle individually, but it would be more efficient to create a selection set of all the circles and then modify them all at the same time. The **Quick Select** dialog box is the most common tool for creating selection sets by specifying object types and property values for selection. See **Figure 15-23**.

NOTE

You can also access the **Quick Select** dialog box by right-clicking in the drawing area and choosing **Quick Select...** or by picking the **Quick Select** button on the **Properties** palette.

PROFESSIONAL TIP

In reference to the example of a part with multiple holes of the same size, you can use parametric tools, explained in Chapter 22 to make all the circles equal in size. Then, when you change the diameter of one circle, all circles change to the new value.

With the **Quick Select** dialog box, you can quickly create a selection set based on specified filtering criteria. One method is to specify an object type (such as text, line, or circle) to select throughout the drawing. Another option is to specify a property (such as a color or layer) that objects must possess in order to be selected. A third option is to

Figure 15-23.
Selection sets can be defined in the **Quick Select** dialog box.

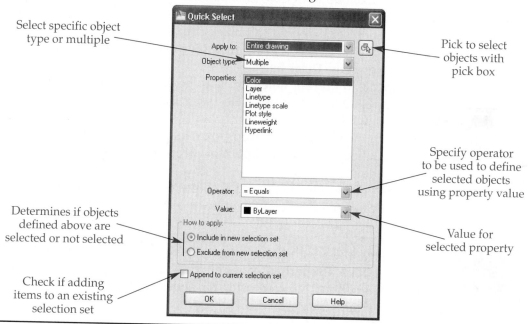

Select specific object type or multiple

Pick to select objects with pick box

Specify operator to be used to define selected objects using property value

Determines if objects defined above are selected or not selected

Value for selected property

Check if adding items to an existing selection set

pick the **Select objects** button and select objects on-screen. Once the selection criteria are defined using one of these techniques, you can use the radio buttons in the **How to apply:** area to include or exclude the selected objects.

Look at **Figure 15-24** as you read the following steps for using the **Quick Select** tool:

1. Open the **Quick Select** dialog box.
2. Select **Entire drawing** from the **Apply to:** drop-down list. (If you access the **Quick Select** dialog box after you select objects, a **Current selection** option allows you to create a subset of the existing set.)
3. Select **Multiple** from the **Object type:** drop-down list. This allows you to select any object type. The drop-down list contains all the object types in the drawing.
4. Select **Color** from the **Properties:** list. The items in the **Properties:** list vary depending on what you specify in the **Object type:** drop-down list.
5. Select **= Equals** from the **Operator:** drop-down list.
6. The **Value:** drop-down list contains values corresponding to the entry in the **Properties:** drop-down list. In this case, color values are listed. Select the color of the right-hand objects in **Figure 15-24A**.
7. Pick the **Include in new selection set** radio button in the **How to apply:** area.
8. Pick the **OK** button to select all objects with the color specified in the **Value:** drop-down list. See **Figure 15-24B**.

Figure 15-24.
Creating selection sets with the **Quick Select** dialog box. A—Objects in the drawing. B—Selection set containing objects with the display color specified. C—Circle object added to the initial selection set.

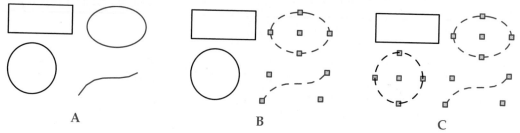

A   B   C

Once you select a set of objects, you can use the **Quick Select** dialog box to refine the selection set. Use the **Exclude from new selection set** option to exclude objects, or use the **Append to current selection set** option to add objects. The following procedure refines the previous selection set to include all black circles in the drawing.

1. While the initial set of objects is selected, right-click in the drawing area and select **Quick Select...** to open the **Quick Select** dialog box.
2. Check the **Append to current selection set** check box at the bottom of the dialog box. AutoCAD automatically selects the **Entire drawing** option in the **Apply to:** drop-down list.
3. Select **Circle** from the **Object type:** drop-down list, **Color** from the **Properties:** drop-down list, **= Equals** from the **Operator:** drop-down list, and **Black** from the **Value:** drop-down list.
4. Pick the **Include in new selection set** radio button in the **How to apply:** area.
5. Pick the **OK** button. The selection now appears as shown in **Figure 15-24C.**

PROFESSIONAL TIP

Use the **Quick Properties** panel or the **Properties** palette to adjust properties of items selected using the **Quick Select** dialog box.

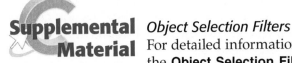

Supplemental Material

Object Selection Filters

For detailed information about selecting multiple objects using the **Object Selection Filters** dialog box, go to the Student Web site (www.g-wlearning.com/CAD), select this chapter, and select **Object Selection Filters**.

Supplemental Material

Creating Object Groups

For information about creating object groups, go to the Student Web site (www.g-wlearning.com/CAD), select this chapter, and select **Creating Object Groups**.

Express Tools

Chapter 15

AutoCAD includes additional tools for improved functionality and productivity during the drawing process. The following Express Tools apply to creating selection sets. For information about these tools, go to the Student Web site (www.g-wlearning.com/CAD), select this chapter, and select **Using Selection Express Tools.**

Get Selection Set
Fast Select

Chapter Test

Answer the following questions. Write your answers on a separate sheet of paper or go to the Student Web site (www.g-wlearning.com/CAD) and complete the electronic chapter test.

1. Name the editing tools that can be accessed automatically using grips.
2. How can you select a grip tool other than the default **STRETCH**?
3. What is the purpose of the **Base Point** option in the grip tools?
4. Explain the function of the **Undo** option in the grip tools.
5. What happens when you choose the **eXit** option from the grips shortcut menu?
6. Which option of the **ROTATE** grip tool option would you use to rotate an object from an existing 60° angle to a new 25° angle?
7. What scale factor would you use to scale an object to become three-quarters of its original size?
8. Describe the options for editing object properties.
9. Where does the **Quick Properties** panel appear by default when an object is selected?
10. By default, how many properties are shown in the **Quick Properties** panel?
11. Describe three items that might be displayed when you pick a property from a **Quick Properties** panel or **Properties** palette row.
12. Identify at least two ways to access the **Properties** palette.
13. Explain how you would change the radius of a circle from 1.375 to 1.875 using the **Properties** palette.
14. How can you change the linetype of an object using the **Properties** palette?
15. For most applications, what value should you use for the color, linetype, and lineweight of objects?
16. What tool is used to change the properties of objects to match the properties of a different object?
17. Briefly explain how the Windows copy and paste function works to copy an object from one drawing to another.
18. Name the paste option that joins a group of objects as a block when they are pasted.
19. When you use the option described in Question 18, how do you separate the objects back into individual objects?
20. Identify three ways to open the **Quick Select** dialog box.

Drawing Problems

Follow these instructions to complete the drawing problems for this chapter:

- *Start AutoCAD if it is not already started.*
- *Start a new drawing using an appropriate template of your choice. The template should include layers, text styles, and table styles necessary for drawing the given objects.*
- *Add layers, text styles, and table styles as needed. Draw all objects using appropriate layers and styles, justification, and format.*
- *Use grips and the associated editing tools or other editing techniques described in this chapter.*
- *Follow the specific instructions for each problem. Do not draw dimensions. Use your own judgment and approximate dimensions when necessary.*

▼ Basic

1. Draw the objects labeled A. Then use the grips **STRETCH** tool to make them look like the objects labeled B. Save the drawing as P15-1.

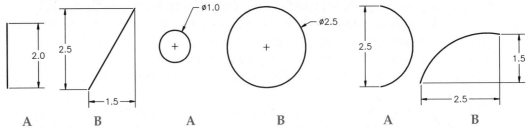

2. Draw the object labeled A. Using the **Copy** option of the grips **MOVE** tool, copy the object to the position labeled B. Edit Object A so it resembles Object C. Edit Object B so it looks like Object D. Save the drawing as P15-2.

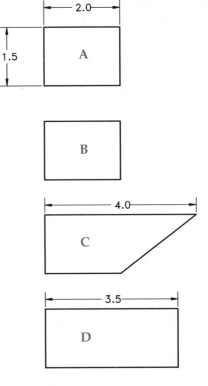

3. Draw the object labeled A. Copy the object, without rotating it, to a position below, as indicated by the dashed lines. Rotate the object 45°. Copy the rotated object labeled B to a position below, as indicated by the dashed lines. Use the **Reference** option to rotate the object labeled C to 25°, as shown. Save the drawing as P15-3.

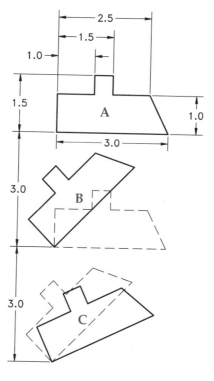

▼ Intermediate

4. Draw the individual objects (vertical line, horizontal line, circle, arc, and C shape) in A using the dimensions given. Use grips and the editing tools to create the object shown in B. Save the drawing as P15-4.

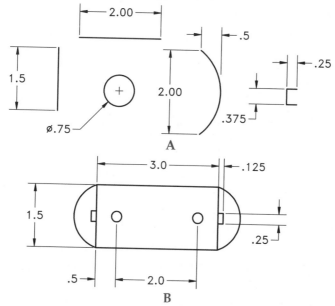

5. Use the completed drawing from Problem 15-4. Erase everything except the completed object and move it to a position similar to that shown in A. Copy the object two times to positions B and C. Use the **SCALE** grip tool to scale the object in position B to 50 percent of its original size. Use the **Reference** option of the **SCALE** tool to enlarge the object in position C from the existing 3.0 length to a 4.5 length, as shown in C. Save the drawing as P15-5.

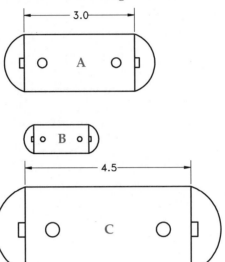

6. Draw the dimensioned partial object shown in A. Mirror the drawing to complete the four quadrants, as shown in B. Change the color of the horizontal and vertical parting lines to Red and the linetype to CENTER. Save the drawing as P15-6.

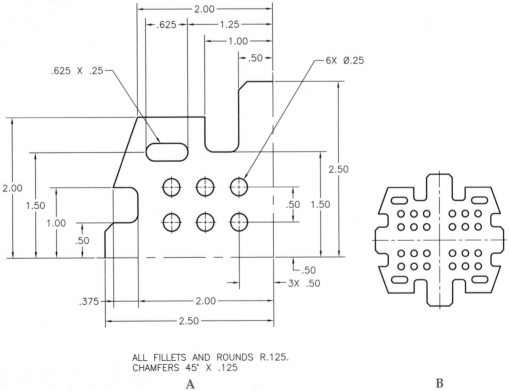

ALL FILLETS AND ROUNDS R.125.
CHAMFERS 45° X .125

A   B

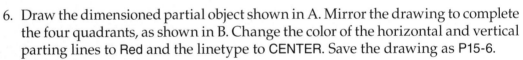

AutoCAD and Its Applications—Basics

7. Load the final drawing you created in Problem 15-6. Use the **Properties** palette to change the diameters of the circles from .25 to .125. Change the linetype of the slots to PHANTOM. Be sure the linetype scale allows the linetypes to be displayed. Save the drawing as P15-7.

8. Use the editing tools described in this chapter to assist you in drawing the following object. Draw the object within the boundaries of the given dimensions. All other dimensions are flexible. Save the drawing as P15-8.

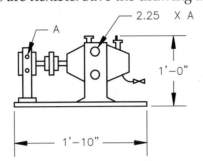

9. Draw the gasket half shown below. Mirror the drawing to complete the other half of the gasket. Save the drawing as P15-9.

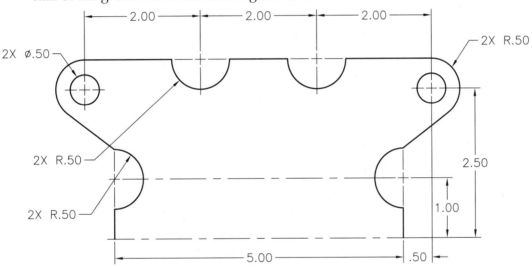

▼ Advanced

10. Draw the following object within the boundaries of the given dimensions. All other dimensions are flexible. After drawing the object, create a page for a vendor catalog, as follows:
    - All labels should be ROMAND text, centered directly below the view. Use a text height of .125".
    - Label the drawing ONE-GALLON TANK WITH HORIZONTAL VALVE.
    - Keep the valve the same scale as the original drawing in each copy.
    - Copy the original tank to a new location and scale it so it is 2 times its original size. Rotate the valve 45°. Label this tank TWO-GALLON TANK WITH 45° VALVE.
    - Copy the original tank to another location and scale it to 2.5 times the size of the original. Rotate the valve 90°. Label this tank TWO-AND-ONE-HALF GALLON TANK WITH 90° VALVE.
    - Copy the two-gallon tank to a new position and scale it so it is 2 times this size. Rotate the valve to 22°30'. Label this tank FOUR-GALLON TANK WITH 22°30' VALVE.

- Left-justify this note at the bottom of the page: Combinations of tank size and valve orientation are available upon request.
- Use the **Properties** palette to change all tank labels to ROMANC, .25" high.
- Change the note at the bottom of the sheet to ROMANS, centered on the sheet, using uppercase letters.
- Save the drawing as P15-10.

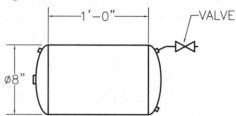

11. Create the interior finish schedule shown below using the **TABLE** tool. Make the measurements for the rows and columns approximately the same as in the given table. Use the **Properties** palette to assist you in constructing the schedule. Save the drawing as P15-11.

INTERIOR FINISH SCHEDULE

ROOM	FLOOR					WALLS				CEILING		
	VINYL	CARPET	TILE	HARDWOOD	CONCRETE	PAINT	PAPER	TEXTURE	SPRAY	SMOOTH	BROCADE	PAINT
ENTRY					•							
FOYER			•			•				•		•
KITCHEN			•					•		•		•
DINING				•		•				•	•	•
FAMILY		•				•				•	•	•
LIVING		•				•			•		•	•
MSTR. BATH			•			•				•		•
BATH #2			•			•		•	•			•
MSTR. BED		•				•			•		•	•
BED #2		•				•				•	•	•
BED #3		•				•				•	•	•
UTILITY	•					•				•	•	•

12. Draw the three views of a sports car, using **SPLINE** to create the curved shapes. Save the drawing as P15-12.

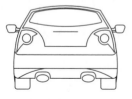
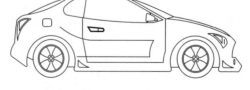

CHAPTER 16

Obtaining Drawing Information

Learning Objectives

After completing this chapter, you will be able to do the following:

✓ Identify a point location and basic object dimensions.
✓ Find the distance between points.
✓ Measure radii, diameters, and angles.
✓ Calculate the area of objects.
✓ List data related to a single point, an object, a group of objects, or an entire drawing.
✓ Determine the drawing status.
✓ Determine the amount of time spent in a drawing session.
✓ Use fields to display object properties in a drawing.
✓ Perform basic and advanced calculations using the **QuickCalc** calculator.

This chapter describes tools that allow you to retrieve geometric values such as distances, angles, and areas. You will also learn how to access and use additional drawing data, such as object properties and overall drawing status. You can even check to see how much time you spend working on a drawing. This chapter also describes how to use the **QuickCalc** tool to calculate values while you work.

Taking Measurements

Taking measurements from your drawing is a common requirement of designing and drafting. In mechanical drafting, for example, you may have to confirm the size of a hole, or identify the angle between two surfaces. In architectural drafting, you often need to check the dimensions of a room, or calculate building square footage.

You can use grips to view basic geometry dimensions. To identify the location of a point that corresponds to an object grip, confirm that the coordinate display field in the status bar is on. Then pick the object to activate grips and hover over a grip. The exact coordinates of the point appear in the coordinate display field. Dynamic input does not have to be active to identify the coordinates of a grip point, but it must be active to view relevant dimensions between grips. Pick the object to activate grips and hover over a grip to display dimensions. The information that appears varies depending on the object type and the selected grip. See **Figure 16-1.**

Figure 16-1.
Examples of hovering over grips to display geometry dimensions.

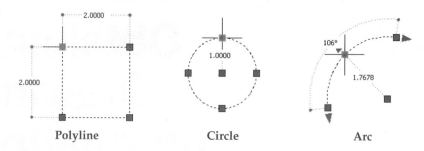

Polyline   Circle   Arc

Exercise 16-1

Access the Student Web site (www.g-wlearning.com/CAD) and complete Exercise 16-1.

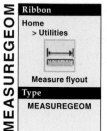

Using the MEASUREGEOM Tool

The **MEASUREGEOM** tool allows you to take a variety of common measurements, including distance, radius, angle, area, perimeter, and volume. The ribbon is an effective way to access **MEASUREGEOM** tool options. See **Figure 16-2**. When you access the **MEASUREGEOM** tool by typing, you must activate a measurement option before you begin. You will notice that the **MEASUREGEOM** tool remains active after you take measurements using the appropriate option. This allows you to continue measuring without reselecting the tool. Select the **eXit** option or press [Esc] to exit the tool.

Measuring Distance

Use the **Distance** option of the **MEASUREGEOM** tool to find the distance between points. Specify the first point followed by the second point. The linear distance between the points, angle *in* the XY plane, and delta X and Y values appear on-screen, as shown in **Figure 16-3**, and at the command line. In a 2D drawing, the angle *from* the XY plane and delta Z values are always 0, as indicated at the command line. As shown in **Figure 16-3**, the first point you specify defines the vertex of the angular dimension.

Once you specify the first point, you can choose the **Multiple points** function to measure the distance between multiple points. AutoCAD calculates the distance between each point and displays the value at the command line during selection.

Figure 16-2.
Use the flyout in the **Utilities** panel of the **Home** ribbon tab to access specific **MEASUREGEOM** tool options.

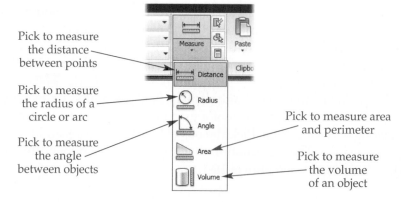

Pick to measure the distance between points

Pick to measure the radius of a circle or arc

Pick to measure the angle between objects

Pick to measure area and perimeter

Pick to measure the volume of an object

Figure 16-3.
The data provided by the **Distance** option. Notice that the first point defines the vertex of the angular value.

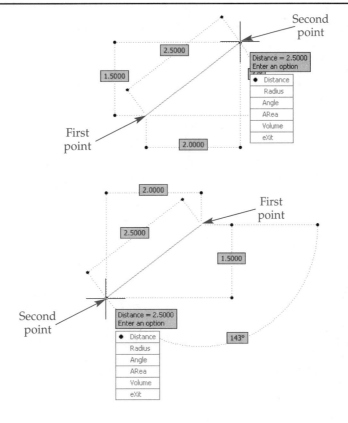

Several options are available for picking multiple points, as described in **Figure 16-4.** When you are finished, use the **Total** or **Close** option to display the distance between all points. **Figure 16-5** shows an example of using the **Multiple points** function to calculate the perimeter of a shape.

> **NOTE**
>
> Use coordinate entry, object snap modes, and other drawing aids to pick points when you use the **MEASUREGEOM** tool.

Figure 16-4.
Options available for the **Multiple points** function of the **Distance** option.

Option	Description
Arc	Measures the length of an arc; includes the same functions available for drawing arcs. Choose the **Line** function to return to measuring the distance between linear points.
Length	Measures the specified length of a line.
Undo	Cancels the effects of an unwanted selection, returning to the previous measurement point.
Total	Finishes multiple point selection and calculates the total distance between points.
Close	Connects the current point to the first point; finishes multiple point selection and calculates the total distance between points.

Figure 16-5.
An example of using the **Multiple points** function of the **Distance** option to calculate the total distance between several points along lines and an arc.

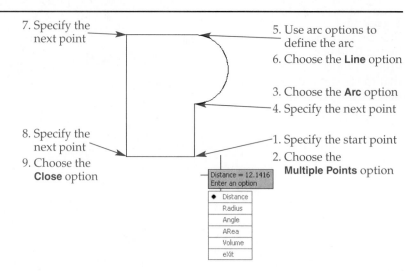

7. Specify the next point

5. Use arc options to define the arc
6. Choose the **Line** option

3. Choose the **Arc** option
4. Specify the next point

8. Specify the next point
9. Choose the **Close** option

1. Specify the start point
2. Choose the **Multiple Points** option

Measuring Radius and Diameter

Specify the **Radius** option of the **MEASUREGEOM** tool and pick an arc or circle to find its radius and diameter. The dimensions appear on-screen, as shown in **Figure 16-6**, and at the command line.

Measuring Angle

Use the **Angle** option of the **MEASUREGEOM** tool to find the angle between lines or points. Measure the angle between lines by selecting the first line followed by the second line. The dimension appears on-screen, as shown in **Figure 16-7A**, and at the command line. Select an arc to measure the angle between arc endpoints. See **Figure 16-7B**.

Figure 16-6.
The data provided by the **Radius** option. You can use the same option to measure an arc.

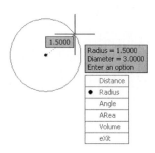

Figure 16-7.
The data provided by the **Angle** option. A—Selecting two lines. B—Picking an arc.

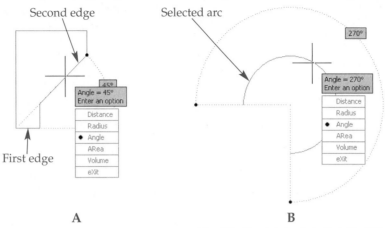

When selecting objects is not appropriate, measure the angle between points by choosing the **Specify vertex** option instead of picking objects. Then select the angle vertex, followed by the first angle endpoint, and finally the second angle endpoint. See **Figure 16-8.**

Exercise 16-2

Access the Student Web site (www.g-wlearning.com/CAD) and complete Exercise 16-2.

Measuring Area

Use the **Area** option of the **MEASUREGEOM** tool to find the area of an object or the area encompassed by selected points. Specify the first corner of the area to measure, followed by all other perimeter corners. Use the **Arc, Length,** or **Undo** functions as needed. These options work the same as they do for measuring distance using the **Multiple points** function of the **Distance** option. A default green (color 100) background fills the area to help you visualize the area. When you are finished specifying perimeter corners, use the **Total** or **Close** option to display the area encompassed by all the points. The area and perimeter appear on-screen, as shown in **Figure 16-9,** and at the command line.

Figure 16-8.
Examples of when it is more appropriate to specify a vertex to measure an angle.

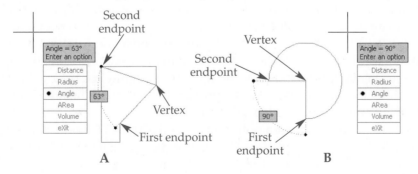

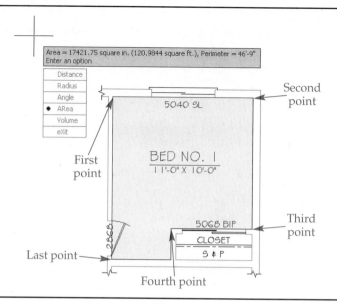

Figure 16-9.
Using the **Area** option of the **MEASUREGEOM** tool to calculate the area encompassed by selected corners.

You can find the area of a polyline object, circle, or spline without picking corners by using the **Object** function of the **Area** option. Once you access the **Area** option, activate the **Object** function instead of picking vertices. Then select the object to display the area. AutoCAD displays the area of the object and a second value. The second value returned by the **Object** function varies, depending on the object type, as shown in the following table:

Object	Values Returned
Polyline	Area and length or perimeter
Circle	Area and circumference
Spline	Area and length or perimeter
Rectangle	Area and perimeter

NOTE

You can calculate the area of an open polyline or spline object, but only if the endpoints show an apparent closure.

The **Area** option includes functions that allow you to calculate the sum of multiple different areas during a single operation. Before selecting corners or an object, activate the **Add area** option and define the first area by picking corners or using the **Object** function. Continue adding areas as needed, or use the **Subtract area** option to remove areas from the selection set. As you add or remove areas, a running total of the area automatically calculates. The **Area** option remains in effect until you exit.

Figure 16-10 shows an example of using the **Add area** and **Subtract area** functions of the **Area** tool in the same operation. In this example, select the **Add area** option, and then select the **Object** option and pick the rectangle. Right-click or press [Enter] or the space bar at the (ADD mode) select objects: prompt to continue. The area and perimeter of the rectangle appear, along with a total area.

Then choose the **Subtract area** option to enter **SUBTRACT** mode. Select the **Object** option and pick the circles to subtract the area of the circles from the area of the rectangle. The area and circumference of each circle appear, along with the total area of the rectangle, minus the total area of the subtracted circles. Right-click or press [Enter]

Figure 16-10.
To calculate the area of a rectangle drawn with the **RECTANGLE** tool, first select the outer boundary of the rectangle using the **Add area** function of the **Area** option. Select the inner circle boundaries using the **Subtract area** option. The total calculation is shown.

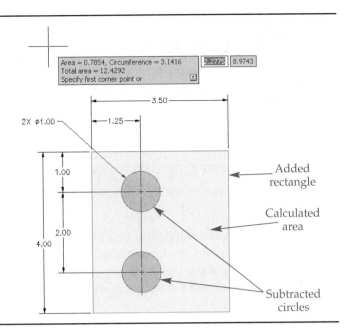

Area = 0.7854, Circumference = 3.1416
Total area = 12.4292
Specify first corner point or

2.2775   8.9743

3.50
2X ⌀1.00
1.25
1.00
2.00
4.00

Added rectangle

Calculated area

Subtracted circles

or the space bar at the (SUBTRACT mode) select objects: prompt to continue. Press [Enter] or the space bar, or right-click and choose **Enter** to display the total area and return to the **MEASUREGEOM** tool prompt.

PROFESSIONAL TIP

Calculating area, circumference, and perimeter values of shapes drawn with the **LINE** tool can be time-consuming, because you must specify each vertex. If you need to calculate areas, it is best to create lines and arcs with the **PLINE** tool. Use the **Object** function of the **Area** option to add or subtract objects.

NOTE

You can use the **Volume** option of the **MEASUREGEOM** tool to measure the volume of a basic drawing in 2D multiview format, but the option is most appropriate for measuring 3D objects. The **Region/Mass Properties** tool provides data related to the properties of a 2D region or 3D solid. *AutoCAD and Its Applications—Advanced* describes the **Volume** option of the **MEASUREGEOM** tool and the **Region/Mass Properties** tool.

NOTE

Traditional *inquiry* tools are available by typing command names. The **ID** tool allows you to display the coordinates of a single selected point. You can use the **DIST** tool to find the distance between points, and the **AREA** tool to calculate areas. You will find that these tools function much like the options available with the **MEASUREGEOM** tool, but they are not as interactive.

Exercise 16-3

Access the Student Web site (www.g-wlearning.com/CAD) and complete Exercise 16-3.

Listing Drawing Data

The **LIST** tool displays a variety of data about any AutoCAD object. Access the **LIST** tool, select the objects to list, and right-click or press [Enter] or the space bar. The data for each object displays at the command line and in the text window. **Figure 16-11A** shows an example of the text window displayed when you list a line. The Delta X and Delta Y values indicate the horizontal and vertical distances between the *from point* and *to point* of the line. These two values, along with the length and angle, provide you with four measurements for a single line. See **Figure 16-11B.**

Figure 16-12 shows an example of the text window displayed with a selection set of multiple objects, including a circle, and a rectangle, and multiline text. When you select multiple objects and not all of the information fits in the window, AutoCAD prompts you to press [Enter] to display additional information.

Figure 16-11.
A—An example of the text window displayed when you use the **LIST** tool to list the properties of a line. B—The various data and measurements of a line provided by the **LIST** tool.

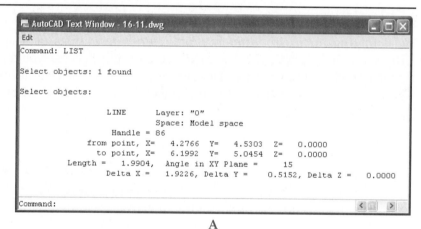

A

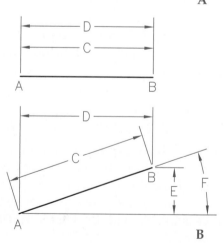

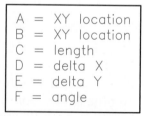

A = XY location
B = XY location
C = length
D = delta X
E = delta Y
F = angle

B

Figure 16-12.
Using the **LIST** tool to list the properties of multiline text, a circle, and a rectangle.

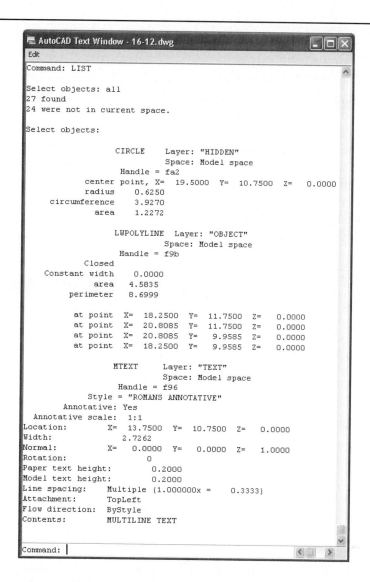

```
AutoCAD Text Window - 16-12.dwg
Edit
Command: LIST

Select objects: all
27 found
24 were not in current space.

Select objects:

                    CIRCLE     Layer: "HIDDEN"
                               Space: Model space
                    Handle = fa2
          center point, X=  19.5000  Y=  10.7500  Z=   0.0000
           radius    0.6250
      circumference    3.9270
           area    1.2272


                    LWPOLYLINE  Layer: "OBJECT"
                               Space: Model space
                    Handle = f9b
          Closed
  Constant width    0.0000
           area    4.5835
       perimeter    8.6999

          at point  X=  18.2500  Y=  11.7500  Z=   0.0000
          at point  X=  20.8085  Y=  11.7500  Z=   0.0000
          at point  X=  20.8085  Y=   9.9585  Z=   0.0000
          at point  X=  18.2500  Y=   9.9585  Z=   0.0000

                    MTEXT      Layer: "TEXT"
                               Space: Model space
                    Handle = f96
           Style = "ROMANS ANNOTATIVE"
        Annotative: Yes
   Annotative scale:  1:1
Location:        X=  13.7500  Y=  10.7500  Z=   0.0000
Width:              2.7262
Normal:          X=   0.0000  Y=   0.0000  Z=   1.0000
Rotation:           0
Paper text height:      0.2000
Model text height:      0.2000
Line spacing:    Multiple (1.000000x =    0.3333)
Attachment:      TopLeft
Flow direction:  ByStyle
Contents:        MULTILINE TEXT

Command:
```

NOTE

The **DBLIST** (database list) tool lists all data for every object in the current drawing. To use this tool, enter DBLIST. The information appears in the same format used by the **LIST** tool, although the text window does not appear automatically.

PROFESSIONAL TIP

The **LIST** tool provides most information about an object, including the area and perimeter of polylines. The **LIST** tool also reports object color and linetype, unless both are BYLAYER.

Exercise 16-4

Access the Student Web site (www.g-wlearning.com/CAD) and complete Exercise 16-4.

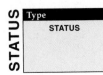
Reviewing the Drawing Status

The **STATUS** tool provides a method to display a variety of drawing information at the command line and in the text window. **Figure 16-13** shows an example of drawing status data in the text window. The number of objects in a drawing refers to the total number of objects—both erased and existing. Free dwg disk (C:) space: represents the space left on the drive containing the drawing file. Drawing aid settings appear, along with the current settings for layer, linetype, and color. Press [Enter] if necessary to proceed to additional information. When you finish reviewing the information, press [F2] to close the text window. You can also switch to the drawing window without closing the text window by picking anywhere inside the drawing window or using the Windows [Alt]+[Tab] feature.

Figure 16-13.
An example of drawing information in the text window when you use the **STATUS** tool.

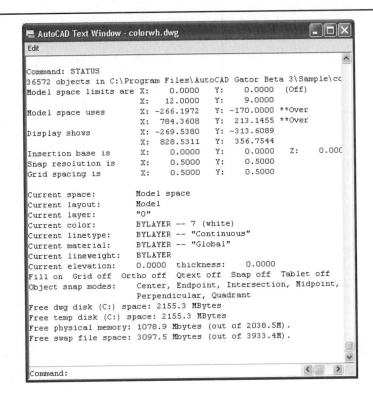

Checking the Time

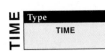
The **TIME** tool allows you to display the current time and time related to the current drawing session. **Figure 16-14** shows an example of drawing time data in the text window. The drawing creation time starts when you begin a new drawing, not when you first save a new drawing. The **SAVE** tool affects the Last updated: time. However, all drawing session time erases when you exit AutoCAD and do not save the drawing.

You can time a specific drawing task using the **Reset** option of the **TIME** tool to reset the elapsed timer. The timer is on by default when you enter the drawing area. Use the **OFF** option to stop the timer. If the timer is off, use the **ON** option to start it again. Time information is static, which means the times you view may be old. Use the **Display** option to request an update.

Figure 16-14.
An example of
the text window
displayed when you
use the **TIME** tool.

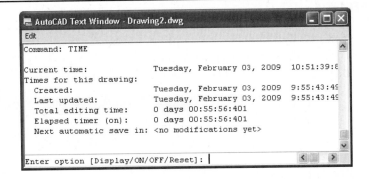

```
AutoCAD Text Window - Drawing2.dwg                          _ □ X
Edit
Command: TIME

Current time:             Tuesday, February 03, 2009   10:51:39:8
Times for this drawing:
  Created:                Tuesday, February 03, 2009    9:55:43:49
  Last updated:           Tuesday, February 03, 2009    9:55:43:49
  Total editing time:     0 days 00:55:56:401
  Elapsed timer (on):     0 days 00:55:56:401
  Next automatic save in: <no modifications yet>

Enter option [Display/ON/OFF/Reset]: |
```

NOTE

The Windows operating system maintains the date and time settings for the computer. You can change these settings in the Windows Control Panel. To access the Control Panel, pick Settings and then Control Panel from the Start menu.

Exercise 16-5

Access the Student Web site (www.g-wlearning.com/CAD) and complete Exercise 16-5.

Displaying Information with Fields

You can use *fields* to list a variety of object properties and drawing information. Each object type has different properties that you can display in a field. For example, you can use a field in an mtext or text object near a circle to list the area and circumference of the circle.

Use the **FIELD** tool and **Field** dialog box, shown in **Figure 16-15**, to add fields to mtext or text objects. To insert a field in an active mtext editor, pick the **Field** button from the **Insert** panel of the **Text Editor** ribbon tab, pick the **Insert Field** option available from the shortcut menu, or press [Ctrl]+[F]. To insert a field in an active single-line text editor, right-click and select **Insert Field...** or press [Ctrl]+[F].

In the **Field** dialog box, pick Objects from the **Field category:** drop-down list, and then pick Object in the **Field names:** list box. Pick the **Select object** button to return to the drawing window and pick an object. When you select an object, the **Field** dialog box reappears with the available properties listed. See **Figure 16-16**. Pick the property, select the format, and pick the **OK** button to insert the field. Once you create the field, whenever you modify the object, the value displayed in the field updates to reflect the new value. In addition to object property settings such as layer, linetype, lineweight, and plot style, you can include many dimensional properties in a field.

field: Text object that displays a property, setting, or value for an object, drawing, or computer system.

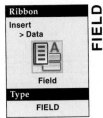

Ribbon
Insert
> Data
Field

Type
FIELD

FIELD

Exercise 16-6

Access the Student Web site (www.g-wlearning.com/CAD) and complete Exercise 16-6.

Figure 16-15.
Pick the Object field to add a property for a specific object to a field. Pick the **Select object** button to select the object.

Select category

Pick to list properties of a specific object

Pick to select object

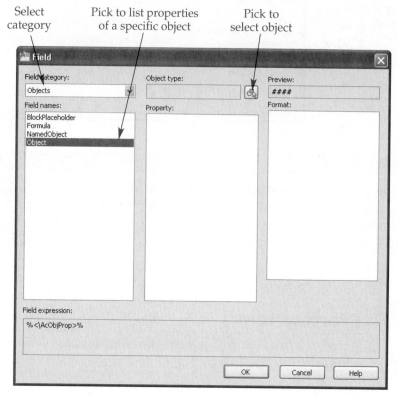

Figure 16-16.
After you pick the object, properties specific to the object type are listed. Select the property and format for the field.

Properties available to field

Value of selected property

Format options are property specific

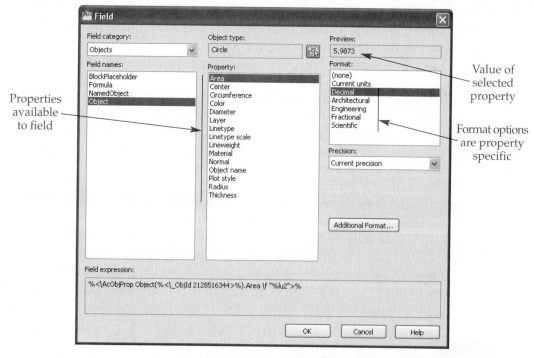

Using QuickCalc

Ribbon
View
> Palettes

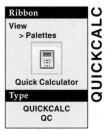

Quick Calculator
Type
QUICKCALC
QC

Most drafting projects require you to make calculations. For example, when working from a sketch with missing dimensions, you may need to calculate a distance or angle, or you may need to double-check dimensions. Often drafters make calculations using a handheld calculator. An alternative is to use **QuickCalc**, which is a palette containing a basic calculator, scientific calculator, units converter, and variables feature. See **Figure 16-17.** You can use **QuickCalc** as you would a handheld calculator, but you can also use it while a tool is active to paste calculations when a prompt asks for a specific value.

NOTE

You can also access the **QuickCalc** palette by right-clicking in the drawing window and selecting **QuickCalc**, or from specific areas such as the button that sometimes appears next to a text box in the **Properties** palette.

Entering Expressions

The basic math functions used in numeric expressions include addition, subtraction, multiplication, division, and exponential notation. You can enter grouped expressions by using parentheses to break up the expressions that calculate separately. For example, to calculate 6 + 2 and then multiply the sum by 4, enter (6+2)*4. The result will be wrong if you do not add the parentheses. The following table shows the symbols used for basic math operators.

Symbol	Function	Example
+	Addition	3+26
–	Subtraction	270–15.3
*	Multiplication	4*156
/	Division	256/16
^	Exponent	22.6^3
( )	Grouped expressions	2*(16+2^3)

Figure 16-17.
The **QuickCalc** palette allows you to perform a variety of calculations.

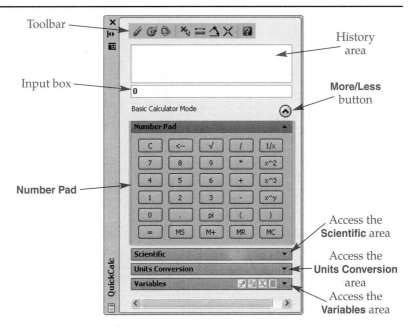

Figure 16-18.
The basic **Number Pad** area contains additional options that you cannot access using the keyboard.

To get the square root of a value

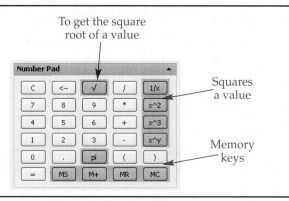

Squares a value

Memory keys

You can add expressions in the input box by picking buttons on the **Number Pad** or by pressing keyboard keys. After creating the expression, pick the equal (=) button on the number pad or press [Enter] to evaluate the expression. **Figure 16-18** shows options found on the basic **Number Pad** that are not available from the keyboard.

The result of an evaluated expression appears in the input box, and the expression moves to the history area. **Figure 16-19** displays the **QuickCalc** palette after calculating 96.27 + 23.58. When you are using only the input box of **QuickCalc**, pick the **More/Less** button below the input box to hide the additional sections, saving valuable drawing space. When the **QuickCalc** palette displays all areas, the button is an up arrow and its tooltip reads **Less**. Pick the button again to display hidden areas.

NOTE

If you move the cursor outside of the **QuickCalc** palette, the drawing area automatically becomes active. Pick anywhere inside the **Quick-Calc** palette to reactivate it.

PROFESSIONAL TIP

If you make a mistake in the input box, you do not need to clear the input box and start over again. Use the left and right arrow keys to move through the field. Right-click in the input box to display a shortcut menu that includes useful options for copying and pasting.

Figure 16-19.
A—Add an expression into the input box.
B—After creating the expression, press [Enter] to evaluate the expression. The history area stores the expression and result.

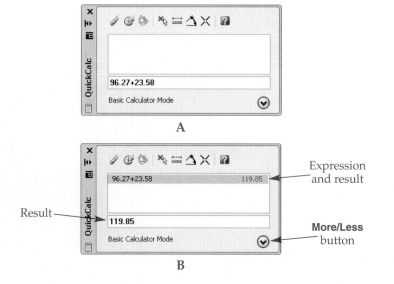

Expression and result

Result

More/Less button

Clearing the Input and History Areas

After picking the equal (=) button or pressing [Enter] to evaluate an expression, you can create a new expression without clearing the last result. AutoCAD automatically starts a new expression. You can clear the input box manually by placing the cursor in the input box and pressing [Backspace] or [Delete], or by picking the **Clear** button from the **QuickCalc** toolbar. To clear the history area, pick the **Clear History** button. See **Figure 16-20.**

CAUTION

If you enter an expression that **QuickCalc** cannot evaluate, AutoCAD displays an **Error in Expression** dialog box. Pick the **OK** button, correct the error, and try it again.

Exercise 16-7

Access the Student Web site (www.g-wlearning.com/CAD) and complete Exercise 16-7.

Scientific Calculations

The **Scientific** area of **QuickCalc** includes trigonometry, exponential, and some geometric functions. See **Figure 16-21.** To use one of the functions, add a value to the input box, pick the appropriate function button, and pick the equal (=) button or press [Enter]. When you pick a function button, the input box value appears in parentheses after the expression. For example, to get the *sine* of 14, clear the input box, type 14 in the input box and pick the **sin** button. The input box now reads sin(14). Pick the equal (=) button or press [Enter] to view the result.

NOTE

You can pick a function button first, but doing so puts a default value of 0 in parentheses. You can then place the cursor in the input box to type a different number in the parentheses if needed.

Figure 16-20.
You can use the buttons on the **QuickCalc** toolbar to clear the input box and history areas.

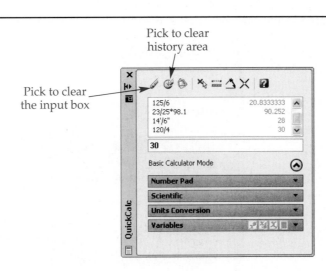

Pick to clear history area

Pick to clear the input box

Figure 16-21.
The scientific functions available in **QuickCalc**.

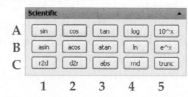

	1	2	3	4	5
A	Sine	Cosine	Tangent	Base–10 Log	Base–10 Exponent
B	Arcsine	Arccosine	Arctangent	Natural Log	Natural Exponent
C	Convert Radians to Degrees	Convert Degrees to Radians	Absolute Value	Round	Truncate

Converting Units

The **Units Conversion** area allows you to convert one unit type to another. The unit types available are **Length**, **Area**, **Volume**, and **Angular**. For example, to use the unit converter to convert 23 centimeters to inches, pick in the **Units type** field to display the drop-down list. See **Figure 16-22.** Pick the drop-down list button to display the different unit types and select Length. Activate the **Convert from** field and select Centimeters from the drop-down list. Activate the **Convert to** field and select Inches from the drop-down list. Type 23 in the **Value to convert** field and pick the equal (=) button or press [Enter]. The **Converted value** field displays the converted units.

To pass the converted value to the input box for use in an expression, pick the **Return Conversion to Calculator Input Area** button. See **Figure 16-23.** If the button is not visible, pick once on the converted units in the **Converted value** field.

Figure 16-22.
Picking the current unit type activates the field and displays the drop-down list button.

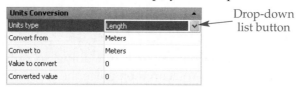

Figure 16-23.
After you convert a value, pass the value to the input box for use in an expression.

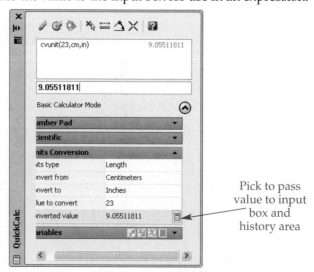

Exercise 16-8

Access the Student Web site (www.g-wlearning.com/CAD) and complete Exercise 16-8.

Using Variables

If you use an expression or value frequently, you can save it as a *variable*. Use the **Variables** area, shown in **Figure 16-24**, to create, edit, delete, and pass variables to the input box. The **Variables** area of **QuickCalc** includes two types of predefined variables: *constant* and *function*.

To create a new variable, select the **New Variable...** button to open the **Variable Definition** dialog box shown in **Figure 16-25**. Type a name for the variable in the **Name:** field. Select a group to contain the variable using the **Group with:** field. Type the value or the expression for the variable in the **Value or expression:** field. Give a description for the variable in the **Description** field. Pick the **OK** button to save the variable and display it in the **Variables** area.

To edit a variable, pick the variable to modify and select the **Edit Variable** button to reopen the **Variable Definition** dialog box. Pick the **Delete** button to delete the selected variable. Pick the **Return Variable to Input Area** button to pass the selected variable to the input box. You can also pass the variable to the input box by double-clicking the variable name.

You can also access variable tools by right-clicking in the **Variable** area to display a shortcut menu. Two additional options are available from the menu. Pick the **New Category** option to create a new category for saving variables. Select the **Rename** option to rename the selected variable. You can also rename a variable by selecting it, pausing, and then selecting it again to activate the name for editing.

variable: Text item that represents another value and can be accessed later as needed.

constant: Expression or value that stays the same.

function: Expression or value that asks.for user input to get values that can be passed to the expression.

Figure 16-24.
The **Variables** area of **QuickCalc** allows you to store values and expressions for later use.

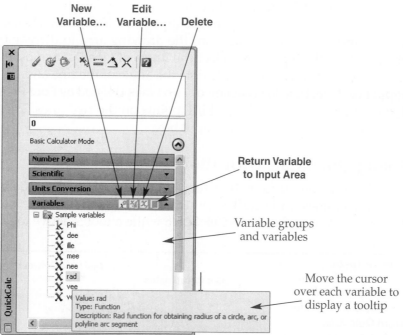

Figure 16-25.
Use the **Variable Definition** dialog box to define new variables.

Choose the type of variable

Group the new variable with the samples or create a new category

Enter the description you want to appear in the tooltip

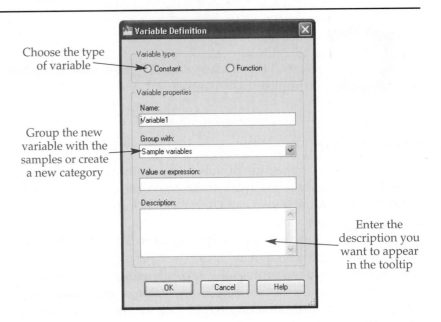

Using Drawing Values

Tools available from the **QuickCalc** toolbar allow you to pass values from the drawing to the **QuickCalc** input box, and from the input box to the drawing. When you select any of the buttons shown in **Figure 16-26**, the **QuickCalc** palette temporarily hides so that you can select points from the drawing window.

Pick the **Get Coordinates** button to select a point from the drawing window and display the X,Y,Z coordinates in the input box. Pick the **Distance Between Two Points** button and select two points from the drawing area to display the distance between the points in the input box. Select the **Angle of Line Defined by Two Points** button and pick two points on a line to calculate the angle of the line and display the angle in the input box. Select the **Intersection of Two Lines Defined by Four Points** button to find the intersection of two lines by picking points on the two lines. The X,Y,Z coordinates of the intersection appears in the input box.

Using QuickCalc with Tools

The previous information focuses on using the **QuickCalc** palette to calculate unknown values while drafting, much like using a handheld calculator or the Windows Calculator. You can also use **QuickCalc** while a tool is active to *pass* a calculated value

Figure 16-26.
You can use buttons on the **QuickCalc** toolbar to pass values from **QuickCalc** to AutoCAD and retrieve values from AutoCAD to pass to **QuickCalc**.

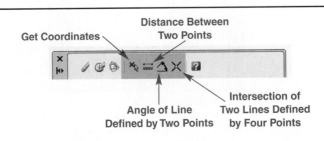

Get Coordinates

Distance Between Two Points

Angle of Line Defined by Two Points

Intersection of Two Lines Defined by Four Points

to the command line as a response to a prompt. There are a few alternatives for using **QuickCalc** while a tool is active.

If the **QuickCalc** palette is active when you access a tool, when the prompt requesting an unknown value appears, calculate the value using the **QuickCalc** palette and then press the **Paste value to command line** button to pass the value to the command line. For example, to draw a line a distance of 14′8″ + 26′3″ horizontally from a start point, activate the **QuickCalc** palette, access the **LINE** tool, and pick a start point. Then use polar tracking or **Ortho** mode to move the crosshairs to the right or left of the start point so the line is at a 0° angle. At the Specify next point or [Undo]: prompt, enter 14′8″ + 26′3″ in the **QuickCalc** palette input box and pick the equal (=) button or press [Enter]. The result in the input box is 40′11″. Pick the **Paste value to command line** button to make 40′11″ appear at the command line. Press [Enter] or the space bar or right-click and select **Enter** to draw the 40′11″ line.

If the **QuickCalc** palette is not active while you are using a tool, you can still calculate and use a value. When the prompt requesting an unknown value appears, access **QuickCalc** by right-clicking and selecting **QuickCalc** or typing 'QC. A **QuickCalc** *window*, which is not the same as the **QuickCalc** palette, opens in command calculation mode. See **Figure 16-27**. Use the necessary tools to evaluate an expression. Then pick the **Apply** button to pass the value back to the command line, and close the **QuickCalc** window.

PROFESSIONAL TIP

The units you use in **QuickCalc** must match the drawing units. In the preceding example, which uses architectural units, the drawing units must also be architectural. If needed, use the **Drawing Units** dialog box to change the drawing units to Architectural.

Exercise 16-9

Access the Student Web site (www.g-wlearning.com/CAD) and complete Exercise 16-9.

Figure 16-27.
The **QuickCalc** window that appears when you access **QuickCalc** while a tool is active. The **Apply** and **Close** buttons are available at the bottom of the window.

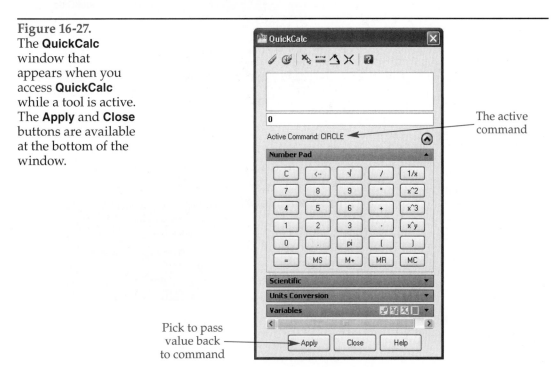

Using QuickCalc with Object Properties

QuickCalc also allows you to calculate expressions for an object while using the **Properties** palette. Pick a field that contains a numeric value to display the calculator icon. **Figure 16-28** shows a selected circle and the active **Radius** field in the **Properties** palette. Pick the calculator icon to open the **QuickCalc** window, again not the same item as the **QuickCalc** palette, in property calculation mode. Use expressions and values in the same manner as when using **QuickCalc** at any other time. Once you evaluate the expression in the input box, pick the **Apply** button to pass the value to the property field in the **Properties** palette. The object automatically updates based on the new value.

Additional QuickCalc Options

The history area contains some settings and features that you can only access from a shortcut menu. This menu, shown in **Figure 16-29**, displays when you right-click anywhere in the history area. The following options are available:

- **Expression Font Color.** Changes the color of the expression font.
- **Value Font Color.** Changes the color of the value font.
- **Copy.** Copies the expression and value to the Windows clipboard.
- **Append Expression to Input Area.** Passes the expression to the input box.
- **Append Value to Input Area.** Passes the value to the input box.
- **Clear History.** Clears the history area.
- **Paste to Command Line.** Passes the value to the Command: prompt.

As with other palettes, picking the **Properties** button on the **QuickCalc** palette displays the **Properties** menu. The settings allow you to change the palette appearance, including its ability to dock, hide, or appear transparent.

Figure 16-28.
The calculator icon appears when you select a numeric field in the **Properties** palette.

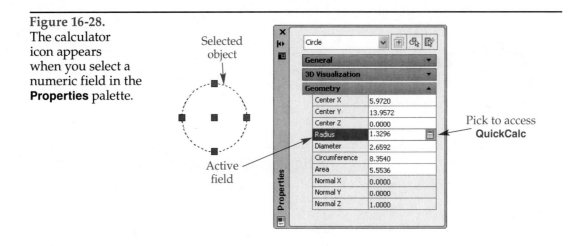

Selected object

Active field

Pick to access **QuickCalc**

Figure 16-29.
The history area shortcut menu contains additional functions and settings.

Shortcut menu

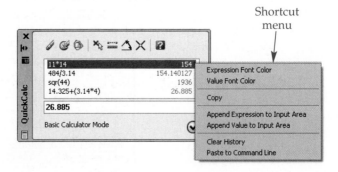

Chapter Test

Answer the following questions. Write your answers on a separate sheet of paper or go to the Student Web site (www.g-wlearning.com/CAD) and complete the electronic chapter test.

1. Explain how to use grips to identify the location of a point, and dimensions of an object.
2. What types of information does the **Distance** option of the **MEASUREGEOM** tool provide?
3. What information does the **Area** option of the **MEASUREGEOM** tool provide?
4. To add the areas of several objects when using the **Area** option of the **MEASUREGEOM** tool, when do you select the **Add area** function?
5. Explain how picking a polyline when using the **Area** option of the **MEASUREGEOM** tool is different from measuring the area of an object drawn with the **LINE** tool.
6. What is the purpose of the **LIST** tool?
7. Describe the meanings of *delta X* and *delta Y*.
8. What is the function of the **DBLIST** tool?
9. What tool, other than **MEASUREGEOM** and **AREA**, provides the area and perimeter of an object?
10. Which tool allows you to list drawing aid settings for the current drawing?
11. What information is provided by the **TIME** tool?
12. When does the drawing creation time start?
13. What term describes a text object that displays a set property, setting, or value for an object?
14. List three ways to open the **QuickCalc** palette.
15. Name the four sections of the **QuickCalc** palette.
16. Give the proper symbol to use for the following math functions:
    A. Addition
    B. Subtraction
    C. Multiplication
    D. Division
    E. Exponent
    F. Grouped expressions
17. Under which section of the **QuickCalc** palette can you find the square root function?
18. Under which section of the **QuickCalc** palette can you find the arccosine function?
19. When using one of the scientific functions, which should you do first: pick the scientific function button or type in the value to be used in the input box? Why?
20. Name the four types of units that can be converted using **QuickCalc**.
21. What term describes a text item that represents another value and can be accessed later as needed?
22. Which tool button is used to pass the value in the **QuickCalc** input box to the command line?
23. How can you start **QuickCalc** while a command is active?
24. When you are using **QuickCalc** while a tool is active, how do you pass the value to the command line?
25. When the **Properties** palette is open, what do you need to do first to see the calculator icon so that **QuickCalc** can be used?

Drawing Problems

Follow these instructions to complete the drawing problems for this chapter:

- *Start AutoCAD if it is not already started.*
- *Start a new drawing using an appropriate template of your choice. The template should include layers and text styles necessary for drawing the given objects.*
- *Add layers and text styles as needed. Draw all objects using appropriate layers and styles, justification, and format.*
- *Follow the specific instructions for each problem. Do not draw dimensions. Use your own judgment and approximate dimensions when necessary.*

▼ Basic

1. Use **QuickCalc** to calculate the result of the following equations.
   - A. 27.375 + 15.875
   - B. 16.0625 − 7.1250
   - C. 5 × 17′-8″
   - D. 48′-0″ ÷ 16
   - E. (12.625 + 3.063) + (18.250 − 4.375) − (2.625 − 1.188)
   - F. 7.25²

2. Convert 4.625″ to millimeters.

3. Convert 26 mm to inches.

4. Convert 65 miles to kilometers.

5. Convert 5 gallons to liters.

6. Find the square root of 360.

7. Calculate 3.25 squared.

▼ Intermediate

8. Show the calculation and answer that would be used with the **LINE** tool to make an 8″ line .006 in./in. longer in a pattern to allow for shrinkage in the final casting. Show only the expression and answer.

9. Solve for the deflection of a structural member. The formula is written as $PL^3/48EI$, where P = pounds of force, L = length of beam, E = modulus of elasticity, and I = moment of inertia. The values to be used are P = 4000 lbs, L = 240″, and E = 1,000,000 lbs/in². The value for I is the result of the beam (width × height³)/12, where width = 6.75″ and height = 13.5″.

10. Calculate the coordinate located at 4,4,0 + 3<30.

11. Calculate the coordinate located at (3 + 5,1 + 1.25,0) + (2.375,1.625,0).

12. Draw the object shown below using the dimensions given. Check the time when you start the drawing. Draw all the features using the **PLINE** and **CIRCLE** tools. Use the **Area** option of the **MEASUREGEOM** tool and the **Object**, **Add area**, and **Subtract area** functions to calculate the following measurements:

   A. The area and perimeter of Object A.
   B. The area and perimeter of area B. The slot ends are full radius.
   C. The area and circumference of one of the circles.
   D. The area of Object A, minus the area of Object B.
   E. The area of Object A, minus the areas of the other three features.
   Enter the **TIME** tool and note the editing time spent on your drawing. Save the drawing as P16-12.

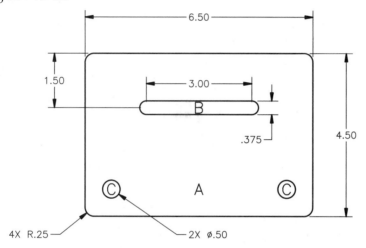

13. Draw the deck shown using the **PLINE** tool. Using the **POLYGON** tool, draw the hot tub. Use the following guidelines to complete this problem:

   A. Specify architectural units for your drawing. Use 1/2″ fractions and decimal degrees. Leave the remaining settings for the drawing units at the default values.
   B. Set the limits to 100′,80′ and use the **All** option of the **ZOOM** tool.
   C. Set the grid spacing to 2′ and the snap spacing to 1′.
   D. Calculate the measurements listed below.
      a. The area and perimeter of the deck.
      b. The area and perimeter of the hot tub.
      c. The area of the deck minus the area of the hot tub.
      d. The distance between Point C and Point D.
      e. The distance between Point E and Point C.
      f. The coordinates of Points C, D, and F.
   E. Enter the **DBLIST** tool and check the information listed for your drawing.
   F. Enter the **TIME** tool and note the total editing time spent on your drawing.
   G. Save the drawing as P16-13.

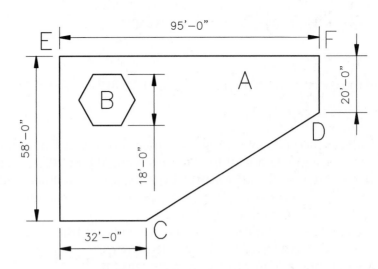

14. The drawing shown below is a side view of a pyramid. The pyramid has four sides. Create an auxiliary view showing the true size of a pyramid face. Save the drawing as P16-14. Using inquiry techniques, calculate the following:
    A. The area of one side.
    B. The perimeter of one side.
    C. The area of all four sides.
    D. The area of the base.
    E. The true length (distance) from the midpoint of the base on one side to the apex.

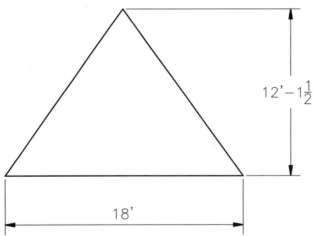

▼ Advanced

15. Given the following right triangle, make the required trigonometry calculations.
    A. Length of side c (hypotenuse).
    B. Sine of angle A.
    C. Sine of angle B.
    D. Cosine of angle A.
    E. Tangent of angle A.
    F. Tangent of angle B.

AutoCAD and Its Applications—Basics

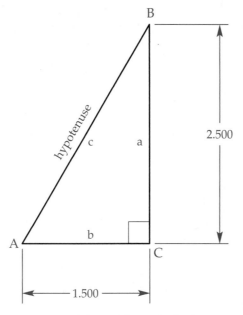

16. The drawing below is a view of the gable end of a house. Draw the house using the dimensions given, and draw the windows as single lines only (the location of the windows is not important). The spacing between each of the second-floor windows is 3″. The width of this end of the house is 16′-6″. The length of the roof is 40′. You may want to use the **PLINE** tool to assist in creating specific shapes in this drawing, except as noted above. Save the drawing as **P16-16**. Calculate the following:

A. The total area of the roof.
B. The diagonal distance from one corner of the roof to the other.
C. The area of the first-floor window.
D. The total area of all second-floor windows, including the 3″ spaces between each of them.
E. Siding will cover the house. What is the total area of siding for this end?

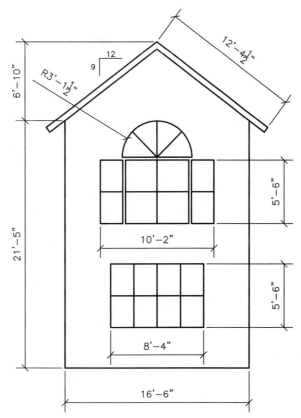

17. Draw the property plat shown below. Label property line bearings and distances only if required by your instructor. Calculate the area of the property plat in square feet and convert to acres. Save the drawing as P16-17.

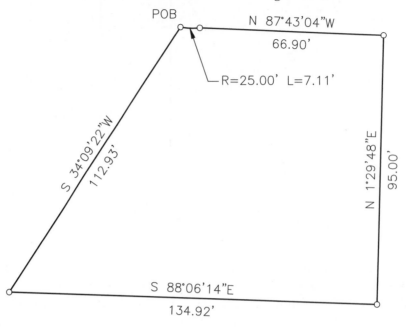

18. Draw the subdivision plat shown below. Label the drawing as shown. Calculate the acreage of each lot and record each value as a label inside the corresponding lot (for example, .249 AC). Save the drawing as P16-18.

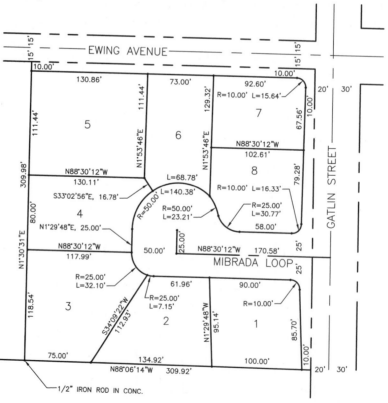

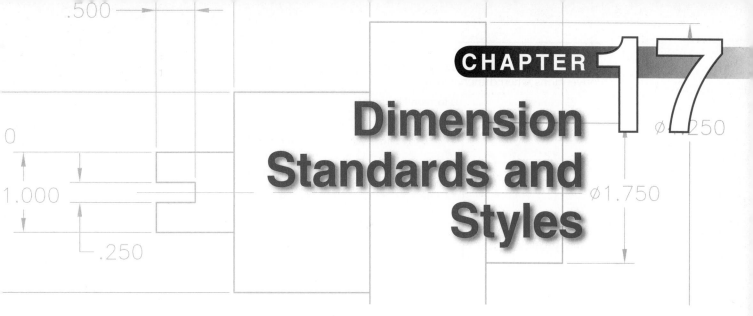

CHAPTER 17

Dimension Standards and Styles

Learning Objectives

After completing this chapter, you will be able to do the following:

✓ Describe common dimension standards and practices.
✓ Create dimension styles.
✓ Manage dimension styles.
✓ Set a dimension style current.

A *dimension* can consist of numerical values, lines, symbols, and notes. **Figure 17-1** shows typical dimension elements and dimensioning applications. Use dimension tools to dimension the size and location of features and objects. Dimension styles control the appearance of dimension elements. Dimensional constraint tools, explained in Chapter 22, allow you to use dimensions to control object size and location.

> **dimension:** A description of the size, shape, or location of features on an object or structure.

Dimension Standards and Practices

Dimensions communicate drawing information. Each drafting field uses different dimensioning practices. Dimensioning practices often depend on product requirements, manufacturing or construction accuracy, standards, and tradition. It is important for you to place dimensions in accordance with industry and company standards and produce drawings that are as clear and easy to read as possible. Dimension standards help to ensure that product manufacturing or construction is accurate.

The standard emphasized in this textbook is ASME Y14.5M-1994, *Dimensioning and Tolerancing,* published by the American Society of Mechanical Engineers (ASME). The *M* in Y14.5M indicates metric numeric values for dimensions. This textbook describes the correct application of inch and metric dimensioning, as well as additional, discipline-specific dimensioning standards and applications.

Unidirectional Dimensioning

Unidirectional dimensioning is common in mechanical drafting. The term *unidirectional* means "in one direction." This type of dimensioning allows you to read all dimensions from the bottom of the sheet. Unidirectional dimensions normally have arrowheads on the ends of dimension lines. The dimension value is usually centered in a break near the center of the dimension line. See **Figure 17-2**.

> **unidirectional dimensioning:** A dimensioning system in which all dimension values display horizontally on the drawing.

Figure 17-1.
Dimensions describe the size and location of objects and features. Follow accepted conventions when dimensioning.

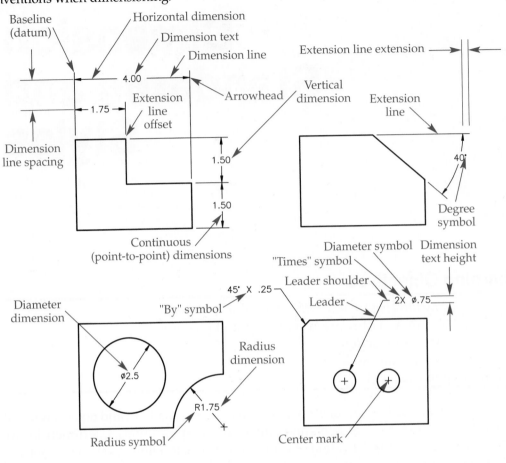

Figure 17-2.
When unidirectional dimensions are used, all dimension numbers and notes are horizontal on the drawing.

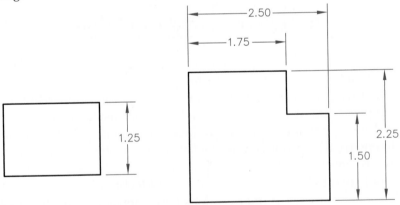

Aligned Dimensioning

aligned dimensioning:
A dimensioning system in which the dimension values align with the dimension lines.

Aligned dimensions are common on architectural and structural drawings. Text for dimensions reads at the same angle as the dimension line. Text applied to horizontal dimensions reads horizontally, and text applied to vertical dimensions rotates 90° to read from the right side of the sheet. Notes usually read horizontally. Tick marks, dots, or arrowheads may terminate aligned dimension lines. In architectural drafting, you generally place the dimension number above the dimension line and use tick mark terminators. See **Figure 17-3.**

Figure 17-3.
An example of aligned dimensioning in architectural drafting. Notice the tick marks used instead of arrowheads and the placement of the dimensions above the dimension line.

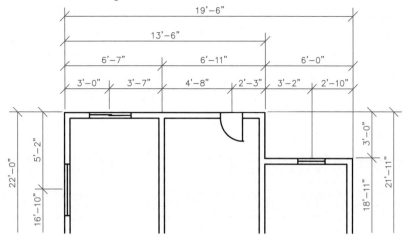

Size and Location Dimensions

Size dimensions provide the size of physical geometric *features*. See **Figure 17-4.** *Location dimensions* locate features on an object. See **Figure 17-5.** Dimension circular features, such as holes and arcs, to their centers in the view in which they appear

Figure 17-4.
Size dimensions.

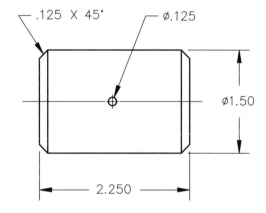

Figure 17-5.
Using location dimensions to locate circular and rectangular features.

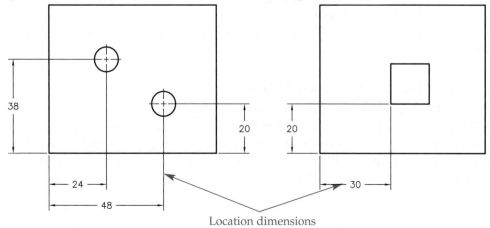

Location dimensions

Figure 17-6.
A—Rectangular coordinate location dimensions. B—Polar coordinate location dimensions.

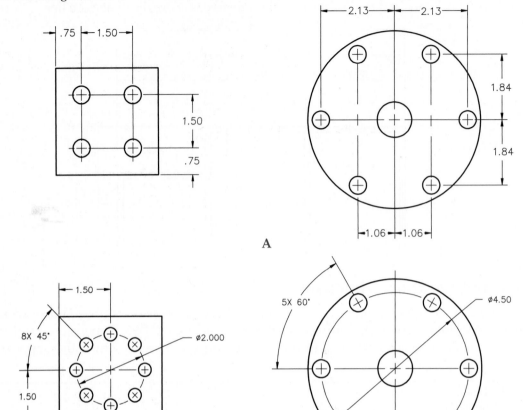

circular. Dimension rectangular features to their edges. An example of location dimensions used in architectural drafting is the dimensioning of windows and doors, usually to their centers, on a floor plan. The *rectangular coordinate system* and the *polar coordinate system* are the two basic systems for creating location dimensions. See **Figure 17-6.**

Notes

rectangular coordinate system: A system for locating dimensions from surfaces, centerlines, or center planes using linear dimensions.

polar coordinate system: A coordinate system in which angular dimensions locate features from surfaces, centerlines, or center planes.

specific notes: Notes that relate to individual or specific features on the drawing.

general notes: Notes that apply to the entire drawing.

Specific notes and *general notes* are another way to describe feature size, location, or additional information. See **Figure 17-7.** Specific notes attach to the dimensioned feature using a leader line. Place general notes in the lower-left corner, upper-left corner, or above or next to the title block, depending on sheet size and industry, company, or school practice.

Dimensioning Features and Objects

In mechanical drafting, you dimension flat surfaces using measurements for each feature. If you provide an overall dimension, you should omit one dimension, as the overall dimension controls the omitted value. See **Figure 17-8A.** In architectural drafting, it is common to place all dimensions without omitting any to help make construction easier. See **Figure 17-8B.**

Figure 17-7.
A—An example of a
specific note. B—An
example of general
notes.

(3) 24" X 76" SOLAR PANELS
SUPPLIED BY OWNER
INSTALLED BY PLUMBING
CONTRACTOR

A

GENERAL NOTES:
1.   PROVIDE SCREENED VENTS @ EA. 3RD. JOIST SPACE @ ALL ATTIC EAVES.
2.   PROVIDE SCREENED ROOF VENTS @ 10'-0" O.C. (1/300 VENT TO ATTIC SPACE).
3.   USE 1/2" CCX PLY. @ ALL EXPOSED EAVES.
4.   USE 300# COMPOSITION SHINGLES OVER 15# FELT.

B

Figure 17-8.
A—Dimensioning
flat surfaces.
B—Dimensioning
architectural
features.

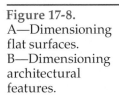

A

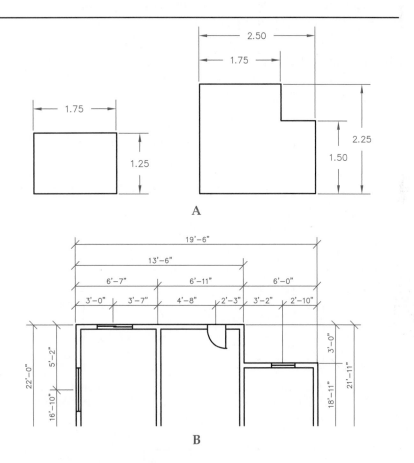

B

Dimensioning Cylindrical Shapes

You typically dimension the diameter and the length of a cylindrical shape in the view in which the cylinder appears rectangular. See **Figure 17-9.** The diameter symbol next to the dimension indicates that the part is a cylinder. This allows you to omit the view in which the cylinder appears as a circle.

Figure 17-9.
Dimensioning cylindrical shapes.

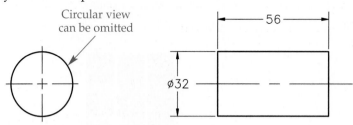

Dimensioning Square and Rectangular Features

You usually dimension square and rectangular features in the views that show the length and height. If appropriate, add a square symbol preceding the dimension for a square feature, to eliminate the need for an additional view. See **Figure 17-10.**

Dimensioning Cones and Regular Polygons

One method to dimension a conical shape is to dimension the length and the diameters at both ends. An alternative is to dimension the taper angle and the length. You usually dimension regular polygons that have an even number of sides by giving the distance across the flats and the length. **Figure 17-11** shows examples of dimensioned cones and regular polygons.

Figure 17-10.
Dimensioning square and rectangular features.

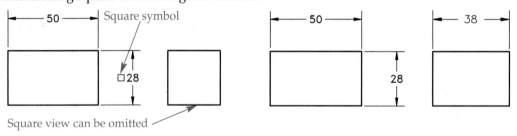

Figure 17-11.
Dimensioning cones and hexagonal cylinders.

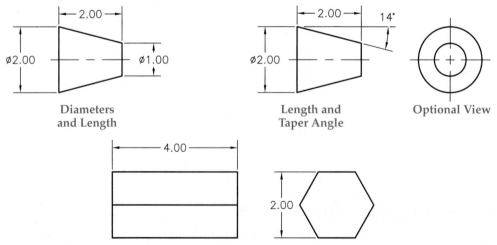

Ideally, you should determine drawing scale, scale factors, and dimension size characteristics before you begin drawing. Incorporate these settings into your drawing template files, and make changes when necessary. The drawing scale factor is important because it determines how dimensions appear on-screen and plot.

To help understand the concept of drawing scale, look at the portion of a floor plan shown in **Figure 17-12.** You should draw everything in model space at full scale. This means that the bathtub, for example, is actually drawn 5' long. However, at this scale, dimension appearance becomes an issue. Full-scale dimension characteristics, such as 1/8" high dimension text, are extremely small compared to the other full-scale objects. See **Figure 17-12A.** To see the dimensions clearly, you must adjust the size of dimension characteristics according to the drawing scale. See **Figure 17-12B.** You can calculate the scale factor manually and apply it to dimensions, or you can allow AutoCAD to calculate the scale factor using annotative dimensions.

Scaling Dimensions Manually

To adjust the size of dimension elements manually according to a specific drawing scale, you must first calculate the drawing scale factor. Once you determine the scale factor, you then multiply the scale factor by the desired plotted dimension size to get the model space dimension size. You can apply this calculation to all dimension elements by entering the scale factor in the **Fit** tab of the **New** (or **Modify**) **Dimension Style** dialog box, described later in this chapter. Refer to Chapter 9 for information on determining the drawing scale factor.

Annotative Dimensions

AutoCAD scales annotative dimensions according to the annotation scale you select, which eliminates the need for you to calculate the scale factor. Once you choose an annotation scale, AutoCAD determines the scale factor and applies it to annotative dimensions and all other annotative objects. For example, if you scale dimensions manually at a drawing scale of 1/4" = 1'-0", or a scale factor of 48, you must enter 48 in the **Fit** tab of the **New** (or **Modify**) **Dimension Style** dialog box. When you place annotative dimensions, using this example, you set an annotation scale of 1/4" = 1'-0". Then, when you add annotative dimensions, AutoCAD scales them automatically according to the 1/4" = 1'-0" annotation scale.

Figure 17-12.
An example of a portion of a floor plan drawn at full scale in model space. If you draw dimensions at full scale, as shown in A, the dimensions are very small compared to the large objects. You must scale the dimensions, as shown in B, in order to see and plot them correctly.

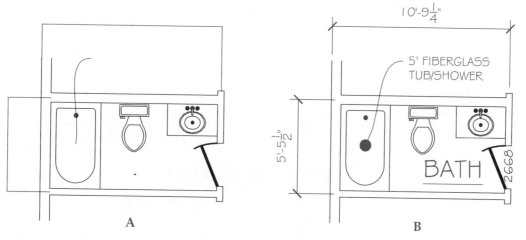

Annotative dimensions offer several advantages over manually scaled dimensions, including the ability to control dimension appearance based on scale, not scale factor. Annotative dimensions are especially effective when the drawing scale changes or when a single sheet includes objects viewed at different scales.

PROFESSIONAL TIP

If you anticipate preparing scaled drawings, you should use annotative dimensions and other annotative objects instead of traditional manual scaling. However, scale factor does influence non-annotative items and is still an important value to identify and use throughout the drawing process.

Setting Annotation Scale

You should usually set annotation scale before you begin adding dimensions so that dimension characteristics scale automatically. However, this is not always possible. It may be necessary to adjust the annotation scale throughout the drawing process, especially if you prepare multiple drawings with different scales on one sheet. This textbook approaches annotation scaling in model space only, using the process of selecting the appropriate annotation scale before placing dimensions. To draw dimensions at another scale, pick the new annotation scale and then place the dimensions.

When you access a dimension tool and an annotative dimension style is current, the **Select Annotation Scale** dialog box appears. This is a very convenient way to set annotation scale before adding dimensions. Dimension styles are described later in this chapter. You can also select the annotation scale from the **Annotation Scale** flyout located on the status bar. See **Figure 17-13.** Remember that the annotation scale is typically the same as the drawing scale.

Figure 17-13.
The status bar provides access to annotation scale options.

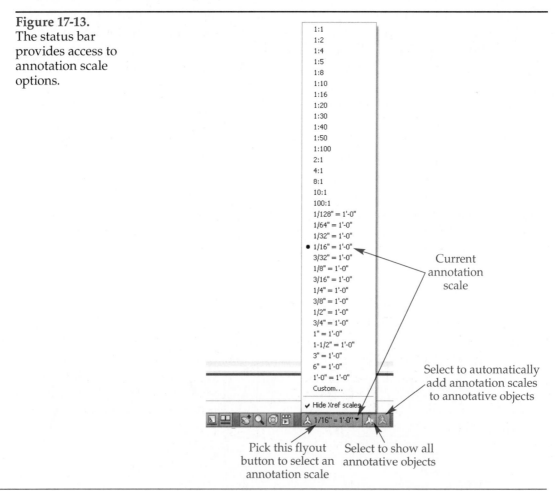

Editing Annotation Scales

If a certain scale is not available, or to change existing scales, pick the **Annotation Scale** flyout in the status bar and choose the **Custom...** option to access the **Edit Scale List** dialog box. From this dialog box, you can move the highlighted scale up or down in the list by picking the **Move Up** or **Move Down** button. To remove the highlighted scale from the list, pick the **Delete** button.

Select the **Edit...** button to open the **Edit Scale** dialog box. Here you can change the name of the scale and adjust the scale by entering the paper and drawing units. For example, a scale of 1/4″ = 1′-0″ uses a paper units value of .25 or 1 and a drawing units value of 12 or 48.

To create a new annotation scale, pick the **Add...** button to display the **Add Scale** dialog box, which functions the same as the **Edit Scale** dialog box. Pick the **Reset** button to restore the default annotation scale. When the correct annotation scale is set current, you are ready to place dimensions that automatically appear at the correct size according to the drawing scale.

> **NOTE**
>
> This textbook describes many additional annotative object tools. Some of these tools are more appropriate for working with layouts, as described later in this textbook.

Dimension Styles

Dimension styles control many dimension appearance characteristics. You might think of a dimension style as a grouping of dimensioning standards. Dimension styles usually apply to a specific type of drafting field or dimensioning application and correspond to appropriate drafting standards. For example, a dimension style used for mechanical drafting may use unidirectional dimensions, the Romans text font placed in a break in the dimension line, and dimension lines terminated with arrowheads. Refer again to **Figure 17-2.** A dimension style for architectural drafting may use aligned dimensions, the Stylus BT text font placed above the dimension line, and dimension lines terminated by tick marks, as shown in **Figure 17-3.**

dimension style: A saved configuration of dimension appearance settings.

Some drawings only require a single dimension style. However, you may need multiple dimension styles, depending on the variety of dimensions you apply and different dimension characteristics. You should generally create a dimension style for each unique dimensioning requirement. You can also override dimension appearance settings for individual dimensions. Add dimension styles to drawing templates for repeated use.

Working with Dimension Styles

Create, modify, and delete dimension styles using the **Dimension Style Manager** dialog box. See **Figure 17-14.** The **Styles:** list box displays existing dimension styles. Standard and Annotative dimension styles are available by default. The Annotative dimension style is preset to create annotative dimensions, as indicated by the icon to the left of the style name. The Standard dimension style does not use the annotative function. To make a dimension style current, double-click the style name, right-click on the name and select **Set current**, or pick the name and selecting the **Current** button.

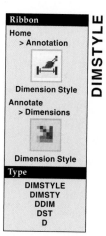

Figure 17-14.
The **Dimension Style Manager** dialog box. The non-annotative dimension style Standard and the annotative dimension style Annotative are available by default.

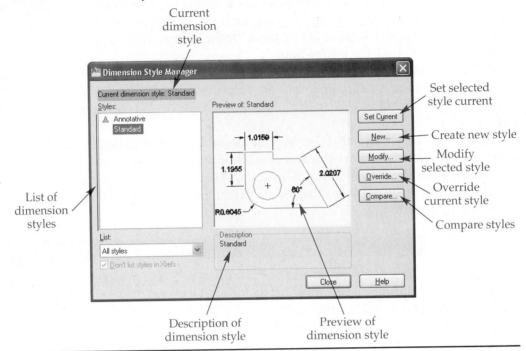

Current dimension style

Set selected style current

Create new style

Modify selected style

Override current style

Compare styles

List of dimension styles

Description of dimension style

Preview of dimension style

The **List:** drop-down list allows you to control whether all styles or only the styles in use appear in the **Styles:** list box. If the current drawing contains external references drawings (xrefs), you can use the **Don't list styles in Xrefs** box to eliminate xref-dependent dimension styles from the **Styles:** list box. This is often valuable because you cannot set xref dimension styles current or use them to create new dimensions. External references are described later in this textbook.

The **Description** area and **Preview of:** image provide information about the selected dimension style. If you change any of the default dimension settings without first creating a new dimension style, the changes are automatically stored as a dimension style override.

NOTE

The **Preview of:** image displays a representation of the dimension style and changes according to the selections you make.

Creating New Dimension Styles

To create a new dimension style, first select an existing dimension style from the **Styles:** list box to use as a base for formatting the new dimension style. Then pick the **New...** button in the **Dimension Style Manager** to open the **Create New Dimension Style** dialog box. See **Figure 17-15.**

Enter a descriptive name for the new dimension style, such as Architectural or Mechanical, in the **New Style Name** text box. If necessary, select a different style from the **Start With** drop-down list from which to base the new style. Pick the **Annotative** check box to make the dimension style annotative. You can also make the dimension style annotative by selecting the **Annotative** check box in the **Fit** tab of the **New (or Modify) Dimension Style** dialog box, described later in this chapter.

AutoCAD and Its Applications—Basics

Figure 17-15.
The **Create New Dimension Style** dialog box.

Enter name of
new style

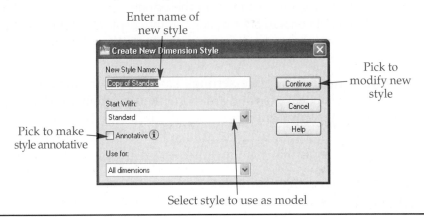

Pick to
modify new
style

Pick to make
style annotative

Select style to use as model

The Use for drop-down list specifies the type of dimensions to which the new style applies. Use the All dimensions option to create a new dimension style for all types of dimensions. If you select the Linear dimensions, Angular dimensions, Radius dimensions, Diameter dimensions, Ordinate dimensions, or Leaders and Tolerances option, you create a sub-style of the dimension style specified in the Start With: text box.

Pick the **Continue** button to access the **New Dimension Style** dialog box, shown in **Figure 17-16,** and adjust dimension style characteristics. The **Lines**, **Symbols and Arrows**, **Text**, **Fit**, **Primary Units**, **Alternate Units**, and **Tolerances** tabs display groups of settings for specifying dimension appearance. The next sections of this chapter describe each tab. After completing the style definition, pick the **OK** button to return to the **Dimension Style Manager** dialog box.

Figure 17-16.
The **Lines** tab of the **New** (or **Modify**) **Dimension Style** dialog box.

Select tab to change
dimension style settings

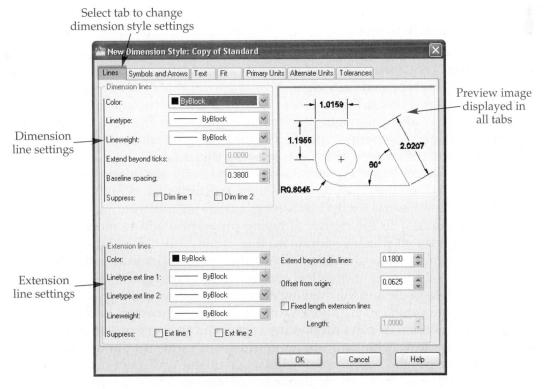

Preview image
displayed in
all tabs

Dimension
line settings

Extension
line settings

NOTE

The preview image shown in the upper-right corner of each **New** (or **Modify**) **Dimension Style** dialog box tab displays a representation of the dimension style and changes according to the selections you make.

AutoCAD stores dimension style settings as *dimension variables*. Dimension variables have limited practical uses and are more likely to apply to advanced applications such as scripting and customizing.

Reference Material

Dimension Variables

For a list of dimension variables, go to the **Reference Material** section of the Student Web site (www.g-wlearning.com/CAD) and select **Dimension Variables**.

CAUTION

Changing dimension variables by typing the variable name is not a recommended method for changing dimension style settings. Changes made in this manner can introduce inconsistencies with other dimensions. You should make changes to dimensions by redefining styles or performing style overrides.

Using the Lines Tab

The options in the **Lines** tab of the **New** (or **Modify**) **Dimension Style** dialog box are shown in **Figure 17-16**. These options allow you to control dimension and extension line display settings.

Dimension Line Settings

The **Dimension lines** area of the **Lines** tab allows you to set dimension line format. **Color**, **Linetype**, and **Lineweight** drop-down lists are available for changing the dimension line color, linetype, and lineweight. All *associative dimensions* are block objects, as further explained later in this textbook. When you assign the default ByBlock setting to color, linetype, and lineweight, the dimension takes on the drawing color, lineweight, and linetype properties, specified in the **Properties** panel of the **Home** ribbon tab, regardless of the layer on which you draw the dimension.

Using the ByBlock setting is noticeable only if you assign absolute values to drawing color, lineweight, and linetype properties in the **Properties** panel of the **Home** ribbon tab. If these properties use the ByLayer setting, as they should, the dimension acquires the settings assigned to the current drawing properties, which adopt the settings applied to the layer on which you draw the dimension.

If you assign the ByLayer setting to color, linetype, and lineweight, the dimension takes on the color, lineweight, and linetype properties of the layer on which you draw the dimension, regardless of the drawing color, lineweight, and linetype properties. If you use absolute values, such as a Blue color, a Continuous linetype, or a 0.05mm lineweight, the dimension displays the specified absolute values regardless of the properties assigned to the drawing or the layer on which you create the dimension.

The **Extend beyond ticks** text box is inactive unless you select oblique or architectural tick terminators from the **Symbols and Arrows** tab of the **New** (or **Modify**) **Dimension Style** dialog box. Architectural tick marks or oblique arrowheads often

Figure 17-17.
Using the **Extend
beyond ticks**
setting to allow
the dimension line
to extend past the
extension line. With
the default value of 0,
the dimension line
does not extend.

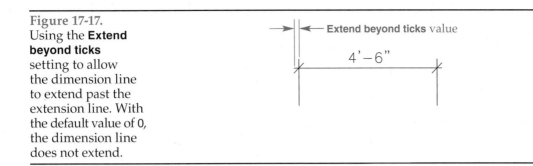

terminate dimension lines on architectural drawings. In this style of dimensioning, dimension lines often extend past extension lines, as shown in **Figure 17-17.** The 0.00 default draws dimension lines that do not extend past extension lines.

The **Baseline spacing** text box allows you to change the spacing between the dimension lines of baseline dimensions created with the **DIMBASELINE** tool. The default spacing is too close for most drawings, as shown in **Figure 17-18.** The ASME minimum dimension line spacing for baseline dimensioning is .375 (10mm). A value of .5 (12mm) or .75 (19mm) is usually more appropriate.

The **Suppress** feature has two toggles that prevent the display of the first, second, or both dimension lines and their arrowheads. The **Dim line 1** and **Dim line 2** check boxes refer to the first and second points picked when you create a dimension. Both dimension lines appear by default. **Figure 17-19** shows the results of using dimension line suppression options.

Extension Line Settings

The **Extension lines** area of the **Lines** tab allows you to set extension line format. **Color, Linetype ext line 1, Linetype ext line 2**, and **Lineweight** drop-down lists are available for changing the extension line color, linetype, and lineweight from the default ByBlock setting, if necessary. You can use the **Linetype ext line 1** and **Linetype ext line 2** drop-down lists to specify the linetype applied to each extension line. Lines 1 and 2 correspond to the first and second points you pick when drawing a dimension.

The **Extend beyond dim lines** option allows you to set the distance the extension line runs past the dimension line. See **Figure 17-20.** An extension line extension of .125 (3mm) is recommended by ASME standards. The **Offset from origin** option specifies the distance between the object and the beginning of the extension line. Most applications require this small offset. ASME standards recommend an extension line offset distance of .063 (1.5mm). When an extension line meets a centerline, however, use a setting of 0.0 to prevent a gap.

Figure 17-18.
The **Baseline
spacing** setting
controls the spacing
between dimension
lines.

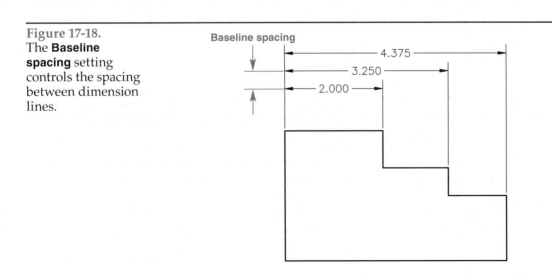

Figure 17-19.
Using the **Dim line 1** and **Dim line 2** dimensioning settings. "Off" is equivalent to an unchecked **Suppress** check box in the **Lines** tab.

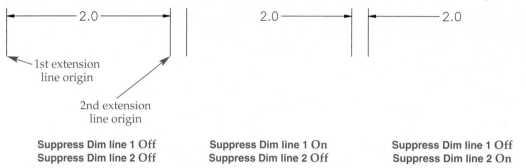

Suppress Dim line 1 Off   Suppress Dim line 1 On   Suppress Dim line 1 Off
Suppress Dim line 2 Off   Suppress Dim line 2 Off   Suppress Dim line 2 On

Figure 17-20.
The extension line extension and extension line offset settings.

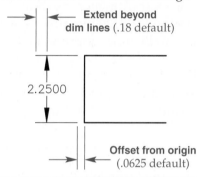

The **Fixed length extension lines** check box sets a given length for extension lines. When this box is checked, the **Length** text box becomes active. The value in the **Length** text box sets a restricted length for extension lines, measured from the dimension line toward the extension line origin.

Extension lines display by default. Use the **Ext line 1** and **Ext line 2** check boxes to suppress extension lines. Though extension line suppression is typically applied to individual dimensions, not a dimension style, you might suppress an extension line, for example, if it coincides with an object line. See **Figure 17-21.**

Using the Symbols and Arrows Tab

The options in the **Symbols and Arrows** tab of the **New** (or **Modify**) **Dimension Style** dialog box are shown in **Figure 17-22.** These options allow you to control the appearance of arrowheads, center marks, and other symbol components of dimensions.

Arrowhead Settings

Use the appropriate drop-down list in **Arrowheads** area to select the arrowhead to use for the first, second, and leader arrowheads. The default arrowhead is **Closed filled**, which is recommended by ASME standards, although **Closed blank**, **Closed**, or

Figure 17-21.
Suppressing extension lines.

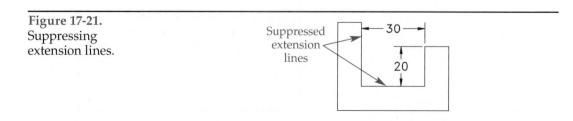

Figure 17-22.
The **Symbols and Arrows** tab of the **New** (or **Modify**) **Dimension Style** dialog box.

Select tab to specify arrow style

Arrowhead properties

Center mark properties

Dimension break size

Arc length dimension settings

Jog symbol angle setting

Jog text height setting

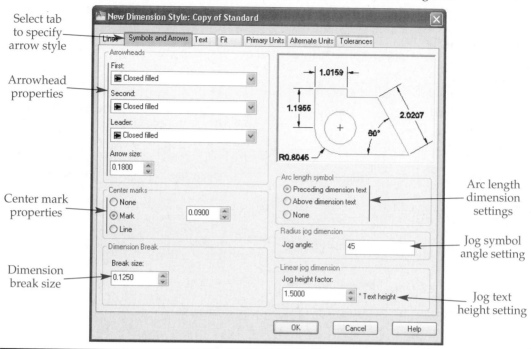

Open arrowheads are sometimes used. A leader pointing to a surface uses a small dot. **Figure 17-23** shows arrowhead styles. If you pick a new arrowhead in the **First:** drop-down list, AutoCAD automatically makes the same selection for the **Second:** drop-down list. When you select the **Oblique** or **Architectural tick** arrowhead, the **Extend beyond ticks:** text box in the **Lines** tab activates.

Notice that **Figure 17-23** does not contain an example of a user arrow. This option allows you to access an arrowhead of your own design. For this to work, you must first design an arrowhead that fits inside a 1 unit square (unit block) with a dimension line "tail" of 1 unit in length, and save the arrowhead as a block. Blocks are described later in this textbook. The **Select Custom Arrow Block** dialog box appears when you pick **User Arrow...** from an **Arrowheads** drop-down list. Type the name of the custom arrow block in the **Select from Drawing Blocks:** text box or pick a block from the drop-down list and then pick **OK** to apply the arrow to the style.

The **Arrow size:** text box allows you to change arrowhead size. A .125" (3 mm) arrowhead size is common on mechanical drawings. See **Figure 17-24.**

Center Mark Settings

The **Center marks** area allows you to select the way center marks appear in circles and arcs when you use circular feature dimensioning tools. Fillets and rounds generally have no center marks. The **None** option provides for no center marks to occur in circles and arcs. The **Mark** option places center marks without centerlines. The **Line** option places center marks and centerlines. Use the **Size:** text box in the **Center marks** area to change the size of the center mark and centerline. The size defines half the length of a centerline dash and the distance that the centerline extends past the object. A value of .0625" (1.5 mm) is appropriate for the centerline dash half-length, but does not provide for the preferred .125" (3 mm) extension past the object. **Figure 17-25** shows the results of specifying center marks and centerlines.

Figure 17-23.
Examples of dimensions drawn using the options found in the **Arrowheads** drop-down lists.

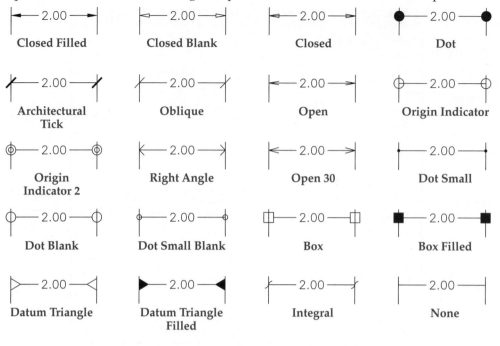

Closed Filled	Closed Blank	Closed	Dot
Architectural Tick	Oblique	Open	Origin Indicator
Origin Indicator 2	Right Angle	Open 30	Dot Small
Dot Blank	Dot Small Blank	Box	Box Filled
Datum Triangle	Datum Triangle Filled	Integral	None

Figure 17-24.
ASME standards specify an arrowhead size of .125″. The **Closed filled**, **Closed blank**, and **Closed** arrowhead styles adhere to the standard 3:1 ratio of length to width.

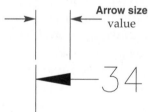

Figure 17-25.
Arcs and circles displayed with center marks and centerlines.

Center Marks	Centerlines
Center mark size	Centerline size

Adjusting Break Size

The **Dimension Break** area controls the amount of extension line removed when you use the **DIMBREAK** tool. Specify a value in the **Break size:** text box to set the total break length. **Figure 17-26** shows an example of a 3 mm extension line break. The default size is .125″ (3 mm). ASME standards do not recommend breaking extension lines.

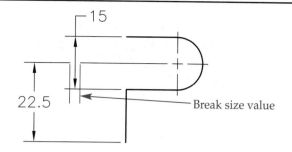

Figure 17-26.
Use the **Break size** setting to specify the length of the break created using the **DIMBREAK** tool.

15

22.5

Break size value

Adding an Arc Length Symbol

The **Arc Length Symbol** area controls the placement of the arc length symbol when you use the **DIMARC** tool. The default **Preceding dimension text** option places the symbol in front of the dimension value. Select the **Above dimension text** radio button to place the arc length symbol over the length value. See **Figure 17-27.** Pick the **None** radio button to suppress the symbol so that it does not show.

Adjusting Jog Angle

The **Jog angle** setting in the **Radius jog dimension** area controls the appearance of the break line applied to the jog symbol when you use the **DIMJOGGED** tool. This value sets the incline formed by the line connecting the extension line and dimension line. The default angle is 45°.

Setting Jog Height

The **Jog height factor** setting in the **Linear jog dimension** area controls the size of the break symbol created using the **DIMJOGLINE** tool. This value sets the height of the break symbol based on a multiple of the text height. For example, the default value of 1.5 creates a break symbol that is .18″ tall if the text height is .12″. The default angle is 45°.

> **NOTE**
>
> Chapter 18 describes the **DIMJOGLINE** tool in more detail, and Chapter 19 covers the **DIMJOGGED**, **DIMARC**, and **DIMBREAK** tools.

Exercise 17-1

Access the Student Web site (www.g-wlearning.com/CAD) and complete Exercise 17-1.

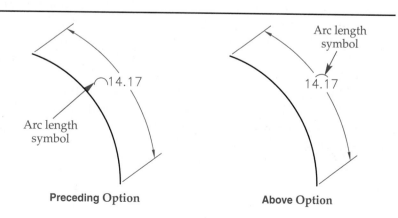

Figure 17-27.
You can place the arc length symbol in front of or above the arc dimension text.

Arc length symbol

14.17

Arc length symbol

14.17

Arc length symbol

Preceding Option

Above Option

Figure 17-28.
The **Text** tab of the **New** (or **Modify**) **Dimension Style** dialog box.

Select tab to set up dimension text

Set appearance of the text

Set location of text relative to dimension line

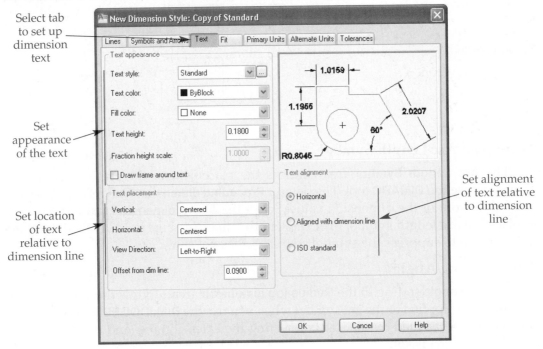

Set alignment of text relative to dimension line

Using the Text Tab

The **Text** tab of the **New** (or **Modify**) **Dimension Style** dialog box is shown in **Figure 17-28.** Use this tab to control the display of dimension values.

Text Appearance Settings

Use the **Text appearance** area to set the dimension text style, color, height, and frame. A text style must be loaded in the current drawing before it is available for use in dimension text. Pick the desired text style from the **Text style** drop-down list. To create or modify an existing text style, pick the ellipsis (…) button next to the drop-down list to launch the **Text Style** dialog box. Use the **Text color** drop-down list to specify the appropriate text color, which should be ByBlock for typical applications.

Use the **Text height** text box to specify the dimension text height. Dimension text height is commonly the same as the text height used for most other drawing text, except for titles, which are often larger. The default dimension text height of .18″ (2.5 mm) is an acceptable standard. Many companies use a text height of .125″ (3 mm). The ASME standard recommends a .12″ (3 mm) text height. The text height for titles and labels is usually .24 (6 mm).

The **Fraction height scale** setting controls the height of fractions for architectural and fractional unit dimensions. The value in the **Fraction height scale** box is multiplied by the text height value to determine the height of the fraction. A value of 1.0 creates fractions that are the same text height as regular (nonfractional) text, which is the normally accepted standard. A value less than 1.0 makes the fraction smaller than the regular text height.

Select the **Draw frame around text** check box to draw a rectangle around the dimension text. A rectangle is most often used to describe a *basic dimension*. The setting for the **Offset from dim line** value, explained later in this section, determines the distance between the text and the frame.

basic dimension:
A theoretically perfect dimension used to describe the exact size, profile, orientation, and location of a feature.

Figure 17-29.
Dimension text justification options. A—Vertical justification options, with the horizontal Centered justification. B—Horizontal justification options, with the vertical Centered justification.

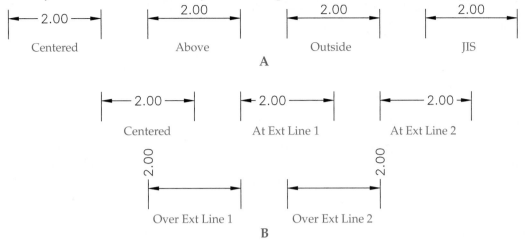

Text Placement Settings

The **Text placement** area controls text placement relative to the dimension line. See **Figure 17-29**. The **Vertical:** drop-down list provides vertical justification options. Use the default **Centered** option to place dimension text centered in a gap provided in the dimension line. This is the most common dimensioning practice in mechanical drafting and many other fields.

Select the **Above** option to place the dimension text horizontally and above horizontal dimension lines. For vertical and angled dimension lines, the text appears in a gap provided in the dimension line. This option is generally used for architectural drafting and building construction. Architectural drafting commonly uses aligned dimensioning, in which the dimension text aligns with the dimension lines and all text reads from either the bottom or the right side of the sheet.

Pick the **Outside** option to place the dimension text outside the dimension line and either above or below a horizontal dimension line or to the right or left of a vertical dimension line. The direction you move the cursor determines the above/below and left/right placement. Choose the **JIS** option to align the text according to the Japanese Industrial Standard (JIS).

The **Horizontal:** drop-down list provides options for controlling the horizontal placement of dimension text. Pick the default **Centered** option to place dimension text centered between the extension lines. Select the **At Ext Line 1** option to locate the text next to the extension line placed first, or choose **At Ext Line 2** to locate the text next to the extension line placed second. Pick **Over Ext Line 1** to place the text aligned with and over the first extension line, or select **Over Ext Line 2** to place the text aligned with and over the second extension line. Placing text aligned with and over an extension line is not common practice.

The **View Direction:** drop-down list determines how you read dimension text. Use the default **Left-to-Right** option to read text from left-to-right or bottom-to-top, depending on the text placement and alignment. Choose the **Right-to-Left** option to flip dimension text. Text may appear inverted and reads from right-to-left or top-to-bottom, depending on the text placement and alignment. Changing text view direction to right-to-left is not common practice.

The **Offset from dim line:** text box sets the gap between the dimension line and dimension text, the distance between the leader shoulder and text, and the space between text and the rectangle drawn around it. The gap should be set to half the text height for most applications. **Figure 17-30** shows the gap in linear and leader dimensions.

Figure 17-30.
The gap (offset) used for text in a linear dimension and a leader dimension.

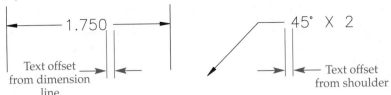

Text Alignment Settings

Use the **Text alignment** area to specify unidirectional or aligned dimensions. The **Horizontal** option draws the unidirectional dimensions commonly used for mechanical drafting applications. The **Aligned with dimension line** option creates aligned dimensions, typically used for architectural drafting applications. The **ISO Standard** option creates aligned dimensions when the text falls between the extension lines and horizontal dimensions when the text falls outside the extension lines.

Using the Fit Tab

The **Fit** tab of the **New** (or **Modify**) **Dimension Style** dialog box is shown in **Figure 17-31**. The settings on this tab allow you to establish dimension *fit format*.

fit format: The arrangement of dimension text and arrowheads on a drawing.

Fit Options

The **Fit options** area controls how text, dimension lines, and arrows behave when there is not enough room between extension lines to accommodate all of the items. The amount of space between the extension lines and the size of the dimension value, offset, and arrowheads, influence fit performance. All fit options place text and dimension lines with arrowheads inside the extension lines if space is available. All except the **Always keep text between ext lines** option place arrowheads, dimension lines, and text outside of the extension lines when space is limited.

Figure 17-31.
The **Fit** tab of the **New** (or **Modify**) **Dimension Style** dialog box.

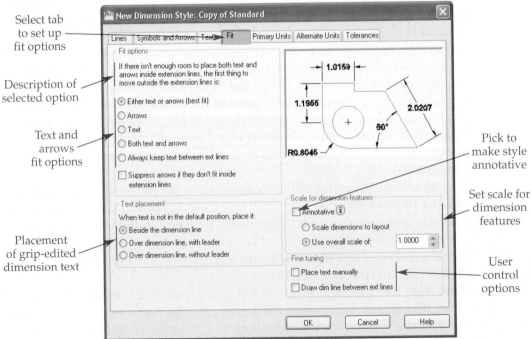

Choose the default **Either text or arrows (best fit)** radio button to move either the dimension value or the arrows outside of the extension lines first. Pick the **Arrows** radio button to attempt to place arrowheads outside of the extension lines first, followed by text. Pick the **Text** radio button to attempt to place text outside of the extension lines first, followed by arrowheads. Choose the **Both text and arrows** radio button to move both text and arrowheads outside of the extension lines.

Select the **Always keep text between ext lines** radio button to place the dimension value between the extension lines. This option typically causes interference between the dimension value and extension lines when there is limited space between extension lines. Pick the **Suppress arrows if they don't fit inside extension lines** radio button to remove the arrowheads if they do not fit inside the extension lines. Use this option with caution, because it can create dimensions that violate standards.

Text Placement Settings

Sometimes it becomes necessary to move the dimension value from its default position. You can use grips to move the value independently of the dimension. The options in the **Text placement** area specify how these grip-editing situations function.

Select the **Beside the dimension line** radio button to restrict dimension text movement. You can grip-move the text with the dimension line, but only within the same plane as the dimension line. If you pick the **Over dimension line, with leader** radio button, you can grip-move the dimension text in any direction away from the dimension line. A leader line forms connecting the text to the dimension line. Choose the **Over dimension line, without leader** radio button to have the ability to move the dimension text in any direction away from the dimension line without a connecting leader.

PROFESSIONAL TIP

To return the dimension text to its default position, select the dimension, right-click and select **Home text** from the **Dim Text position** cascading submenu.

Text Scale Options

Use the **Scale for dimension features** area to set the dimension scale factor. Select the **Annotative** check box to create an annotative dimension style. The **Annotative** check box is already set when you modify the default Annotative dimension style or pick the **Annotative** check box in the **Create New Dimension Style** dialog box.

You can select the **Scale dimensions to layout** radio button to dimension in a floating viewport in a paper space layout. You must add dimensions to the model in a floating viewport in order for this option to function. Scaling dimensions to the layout allows the overall scale to adjust according to the active floating viewport by setting the overall scale equal to the viewport scale factor.

Pick the **Use overall scale of** radio button to scale a drawing manually, and enter the drawing scale factor to be applied to all dimension settings. The scale factor is multiplied by the desired plotted dimension size to get the model space dimension size. For example, if the height of dimension text is set to .12 and the value for the overall scale is set to 2 for a half scale drawing, then the dimension text measures .24 units (2 × .12 = .24). If you then plot the drawing at a plot scale of 1:2 (half), the size of the dimension text on the paper measures .12 units.

You can draw dimensions in either model space or layout space. Model space dimension scale is set by the drawing scale factor to achieve the correct dimension appearance. Associative paper space dimensions automatically adjust to model modifications and do not require scaling. In addition, if you dimension in paper space, you can dimension the model differently in two viewports. However, paper space dimensions are not visible when you work in model space, so you must be careful not to move a model space object into a paper space dimension. Avoid using nonassociative paper space dimensions.

Fine Tuning Settings

The **Fine tuning** area provides flexibility in controlling the placement of dimension text. Select the **Place text manually** check box to have the ability to place text where you want it, such as to the side within the extension lines or outside of the extension lines. However, this feature can make equally offsetting dimension lines somewhat more cumbersome, and it is not necessary for standard dimensioning practices.

The **Draw dim line between ext lines** option forces AutoCAD to place the dimension line inside the extension lines, even when the text and arrowheads are outside. The default application is to place the dimension line and arrowheads outside the extension lines. See **Figure 17-32.** Though some companies prefer the appearance, forcing the dimension line inside the extension lines is not an ASME standard.

Exercise 17-2

Access the Student Web site (www.g-wlearning.com/CAD) and complete Exercise 17-2.

Using the Primary Units Tab

The **Primary Units** tab of the **New** (or **Modify**) **Dimension Style** dialog box is shown in **Figure 17-33.** This tab controls linear and angular dimension units.

Linear Dimension Settings

The **Linear dimensions** area allows you to specify settings for primary linear dimensions. The options from the **Unit format** drop-down list are the same as those in the **Length** area of the **Drawing Units** dialog box. Typically, primary linear dimension unit format is the same as the corresponding drawing units.

The **Precision** drop-down list allows you to specify the precision applied to dimensions, which may be the same as the related drawing units precision. A variety of dimension precisions are often found on the same drawing. When you are using decimal units, precision determines how many zeros follow the decimal place. Precision settings in mechanical drafting depend on the accuracy required to manufacture specific features. Some features require greater precision, generally due to fits

Figure 17-32.
The effect of the **Draw dim line between ext lines** option in the **Fine tuning** area of the **Fit** tab.

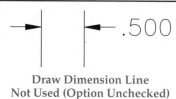

Draw Dimension Line
Not Used (Option Unchecked)

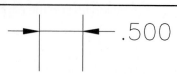

Draw Dimension Line
Enabled (Option Checked)

Figure 17-33.
The **Primary Units** tab of the **New** (or **Modify**) **Dimension Style** dialog box.

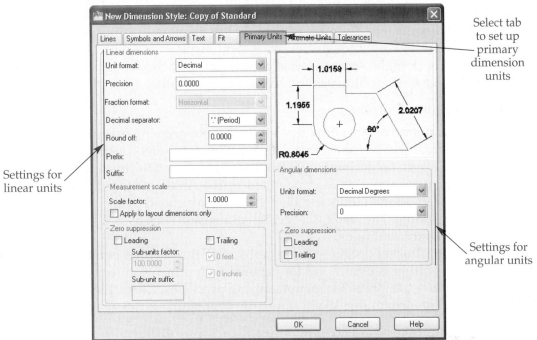

Select tab to set up primary dimension units

Settings for linear units

Settings for angular units

between mating parts. For example, a precision setting of 0.00 represents less exactness than a setting of 0.0000. Chapter 20 further explains this concept. Common precisions in mechanical drafting include 0.0000, 0.000, and 0.00. When you specify fractional units, precision values identify the smallest desired fractional denominator. Precisions of 1/16 to 1/64 are common, but you can choose other options, ranging from 1/256 to 1/2; 0 displays no fractional values.

Use the **Decimal separator** drop-down list to specify commas, periods, or spaces as separators for decimal numbers. The '.' **(Period)** option is default and is appropriate for typical applications. The **Decimal separator** option is not available if the unit format is **Architectural** or **Fractional**. The **Fraction format** drop-down list is available if the unit format is **Architectural** or **Fractional**. The options for controlling the display of fractions are **Diagonal**, **Horizontal**, and **Not Stacked**.

The **Round off** text box specifies the accuracy of rounding for dimension numbers. The default is zero, which means that no rounding takes place and all dimensions specify the value exactly as measured. No rounding is appropriate for most applications. If you enter a value of .1, all dimensions are rounded to the closest .1 unit. For example, an actual measurement of 1.188 is rounded to 1.2.

Add a *prefix* to a dimension by entering a value in the **Prefix** text box. A typical application for a prefix is SR3.5, where SR means "spherical radius." When a prefix is used on a diameter or radius dimension, the prefix replaces the ∅ or R symbol. Add a *suffix* to a dimension by entering a value in the **Suffix** text box. A typical application for a suffix is 3.5 MAX, where MAX is the abbreviation for "maximum." The abbreviation in could be used when one or more inch dimensions are placed on a metric-dimensioned drawing. Conversely, a suffix of mm could be used on one or more millimeter dimensions placed on an inch drawing.

prefixes: Special notes or applications placed in front of the dimension text.

suffixes: Special notes or applications placed after the dimension text.

PROFESSIONAL TIP

A prefix or suffix is normally a special specification, used in only a few cases on a drawing. As a result, often it is easiest to enter a prefix or suffix using the **MText** or **Text** option of the related dimensioning tool.

Set the scale factor of linear dimensions in the **Scale factor:** text box of the **Measurement scale** area. If you set a scale factor of 1, dimension values display the same as they measure. If the scale factor is 2, dimension values are twice as much as the measured amount. For example, an actual measurement of 2 inches displays as 2 with a scale factor of 1, but the same measurement displays as 4 when the scale factor is 2. Placing a check in the **Apply to layout dimensions only** check box makes the linear scale factor active only for dimensions created in paper space.

Zero Suppression Options

The **Zero suppression** area provides options for suppressing primary unit leading and trailing zeros, and for controlling the function sub-units. Uncheck the **Leading** option to leave a zero on decimal units less than 1, such as 0.5. This option is suitable to create metric dimensions as recommended by the ASME standard. Check the box to remove the 0 on decimal units less than 1, as recommended by the ASME standard for inch dimensioning. The result is a decimal dimension such as .5. This option is not available for architectural units.

Uncheck the **Trailing** option to leave zeros after the decimal point based on the precision. This setting is usually off for decimal inch dimensioning because trailing zeros often control tolerances for manufacturing processes. Check the box for metric dimensions to conform to the ASME standard. This option is not available for architectural units.

The **0 feet** check box is enabled for architectural and engineering units. Check the box to remove the zero in dimensions given in feet and inches when there are zero feet. For example, when this box is checked, a dimension reads 11″. When **0 feet** is unchecked, however, the same dimension reads 0′-11″.

The **0 inches** check box is also enabled for architectural and engineering units. Check the box to remove the zero when the inch portion of dimensions displayed in feet and inches is less than one inch, such as 12′-7/8″. If this box is unchecked, the same dimension reads 12′-0 7/8″. In addition, this option removes the zero from a dimension with no inch value; for example, 12′ appears instead of 12′-0″.

sub-units: Unit formats that are smaller than the primary unit format. For example, centimeters can be defined as a sub-unit of meters.

The **Sub-units factor** and **Sub-unit suffix** text boxes become enabled when you use decimal units and select the **Leading** check box. Most drawings use a single format for all dimension values. For example, all dimensions on a decimal inch drawing measure in inches, or decimals of an inch. *Sub-units* allow you to apply a different unit format to dimensions that are smaller than the primary unit format, without using decimals. For example, if you use meters to dimension most objects on a metric civil engineering drawing, you can use a **Sub-units factor** value of 100 (100 cm/m) and a **Sub-unit suffix** of cm to dimension objects smaller than one meter using centimeters, instead of decimals of a meter. Now when you dimension an object that is 0.5 meters, the dimension reads 500 cm.

> **NOTE**
>
> For drawings that do not require sub-units, but do suppress leading zeros, specify no sub-unit suffix. As long as you do not add a suffix, there is no need to change the sub-unit factor, though a factor of 0 also disables sub-units.

Angular Dimension Settings

The **Angular dimensions** area allows you to specify settings for primary angular dimensions. The options from the **Units format** drop-down list are the same as those in the **Angle** area of the **Drawing Units** dialog box. Typically, the primary angular dimension unit format is the same as the corresponding drawing units. Use the **Precision** drop-down list to set the appropriate angular dimension value precision. The **Zero**

suppression area has check boxes for suppressing angular dimension **Leading** and **Trailing** zeros. Zero suppression for angular units is usually the same as applied to linear dimensions.

Using the Alternate Units Tab

The **Alternate Units** tab of the **New** (or **Modify**) **Dimension Style** dialog box, shown in **Figure 17-34,** allows you to set *alternate units*, or *dual dimensioning units*. Dual dimensioning practices are no longer a recommended ASME standard. ASME recommends that drawings be dimensioned using inch or metric units only. However, many other applications do use alternate units.

To use alternate units, select the **Display alternate units** check box to enable the settings. The **Alternate Units** tab includes most of the same settings found in the **Primary Units** tab. The **Multiplier for alt units** setting is multiplied by the primary unit to establish the value for the alternate unit. The default 25.4 allows you to use millimeters as alternate units on an inch unit drawing. The **Placement** area controls the location of the alternate-unit dimension. You can choose to place the alternate-unit dimension after or below the primary value.

alternate units (dual dimensioning units): Dimensions in which measurements in one system, such as inches, are followed by bracketed measurements in another system, such as millimeters.

NOTE

Chapter 20 describes the **Tolerances** tab found in the **New** (or **Modify**) **Dimension Style** dialog box.

Exercise 17-3

Access the Student Web site (www.g-wlearning.com/CAD) and complete Exercise 17-3.

Figure 17-34.
The **Alternate Units** tab of the **New** (or **Modify**) **Dimension Style** dialog box.

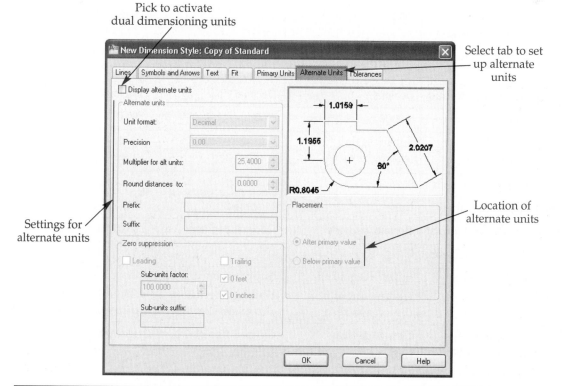

Pick to activate dual dimensioning units

Select tab to set up alternate units

Settings for alternate units

Location of alternate units

Making Your Own Dimension Styles

Creating and using dimension styles is an important element of drafting with AutoCAD. Carefully evaluate the characteristics of the dimensions you will add to drawings. Check school, company, and national standards to verify the accuracy of the dimension settings you plan to use. When you are ready, use the **Dimension Style Manager** dialog box to establish appropriate dimension styles. **Figure 17-35** provides possible settings for three common dimension styles. Use the AutoCAD default values for settings not listed.

Figure 17-35.
This chart shows dimension settings for typical mechanical and architectural drawings.

Setting	Mechanical—Inch	Mechanical—Metric (mm)	Architectural—U.S. Customary
Baseline spacing	.5	12	1/2″
Extend beyond dimension lines	.125	3	1/8″
Offset from origin	.063	1.5	1/16″ or 3/32″
Arrowhead options	Closed filled, Closed, or Open	Closed filled, Closed, or Open	Architectural tick, Dot, Closed filled, Oblique, or Right angle
Arrow size	.125	3	1/8″
Center marks	Line	Line	Mark
Center mark size	.0625	1.5	1/16″
Text style	Romans	Romans	Stylus BT
Text height	.12	3	1/8″
Vertical and horizontal text placement	Centered	Centered	Vertical: Above Horizontal: Centered
View direction	Left-to-right	Left-to-right	Left-to-right
Offset from dimension line	.063	1.5	1/16″
Text alignment	Horizontal	Horizontal	Aligned with dimension line
Linear unit format	Decimal	Decimal	Architectural
Linear precision	0.0000	0.00	1/16″
Linear zero suppression	Suppress only the leading zero	Suppress only the trailing zero	Suppress only the 0 feet zero
Sub-units factor	0	Disabled	Disabled
Sub-units suffix	None	Disabled	Disabled
Angular unit format	Decimal degrees	Decimal degrees	Decimal degrees
Angular precision	0	0	0
Angular zero suppression	Suppress only the leading angular dimension zero	Suppress only the trailing angular dimension zero	Suppress only the leading angular dimension zero
Alternate units	Do not display	Do not display	Do not display
Tolerances	By application	By application	None

Exercise 17-4

Access the Student Web site (www.g-wlearning.com/CAD) and complete Exercise 17-4.

Changing Dimension Styles

Use the **Dimension Style Manager** to change the characteristics of an existing dimension style. Pick the **Modify** button to open the **Modify Dimension Style** dialog box, which allows you to make changes to the selected style. When you make changes to a dimension style, such as selecting a different text style or linear precision, all existing dimensions drawn using the modified dimension style update to reflect the changes. Use a different dimension style with unique characteristics when appropriate.

To *override* a dimension style, pick the **Override** button in the **Dimension Style Manager** to open the **Override Current Style** dialog box. An example of an override is including a text prefix for a few of the dimensions in a drawing. The **Override** button is only available for the current style. Once you create an override, it is current and appears as a branch, called the *child*, of the *parent* style. The override settings are lost when any other style, including the parent, is set current.

Sometimes it is useful to view the details of two styles to determine their differences. Select the **Compare...** button in the **Dimension Style Manager** to display the **Compare Dimension Styles** dialog box. Here you can compare two styles by selecting the name of one style from the **Compare:** drop-down list and the name of the other in the **With:** drop-down list. The differences between the selected styles display in the dialog box.

override: A temporary change to the current style settings; the process of changing a current style temporarily.

child: A style override.

parent: The dimension style from which a style override is created.

NOTE

The **New Dimension Style**, **Modify Dimension Style**, and **Override Current Style** dialog boxes have the same tabs.

Renaming and Deleting Dimension Styles

To rename a dimension style using the **Dimension Style Manager**, slowly double-click on the name or right-click the name and select **Rename**. To delete a dimension style using the **Dimension Style Manager**, right-click the name and select **Delete**. You cannot delete a dimension style that is assigned to dimensions. To delete a style that is in use, assign a different style to the dimensions that reference the style to be deleted.

NOTE

You can also rename styles using the **Rename** dialog box. Select **Dimension styles** in the **Named Objects** list to rename a dimension style.

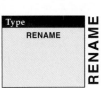

Type
RENAME

RENAME

Setting a Dimension Style Current

You can set a dimension style current using the **Dimension Style Manager** by double-clicking the style in the **Styles** list box, right-clicking on the name and selecting **Set current**, or picking the style and selecting the **Set current** button. To set a text style current without opening the **Dimension Style Manager** dialog box, use the **Dimension Style** flyout located in the expanded **Annotation** panel of the **Home** ribbon tab and on the **Dimensions** panel of the **Annotate** ribbon tab.

PROFESSIONAL TIP

You can import dimension styles from existing drawings using **DesignCenter**. See Chapter 5 for more information about using **DesignCenter** to import file content.

Exercise 17-5

Access the Student Web site (www.g-wlearning.com/CAD) and complete Exercise 17-5.

Template Development

Chapter 9

For instructions on adding dimension styles to each drawing template, go to the Student Web site (www.g-wlearning.com/CAD), select this chapter, and select **Template Development**.

Chapter Test

Answer the following questions. Write your answers on a separate sheet of paper or go to the Student Web site (www.g-wlearning.com/CAD) and complete the electronic chapter test.

1. List at least three factors that influence company dimensioning practices.
2. What does the *M* mean in the title of the standard ASME Y14.5M-1994?
3. Name two basic coordinate systems that are used to create location dimensions.
4. Define the term *general notes*.
5. Briefly explain the difference between placing specific and general notes on a drawing.
6. Explain how to dimension a cylinder using only one view.
7. Describe two ways to dimension a cone.
8. When is the best time to determine the drawing scale and scale factors for a drawing?
9. Explain how to add a scale to the **Annotation Scale** flyout in the status bar.
10. Define *dimension style*.
11. Name the dialog box that is used to create dimension styles.
12. Identify two ways to access the dialog box identified in Question 11.
13. Name the dialog box tab used to control the appearance of dimension lines and extension lines.

14. Name at least four arrowhead types that are available in the **Symbols and Arrows** tab for common use on architectural drawings.
15. Name the dialog box tab used to control the settings that display the dimension text.
16. What has to happen before a text style can be accessed for use in dimension text?
17. What is the ASME recommended height for dimension numbers and notes on drawings?
18. Name the dialog box tab used to control settings that adjust the location of dimension lines, dimension text, arrowheads, and leader lines.
19. How can you delete a dimension style from a drawing?
20. How do you set a dimension style current?

Drawing Problems

Start AutoCAD if it is not already started. Start a new drawing for each problem using an appropriate template of your choice. Follow the specific instructions for each problem.

▼ Basic

1. Start a new drawing using one of your templates and create a RomanS text style using the romans font. Create the Mechanical (Inch) dimension style shown in **Figure 17-35.** Use the default AutoCAD settings for the dimension style settings not listed. Save the drawing as P17-1.

2. Start a new drawing using one of your templates and create a RomanS text style using the romans font. Create the Mechanical (Metric) dimension style shown in **Figure 17-35.** Use the default AutoCAD settings for the dimension style settings not listed. Save the drawing as P17-2.

3. Start a new drawing using one of your templates and create a Stylus BT text style using the Stylus BT font. Create the Architectural dimension style shown in **Figure 17-35.** Use the default AutoCAD settings for the dimension style settings not listed. Save the drawing as P17-3.

4. Write a short report explaining the difference between unidirectional and aligned dimensioning. Use a word processor and include sketches giving examples of each method.

5. Write a short report explaining the difference between size and location dimensions. Use a word processor and include sketches giving examples of each method.

6. Write a short report describing the basic difference between dimensioning for mechanical drafting (drafting for manufacturing) and architectural drafting. Use a word processor and include sketches giving examples of each method.

7. Make sketches showing the standard practice for dimensioning a cylindrical object, a square object, and a conical object.

8. Make sketches showing the standard practice for dimensioning angles. Make one sketch showing coordinate dimensioning and another showing angular dimensioning.

9. Find a copy of the ASME Y14.5M, *Dimensioning and Tolerancing* standard and write a report of approximately 350 words explaining the importance and basic content of this standard.

10. Interview your drafting instructor or supervisor and determine what dimension standards exist at your school or company. Write them down and keep them with you as you learn AutoCAD. Make notes as you progress through this textbook on how you use these standards. Also, note how the standards could be changed to better match the capabilities of AutoCAD.

▼ Advanced

11. Create a freehand sketch of **Figure 17-1.** Label each of the dimension items. To the side of the sketch, write a short description of each item.

12. Research civil drafting and create a template establishing the dimension styles for a civil drawing.

13. Visit at least three local manufacturing companies where design drafting work is done as part of their business. Write a report with sketched examples identifying the standards used at each company.

14. Find a local manufacturing company where design drafting work is done as part of their business. Write a report with sketched examples identifying the standards used at the company.

15. Find and visit two local companies, one architectural and one civil, where design drafting work is done as part of their business. Write a report with sketched examples identifying the standards used at each company.

Linear and Angular Dimensioning

Learning Objectives

After completing this chapter, you will be able to do the following:

✓ Add linear dimensions to a drawing.
✓ Add angular dimensions to a drawing.
✓ Draw datum and chain dimensions.
✓ Add dimensions for multiple items using the **QDIM** tool.

A drawing often requires a variety of dimensions to describe the size and shape of features and objects. Linear and angular dimensions are two of the most common. This chapter covers the process of adding linear and angular dimensions to a drawing using several dimensioning tools. You will also learn how to add a break symbol to a dimension line and use the **QDIM** tool.

Placing Linear Dimensions

Linear dimensions usually measure straight distances, such as distances between horizontal, vertical, or slanted surfaces. The **DIMLINEAR** tool allows you to place linear dimensions.

Dimension tools reference the current dimension style and the points or objects you select to create a single dimension object. When you use the **DIMLINEAR** tool, for example, you create a dimension object that includes all related dimension style characteristics, dimension and extension lines, arrowheads, and a dimension value associated with the distance between selected points.

Once you access the **DIMLINEAR** tool, pick a point to locate origin of the first extension line, and then pick a point to locate the origin of the second extension line. See **Figure 18-1.** Use object snap modes and other drawing aids to pick the exact points where extension lines begin. Once you establish the extension line origins, you can select from several options that appear at the Specify dimension line location or [Mtext/ Text/Angle/Horizontal/Vertical/Rotated] prompt. To apply the default option and create a linear dimension, move the dimension line to the desired location and pick. See **Figure 18-2.**

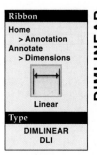

Ribbon
Home
> Annotation
Annotate
> Dimensions

Linear

Type
DIMLINEAR
DLI

DIMLINEAR

Figure 18-1.
Establishing extension line origins. The **Endpoint** and **Intersection** object snap modes are useful for locating origins accurately.

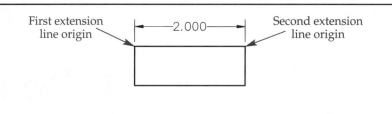

First extension line origin — 2.000 — Second extension line origin

Figure 18-2.
Establishing the location of a dimension line.

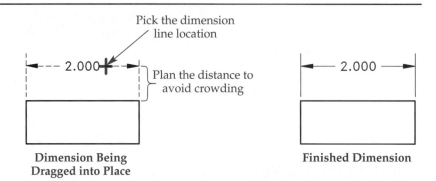

Pick the dimension line location

Plan the distance to avoid crowding

2.000

Dimension Being Dragged into Place

2.000

Finished Dimension

NOTE

When you dimension objects with AutoCAD, the objects automatically measure exactly as you have drawn them. This makes it important for you to draw objects and features accurately and to select the origins of the extension lines accurately.

PROFESSIONAL TIP

Use preliminary plan sheets and sketches to help you determine proper dimension line location and distances between dimension lines to avoid crowding.

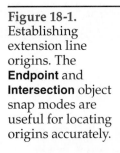

Exercise 18-1

Access the Student Web site (www.g-wlearning.com/CAD) and complete Exercise 18-1.

Selecting an Object to Dimension

An alternative method for locating extension line origins involves picking a single line, circle, or arc to dimension. You can use this option whenever you see the Specify first extension line origin or <select object>: prompt. Press [Enter] or the space bar or right-click and then pick the object to dimension. When you select a line or arc, extension lines begin from endpoints. When you pick a circle, extension lines begin from the closest quadrant and its opposite quadrant. See **Figure 18-3.**

Figure 18-3.
AutoCAD can determine the extension line origins automatically when you select a line, arc, or circle.

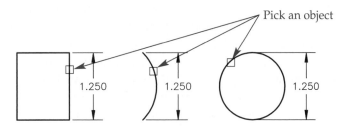

Adjusting Dimension Text

The value attached to the dimension corresponds to the distance between extension lines. Use the **Mtext** option to access the multiline text editor to adjust the dimension value. See **Figure 18-4.** The highlighted value represents the current dimension value. Add to or modify the dimension text and then close the text editor. The tool continues, allowing you to pick the dimension line location.

The **Text** option allows you to use the single-line text editor to change dimension text, even though the final dimension value is an mtext object. The current dimension value appears in brackets. Add to or modify the value as necessary, and then press [Enter] to exit the option. The tool continues, allowing you to pick the dimension line location.

NOTE

Dimension values are horizontal or aligned with the dimension line, according to the current dimension style format. The **Angle** option has limited applications, but allows you to rotate the dimension text. Enter the desired angle at the Specify angle of dimension text: prompt to use this option.

Figure 18-4.
When you use the **Mtext** option, the **Text Editor** ribbon tab appears, and AutoCAD's calculated dimension value appears in a text box for editing.

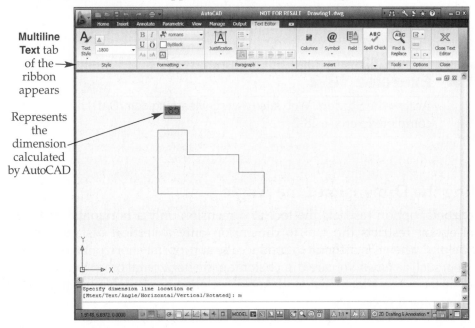

Multiline **Text** tab of the ribbon appears

Represents the dimension calculated by AutoCAD

Including Symbols with Dimension Text

With some dimension tools, AutoCAD automatically places appropriate symbols with the dimension value. For example, when you dimension an arc using the **DIMRADIUS** tool, an R appears before the dimension value. When you dimension a circle using the **DIMDIAMETER** tool, a ∅ symbol appears before the dimension value. The ASME standard recommends these symbols. However, dimension tools such as **DIMLINEAR** do not automatically place certain symbols or add necessary characters.

One option is to use the **Mtext** option to activate the multiline text editor. Place the cursor at the correct location and use options from the **Symbol** flyout or type characters to add information. For example, pick the **Diameter** symbol from the **Symbol** flyout to add ∅. Another example is enclosing a reference dimension in parentheses, as recommended by the ASME standard. To create a reference dimension, type open and close parentheses around the highlighted value.

You can also add content using the **Text** option and the single-line text editor. Place the cursor at the correct location and use control codes or characters to add content. For example, type %%C to display ∅ or type parentheses around the value to create a reference dimension.

Still another way to place symbols with dimension text is to create a dimension style that references a text style using the gdt.shx font. A text style with the gdt.shx font allows you to place common dimension symbols, including geometric dimensioning and tolerancing (GD&T) symbols, using the lowercase letter keys.

PROFESSIONAL TIP

Although you can add a prefix and suffix to a dimension style, usually it is more appropriate to adjust the limited number of dimensions that require a prefix or suffix.

Reference Material *Drafting Symbols*

For more information about common drafting symbols and the gdt.shx font, go to the **Reference Material** section of the Student Web site (www.g-wlearning.com/CAD) and select **Drafting Symbols** in the list.

Exercise 18-2

Access the Student Web site (www.g-wlearning.com/CAD) and complete Exercise 18-2.

Controlling the Dimension Line Angle

The **Horizontal** option restricts the tool to dimension only a horizontal distance. The **Vertical** option restricts the tool to dimension only a vertical distance. These options are helpful when it is difficult to produce the appropriate horizontal or vertical dimension line, such as when you are dimensioning the horizontal or vertical distance of a slanted surface. The **Mtext**, **Text**, and **Angle** options are available to change the dimension text value if necessary.

The **Rotated** option allows you to specify a dimension line angle. A practical application is dimensioning to angled surfaces and auxiliary views. This technique is different from other dimensioning tools because you provide a dimension line angle. See **Figure 18-5**. At the Specify angle of dimension line <0>: prompt, enter a value or pick two points on the line to dimension.

PROFESSIONAL TIP

AutoCAD dimensioning should be as accurate and neat as possible. You can achieve consistent, professional results by using the following guidelines:

- Never truncate, or round off, decimal values when entering locations, distances, or angles. For example, enter .4375 for 7/16, rather than .44.
- Set the precision to the most common precision level in the drawing before adding dimensions. Most drawings have varying levels of precision for specific drawing features, so adjust the precision as needed for each dimension.
- Always use precision drawing aids, such as object snaps, to ensure the accuracy of dimensions.
- Never type a different dimension value from what appears highlighted or in <> brackets. To change a dimension, revise the drawing or dimension settings. Only adjust dimension text when it is necessary to add prefixes and suffixes, or use a different text format.

Exercise 18-3

Access the Student Web site (www.g-wlearning.com/CAD) and complete Exercise 18-3.

Dimensioning Angled Surfaces and Auxiliary Views

When you dimension a surface drawn at an angle, such as an auxiliary view, it is often necessary to align the dimension line with the surface, with extension lines perpendicular to the surface. In order to dimension these features properly, use the **DIMALIGNED** tool or the **Rotated** option of the **DIMLINEAR** tool.

Figure 18-5.
Rotating a dimension for an angled view.

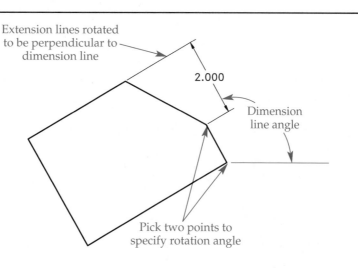

Extension lines rotated to be perpendicular to dimension line

2.000

Dimension line angle

Pick two points to specify rotation angle

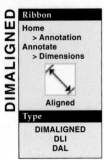
DIMALIGNED

Ribbon

Home
> Annotation
Annotate
> Dimensions

Aligned

Type

DIMALIGNED
DLI
DAL

Figure 18-6 shows the results of using the **DIMALIGNED** tool. Notice the difference between the aligned dimension in this figure and the rotated dimension in **Figure 18-5**. You can usually use the **DIMALIGNED** tool when the length of the extension lines is equal. The **Rotated** option of the **DIMLINEAR** tool is often necessary when extension lines are unequal.

Exercise 18-4

Access the Student Web site (www.g-wlearning.com/CAD) and complete Exercise 18-4.

Dimensioning Long Objects

When you create a drawing of a long part that has a constant shape, the view may not fit on the desired sheet size, or it may look strange compared to the rest of the drawing. To overcome this problem, use a *conventional break* (or *break*) to shorten the view. **Figure 18-7** shows examples of standard break lines. For many long parts, a conventional break is required to display views or increase the view scale without increasing the sheet size. Dimensions added to conventional breaks describe the actual length of the product in its unbroken form. The dimension line often includes a break symbol to indicate that the drawing view is broken and that the feature is longer than it appears. See **Figure 18-8**.

conventional break (break): Removal of a portion of a long, constant-shaped object that has been removed from the drawing to make the object fit better on the drawing sheet.

You can use the **DIMJOGLINE** tool to add a break symbol to dimension lines created using the **DIMLINEAR** or **DIMALIGNED** tools. Once you access the **DIMJOGLINE** tool, pick a linear or aligned dimension line. Then pick a location on the dimension line to place the break symbol, as shown in **Figure 18-8**. An alternative to selecting the

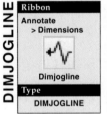
DIMJOGLINE

Ribbon

Annotate
> Dimensions

Dimjogline

Type

DIMJOGLINE

Figure 18-6.
The **DIMALIGNED** tool allows you to place dimension lines parallel to angled features.

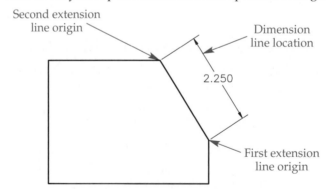

Figure 18-7.
Standard break lines.

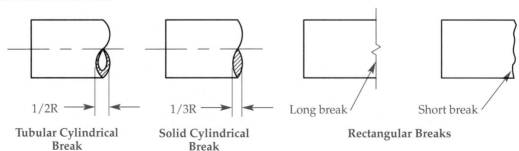

Tubular Cylindrical Break   Solid Cylindrical Break   Rectangular Breaks

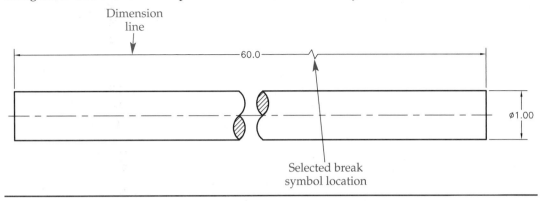

Dimension line

60.0

ø1.00

Selected break symbol location

location of the break symbol is to press [Enter] to accept the default location. You can move the break later using grip editing or by reusing the **DIMJOGLINE** tool to select a different location. To remove the break symbol, access the **DIMJOGLINE** tool and select the **Remove** option.

> **NOTE**
>
> You can add only one break symbol to a dimension line.

Exercise 18-5

Access the Student Web site (www.g-wlearning.com/CAD) and complete Exercise 18-5.

Dimensioning Angles

Coordinate and angular dimensioning are both accepted methods for dimensioning angles. **Figure 18-9** shows an example of *coordinate dimensioning* using the **DIMLINEAR** tool.

Figure 18-10 shows an example of *angular dimensioning* using the **DIMANGULAR** tool. You can dimension the angle between any two nonparallel lines from the *vertex* of the angle. AutoCAD automatically draws extension lines if needed.

Once you access the **DIMANGULAR** tool, pick the first leg of the angle to dimension, and then pick the second leg of the angle. The last prompt asks you to pick the location of the dimension line arc. **Figure 18-11** shows examples of angular dimensions and the effect that limited space may have on dimension fit and placement. Fit characteristics apply to most dimensions.

coordinate dimensioning: A method of dimensioning angles in which dimensions locate the corner of the angle.

angular dimensioning: A method of dimensioning angles in which one corner of an angle is located with a dimension and the value of the angle is provided in degrees.

vertex: The point at which the two lines that form an angle meet.

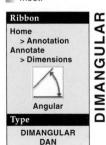

Ribbon

Home
> Annotation
Annotate
> Dimensions

Angular

Type
DIMANGULAR
DAN

DIMANGULAR

Figure 18-9.
Coordinate
dimensioning of
angles.

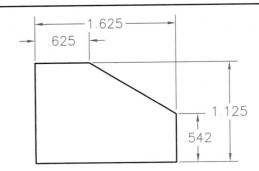

Figure 18-10.
Two examples of
drawing angular
dimensions.

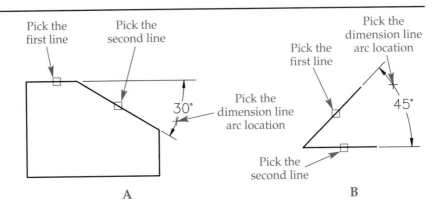

Pick the
first line

Pick the
second line

Pick the
dimension line
arc location

Pick the
first line

Pick the
dimension line
arc location

30°

45°

Pick the
dimension line
arc location

Pick the
second line

A

B

Figure 18-11.
The dimension
line arc location
determines where
the dimension line
arc, text, and arrows
display.

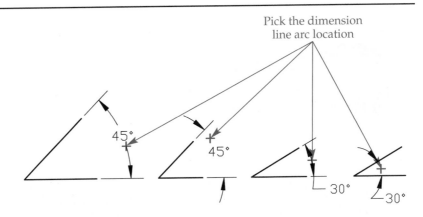

Pick the dimension
line arc location

45°

45°

30°

30°

Dimensioning Angles on Arcs and Circles

You can use the **DIMANGULAR** tool to dimension the included angle of an arc or a
portion of a circle. When you dimension an arc, the center point becomes the angle vertex,
and the two arc endpoints establish the extension line origins. See **Figure 18-12.**

When you dimension a circle using **DIMANGULAR**, the center point becomes the
angle vertex and two picked points specify the extension line origins. See **Figure 18-13.**
The point you pick to select the circle locates the origin of the first extension line. You
then select the second angle endpoint, which locates the origin of the second extension line.

Figure 18-12.
Placing angular
dimensions on arcs.

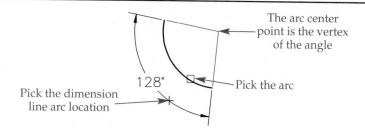

The arc center
point is the vertex
of the angle

128°

Pick the arc

Pick the dimension
line arc location

Figure 18-13.
Placing angular
dimensions on
circles.

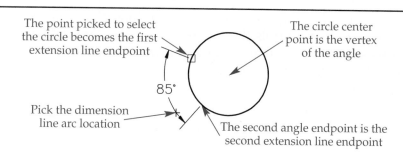

The point picked to select
the circle becomes the first
extension line endpoint

The circle center
point is the vertex
of the angle

85°

Pick the dimension
line arc location

The second angle endpoint is the
second extension line endpoint

PROFESSIONAL TIP

Using angular dimensioning for circles increases the number of possible solutions for a given dimensioning requirement, but the actual uses are limited. One application is dimensioning an angle from a quadrant point to a particular feature without first drawing a line to dimension. Another benefit of this option is the ability to specify angles that exceed 180°.

Angular Dimensioning through Three Points

You can also establish an angular dimension through three points. The points are the angle vertex and two angle line endpoints. See **Figure 18-14.** To apply this technique, press [Enter] or the space bar, or right-click after the first prompt. Then pick the vertex, followed by the two points. This method also dimensions angles over 180°.

Exercise 18-6

Access the Student Web site (www.g-wlearning.com/CAD) and complete Exercise 18-6.

Figure 18-14.
Placing angular
dimensions using
three points.

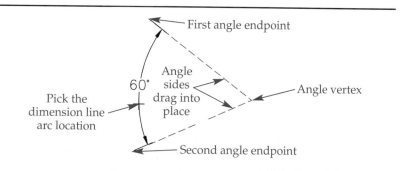

First angle endpoint

Angle
sides
drag into
place

Angle vertex

60°

Pick the
dimension line
arc location

Second angle endpoint

Datum and Chain Dimensioning

Mechanical drafting often requires *datum dimensioning* because each dimension is independent of the others, and references a *datum*. This achieves more accuracy in manufacturing. **Figure 18-15** shows an object dimensioned with surface datums.

Mechanical drafting sometimes uses *chain dimensioning*, also called *point-to-point dimensioning*. However, this method provides less accuracy than datum dimensioning because each dimension is dependent on other dimensions in the chain. In mechanical drafting, it is common to leave one dimension out and provide an overall dimension that controls the missing value. See **Figure 18-16.** Architectural drafting uses chain dimensioning in most applications to reduce the need to calculate or find dimension values during construction. Architectural drafting practices usually show dimensions for all features, plus an overall dimension.

Figure 18-15.
Datum dimensioning.

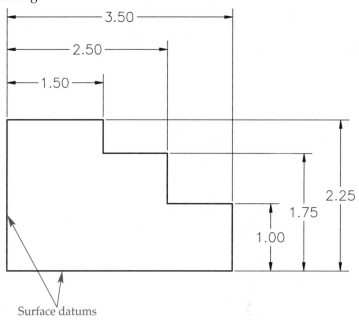

Figure 18-16.
Chain dimensioning. The example on the right is most common, but either technique is acceptable, depending on the design and dimensioning requirement.

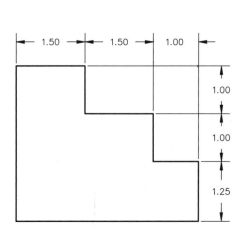
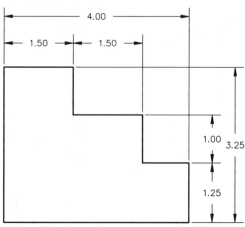

Placing Datum Dimensions

AutoCAD refers to datum dimensioning as *baseline dimensioning*. The **DIMBASELINE** tool controls datum dimensioning and allows you to select several points to define a series of datum dimensions. You can create baseline dimensions with linear, angular, and ordinate dimensions. Chapter 19 describes ordinate dimensions.

The **DIMBASELINE** tool continues from an existing dimension. Therefore, you must create the first dimension using the **DIMLINEAR** tool. The first point you select when drawing the linear dimension defines the datum. Then access the **DIMBASELINE** tool and pick the next second extension line origin. Continue picking extension line origins until you have dimensioned all of the features. Press [Enter] or the space bar or right-click twice, once at the Specify a second extension line origin or: prompt, and again at the Select base dimension: prompt, to create the dimensions and exit the tool. Notice that as you pick additional extension line origins, AutoCAD automatically places the dimension text; you do not specify a location. **Figure 18-17** shows an example of using the **DIMBASELINE** tool to pick two additional extension line origins to an existing linear dimension.

AutoCAD automatically selects the most recently drawn dimension as the base dimension unless you specify a different dimension. To add datum dimensions to an existing dimension other than the most recently drawn dimension, use the **Select** option by pressing [Enter] or the space bar, or by right-clicking and selecting **Enter,** at the first prompt. At the Select base dimension: prompt, pick the dimension to serve as the base. The extension line nearest the point where you select the dimension establishes the datum. Then select the new second extension line origins as previously described.

You can also draw baseline dimensions to angular features. First, draw an angular dimension. Then enter the **DIMBASELINE** tool. **Figure 18-18** shows angular baseline dimensions. As with linear dimensions, you can pick an existing angular dimension other than the dimension most recently drawn.

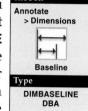

baseline dimensioning: The AutoCAD term for datum dimensioning.

Ribbon
Annotate
> Dimensions

Baseline

Type
DIMBASELINE
DBA

DIMBASELINE

Placing Chain Dimensions

AutoCAD refers to chain dimensioning as *continued dimensioning*. The **DIMCONTINUE** tool controls chain dimensioning and allows you to select several points to define a series of chain dimensions. **Figure 18-19** shows chain dimensioning.

Use the **DIMBASELINE** and **DIMCONTINUE** tools in the same manner. When creating chain dimensions, you will see the same prompts and options you see when creating datum dimensions. Like datum dimensions, you can create continued dimensions with linear, angular, and ordinate dimensions.

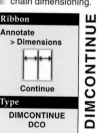

continued dimensioning: The AutoCAD term for chain dimensioning.

Ribbon
Annotate
> Dimensions

Continue

Type
DIMCONTINUE
DCO

DIMCONTINUE

Figure 18-17.
Using the **DIMBASELINE** tool. AutoCAD automatically places the extension lines, dimension lines, arrowheads, and text.

First dimension line location

Dimensions are automatically drawn

4.375

3.250

2.000

First extension line origin

Second extension line origin

Next second extension line origin

Next second extension line origin

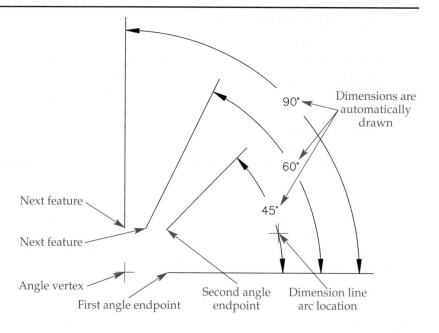

Figure 18-18.
Using the **DIMBASELINE** tool to add datum dimensions to angular features.

90°
60°
45°

Dimensions are automatically drawn

Next feature
Next feature
Angle vertex
First angle endpoint
Second angle endpoint
Dimension line arc location

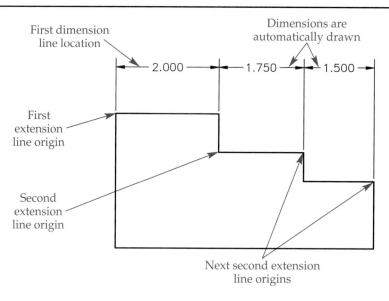

Figure 18-19.
Using the **DIMCONTINUE** tool to create chain dimensions.

First dimension line location

Dimensions are automatically drawn

2.000
1.750
1.500

First extension line origin

Second extension line origin

Next second extension line origins

NOTE

Use the **Undo** option in the **DIMBASELINE** and **DIMCONTINUE** tools to undo previously drawn dimensions.

PROFESSIONAL TIP

You do not have to use **DIMBASELINE** or **DIMCONTINUE** immediately after you create the base or chain dimension. Use the **Select** option later during the drawing session to pick the datum or chain dimension.

Exercise 18-7

Access the Student Web site (www.g-wlearning.com/CAD) and complete Exercise 18-7.

Using QDIM to Dimension

The **QDIM**, or quick dimension, tool makes chain and datum dimensioning easier by eliminating the need to define exact dimension points. Often, the points you select for dimensioning are the endpoints of lines or the center points of arcs. AutoCAD automates the process of point selection in the **QDIM** tool by finding those points for you.

The type of geometry you select affects the **QDIM** output. If you choose a single polyline, **QDIM** attempts to draw linear dimensions to every vertex. If you pick a single arc or circle, **QDIM** draws a radius or diameter dimension. If you select multiple objects, linear dimensions occur from the vertex of every line or polyline and to the center of every arc or circle. In each case, AutoCAD finds the points automatically.

Once you access the **QDIM** tool, pick several lines, polylines, arcs, and/or circles and press [Enter] or the space bar, or right-click. Then pick a position for the dimension lines to create the dimensions and exit the tool. **Figure 18-20** shows examples of using the **QDIM** tool to dimension different types of objects. To create the upper dimensions, select each object separately. To create the lower dimensions, select all of the objects at once.

The **Continuous** option of the **QDIM** tool creates chain dimensions. The **Baseline** option creates datum dimensions. The **Staggered** option creates staggered (noncontinuous) dimensions. The **Ordinate, Radius,** and **Diameter** options provide methods of adding ordinate, radius, and diameter dimensions. Chapter 19 describes these types of dimensions.

To dimension as shown in **Figure 18-20A**, access the **QDIM** tool and select the polyline. Then issue the **Baseline** option and select a position for the dimension line above the polyline to create the dimension and exit the tool. To create the dimensions at the bottom of **Figure 18-20**, use the **Continuous** option of the **QDIM** tool and select all three objects.

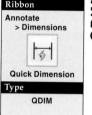

Ribbon
Annotate
> Dimensions

Quick Dimension

Type

QDIM

QDIM

Figure 18-20.
The **QDIM** tool can dimension multiple features or objects at the same time.

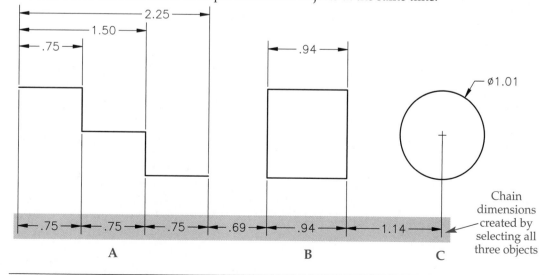

A   B   C

You can use the **datumPoint** option to change the datum point for datum or chain dimensions. The **Settings** option allows you to set the object snap mode for establishing the extension line origins to **Endpoint** or **Intersection**.

NOTE

You can also use the **QDIM** tool to edit any existing associative dimension. Chapter 21 describes editing dimensions and using the **Edit** option of the **QDIM** tool.

Chapter Test

Answer the following questions. Write your answers on a separate sheet of paper or go to the Student Web site (www.g-wlearning.com/CAD) and complete the electronic chapter test.

1. Name the two **DIMLINEAR** options that allow you to change dimension text.
2. Give two examples of symbols that automatically appear with some dimensions.
3. What is the purpose of AutoCAD's gdt.shx font?
4. Name the two dimensioning tools that provide linear dimensions for angled surfaces.
5. Which tool allows you to place a break symbol in a dimension line?
6. Name the tool used to dimension angles in degrees.
7. Describe a way to specify an angle in degrees if the angle is greater than 180°.
8. Which type of dimensioning is generally preferred for manufacturing because of its accuracy?
9. What is the AutoCAD term for datum dimensioning?
10. How do you place a datum dimension from the origin of the previously drawn dimension?
11. How do you place a datum dimension from the origin of a dimension drawn during a previous drawing session?
12. What is the conventional term for the type of dimensioning AutoCAD refers to as continuous dimensioning?
13. Which type of dimensions are created when you select multiple objects in the **QDIM** tool?
14. Which tool other than **DIMBASELINE** is used to create baseline dimensions?
15. Name at least three modes of dimensioning available through the **QDIM** tool.

Drawing Problems

- *Start AutoCAD if it is not already started.*
- *Start a new drawing using an appropriate template of your choice. The template should include layers, text styles, and dimension styles appropriate for drawing the given objects.*
- *Add layers, text styles, and dimension styles as needed. Draw all objects using appropriate layers, text styles, dimension styles, justification, and format.*
- *Follow the specific instructions for each problem. Use your own judgment and approximate dimensions when necessary.*
- *Apply dimensions accurately using ASME or appropriate industry standards. Use object snap modes to your best advantage.*
- *For mechanical drawings, place the following general notes 1/2" from the lower-left corner:*

> NOTES:
> 1. INTERPRET DIMENSIONS AND TOLERANCES PER ASME Y14.5M-1994.
> 2. REMOVE ALL BURRS AND SHARP EDGES.
> 3. UNLESS OTHERWISE SPECIFIED, ALL DIMENSIONS ARE IN INCHES (or MILLIMETERS as applicable).

▼ Basic

1. Open a new drawing and save the file as P18-1. Open P3-6 and copy one instance of Object A and Object B to the P18-1 drawing. The P18-1 file should be active. Dimension the two objects as shown.

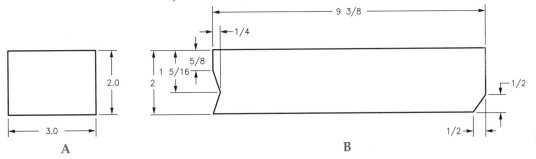

2. Open P3-4 and save the file as P18-2. The P18-2 file should be active. Dimension the drawing as shown using datum dimensions.

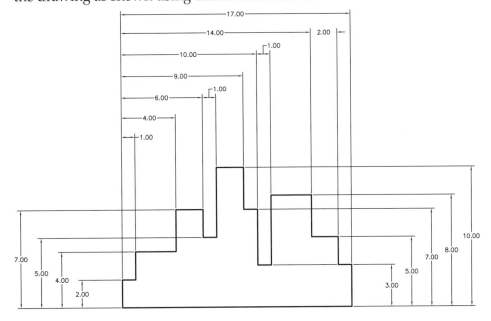

Chapter 18   Linear and Angular Dimensioning

3. Open P3-9 and save the file as P18-3. The P18-3 file should be active. Dimension both objects as shown.

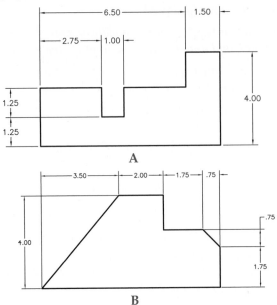

4. Open P3-10 and save the file as P18-4. The P18-4 file should be active. Dimension the drawing as shown.

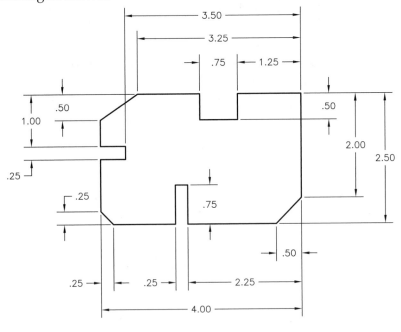

5. Open P3-11 and save the file as P18-5. The P18-5 file should be active. Dimension the drawing as shown. Note that this is a metric drawing.

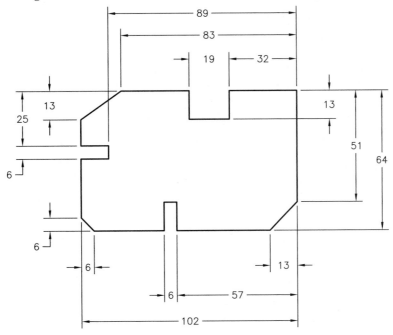

6. Open P3-7 and save the file as P18-6. The P18-6 file should be active. Dimension the drawing as shown.

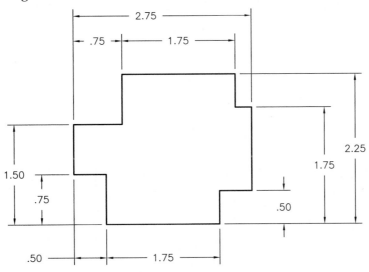

7. Open P3-8 and save the file as P18-7. The P18-7 file should be active. Dimension the drawing as shown.

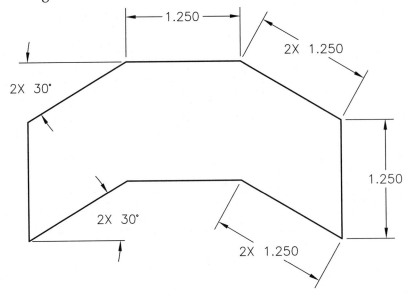

8. Write a report explaining the difference between datum and chain dimensioning. Use a word processor and include sketches giving examples of each method.

▼ Intermediate

9. Create the views of a shaft and dimension as shown. Save the drawing as P18-9.

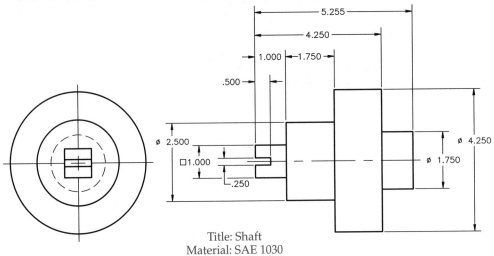

Title: Shaft
Material: SAE 1030

10. Open P3-12 and save the file as P18-10. The P18-10 file should be active. Dimension the drawing as shown.

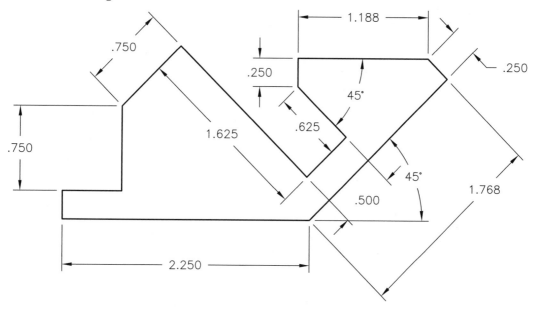

11. Create the partial floor plan shown below and dimension as shown. Save the drawing as P18-11.

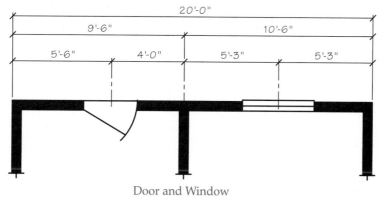

Door and Window

12. Create the partial floor plan and dimension as shown. Save the drawing as P18-12.

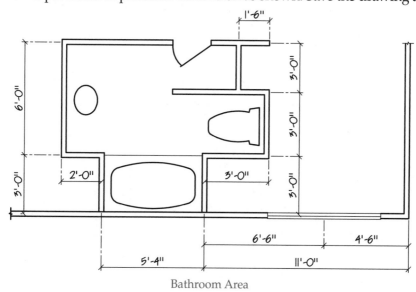

Bathroom Area

13. Create the object and dimension as shown. Save the drawing as P18-13.

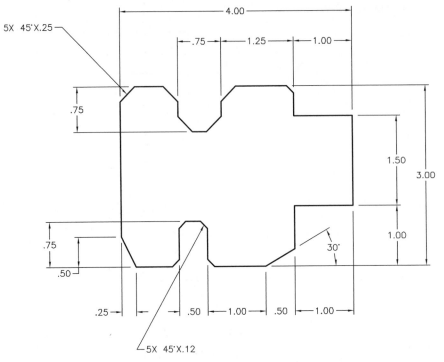

14. Create the object and dimension as shown. Save the drawing as P18-14.

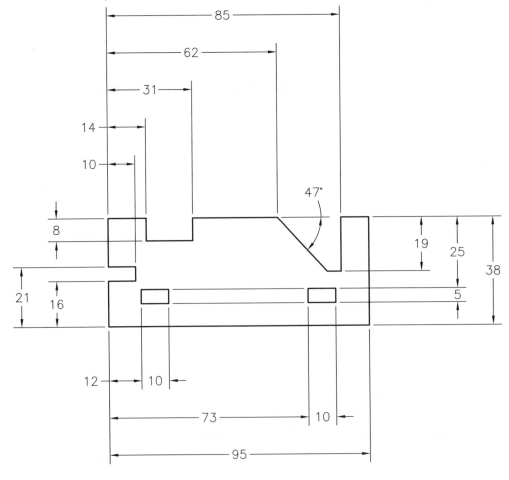

AutoCAD and Its Applications—Basics

▼ Advanced

15. Open P3-3 as shown below and save the file as P18-15. The P18-15 file should be active. Make one copy of the object at a new location to the right of the original object. Dimension the object on the left using datum dimensioning. Dimension the object on the right using chain dimensioning.

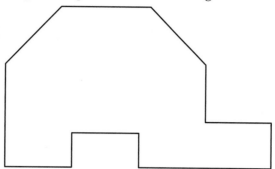

16. Open P8-5 and save the file as P18-16. The P18-16 file should be active. Dimension the drawing as shown.

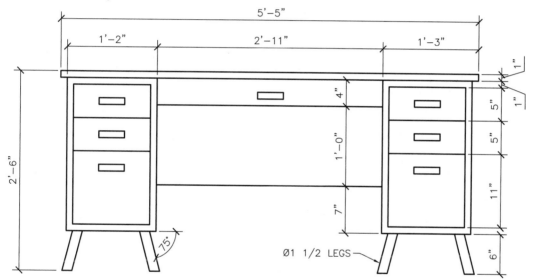

17. Draw this floor plan. Size the windows and doors to your own specifications. Dimension the drawing as shown. Save the drawing as P18-17.

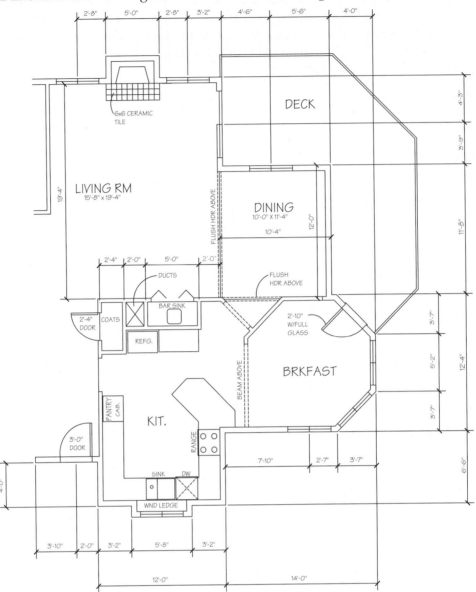

Dimensioning Features and Alternate Practices

Learning Objectives

After completing this chapter, you will be able to do the following:

✓ Dimension circles and arcs.
✓ Create and use multileader styles.
✓ Draw leaders using the **MLEADER** tool.
✓ Apply alternate dimensioning practices.
✓ Dimension using the **DIMORDINATE** tool.
✓ Mark up a drawing using the **REVCLOUD** and **WIPEOUT** tools.

A drawing must describe the size and location of all features for manufacturing or construction. This chapter describes dimensioning practices for object features and introduces alternate mechanical drafting dimensioning practices. This chapter also introduces basic redlining techniques using the **Revision Cloud** and **Wipeout** tools.

Dimensioning Circles

Diameter usually describes the size of circles. The ASME standard for dimensioning arcs is to give the radius. However, AutoCAD allows you to dimension a circle or an arc with a diameter using the **DIMDIAMETER** tool. Access the **DIMDIAMETER** tool and select a circle or arc to display a leader line and a diameter dimension value attached to the crosshairs. Move the leader to the desired location and pick to place the dimension. See **Figure 19-1.**

Like other dimension tools, the **DIMDIAMETER** tool references the current dimension style and the object you select to create a single dimension object. The dimension you create includes all related dimension style characteristics, centerlines, leader, arrowheads, and a dimension value associated with the diameter. The leader points to the center of the circle or arc, as recommended by the ASME standard.

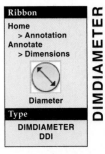

Ribbon
Home
    > Annotation
Annotate
    > Dimensions

Diameter

Type
DIMDIAMETER
DDI

DIMDIAMETER

Figure 19-1.
Using the **DIMDIAMETER** tool.

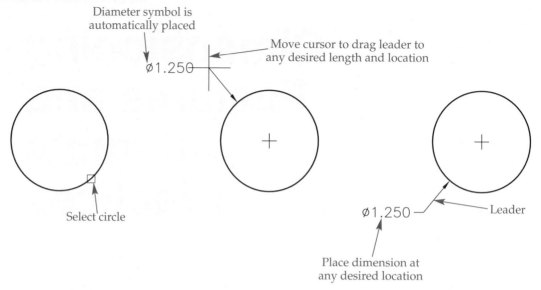

Figure 19-1.
Using the **DIMDIAMETER** tool.

Diameter symbol is automatically placed

Ø1.250

Move cursor to drag leader to any desired length and location

Select circle

Ø1.250

Leader

Place dimension at any desired location

NOTE

The **DIMDIAMETER** tool includes the **Mtext**, **Text**, and **Angle** options. Use the **Mtext** or **Text** option to add information to or change the dimension value. The **Angle** option changes the dimension text angle, although this practice is not common.

Exercise 19-1

Access the Student Web site (www.g-wlearning.com/CAD) and complete Exercise 19-1.

Dimensioning Holes

Dimension holes in the view in which they appear as circles. Give location dimensions to the center and a leader showing the diameter. The **DIMDIAMETER** tool is effective for dimensioning holes. To note multiple holes of the same size, dimension the size of one hole using the **DIMDIAMETER** tool and the **Mtext** or **Text** option. Precede the diameter with the number of holes followed by X and then a space. The 2X Ø.50 dimension in **Figure 19-2** is an example of this practice.

PROFESSIONAL TIP

The ASME standard recommends a small space between the object and the extension line. To specify this space, adjust the **Offset from origin** setting in the dimension style to an appropriate positive value, such as .063 (1.5 mm). This is very useful *except* when you dimension to centerlines to locate holes. When you pick the endpoint of the centerline, a positive value leaves a space between the centerline and the beginning of the extension line. This is not a preferred practice. Change the **Offset from origin** setting to 0 to remove the gap. Be sure to change back to the positive setting before dimensioning other objects.

Figure 19-2.
Dimensioning holes.

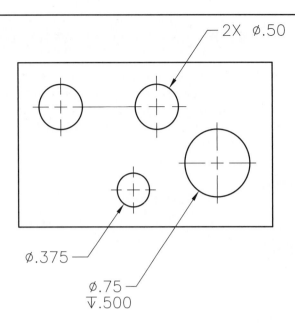

Dimensioning for Manufacturing Processes

Counterbore, *spotface*, and *countersink* manufacturing processes are examples of features dimensioned using symbols. Dimension these processes in the circular view, like holes, with a leader providing machining information in a note. See **Figure 19-3**. The **Mtext** option of the **DIMDIAMETER** tool provides a convenient method for dimensioning manufacturing processes. Many symbols are available from the gdt.shx font. You can also create custom symbols as blocks. Blocks are described later in this textbook.

Drafting Symbols

For more information about common drafting symbols and the gdt.shx font, go to the **Reference Material** section of the Student Web site (www.g-wlearning.com/CAD) and select **Drafting Symbols** in the list.

counterbore: A larger-diameter hole machined at one end of a smaller hole that provides a place for the screw head.

spotface: A larger-diameter hole machined at one end of a smaller hole that provides a smooth, recessed surface for a washer; similar to a counterbore, but not as deep.

countersink: A cone-shaped recess at one end of a hole that provides a mating surface for a screw head of the same shape.

Figure 19-3.
Dimension notes for machining processes. You can insert symbols as blocks or use lowercase letters in the gdt.shx font.

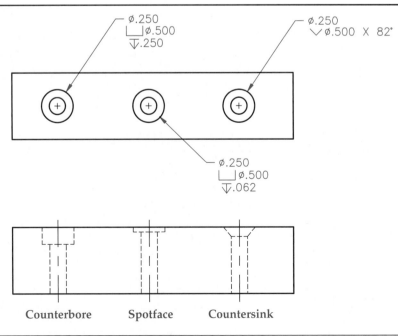

repetitive features:
Many features having the same shape and size.

For *repetitive features*, an X, a space, and the size dimension follow the number of repetitions. The dimension connects to the feature with a leader. **Figure 19-4** shows how the **Mtext** or **Text** option, available with several dimensioning tools, allows you to dimension repetitive features. Use the **MLEADER** tool to create the 8X note shown, as described later in this chapter.

Exercise 19-2

Access the Student Web site (www.g-wlearning.com/CAD) and complete Exercise 19-2.

Figure 19-4.
Dimensioning repetitive features (shown in color).

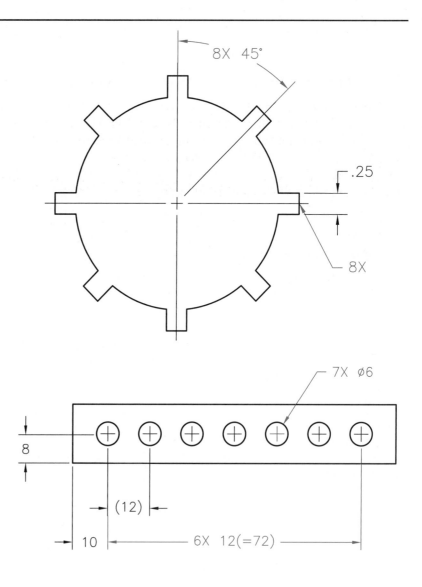

Dimensioning Arcs

The standard for dimensioning arcs is a radius dimension, which you can create using the **DIMRADIUS** tool. Access the **DIMRADIUS** tool and select an arc or circle to display a leader line and radius dimension value attached to the crosshairs. Move the leader to the desired location, and pick to place the dimension. See **Figure 19-5.** The resulting leader points to the center of the arc or circle, as recommended by the ASME standard. Centerlines or center marks appear, depending on the current dimensions style setting.

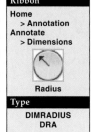

Ribbon
Home > Annotation Annotate > Dimensions
Radius
Type
DIMRADIUS DRA

DIMRADIUS

> **NOTE**
>
> The **DIMRADIUS** tool includes the **Mtext**, **Text**, and **Angle** options. Use the **Mtext** or **Text** option to add information to or change the dimension value. The **Angle** option changes the dimension text angle, although this practice is not common.

Dimensioning Arc Length

You can use the **DIMARC** tool to dimension the length of an arc. The length measures the distance along the arc segment. Access the **DIMARC** tool and select an arc or polyline arc segment to display the arc length symbol and dimension value attached to the crosshairs. Move the text to the desired location and pick. By default, the symbol occurs before the text. The ASME standard recommends placing the symbol over the text, as shown in **Figure 19-6.** The dimension style controls the symbol placement.

Before placing the arc length dimension, add information to or change the dimension value using the **Mtext** or **Text** option. The **Angle** option is available to change the text angle. Use the **Partial** option to dimension a portion of the arc length. Select a first point on the arc followed by a second point to dimension the length between the points. When the arc is greater than 90°, the **Leader** option is available. This option allows you to add a leader pointing to the arc you are dimensioning.

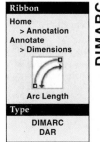

Ribbon
Home > Annotation Annotate > Dimensions
Arc Length
Type
DIMARC DAR

DIMARC

Dimensioning Large Circles and Arcs

When a circle or arc is so large that the center point cannot appear on the layout, use the **DIMJOGGED** tool to create the dimension. Access the **DIMJOGGED** tool and pick an arc or circle. Then pick a location for the origin of the center. This is the point

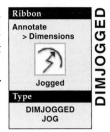

Ribbon
Annotate > Dimensions
Jogged
Type
DIMJOGGED JOG

DIMJOGGED

Figure 19-5.
Using the **DIMRADIUS** tool to dimension arcs.

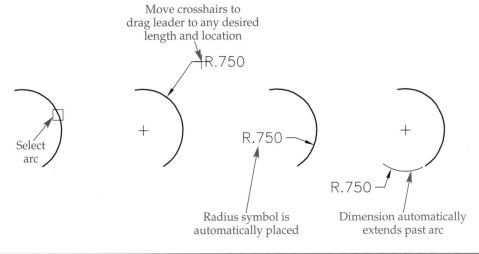

Move crosshairs to drag leader to any desired length and location

R.750

Select arc

R.750

Radius symbol is automatically placed

R.750

Dimension automatically extends past arc

Figure 19-6.
Using the **DIMARC** tool to dimension the length of an arc.

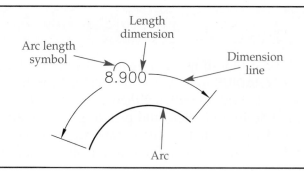

that represents, or overrides, the center of the arc or circle. The associated radius value does not change. Select a location for the dimension line and then pick a location for the break symbol. See **Figure 19-7.** You can move the components of the dimension by grip editing after you place the dimension.

Dimensioning Fillets and Rounds

fillets: Small inside arcs designed to strengthen inside corners.

rounds: Small arcs on outside corners used to relieve sharp corners.

You can dimension *fillets* and *rounds* individually as arcs, using the **DIMRADIUS** tool, or collectively in a general note. See **Figure 19-8.** On mechanical drawings, it is common to include a general note such as ALL FILLETS AND ROUNDS R.125 UNLESS OTHERWISE SPECIFIED on the drawing.

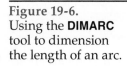

Exercise 19-3

Access the Student Web site (www.g-wlearning.com/CAD) and complete Exercise 19-3.

Figure 19-7.
Using the **DIMJOGGED** tool to place a radius dimension for a large arc.

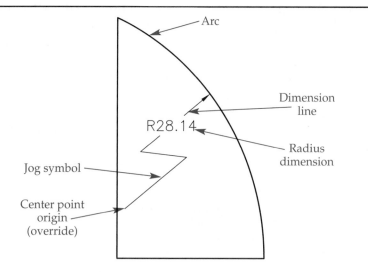

Figure 19-8.
Dimensioning fillets and rounds.

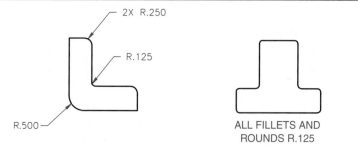

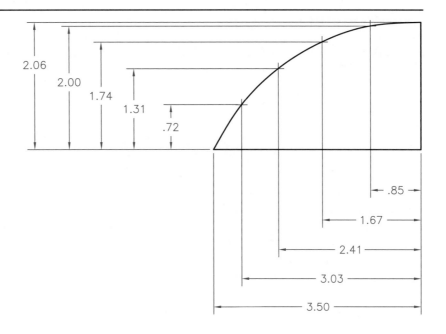

Figure 19-9.
Dimensioning curves that do not have a constant radius.

Dimensioning Curves

Dimension curves as arcs when possible. When an arc is not in the shape of a constant-radius arc, you should dimension to points along the curve using the **DIMLINEAR** tool. See **Figure 19-9.**

Adding Center Dashes and Centerlines

Depending on the current dimension style setting, when you use the **DIMDIAMETER** and **DIMRADIUS** tools, small circles or arcs automatically receive center dashes, and large circles or arcs display center dashes or centerlines. Use the **DIMCENTER** tool to add center dashes or centerlines to objects that are not dimensioned using the **DIMDIAMETER** or **DIMRADIUS** tools. Access the **DIMCENTER** tool and pick a circle or an arc to display center marks.

The **DIMCENTER** tool references the current dimension style and the size of the circle or arc to place center dashes, centerlines, or no symbol. The ASME standard refers to AutoCAD center marks as the center dashes of centerlines. Center marks are typically applied to circular objects that are too small to receive centerlines. Center marks are also common for rectangular coordinate dimensioning without dimension lines, regardless of circular object size, as described later in this chapter.

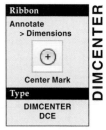

Ribbon
Annotate
> Dimensions

Center Mark

Type
DIMCENTER
DCE

DIMCENTER

Drawing Leader Lines

The **DIMDIAMETER** and **DIMRADIUS** tools automatically place *leader lines* when you dimension circles and arcs. AutoCAD multileaders created using the **MLEADER** tool allow you to begin and end a leader line at a specific location. Multileaders consist of single or multiple lines of *annotations*, including symbols, with the leader. Multileader styles control multileader characteristics, such as leader format, annotation style, and arrowhead size. You can create multiple-segment leaders and align and group separate leaders. Chapter 21 describes adding and removing multiple leader lines and aligning leaders.

leader line: A line that connects note text to a specific feature or location on a drawing.

annotation: Text or similar information on a drawing, such as a note or dimension.

Multileader Styles

multileader styles: Saved configurations for the appearance of leaders.

shoulder: A short horizontal line usually added to the end of straight leader lines.

MLEADERSTYLE

| Ribbon |
| Home
> Annotation |
| Multileader Style |
| Annotate
> Leaders |
| Multileader Style |
| Type |
| MLEADERSTYLE
MLS |

Multileader styles allow you to control multileader settings as needed to achieve the desired leader appearance. This process is similar to using a dimension style. In mechanical drafting, properly drawn leaders have one straight segment extending from the feature to a horizontal *shoulder* that is 1/8"–1/4" (3 mm–6 mm) long. While most other fields also use straight leaders, AutoCAD provides the option of drawing curved leaders, which are common in architectural drafting. See **Figure 19-10**.

Working with Multileader Styles

Create, modify, and delete multileader styles using the **Multileader Style Manager** dialog box. See **Figure 19-11**. The **Styles:** list box displays existing multileader styles. Standard and Annotative multileader styles are available by default in some templates. The Annotative multileader style is preset to create annotative leaders, as indicated by the icon to the left of the style name. The Standard multileader style does not use the annotative function. The **List:** drop-down list allows you to control whether all styles or only the styles in use appear in the **Styles:** list box. To make a multileader style current, double-click the style name, right-click on the name and select **Set current**, or pick the name and select the **Current** button.

Figure 19-10.
Multileader styles control the display of leaders created using the **MLEADER** tool.

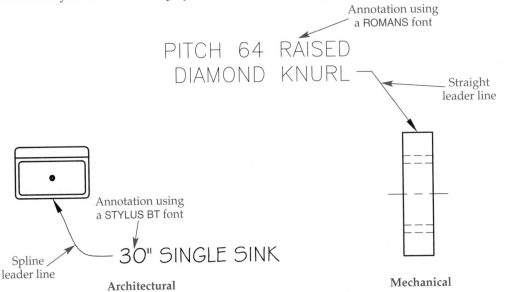

Figure 19-11.
The **Multileader Style Manager** dialog box. The Standard multileader style is the AutoCAD default.

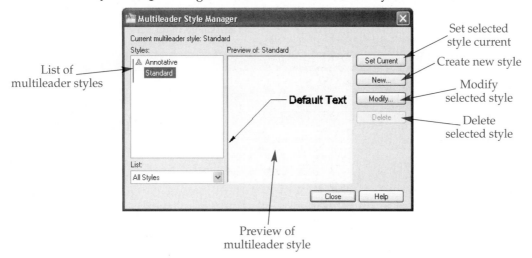

List of multileader styles

Set selected style current

Create new style

Modify selected style

Delete selected style

Preview of multileader style

Creating a New Multileader Style

To create a new multileader style, first select an existing multileader style from the **Styles:** list box to use as a base for formatting the new multileader style. Then pick the **New...** button in the **Multileader Style Manager** to open the **Create New Multileader Style** dialog box. See **Figure 19-12.**

Enter a descriptive name for the new multileader style, such as Architectural, Mechanical, Straight, or Spline, in the **New Style Name** text box. If necessary, select a different style from the **Start With** drop-down list on which to base the new style. Pick the **Annotative** check box to make the multileader style annotative.

Pick the **Continue** button to access the **Modify Multileader Style** dialog box, shown in **Figure 19-13.** The **Leader Format**, **Leader Structure**, and **Content** tabs display groups of settings for specifying leader appearance. The next sections of this chapter describe each tab. After completing the style definition, pick the **OK** button to return to the **Multileader Style Manager** dialog box.

> **NOTE**
>
>
> The preview image in the upper-right corner of each **Modify Multileader Style** dialog box tab displays a representation of the multileader style and changes according to the selections you make.

Figure 19-12.
The **Create New Multileader Style** dialog box.

Name new style

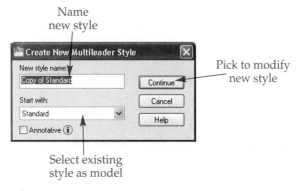

Pick to modify new style

Select existing style as model

Figure 19-13.
The **Leader Format** tab of the **Modify Multileader Style** dialog box.

Preview image is
displayed in all tabs

General
multileader
format
settings

Multileader
arrowhead
settings

Multileader
break size
setting

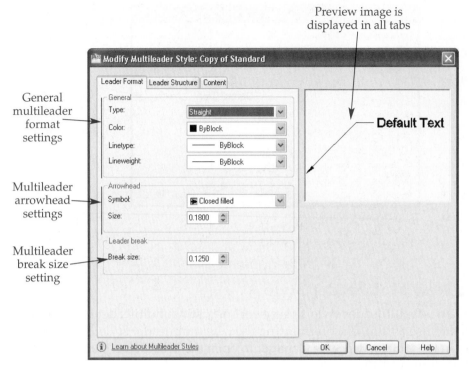

Leader Format Settings

The **Leader Format** tab of the **Modify Multileader Style** dialog box, shown in **Figure 19-13**, controls leader line settings. It allows you to set the appearance of the leader line and arrowhead.

General Leader Format Settings

The **General** area contains a **Type** drop-down list that that you can use to specify the leader line shape. The **Straight** option produces leaders with straight-line segments. The **Spline** option produces the curved leader lines common in architectural drafting. Pick the **None** option to create a multileader style that does not use a leader line. Use this option to create a leader that you can associate with other leaders using the **MLEADERALIGN** and **MLEADERCOLLECT** tools, described in Chapter 21.

Color, **Linetype**, and **Lineweight** drop-down lists are available for changing the multileader color, linetype, and lineweight. These options function the same as they do for adjusting dimension elements.

Arrowhead Settings

The **Arrowhead** area sets the leader arrowhead style and size. Select the arrowhead style from the **Symbol:** drop-down list. The arrowhead symbol options are the same as those for dimension style arrowheads. Set the arrowhead size using the **Size:** text box. A common arrowhead size is .125″ (3 mm). Leader arrowheads are typically the same size as dimension arrowheads.

Adjusting Break Size

The **Leader Break** area controls the amount of leader line removed by the **DIMBREAK** tool. Specify a value in the **Break size:** text box to set the total length of the break. The default size is .125″ (3 mm). ASME standards do not recommend breaking leader lines.

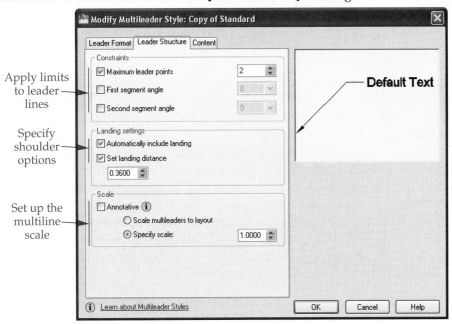

Apply limits to leader lines

Specify shoulder options

Set up the multiline scale

Leader Structure Settings

The **Leader Structure** tab of the **Modify Multileader Style** dialog box is shown in **Figure 19-14.** This tab contains settings that control leader construction and size.

Setting Constraints

The **Constraints** area restricts the number of points you can select to create a leader, as well as the leader line angle. Pick the **Maximum leader points** check box to set a maximum number of vertices on the leader line. The multileader automatically forms once you pick the maximum number of points. To use fewer than the maximum number of points, press [Enter] at the Specify next point: prompt. Deselect the **Maximum leader points** check box to allow an unlimited number of vertices.

You can use the **First segment angle** and **Second segment angle** check boxes to restrict the first two leader line segments to certain angles. Deselect the check boxes to draw leader lines at any angle. Select the appropriate check boxes and pick a value from the drop-down list to restrict the angle of the leader segment according to the selected value. The **Ortho** mode setting overrides the angle constraints, so it is advisable to turn **Ortho** mode off while you are placing leaders.

PROFESSIONAL TIP

The ASME standard for leaders recommends that leader lines have angles not less than 15° and not greater than 75° from horizontal. Use the **First segment angle** and **Second segment angle** settings to help maintain this standard.

Landing Settings

The **Landing settings** area controls the display and size of the *landing* and is only available with straight multileader styles. Select the **Automatically include landing** check box to display a shoulder automatically when you select the second leader line point. This is the preferred method for creating straight leader lines. Deselect the check box to create leaders without shoulders, or to pick a third point to manually draw

landing: The AutoCAD term for a leader shoulder.

the shoulder. When you check **Automatically include landing**, the **Set landing distance** check box enables. Pick the **Set landing distance** check box to define a specific shoulder length, typically 1/8"–1/4" (3 mm–6 mm), in the text box. If you deselect the text box, a prompt asks you for the shoulder length when you place a leader.

Scale Options

The **Scale** area sets the multileader scale factor. Select the **Annotative** check box to create an annotative multileader style. The **Annotative** check box is already set when you modify the default Annotative multileader style or pick the **Annotative** check box in the **Create New Multileader Style** dialog box.

You can select the **Scale dimensions to layout** radio button to add leaders in a floating viewport in a paper space layout. You must add leaders to the model in a floating viewport in order for this option to function. Scaling leaders to the layout allows the overall scale to adjust according to the active floating viewport by setting the overall scale equal to the viewport scale factor. Pick the **Use overall scale of** radio button to manually scale a drawing, and enter the drawing scale factor applied to all leader settings. The scale factor is multiplied by the desired plotted leader size to get the size of the leader in model space. Layouts are described later in this textbook.

Content Settings

The **Content** tab of the **Modify Multileader Style** dialog box, shown in **Figure 19-15**, controls the display of text or a block with the leader line. Use the **Multileader type:** drop-down list to select the type of object to attach to the end of the leader line or shoulder. **Figure 19-16** shows an example of a leader drawn with each content option.

Attaching Mtext

Pick the **Mtext** option from the **Multileader type:** drop-down list, as shown in **Figure 19-15**, to attach a multiline text object to the leader. The **Text options** and **Leader connection** areas of the **Content** tab appear when you select the **Mtext** content type. The **Default text** option allows you to specify a value to attach leaders during leader

Figure 19-15.
The **Content** tab of the **Modify Multileader Style** dialog box with the **Mtext** multileader type selected.

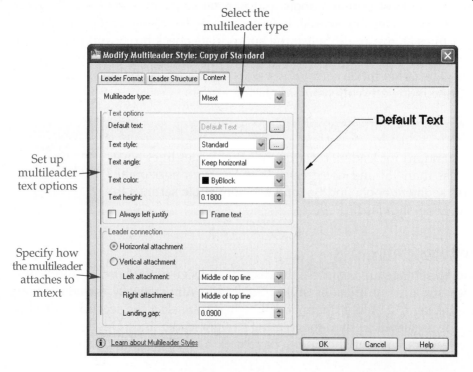

Figure 19-16.
Examples of each multileader content type.

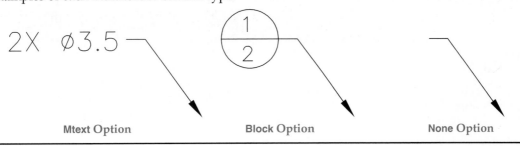

Figure 19-16.
Examples of each multileader content type.

Mtext Option   Block Option   None Option

placement. This is useful when the same note or symbol is required throughout a drawing. Pick the ellipsis (...) button to return to the drawing window and use the multiline text editor to enter the default text value. Close the text editor to return to the **Modify Multileader Style** dialog box.

A text style must be loaded in the current drawing before it is available for use in leader text. Pick the desired text style from the **Text style** drop-down list. To create or modify an existing text style, pick the ellipsis (…) button next to the drop-down list to launch the **Text Style** dialog box.

Select an option from the **Text angle** drop-down list to control the angle at which text appears in reference to the angle of the leader line or shoulder. Figure 19-17 shows the effects of applying each text angle option to the same leader. Use the **Text color** drop-down list to specify the text color, which should be ByBlock for typical applications. Use the **Text height** text box to specify the leader text height. Leader text height is commonly the same as the text height used for dimensions.

The **Always left justify** option forces leader text to left-justify, regardless of the leader line direction. Pick the **Frame text** check box to create a box around the multiline text box. The current multileader style settings control the default properties of the frame.

The **Leader connection** area contains options that determine how the mtext object positions relative to the endpoint of the leader line or shoulder. Most drawings require leaders that use the **Horizontal attachment** option. Use the **Left attachment:** drop-down list to define how multiple lines of text are positioned when the leader is on the left side

Figure 19-17.
Text angle options available for mtext.

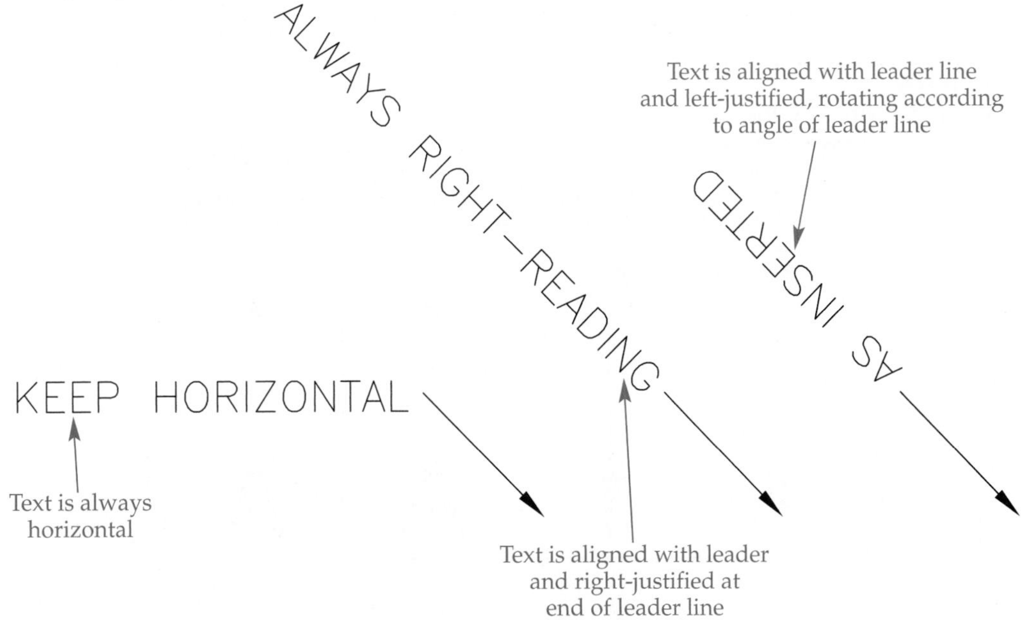

Text is aligned with leader line and left-justified, rotating according to angle of leader line

ALWAYS RIGHT-READING

AS INSERTED

KEEP HORIZONTAL

Text is always horizontal

Text is aligned with leader and right-justified at end of leader line

Figure 19-18.
Horizontal alignment options. The shaded examples are the recommended ASME standards.

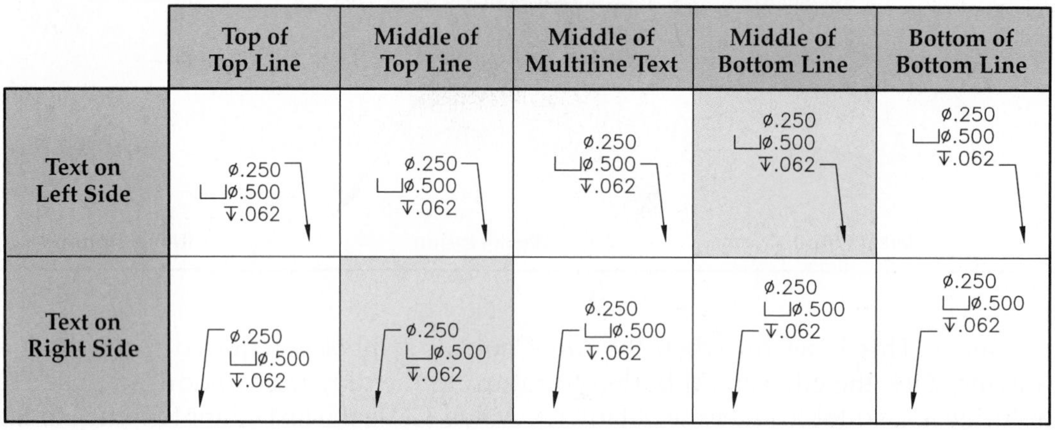

	Top of Top Line	Middle of Top Line	Middle of Multiline Text	Middle of Bottom Line	Bottom of Bottom Line
Text on Left Side	∅.250 ⌴∅.500 ▽.062	∅.250 ⌴∅.500 ▽.062	∅.250 ⌴∅.500 ▽.062	∅.250 ⌴∅.500 ▽.062	∅.250 ⌴∅.500 ▽.062
Text on Right Side	∅.250 ⌴∅.500 ▽.062	∅.250 ⌴∅.500 ▽.062	∅.250 ⌴∅.500 ▽.062	∅.250 ⌴∅.500 ▽.062	∅.250 ⌴∅.500 ▽.062

of the text. Use the **Right attachment:** drop-down list to define how multiple lines of text are positioned when the leader is on the right side of the text. **Figure 19-18** shows typical selections. The **Underline bottom line** option draws a line along the bottom of the multiline text box. The **Underline all text** option underlines each line of leader text. The **Leading gap** text box specifies the space between the leader line or shoulder and the text. The default is .09, but .063 (1.5 mm) is standard.

PROFESSIONAL TIP

Common drafting practice is to use the **Middle of bottom line** option for left attachment and the **Middle of top line** option for right attachment.

For some applications, you may find it necessary to select the **Vertical attachment** radio button, although this is not common. This option eliminates the possible use of a shoulder and connects the leader endpoint to the top center or bottom center of the text, depending on the leader line position. Use the **Top attachment:** drop-down list to define how text is positioned when the leader is above the text. Use the **Bottom attachment:** drop-down list to define how text is positioned when the leader is below the text. **Figure 19-19** shows each option.

Figure 19-19.
Vertical alignment options.

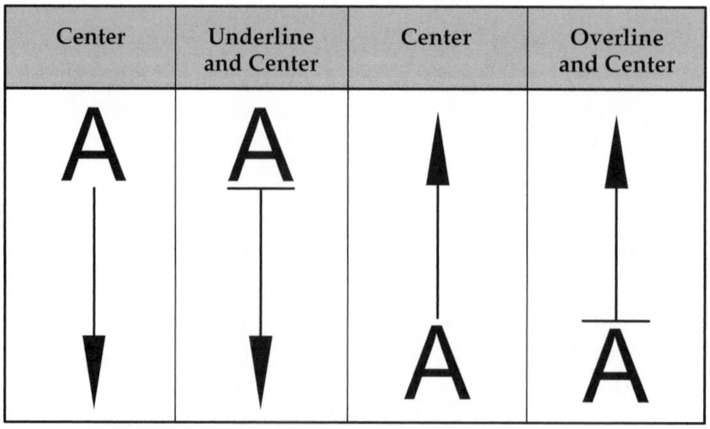

Center	Underline and Center	Center	Overline and Center
A ↓	A ↓	↑ A	↑ A

Figure 19-20.
The **Content** tab of the **Modify Multileader Style** dialog box with the **Block** multileader type selected.

Select the multileader type

Set the options for the block

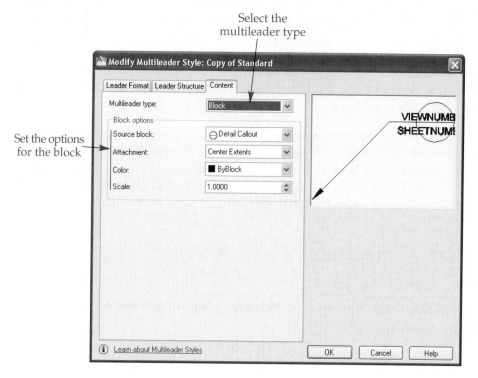

Attaching a Symbol

Pick the **Block** option from the **Multileader type:** drop-down list, as shown in Figure 19-20, to attach a *block* to the leader. Chapters 25 through 28 describe blocks in detail. Several blocks are available by default from the **Source block:** drop-down list. You also have the option of picking the **User Block...** option to select your own saved block. The **Select Custom Content Block** dialog box appears when you pick the **User Block...** option. Pick a block in the current drawing from the **Select from Drawing Blocks:** drop-down list and then pick the **OK** button.

Use the **Attachment:** drop-down list to specify how to attach the block to the leader. Pick the **Insertion point** option to attach the block to the leader according to the block insertion point, or base point. Choose the **Center extents** option to attach the block directly to the leader, aligned to the center of the block, even if the block insertion point is not on the block itself. See **Figure 19-21**.

block: A symbol that was previously created and saved for reuse.

Figure 19-21.
Adjusting multileader block attachment.

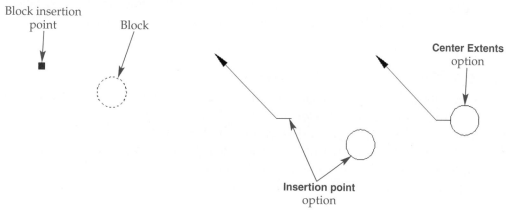

Use the **Color** drop-down list to specify the appropriate block color, which should be ByBlock for typical applications. Use the **Scale** text box to proportionately increase or decrease the block size. Scale does not affect the appearance of the leader line, arrowhead, or shoulder, or the scale applied to the multileader object.

Using No Content

Select the **None** option from the **Multileader type:** drop-down list to end the leader with no annotation. You can use the **None** option whenever there is a need to create only a leader, without text or a symbol attached to the leader line or shoulder.

> **NOTE**
>
> You can add leaders to existing multileaders using the **Add Leader** tool. This eliminates the need to create a separate multileader style that uses the **None** multileader content type for most applications.

Exercise 19-4

Access the Student Web site (www.g-wlearning.com/CAD) and complete Exercise 19-4.

Modifying Multileader Styles

Use the **Multileader Style Manager** to change the characteristics of an existing multileader style. Pick the **Modify** button to open the **Modify Multileader Style** dialog box, which allows you to make changes to the selected style. If you make changes to a multileader style, such as selecting a different text or arrowhead style, existing leaders you drew using the modified multileader style update to reflect the changes. Use a different multileader style with unique characteristics when appropriate.

Renaming and Deleting Multileader Styles

To rename a multileader style using the **Multileader Style Manager**, slowly double-click on the name or right-click the name and select **Rename**. To delete a multileader style using the **Multileader Style Manager**, right-click the name and select **Delete**. You cannot delete a multileader style that is assigned to leaders. To delete a style that is in use, assign a different style to the leaders that reference the style to be deleted.

> **NOTE**
>
> You can also rename styles using the **Rename** dialog box. Select **Multileader styles** in the **Named Objects** list to rename the style.

Setting a Multileader Style Current

You can set a multileader style current using the **Multileader Style Manager** by double-clicking the style in the **Styles** list box, right-clicking on the name and selecting **Set current**, or picking the style and selecting the **Set Current** button. To set a multileader style current without opening the **Multileader Style Manager**, use the **Multileader Style** drop-down list located in the expanded **Annotation** panel on the **Home** ribbon tab and on the **Leaders** panel on the **Annotate** ribbon tab.

You can import multileader styles from existing drawings using **DesignCenter**. See Chapter 5 for more information about using **DesignCenter** to import file content.

Inserting Multileaders

Once you develop a multileader style and make the style current, you are ready to insert leaders using the **MLEADER** tool. How you insert a leader depends on the current multileader style settings and the option you choose to construct the leader. In general, there are three methods for inserting a multileader, depending on what portion of the leader you locate first. Review the components of a leader, shown in **Figure 19-22**, before reading the options for creating a multileader.

The first option for inserting a leader is **Specify leader arrowhead location**. To use this option, first select the location at which you want the arrowhead to point. Then choose where the leader ends and the shoulder begins. If the **Mtext** option is active, enter leader text using the multiline text editor.

The second method uses the **leader Landing first** option. To use this technique, first select where the leader ends and the shoulder begins. Then choose the location where the arrowhead points. If the **Mtext** option is active, enter leader text using the multiline text editor.

The third method involves using the **Content first** option. To use this technique, first define the leader content. The **Mtext** option allows you to type text using the multiline text editor. Then you can select the location where the arrowhead points.

Select **Options** to access a list of options that allow you to override the current multileader style characteristics. These options are the same as those found in the **Modify Multileader Style** dialog box.

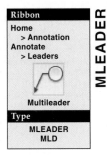

Ribbon

Home
  > Annotation
Annotate
  > Leaders

Multileader

Type

MLEADER
MLD

MLEADER

Figure 19-22.
Examples of leaders created using the **MLEADER** tool. A—An architectural leader created using a spline leader line, the **Specify leader arrowhead location** option, and three leader points. B—A mechanical leader created using a straight leader line, the **leader Landing first** option, and two leader points.

Exercise 19-5

Access the Student Web site (www.g-wlearning.com/CAD) and complete Exercise 19-5.

Dimensioning Chamfers

chamfer: An angled surface used to relieve sharp corners.

Dimension *chamfers* of 45° either with a leader giving the angle and linear dimension, or with two linear dimensions. See **Figure 19-23.** Place the leader using the **MLEADER** tool. Chamfers other than 45° must include either the angle and a linear dimension or two linear dimensions. See **Figure 19-24.** Use the **DIMLINEAR** and **DIMANGULAR** tools for this purpose.

Exercise 19-6

Access the Student Web site (www.g-wlearning.com/CAD) and complete Exercise 19-6.

Thread Drawings and Notes

Figure 19-25 shows the elements of a screw thread. However, threads commonly appear on a drawing as a simplified representation in which a hidden line indicates thread depth. See **Figure 19-26.** Both external and internal threads use this method. An externally threaded part often includes a chamfer to help engage the mating thread.

Thread representations show the reader that a thread exists, but the thread note gives the exact specifications. The thread note typically connects to the thread with a leader. See **Figure 19-27.** The most common thread forms are the Unified and metric screw threads, but a variety of other thread forms are required for specific applications.

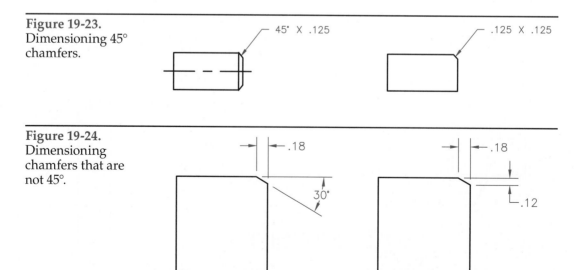

Figure 19-23.
Dimensioning 45° chamfers.

Figure 19-24.
Dimensioning chamfers that are not 45°.

Figure 19-25.
Features of a screw thread.

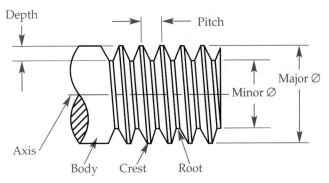

Figure 19-26.
Simplified thread representations.

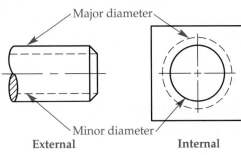

Figure 19-27.
Displaying the thread note with a leader.

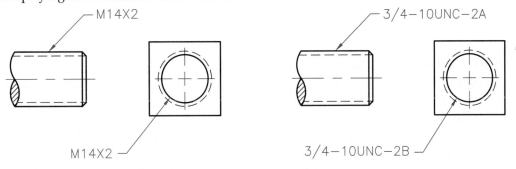

The following format specifies the thread note for Unified screw threads.

3/4-   10UNC-   2A
 (1)   (2)   (3)   (4)(5)

(1) Major diameter of thread, given as a fraction or number.
(2) Number of threads per inch.
(3) Thread series: UNC = Unified National Coarse, UNF = Unified National Fine.
(4) Class of fit: 1 = large tolerance, 2 = general-purpose tolerance, 3 = tight tolerance.
(5) Thread type: A = external thread. B = internal thread.

The following format specifies the thread note for metric threads.

M   14X2
(1)   (2)   (3)

(1) M = metric thread.
(2) Major diameter in millimeters.
(3) Pitch in millimeters.

There are too many screw threads to describe in detail in this textbook. Refer to the *Machinery's Handbook,* published by Industrial Press Inc., or a comprehensive mechanical drafting text for more information.

Exercise 19-7

Access the Student Web site (www.g-wlearning.com/CAD) and complete Exercise 19-7.

Alternate Dimensioning Practices

Dimension lines are often omitted on drawings in industries in which computer-controlled machining processes are used, and unconventional dimensioning practices are sometimes required because of product features. Rectangular coordinate dimensioning without dimension lines, tabular, and chart dimensioning are three examples of dimensioning methods that omit dimension lines.

Dimensioning without Dimension Lines

rectangular coordinate dimensioning without dimension lines (arrowless dimensioning): A type of dimensioning that includes only extension lines and text aligned with the extension lines.

Rectangular coordinate dimensioning without dimension lines, traditionally known as *arrowless dimensioning,* is popular in mechanical drafting for specific applications. Common applications include precision sheet metal part drawings and electronics drafting, especially for chassis layout. Each dimension represents a measurement originating from a *datum.* Identification letters label holes or similar features. Often a table, keyed to the identification letters, indicates feature size or specifications. See **Figure 19-28.**

datum: The 0 dimension, baseline, or common point from which all measurements are made while dimensioning.

Tabular Dimensioning

tabular dimensioning: A form of rectangular coordinate dimensioning without dimension lines in which dimensions are shown in a table.

In *tabular dimensioning,* each feature receives a label with a letter or number that correlates to a table. The table gives the location of features from the X and Y axes. See **Figure 19-29.** The table also provides the depth of features from the Z axis and other specifications when appropriate.

Figure 19-28. Rectangular coordinate dimensioning without dimension lines, or arrowless dimensioning.

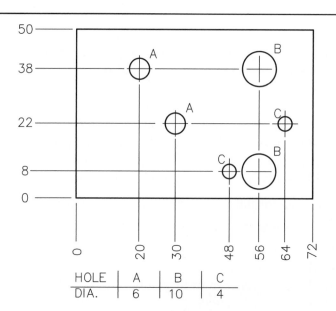

HOLE	A	B	C
DIA.	6	10	4

Figure 19-29.
Tabular dimensioning. (Doug Major)

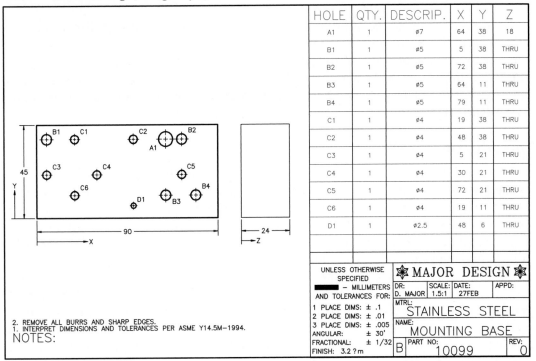

HOLE	QTY.	DESCRIP.	X	Y	Z
A1	1	ø7	64	38	18
B1	1	ø5	5	38	THRU
B2	1	ø5	72	38	THRU
B3	1	ø5	64	11	THRU
B4	1	ø5	79	11	THRU
C1	1	ø4	19	38	THRU
C2	1	ø4	48	38	THRU
C3	1	ø4	5	21	THRU
C4	1	ø4	30	21	THRU
C5	1	ø4	72	21	THRU
C6	1	ø4	19	11	THRU
D1	1	ø2.5	48	6	THRU

UNLESS OTHERWISE SPECIFIED
▬ — MILLIMETERS AND TOLERANCES FOR:
1 PLACE DIMS: ± .1
2 PLACE DIMS: ± .01
3 PLACE DIMS: ± .005
ANGULAR: ± 30'
FRACTIONAL: ± 1/32
FINISH: 3.2 ? m

❋ MAJOR DESIGN ❋
DR: D. MAJOR | SCALE: 1.5:1 | DATE: 27FEB | APPD:
MTRL: STAINLESS STEEL
NAME: MOUNTING BASE
B | PART NO: 10099 | REV: 0

NOTES:
2. REMOVE ALL BURRS AND SHARP EDGES.
1. INTERPRET DIMENSIONS AND TOLERANCES PER ASME Y14.5M−1994.

Chart Dimensioning

Chart dimensioning may take the form of unidirectional, aligned, rectangular coordinate dimensioning without dimension lines, or tabular dimensioning. Chart dimensioning provides flexibility when dimensions change as requirements of the product change. See **Figure 19-30**.

chart dimensioning: A type of dimensioning in which the variable dimensions are shown with letters that correlate to a chart where the possible dimensions are given.

Figure 19-30.
Chart dimensioning.

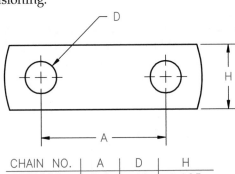

CHAIN NO.	A	D	H
SST1000	2.6	.44	1.125
SST1001	3.0	.48	1.525
SST1002	3.5	.95	2.125

NOTE:
OVERALL LENGTH IS 1.5 X A
END RADII ARE .9 X A

Creating Ordinate Dimension Objects

AutoCAD refers to rectangular coordinate dimensioning without dimension lines as *ordinate dimensioning*. In order to create ordinate dimension objects accurately, you must move the default origin (0,0,0 coordinate) to the object datum. This involves understanding AutoCAD's world coordinate system and user coordinate systems, as described below. Once you establish the datum by temporarily moving the origin, use the **DIMORDINATE** tool to place ordinate dimension objects.

Introduction to WCS and UCS

The origin (0,0,0 coordinate) of the *world coordinate system (WCS)* has been in the lower-left corner of the drawing window for the drawings you have created throughout this textbook. In most cases, this is appropriate. However, when you dimension without dimension lines, it is best to have the dimensions originate from a primary datum, which is often a corner of the object. Depending on how you draw the object, this point may or may not align with the WCS origin.

While the WCS is fixed, a *user coordinate system (UCS)* can move to any orientation. The UCS is described in detail in *AutoCAD and Its Applications—Advanced*. In general, a UCS allows you to set your own coordinate system and origin. Measurements made with the **DIMORDINATE** tool originate from the current UCS origin. By default, this is the 0,0 origin. One method for relocating the origin is to pick the **Origin** button from the **Coordinates** panel of the **View** ribbon tab. Then specify a new origin point, such as the corner of an object or another appropriate datum. See **Figure 19-31**.

When you finish drawing dimensions from a datum, you can leave the UCS origin at the datum or move it back to the WCS origin. To return to the WCS, pick the **World** button from the **Coordinates** panel of the **View** ribbon tab.

Using the DIMORDINATE Tool

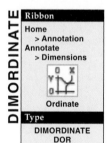

DIMORDINATE

Ribbon
Home
> Annotation
Annotate
> Dimensions

Ordinate

Type
DIMORDINATE
DOR

When you use the **DIMORDINATE** tool, AutoCAD automatically places an extension line and a dimension at the location you pick. The dimension is measured as an X or Y coordinate distance from the UCS origin.

Since you are working in the XY plane, you may want to set vertical and horizontal polar tracking or turn **Ortho** mode on before using the **DIMORDINATE** tool. Also, if the drawing includes circular features, use the **DIMCENTER** tool to place center dashes as shown in **Figure 19-31**. This conforms to ASME standards and provides something to pick when you dimension circular features.

Figure 19-31.
Before you use the **DIMORDINATE** tool, move the UCS origin to the appropriate datum location. Also, add center marks to circular features that you plan to dimension.

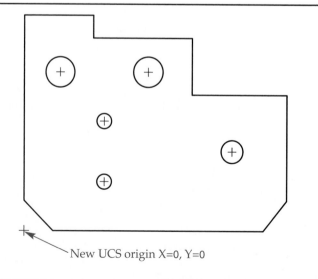

New UCS origin X=0, Y=0

Now you are ready to start placing ordinate dimensions. Access the **DIMORDINATE** tool. When the Specify feature location: prompt appears, pick a point to locate the origin of the extension line. If the feature is the corner of the object, pick an endpoint. If the feature is a circle, pick the *end* of the center dash, as shown in **Figure 19-32,** not the center of the object. This leaves the required space between the center mark and the extension line. Zoom in if needed and use object snap modes when necessary. The next prompt asks for the leader endpoint. This actually refers to the extension line endpoint, so pick the endpoint of the extension line.

If the X axis or Y axis distance between the feature and the extension line endpoint is large, the axis AutoCAD uses for the dimension by default may not be correct. When this happens, use the **Xdatum** or **Ydatum** option to specify the axis from which the dimension originates. The **Mtext**, **Text**, and **Angle** options are identical to the options available with other dimensioning tools. Pick the extension line endpoint to complete the process.

Figure 19-33A shows ordinate dimensions placed on the object. Notice that the dimension text aligns with the extension lines. Aligned dimensioning is standard with ordinate dimensioning. Finally, complete the drawing by adding any missing lines, such as centerlines or fold lines. Identify the holes with letters and create a correlated dimensioning table if appropriate. See **Figure 19-33B.**

PROFESSIONAL TIP

Most ordinate dimensioning tasks work best with polar tracking or **Ortho** mode on. However, when the extension line is too close to an adjacent dimension, it is best to stagger the extension line as shown in the following illustration. With **Ortho** mode off, the extension line automatically staggers when you pick the second extension line point, as shown here.

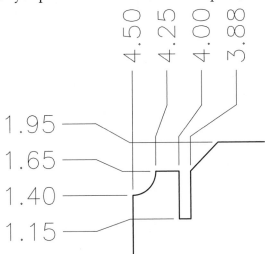

Figure 19-32.
Pick the endpoints of center marks to establish the correct offset, or develop a specific dimension style with an extension line origin offset of 0 for placing dimensions from center marks.

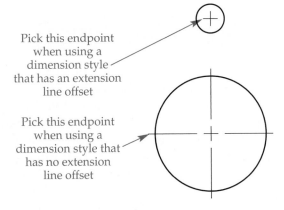

Pick this endpoint when using a dimension style that has an extension line offset

Pick this endpoint when using a dimension style that has no extension line offset

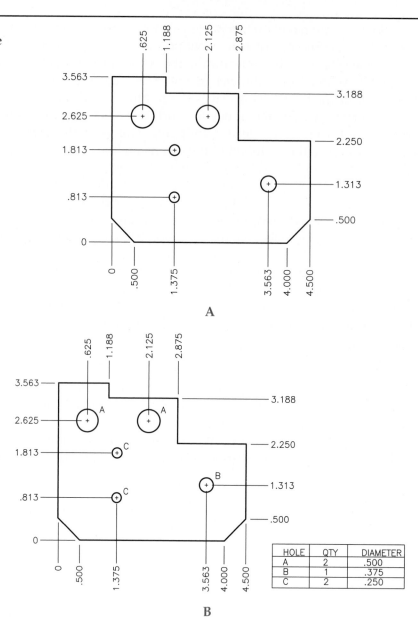

Figure 19-33.
A—Placing ordinate dimensions.
B—Completing the drawing.

HOLE	QTY	DIAMETER
A	2	.500
B	1	.375
C	2	.250

Exercise 19-8

Access the Student Web site (www.g-wlearning.com/CAD) and complete Exercise 19-8.

Marking Up Drawings

marking up (redlining): The process of reviewing a drawing and marking required changes.

Marking up, or *redlining,* is not a dimensioning practice, but it is similar to dimensioning in that it explains information to the reader. Redlines are typically added directly to a final drawing by someone who reviews the drawing for accuracy and design changes. You might experience this process with your instructor or supervisor. Common mark-up techniques include redlining a plot with a red pen, using separate mark-up software to review exported drawings, or redlining directly in the drawing file. Redlines drawn in AutoCAD sometimes become part of the drawing to document revision history.

You can redline a drawing using any appropriate AutoCAD tools, typically using a separate layer. Redlining often includes basic objects, text, and leaders. In some cases, you may add redline dimensions and even an entire drawing or detail. The **REVCLOUD** and **WIPEOUT** tools are also common mark-up tools.

Creating Revision Clouds

Ribbon
Home
> Draw
Annotate
> Markup

Revision Cloud

Type
REVCLOUD

A *revision cloud* is a polyline of sequential arcs forming a cloud-shaped object. **Figure 19-34** shows a revision cloud with a leader and note attached. The revision cloud points the drafter to a specific portion of the drawing that may require an edit.

Drawing a revision cloud using the **REVCLOUD** tool is somewhat different than drawing most other objects, because a single pick is all that is required. To begin drawing the revision cloud, pick a start point in the drawing, and then move the crosshairs around the objects to enclose until you return close to the start point. AutoCAD closes the cloud automatically and exits the tool. Options are available before you pick the start point.

revision cloud: A polyline of sequential arcs used to form a cloud shape around changes in a drawing.

Defining Arc Length

Use the **Arc length** option to specify the size of revision cloud arcs. The value measures the length of an arc from the arc start point to the arc endpoint. AutoCAD prompts for the minimum arc length and then for the maximum arc length. Specifying different minimum and maximum values causes the revision cloud to have an uneven, hand-drawn appearance.

Converting Objects to Revision Clouds

Use the **Object** option to convert a circle, closed polyline, ellipse, polygon, or rectangle to a revision cloud. Pick the object to convert to a revision cloud. Enter the **No** option at the Reverse direction: prompt, or use the **Yes** option to reverse the direction of the cloud arcs.

Figure 19-34.
An example of a revision cloud identifying a modified area of a drawing. Notice the leader describing the change.

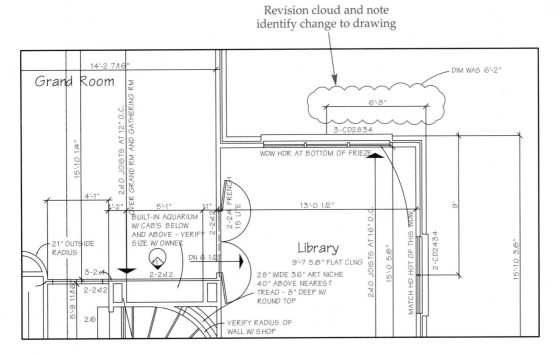

Changing Revision Cloud Style

The **Style** option offers two style choices: **Normal** and **Calligraphy**. The default style is **Normal**, which displays arcs with a consistent width. When you specify the **Calligraphy** style, the start and end widths of the individual arcs are different, creating a more stylized revision cloud. See **Figure 19-35.**

Exercise 19-9

Access the Student Web site (www.g-wlearning.com/CAD) and complete Exercise 19-9.

Using the WIPEOUT Tool

WIPEOUT

Ribbon
Home
  > Draw
Annotate
  > Markup

Wipeout

Type
WIPEOUT

The **WIPEOUT** tool allows you to clear a portion of the drawing without erasing objects. This tool is sometimes appropriate for applications similar to those for **REVCLOUD**, most often redlining. **Figure 19-36** shows an example of a wipeout used to lay out the location of a proposed building sight.

Specify the first corner of the wipeout, followed by all other perimeter corners. Use the **Undo** option as needed to reverse the effects of an incorrect selection. When you are finished selecting points, use the **Close** option, press [Enter] or the space bar, or right-click and choose **Enter** to create the wipeout.

An alternative to picking points is to use the **Polyline** option and select a closed polyline object to convert to a wipeout. Use the **Frames** option to turn the display of all wipeout boundaries on or off. You may need to regenerate the display to observe the effects of changing the frame setting. To reveal objects hidden by a wipeout, freeze or turn off the wipeout layer, use draw order tools, or erase the wipeout if it is no longer needed.

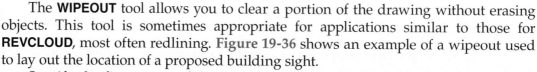

Template Development

Chapter 19

For detailed instructions on adding multileader styles to each drawing template, go to the Student Web site (www.g-wlearning.com/CAD), select this chapter, and select **Template Development**.

Figure 19-35.
You can create revision clouds in two different styles: the **Normal** style and the **Calligraphy** style.

Normal Style

Calligraphy Style

Figure 19-36.
An example of using the **WIPEOUT** tool to clear a portion of a drawing. Objects below the wipeout still exist. Additional information is added to the wipeout in this example.

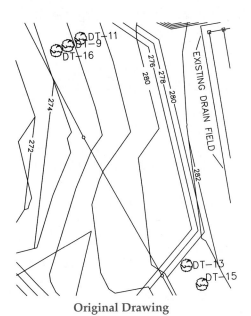

Original Drawing

Pick points to create the wipeout

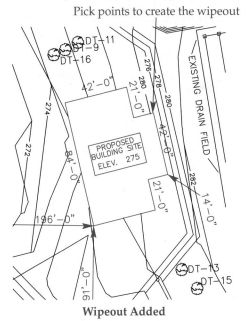

Wipeout Added

Chapter Test

Answer the following questions. Write your answers on a separate sheet of paper or go to the Student Web site (www.g-wlearning.com/CAD) and complete the electronic chapter test.

1. Which tool provides diameter dimensions for circles?
2. Which tool provides radius dimensions for arcs?
3. Explain how to add a center mark to a circle without using the **DIMDIAMETER** or **DIMRADIUS** tool.
4. What is the most common size for leader arrowheads?
5. What angle constraints should you use for leaders to maintain the ASME standard?
6. What is the usual length for the shoulder of a leader in mechanical drafting?
7. Describe two ways to dimension a 45° chamfer.
8. Identify the elements of this Unified screw thread note: 1/2-13UNC-2B.
   A. 1/2
   B. 13
   C. UNC
   D. 2
   E. B
9. Identify the elements of this metric screw thread note: M 14 X 2.
   A. M
   B. 14
   C. 2
10. Define *rectangular coordinate dimensioning without dimension lines.*
11. What term does AutoCAD use to refer to rectangular coordinate dimensioning without dimension lines?
12. Explain the importance of the user coordinate system (UCS) for drawing ordinate dimension objects.
13. What is the purpose of a revision cloud?
14. How do you close a revision cloud?
15. What is the purpose of the **WIPEOUT** tool?

Drawing Problems

- *Start AutoCAD if it is not already started.*
- *Start a new drawing using an appropriate template of your choice. The template should include layers, text styles, dimension styles, and multileader styles appropriate for drawing the given objects.*
- *Add layers, text styles, dimension styles, and multileader styles as needed. Draw all objects using appropriate layers, text styles, dimension styles, multileader styles, justification, and format.*
- *Follow the specific instructions for each problem. Use your own judgment and approximate dimensions when necessary.*
- *Apply dimensions accurately using ASME or appropriate industry standards. Use object snap modes to your best advantage.*
- *For mechanical drawings, place the following general notes 1/2" from the lower-left corner:*

> NOTES:
>
> 1. INTERPRET DIMENSIONS AND TOLERANCES PER ASME Y14.5M-1994.
> 2. REMOVE ALL BURRS AND SHARP EDGES.
> 3. UNLESS OTHERWISE SPECIFIED, ALL DIMENSIONS ARE IN INCHES (or MILLIMETERS as applicable).

▼ Basic

1. Draw the view and add dimensions as shown. Save the drawing as P19-1.

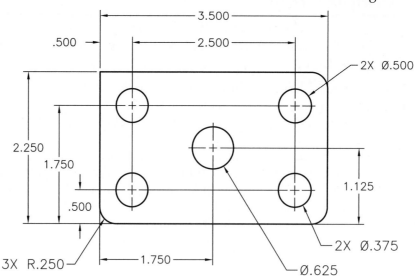

2. Draw the views and add dimensions as shown. Save the drawing as P19-2.

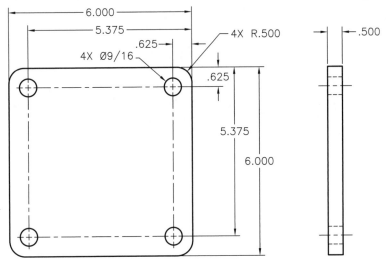

3. Draw the view and add dimensions as shown. Save the drawing as P19-3.

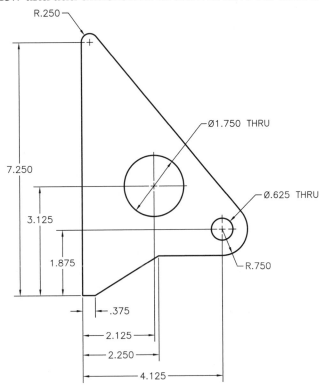

4. Draw the pin and add dimensions as shown. Save the drawing as P19-4.

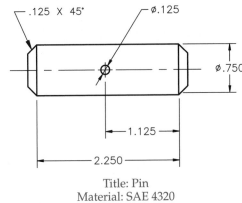

Title: Pin
Material: SAE 4320

5. Draw the spline and add dimensions as shown. Save the drawing as P19-5.

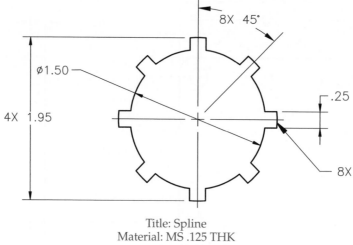

Title: Spline
Material: MS .125 THK

6. Draw the gasket and add dimensions as shown. Save the drawing as P19-6.

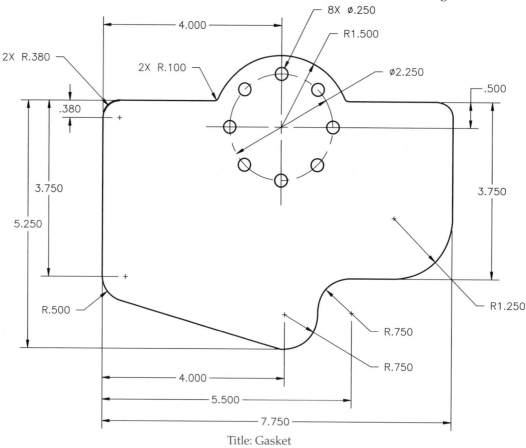

Title: Gasket
Material: 00 Phosphor Bronze

7. Draw the chain link and add dimensions as shown. Save the drawing as P19-7.

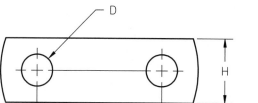

CHAIN NO.	A	D	H
SST1000	2.6	.44	1.125
SST1001	3.0	.48	1.525
SST1002	3.5	.95	2.125

Note:
Overall Length is 1.5XA
end radii are .9XA

Title: Chain Link
Material: Steel

8. Draw the view and add dimensions as shown. Save the drawing as P19-8.

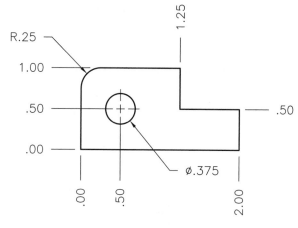

9. Draw the chassis spacer and add dimensions as shown. Save the drawing as P19-9.

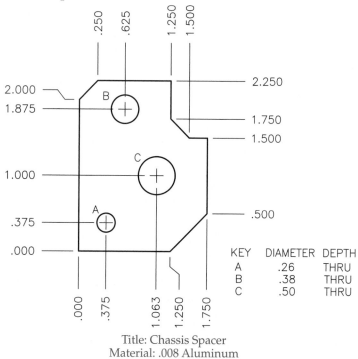

KEY	DIAMETER	DEPTH
A	.26	THRU
B	.38	THRU
C	.50	THRU

Title: Chassis Spacer
Material: .008 Aluminum

Drawing Problems - Chapter 19

10. Draw the chassis and add dimensions as shown. Save the drawing as P19-10.

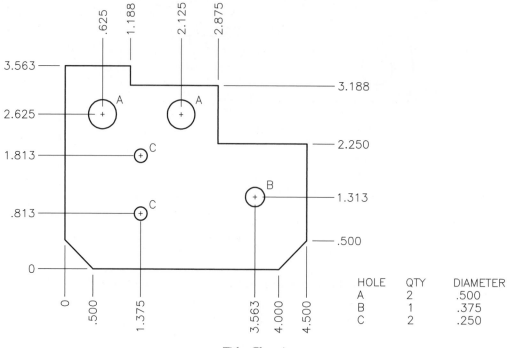

HOLE	QTY	DIAMETER
A	2	.500
B	1	.375
C	2	.250

Title: Chassis
Material: Aluminum .100 THK

11. Draw the view and add dimensions as shown. Save the drawing as P19-11.

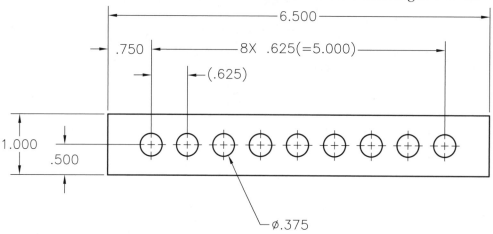

▼ Intermediate

12. Convert the given drawing to a drawing with the holes located using the **DIMORDINATE** tool based on the X and Y coordinates given in the table. Place a table above your title block with columns for Hole (identification), Quantity, Description, and Depth (Z axis). Save the drawing as P19-12.

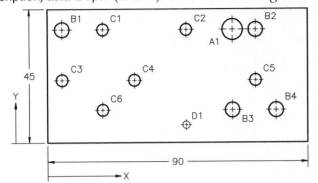

HOLE	QTY	DESC	X	Y	Z
A1	1	ø7	64	38	18
B1	1	ø5	5	38	THRU
B2	1	ø5	72	38	THRU
B3	1	ø5	64	11	THRU
B4	1	ø5	79	11	THRU
C1	1	ø4	19	38	THRU
C2	1	ø4	48	38	THRU
C3	1	ø4	5	21	THRU
C4	1	ø4	30	21	THRU
C5	1	ø4	72	21	THRU
C6	1	ø4	19	11	THRU
D1	1	ø2.5	48	6	THRU

Title: Base
Material: Bronze

For Problems 13-15, use the isometric drawing provided to create a multiview orthographic drawing for the part. Include only the views necessary to fully describe the object. Add all required dimensions according to the ASME standards.

13. Save the drawing as P19-13.

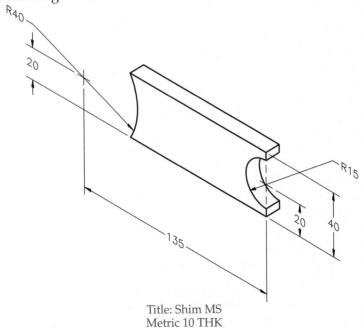

Title: Shim MS
Metric 10 THK

Chapter 19 Dimensioning Features and Alternate Practices

14. Half of the object is removed for clarity. The entire object should be drawn. Save the drawing as P19-14.

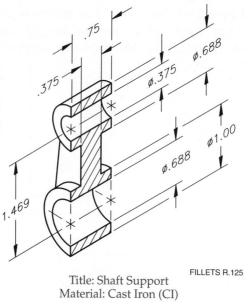

FILLETS R.125

Title: Shaft Support
Material: Cast Iron (CI)

15. Half of the object is removed for clarity. The entire object should be drawn. Save the drawing as P19-15.

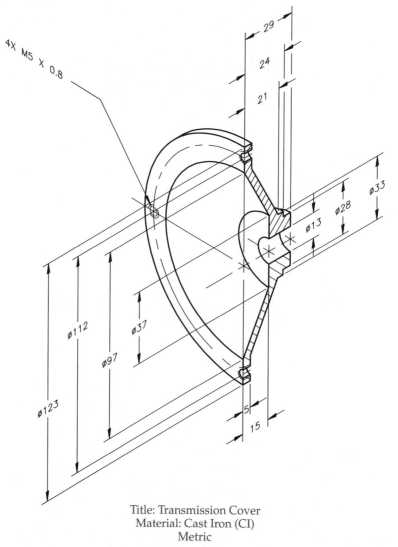

Title: Transmission Cover
Material: Cast Iron (CI)
Metric

16. Create the drawing as shown and save it as P19-16.

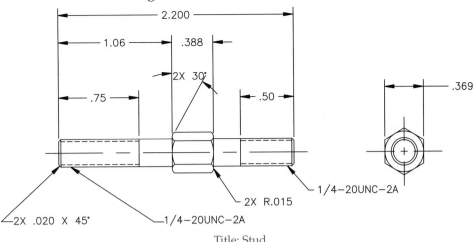

Title: Stud
Material: Stainless Steel

17. Create the drawing as shown and save it as P19-17.

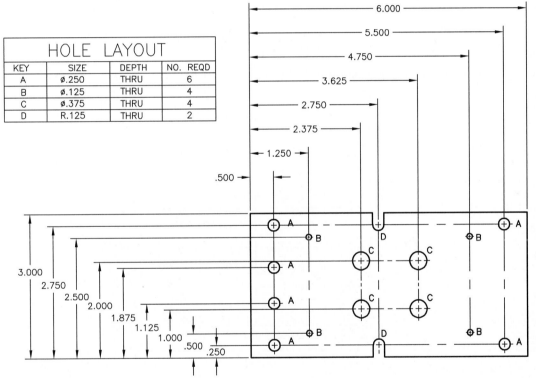

HOLE LAYOUT			
KEY	SIZE	DEPTH	NO. REQD
A	ø.250	THRU	6
B	ø.125	THRU	4
C	ø.375	THRU	4
D	R.125	THRU	2

Title: Chassis Base (datum dimensioning)
Material: 12 gage Aluminum

18. Create the drawing as shown and save it as **P19-18**.

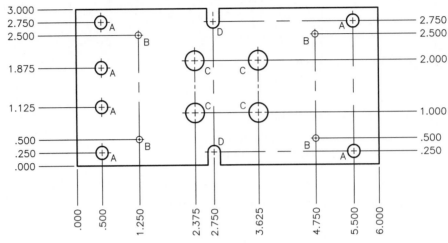

HOLE LAYOUT			
KEY	SIZE	DEPTH	NO. REQD
A	⌀.250	THRU	6
B	⌀.125	THRU	4
C	⌀.375	THRU	4
D	R.125	THRU	2

Title: Chassis Base (arrowless dimensioning)
Material: 12 gage Aluminum

19. Create the drawing as shown and save it as **P19-19**.

HOLE LAYOUT				
KEY	X	Y	SIZE	TOL
A1	.500	2.750	⌀.250	±.002
A2	.500	1.875	⌀.250	±.002
A3	.500	1.125	⌀.250	±.002
A4	.500	.250	⌀.250	±.002
A5	5.500	2.750	⌀.250	±.002
A6	5.500	.250	⌀.250	±.002
B1	1.250	2.500	⌀.125	±.001
B2	1.250	.500	⌀.125	±.001
B3	4.750	2.500	⌀.125	±.001
B4	4.750	.500	⌀.125	±.001
C1	2.375	2.000	⌀.375	±.005
C2	2.375	1.000	⌀.375	±.005
C3	3.625	2.000	⌀.375	±.005
C4	3.625	1.000	⌀.375	±.005
D1	2.750	2.750	R.125	±.002
D2	2.750	.250	R.125	±.002

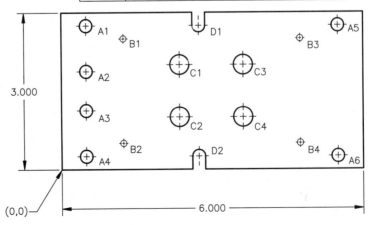

Title: Chassis Base (arrowless tabular dimensioning)
Material: 12 gage Aluminum

▼ Advanced

20. Create the drawing as shown and save it as P19-20.

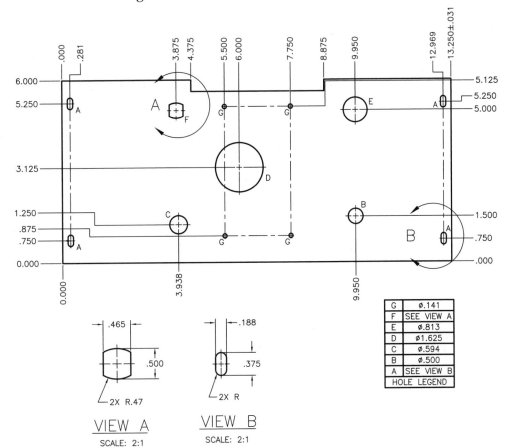

VIEW A
SCALE: 2:1

VIEW B
SCALE: 2:1

G	ø.141
F	SEE VIEW A
E	ø.813
D	ø1.625
C	ø.594
B	ø.500
A	SEE VIEW B
HOLE LEGEND	

21. Create the drawing as shown and save it as P19-21.

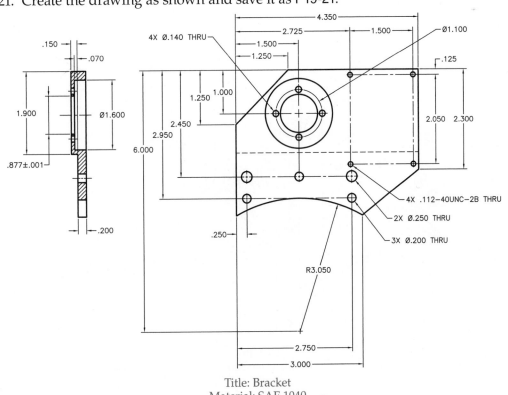

Title: Bracket
Material: SAE 1040

22. Create the drawing as shown and save it as P19-22.

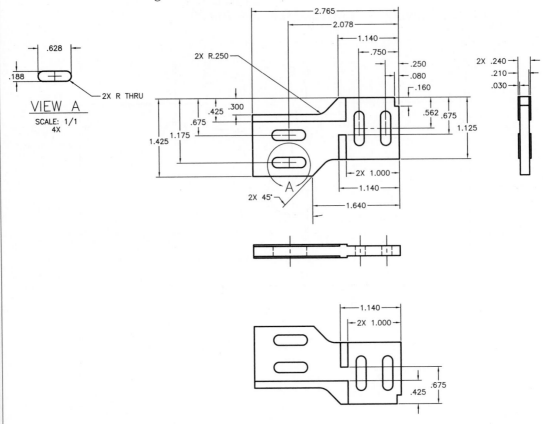

VIEW A

SCALE: 1/1
4X

For Problems 23 and 24, use the isometric drawing provided to create a multiview orthographic drawing for the part. Include only the views necessary to fully describe the object. Add all required dimensions according to the ASME standards.

23. Create the drawing as shown and save it as P19-23.

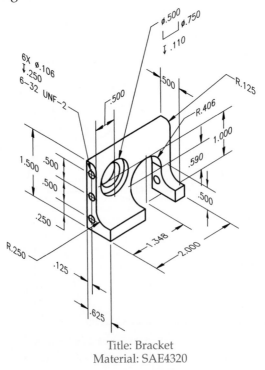

Title: Bracket
Material: SAE4320

24. Carefully evaluate the given problem before beginning. Many given dimensions are provided to the inside surfaces of the bracket. This application is incorrect. Calculate the dimensions as needed to place datum dimensioning from the surfaces labeled A and B. Do not place the A and B on your final drawing. Save the drawing as P19-24.

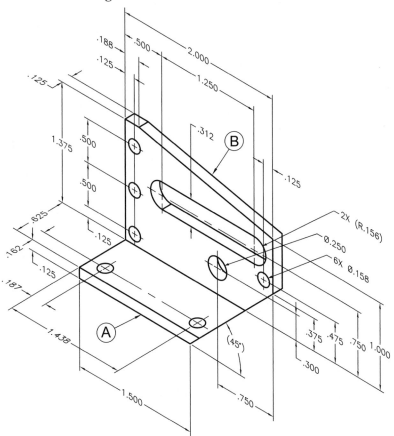

25. Open P18-17 and save the file as P19-25. The P19-25 file should be active. Add the client-requested redlines to the floor plan as shown.

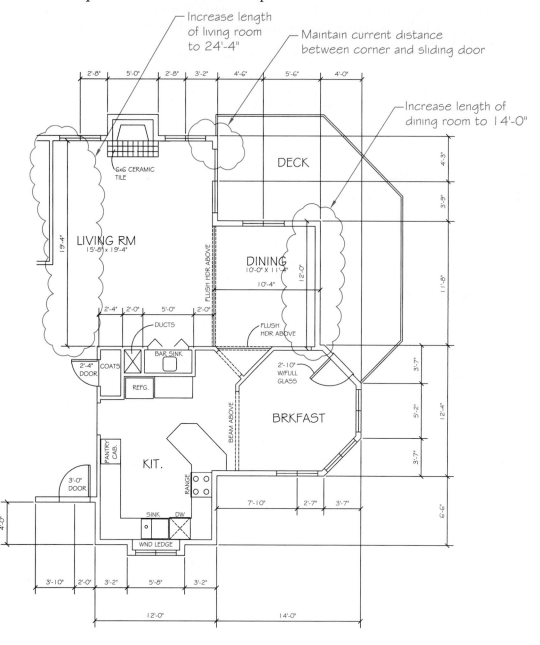

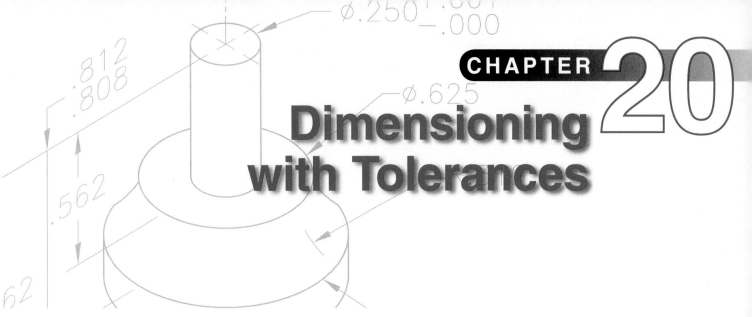

CHAPTER 20

Dimensioning with Tolerances

Learning Objectives

After completing this chapter, you will be able to do the following:
✓ Define and use dimensioning and tolerancing terminology.
✓ Set the precision for dimensions and tolerances.
✓ Set up the primary units for use with inch or metric dimensions.
✓ Create and use dimension styles with various tolerance settings.
✓ Explain the purpose of geometric dimensioning and tolerancing (GD&T).

This chapter introduces tolerancing and explains how to prepare dimensions with tolerances for mechanical manufacturing drawings. This chapter also introduces drawing geometric dimensioning and tolerancing (GD&T) symbols and offers information on how you can learn more about GD&T.

Tolerancing Fundamentals

Methods of specifying *tolerance* include direct placement with a dimension, a general note, and specifications in the drawing title block. See **Figure 20-1**. The dimension stated as 12.50±.25 in **Figure 20-2A** is in a style known as *plus-minus dimensioning*. The tolerance of this dimension is the difference between the maximum and minimum *limits*. This tolerance style applies when the variance is the same in the positive and negative directions. In this case, the upper limit is 12.75 (12.50 +.25 = 12.75), and the lower limit is 12.25 (12.50 − .25 = 12.25). To find the tolerance, subtract the lower limit from the upper limit. The tolerance in this example is .50 (12.75 − 12.25 = .50). The *specified dimension* of the feature shown in **Figure 20-2** is 12.50.

Limits dimensioning, shown in **Figure 20-2B**, is another common method of showing and calculating tolerance. Many schools and companies prefer this method because it does not require calculating limits. Additional tolerance methods are also common, depending on the design requirement. **Figure 20-3** shows examples of equal and unequal *bilateral tolerances*. **Figure 20-4** shows an example of a *unilateral tolerance*.

tolerance: The total amount by which a specific dimension is permitted to vary.

plus-minus dimensioning: A tolerance style in which the positive and negative variance is equal and is preceded by a ± symbol.

limits: The largest and smallest numerical values the feature can have.

specified dimension: The part of the dimension from which the limits are calculated.

limits dimensioning: Method in which the upper and lower limits are given, instead of the specified dimension and tolerance.

bilateral tolerance: A tolerance style that permits variance in both the positive and negative directions from the specified dimension.

unilateral tolerance: A tolerance style that permits a variation in only one direction from the specified dimension.

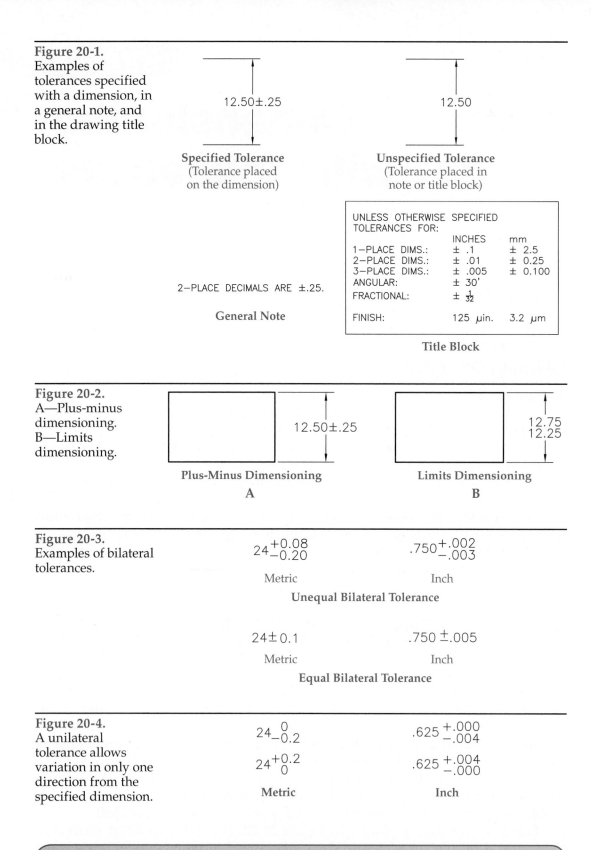

Figure 20-1.
Examples of tolerances specified with a dimension, in a general note, and in the drawing title block.

12.50±.25

Specified Tolerance
(Tolerance placed on the dimension)

12.50

Unspecified Tolerance
(Tolerance placed in note or title block)

2—PLACE DECIMALS ARE ±.25.

General Note

UNLESS OTHERWISE SPECIFIED
TOLERANCES FOR:

	INCHES	mm
1—PLACE DIMS.:	± .1	± 2.5
2—PLACE DIMS.:	± .01	± 0.25
3—PLACE DIMS.:	± .005	± 0.100
ANGULAR:	± 30'	
FRACTIONAL:	± $\frac{1}{32}$	
FINISH:	125 μin.	3.2 μm

Title Block

Figure 20-2.
A—Plus-minus dimensioning.
B—Limits dimensioning.

12.50±.25

Plus-Minus Dimensioning
A

12.75
12.25

Limits Dimensioning
B

Figure 20-3.
Examples of bilateral tolerances.

$24^{+0.08}_{-0.20}$

Metric

$.750^{+.002}_{-.003}$

Inch

Unequal Bilateral Tolerance

24 ± 0.1

Metric

$.750\pm.005$

Inch

Equal Bilateral Tolerance

Figure 20-4.
A unilateral tolerance allows variation in only one direction from the specified dimension.

$24^{0}_{-0.2}$

$24^{+0.2}_{0}$

Metric

$.625^{+.000}_{-.004}$

$.625^{+.004}_{-.000}$

Inch

Assigning Decimal Places

The ASME Y14.5M *Dimensioning and Tolerancing* standard has separate recommendations for the display of decimal places in inch and metric dimensions. **Figure 20-3** and **Figure 20-4** show examples of decimal dimension values in inches and metric units.

Inch Dimensioning

A specified inch dimension uses the same number of decimal places as its tolerance. Add zeros to the right of the decimal point if needed. For example, the inch dimension .250±.005 has an additional zero added to the .25 to match the three-decimal tolerance. Similarly, the dimensions 2.000±.005 and 2.500±.005 have zeros added to match the tolerance.

Both of the values in a plus-minus tolerance for an inch dimension have the same number of decimal places. Add zeros to fill in where needed. For example:

$$\begin{array}{ccc} +.005 & & +.005 \\ -.010 & not & -.01 \end{array}$$

Metric Dimensioning

Omit the decimal point and zeros from the specified dimension of a metric whole number. For example, the metric dimension 12 has no decimal point followed by a zero. This rule is true unless the drawing displays tolerance values. When a metric dimension includes a decimal portion, a zero does not follow the last digit to the right of the decimal point. For example, the metric dimension 12.5 has no zero to the right of the 5. This rule is true unless the drawing displays tolerance values.

Both values in a bilateral tolerance for a metric dimension have the same number of decimal places. Add zeros to fill in where needed. Typically, no additional zeros appear after the specified dimension to match the tolerance. For example, both 24±0.25 and 24.5±0.25 are correct. However, some companies prefer to add zeros after the specified dimension to match the tolerance, in which case 24.00±0.25 and 24.50±0.25 are both correct.

Setting Primary Units

The dimension style controls the appearance of dimensions, including dimension values and tolerance. The initial phase of dimensioning with tolerances involves setting the appropriate values for the primary units of the dimension style you plan to use. Use the **Primary Units** tab of the **New** (or **Modify**) **Dimension Style** dialog box, shown in **Figure 20-5**, to set the dimension units and precision.

The **Precision** drop-down list in the **Linear dimensions** area allows you to specify the number of zeros displayed after the decimal point of the specified dimension. The ASME standard recommends that the precision for the dimension and the tolerance be the same for inch dimensions, but it may be different for metric values, as previously described.

The **Zero suppression** settings control the display of zeros before and after the decimal point. For inch dimensions, the **Leading** options should be on, and the **Trailing** options should be off. For typical metric dimensions, without using sub-units, the **Leading** options should be off, and the **Trailing** options should be on.

Figure 20-5.
The **Primary Units** tab of the **New** (or **Modify**) **Dimension Style** dialog box sets the unit format and precision of linear dimensions.

Set the precision for specified dimensions

Set the zero suppression

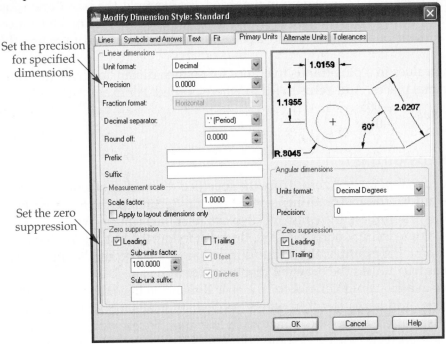

Setting Tolerance Methods

The **Tolerances** tab of the **New** (or **Modify**) **Dimension Style** dialog box, shown in **Figure 20-6,** allows you to apply a tolerance method to your drawing. The default option in the **Method:** drop-down list is None. This means dimensions do not include

Figure 20-6.
The **Tolerances** tab of the **New** (or **Modify**) **Dimension Style** dialog box contains formatting settings for tolerance dimensions.

Select a tolerance method

Set the precision for tolerance dimensions

Settings should match the **Zero suppression** linear dimension settings in the **Primary Units** tab

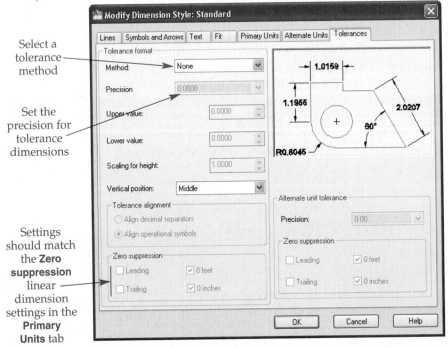

Figure 20-7.
Select a tolerance
dimensioning
method from the
Method: drop-down
list, in the **Tolerance
format** area.

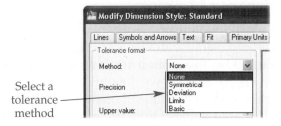

Select a
tolerance
method

a tolerance method, or specific tolerance. As a result, most of the options in the **Tolerances** tab are disabled. When you pick a tolerance method from the drop-down list, appropriate options enable, and the image in the tab reflects the selected method. **Figure 20-7** shows the drop-down list options.

> **NOTE**
>
> The following information describes tolerance methods and the settings unique to each. General settings, including tolerance precision, height, vertical position, alignment, and zero suppression, are explained later in this chapter.

Symmetrical Tolerance Method

Select the **Symmetrical** option from the **Method:** drop-down list to create a *symmetrical tolerance*. Use this option to draw dimensions that display an equal bilateral tolerance in the plus-minus format. **Figure 20-8** shows the options that enable symmetrical tolerances, the preview that appears, and an example of using the **Symmetrical** option. Enter a tolerance value in the **Upper value:** text box. Although it is disabled, you can see that the value in the **Lower value:** text box matches the value in the **Upper value:** text box.

symmetrical tolerance: AutoCAD's term for an equal bilateral tolerance.

Exercise 20-1

Access the Student Web site (www.g-wlearning.com/CAD) and complete Exercise 20-1.

Figure 20-8.
Setting the
Symmetrical
tolerance method
option current, with
an equal bilateral
tolerance value of
.005.

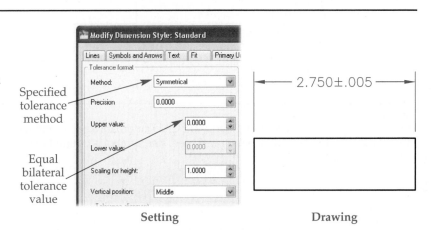

Specified
tolerance
method

Equal
bilateral
tolerance
value

Setting   Drawing

Deviation Tolerance Method

deviation tolerance: AutoCAD's term for an unequal bilateral tolerance.

Pick the **Deviation** option from the **Method:** drop-down list to create a *deviation tolerance*. A deviation tolerance deviates (varies) from the specified dimension with two different values. Use this option to draw dimensions that display an unequal bilateral tolerance. **Figure 20-9** shows the options that enable deviation tolerances, the preview that appears, and an example of using the **Deviation** option. Enter the desired upper and lower tolerance values in the **Upper value:** and **Lower value:** text boxes.

You can also use the deviation option to draw a unilateral tolerance by entering 0 for either the **Upper value:** or **Lower value:** setting. If you are using inch units, AutoCAD includes the plus or minus sign before the zero tolerance. When you use metric units, AutoCAD omits the sign for the zero tolerance. See **Figure 20-10**.

Exercise 20-2

Access the Student Web site (www.g-wlearning.com/CAD) and complete Exercise 20-2.

Limits Tolerance Method

Select the **Limits** option from the **Method:** drop-down list to apply the limits tolerance method. In limits dimensioning, the tolerance limits are given, and no calculations from the specified dimension are required (unlike plus-minus dimensioning). **Figure 20-11** shows the options that enable the limits method, the preview that appears, and an example of using the **Limits** option. Use the **Upper value:** and **Lower value:** text

Figure 20-9.
Setting the **Deviation** tolerance method option current, with unequal bilateral tolerance values.

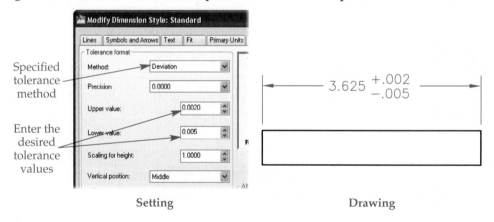

Figure 20-10.
When a unilateral tolerance is specified, AutoCAD automatically places the plus or minus symbol in front of the zero tolerance, if inch units are used. The symbol is omitted with metric units.

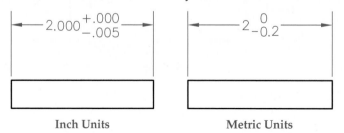

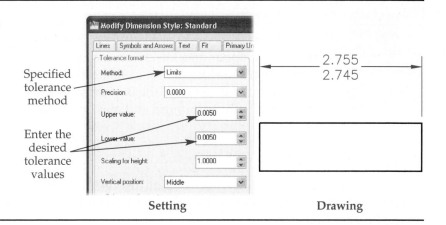

Figure 20-11.
Selecting the limits tolerance method and setting limit values.

Specified tolerance method

Enter the desired tolerance values

Setting   Drawing

boxes to enter the upper and lower tolerance values to add and subtract from the specified dimension. The upper and lower values can be equal or different.

Exercise 20-3

Access the Student Web site (www.g-wlearning.com/CAD) and complete Exercise 20-3.

Basic Tolerance Method

Pick the **Basic** option from the **Method:** drop-down list to draw *basic dimensions*. **Figure 20-12** shows the options that enable the basic method, the preview that appears, and an example of using the **Basic** option. Few options are enabled because a basic dimension has no tolerance. A rectangle placed around the dimension number distinguishes a basic dimension from other dimensions.

basic dimension: A theoretically perfect dimension used in geometric dimensioning and tolerancing.

NOTE

Picking the **Draw frame around text** check box in the **Text** tab of the **New** (or **Modify**) **Dimension Style** dialog box also activates the basic tolerance method.

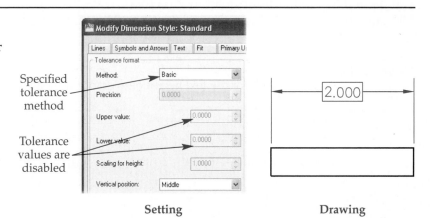

Figure 20-12.
Use the basic tolerance method for basic dimensioning. The dimension text for a basic dimension appears inside a rectangle.

Specified tolerance method

Tolerance values are disabled

Setting   Drawing

Tolerance Precision

After you specify the tolerance method using the **Method:** drop-down list, you can adjust the tolerance precision. By default, when you set the primary unit precision on the **Primary Units** tab, AutoCAD automatically makes the tolerance precision in the **Tolerances** tab the same unit precision. If the setting does not reflect the correct level of precision, change it using the **Precision** drop-down list in the **Tolerance format** area.

Tolerance Height

You can set the text height of the tolerance dimension in relation to the text height of the specified dimension using the **Scaling for height:** text box in the **Tolerance format** area. The default of 1.0000 makes the tolerance dimension text the same height as the specified dimension text. This is recommended in the ASME standard.

To make the tolerance dimension height three-quarters as high as the specified dimension height, type .75 in the **Scaling for height:** text box. Some companies prefer this practice to keep the tolerance part of the dimension from taking up additional space. **Figure 20-13** shows examples of tolerance dimensions with different text heights.

Vertical Position

Use the options in the **Vertical position:** drop-down list in the **Tolerance format** area to control the alignment, or justification, of deviation tolerance dimensions. The **Middle** option, which is the default, centers the tolerance with the specified dimension. This is the recommended ASME practice. The other justification options are **Top** and **Bottom**. **Figure 20-14** displays deviation tolerance dimensions with each of the justification options.

Tolerance Alignment

The options in the **Tolerance alignment** area become available for selection when you use a deviation or limits tolerance method. The setting controls the left and right tolerance justification. When using a deviation tolerance method, pick the **Align decimal**

Figure 20-13.
Using different scale settings for the text height of tolerance dimensions.

$3.250 \begin{array}{c} +.005 \\ -.002 \end{array}$

Tolerance Scale Setting = 1

$3.250 \begin{array}{c} +.005 \\ -.002 \end{array}$

Tolerance Scale Setting = .75

3.255
3.248

Tolerance Scale Setting = 1

3.255
3.248

Tolerance Scale Setting = .75

Figure 20-14.
Examples of the tolerance justification options for deviation tolerance dimensions.

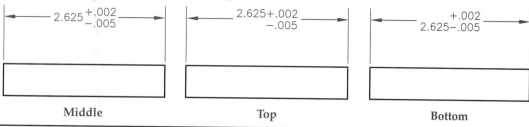

Middle   Top   Bottom

separators radio button to align the upper and lower tolerance value decimal points vertically. Select the **Align operational symbols** radio button to align the upper and lower tolerance plus and minus symbols vertically. See **Figure 20-15.** When using the limits tolerance method, pick the **Align decimal separators** radio button to align the upper and lower limit decimal points vertically. Select the **Align operational symbols** radio button to left-justify the upper and lower limits. See **Figure 20-16.**

Zero Suppression

You must select a tolerance method to enable the options in the **Zero suppression** area. The suppression settings for linear dimensions in the **Tolerances** tab should be the same as the **Zero suppression** tolerance format settings in the **Primary Units** tab. AutoCAD does not automatically match the tolerance setting to the primary units setting.

Select the **Leading** check box in the **Zero suppression** area of the **Tolerances** tab when you are drawing inch tolerance dimensions. Activate the same option for linear dimensions in the **Primary Units** tab. You can then draw inch tolerance dimensions without placing the zero before the decimal point, as recommended by ASME standards. These settings allow you to draw a tolerance dimension such as .625±.005.

Deselect the **Leading** check box in the **Zero suppression** area of the **Tolerances** tab when drawing metric tolerance dimensions. Deactivate the same option for linear dimensions in the **Primary Units** tab. This allows you to place a metric tolerance dimension with the zero before the decimal point, such as 12±0.2, as recommended by ASME standards.

Figure 20-15.
Changing tolerance alignment for use with a deviation tolerance method.

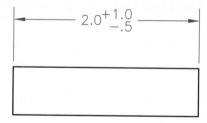

Aligned on Decimal Separator

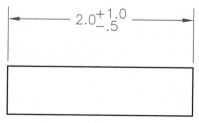

Aligned on Operational Symbols

Figure 20-16.
Changing tolerance
alignment for
use with a limits
tolerance method.

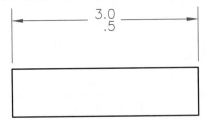

3.0
.5

Aligned on Decimal Separator

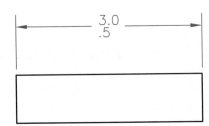

3.0
.5

Aligned on Operational Symbols

NOTE

The options in the **Alternate unit tolerance** area become enabled when you pick the **Display alternate units** check box in the **Alternate Units** tab of the **New** (or **Modify**) **Dimension Style** dialog box. Use the **Alternate unit tolerance** area to set specific tolerances for alternate units.

Exercise 20-4

Access the Student Web site (www.g-wlearning.com/CAD) and complete Exercise 20-4.

Introduction to GD&T Symbols

geometric dimensioning and tolerancing (GD&T): The dimensioning and tolerancing of individual features of a part where the permissible variations relate to characteristics of form, profile, orientation, runout, or the relationship between features.

Geometric dimensioning and tolerancing (GD&T) is the dimensioning and tolerancing of individual features of a part where the permissible variations relate to characteristics of form, profile, orientation, runout, or the relationship between features. For complete coverage of GD&T, refer to *Geometric Dimensioning and Tolerancing* by David A. Madsen, published by Goodheart-Willcox Company, Inc.

Reference Material

Drafting Symbols
For the names and examples of GD&T symbols and symbol applications, go to the **Reference Material** section of the Student Web site (www.g-wlearning.com/CAD) and select **Drafting Symbols** in the list.

Chapter Test

Answer the following questions. Write your answers on a separate sheet of paper or go to the Student Web site (www.g-wlearning.com/CAD) and complete the electronic chapter test.

1. Define the term *tolerance.*
2. What are the limits of the tolerance dimension 3.625±.005?
3. Give an example of an equal bilateral tolerance in inches and in metric units.
4. Give an example of an unequal bilateral tolerance in inches and in metric units.
5. Give an example of a unilateral tolerance in inches and in metric units.
6. What is the purpose of the **Symmetrical** tolerance method option?
7. What is the purpose of the **Deviation** tolerance method option?
8. What is the purpose of the **Limits** tolerance method option?
9. How do you set the number of zeros displayed after the decimal point for a tolerance dimension?
10. Explain the result of setting the **Scaling for height:** option to 1 in the **Tolerances** tab.
11. What setting would you use for the **Scaling for height:** option if you wanted the tolerance dimension height to be three-quarters of the specified dimension height?
12. Name the tolerance dimension justification option recommended by the ASME standards.
13. Which **Zero suppression** settings should you specify for linear and tolerance dimensions when you are using inch units?
14. Which **Zero suppression** settings should you specify for linear and tolerance dimensions when you are using metric units?
15. What is the purpose of geometric dimensioning and tolerancing?

Drawing Problems

- *Start AutoCAD if it is not already started.*

- *Start a new drawing using an appropriate template of your choice. The template should include layers, text styles, dimension styles, and multileader styles appropriate for drawing the given objects.*

- *Add layers, text styles, dimension styles, and multileader styles as needed. Draw all objects using appropriate layers, text styles, dimension styles, multileader styles, justification, and format.*

- *Follow the specific instructions for each problem. Use your own judgment and approximate dimensions when necessary.*

- *Apply dimensions accurately using ASME or appropriate industry standards. Use object snap modes to your best advantage.*

- *For mechanical drawings, place the following general notes 1/2" from the lower-left corner:*

> NOTES:
> 1. INTERPRET DIMENSIONS AND TOLERANCES PER ASME Y14.5M-1994.
> 2. REMOVE ALL BURRS AND SHARP EDGES.
> 3. UNLESS OTHERWISE SPECIFIED, ALL DIMENSIONS ARE IN INCHES
>   (or MILLIMETERS as applicable).

For Problems 1-7, use the isometric drawing provided to create a multiview orthographic drawing for the part. Include only the views necessary to fully describe the object. Add all required dimensions according to the ASME standards.

▼ Basic

1. Create the drawing as shown and save it as P20-1.

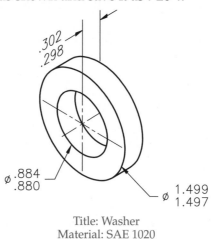

Title: Washer
Material: SAE 1020
Inch

2. Create the drawing as shown and save it as P20-2.

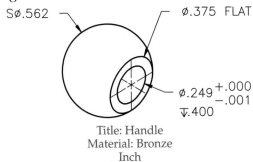

SØ.562   Ø.375 FLAT

Ø.249 +.000 −.001

▽.400

Title: Handle
Material: Bronze
Inch

3. Create the drawing as shown and save it as P20-3.

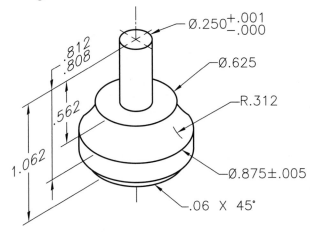

Ø.250 +.001 −.000

Ø.625

R.312

Ø.875±.005

.06 X 45°

.812
.808

.562

1.062

ALL OTHER THREE PLACE DECIMALS ±.010

4. Create the drawing as shown and save it as P20-4.

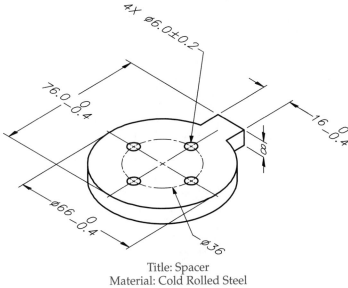

4X Ø6.0±0.2

76.0 −0.4

16 −0.4

8

Ø66 −0.4

Ø36

Title: Spacer
Material: Cold Rolled Steel
Metric

▼ Intermediate

5. Create the drawing as shown and save it as P20-5.

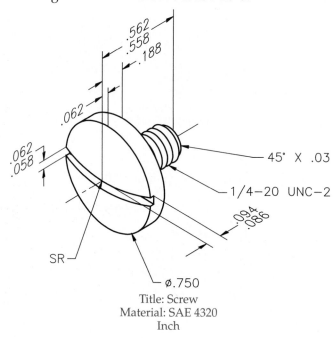

.562
.558
.188
.062
.062
.058
45° X .03
1/4–20 UNC–2
.094
.086
SR
Ø.750

Title: Screw
Material: SAE 4320
Inch

6. This object is shown as a section for clarity. Do not draw a section. Save the drawing as P20-6.

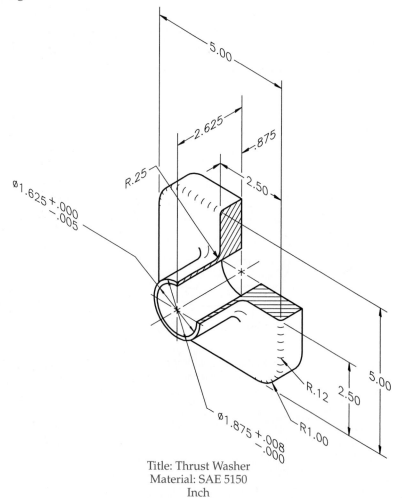

5.00
2.625
.875
R.25
2.50
Ø1.625 +.000 −.005
R.12
5.00
2.50
Ø1.875 +.008 −.000
R1.00

Title: Thrust Washer
Material: SAE 5150
Inch

▼ Advanced

7. Create the drawing as shown and save it as P20-7.

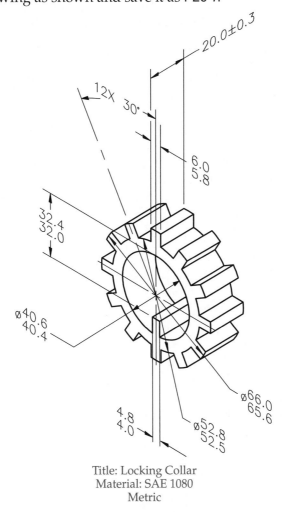

Title: Locking Collar
Material: SAE 1080
Metric

8. Draw the vise clamp and add dimensions as shown. Save the drawing as P20-8.

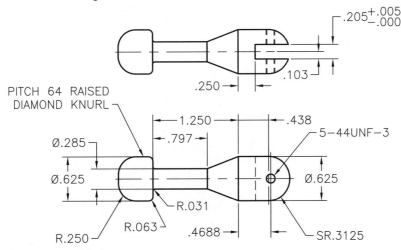

For Problems 9-15, use the isometric drawing provided to create a multiview orthographic drawing for the part. Include only the views necessary to fully describe the objects. Add all required dimensions according to the ASME standards. Use GD&T tools and practices as described in the GD&T with AutoCAD supplement available in the Supplemental Material for this chapter on the Student Web site.

9. Create the drawing as shown. Untoleranced dimensions are ±0.3. Save the drawing as P20-9.

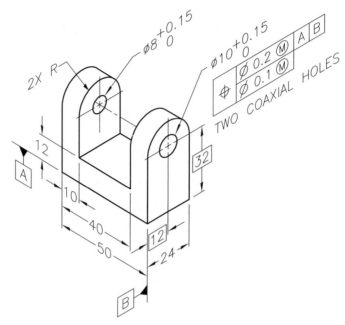

10. Open P20-6 and save the file as P20-10. The P20-10 file should be active. Add the geometric tolerancing applications shown. Untoleranced dimensions are ±.02 for two-place decimal precision and ±.005 for three-place decimal precision.

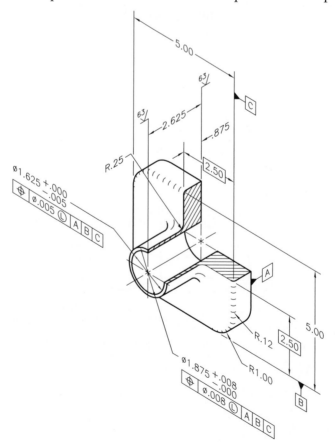

11. Open P20-4 and save the file as P20-11. The P20-11 file should be active. Add the geometric tolerancing applications shown. Untoleranced dimensions are ±0.5.

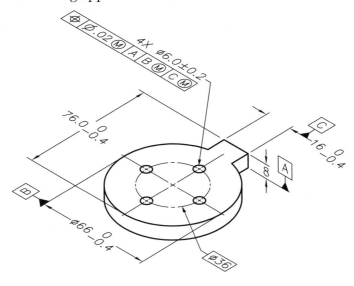

12. Open P20-7 and save the file as P20-12. The P20-12 file should be active. Add the geometric tolerancing applications shown.

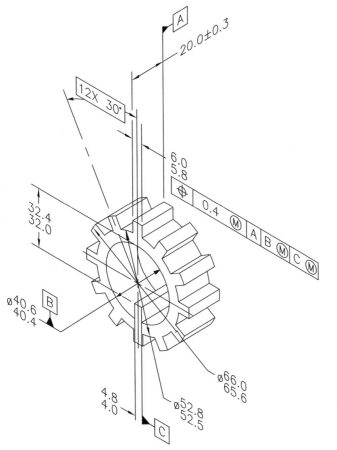

13. This problem is shown with a full section for clarity. You do not need to draw a section. Untoleranced dimensions are ±.010. Save the drawing as P20-13.

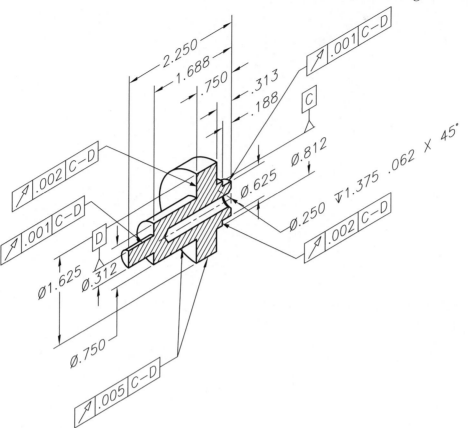

14. Open P19-21 and save the file as P20-14. The P20-14 file should be active. Add the geometric tolerancing applications shown.

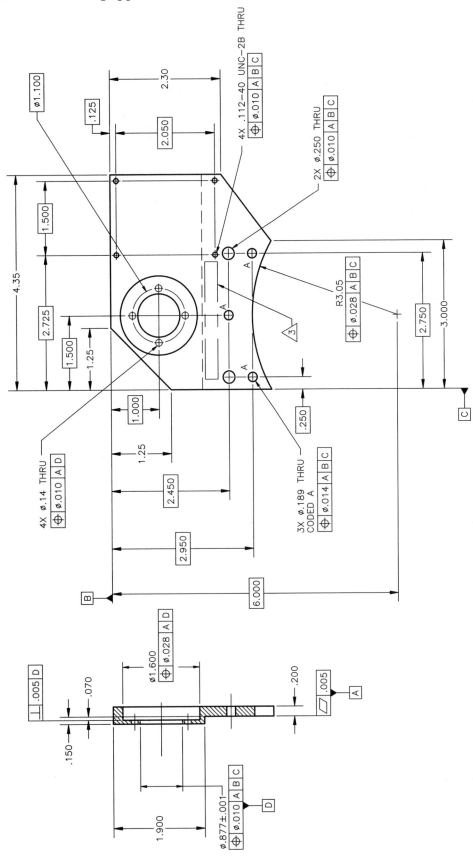

15. The problem is shown with a half section for clarity. You do not need to draw a section. Untoleranced dimensions are ±.010. Save the drawing as P20-15.

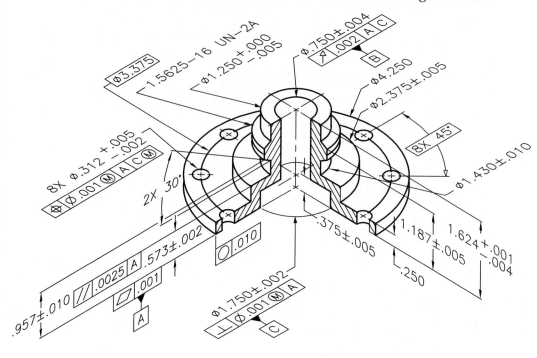

Editing Dimensions

25

Learning Objectives

After completing this chapter, you will be able to do the following:

- ✓ Describe and control associative dimensions.
- ✓ Control the appearance of existing dimensions and dimension text.
- ✓ Update dimensions to reflect the current dimension style.
- ✓ Override dimension style settings and match dimension properties.
- ✓ Change dimension line spacing and alignment.
- ✓ Break dimension, extension, and leader lines.
- ✓ Create inspection dimensions.
- ✓ Edit existing multileaders.

You can use most typical object editing tools to modify and arrange dimensions. In addition to these modification techniques, there are specific considerations and tools for adjusting dimension objects. This chapter describes a variety of useful techniques for editing dimension placement, value, and appearance.

Associative Dimensioning

A dimension is a group of elements treated as a single object. When you select a dimension to edit, all dimension elements highlight. If you use the **ERASE** tool, for example, you pick the dimension as a single object and erase all elements at once. Additionally, dimensions reference objects or points. This allows you to use editing tools such as **STRETCH**, **MOVE**, **ROTATE**, and **SCALE** to make changes to the drawing that also affect dimensions. See **Figure 21-1**.

An *associative dimension* forms by default when you create a dimension using object selection, or by picking points using object snaps. For example, when you dimension the ∅1.0 circle in **Figure 21-1** using the **DIMDIAMETER** tool, and then change size of the circle to ∅2.00, the diameter dimension adapts to correctly reflect the size of the modified circle. Create associative dimensions when possible and practical by selecting objects or using appropriate object snaps. This allows dimensions to relate best to object size, and often makes revisions easier.

associative dimension: Dimension associated with an object. The dimension value updates automatically when the object changes.

Figure 21-1.
An example of typical drawing revisions. The dimensions adjust with the changes, and dimension values update to reflect the size and location of the modified geometry.

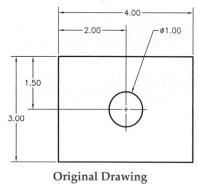

Original Drawing

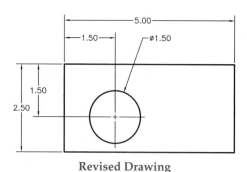

Revised Drawing

nonassociative dimension: A dimension linked to point locations, not an object; does not update when the object changes.

A *nonassociative dimension* forms when you create a dimension by selecting points without using object snaps. A nonassociative dimension is still a single object that updates when you make changes to the dimension, such as stretching the extension line origin. Nonassociative dimensions are appropriate for applications in which using associative dimensions would result in dimensioning difficulty or unacceptable standards. When using nonassociative dimensions, remember to edit the dimension with the object it dimensions, or adjust the dimension after the object changes.

NOTE

Refer to the **Associative** property in the **General** area of the **Properties** palette to determine whether a dimension is associative.

PROFESSIONAL TIP

Dimension tools allow you to dimension a drawing, but they do not control object size and location. Chapter 22 explains how to use dimensional constraint tools to control object size and location. If you anticipate creating a drawing with features that may require significant or constant change, you may want to use dimensional constraints instead of, or in addition to, traditional dimensioning tools.

Associating Dimensions with Objects

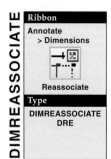

Ribbon
Annotate
> Dimensions

Reassociate

Type
DIMREASSOCIATE
DRE

DIMREASSOCIATE

Dimensions associate with objects by default when you use object selection or pick points using object snaps. To deactivate associative dimensioning for new objects, access the **Options** dialog box, and deselect the **Make new dimensions associative** check box, in the **Associative Dimensioning** area of the **User Preferences** tab.

Often the easiest way to convert a nonassociative dimension to an associative dimension is to select the dimension for grip editing and stretch the appropriate grip to the corresponding object snap point. You can also make the conversion using the **DIMREASSOCIATE** tool. Select the dimension to associate with an object. An X marker appears at a dimension origin, such as the origin of a linear dimension extension line or the center of a radial dimension. Select a point on an object to associate with the marker location. Repeat the process to locate the second object point for the first extension line, if required.

AutoCAD and Its Applications—Basics

Use the **Next** option to advance to the next definition point. Use the **Select object** option to select an object to associate with the dimension. The extension line endpoints automatically associate with the object endpoints.

Type
DIMDISASSOCIATE

NOTE

To disassociate a dimension from an object, use grip editing to stretch an appropriate grip point away from the associated object, or use the **DIMDISASSOCIATE** tool.

Dimension Definition Points

Definition points, or *defpoints*, occur as a dimension element whenever you create a dimension. Use the **Node** object snap to snap to a definition point. If you select an object to edit and want to include dimensions in the edit, you must include the definition points in the selection set. AutoCAD automatically creates a Defpoints layer and places definition points on the layer. By default, the Defpoints layer does not plot. You can only plot definition points if you rename and then set the Defpoints layer to plot. Definition points display even if you turn off or freeze the Defpoints layer.

definition points (defpoints):
The points used to specify the dimension location and the center point of the dimension text.

Exercise 21-1

Access the Student Web site (www.g-wlearning.com/CAD) and complete Exercise 21-1.

CAUTION

A dimension is a single object even though it consists of extension lines, a dimension line, arrowheads, and text. You may be tempted to explode the dimension using the **EXPLODE** tool to modify individual dimension elements. You should rarely, or never, explode dimensions. Exploded dimensions lose their layer assignments and their association to related features and dimension styles.

PROFESSIONAL TIP

You can edit individual dimension properties without exploding a dimension by using dimension shortcut menu options or the **Properties** palette to create a dimension style override.

Dimension Editing Tools

Attention to dimension style settings and careful placement of dimensions allow you to dimension a drawing according to specific drafting standards. As a drawing process evolves and design changes occur, you will find it necessary to make changes to dimensioned objects and dimensions. Dimension-specific editing tools and techniques are available to help you adjust dimensions as necessary.

Figure 21-2.
Select a dimension and then right-click to access this shortcut menu.

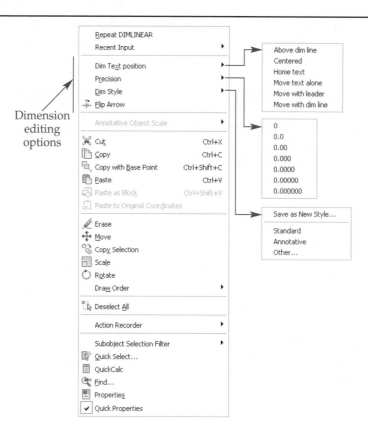

Dimension editing options

Dimension Shortcut Menu Options

Select a dimension and then right-click to display the shortcut menu shown in **Figure 21-2.** The **Dim Text position** cascading menu provides options for adjusting the dimension value location. Pick **Above dim line** to move the dimension text above the dimension line. Select **Centered** to center the dimension text on the dimension line. Pick **Home text** to reposition the text at its original position. **Move text alone** allows you to move the text away from the dimension line. **Move with leader** allows you to move the text away from the dimension line with a leader attaching the text to the dimension line. **Move with dimension line** allows you to move the text, but maintain its alignment with the dimension line.

The options in the **Precision** cascading menu allow you to adjust the number of decimal places displayed in a dimension text value. This is often the easiest way to specify an alternative tolerance. The **Dim Style** cascading menu allows you to create a new dimension style based on the properties of the selected dimension. You can also apply a different dimension style to the dimension.

The **Flip Arrow** option allows you to flip the direction of a dimension arrowhead to the opposite side of the extension line or object that the arrow touches. For example, if the arrowheads and dimension value are crowded inside the extension lines, you can flip the arrowheads to the outside of the extension lines to make the dimension easier to read. If the selected dimension includes two arrowheads, only one of the arrowheads flips, allowing you to control the arrowheads independently. The arrowhead that flips is the one closest to the point you pick when you select the dimension (not the right-click point).

Assigning a Different Dimension Style

To assign a different dimension style to existing dimensions, you can use the options from the **Dim Style** cascading menu of the dimension shortcut menu. A second option is to pick the dimensions to change and select a different dimension style from the **Dimension Style** drop-down list on the **Home** and **Annotation** ribbon tabs. A third option is to select the dimensions to change and choose a new dimension style from the **Quick Properties** panel or the **Properties** palette.

Another technique for changing the dimension style of existing dimensions is to use the **Update** dimension tool. Before you access the **Update** dimension tool, be sure the current dimension style is the dimension style you want to assign to existing dimensions. Then access the **Update** dimension tool and pick the dimensions to change to the current style.

Ribbon
Annotate
> Dimensions

Update

Editing the Dimension Value

The **DDEDIT** tool allows you to add a prefix or suffix to the text or edit the dimension text format. For example, use the **DDEDIT** tool to dimension a linear diameter if you forget to use the **Mtext** or **Text** option of the **DIMLINEAR** tool to add a diameter symbol to the value. See **Figure 21-3**. Access the **DDEDIT** tool and select a dimension to enter the multiline text editor with the current dimension value highlighted. Add to or modify the dimension text and then close the text editor. The tool continues, allowing you to edit other text if necessary.

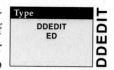

Type
DDEDIT
ED

DDEDIT

> ### CAUTION
>
> You can replace the highlighted text, which represents the dimension value, with numeric values. However, this action disassociates the dimension value with the object or points it dimensions. Therefore, leave the default value intact whenever possible.

Exercise 21-2

Access the Student Web site (www.g-wlearning.com/CAD) and complete Exercise 21-2.

Editing Dimension Text Placement

Proper dimensioning practice requires dimensions that are clear and easy to read. This sometimes involves moving the text of adjacent dimensions to separate the text elements. See **Figure 21-4**. You can use the dimension shortcut menu to adjust dimension text position, but often the quickest method is to use grips. Select the dimension, pick the dimension text grip, and stretch the text to the new location. AutoCAD automatically reestablishes the break in the dimension line when you pick the new location.

Using the DIMTEDIT Tool

The **DIMTEDIT** tool allows you to change the placement and orientation of existing dimension text. Access the **DIMTEDIT** tool and select the dimension to alter. Drag the dimension text to a new location and pick to move the text and automatically reestablish the break in the dimension line.

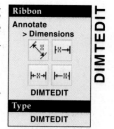

Ribbon
Annotate
> Dimensions

DIMTEDIT

Type
DIMTEDIT

DIMTEDIT

Figure 21-3.
Using the **DDEDIT** tool to add a diameter symbol to an existing dimension.

— 3.250 —

Original

— ⌀3.250 —

Diameter Symbol Added

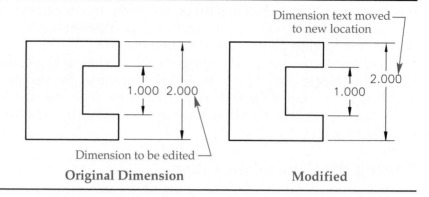

Figure 21-4.
Staggering dimension text for improved readability.

Dimension text moved to new location

2.000

1.000

1.000 2.000

Dimension to be edited

Original Dimension   **Modified**

The **DIMTEDIT** tool also provides options for relocating dimension text to a specific location and for rotating the text. However, it is usually quicker to select the appropriate button from the expanded **Dimensions** panel of the **Annotation** ribbon tab or select a similar option from the dimension shortcut menu. Select **Text Angle (Angle)** to rotate the dimension text. **Left Justify (Left)** moves horizontal text to the left and vertical text down. **Center Justify (Center)** centers the dimension text on the dimension line. **Right Justify (Right)** moves horizontal text to the right and vertical text up. Select the **Home** option to relocate text back to its original position. **Figure 21-5** shows the result of using each **DIMTEDIT** tool option.

PROFESSIONAL TIP

You may want to activate the **Place text manually** check box in the **Fit** tab of the **New** (or **Modify**) **Dimension Style** dialog box to provide greater flexibility for the initial placement of dimensions.

Exercise 21-3

Access the Student Web site (www.g-wlearning.com/CAD) and complete Exercise 21-3.

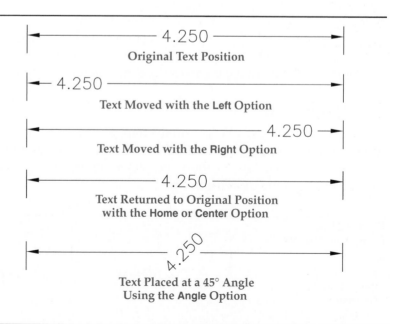

Figure 21-5.
A comparison of the **DIMTEDIT** tool options.

4.250
Original Text Position

4.250
Text Moved with the **Left** Option

4.250
Text Moved with the **Right** Option

4.250
Text Returned to Original Position with the **Home** or **Center** Option

4.250
Text Placed at a 45° Angle Using the **Angle** Option

Using the DIMEDIT Tool

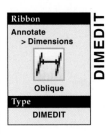

The **DIMEDIT** tool, not to be confused with the **DIMTEDIT** tool, provides **Home** and **Rotate** options that function the same as the **Home** and **Angle** options of the **DIMTEDIT** tool. The **New** option is similar to using the **DDEDIT** tool to edit dimension text values. When you activate the **New** option, the multiline text editor displays with a 0.0000 (depending on units and precision) value highlighted. The highlighted 0.0000 value represents the associated dimension value. Add to or modify the dimension text and then close the text editor.

Creating Oblique Extension Lines

The **Oblique** option is unique to the **DIMEDIT** tool and allows you to change the extension line angle without affecting the associated dimension value. In **Figure 21-6A**, for example, the curve on the left is dimensioned using linear dimensions. On the right, the extension lines of two of the linear dimensions have been made oblique to adjust the placement of dimensions when space is limited. **Figure 21-6B** shows an example of oblique extension lines used to orient extension lines properly with the angle of the stairs in a stair section. Notice that the associated values and orientation of the dimension lines in these examples do not change.

To create oblique extension lines, first dimension the object using the **DIMLINEAR** tool as appropriate, even if dimensions are crowded or overlap. Then access the **Oblique** option of the **DIMEDIT** tool. The quickest way to access the **Oblique** option is to pick the corresponding button from the expanded **Dimensions** panel of the **Annotation** ribbon

Figure 21-6.
Drawing dimensions with oblique extension lines.

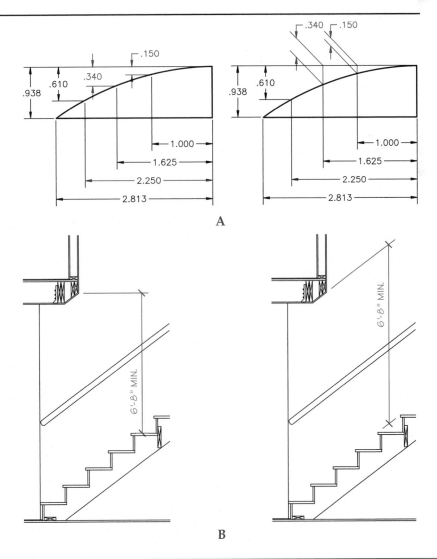

tab. Once you activate the **Oblique** option, pick the dimensions to redraw at an oblique angle. Next, specify the obliquing angle. Plan carefully to make sure you enter the correct obliquing angle. Obliquing angles originate from 0° East and revolve counter-clockwise. Enter a specific value or pick two points to define the obliquing angle.

NOTE

You can use the **Oblique** option of the **DIMEDIT** tool to dimension oblique and isometric drawings, as described in Chapter 24.

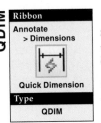

Exercise 21-4

Access the Student Web site (www.g-wlearning.com/CAD) and complete Exercise 21-4.

Editing Dimensions with the QDIM Tool

The **QDIM** tool includes options for changing the arrangement of existing dimensions, adding a dimension, and removing a dimension. The **Continuous** option allows you to change a selected group of dimensions to chain dimensions. See **Figure 21-7A**. The **Baseline** option allows you to change a selected group of dimensions to datum dimensions. **Figure 21-7B** shows an example of using the **Baseline** option to change the dimensioning arrangement from chain to datum.

You can use the **Edit** option to add dimensions to, or remove dimensions from, a selected group of existing dimensions and automatically reorder the group. Use the **Add** function of the **Edit** option to add a dimension, or use the **Remove** function to remove a dimension. For example, to add the dimension shown in **Figure 21-7C**, access the **QDIM** tool, select all of the dimensions in the group to change, and right-click or press [Enter] or the space bar. Then activate the **Edit** option and the **Add** function. Pick the location or feature where the new dimension is to originate and right-click or press [Enter] or the space bar. Finally, pick a location for the baseline dimension arrangement.

The dimensions automatically realign after you pick a location for the arrangement. You do not have to pick all of the dimensions in the group. However, if you do not pick the entire group, you must carefully select the location. In addition, the spacing for the edited dimension and the dimensions in the group that you do not select may be inconsistent.

Figure 21-7.
You can use the **QDIM** tool to change existing dimension arrangements and add or remove dimensions.

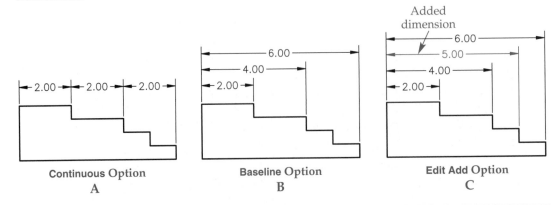

Exercise 21-5

Access the Student Web site (www.g-wlearning.com/CAD) and complete Exercise 21-5.

Overriding Existing Dimension Style Settings

Generally, you should set up one or more dimension styles to perform specific dimensioning tasks. However, in some situations, a few dimensions require specific settings that an existing dimension style cannot provide. These situations may be too few to merit creating a new style. For example, assume you have the value for **Offset from origin** set at .063, which conforms to ASME standards. However, three dimensions in your final drawing require a 0 **Offset from origin** setting. For these dimensions, you can perform a *dimension style override*.

dimension style override: A temporary alteration of settings for the dimension style that does not actually modify the style.

Dimension Style Overrides for Existing Dimensions

The **Properties** palette is the most effective tool to use for overriding the dimension style settings of existing dimensions. The **Properties** palette divides dimension properties into several categories. See **Figure 21-8**. To change a property, access the proper category, pick the property to highlight, and adjust the corresponding value. Most changes made using the **Properties** palette are overrides to the dimension style for the selected dimension. The changes do not alter the original dimension style and do not apply to new dimensions.

NOTE

The **Quick Properties** panel provides a limited number of dimension properties and style overrides.

Dimension Style Overrides for New Dimensions

Use the **Dimension Style Manager** to override the dimension style for dimensions you are about to create. An example of an override is including a text prefix for a few of the dimensions in a drawing. Select the dimension style to override from the **Styles** list and then pick the **Override** button to open the **Override Current Style** dialog box. The **Override** button is only available for the current style. The **Override Current Style** dialog box includes the same tabs as the **New Dimension Style** and **Modify Dimension Style** dialog boxes. Make the necessary changes and pick the **OK** button. Once you create an override, it is current and appears as a branch under the original style labeled **<style overrides>**. Close the **Dimension Style Manager** and draw the needed dimensions.

To clear style overrides, return to the **Dimension Style Manager** and set any other style current. The override settings are lost when any other style, including the parent, is set current. To incorporate the overrides into the overridden style, right-click on the **<style overrides>** name and select **Save to current style**. To save the changes to a new style, pick the **New...** button. Then select **<style overrides>** in the **Start With** drop-down list in the **Create New Dimension Style** dialog box. In the **New Dimension Style** dialog box, pick **OK** to save the overrides as a new style.

Figure 21-8.
The **Properties** palette allows you to edit dimension properties and create a dimension style override.

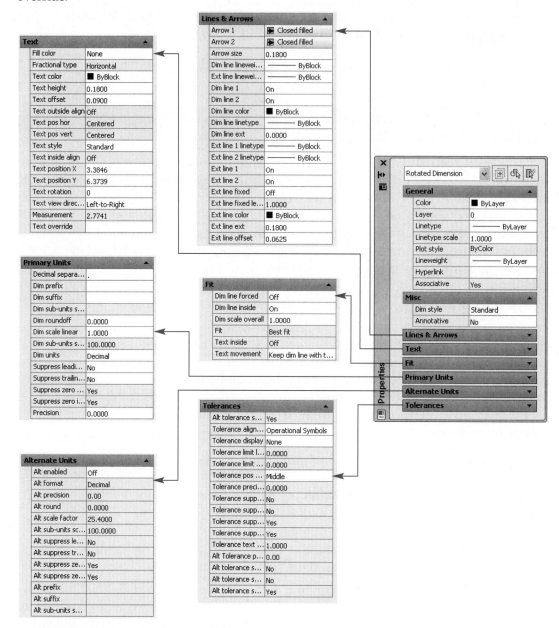

Carefully evaluate the dimensioning requirements in a drawing before performing a style override. It may be better to create a new style. For example, if a number of the dimensions in the current drawing require the same overrides, generating a new dimension style is a good idea. If only one or two dimensions need the same changes, performing an override may be more productive.

Exercise 21-6

Access the Student Web site (www.g-wlearning.com/CAD) and complete Exercise 21-6.

Using the MATCHPROP Tool

The **MATCHPROP** tool provides an effective method of applying the properties of a selected dimension to existing dimensions in the same drawing or a different drawing. Access the **MATCHPROP** tool, pick the source dimension that has the desired properties, and then pick the destination dimensions to change. Press [Enter] to update all of the destination dimensions to reflect the properties of the source dimension.

For the **MATCHPROP** tool to work with dimensions, the **Dimension** setting must be active. You can check this after you select the source object. When the Current active settings: prompt appears, Dim should appear with the other settings. If Dim does not appear when you are prompted to select a destination object, use the **Settings** option to display the **Property Settings** dialog box. Activate the **Dimension** check box in the **Special Properties** area and pick **OK**. Then select the destination dimensions.

Exercise 21-7

Access the Student Web site (www.g-wlearning.com/CAD) and complete Exercise 21-7.

Using the DIMSPACE Tool

The amount of space between a drawing view and the first dimension line, and the space between dimension lines, varies depending on the drawing and industry or company standard. ASME standards recommend a minimum spacing of .375″ (10 mm) from a drawing feature to the first dimension line and a minimum spacing of .25″ (6 mm) between dimension lines. A minimum spacing of 3/8″ is common for architectural drawings. These minimum recommendations are generally less than the spacing required by actual company or school standards.

Typically, the spacing between dimension lines is equal, and chain dimensions align. See **Figure 21-9.** As a result, it is important to determine the correct location and spacing of dimension lines. However, you can adjust dimension line spacing and alignment after you place dimensions. This is a common requirement when there is a need to increase or decrease the space between dimension lines, such as when the drawing scale changes, or when dimensions are spaced unequally or misaligned.

The **STRETCH** and **DIMTEDIT** tools or grips are common methods for adjusting the location and alignment of dimension lines. You must determine the exact location or amount of stretch applied to each dimension line before using these tools. An alternative is to use the **DIMSPACE** tool, which allows you to adjust the space equally between dimension lines or align dimension lines.

Figure 21-9.
Correct drafting practice requires equal space and alignment between dimension lines for readability.

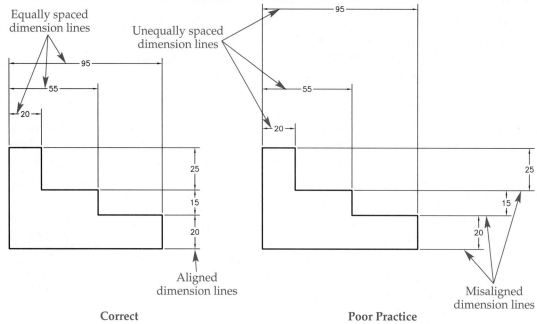

Correct

Poor Practice

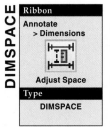
Access the **DIMSPACE** tool and select the *base dimension*, followed by each dimension to space. Right-click or press [Enter] or the space bar to display the Enter value or [Auto]: prompt. Enter a value to space the dimension lines equally. For example, enter .5 to space the selected dimension lines .5″ apart. Enter a value of 0 to align the dimensions. See **Figure 21-10.** Use the **Auto** option to space dimension lines using a value that is twice the height of the dimension text.

base dimension:
The dimension line that remains in the same location, with which other dimension lines are spaced or aligned.

NOTE

You can use the **DIMSPACE** tool to space and align linear and angular dimensions.

Figure 21-10.
Using the **DIMSPACE** tool to space and align dimension lines correctly.

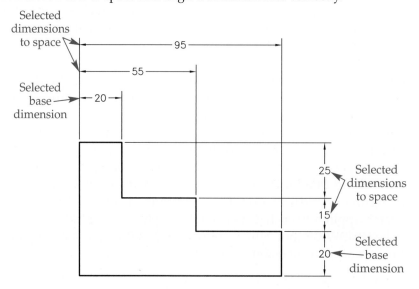

Exercise 21-8

Access the Student Web site (www.g-wlearning.com/CAD) and complete Exercise 21-8.

Using the DIMBREAK Tool

Drafting standards state that when dimension, extension, or leader lines cross a drawing feature or another dimension, neither line is broken at the intersection. See **Figure 21-11.** However, you can use the **DIMBREAK** tool to create breaks if desired.

Access the **DIMBREAK** tool and select the dimension to break. This is the dimension that contains the dimension, extension, or leader line that you want to break across an object. If you pick a single dimension to break, the Select object to break dimension or [Auto/Restore/Manual]: prompt appears.

The **Auto** option of the **DIMBREAK** tool, which is the default, breaks the dimension, extension, or leader line at the selected object. The **Dimension Break** setting of the current dimension style controls the break size. You can pick additional objects if necessary to break the dimension at additional locations. See **Figure 21-12.** Use the **Manual** option to define the size of the break by selecting two points along the dimension, extension, or leader line, instead of using the break size set in the current dimension style. Activate the **Restore** option to remove an existing break created using the **DIMBREAK** tool.

Another technique is to use the **Multiple** option to select more than one dimension. The **Break** and **Restore** options are available when you use the **Multiple** option to select multiple dimensions. Right-click or press [Enter] or the space bar after you select the dimensions to display the Enter and option [Break/Restore]: prompt. Select the **Break** option to break the selected dimension, extension, or leader lines everywhere they intersect another object. Use the **Restore** option to remove any existing breaks added to the selected dimensions using the **DIMBREAK** tool.

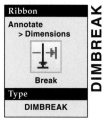

Ribbon
Annotate
> Dimensions

Break

Type
DIMBREAK

DIMBREAK

Figure 21-11.
Drafting standards state that when dimension, extension, or leader lines cross a drawing feature or another dimension, the line is not broken at the intersection.

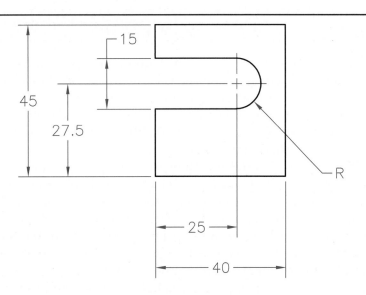

Figure 21-12.
Use the **DIMBREAK** tool to break dimension, extension, or leader lines when they cross an object. *Caution:* This example violates ASME standards and is for reference only. Extension and leader lines do not break over object lines, but some drafters prefer to break an extension line when it crosses a dimension line.

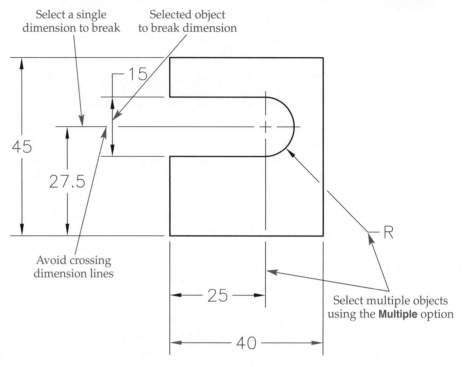

Select a single dimension to break

Selected object to break dimension

Avoid crossing dimension lines

Select multiple objects using the **Multiple** option

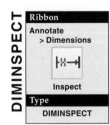

Exercise 21-9

Access the Student Web site (www.g-wlearning.com/CAD) and complete Exercise 21-9.

Creating Inspection Dimensions

Inspections and tests occur throughout the design and manufacturing of a product. These tests help ensure the correct size and location of product features. In some cases, size and location dimensions include information about how frequently a test on the dimension occurs for consistency and tolerance during the manufacturing process. See **Figure 21-13.** You can use the **DIMINSPECT** tool to add inspection information to most types of existing dimensions.

Access the **DIMINSPECT** tool to display the **Inspection Dimension** dialog box, shown in **Figure 21-14.** Pick the **Select dimensions** button and choose the dimensions to receive inspection information. You can select multiple dimensions, although the same inspection specifications apply to each. Define the shape of the inspection dimension frame by picking the appropriate radio button in the **Shape** area. The inspection dimension contains the inspection label, the dimension value, and the inspection rate. Select the **None** option to omit frames around values.

To include a label, pick the **Label** check box and type the label in the text box. The label is located on the left side of the inspection dimension and identifies the specific dimension. The inspection dimension shown in **Figure 21-13** is labeled A. The dimension frame houses the dimension value specified when you create the

DIMINSPECT

Ribbon
Annotate
> Dimensions

Inspect

Type
DIMINSPECT

Figure 21-13.
An example of an inspection dimension added to a part drawing. This example shows an angular shape with a label, dimension, and inspection rate frame.

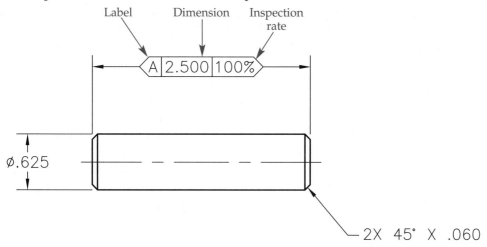

Figure 21-14.
The **Inspection Dimension** dialog box allows you to add inspection information to existing dimensions.

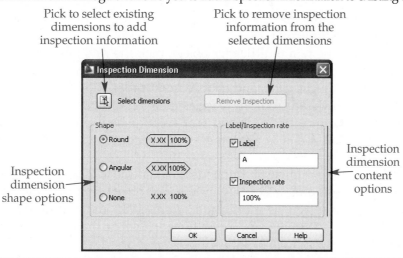

dimension. The length of the part shown in **Figure 21-13** is 2.500, as specified when using the **DIMLINEAR** tool. The **Inspection rate** check box is selected by default. Enter a value in the text box to indicate how often a test should be performed on the dimension. The inspection rate for the dimension shown in **Figure 21-13** is 100%. This rate can have different meanings depending on the application. In this example, the inspection rate of 100% means that the length of the part must be checked for tolerance every time the part is added to an assembly.

To remove an inspection dimension, access the **DIMINSPECT** tool, pick the **Select dimensions** button in the **Inspection Dimension** dialog box, and choose the dimensions that contain the inspection information to remove. Right-click or press [Enter] or the space bar to return to the **Inspection Dimension** dialog box, and pick the **Remove Inspection** button to return the dimension to its condition prior to adding the inspection content.

Exercise 21-10

Access the Student Web site (www.g-wlearning.com/CAD) and complete Exercise 21-10.

Editing Multileaders

The methods to edit multileaders are similar to those you use to edit dimensions. Use editing tools such as **STRETCH, MOVE, ROTATE**, and **SCALE** as needed. Grips are particularly effective for adjusting the location of leader elements. Use the grips at the end of leader lines to relocate the arrowhead. Use the grips at each end of a landing to stretch the landing, but be careful not to violate drafting standards. Use the grips positioned at the middle of a landing or with leader content to relocate content.

To make changes to multileader text, double-click on the text to re-enter the multiline text editor. Use the **Properties** palette or **Quick Properties** panel to override specific multileader properties. You can also use the **MATCHPROP** tool. In addition to these general multileader editing techniques, specific tools allow you to add and remove leader lines, and space, align, and group multileader objects. Select a multileader and right-click to display a shortcut menu with specific options for adjusting multileaders and assigning a different multileader style.

Adding and Removing Multiple Leader Lines

The **MLEADEREDIT** tool provides options for adding leader lines to, and removing leader lines from, an existing multileader object. Multiple leaders are not a recommended ASME standard, but they are appropriate for some applications, such as welding symbols. See **Figure 21-15.**

To add a leader line to a multileader object, pick the **Add Leader** button from the ribbon and select the multileader that will receive the new leader line. You can also select the multileader, right-click, and choose **Add Leader**. Pick a location for the additional leader line arrowhead. You can place as many additional leader lines as needed without accessing the tool again. When you are finished, press [Enter], [Esc] or the space bar or right-click and select **Enter.** All of the leader lines group to form a single object.

The quickest way to remove an unneeded leader line is to pick the **Remove Leader** button from the ribbon and select the multileader object that includes the leader to remove. You can also select the multileader, right-click, and choose **Remove Leader.** Select the leader lines to remove and press [Enter], [Esc] or the space bar or right-click and select **Enter.**

> **NOTE**
>
> If you type **MLEADEREDIT** to access the tool, you must activate the **Remove leaders** option to remove leader lines.

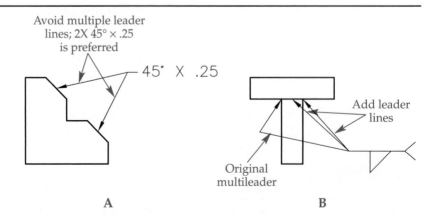

Figure 21-15. Applications of multiple leader lines. A—Do not use multiple leader lines in mechanical applications. B— Welding applications often use multiple leader lines.

To adjust the properties of a specific leader line in a group of leaders attached to the same content, hold down [Ctrl] and pick the leader to modify. Then access the **Properties** palette. Options specific to the selected leader appear, and all other properties are filtered out.

Exercise 21-11

Access the Student Web site (www.g-wlearning.com/CAD) and complete Exercise 21-11.

Aligning Multileaders

An advantage of using multileaders is the ability to space and align leaders in an easy-to-read pattern. It is good practice to determine the correct location and spacing of the leaders before placement, but you can also adjust leader spacing and alignment after you add leaders. This is a common requirement when there is a need to increase or decrease the space between leader lines, such as when the drawing scale changes, or when leaders are spaced unequally or misaligned. See **Figure 21-16.**

The **STRETCH** tool or grips are common methods for adjusting the location and alignment of leaders. You must determine the exact location of or amount of stretch applied to each leader before using these tools. An alternative is to use the **MLEADERALIGN** tool, which allows you to align and adjust the space between leaders.

Access the **MLEADERALIGN** tool and select the leaders to space and align. You can use the **MLEADERALIGN** tool to adjust the location of a single leader in reference to another leader, but for most applications, you should select several leaders. Select each

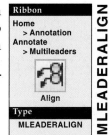

Ribbon
Home
> Annotation
Annotate
> Multileaders
Align
Type
MLEADERALIGN

MLEADERALIGN

Figure 21-16.
Leaders that are equally spaced and aligned improve drawing readability.

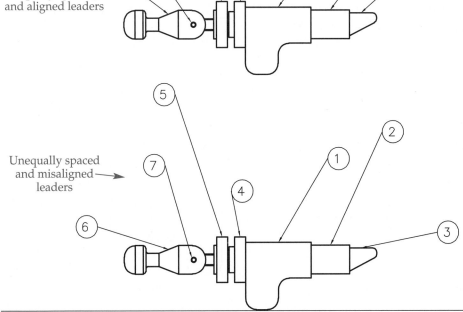

leader to space or align and right-click or press [Enter] or the space bar. When a prompt asks you to select the multileader to align to, you can activate **Options** to change the multileader alignment.

Using the Current Leader Spacing

Apply the **Use current spacing** option to align and space the selected leaders equally according to the distance between one of the selected leaders and the next closest leader. Select the multileader with which all other leaders are aligned and spaced. Then specify the direction of the leader arrangement by entering or picking a point. The space between leaders is maintained if possible, depending on the selected direction. See **Figure 21-17**.

Using the Distribute Option

Select the **Distribute** option to align and distribute the leaders, or place them at equally spaced locations between two points. The first point you pick identifies the location of one of the leaders and determines where distribution begins. The second point you pick identifies the location of the last leader. All other leaders distribute equally between the two points. Leaders align with the first point. See **Figure 21-18**.

Figure 21-17.
Using the **Use current spacing** option to align and equally space leaders.

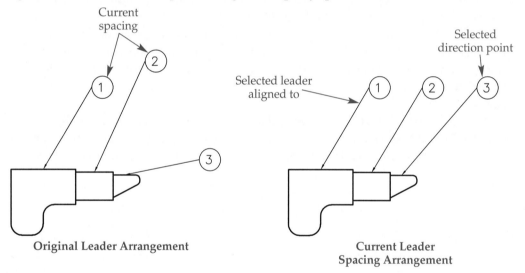

Figure 21-18.
Using the **Distribute** option to align and equally space leaders. In this example, horizontally aligned points are used.

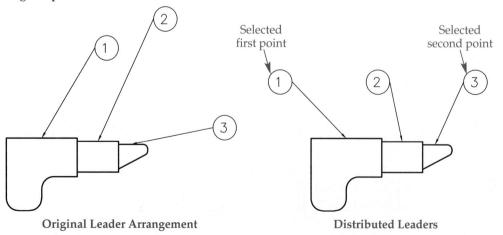

Making Leader Segments Parallel

Use the **make leader segments Parallel** option to make all the selected leader lines parallel to one of the selected leader lines. Select the existing leader to keep in the same location and at the same angle. All other leaders become parallel to this selection. The length of each leader line, except for the leader aligned to, increases or decreases in order to become parallel with the first leader. See **Figure 21-19**.

Specifying the Leader Spacing

Choose the **Specify spacing** option to align and equally space the selected leaders according to the distance, or clear space, between the extents of the content of each leader. Select the multileader with which all other leaders align and space. Finally, specify the direction of the leader arrangement by entering or picking a point. See **Figure 21-20**.

Exercise 21-12

Access the Student Web site (www.g-wlearning.com/CAD) and complete Exercise 21-12.

Figure 21-19.
Using the **make leader segments Parallel** option to make leader lines parallel to each other.

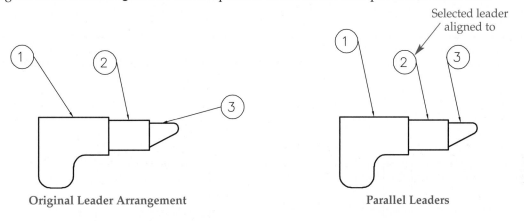

Figure 21-20.
Using the **Specify spacing** option to align and equally space leaders.

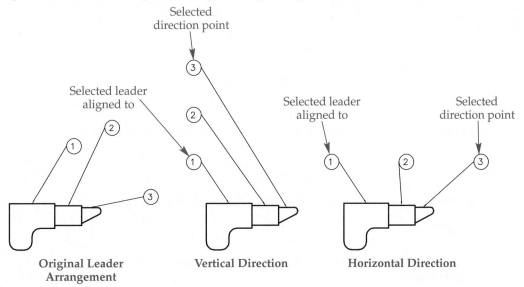

Grouping Multileaders

MLEADERCOLLECT

Ribbon
Home
> Annotation
Annotate
> Multileaders

Collect
Type
MLEADERCOLLECT

You can group separate multileaders created using a **Block** multileader content style using a single leader line. This practice is common when adding *balloons* to assembly drawings. *Grouped balloons* allow you to identify closely related clusters of assembly components, such as a bolt, washer, and nut. See **Figure 21-21.** Use the **MLEADERCOLLECT** tool to group multiple existing leaders together using a single leader line.

Access the **MLEADERCOLLECT** tool and select the leaders to group. The order in which you select the leaders determines how the leaders group. Select leaders in a sequential order, ending with the leader line to keep. The options illustrated in **Figure 21-22** become available after you select the leaders.

Select the **Horizontal** option to align the grouped content horizontally or the **Vertical** option to align the grouped content vertically. Pick a point to locate the grouped leader. Select the **Wrap** option to wrap the grouped content to additional lines as needed when the number of items exceeds a specified width or quantity. Enter the width at the Specify width prompt, or use the **Number** option to enter a quantity not to exceed before the grouped leaders wrap. Then pick a point to locate the grouped leader.

balloons: Circles that contain a number or letter to identify the part and correlate it to a parts list or bill of materials. Balloons connect to a part with a leader line.

grouped balloons: Balloons that share the same leader, which typically connects to the most obviously displayed component.

> **NOTE**
>
> You can only use the **MLEADERCOLLECT** tool to group symbols attached to leaders created using the **Block** content style.

Figure 21-21.
An example of grouped balloons identifying closely related parts. Often some parts or features are hidden.

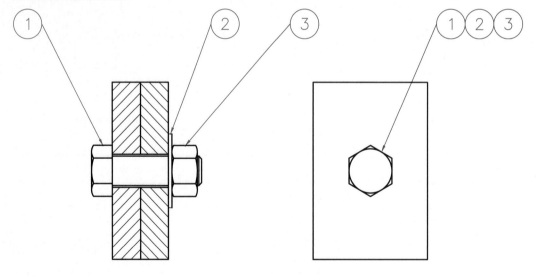

Figure 21-22.
Examples of options for grouping leaders using the **MLEADERCOLLECT** tool.

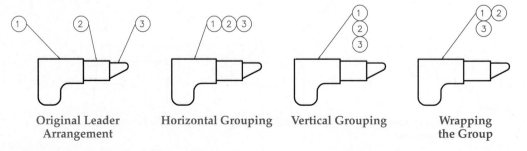

Original Leader Arrangement   Horizontal Grouping   Vertical Grouping   Wrapping the Group

Chapter Test

Answer the following questions. Write your answers on a separate sheet of paper or go to the Student Web site (www.g-wlearning.com/CAD) and complete the electronic chapter test.

1. Define *associative dimension*.
2. Why is it important to have associative dimensions for editing objects?
3. Which **Options** dialog box setting controls associative dimensioning?
4. Which tool allows you to convert nonassociative dimensions to associative dimensions?
5. Which tool allows you to convert associative dimensions to nonassociative dimensions?
6. What are definition points?
7. Which four tool options related to dimension editing appear in the shortcut menu accessed when you select a dimension?
8. Name three methods of changing the dimension style of a dimension.
9. How does the **Dimension Update** tool affect selected dimensions?
10. Explain how to add a diameter symbol to a dimension text value using the **DDEDIT** tool.
11. Name the tool that allows you to control the placement and orientation of an existing associative dimension text value.
12. Name two applications in which you might need to create oblique extension lines.
13. Which tool and option can you use to add a new baseline dimension to an existing set of baseline dimensions?
14. When you use the **Properties** palette to edit a dimension, what is the effect on the dimension style?
15. How do you access the **Property Settings** dialog box?
16. Which tool can you use to adjust the space equally between dimension lines or align dimension lines without having to determine the exact location or amount of stretch needed?
17. What two options are available when you use the **Multiple** option of the **DIMBREAK** tool?
18. What tool allows you to add information about how frequently the dimension should be tested for consistency and tolerance during the manufacturing of a product?
19. Name an application in which leaders with multiple leader lines are commonly used.
20. Identify the four options available to change the multileader alignment.

Drawing Problems

- *Start AutoCAD if it is not already started.*
- *Start a new drawing using an appropriate template of your choice. The template should include layers, text styles, dimension styles, and multileader styles appropriate for drawing the given objects.*
- *Add layers, text styles, dimension styles, and multileader styles as needed. Draw all objects using appropriate layers, text styles, dimension styles, multileader styles, justification, and format.*
- *Follow the specific instructions for each problem. Use your own judgment and approximate dimensions when necessary.*
- *Apply dimensions accurately using ASME or appropriate industry standards. Use object snap modes to your best advantage.*
- *For mechanical drawings, place the following general notes 1/2" from the lower-left corner:*
  NOTES:
  1. INTERPRET DIMENSIONS AND TOLERANCES PER ASME Y14.5M-1994.
  2. REMOVE ALL BURRS AND SHARP EDGES.
  3. UNLESS OTHERWISE SPECIFIED, ALL DIMENSIONS ARE IN INCHES (or MILLIMETERS as applicable).

▼ Basic

1. Open P18-9 and save the file as P21-1. The P21-1 file should be active. Edit the drawing as follows:
   A. Erase the front (circular) view.
   B. Stretch the vertical dimensions to provide more space between dimension lines. Be sure the space you create is the same between all vertical dimensions.
   C. Stagger the existing vertical dimension text numbers if they are not staggered as shown in the original problem.
   D. Erase the 1.750 horizontal dimension and then stretch the 5.255 and 4.250 dimensions to make room for a new datum dimension from the baseline to where the 1.750 dimension was located. This should result in a new baseline dimension that equals 2.750. Be sure all horizontal dimension lines are equally spaced.

2. Open P19-1 and save the file as P21-2. The P21-2 file should be active. Edit the drawing as follows:
   A. Stretch the total length from 3.500 to 4.000, leaving the holes the same distance from the edges.
   B. Fillet the upper-left corner. Modify the 3X R.250 dimension accordingly.

3. Open P18-12 and save the file as P21-13. The P21-3 file should be active. Edit the drawing as follows:
   A. Make the bathroom 8'-0" wide by stretching the walls and vanity that are currently 6'-0" wide to 8'-0". Do this without increasing the size of the water closet compartment. Provide two equally spaced oval sinks where there is currently one.

4. Open P19-18 and save the file as P21-4. The P21-4 file should be active. Edit the drawing as follows:
   A. Lengthen the part .250 on each side for a new overall dimension of 6.500.
   B. Change the width of the part from 3.000 to 3.500 by widening an equal amount on each side.

5. Open P19-16 and save the file as P21-5. The P21-5 file should be active. Edit the drawing as follows:
   A. Shorten the .75 thread on the left side to .50.
   B. Shorten the .388 hexagon length to .300.

6. Draw the shim as shown at A. Then edit the .150 and .340 values using oblique dimensions as shown at B. Save the drawing as P21-6.

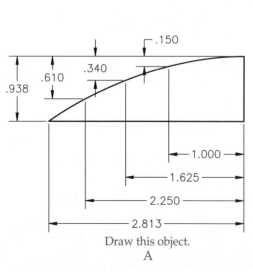

Draw this object.
A

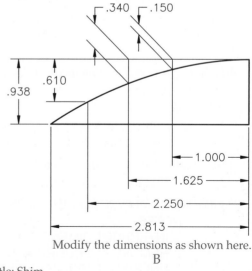

Modify the dimensions as shown here.
B

Title: Shim

▼ Intermediate

7. Open P19-4 and save the file as P21-7. The P21-7 file should be active. Edit the drawing as follows:
   A. Use the existing drawing as the model and make four copies.
   B. Leave the original drawing as it is and edit the other four pins in the following manner, keeping the ∅.125 hole exactly in the center of each pin.
   C. Give one pin a total length of 1.500.
   D. Create the next pin with a total length of 2.000.
   E. Edit the third pin to a length of 2.500.
   F. Change the last pin to a length of 3.000.
   G. Organize the pins on your drawing in a vertical row ranging in length from the smallest to the largest. You may need to change the drawing limits.

8. Open P19-5 and save the file as P21-8. The P21-8 file should be active. Edit the drawing as follows:
   A. Modify the spline to have twelve projections, rather than eight.
   B. Change the angular dimension, linear dimension, and 8X dimension to reflect the modification.

9. Open P19-11 and save the file as P21-9. The P21-9 file should be active. Edit the drawing as follows:
   A. Stretch the total length from 6.500 to 7.750.
   B. Add two more holes that continue the equally spaced pattern of .625 apart.
   C. Change the 8X .625(=5.00) dimension to read 10X .625(=6.250).

▼ Advanced

10. Draw the stairs cross section using the dimensions and notes provided. Use oblique dimensions where necessary. Save the drawing as P21-10.

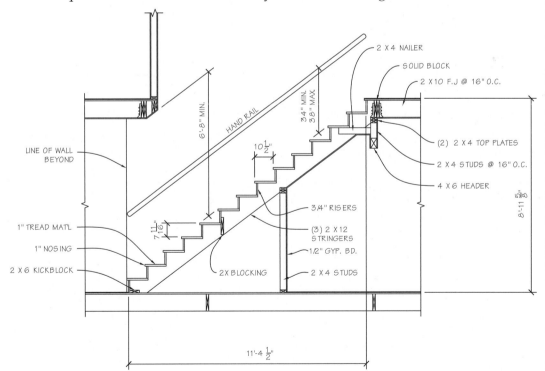

11. Use a word processor to write a report of at least 250 words explaining the importance of associative dimensioning. Site at least three examples from actual industry applications. Show at least four drawings illustrating your report.

12. Open P18-17 and save the file as P21-12. The P21-12 file should be active. Make the client-requested revisions to the floor plan as shown. Make sure the dimensions reflect the changes.

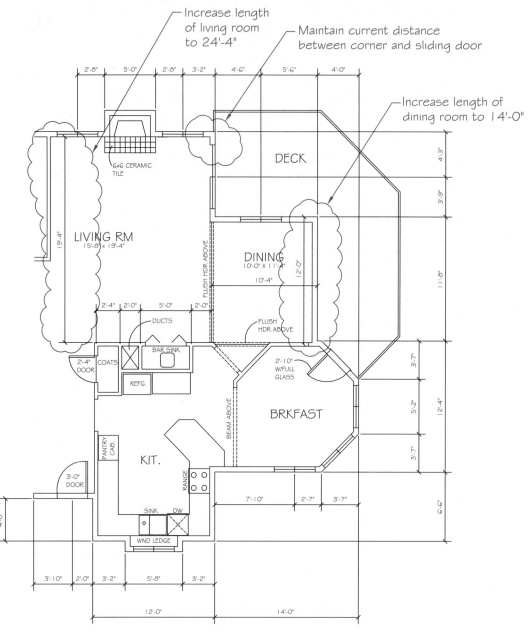

Parametric Drafting

22

Learning Objectives

After completing this chapter, you will be able to do the following:

✓ Explain parametric drafting processes and applications.
✓ Create and edit parametric drawings.
✓ Add and manage geometric constraints.
✓ Add and manage dimensional constraints.
✓ Convert dimensional constraints.

parametric drafting: A form of drafting in which parameters and constraints drive object size and location to produce drawings with features that adapt to changes made to other features.

parameters (constraints): Geometric characteristics and dimensions that control the size, shape, and position of drawing geometry.

Parametric drafting tools allow you to assign *parameters*, or *constraints*, to objects. The parametric concept, also known as *intelligence*, provides a way to associate objects and limit design changes. You cannot change a constraint so that it conflicts with other parametric geometry. A database stores and allows you to manage all parameters. You typically use parametric tools with standard drafting practices to create a more interactive drawing.

Parametric Fundamentals

AutoCAD
NEW

Parametric drafting allows you to control every aspect of a drawing during and after the design and documentation process. You should always use standard drafting practices and drawing aids to construct initial geometry. When used correctly, these techniques allow you to produce accurate drawings efficiently. You typically use parametric tools as a supplement or drawing aid. First, create a drawing using the same tools and techniques you use to create a non-parametric drawing. Then add parameters using *geometric constraints* and *dimensional constraints*.

geometric constraints: Geometric characteristics applied to restrict the size or location of geometry.

Understanding Constraints

Well-defined constraints allow you to incorporate and preserve specific design intentions and increase revision efficiency. For example, if two holes through a part, drawn as circles, must always be the same size, use a geometric constraint to make the circles equal and add a dimensional constraint to size one of the circles. The size of both circles changes when you modify the dimensional constraint value. See **Figure 22-1.**

dimensional constraints: Measurements that numerically control the size or location of geometry.

Figure 22-1.
An example of a basic parametric relationship. The dimensional constraint controls the size of both circles with the aid of an equal geometric constraint.

Icon indicates dimensional constraint

dia1=1.000

Dimensional constraint added to one circle only

Equal geometric constraints

Original Circles

Value controls circle diameter

Circle diameter changes parametrically

dia1=2.000

Revised Circles

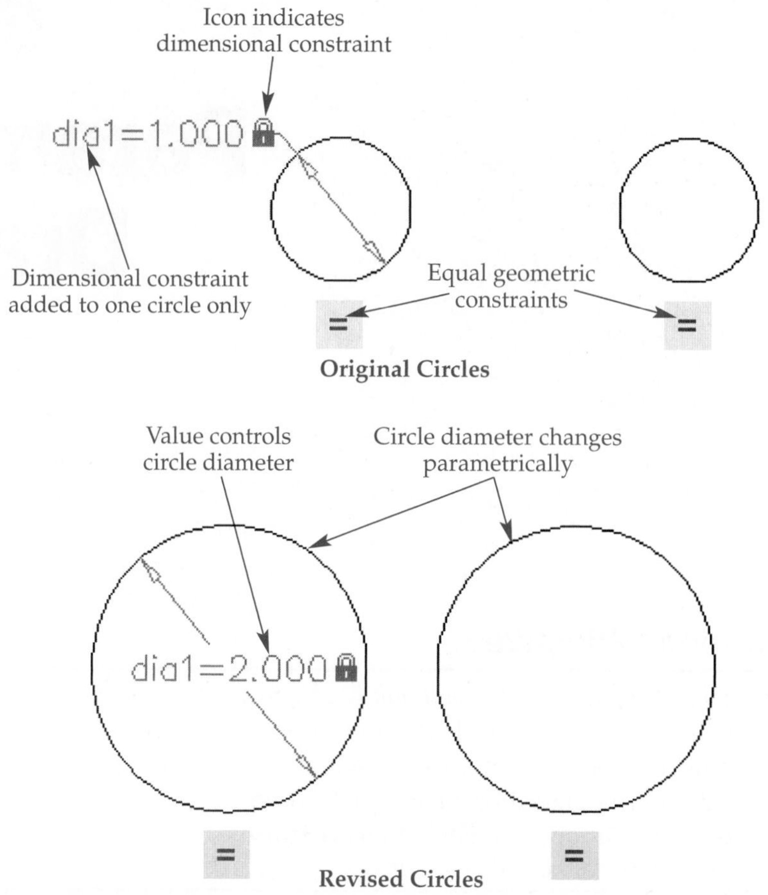

You must add constraints to make an object parametric. Dimensional constraints create parameters that direct object size and location. A traditional associative dimension is associated with an object, but it does not control object size or location. **Figure 22-2** shows an example constraint levels, including *under-constrained*, *fully constrained*, and *over-constrained*. As you progress through the design process, you will often fully or near fully constrain the drawing to ensure that your design is accurate. However, if you attempt to over-constrain the drawing, a message appears allowing you to decide what to do. See **Figure 22-3**. AutoCAD does not allow you to over-constrain a drawing, as shown by the *reference dimension* in **Figure 22-2**.

Figure 22-4 shows an extreme example of constraining, for reference only. Study the figure to help understand how constraints work, and how applying constraints differs from and compliments traditional drafting. Typically, you should prepare initial objects as accurately as possible using the many tools and drawing aids available, and then add constraints.

Parametric Applications

Parametric tools aid the design and revision process, place limits on geometry to preserve design intent, and help form geometric constructions. You may want to consider using constraints to help maintain relationships between objects in a drawing, especially during the design process, when changes are often frequent. However, you must decide if the additional steps required to make a drawing parametric are appropriate and necessary for your applications.

under-constrained: Describes a drawing that includes constraints, but not enough to size and locate all geometry.

fully constrained: Describes a drawing in which objects have no freedom of movement.

over-constrained: Describes a drawing that contains too many constraints.

reference dimension: A dimension used for reference purposes only. Parentheses enclose reference dimensions to differentiate them from other dimensions.

Figure 22-2.
Levels of parametric constraint.

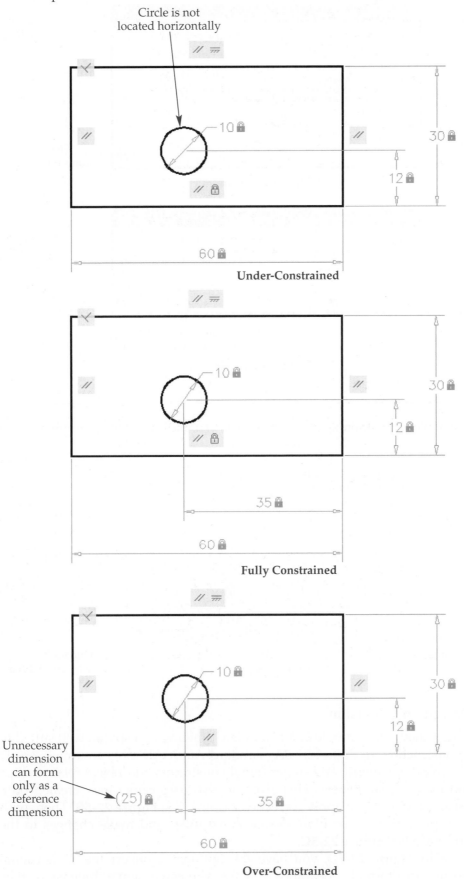

Circle is not located horizontally

10

30

12

60

Under-Constrained

10

30

12

35

60

Fully Constrained

10

30

12

Unnecessary dimension can form only as a reference dimension

(25)

35

60

Over-Constrained

Figure 22-3.
Error messages that appear when you attempt to over-constrain objects.

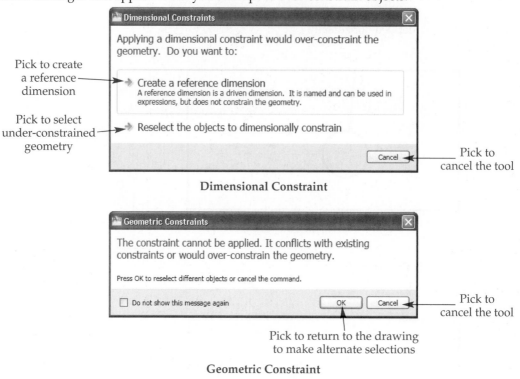

Pick to create a reference dimension

Pick to select under-constrained geometry

Pick to cancel the tool

Dimensional Constraint

Pick to cancel the tool

Pick to return to the drawing to make alternate selections

Geometric Constraint

Figure 22-4.
An extreme example of constraining a drawing to help you understand how you apply constraints.

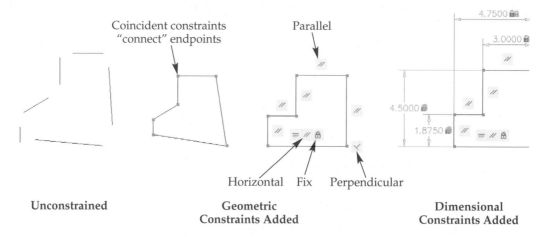

Coincident constraints "connect" endpoints

Parallel

Horizontal Fix Perpendicular

Unconstrained

Geometric Constraints Added

Dimensional Constraints Added

Product Design and Revision

Figure 22-5 shows an example of a front view drawing of a spacer, well suited to parametric construction. In **Figure 22-5A,** standard AutoCAD tools accurately create the geometry. Next, geometric and dimensional constraints add object relationships and size and location parameters. This example also uses centerlines, created on a separate construction layer, to apply correct constrains. See **Figure 22-5B.** Then, with parameters in place, you can explore design alternatives and make changes to the drawing efficiently. See **Figure 22-5C.**

As shown in **Figure 22-5D,** you have the option to convert the dimensional constraint format to a formal appearance to which you can assign a dimension style. You can still use converted dimensions to adjust geometry parametrically. In this

Figure 22-5.
The front view of this spacer is a good candidate for parametric associations.
A—Accurate view geometry constructed using standard AutoCAD practices.
B—Adding geometric and dimensional constraints to constrain the drawing.
C—Changing the values of a few dimensional constraints to update the entire drawing.
D—Reusing dimensional constraints to help prepare a formal drawing.

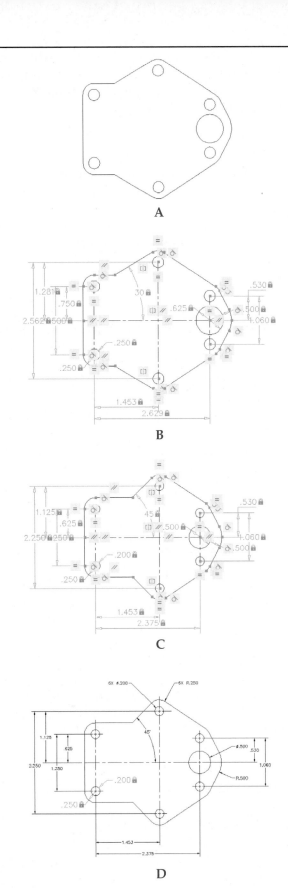

example, all of the dimensions except the .250 radius and diameter dimensions have been converted. Standard associative .250 radial and diameter dimensions allow you to add the 6X prefix and relocate the dimensions. You can then hide constraint information to view the finished drawing.

Geometric Construction

You can use constraints to form geometric constructions in specific situations when standard AutoCAD tools are inefficient or ineffective. For example, suppose you know that the angle of a line is 30°, and you know the line is tangent to a circle. However, you do not know the length of the line or the location of the line endpoints. One option is to position a 30° construction line, using the **Ang** option of the **XLINE** tool, anywhere in the drawing. See **Figure 22-6A.** Use a **Tangent** geometric constraint to form a tangent relationship between the xline and circle. See **Figure 22-6B.** You can then hide or delete the constraint if necessary.

Unsuitable Applications

You may find that parametric drafting is unsuitable or ineffective for some applications. For example, it may be unsuitable to add parameters to a drawing if the drawing is of a finalized product that will not require extensive revision, or if you can easily modify drawing geometry without associating objects.

In addition, if your drawing includes a large number of objects, you may find it cumbersome to add the constraints required to form an intelligent drawing. For instance, you can use constraints to form all desired relationships between objects in a floor plan. See **Figure 22-7.** In this example, dimensional constraints can specify wall thickness, position windows between walls, locate sinks on vanities, and form many other parametric relationships. You can then adjust dimensional constraints as needed to update the drawing.

If you effectively constrain *all* objects shown in **Figure 22-7,** you have the ability to change the 11′-10 1/2″ dimensional constraint to increase the width of the master bedroom, for example. The entire floor plan adjusts to the modified room size. Consider, however, what this process requires. You must constrain all wall endpoints; the points where doors and windows meet walls; the distance between walls and objects, such as cabinets, sinks, and water closets; and form all other geometric and dimensional constraints.

Figure 22-6.
A—A 30° xline placed near an associated circle. B—Using a **Tangent** geometric constraint to form a tangent construction. Notice the appropriate selection process. C—The final drawing.

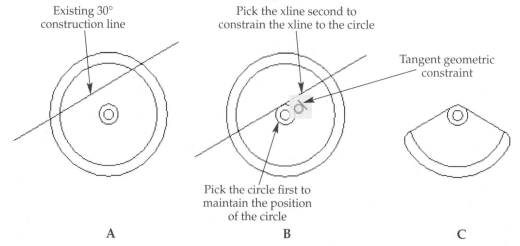

Existing 30° construction line

Pick the xline second to constrain the xline to the circle

Tangent geometric constraint

Pick the circle first to maintain the position of the circle

A   B   C

Figure 22-7.
An architectural floor plan usually includes too many objects to effectively and efficiently constrain.

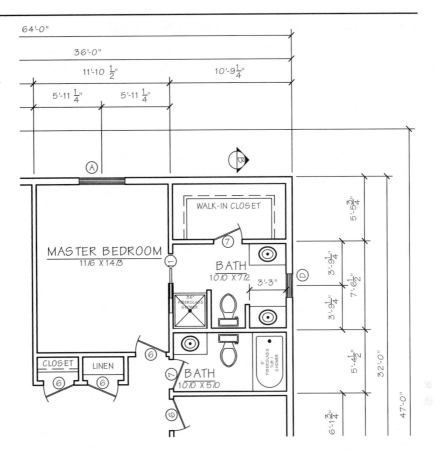

NOTE

You can also use parametric tools when constructing blocks, as described later in this textbook.

PROFESSIONAL TIP

Constraints can prove effective for multiview layout, in order to help maintain alignment between views.

Exercise 22-1

Access the Student Web site (www.g-wlearning.com/CAD) and complete Exercise 22-1.

Adding Geometric Constraints

AutoCAD
2010
NEW

Geometric constraint tools allow you to add the geometric relationships required to build a parametric drawing. You should typically add geometric constraints, or at least a portion of the necessary geometric constraints, before dimensional constraints to help preserve design intent. By default, a constraint-specific icon is visible to indicate the presence of a geometric constraint. Once you have placed geometric constraints, you can view, adjust, and remove them as needed.

Using the GEOMCONSTRAINT Tool

The **GEOMCONSTRAINT** tool allows you to assign specific geometric constraints to objects. Each constraint is a separate **GEOMCONSTRAINT** tool option. The quickest way to add geometric constraints using this tool is to pick the appropriate button from the **Geometric** panel of the **Parametric** ribbon tab. Then follow the prompts to make the required selection(s), form the constraint, and exit the tool.

When using geometric constraint tools that allow you to pick two objects or points, keep in mind that the first object or point you select always remains the same. The second object or point you select changes in relation to the first. For example, to create perpendicular lines using the **Perpendicular** constraint, first select the line that remains in the same position at the same angle. Then select the line to make perpendicular to the first line.

Assigning Coincident Constraints

coincident:
A geometric construction that specifies two points sharing the same position.

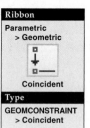

Ribbon

**Parametric
> Geometric**

Coincident

Type

**GEOMCONSTRAINT
> Coincident**

A *coincident* constraint is one of the most common parametric constructions. For example, the endpoints of two lines at the corner of a rectangle coincide. Access the **Coincident** constraint, and move the pick box near a point on an existing object to display a point marker. See **Figure 22-8**. The object associated with a specific point highlights, allowing you to confirm that the point is on the appropriate object. Pick the marked location, and then pick a point on another object to make the two points coincide. **Figure 22-9** shows coincident constraints applied to a basic multiview drawing.

The **Object** function allows you to select an object, followed by a point. This is useful when it is necessary to constrain a point along a curve. The point does not have to contact the curve. The **Autoconstrain** function allows you to select multiple objects to form coincident constraints at every possible coincident intersection, in a single operation. Right-click or press [Enter] or the space bar to use the **Autoconstrain** function.

> **NOTE**
>
> Connected polyline segments already include a form of coincident constraint, although no icon displays.

Figure 22-8.
Examples of objects with selectable constraint points. Study these points, because they are the same points used to add other geometric constraints when point selection is necessary.

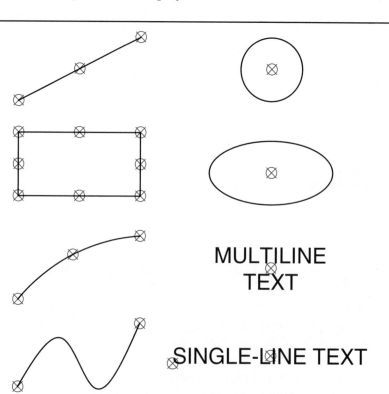

AutoCAD and Its Applications—Basics

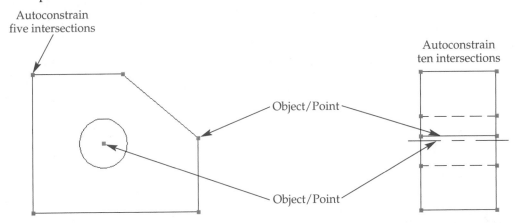

Autoconstrain
five intersections

Autoconstrain
ten intersections

Object/Point

Object/Point

PROFESSIONAL TIP

When you are selecting points, be sure to select the points corresponding to the surface to constrain.

Exercise 22-2

Access the Student Web site (www.g-wlearning.com/CAD) and complete Exercise 22-2.

Assigning Horizontal and Vertical Constraints

The **Horizontal** constraint aligns lines, polylines, or points along the X axis, or horizontally. The **Vertical** constraint aligns lines, polylines, or points along the Y axis, or vertically. These constraints are commonly used to define a horizontal or vertical surface datum or to align points. Access the **Horizontal** or **Vertical** constraint and select the object to constrain. Both options include a **2Points** function that allows you to pick two points to align horizontally or vertically. See **Figure 22-10**.

Assigning Parallel and Perpendicular Constraints

The **Parallel** constraint creates a *parallel* constraint between lines or polylines. The **Perpendicular** constraint creates a *perpendicular* constraint between lines or polylines. These constraints are used for a variety of applications. Access the **Parallel** or **Perpendicular** constraint, and select two lines or polylines, or a line and a polyline. See **Figure 22-11**.

Exercise 22-3

Access the Student Web site (www.g-wlearning.com/CAD) and complete Exercise 22-3.

Ribbon
Parametric > Geometric
Horizontal
Type
GEOMCONSTRAINT > Horizontal

Ribbon
Parametric > Geometric
Vertical
Type
GEOMCONSTRAINT > Vertical

Ribbon
Parametric > Geometric
Parallel
Type
GEOMCONSTRAINT > Parallel

Ribbon
Parametric > Geometric
Perpendicular
Type
GEOMCONSTRAINT > Perpendicular

parallel: A geometric construction that specifies that objects such as lines will never intersect, no matter how long they become.

perpendicular: A geometric construction that defines a 90° angle between objects such as lines.

Figure 22-10.
Examples of vertically and horizontally constrained objects and points.

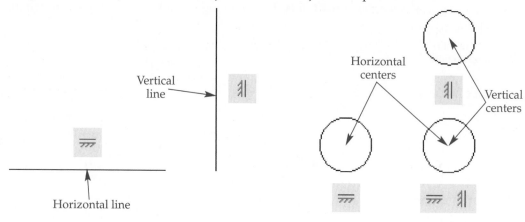

Figure 22-11.
Parallel and perpendicular constraints required for the example multiview drawing.

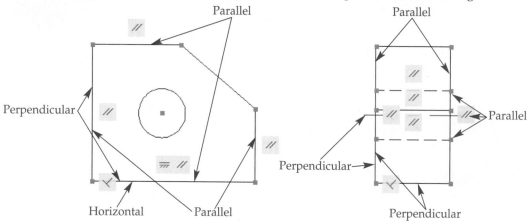

Assigning Collinear and Tangent Constraints

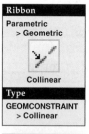
Ribbon
Parametric
> Geometric

Collinear

Type
GEOMCONSTRAINT
> Collinear

The **Collinear** constraint allows you to align two lines or polylines along the same line. A collinear constraint is common for applications such as aligning multiview surfaces. Access the **Collinear** constraint and select two lines or polylines, or activate the **Multiple** function to select multiple lines and/or polylines to align in a single operation. Right-click or press [Enter] or the space bar to complete a multiple constrain.

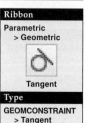
Ribbon
Parametric
> Geometric

Tangent

Type
GEOMCONSTRAINT
> Tangent

Use the **Tangent** constraint option to form a tangent constraint between a line or polyline and an arc, circle, or ellipse. You can also make two circular objects tangent. Access the **Tangent** constraint and select two appropriate objects. **Figure 22-12** shows examples of collinear and tangent constraints.

Exercise 22-4

Access the Student Web site (www.g-wlearning.com/CAD) and complete Exercise 22-4.

Figure 22-12.
A—Collinear and tangent constraints required for the example multiview drawing. B—Example of common tangent constraint applications.

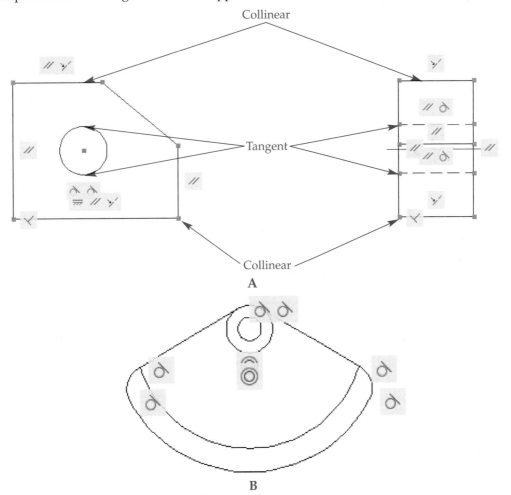

A

B

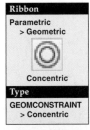

Ribbon
Parametric
> Geometric
Concentric

Type
GEOMCONSTRAINT
> Concentric

Ribbon
Parametric
> Geometric
Equal

Type
GEOMCONSTRAINT
> Equal

concentric: Arcs, circles, and/or ellipses sharing the same center point.

Ribbon
Parametric
> Geometric
Symmetric

Type
GEOMCONSTRAINT
> Symmetric

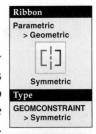

Ribbon
Parametric
> Geometric
Fix

Type
GEOMCONSTRAINT
> Fix

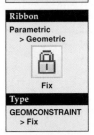

Ribbon
Parametric
> Geometric
Smooth

Type
GEOMCONSTRAINT
> Smooth

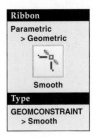

Assigning Concentric and Equal Constraints

To assign a *concentric* constraint, access the **Concentric** constraint and select any combination of arcs, circles, or ellipses. The **Equal** constraint allows you to size objects equally and, in some cases, locate objects. Access the **Equal** constraint and select two objects, or activate the **Multiple** function to select multiple objects to equalize in single operation. Right-click or press [Enter] or the space bar to complete a multiple constrain. **Figure 22-13** shows examples of concentric and equal constraints.

Assigning Symmetric, Fix, and Smooth Constraints

By default, the **Symmetric** constraint allows you to establish symmetry by selecting one object, followed by another, and finally, a line of symmetry. The **2Points** option allows you to pick points followed by the line of symmetry to constrain symmetrical points.

The **Fix** constraint secures a point or object to its current location in space to help preserve design intent. A single fix constraint is often required to fully constrain a drawing. Use the default method to fix a point, or activate the **Object** function to select an object to fix. **Figure 22-14** shows examples of symmetric and fix constraints. Use the **Smooth** constraint to create a curvature-continuous situation, or G2 curve, between a selected spline and a line, second spline, or arc connected to the spline endpoint.

Figure 22-13.
Examples of concentric and equal constraints. All circles are concentric to an arc. All small arcs are equal. All small circles are equal. Use the **Multiple** option to help make multiple objects equal in a single operation.

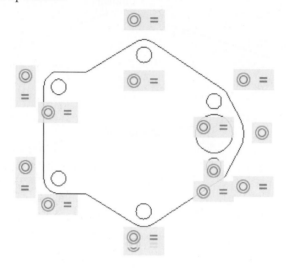

Figure 22-14.
An example of a symmetrical parametric drawing created by adding symmetric constraints to circles and arcs. A fix constraint secures the drawing in space.

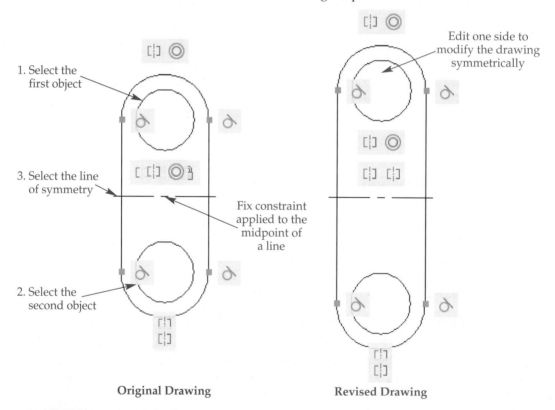

Original Drawing   Revised Drawing

Exercise 22-5

Access the Student Web site (www.g-wlearning.com/CAD) and complete Exercise 22-5.

Using the AUTOCONSTRAIN Tool

You can use the **AUTOCONSTRAIN** tool in an attempt to add all required geometric constraints in a single operation. Before using this tool, access the **Constraint Settings** dialog box and use the options in the **AutoConstrain** tab to specify which geometric constraints to apply. The constraint priority determines which constraints apply first. The higher the priority, the more likely and often the constraint will form if appropriate geometry is available. Select a constraint and use the **Move Up** and **Move Down** buttons to change its priority. Use the corresponding **Apply** check marks and the **Select All** and **Clear All** buttons if there are specific constraints you want to omit during the constraining procedure.

The check boxes determine whether tangent and perpendicular constraints can form if objects do not intersect. See **Figure 22-15**. The **Tolerances** area controls how specific constraints form based on the distance between and angle of objects. A distance less than or equal to the value specified in the **Distance** text box receives constraints. An angle less than or equal to the value specified in the **Angle** text box receives constraints. See **Figure 22-16**.

Once you specify the settings, access **AUTOCONSTRAIN** tool and select the objects to assign geometric constraints. The **Settings** option is available before selection to access the **AutoConstrain** tab of the **Constraint Settings** dialog box. A fix constraint does not occur.

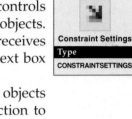

Ribbon
Parametric
> Geometric

Auto Constrain

Type
AUTOCONSTRAIN

Ribbon
Parametric
> Geometric

Constraint Settings

Type
CONSTRAINTSETTINGS

Figure 22-15.
Examples of constraints that will form when you select the **Tangent objects must share an intersection** the **Perpendicular objects must share an intersection** check boxes.

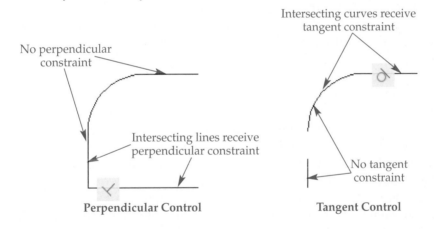

No perpendicular constraint

Intersecting lines receive perpendicular constraint

Perpendicular Control

Intersecting curves receive tangent constraint

No tangent constraint

Tangent Control

Figure 22-16.
Examples of constraints that form based on **Distance** and **Angle** tolerance values.

	Distance between Objects > Distance Tolerance	Distance between Objects ≥ Distance Tolerance	Angle > Angle tolerance	Angle ≤ Angle tolerance
Before				
After				

Managing Geometric Constraints

constraint bars:
Toolbars that allow you to view and remove geometric constraints.

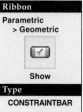

Ribbon
Parametric > Geometric
Show
Type
CONSTRAINTBAR

Once you apply geometric constraints, geometric *constraint bars* appear by default. See **Figure 22-17.** The **CONSTRAINTBAR** tool includes options to show specific constraint bars, show all constraint bars, or hide all constraint bars. The quickest way to access these options is to pick the appropriate button from the **Geometric** panel of the **Parametric** ribbon tab. Select the **Show** button, and then pick objects to display hidden constraint bars. Choose the **Show All** button to display all constraint bars, or choose the **Hide All** button to hide all constraint bars. Hiding constraint bars does not remove geometric constraints.

Use the **Geometric** tab of the **Constraint Settings** dialog box to specify the geometric bars that appear and other geometric bar characteristics. Select the check boxes for individual constraints to display them in constraint bars. Use the **Select All** and **Clear All** buttons to select or deselect all constraint type check boxes. By default, all constraint types show. Limiting constraint bar visibility to specific constraint types often helps to locate and adjust constraints. Use the **Constraint bar transparency** slider or text box to increase or decease the constraint bar transparency.

A coincident constraint appears as a dot that, when hovered over, shows the coincident constraint bar. All other constraints appear as constraint bar icons. When you hover over or select a constraint bar icon, the corresponding constrained objects highlight and markers identify constrained points, as shown in **Figure 22-17.** This allows you to recognize which objects and points are associated with the constraint.

If a constraint bar blocks your view, drag it to a new location. To close a constraint bar, pick the **Hide Constraint Bar** button located to the right of the symbols, or right-click and choose **Hide**. The right-click shortcut menu also includes options for hiding all constraint bars, and for accessing the **Geometric** tab of the **Constraint Settings** dialog box. In some cases, you may need to delete existing constraints in order to apply

Figure 22-17.
Use geometric constraint bars to view and delete geometric constraints.

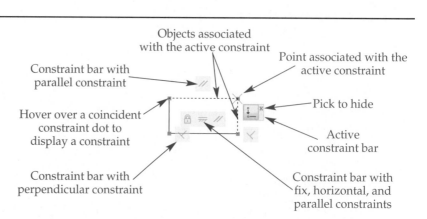

Objects associated with the active constraint

Point associated with the active constraint

Constraint bar with parallel constraint

Hover over a coincident constraint dot to display a constraint

Pick to hide

Active constraint bar

Constraint bar with perpendicular constraint

Constraint bar with fix, horizontal, and parallel constraints

different constructions. To delete constraints, hover over an icon in the constraint bar and press [Delete], or right-click and select **Delete**.

Exercise 22-6

Access the Student Web site (www.g-wlearning.com/CAD) and complete Exercise 22-6.

Adding Dimensional Constraints

AutoCAD 2010
NEW

Dimensional constraints establish size and location parameters. You must include dimensional constraints to create a truly parametric drawing. Dimensional constraints use a *dynamic format* by default and display a unique appearance. You cannot modify how dimensional constraints appear, but you can convert them to an *annotational format*. Once you have placed dimensional constraints, you can view, adjust, and remove them as needed.

dynamic format: A dimensional constraint format specifically for controlling the size or location of geometry.

annotational format: A dimensional constraint format in which the constraints look like traditional dimensions, using a dimension style. Constraints displayed in this format can still control the size or location of geometry.

Using the DIMCONSTRAINT Tool

The **DIMCONSTRAINT** tool allows you to assign linear, diameter, radius, and angular dimensional constraints. You can also use the tool to convert dimensional constraints and associative dimensions. Each dimensional constraint is a separate **DIMCONSTRAINT** tool option. The quickest way to add or convert dimensional constraints using this tool is to pick the appropriate button from the **Dimensional** panel of the **Parametric** ribbon tab.

Dimensional Constraint Procedures

To create a dimensional constraint, follow the prompts to make the required selections, pick a location for the dimension line, enter a value to form the constraint, and exit the tool. The process of selecting points or an object to create a dimensional constraint is the same as that for adding geometric constraints. When a dimensional constraint tool requires you to pick two points or objects, the first point or object you select always remains the same. In some cases, you should consider the first point or object to be the datum. The second object or point you select changes in relation to the first. When selecting points, be sure to select the points corresponding to the surface you want to constrain.

A text editor appears after you select the location for the dimension line, allowing you to specify the dimension value. See **Figure 22-18A.** Each dimensional constraint is a parameter with a specific name, expression, and value. By default, linear dimensions receive d names, angular dimensions receive ang names, diameter dimensions receive dia names, and radial dimensions receive rad names. The name of the first of each type of dimension includes a 1, such as d1. The next dimension includes a 2, such as d2, and so on. You can use the text editor to enter a more descriptive name, such as Length, Width, or Diameter. Every parameter must have a unique name. Follow the name with the = symbol and then the dimension value. Press [Enter] or pick outside of the text editor to form the constraint. See **Figure 22-18B.**

You can specify the dimension value for a dimensional constraint in different ways. The most basic option is to type a value in the text editor. Dimensional constraint units reflect the current work environment and unit settings, including length, angle type, and precision. If the drawing is accurate, you should be able to accept the current value. Enter a different value if the drawing is inaccurate, or to change object size.

Another option is to enter an expression in the text editor. You can enter an expression if you do not know an exact value, much like using a calculator. Usually, however, expressions include parameters to associate dimensional constraints. This enables the drawing to adapt according to dimension changes. In the active text editor, include the parameter name in the expression. **Figure 22-19** shows a basic example of using an existing parameter in an expression to control the size of an object. In this example, the design requires that the height of the object always be half the value of the length.

> **NOTE**
>
> If you enter an existing parameter in the text editor without making a calculation, such as d1 = d2, the dimensional constraint references another dimensional constraint value. This is necessary for many applications. Use an equal geometric constraint when possible.

Applying Linear Dimensional Constraints

The **Linear** option allows you to place horizontal or vertical linear dimensional constraints. The **Horizontal** option sets the tool to constrain only a horizontal distance. The **Vertical** option sets the tool to constrain only a vertical distance. The **Horizontal** and **Vertical** options are helpful when it is difficult to produce the appropriate linear

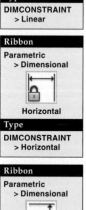

Ribbon
Parametric > Dimensional

Linear

Type
DIMCONSTRAINT > Linear

Ribbon
Parametric > Dimensional

Horizontal

Type
DIMCONSTRAINT > Horizontal

Ribbon
Parametric > Dimensional

Vertical

Type
DIMCONSTRAINT > Vertical

Figure 22-18.
A—Use the text editor that appears after you establish the location of the dimension line to specify a parametric dimension. B—The modified parameter name and value. Notice that the dimension controls the object size.

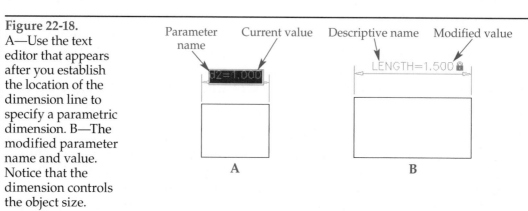

Figure 22-19.
A—An example of a modified parameter name and expression that references another parameter.
B—The dimension that includes a parameter references the parameter when changes occur.

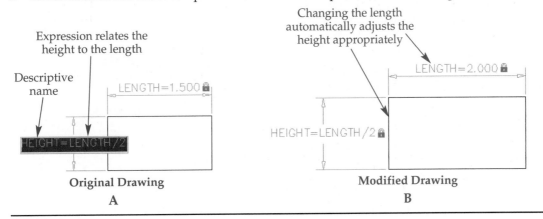

Original Drawing
A

Modified Drawing
B

dimensional constraint, such as when dimensioning the horizontal or vertical distance of an angled surface.

Access the appropriate linear dimensional constraint option and pick two points to specify the origin of the dimensional constraint. The **Object** function allows you to select a line, polyline, or arc to constrain, instead of two points. After you select points or an object, move the dimension line to an appropriate location and pick. Specify the dimension value and adjust the parameter name if desired. Press [Enter] or pick outside of the text editor to form the constraint. **Figure 22-18** and **Figure 22-19** show examples of linear dimensions.

Exercise 22-7

Access the Student Web site (www.g-wlearning.com/CAD) and complete Exercise 22-7.

Applying Aligned Dimensional Constraints

The **Aligned** option allows you to place a linear dimensional constraint with a dimension line aligned with, and extension lines perpendicular to, an angled surface. Access the **Aligned** option and pick two points to specify the origin of the dimensional constraint. The **Object** function allows you to select a line, polyline, or arc to constrain, instead of two points. Often when you apply aligned dimensions, such as when you are dimensioning an auxiliary view, it is necessary to pick a point and an aligned surface or two aligned surfaces. Use the **Point & Line** function to select a point and an alignment line. Use the **2Line** function to select two alignment lines.

Once you make the selections, move the dimension line to an appropriate location and pick. Specify the dimension value, and adjust the parameter name if desired. Press [Enter] or pick outside of the text editor to form the constraint. **Figure 22-20** shows examples of aligned dimensional constraints created using each method.

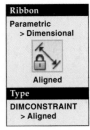

Ribbon
Parametric
> Dimensional

Aligned

Type
DIMCONSTRAINT
> Aligned

Applying Angular Dimensional Constraints

The **Angular** option allows you to place an angular dimension between two objects or three points. Access the **Angular** option, and by default, pick two lines, polylines, or arcs. The **3Point** function allows you to select the angle vertex, followed by two points to locate each side of the angle. After you make the selections, move the dimension

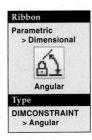

Ribbon
Parametric
> Dimensional

Angular

Type
DIMCONSTRAINT
> Angular

Figure 22-20.
An example of an auxiliary view with full dimensional constraints created using the **Aligned** option.

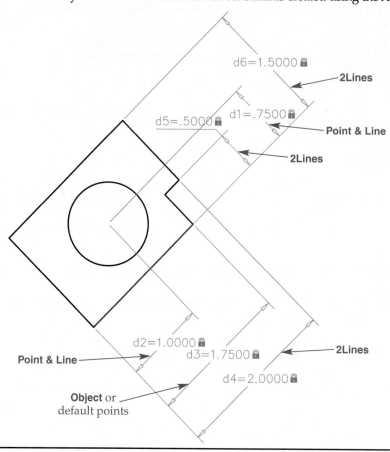

d6=1.5000 🔒
2Lines

d5=.5000🔒 d1=.7500 🔒
Point & Line
2Lines

d2=1.0000 🔒
Point & Line
d3=1.7500 🔒
2Lines
d4=2.0000 🔒

Object or
default points

line to an appropriate location and pick. Specify the dimension value, and adjust the parameter name if desired. Press [Enter] or pick outside of the text editor to form the constraint. **Figure 22-21** shows examples of angular dimensional constraints created using each method.

Exercise 22-8

Access the Student Web site (www.g-wlearning.com/CAD) and complete Exercise 22-8.

Figure 22-21.
Forming angular dimensional constraints.

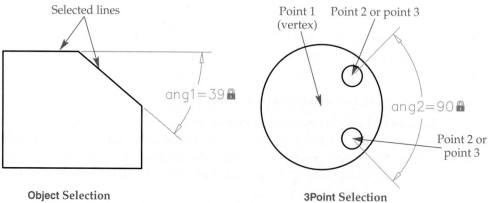

Selected lines

ang1=39 🔒

Point 1
(vertex)   Point 2 or point 3

ang2=90 🔒

Point 2 or
point 3

Object Selection   **3Point Selection**

Adding Diameter and Radial Dimensional Constraints

Use the **Diameter** option to create a diameter dimensional constraint, and use the **Radial** option to form a radial dimensional constraint. You can select a circle or arc when using either tool. In formal drafting, circles receive diameter constraints and arcs receive radial constraints. See **Figure 22-22**. In some parametric applications, however, you may find it appropriate to constrain arcs using a diameter and circles using a radius.

Creating Reference Dimensional Constraints

When you constrain a defined object, the drawing should become over-constrained. However, AutoCAD does not allow over-constraining to occur. You can either cancel the tool without accepting the dimension, or accept the dimension and allow it to become a reference dimension. See **Figure 22-23**. Reference dimensions are sometimes required to form specific parameters or expressions. You cannot edit a reference dimension to change the size of an object, but a reference dimension changes when you modify corresponding dimensions.

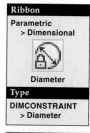

Ribbon
Parametric
> Dimensional

Diameter

Type
DIMCONSTRAINT
> Diameter

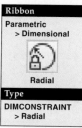

Ribbon
Parametric
> Dimensional

Radial

Type
DIMCONSTRAINT
> Radial

Figure 22-22.
Forming diameter and radial dimensional constraints. In this example, equal geometric constraints control the size of the undimensioned arcs.

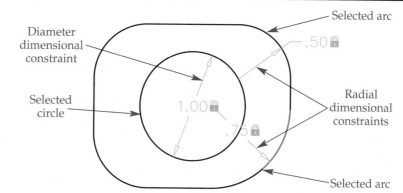

Diameter dimensional constraint

Selected circle

Selected arc

Radial dimensional constraints

Selected arc

Figure 22-23.
You cannot directly modify reference dimensions, which are identified by parentheses.

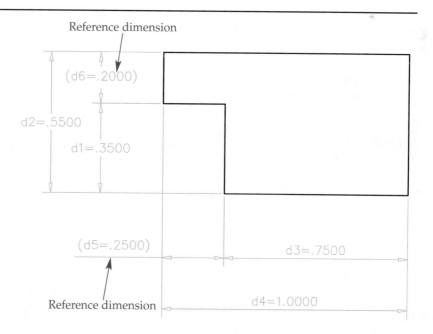

Reference dimension

(d6=.2000)

d2=.5500

d1=.3500

(d5=.2500)

d3=.7500

d4=1.0000

Reference dimension

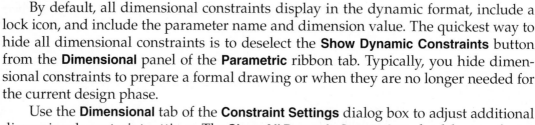

NOTE

Dimensional constraints define and constrain drawing geometry. Dimensional constraints use a unique AutoCAD style and do not comply with ASME standards. Do not be overly concerned about the placement or display characteristics of these dimensions, but if possible, apply dimensional constraints just as you would add dimensions to a drawing, using correct drafting practices. In addition, move and manipulate dimensional constraints so the drawing is as uncluttered as possible. Use grips to make basic adjustments to dimensional constraint position.

Managing Dimensional Constraints

Ribbon
**Parametric
> Dimensional**

**Show Dynamic
Constraints**

Ribbon
**Parametric
> Dimensional**

Constraint Settings

Type
CONSTRAINTSETTINGS

By default, all dimensional constraints display in the dynamic format, include a lock icon, and include the parameter name and dimension value. The quickest way to hide all dimensional constraints is to deselect the **Show Dynamic Constraints** button from the **Dimensional** panel of the **Parametric** ribbon tab. Typically, you hide dimensional constraints to prepare a formal drawing or when they are no longer needed for the current design phase.

Use the **Dimensional** tab of the **Constraint Settings** dialog box to adjust additional dimensional constraint settings. The **Show All Dynamic Constraints** check box performs the same function as the **Show Dynamic Constraints** ribbon button. The **Dimension name format** drop-down list allows you to display dimensional constraints with the parameter name and value, parameter name, or value. Use the **Show lock icon for additional constraints** check box to toggle the lock icon on or off for new dimensional constraints. If you hide dimensional constraints and select the **Show hidden dynamic constraints of selected objects** check box, you can pick an object to show associated dimensional constraints temporarily.

Hiding dimensional constraints does not remove them. In some cases, you may need to delete existing constraints in order to apply different parameters. Use the **ERASE** tool to eliminate specific dimensional constraints.

NOTE

You can also right-click with no objects selected to access a **Parametric** cascading submenu that provides options for displaying and hiding constraints, changing the dimension name format, and accessing the **Constraint Settings** dialog box. The options are also available when you pick a dimensional constraint and then right-click.

Working with Parameters

dimensional constraint parameters: Parameters added when you insert a dimensional constraint.

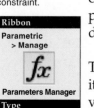

Ribbon
**Parametric
> Manage**

fx

Parameters Manager

Type
PARAMETERS

A *dimensional constraint parameter* automatically forms every time you add a dimensional constraint. You can adjust parameters by changing the dimensional constraint value or by using the options in the **Constraint** category of the **Properties** palette. The **Parameters Manager** allows you to manage all of the parameters in the drawing. See **Figure 22-24**.

You can use the **Filter** flyout to show only those parameters used in expressions. To change a parameter name, pick inside a text box in the **Name** column to activate it, type the new name, and press [Enter] or pick outside of the text box. Enter a new value or expression for the parameter in a text box in the **Expression** column. The value appears in the **Value** column display boxes for reference. Use the **Delete** button or

Figure 22-24.
The **Parameters Manager** is a good resource for reviewing and editing parameters and for creating user parameters.

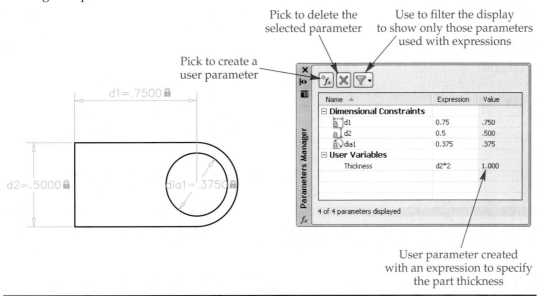

Pick to delete the selected parameter

Use to filter the display to show only those parameters used with expressions

Pick to create a user parameter

User parameter created with an expression to specify the part thickness

right-click on a parameter and select **Delete** to remove the parameter and the corresponding dimensional constraint.

You can also specify your own *user parameters* by picking the **User Parameters** button to display a **User Variables** node. User parameters function like dimensional constraints in the **Parameters Manager**. Create parameters in order to access specific parameters throughout the design process. For example, if you know the thickness of a part will always be twice a certain dimensional constraint, create a user parameter similar to the one in **Figure 22-24** to define the thickness. You can then use the custom parameter for reference and in expressions when you place additional dimensional constraints.

user parameters: Additional parameters you define.

Exercise 22-9

Access the Student Web site (www.g-wlearning.com/CAD) and complete Exercise 22-9.

Converting Dimensional Constraints

The **Convert** option of the **DIMCONSTRAINT** tool allows you to convert an associative dimension to a dimensional constraint. This allows you to prepare a parametric drawing using existing associative dimensions. Access the **Convert** option, pick the associative dimensions to convert, and press [Enter] or pick outside of the text editor. By default, new dimensional constraints and converted dimensions use the dynamic format. See **Figure 22-25**. Recall that this is the primary format for creating parametric geometry, but it is not necessarily associated with formal dimension practices. **Figure 22-26A** shows all of the dynamic dimensions and geometric constraints required to fully constrain the example multiview drawing.

You can use the **Form** option of the **DIMCONSTRAINT** tool to change the dimensional constraint format. As explained earlier, annotational dimensions reference the current dimension style. To convert dynamic dimensions to the annotational format, first make the dimension style and dimension layer to apply to the annotational

Ribbon
Parametric > Dimensional

Convert

Type
DIMCONSTRAINT

Figure 22-25.
An example of converting a dimension drawn using the **DIMDIAMETER** tool to a dimensional constraint. The default dimensional constraint format is dynamic.

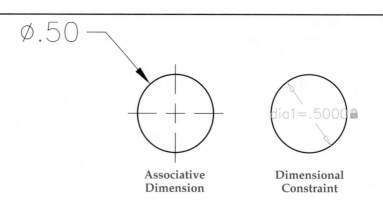

Associative
Dimension

Dimensional
Constraint

dimensions current. Then access the **Form** option, pick the **Annotational** format, select the associative or dynamic dimensions, and press [Enter] or pick outside of the text editor. See **Figure 22-26B.** You can also use the **Constraint Form** drop-down list in the **Constraint** category of the **Properties** palette to change dimensional constraint form. Annotational constraints still control object size and location.

NOTE

The specified **Annotational** or **Dynamic** form remains set and applies to new dimensional constraints until you change to the alternate setting.

Annotational dimensions, especially those converted from dynamic dimensions, may require that you make some format and organizational changes to prepare the final drawing. You may also have to add non-parametric dimensions. Make the following changes to create the final drawing shown in **Figure 22-26C.**
- Hide all geometric constraints and dynamic dimensions.
- Disable the lock icon display.
- Change the dimension name format to **Value**.
- Make basic dimension style overrides and dimension location adjustments if necessary.

Exercise 22-10

Access the Student Web site (www.g-wlearning.com/CAD) and complete Exercise 22-10.

AutoCAD
NEW

Parametric Editing

You can adjust existing parametric drawings in a variety of ways. Drawing objects adds additional unconstrained geometry. You may need to delete or replace existing constraints in order to constrain new objects. Erasing objects and exploding polyline objects removes constraints. If you erase geometry associated with an expression, an alert appears asking if you want to convert the dimensional constraint to a user parameter, maintain the information, or remove the parameter with the dimensional constraint.

Figure 22-26.
A typical
dimensional
constraint
conversion process.
A—A fully
constrained drawing
with dynamic
dimensional
constraints. Notice
that dimensions
apply to all items,
including centerline
extensions and
the distance
between views.
B—Converting
dynamic dimensions
to annotational
dimensions.
Convert only
those dimensions
required for formal
dimensioning.

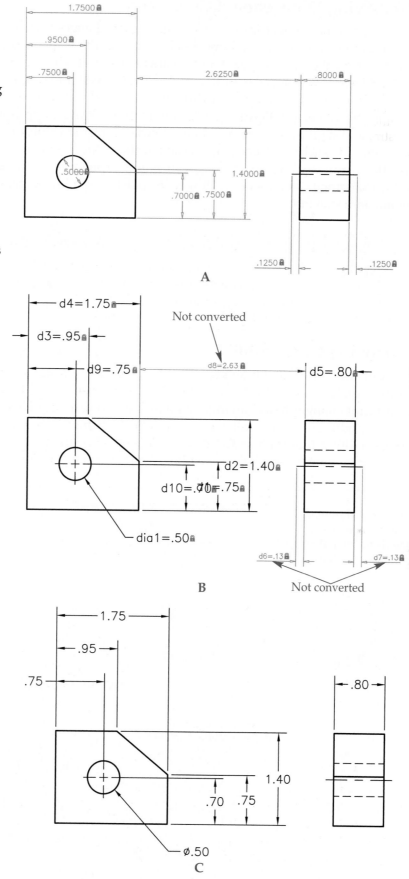

Modifying Dimensional Constraints

Adjust dimensional constraints to make design changes to a constrained drawing. To edit a dimensional constraint value, double-click on the value, or select the dimensional constraint, right-click, and pick **Edit Constraint**. The text editor appears, allowing you to make changes. Press [Enter] or pick outside of the text editor to complete the operation.

Another method is to pick a dimensional constraint and use parameter grips to change the value. See **Figure 22-27.** This method is most appropriate when you are analyzing design options, if you do not know a specific value, or when you are using drawing aids such as object snaps to adjust the value. You can also modify a dimensional constraint value by entering a different expression in the **Expression** text box in the **Constraint** category of the **Properties** palette, or in the **Expression** text box of the **Parameters Manager**.

NOTE

If your drawing includes enough geometric constraints, and the geometric constraints are accurate, changing dimensions should maintain all geometric relationships.

Removing Constraints

Constraints limit the ability to make changes to a drawing. For example, you cannot rotate a horizontally or vertically constrained line. If it is necessary to make significant changes to a drawing, you may have to relax or delete the constraints that limit the edit. Constraints relax using standard editing practices. When you are using a basic editing tool such as **ROTATE**, a message appears asking if you want to relax constraints. To relax constraints while grip stretching, moving, or scaling, you may

Figure 22-27.
Using a parameter grip to change the dimensional constraint value.

Selected dimensional constraint

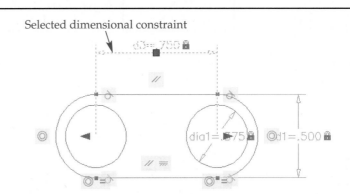

Original Drawing

Value changes according to the second stretch point

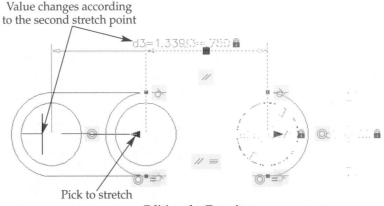

Pick to stretch

Editing the Drawing

need to press [Ctrl] to toggle relaxing on and off. See **Figure 22-28A**. Other edits automatically remove constraints. See **Figure 22-28B**. The only constraints, if any, that are removed are those required to achieve the edit.

In addition to the methods described for deleting individual constraints, a **DELCONSTRAINT** tool is available. Access the tool and select the geometric and dimensional constraints to remove. This tool is especially effective for removing a significant number of constraints. Use the **All** selection option to remove all constraints from the drawing.

Ribbon
Parametric > Manage
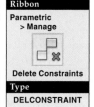
Delete Constraints
Type
DELCONSTRAINT

NOTE

Some editing techniques, such as mirroring or moving, may only require removal of fix constraints, especially when they are applied to an entire drawing.

PROFESSIONAL TIP

Occasionally, constraining geometry causes objects to twist out of shape, making it difficult to control the size and position of the drawing. Use the **UNDO** tool to return to the previous design. To help avoid this situation, consider the following suggestions:
- Construct objects at or close to their finished size using standard drafting practices.
- Add as many geometric constraints as appropriate before dimensioning.
- Dimension the largest objects first.
- Move objects to a more appropriate location, if necessary, and change object size before constraining.

Exercise 22-11

Access the Student Web site (www.g-wlearning.com/CAD) and complete Exercise 22-11.

Figure 22-28.
Examples of relaxing constraints.
A—Access the MOVE tool and then press [Ctrl] to remove the concentric constraint. The process is similar for stretching and scaling. B—Rotating automatically removes, in this example, a vertical constraint.

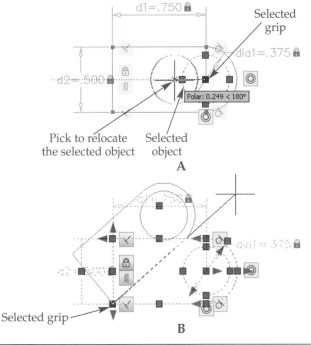

Chapter Test

Answer the following questions. Write your answers on a separate sheet of paper or go to the Student Web site (www.g-wlearning.com/CAD) and complete the electronic chapter test.

1. Give an example demonstrating how you can use constraints to form geometric constructions in specific situations when standard AutoCAD tools are inefficient or ineffective.
2. Describe two applications in which parametric drafting is unsuitable or ineffective.
3. Briefly describe what geometric constraint tools allow you to do, and identify what you see that indicates the presence of a geometric constraint.
4. Name the tool that allows you to assign specific geometric constraints to objects.
5. When you use geometric constraint tools that allow you to pick two objects or points, describe what happens to the first and second objects you select.
6. A coincident constraint is one of the most common parametric constructions. Give an example of objects that coincide.
7. Identify common uses for horizontal and vertical constraints.
8. Name the two basic object types that can form parallel or perpendicular constraints.
9. What does the **Collinear** constraint allow you to do?
10. Name the types of objects you can constrain with the **Tangent** constraint.
11. Explain the basic function of the **Equal** constraint.
12. Describe the default function of the **Symmetric** constraint.
13. Name the tool you can use if you want to attempt to add all required geometric constraints in a single operation.
14. Briefly describe how to specify the appearance and characteristics of geometric bars.
15. Compare the appearance of a coincident constraint with the display of other constraints.
16. Explain how you can determine which objects and points are associated with constraints.
17. What should you do if constraint bars block your view or if you want to hide constraint bars?
18. Name the tool that allows you to assign linear, diameter, radius, and angular dimensional constraints.
19. Describe the basic process used to create a dimensional constraint.
20. What is the most basic method to specify dimension values when you create a dimensional constraint?
21. Name the option that allows you to place horizontal or vertical linear dimensional constraints.
22. Name the option that allows you to place a linear dimensional constraint with a dimension line aligned with, and extension lines perpendicular to, an angled surface.
23. Which option allows you to place an angular dimension between two objects or between three points?
24. Explain the options AutoCAD provides when you try to over-constrain a defined object.
25. What happens every time you add a dimensional constraint?
26. How do you adjust parameters?
27. Name the tool and option that allow you to convert an associative dimension to a dimensional constraint, and give the advantage of using this option.
28. Explain how to edit a dimensional constraint value.
29. Briefly describe how to relax constraints.
30. Which tool provides an efficient method of removing a significant number of constraints in a single operation?

Drawing Problems

- *Start AutoCAD if it is not already started.*
- *Start a new drawing using an appropriate template of your choice. The template should include layers, text styles, dimension styles, and multileader styles appropriate for drawing the given objects.*
- *Add layers, text styles, dimension styles, and multileader styles as needed. Draw all objects using appropriate layers, text styles, dimension styles, multileader styles, justification, and format.*
- *Follow the specific instructions for each problem. Use your own judgment and approximate dimensions when necessary.*
- *Apply formal dimensions accurately using ASME or appropriate industry standards.*

Note: Dimensional constraints shown for reference are created using AutoCAD and may not comply with ASME standards.

▼ Basic

1. Use the **POLYGON** tool to draw the hexagon and fully constrain it as shown. All sides are equal. Edit the d1 parameter to change the distance across the flats to 4.000. Save the drawing as P22-1.

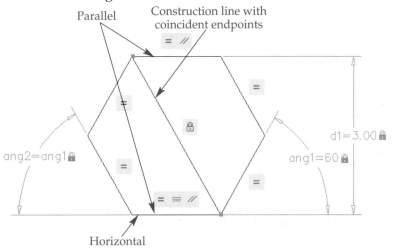

2. Use the **RECTANGLE** and **CIRCLE** tools to draw the view and fully constrain it as shown. The circle is tangent to the rectangle in two locations. Edit the d2 parameter to change the distance to 4.500. Edit the d3 parameter to change the distance to 2.000. Save the drawing as P22-2.

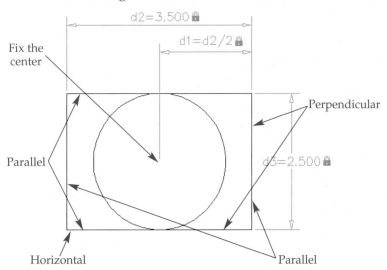

3. Draw and fully constrain the circles shown. Both of the smaller circles are tangent to the larger circle. Edit the dia1 parameter to change the diameter to 2.000. Save the drawing as P22-3.

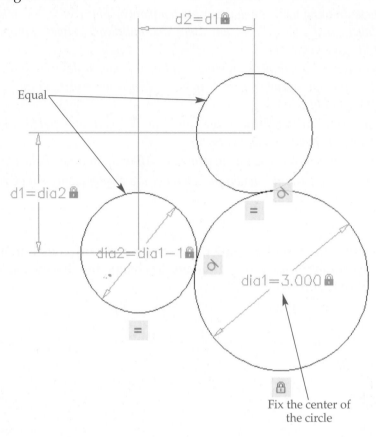

4. Draw and fully constrain the pipe spacer shown. Use the **AUTOCONSTRAIN** tool, with default settings, to apply most of the geometric constraints. Make all circles equal in size. Edit the d1 parameter to change the distance to 2.000. Edit the rad1 parameter to change the radius to 1.250. Save the drawing as P22-4.

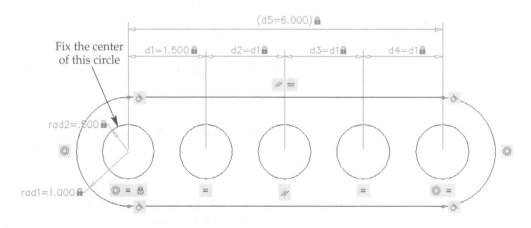

5. Draw and fully constrain the view shown. Follow the guidelines below.
   - Begin by using the **Fillet** option of the **RECTANGLE** tool to create the 4.6250 by 6.3750 rectangle with .6250 rounded corners. Notice that the circles are not concentric to the arcs.
   - Use the **RECTANGLE** tool to create the construction rectangle.
   - Apply geometric constraints in the following order: fix the center of the lower-left circle; use the **Autoconstrain** function of the **Coincident** option to apply all coincident constraints; make all circles equal; make all arcs equal.
   - Use the **AUTOCONSTRAIN** tool with default settings to apply the remaining geometric constraints.
   - Edit the d1 parameter to change the distance to 1.0000.
   - Edit the d5 parameter to change the distance to 5.1250.
   - Save the drawing as P22-5.

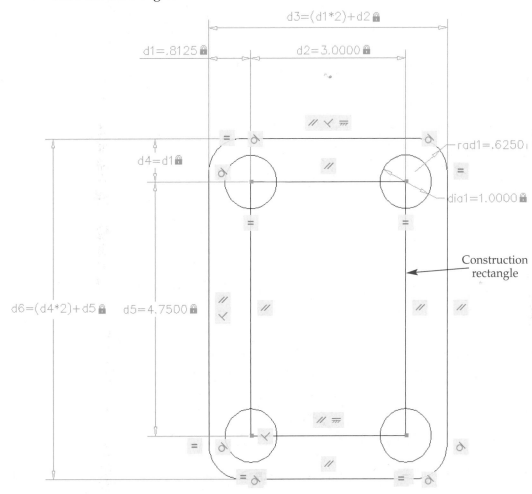

6. Draw and fully constrain the view shown. Apply geometric constraints in the following order: fix the center of the center circle; use the **Autoconstrain** function of the **Coincident** option to apply all coincident constraints; make all circles equal; make all small arcs equal; make all large arcs concentric to the appropriate circle; use the **AUTOCONSTRAIN** tool with default settings to apply the remaining geometric constraints. Edit the d1 parameter to change the distance to 1.750. Save the drawing as P22-6.

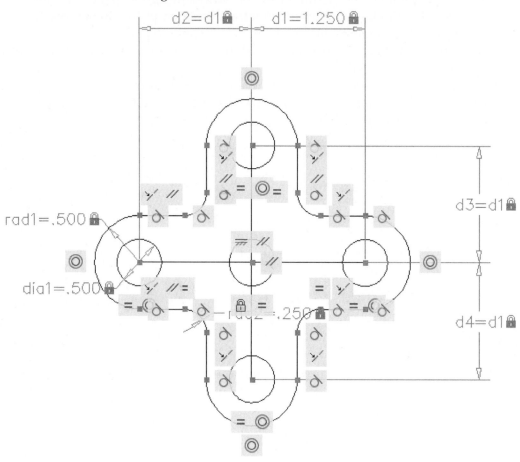

▼ **Advanced**

7. Draw and fully constrain the view shown. Save the drawing as P22-7.

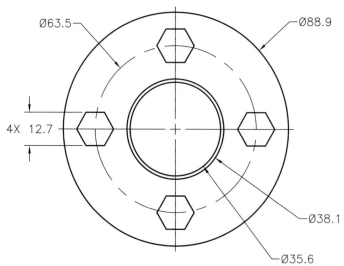

8. Use the information below to draw the non-parametric view shown in A. Then fully constrain the view as shown in B. Edit the view as shown in C. Finish by converting dimensional constraints to create the formal drawing shown in D. Save the drawing as P22-8.

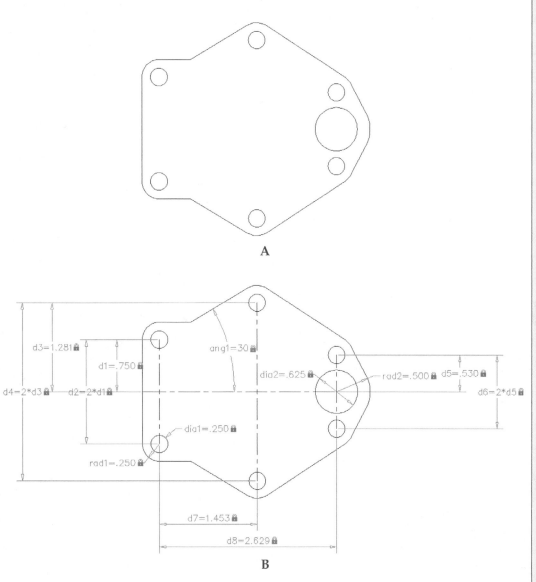

A

B

(Continued)

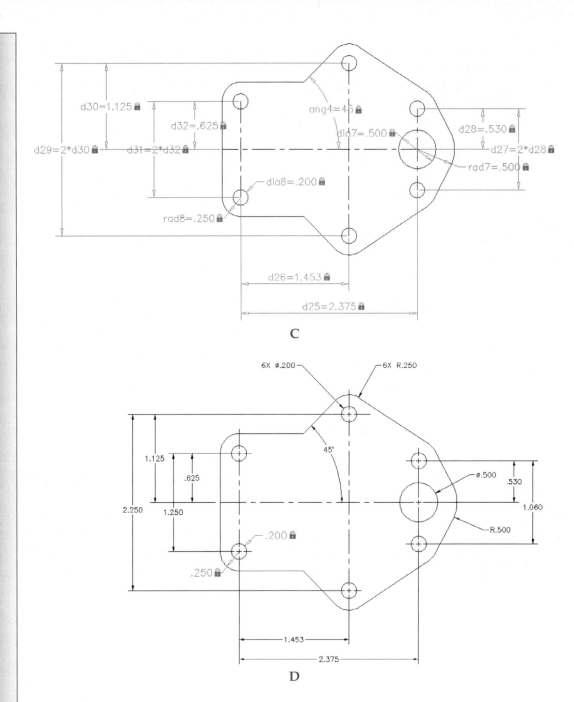

C

D

9. Draw and fully constrain the view shown. Convert dynamic dimensional constraints to annotational dimensional constraints. Adjust the drawing and add associative dimensions as needed to create the formal drawing shown. Save the drawing as P22-9.

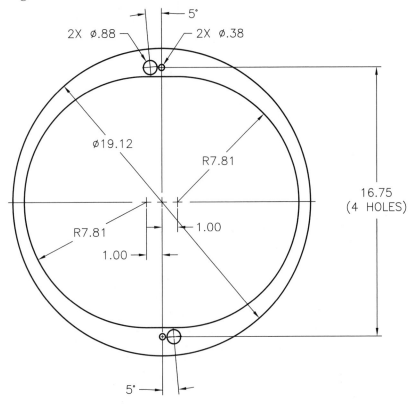

10. Draw and fully constrain the multiview drawing of the support. Convert dynamic dimensional constraints to annotational dimensional constraints. Adjust the drawing and add associative dimensions as needed to create the formal drawing shown. Save the drawing as P22-10.

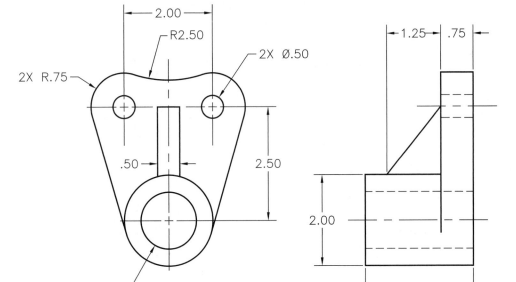

11. Use the isometric drawing provided to create a multiview orthographic drawing for the part. Include only the views necessary to fully describe the object. Draw and fully constrain the drawing. Convert dynamic dimensional constraints to annotational dimensional constraints. Adjust the drawing and add associative dimensions as needed to create a formal drawing. Save the drawing as P22-11.

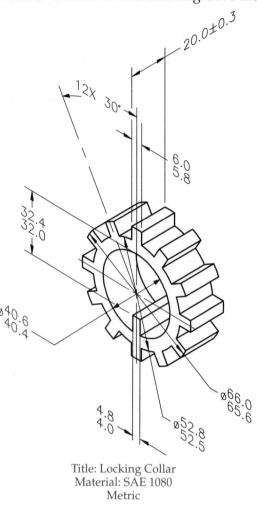

Title: Locking Collar
Material: SAE 1080
Metric

Section Views and Graphic Patterns

Learning Objectives

After completing this chapter, you will be able to do the following:

✓ Identify sectioning techniques.
✓ Add graphic patterns using the **HATCH** tool.
✓ Insert hatch patterns into drawings using **DesignCenter** and tool palettes.
✓ Edit existing hatch patterns with the **HATCHEDIT** tool and grips.

Many drawings require *graphic patterns* to describe specific information. For example, the front elevation of the house shown in **Figure 23-1** contains patterns of lines that create graphic representations of building materials. One of the most common graphic patterns is a group of section lines added to a section view. This chapter focuses on using the **HATCH** tool to draw graphic patterns quickly.

graphic pattern: A patterned arrangement of objects or symbols.

Figure 23-1.
Graphic patterns describe repetitive drawing information, such as the siding, brick, roof, and concrete step materials added to the front elevation of a house. In this illustration, all of the items shown in color were drawn as graphic patterns.

Section Views

It is poor practice, and sometimes not even possible, to dimension internal, hidden features. *Section views*, or *sections*, allow you to clarify hidden features. Typically, you add section views to a multiview drawing to describe both the exterior and interior features of a product. Often, as shown in **Figure 23-2,** a primary view is a section or includes a section.

When a section is included, one of the drawing views contains a *cutting-plane line* to show the location of the cut. The cutting-plane line is drawn with a thick dashed or phantom line in accordance with ASME Y14.2M, *Line Conventions and Lettering*. The arrows on the cutting-plane line indicate the line of sight when you are looking at the section view. Cutting-plane lines often include labels with letters that relate to the proper section view.

A title, such as SECTION A-A, under the section view links the section with the labeled cutting-plane. When more than one section view is drawn, labels continue with B-B through Z-Z. The letters *I, O,* and *Q* are not used because they may be confused with numbers. Labeling section views is necessary for drawings with multiple sections. The label is sometimes omitted when only one section view is present and its location is obvious. *Section lines* help distinguish hidden features from exterior objects.

Other drafting fields, including architectural, structural, and civil, also use sectioning. Cross sections through buildings show the construction methods and materials. See **Figure 23-3.** The cutting-plane lines used in these fields are often composed of letter and number symbols. This helps coordinate the large number of sections found in a set of architectural drawings.

Section Lines

Section views include section line symbols to show where material has been cut to reveal hidden features. In most cases, 45° section lines are standard, unless another angle is required to satisfy the next two rules. Avoid section lines placed at angles greater than 75° or less than 15° from horizontal. Section lines should never be parallel or perpendicular to any other adjacent lines on the drawing. In addition, section lines should not cross object lines.

section view (section): A view that shows internal features as if a portion of the object has been cut away.

cutting-plane line: The line that cuts through the object to expose internal features.

section lines: Lines that show where material has been cut away.

Figure 23-2.
A two-view drawing with a full section view and a broken-out section.

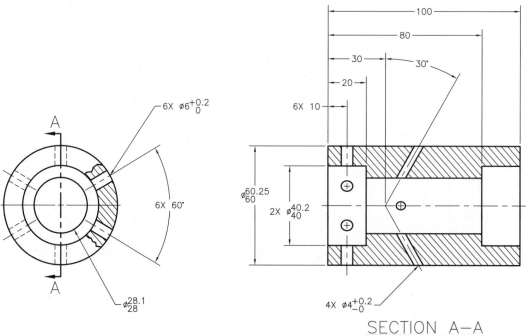

SECTION A—A

Figure 23-3.
An architectural section view. (Alan Mascord Design Associates)

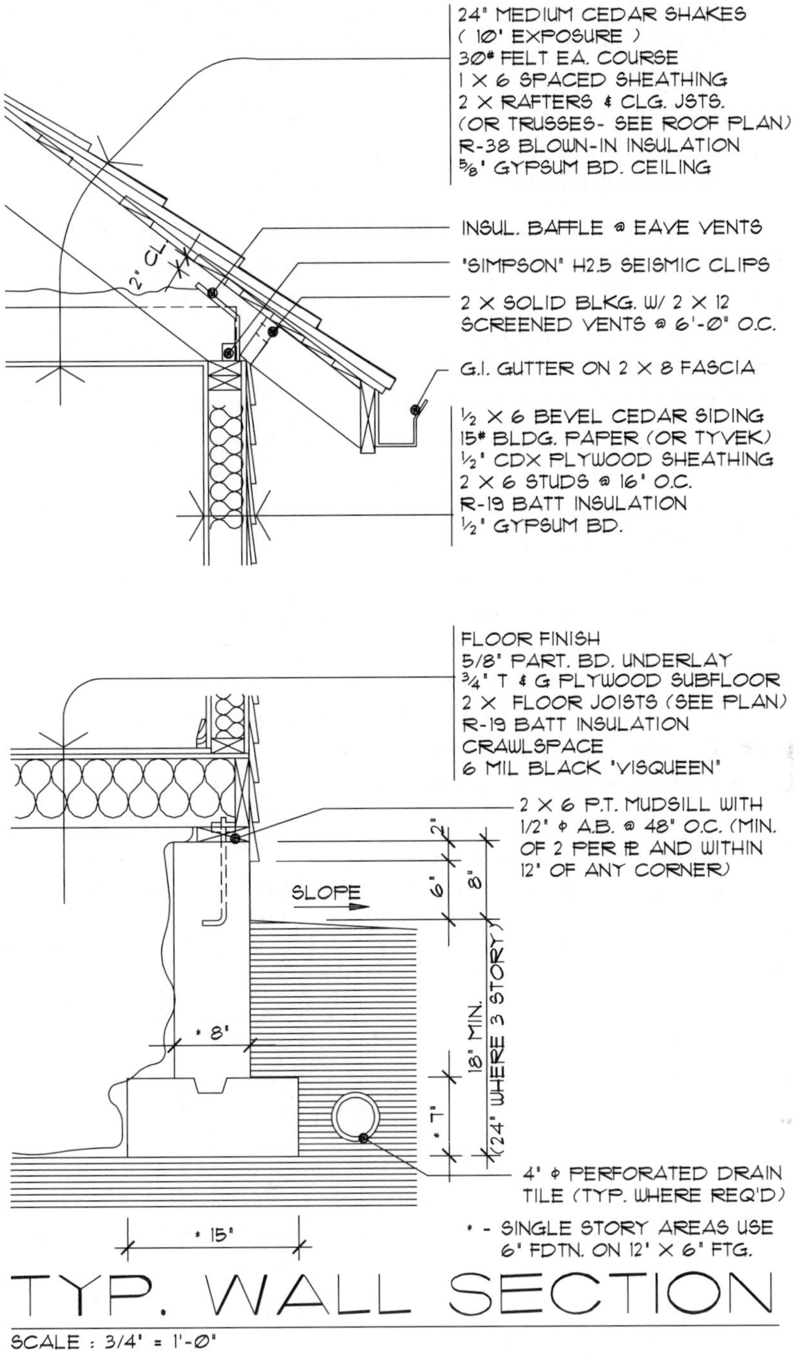

24" MEDIUM CEDAR SHAKES
(10" EXPOSURE)
30# FELT EA. COURSE
1 X 6 SPACED SHEATHING
2 X RAFTERS & CLG. JSTS.
(OR TRUSSES- SEE ROOF PLAN)
R-38 BLOWN-IN INSULATION
⅝" GYPSUM BD. CEILING

INSUL. BAFFLE @ EAVE VENTS

'SIMPSON' H2.5 SEISMIC CLIPS

2 X SOLID BLKG. W/ 2 X 12
SCREENED VENTS @ 6'-0" O.C.

G.I. GUTTER ON 2 X 8 FASCIA

½ X 6 BEVEL CEDAR SIDING
15# BLDG. PAPER (OR TYVEK)
½" CDX PLYWOOD SHEATHING
2 X 6 STUDS @ 16" O.C.
R-19 BATT INSULATION
½" GYPSUM BD.

FLOOR FINISH
⅝" PART. BD. UNDERLAY
¾" T & G PLYWOOD SUBFLOOR
2 X FLOOR JOISTS (SEE PLAN)
R-19 BATT INSULATION
CRAWLSPACE
6 MIL BLACK 'VISQUEEN'

2 X 6 P.T. MUDSILL WITH
½" ⌀ A.B. @ 48" O.C. (MIN.
OF 2 PER 12 AND WITHIN
12" OF ANY CORNER)

SLOPE

4" ⌀ PERFORATED DRAIN
TILE (TYP. WHERE REQ'D)

* - SINGLE STORY AREAS USE
6" FDTN. ON 12" X 6" FTG.

TYP. WALL SECTION

SCALE : 3/4" = 1'-0"

Section lines may be drawn using different patterns to represent specific types of material. Equally spaced section lines like those in **Figure 23-2** represent a general application. This is adequate in most situations. Additional patterns are not necessary if the title block or a note clearly indicates the type of material. However, different section line material symbols are needed when connected parts of different materials are sectioned.

AutoCAD provides standard section line symbols known as *hatch patterns*. The acad.pat file stores standard hatch pattern symbols. The ANSI31 pattern is a general section line symbol and is the default pattern in some templates. The ANSI31 pattern also represents cast iron in a section. The ANSI32 symbol identifies steel in a section. When you change to a different hatch pattern, the new pattern becomes the default in the current drawing until changed.

hatch patterns:
AutoCAD section line symbols and graphic patterns.

When you section very thin objects, section lines can completely blacken or shade the cut material to clarify features. Use the Solid hatch pattern for this application, or whenever it is necessary to create a solid filled area. The ASME Y14.2M standard recommends that you draw very thin sections without section lines or solid fills.

Types of Sections

Many types of sections can be used, depending on the application. Choose the appropriate section according to the features and detail to be sectioned. For example, an object that includes a significant amount of hidden detail may require a section that cuts completely through the object. In contrast, a drawing may require that you only remove a small portion to expose and dimension a single, minor interior feature.

Figure 23-2 shows an example of a *full section*. In this type of section, the cutting plane passes completely through the object along the center plane, as shown by the cutting-plane line. *Offset sections* are similar to full sections, except the cutting plane staggers. This allows you to cut through features that are not in a straight line. See **Figure 23-4**.

Figure 23-5 provides an example of an *aligned section*. The cutting plane cuts through the feature to section, and then *rotates* to align with the center plane before projecting onto the section view. Using an offset section for this application would distort the image.

Figure 23-6 shows an example of a *revolved section*. A revolved section may appear in place within the object, or a portion of the view may be broken away to make dimensioning easier. *Removed sections* serve much the same function as revolved sections. See **Figure 23-7**. A cutting-plane line identifies the location of the section. Multiple removed sections require labeled cutting-plane lines and related views. Drawing only the ends of the cutting-plane lines simplifies the views.

full sections:
Sections in which half the object is removed.

offset sections:
Sections that have a staggered cutting plane.

aligned sections:
Sections used when a feature is out of alignment with the center plane.

revolved sections:
Sections that clarify the contour of objects that have the same shape throughout their length.

removed sections:
Section views that are similar to revolved sections, but are removed from the regular view.

Figure 23-4.
An offset section.

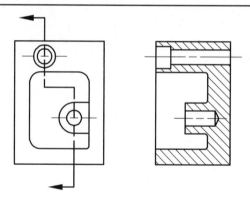

Figure 23-5.
An aligned section.

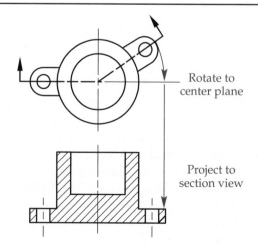

Rotate to center plane

Project to section view

AutoCAD and Its Applications—Basics

Figure 23-6.
A revolved section.

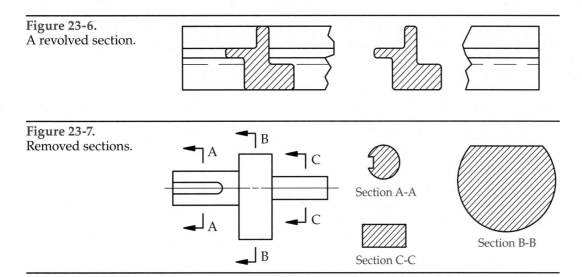

Figure 23-7.
Removed sections.

Figure 23-8 shows an example of a *half section*. The term *half* describes how half of the view appears in section, while the other half remains as an exterior view. Half sections are common when drawing symmetrical objects. A centerline separates the sectioned part of the view from the unsectioned part. You normally omit hidden lines from the unsectioned side. *Broken-out sections* clarify a specific hidden feature(s). See **Figure 23-9.** **Figure 23-2** also shows a broken-out section.

half sections:
Sections that show one-quarter of the object removed.

broken-out sections: Sections that show only a small portion of the object removed.

NOTE

Section lines are a basic application for hatch patterns. Many other drawings also require hatching, such as for material representation on an architectural elevation, shading on a technical illustration, and artistic patterns for graphic layouts.

Figure 23-8.
A half section.

Figure 23-9.
A broken-out
section.

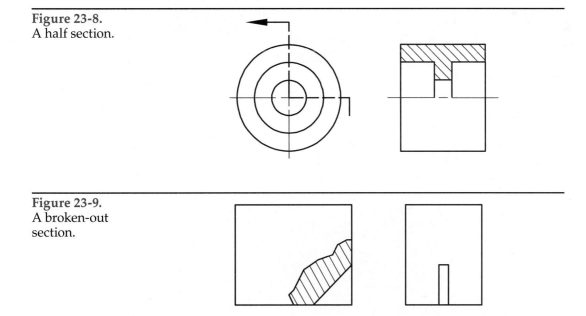

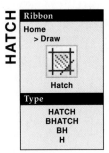

HATCH

Ribbon
Home
> Draw

Hatch

Type
HATCH
BHATCH
BH
H

boundary: The area filled by a hatch pattern.

The **HATCH** tool simplifies the process of creating section lines and graphic patterns. The **Hatch and Gradient** dialog box, shown in **Figure 23-10,** appears when you access the **HATCH** tool and allows you to apply a hatch pattern to a *boundary*. Use the **Hatch** tab to apply common drafting graphic patterns, such as section lines.

Selecting a Hatch Pattern

Use the **Type and pattern** area of the **Hatch** tab to select a hatch pattern. Hatch pattern categories are available in the **Type:** drop-down list. Hatch types include **Predefined**, **User defined**, and **Custom**. Once you select the pattern type, specific options become enabled for selecting a pattern and controlling pattern characteristics.

Predefined Hatch Patterns

The **Predefined** hatch type provides patterns stored in the acad.pat and acadiso.pat files. Use the **Pattern:** drop-down list to select a predefined hatch pattern by name. Alternatively, pick the ellipsis (**...**) button next to the **Pattern:** drop-down list or pick in the **Swatch:** preview box to display the **Hatch Pattern Palette** dialog box. See **Figure 23-11.** The **ANSI, ISO, Other Predefined**, and **Custom** tabs divide the hatch patterns into groups. The name and an image identify each pattern. Select the pattern from the appropriate tab and pick the **OK** button to return to the **Hatch and Gradient** dialog box. The selected pattern appears in the **Pattern:** text box and the **Swatch:** preview box.

Figure 23-10.
The **Hatch** tab of the **Hatch and Gradient** dialog box.

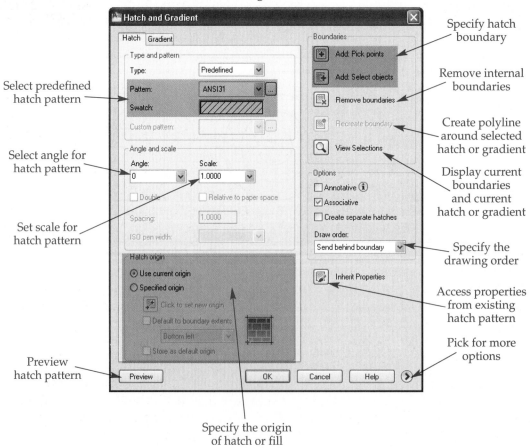

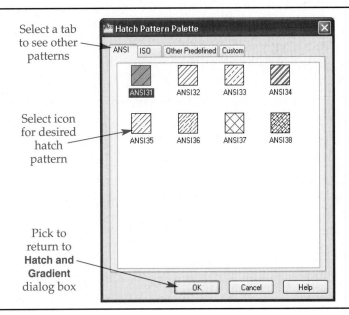

Select a tab to see other patterns

Select icon for desired hatch pattern

Pick to return to **Hatch and Gradient** dialog box

Predefined hatch patterns use specific angle and scale characteristics. You can modify these settings for unique applications using the options in the **Angle and scale** area. Specify a value in the **Angle:** text box to rotate the pattern. Specify a value in the **Scale:** text box to change the pattern size. For example, by default, the ANSI31 hatch is a pattern of 45° lines spaced .125″ (3 mm) apart. If you change the angle to 15° and the scale to 2, a pattern of 60° (45+15=60) lines spaced .25″ (6 mm) apart forms.

NOTE

You can control the ISO pen width for predefined ISO patterns using the **ISO pen width:** drop-down list.

User-Defined Hatch Patterns

The **User defined** hatch type creates a pattern of equally spaced lines for basic hatching applications. The lines use the current linetype. The options in the **Angle and scale** area allow you to form a specific pattern of lines. Specify a value in the **Angle:** text box to rotate the pattern relative to the X axis. Use the **Spacing:** text box to specify the distance between lines in the pattern. Check **Double** to create a pattern of double lines. **Figure 23-12** shows examples of user-defined hatch patterns.

Custom Hatch Patterns

You create and save custom hatch patterns in PAT files. The **Custom** hatch type allows you to specify a pattern defined in any custom PAT file you add to the AutoCAD search path. Use the **Custom pattern:** drop-down list to select a custom pattern name, or pick the ellipsis (...) button or the **Swatch:** preview box to select the pattern from the **Custom** tab of the **Hatch Pattern Palette** dialog box. You can set the angle and scale of custom hatch patterns using the same techniques as for predefined hatch patterns.

Setting the Hatch Pattern Size

Adjust the size of predefined and custom hatch patterns using the **Scale:** text box. The default scale is 1. If the pattern appears too small or large, specify a different scale. See **Figure 23-13**. User-defined hatch pattern size is set using the **Spacing:** text box and defines the exact distance between lines. Refer again to **Figure 23-12**.

Figure 23-12.
Examples of user-defined hatch patterns with different hatch angles and spacing.

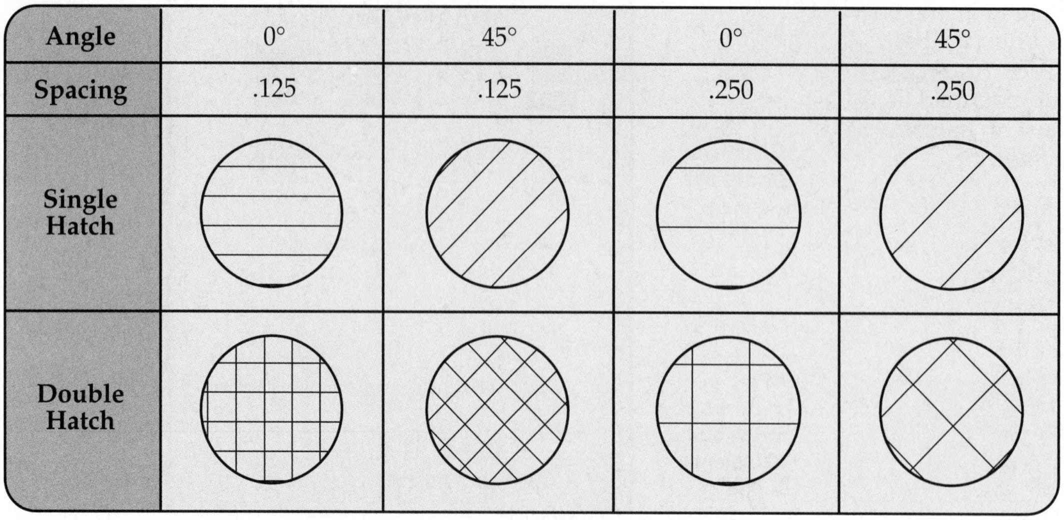

Angle	0°	45°	0°	45°
Spacing	.125	.125	.250	.250
Single Hatch				
Double Hatch				

Figure 23-13.
Hatch pattern scale factors.

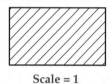 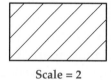 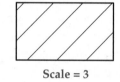

| Scale = 1 | Scale = 2 | Scale = 3 |

PROFESSIONAL TIP

Use a smaller hatch size for small objects and a larger hatch size for larger objects. This makes section lines look appropriate for the drawing scale. Often you must use your best judgment when selecting a hatch size.

The drawing scale is an important consideration when you select the hatch pattern scale or spacing. You must use an appropriate hatch size in order to make sure the hatch pattern appears on-screen and plots correctly. To understand the concept of hatch size, look at the section view shown in **Figure 23-14.** In this example, which uses the ANSI31 hatch pattern, the section line spacing should be the same distance apart regardless of drawing scale. The section lines on the full-scale (1:1) drawing display correctly. However, the section lines are too close on the half-scale (1:2) drawing, and they are too far apart on the double-scale (2:1) drawing. To overcome this issue, you must adjust the hatch pattern according to the drawing scale. You can calculate and manually apply the scale factor to the hatch pattern scale or spacing, or you can allow AutoCAD to calculate the scale factor by using annotative hatch patterns.

Scaling Hatch Patterns Manually

To adjust hatch size manually according to a specific drawing scale, you must calculate the drawing scale factor and then multiply the scale factor by the desired plotted hatch scale or spacing to get the model space hatch scale or spacing. Enter the adjusted scale of predefined or custom hatch patterns in the **Scale:** text box. Enter the adjusted spacing of a user-defined hatch pattern in the **Spacing:** text box. **Figure 23-15** shows examples of adjusting hatch scale according to drawing scale. Refer to Chapter 9 for information on determining the drawing scale factor.

Figure 23-14.
The hatch pattern may appear incorrect if the drawing scale changes.

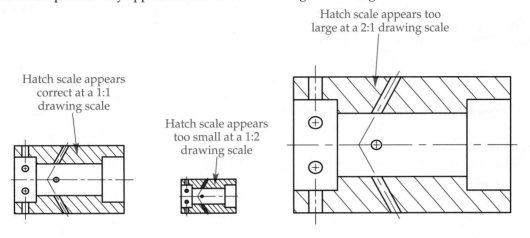

Hatch scale appears too
large at a 2:1 drawing scale

Hatch scale appears
correct at a 1:1
drawing scale

Hatch scale appears
too small at a 1:2
drawing scale

Figure 23-15.
The hatch pattern scale may require adjusting, depending on the drawing scale.

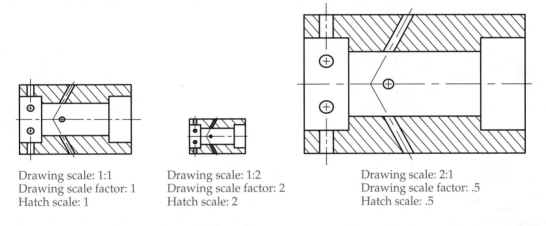

Drawing scale: 1:1
Drawing scale factor: 1
Hatch scale: 1

Drawing scale: 1:2
Drawing scale factor: 2
Hatch scale: 2

Drawing scale: 2:1
Drawing scale factor: .5
Hatch scale: .5

Annotative Hatch Patterns

Pick the **Annotative** check box in the **Options** area to make the hatch pattern annotative. AutoCAD scales annotative hatch patterns according to the annotation scale you select, which eliminates the need for you to calculate the scale factor. When you select an annotation scale from the **Annotation Scale** flyout button on the status bar, AutoCAD determines the scale factor and automatically applies it to annotative hatch patterns, or any annotative object. The result is a hatch pattern that displays the proper size regardless of the drawing scale, much like the example shown in **Figure 23-15,** but without requiring you to change the hatch scale or spacing between different drawing scales. For example, if you enter a value in the **Scale:** text box or **Spacing:** text box that is appropriate for an annotation scale of 1/4″ = 1′-0″, and then change the annotation scale to 1″ = 1′-0″, the appearance of the hatch pattern relative to the drawing scale does not change. It looks the same on the 1/4″ = 1′-0″ scale drawing as it does on the 1″ = 1′-0″ scale drawing.

Scaling Relative to Paper Space

The **Relative to paper space** check box in the **Angle and scale** area allows you to scale the hatch pattern relative to the scale of the active layout viewport. You must enter a floating layout viewport in order to select the **Relative to paper space** check box. The hatch scale automatically adjusts according to the viewport scale. For example, a floating viewport scale set to 4:1 uses a scale factor of .25 (1 ÷ 4 = .25). If you enter a hatch scale of 1, the hatch automatically appears at a scale of .25 (1 × .25 = .25).

Setting the Hatch Origin Point

The **Hatch origin** area includes options that control the position of hatch patterns. The default setting is **Use current origin**, which refers to the current UCS origin, to define the point from which the hatch pattern forms and how the pattern repeats. In some cases, it is important that a hatch pattern align with, or originate from, a specific point. A common example is hatching the representation of bricks.

To specify a different origin point, select the **Specified origin** radio button. See **Figure 23-16**. Pick the **Click to set new origin** button to return to the drawing and select an origin point. **Figure 23-17** shows an example of the difference between applying the **Use current origin** setting and picking a specific origin point. The **Hatch and Gradient** dialog box reappears after you select an origin point.

You can align the hatch origin point with a specific point on the hatch boundary by checking **Default to boundary extents**. Then use the corresponding drop-down list to select **Bottom left**, **Bottom right**, **Top right**, **Top left**, or **Center**. The hatch origin positions at the selected point on the boundary. For example, you could use the **Bottom right** option to create the pattern shown in **Figure 23-17B**. Check **Store as default origin** to save the custom origin point.

Specifying the Hatch Boundary

The **Add: Pick points** button often provides the easiest method of defining the hatch boundary. Pick the **Add: Pick points** button to return to the drawing and pick a point within the area to hatch. AutoCAD automatically defines and highlights the boundary around the selected point. You can select more than one internal point. Press [Enter] or the space bar, or right-click and select **Enter** to return to the **Hatch and Gradient** dialog box. See **Figure 23-18**.

Figure 23-16.
The **Specified origin** setting in the **Hatch origin** area of the **Hatch** tab activates the other hatch origin settings.

Pick to specify a different origin point

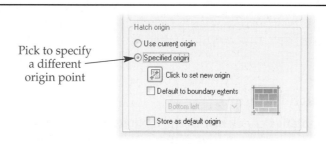

Figure 23-17.
A—The default **Use current origin** setting. B—The **Specified origin** option is selected and the **Click to set new origin** button is used to select the lower-left corner (endpoint) of the rectangle. Notice how the pattern, or in this example the first brick, starts exactly at the corner of the hatched area.

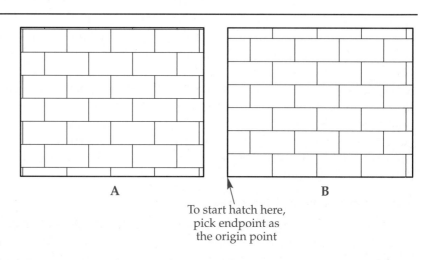

A

B

To start hatch here, pick endpoint as the origin point

Figure 23-18.
Picking a point to
define the hatch
boundary.

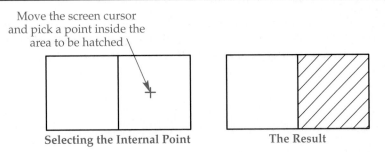

Move the screen cursor
and pick a point inside the
area to be hatched

Selecting the Internal Point   The Result

NOTE

If you pick the wrong area, use the **UNDO** tool at the Select internal point: prompt to undo the selection. You can also undo the hatch pattern after the pattern is drawn. However, to save time, you can preview the hatch before applying it.

Use the **Add: Select objects** button to define the hatch boundary by selecting objects, rather than picking inside an area. See **Figure 23-19.** Remember to select objects within the hatch boundary to exclude from the hatch pattern if necessary. See **Figure 23-20.** Press [Enter] or the space bar, or right-click and select **Enter** to return to the **Hatch and Gradient** dialog box. Selecting objects works especially well if other objects cross the area to hatch, forming several possible boundaries. **Figure 23-21** shows the difference between picking points and objects when hatching a bar graph.

The **Remove boundaries** button is available after you select a point or objects to define a boundary. If necessary, pick the **Remove boundaries** button to return to the drawing and select objects to remove from the hatch boundary. Press [Enter] or the space bar, or right-click and select **Enter** to return to the **Hatch and Gradient** dialog box.

Figure 23-19.
Selecting objects to
hatch.

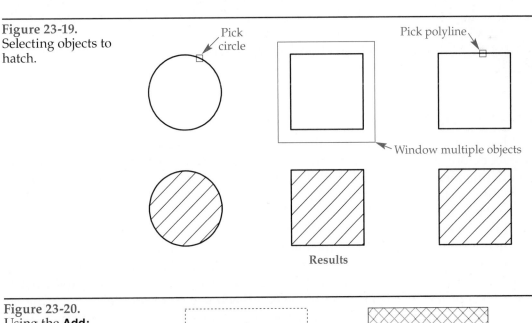

Pick circle

Pick polyline

Window multiple objects

Results

Figure 23-20.
Using the **Add: Select objects** button to exclude an object from the hatch pattern.

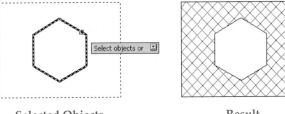

Selected Objects   Result

Figure 23-21.
A—Applying a hatch pattern to objects that cross each other using the **Add: Pick points** button.
B—Applying a hatch pattern to a closed polygon using the **Add: Select objects** button.

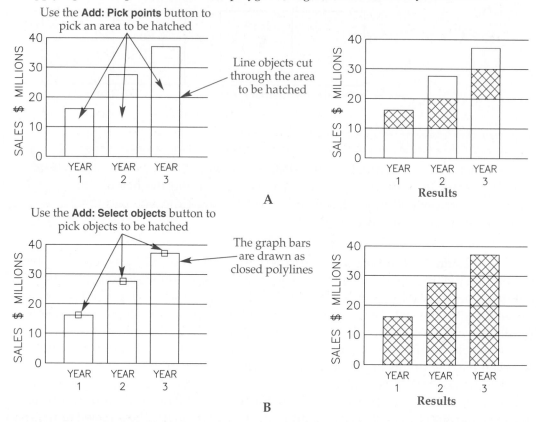

NOTE

Prompt options are available for toggling between point selection, object selection, or object removal while you are defining a boundary.

Previewing the Hatch

Use preview tools to be sure the hatch pattern and hatch boundary settings are correct before applying a hatch pattern. The **View Selections** button is available after you define a hatch boundary. Pick this button to return to the drawing window to view the selected hatch boundary. Press [Enter] or the space bar, or right-click to return to the **Hatch and Gradient** dialog box.

Pick the **Preview** button, located in the lower-left corner of the **Hatch and Gradient** dialog box, to temporarily place the hatch pattern on the drawing. This allows you to see if any changes are required before the hatch is drawn. Press [Enter] or right-click to accept the preview and create the hatch. To make changes after previewing the hatch, press [Esc] or the space bar to return to the **Hatch and Gradient** dialog box. Change hatch pattern settings as needed and preview the hatch again. Apply the hatch when you are satisfied with the preview. Pick the **OK** button in the **Hatch and Gradient** dialog box to create the hatch pattern without previewing, or at any time after you define the boundary.

Exercise 23-1

Access the Student Web site (www.g-wlearning.com/CAD) and complete Exercise 23-1.

Hatch Pattern Composition Options

The **HATCH** tool creates an *associative hatch pattern* by default. To create a *nonassociative hatch pattern*, deselect the **Associative** check box in the **Options** area of the **Hatch and Gradient** dialog box. An associative hatch is appropriate for most applications. If you stretch, scale, or otherwise edit the objects that define the boundary of an associative hatch, the pattern automatically adjusts to and fills the modified boundary. A nonassociative hatch pattern does not respond this way. Instead, nonassociative hatch boundary grips are available for changing the extents of the hatch, separate from the original boundary objects.

You can select multiple points and objects during a single hatch operation. By default, multiple boundaries form a single hatch object. Selecting and editing one of the hatch patterns selects and edits all patterns created during the same operation. If this is not the preferred result, check the **Create separate hatches** check box in the **Options** area before applying the hatch patterns. Individual hatch patterns will then form for each boundary.

associative hatch pattern: Pattern that updates automatically when the associated objects are edited.

nonassociative hatch pattern: A pattern that is independent of objects; it updates when the boundary changes, but not when changes are made to objects.

Controlling the Draw Order

The **Draw order** drop-down list in the **Options** area provides options for controlling the order of display when a hatch pattern overlaps other objects. The **Send behind boundary** option is default and makes the hatch pattern appear behind the boundary. Select the **Bring in front of boundary** option to make the hatch pattern appear on top of the boundary. Select the **Do not assign** option to have no automatic drawing order setting assigned to the hatch.

Use the **Send to back** option to send the hatch pattern behind all other objects in the drawing. Any objects that are in the hatching area appear as if they are on top of the hatch pattern. Use the **Bring to front** option to bring the hatch pattern in front of, or on top of, all other objects in the drawing. Any objects that are in the hatching area appear as if they are behind the hatch pattern.

> **NOTE**
>
> Use the **DRAWORDER** tool to change the draw order setting after creating a hatch pattern.

Hatching Objects with Islands

When defining a hatch boundary using the **Add: Pick points** button, you may need to adjust how AutoCAD treats *islands*, like those shown in **Figure 23-22**. By default, islands do not hatch, as shown in **Figure 23-22B**. One option to hatch islands is to use the **Remove boundaries** button after selecting the internal point. Then pick the islands to remove and press [Enter] or the space bar, or right-click and pick **Enter** to return to the dialog box and create the hatch. See **Figure 23-22C**.

Another method is to adjust island detection in the **Islands** area. Pick the **More Options** button in the lower-right corner of the **Hatch and Gradient** dialog box to expand the dialog box to show the **Islands** area and additional hatch settings. See **Figure 23-23**.

islands: Boundaries inside another boundary.

Figure 23-22.
A—Original objects. B—Using the **Add: Pick points** button to hatch an internal area leaves islands unhatched. C—After picking an internal point, use the **Remove boundaries** button and pick the islands to hatch over the islands.

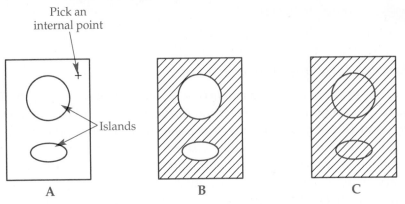

Figure 23-23.
The **Islands** area and additional hatch settings appear when you pick the **More Options** button. The **Boundary retention** and **Boundary set** areas of the **Hatch and Gradient** dialog box provide options to improve hatching efficiency.

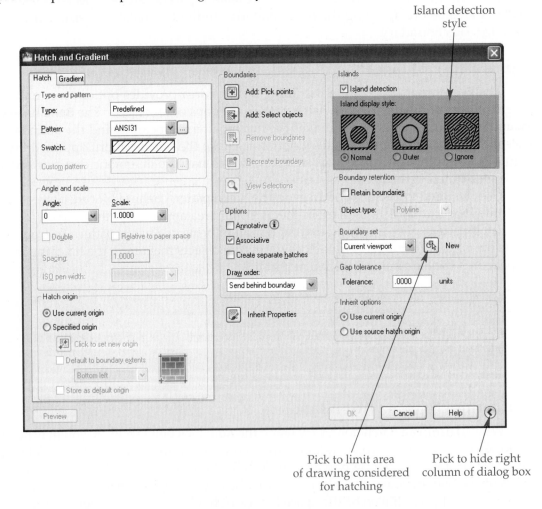

Select the **Island detection** check box to adjust the island display style. The island display style images illustrate the effect of each island detection option.

Pick the **Normal** radio button to hatch every other boundary, stepping inward from the outer boundary. If AutoCAD encounters an island, it turns off hatching until it encounters another island, and then reactivates hatching. Select the **Outer** radio button to hatch only the outermost area inward from the outer boundary. AutoCAD stops hatching when it encounters the first island. Choose the **Ignore** radio button to ignore all islands and hatch everything within the selected boundary.

PROFESSIONAL TIP

Pick the **Outer** island display style to ensure that islands do not unintentionally hatch.

Exercise 23-2

Access the Student Web site (www.g-wlearning.com/CAD) and complete Exercise 23-2.

Improving Boundary Hatching Speed

You can improve the hatching speed and resolve other hatching problems using options in the expanded **Hatch and Gradient** dialog box. By default, a hatch pattern forms according to objects in the drawing. The boundary is essentially temporary. This is appropriate for most applications. To form the boundary as a separate object that overlaps the original objects, pick the **Retain boundaries** check box in the **Boundary retention** area. See **Figure 23-24.** Select the Polyline option from the **Object type:** drop-down list to specify the boundary as a polyline object around the hatch area. Choose the Region option to specify the hatch boundary as a hatched *region*. Retaining a boundary forms additional boundary objects that you can use even if you remove or edit the original objects.

region: A closed two-dimensional area.

The **Boundary set** area specifies the amount of drawing area evaluated during the hatch operation. The default setting is Current viewport. To limit the evaluation area, possibly increasing hatch performance, pick the **New** button and use a window to select the area to evaluate. See **Figure 23-25.** After you define the new boundary set, the **Hatch and Gradient** dialog box returns, and the drop-down list in the **Boundary set** area displays Existing set. **Figure 23-26** shows a new hatch pattern applied to a boundary within the boundary set. You can make as many boundary sets as necessary. The last boundary set remains current until you create another.

Figure 23-24.
You can form a boundary as a polyline or a region. These options are only available if the **Retain boundaries** option is checked.

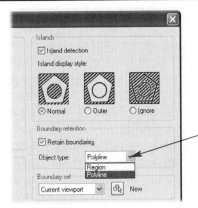

Retain boundary as a region or polyline

Figure 23-25.
The boundary set limits the area that AutoCAD evaluates during a hatching operation.

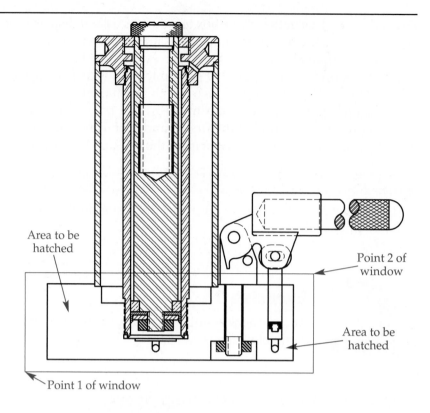

Area to be hatched

Point 2 of window

Area to be hatched

Point 1 of window

Figure 23-26.
Results of hatching the drawing in **Figure 23-25** after selecting a boundary set.

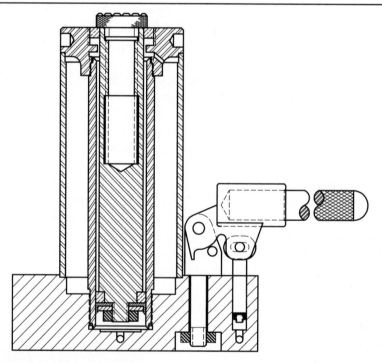

PROFESSIONAL TIP

Apply the following techniques to help save time when hatching, especially large and complex drawings:
- Zoom in on the area to hatch to make it easier for you to define the boundary.
- Preview the hatch before you apply it to make last-minute adjustments.
- Turn off layers that contain lines or text that might interfere with your ability to define hatch boundaries accurately.
- Create boundary sets of small areas within a complex drawing.

AutoCAD and Its Applications—Basics

Hatching Unclosed Areas and Correcting Boundary Errors

The **HATCH** tool works well unless there is a gap in the hatch boundary or you pick a point outside a likely boundary. When you select a point where no boundary can form, an error message states that a valid boundary cannot be determined. Close the message and try again to specify the boundary. When you try to hatch an area that does not close because of a small gap, you will see the error message and circles shown in **Figure 23-27.** Close the message and eliminate the gap to create the hatch.

For most applications, it is best to identify and close a gap. However, you can hatch an unclosed boundary by setting a *gap tolerance* in the **Gap Tolerance** area. AutoCAD ignores any gaps in the boundary less than or equal to the value specified in the **Tolerance:** text box. Before generating the hatch, AutoCAD displays a message allowing you to hatch the unclosed area or return to the **Hatch and Gradient** dialog box.

gap tolerance:
The amount of gap allowed between segments of a boundary to be hatched.

PROFESSIONAL TIP

When you create an associative hatch, it is often best to specify a single internal point per hatch. If you specify more than one internal point in the same operation, AutoCAD creates one hatch object from all points picked. This can cause unexpected results when you try to edit what appears to be a separate hatch object.

Creating Solid and Gradient Fills

The Solid predefined hatch pattern, available from the **Pattern:** drop-down list or **Other Predefined** tab of the **Hatch Pattern Palette** dialog box, is an effective way to fill a boundary with a solid. See **Figure 23-28.** To create a more advanced gradient fill, use the options in the **Gradient** tab of the **Hatch and Gradient** dialog box. See **Figure 23-29.**

Gradient fills are commonly used to simulate color-shaded objects. Gradients create the appearance of a lit surface with a gradual transition from an area of highlight to a filled area. Use two colors to simulate a transition from light to dark between

gradient fill: A shading transition between the tones of one color or two separate colors.

Figure 23-27.
Close a boundary to create a hatch pattern.

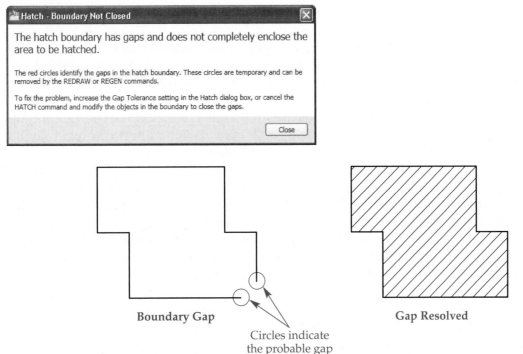

Boundary Gap

Gap Resolved

Circles indicate the probable gap

Figure 23-28.
Using the Solid hatch pattern to make a solid hatch object.

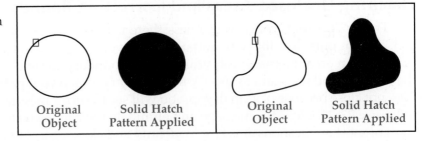

Original Object   Solid Hatch Pattern Applied   Original Object   Solid Hatch Pattern Applied

Figure 23-29.
The **Gradient** tab of the **Hatch and Gradient** dialog box contains options for creating gradient fills.

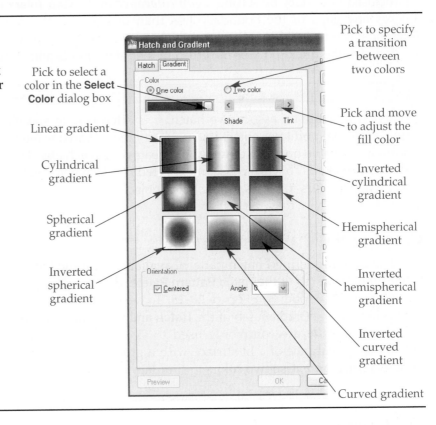

Pick to select a color in the **Select Color** dialog box

Linear gradient

Cylindrical gradient

Spherical gradient

Inverted spherical gradient

Pick to specify a transition between two colors

Pick and move to adjust the fill color

Inverted cylindrical gradient

Hemispherical gradient

Inverted hemispherical gradient

Inverted curved gradient

Curved gradient

the colors. Several different gradient fill patterns are available to create linear sweep, spherical, radial, or curved shading.

The **One color** radio button is the default and creates a fill that has a smooth transition between the darker shades and lighter tints of one color. To select a color, pick the ellipsis (**...**) button next to the color swatch to access the **Select Color** dialog box. When the **One color** option is active, the **Shade** and **Tint** slider appears. Use the slider to specify the *tint* or *shade* of a color used for a one-color gradient fill. Pick the **Two color** radio button to specify a fill using a smooth transition between two colors. A color swatch with an ellipsis (**...**) button is available for each color.

tint: A specific color mixed with white.

shade: A specific color mixed with gray or black.

Pick the **Centered** check box to apply a symmetrical configuration. If you do not select this option, the gradient fill shifts to simulate the projection of a light source from the left of the object. Use the **Angle** text box to specify the gradient fill angle relative to the current UCS. The default angle is 0°. Once you specify gradient characteristics, create gradients using the same boundary selection process, options, and settings you would use to apply a hatch pattern. Gradient fills are associative by default and can be edited using the same methods as other hatch patterns.

The **SOLID** tool allows you to draw basic solid-filled shapes without creating a boundary. Access the **SOLID** tool and pick points in a specific sequence to form different shapes. Using the **HATCH** tool with the Solid predefined hatch pattern is typically a better method of creating solid fills.

Type	
SOLID	SOLID
SO	

Exercise 23-3

Access the Student Web site (www.g-wlearning.com/CAD) and complete Exercise 23-3.

Reusing Existing Hatch Properties

You can specify hatch pattern characteristics by referencing an identical hatch pattern from the drawing. Pick the **Inherit Properties** button and select an existing hatch pattern. The crosshairs appears with a paintbrush icon. Pick a point inside a different area to define the boundary, and then press [Enter] or the space bar, or right-click and pick **Enter** to return to the **Hatch and Gradient** dialog box. The dialog box displays the settings of the selected pattern.

The **Inherit options** area in the **Hatch and Gradient** dialog box controls the hatch origin. The default **Use current origin** option uses the origin point setting specified in the **Hatch origin** area on the **Hatch** tab. Select the **Use source hatch origin** option to originate the new hatch pattern from the origin of the hatch selected with the **Inherit Properties** button.

Exercise 23-4

Access the Student Web site (www.g-wlearning.com/CAD) and complete Exercise 23-4.

Using DesignCenter to Insert Hatch Patterns

To pattern a boundary using **DesignCenter**, locate and select a PAT file to display the contents in the preview pane. See **Figure 23-30A**. Usually the quickest and most effective technique for transferring a hatch pattern from **DesignCenter** into the active drawing is to use a drag-and-drop operation. Pick the hatch pattern from **DesignCenter** and hold down the pick button. When you move the cursor into the active drawing, a hatch pattern symbol appears with the cursor. See **Figure 23-30B**. Place the cursor in the area to hatch and release the pick button to apply the hatch pattern. See **Figure 23-30C**.

An alternative to drag and drop is copy and paste. Right-click on a hatch pattern in **DesignCenter** and pick **Copy**. Move the cursor into the active drawing, right-click, and select **Paste**. The hatch pattern symbol appears with the crosshairs. Pick in the area to hatch to apply the hatch pattern. You can also use **DesignCenter** in combination with the **Hatch and Gradient** dialog box. Right-click on a hatch pattern in **DesignCenter** and select **BHATCH...** to access the **Hatch and Gradient** dialog box with the selected hatch pattern active.

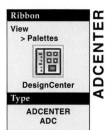

Ribbon	
View	
> Palettes	
DesignCenter	
Type	
ADCENTER	
ADC	

ADCENTER

Figure 23-30.
A—Pick a PAT file in **DesignCenter** to display the available hatch patterns in the preview palette. B—The hatch pattern symbol appears under the cursor during the drag-and-drop and paste operations. C—Pick a point to apply the hatch pattern.

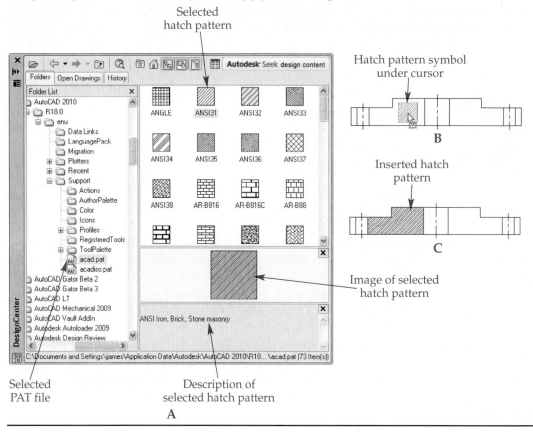

A

When you drag and drop or paste a hatch pattern into an area that is not a closed boundary, the same rules apply as when using the **HATCH** tool. When you insert or reference hatch patterns using **DesignCenter**, the angle, scale, and island detection settings match the settings of the previous hatch pattern. Edit the hatch pattern, as described later in this chapter, to change the settings after you insert the hatch pattern.

> **NOTE**
>
> AutoCAD includes two PAT files: acad.pat and acadiso.pat. These files are located in Program Files/AutoCAD 2010/UserDataCache/ Support. To verify the location of AutoCAD support files, access the **Files** tab in the **Options** dialog box and check the path listed under the Support File Search Path.

Exercise 23-5

Access the Student Web site (www.g-wlearning.com/CAD) and complete Exercise 23-5.

Using Tool Palettes to Insert Hatch Patterns

The **Tool Palettes** palette, shown in **Figure 23-31,** provides an alternative means of storing and inserting hatch patterns. *Tool palettes* can also store and activate many other types of drawing content and tools, such as blocks, images, tables, external reference files, drawing and editing tools, user-defined macros, script files, and AutoLISP routines. The **Command Tools Samples** tool palette contains examples of custom tools. For more information on AutoCAD customization and using tool palettes, refer to *AutoCAD and Its Applications—Advanced.*

Ribbon
View > Palettes

Tool Palettes

Type
TOOLPALETTES TP

TOOLPALETTES

tool palette: A palette that contains tabs to help organize tools and other features.

Locating and Viewing Content

Each tool palette in the **Tool Palettes** palette has its own tab along the side of the window. To view the content in a tool palette, pick the related tab. If the **Tool Palettes** palette contains more palettes than can display on-screen, pick on the edge of the lowest tab to display a selection menu listing the palette tabs. Select the name of the tab to access the related tool palette. If not all of the content of a selected tool palette fits in the window, you can view the remainder using the scroll bar or the scroll hand. The scroll hand appears when you place the cursor in an empty area in the tool palette. Picking and dragging scrolls the tool palette up and down.

> **NOTE**
>
> By default, icons represent the tools in each tool palette. Several tool palette view options are available, including the ability to display a tool as an image of your choice. For more information on adjusting tool palette display, refer to *AutoCAD and Its Applications—Advanced.*

Inserting Hatch Patterns

To insert a hatch pattern from the **Tool Palettes** palette, access the tool palette in which the pattern resides. To drag and drop the pattern, hold down the pick button on the image. When you move the cursor into the active drawing, a hatch pattern symbol appears with the cursor. Place the cursor in the area to hatch and release the

Figure 23-31.
You can use the **Tool Palettes** palette to access and insert hatch patterns.

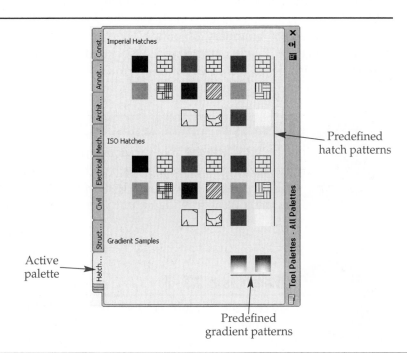

Active palette

Predefined hatch patterns

Predefined gradient patterns

pick button to apply the hatch pattern. An alternative to the drag-and-drop method is to pick once on the hatch image to attach the hatch pattern to the crosshairs, and then pick a boundary in the drawing to apply the hatch pattern. Edit the hatch pattern, as described later in this chapter, to change the settings after inserting the hatch pattern.

NOTE

You can add tool palettes to the **Tool Palettes** window and add tools to tool palettes. For more information on creating and modifying tool palettes, refer to *AutoCAD and Its Applications—Advanced.*

Exercise 23-6

Access the Student Web site (www.g-wlearning.com/CAD) and complete Exercise 23-6.

Editing Hatch Patterns

HATCHEDIT

Ribbon
Home
> Modify

Edit Hatch

Type
HATCHEDIT
HE

A hatch pattern is a single object that you can edit with tools such as **ERASE**, **COPY**, and **MOVE**, grips, or the **Properties** palette. You can also use the **HATCHEDIT** tool to adjust the characteristics of an existing hatch pattern. A quick way to access the **HATCHEDIT** tool is to double-click the hatch pattern to be edited. The **HATCHEDIT** tool opens the **Hatch Edit** dialog box shown in Figure **23-32**. The **Hatch Edit** dialog box is identical to the **Hatch and Gradient** dialog box except for the available **Recreate boundary** and **Select boundary objects** buttons in the **Boundaries** area.

You can use the **Recreate boundary** button to trace boundary objects over the original objects defining the boundary. However, a more practical application is to recreate the geometry if you have erased the original boundary object. See **Figure 23-33.** Pick the button and follow the prompts to create the boundary objects as a region or polyline. You can also specify whether to associate the hatch with the objects. The **Hatch Edit** dialog box reappears after you choose the desired options.

Pick the **Select boundary objects** button to exit the **Hatch and Gradient** dialog box and select the hatch pattern and associated boundary object or nonassociated boundary. This allows you to use grips to make changes to the size and shape of the associated object or the nonassociated boundary.

PROFESSIONAL TIP

The **MATCHPROP** tool provides an alternate method of inheriting the properties of an existing hatch pattern and applying the properties to a different hatch pattern. This tool applies existing hatch patterns to objects in the current drawing file or to objects in other open drawing files.

Exercise 23-7

Access the Student Web site (www.g-wlearning.com/CAD) and complete Exercise 23-7.

Figure 23-32.
The **Hatch Edit** dialog box allows you to edit hatch patterns. Notice that only the options related to hatch characteristics are available.

Pick to recreate boundary

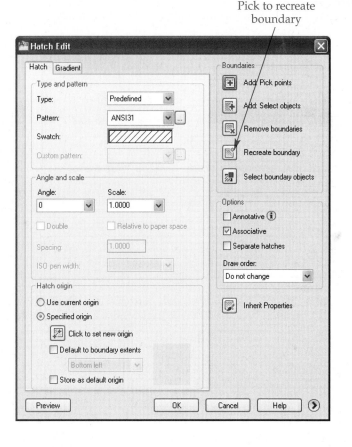

Figure 23-33.
An example of using the **Recreate boundary** button to recreate a lost object associated with the hatch boundary.

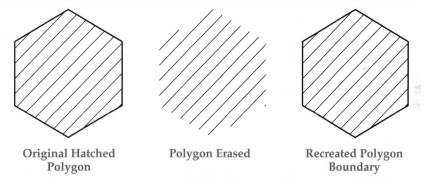

Original Hatched Polygon   Polygon Erased   Recreated Polygon Boundary

Adding and Removing Boundaries

Use the **Add: Pick points**, **Add: Select objects**, and **Remove boundaries** buttons in the **Hatch Edit** dialog box to add boundaries to and remove them from existing associative and nonassociative hatch patterns. In **Figure 23-34**, for example, a rectangle is drawn to create a window. To add the window as an island in the boundary, double-click the hatch pattern to open the **Hatch Edit** dialog box. Pick the **Add: Select objects** button to return to the drawing and pick the rectangle. Return to the **Hatch Edit** dialog box and complete the operation.

Editing Associative Hatch Patterns

When you edit an object associated with a hatch pattern, the hatch pattern changes to adapt to the edit. **Figure 23-35** shows examples of stretching an associated object and removing an island from an associative boundary. As long as you edit the objects associated with the boundary, the hatch pattern will update.

Figure 23-34.
Adding an object to an existing hatch pattern boundary.

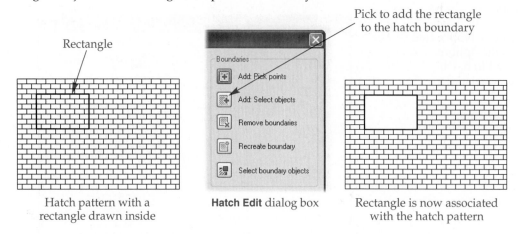

Rectangle

Pick to add the rectangle
to the hatch boundary

Hatch pattern with a
rectangle drawn inside

Hatch Edit dialog box

Rectangle is now associated
with the hatch pattern

Figure 23-35.
Editing objects with associative hatch patterns. A—The hatch pattern stretches with the object. B—The hatch pattern revises to fill the area of an erased island.

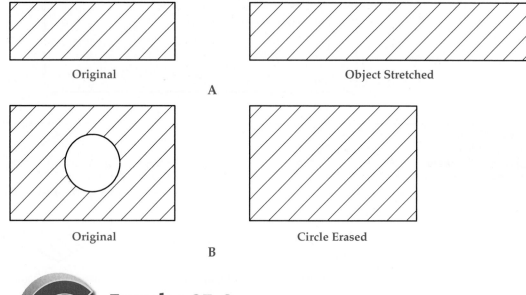

Original

Object Stretched

A

Original

Circle Erased

B

Exercise 23-8

Access the Student Web site (www.g-wlearning.com/CAD) and complete Exercise 23-8.

AutoCAD
2010
NEW

Editing Nonassociative Hatch Patterns

You can create a nonassociative hatch pattern using the **HATCH** or **HATCHEDIT** tool, or by moving an associative hatch pattern away from the associated objects. The objects you reference to create a nonassociative hatch pattern cannot control the size and shape of the hatch. However, you can edit nonassociative hatch patterns using many tools, such as **ROTATE**, **COPY**, and **MOVE**. In addition, special grips allow you to make changes to the boundary. See **Figure 23-36.** Standard grip editing tools and other editing options are available, depending on the selected grip and boundary geometry.

Figure 23-36.
Using the grips that appear when you select a nonassociative hatch pattern. A—Original nonassociative hatch. B—Moving an island out of the boundary and adjusting edge grips. C—Adjusting edge and point grips. D—Moving an island back into the boundary and adjusting edge and point grips.

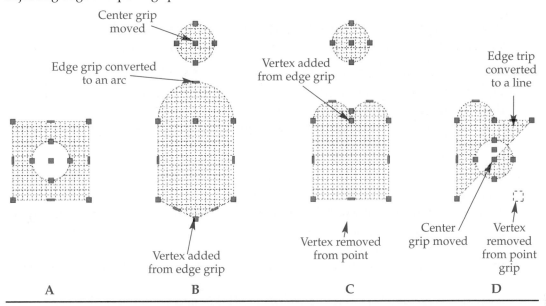

Once you select a grip, press [Ctrl] to cycle though unique editing functions. You may be able to add a vertex to create a new line or arc, remove a vertex to eliminate a line or arc, or convert a line to an arc or an arc to a line. **Figure 23-36** shows the process of making several changes to a boundary using grip editing techniques.

PROFESSIONAL TIP

When working with associative and nonassociative hatch patterns, remember that associative hatch patterns are associated with objects. The objects define the hatch boundary. Nonassociative hatch patterns are not associated with objects, but they do show association with the hatch boundary.

Exercise 23-9

Access the Student Web site (www.g-wlearning.com/CAD) and complete Exercise 23-9.

Express Tools
Chapter 23

The *Express Tools* ribbon tab includes additional tools for improved functionality and productivity during the drawing processes. The following Express Tool is a hatch express tool. For information about this tool, go to the Student Web site (www.g-wlearning.com/CAD), select this chapter, and select *Inserting Hatch Patterns with the SUPERHATCH Tool.*
Super Hatch

Chapter Test

1. What is AutoCAD's term for standard section line symbols?
2. Which AutoCAD hatch pattern is used as a general section line symbol?

For Questions 3–8, name the type of section identified in each of the following statements:

3. Half of the object is removed; the cutting-plane line generally cuts completely through along the center plane.
4. The cutting-plane line is staggered through features that do not lie in a straight line.
5. The section is turned in place to clarify the contour of the object.
6. The section is rotated and located away from the object. The location of the section is normally identified with a cutting-plane line.
7. The cutting-plane line cuts through one-quarter of the object; used primarily on symmetrical objects.
8. A small portion of the view is removed to clarify an internal feature.
9. Identify two ways to select a predefined hatch pattern in the **Hatch and Gradient** dialog box.
10. How do you change the hatch angle in the **Hatch and Gradient** dialog box?
11. Describe the fundamental difference between using the **Add: Pick points** and the **Add: Select objects** buttons in the **Hatch and Gradient** dialog box.
12. Define *associative hatch pattern*.
13. What is the result of stretching an object that is hatched with an associative hatch pattern?
14. If you use the **Add: Pick points** button inside the **Hatch and Gradient** dialog box to hatch an area, how do you hatch around an island inside the area to be hatched?
15. Explain the three island detection style options.
16. How do you limit AutoCAD hatch evaluation to a specific area of the drawing?
17. What is the purpose of the **Gap Tolerance** setting in the **Hatch and Gradient** dialog box?
18. What are gradient fill hatch patterns? How are they created with the **HATCH** tool?
19. Explain how to use an existing hatch pattern on a drawing as the pattern for your next hatch.
20. Explain how to use drag and drop to insert a hatch pattern from **DesignCenter** into an active drawing.
21. Name the two files that contain hatch patterns that can be copied from **DesignCenter**.
22. Explain two ways to use drag and drop for inserting a hatch pattern from a tool palette into the drawing.
23. Name the tool used to edit existing associative hatch patterns.
24. How does the **Hatch Edit** dialog box compare to the **Hatch and Gradient** dialog box?
25. What happens if you erase an island inside an associative hatch pattern?

Drawing Problems

Start AutoCAD if it is not already started. Start a new drawing using an appropriate template of your choice. The template should include layers, text styles, dimension styles, and multileader styles appropriate for drawing the given objects. Add layers, text styles, dimension styles, and multileader styles as needed. Follow the specific instructions for each problem. Use your own judgment and approximate dimensions when necessary. Draw all objects using appropriate layers, text styles, dimension styles, multileader styles, justification, and format.

▼ Basic

1. Draw the game board as shown. Save the drawing as P23-1.

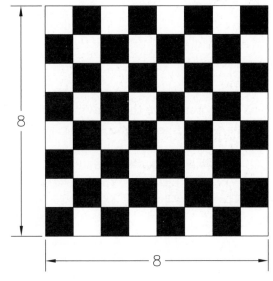

2. Draw the bar graph as shown. Save the drawing as P23-2.

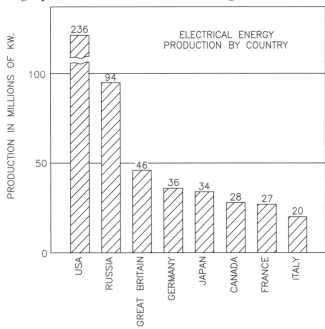

3. Draw the component layout as shown. Save the drawing as P23-3.

COMPONENT LAYOUT

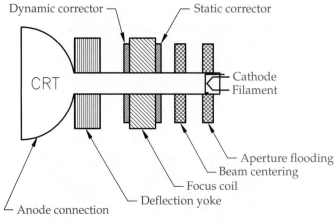

4. Draw the bar graphs as shown. Save the drawing as P23-4.

SOLOMAN SHOE COMPANY

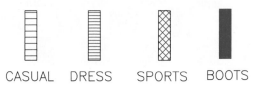

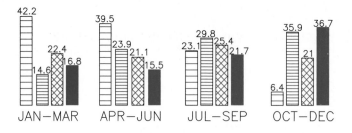

5. Draw the pie chart as shown. Save the drawing as P23-5.

DIAL TECHNOLOGIES
EXPENSE BUDGET
FISCAL YEAR

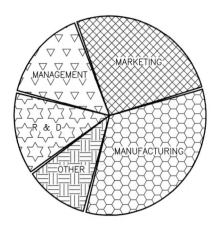

Drawing Problems - Chapter 23

6. Draw the bar graph as shown. Save the drawing as P23-6.

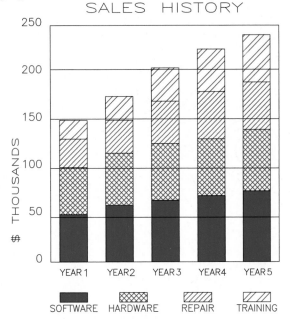

For Problems 7–15, use the following additional guidelines:

- *Apply dimensions accurately using ASME or appropriate industry standards. Use object snap modes to your best advantage.*
- *For mechanical drawing problems in which no notes are specified, place the following notes 1/2" from the lower-left corner:*

   NOTES:
   1. INTERPRET DIMENSIONS AND TOLERANCES PER ASME Y14.5M-1994.
   2. REMOVE ALL BURRS AND SHARP EDGES.
   3. UNLESS OTHERWISE SPECIFIED, ALL DIMENSIONS ARE IN INCHES (or MILLIMETERS as applicable).

▼ Intermediate

7. Draw and dimension the views given, which include aligned and broken-out sections. Add the following notes: FINISH ALL OVER 1.63 mm UNLESS OTHERWISE SPECIFIED and ALL DIMENSIONS ARE IN MILLIMETERS. All arc and circle contours are tangent. Save the drawing as P23-7.

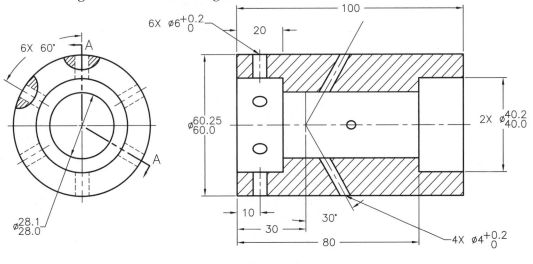

SECTION A—A

Name: Nozzle
Material: Phosphor Bronze

Drawing Problems - Chapter 23

8. Draw and dimension the given views. Save the drawing as P23-8.

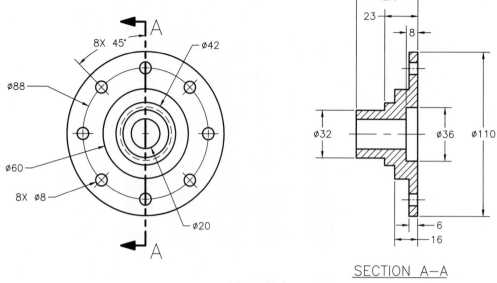

SECTION A—A

Name: Hub
Material: Cast Iron

9. Draw and dimension the given views, including the aligned section shown on the right. Add the following notes: FINISH ALL OVER 1.63 mm UNLESS OTHERWISE SPECIFIED and ALL DIMENSIONS ARE IN MILLIMETERS. Save the drawing as P23-9.

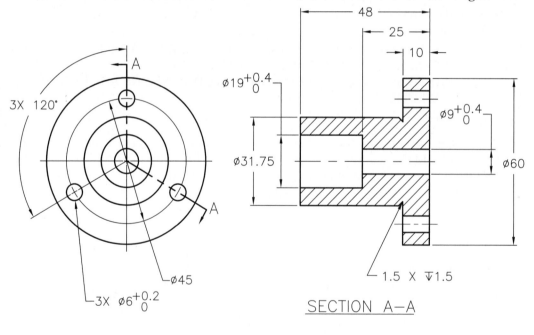

SECTION A—A

10. Draw and dimension the views of the chain guide as shown. Save the drawing as P23-10.

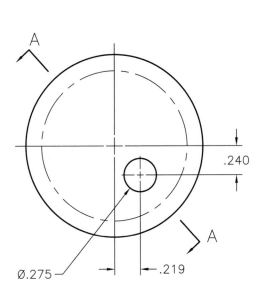

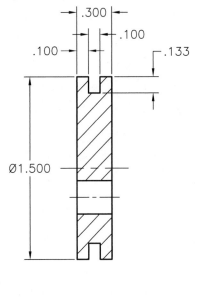

SECTION A—A

11. Draw and dimension the views of the sleeve and add the notes as shown. Save the drawing as P23-11.

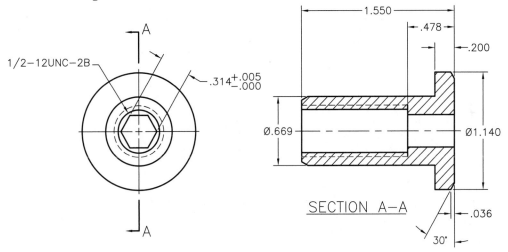

SECTION A—A

4. PAINT ACE GLOSS BLACK ALL OVER.
3. CASE HARDEN 45—50 ROCKWELL.
2. REMOVE ALL BURRS AND SHARP EDGES.
1. INTERPRET ALL DIMENSIONS AND
   TOLERANCES PER ASME Y14.5M—1994.

12. Draw and dimension the given views. Add the following notes: OIL QUENCH 40-45C, CASE HARDEN .020 DEEP, and 59-60 ROCKWELL C SCALE. Save the drawing as P23-12.

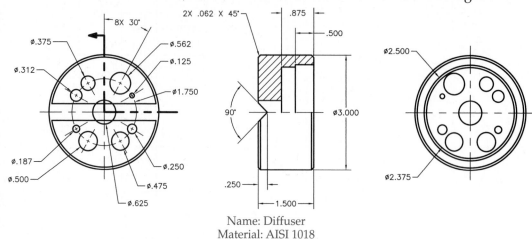

Name: Diffuser
Material: AISI 1018

13. Draw and dimension the views of the tow hook as shown. Save the drawing as P23-13.

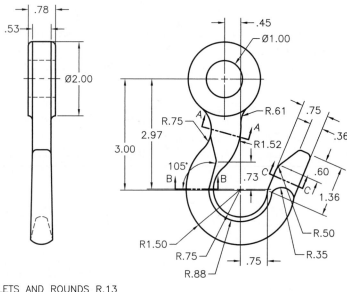

ALL FILLETS AND ROUNDS R.13
UNLESS OTHERWISE SPECIFIED

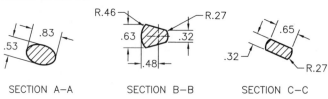

SECTION A–A   SECTION B–B   SECTION C–C

14. Draw the foundation detail as shown. Save the drawing as P23-14.

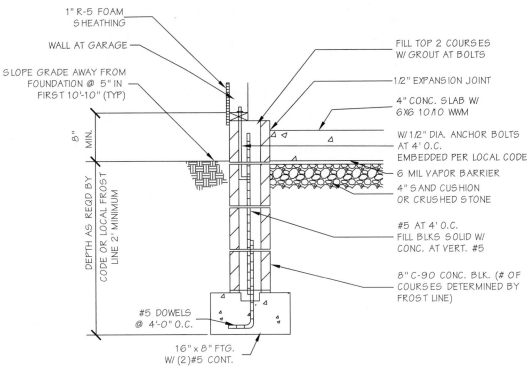

1" R-5 FOAM SHEATHING

WALL AT GARAGE

SLOPE GRADE AWAY FROM FOUNDATION @ 5" IN FIRST 10'-10" (TYP)

8" MIN.

DEPTH AS REQD BY CODE OR LOCAL FROST LINE 2' MINIMUM

#5 DOWELS @ 4'-0" O.C.

16" x 8" FTG. W/ (2)#5 CONT.

FILL TOP 2 COURSES W/ GROUT AT BOLTS

1/2" EXPANSION JOINT

4" CONC. SLAB W/ 6X6 10/10 WWM

W/ 1/2" DIA. ANCHOR BOLTS AT 4' O.C. EMBEDDED PER LOCAL CODE

6 MIL VAPOR BARRIER

4" SAND CUSHION OR CRUSHED STONE

#5 AT 4' O.C. FILL BLKS SOLID W/ CONC. AT VERT. #5

8" C-90 CONC. BLK. (# OF COURSES DETERMINED BY FROST LINE)

15. Draw the stair detail as shown. Save the drawing as P23-15.

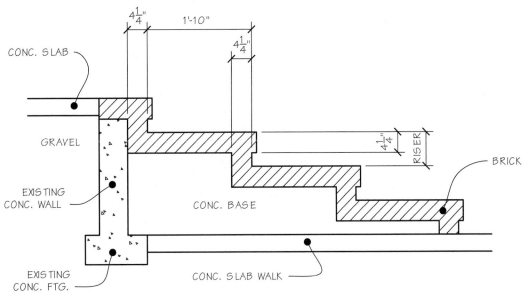

CONC. SLAB

GRAVEL

EXISTING CONC. WALL

EXISTING CONC. FTG.

$4\frac{1}{4}$"

1'-10"

$4\frac{1}{4}$"

$4\frac{1}{4}$" RISER

BRICK

CONC. BASE

CONC. SLAB WALK

16. Draw the front elevation as shown. Save the drawing as P23-16.

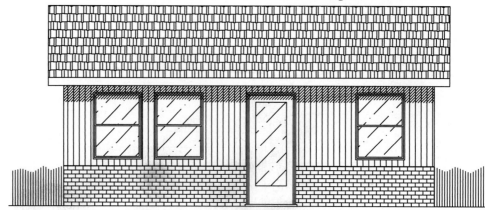

17. Draw the plan as shown. Save the drawing as P23-17.

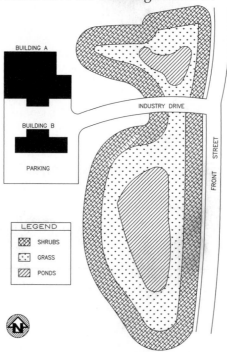

BUILDING A

BUILDING B

PARKING

INDUSTRY DRIVE

FRONT STREET

LEGEND

SHRUBS

GRASS

PONDS

18. Draw the plan as shown. Save the drawing as P23-18.

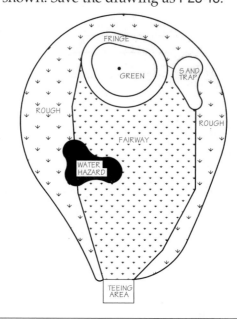

FRINGE

GREEN

SAND TRAP

ROUGH

ROUGH

FAIRWAY

WATER HAZARD

TEEING AREA

19. Draw the plan as shown. Save the drawing as P23-19.

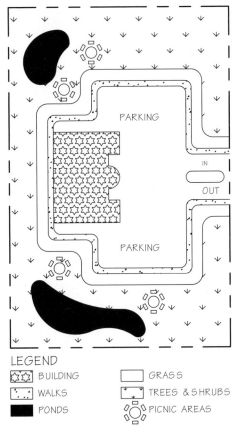

PARKING

IN

OUT

PARKING

LEGEND

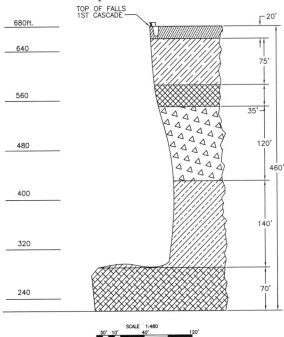

BUILDING

WALKS

PONDS

GRASS

TREES & SHRUBS

PICNIC AREAS

20. Draw the profile as shown. Save the drawing as P23-20.

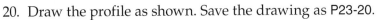

TOP OF FALLS
1ST CASCADE

680ft.

640

560

480

400

320

240

20'

75'

35'

120'

460'

140'

70'

PROFILE OF MULTNOMAH FALLS
&
GEOLOGIC INFORMATION

20' COLLONADE OF AN 80-FOOT THICK FLOW, NOTCHED BY MULTNOMAH CREEK.

75' PILLOW LAVA

35' A GLASSY FLOW, WITH WELL-FORMED ENTABLATURE AND COLLONADE.

120' CONSISTING OF TWO TIERS OF HACKLY-JOINTED BASALT, WITH NO COLLONADE.

140' ENTABLATURE WITH THIN COLUMNS, TOPPED BY A VESICULAR ZONE.

70' OF ENTABLATURE BENEATH THE LOWER FALLS.

BRIEF DESCRIPTION OF TERMS.
COLONNADE: THE LOWER PORTION OF A LAVA FLOW OF COLUMNAR-JOINTED BASALT.
ENTABLATURE: THE UPPER MASSIVE OF A LAVA FLOW OF HACKLY-JOINTED BASALT.

* INFORMATION TAKEN FROM:
"THE MAGNIFICENT GATEWAY"
AUTHOR: JOHN ELIOT ALLEN
PAGES: 89-91

SCALE 1:480

30' 10'   40'   120'
40' 20' 0'   80'

21. Draw the drop cleanout detail as shown. Save the drawing as P23-21.

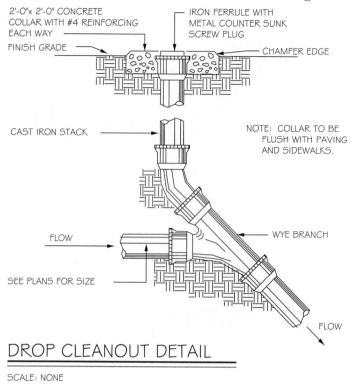

2'-0"x 2'-0" CONCRETE
COLLAR WITH #4 REINFORCING
EACH WAY

FINISH GRADE

IRON FERRULE WITH
METAL COUNTER SUNK
SCREW PLUG

CHAMFER EDGE

CAST IRON STACK

NOTE: COLLAR TO BE
FLUSH WITH PAVING
AND SIDEWALKS.

FLOW

WYE BRANCH

SEE PLANS FOR SIZE

FLOW

DROP CLEANOUT DETAIL

SCALE: NONE

22. Draw the map shown below, using **SPLINE** to create the curved shapes. Try to make the shapes as similar to those on the map as possible. Save the drawing as P23-22.

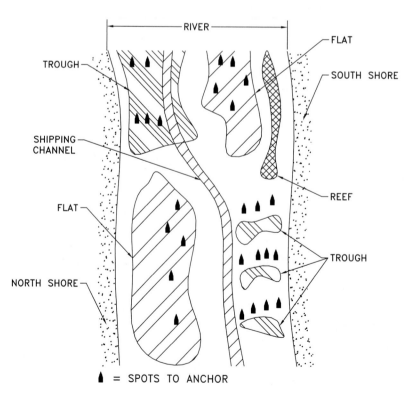

CATFISHING ANCHOR LOCATIONS

RIVER

TROUGH

FLAT

SOUTH SHORE

SHIPPING
CHANNEL

REEF

FLAT

TROUGH

NORTH SHORE

▲ = SPOTS TO ANCHOR

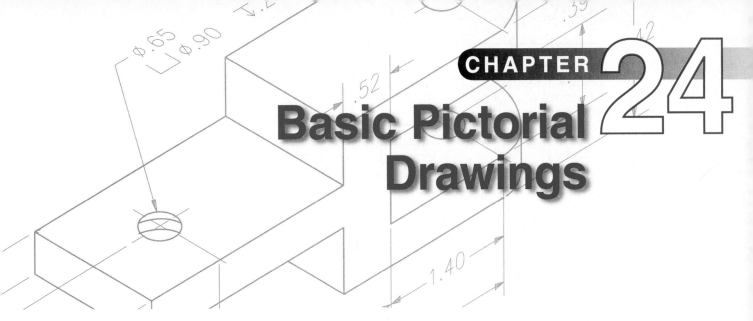

Learning Objectives

After completing this chapter, you will be able to do the following:
- ✓ Describe the three basic types of pictorial drawings.
- ✓ Construct accurate isometric drawings.
- ✓ Dimension isometric drawings.

Being able to visualize and draw three-dimensional shapes is a skill that every drafter, designer, and engineer should possess. This is especially important in 3D modeling. However, there is a distinct difference between drawing a view that *looks* three-dimensional and creating a *true* 3D model.

A 3D model can be rotated on the display screen and viewed from any angle. The computer calculates the points, lines, and surfaces of the objects in space. For information on 3D modeling, viewing, and visualization techniques, see *AutoCAD and Its Applications—Advanced*. This chapter is provided as background information on the classic drawing techniques used in pictorial drawing. The focus of this chapter is creating views that *look* three-dimensional using some special AutoCAD functions and two-dimensional coordinates and objects.

Pictorial Drawing Overview

The word *pictorial* means "like a picture." It refers to any form of 2D drawing that illustrates height, width, and depth. Several forms of pictorial drawings are used in industry today. The least realistic is oblique. However, this is the simplest type. The most realistic, but also the most complex, is perspective. The realism and complexity of isometric drawing falls midway between the two.

pictorial: A 2D drawing that shows height, width, and depth; similar to a picture.

Oblique Drawings

An *oblique drawing* shows objects with one or more parallel faces at their true shape and size. A scale is selected for the orthographic, or front, faces. Then an angle for the depth (receding axis) is chosen. The three types of oblique drawings are cavalier, cabinet, and general. See **Figure 24-1**. These types vary in the scale of the receding axis. The receding axis is drawn at full scale for a cavalier view and at half scale for a

oblique drawing: A drawing that shows objects with one or more parallel faces having true shape and size.

Figure 24-1.
The scale of the
receding axis differs
in the three types of
oblique drawings.

Cavalier   Cabinet   General

cabinet view. The receding axis on a general oblique view is normally drawn with a 3/4 scale. Oblique drawings were traditionally used by cabinetmakers, but this type of pictorial drawing is rarely used in the CAD environment.

Axonometric Drawings

The general term for drawings that display three dimensions on a two-dimensional surface such as a drawing sheet is *axonometric drawings*. The three types of axonometric drawings are isometric, dimetric, and trimetric.

Isometric drawings are more realistic than oblique drawings. The entire object appears as if it is tilted toward the viewer. The word *isometric* means "equal measure." This equal measure refers to the angle between the three axes (120°) after the object has been tilted. The tilt angle is 35°16'. This is shown in **Figure 24-2**. The 120° angle corresponds to an angle of 30° from horizontal. In the construction of isometric drawings, lines that are parallel in the orthogonal views must be parallel in the isometric view.

The most appealing aspect of isometric drawing is that all three axis lines can be measured using the same scale. This saves time, while still producing a pleasing pictorial of the object.

Dimetric and trimetric drawings are closely related to isometric drawing. These forms of pictorial drawing differ from isometric in the scales used to measure the three axes. *Dimetric* drawing uses two different scales and *trimetric* uses three scales. Using different scales is an attempt to create *foreshortening*. The relationship between isometric, dimetric, and trimetric drawings is illustrated in **Figure 24-3**.

axonometric drawings: Drawings in which a 3D object is rotated for display on a 2D drawing sheet so all three dimensions can be seen.

isometric drawings: Drawings in which the three axes are equally spaced at 120°.

dimetric: An axonometric drawing in which two different scales are used to measure the three axes.

trimetric: An axonometric drawing in which three different scales are used to measure the three axes.

foreshortening: Property of a drawing in which objects appear to recede in the distance.

Figure 24-2.
An object is tilted
35°16' to achieve
an isometric
view having 120°
between the three
axes. Notice the
highlighted face in
each view.

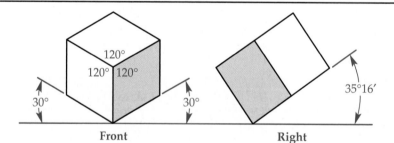

Front   Right

Figure 24-3.
Isometric, dimetric,
and trimetric
drawings differ in the
scales used to draw
the three axes. In the
isometric shown here,
all three sides are
drawn at full scale,
or 1. You can see how
the dimetric and
trimetric scales vary.

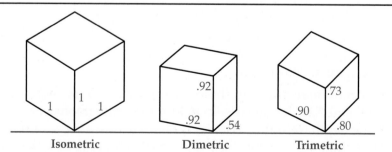

Isometric   Dimetric   Trimetric

Perspective Drawings

The most realistic form of pictorial drawing is a *perspective drawing*. The eye naturally sees objects in perspective. Look down a long hall and notice that the wall and floor lines seem to converge in the distance. The point at which they converge is called the *vanishing point*. The most common types of perspective drawing are one-point and two-point perspectives. These forms of pictorial drawing are often used in architecture. They are also used in the automotive and aircraft industries. Examples of one-point and two-point perspectives are shown in **Figure 24-4.** A true perspective of a 3D model can be produced in AutoCAD using the **ViewCube**. See *AutoCAD and Its Applications—Advanced* for complete coverage of the **ViewCube** and 3D modeling.

perspective drawing: The most realistic form of pictorial drawing, in which receding objects meet at one or more vanishing points on the horizon.

vanishing point: The point at which objects seem to converge in the distance.

Figure 24-4.
An example of one-point perspective and two-point perspective.

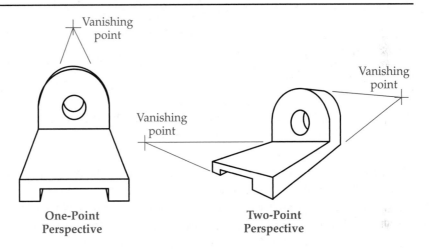

One-Point Perspective

Two-Point Perspective

Isometric Drawing

The method of pictorial drawing that is used most commonly in industry is isometric. Isometric drawings provide a single view showing three sides that can be measured using the same scale. An isometric view has no perspective and may appear somewhat distorted. Two of the isometric axes are drawn at 30° to horizontal, and the third is drawn at 90°. See **Figure 24-5.**

The three axes shown in **Figure 24-5** represent the width, height, and depth of the object. Lines that appear horizontal in an orthographic view are placed at a 30° angle. Lines that are vertical in an orthographic view are placed vertically. These lines are parallel to the axes. Any line parallel to an axis can be measured and is called an *isometric line*. Lines that are not parallel to the axes are called *nonisometric lines* and cannot be measured. Note the two nonisometric lines in **Figure 24-5.**

Circles appear as ellipses in an isometric drawing. Circular features shown on isometric objects must be oriented properly or they appear distorted. The correct orientation of isometric circles on the three principal planes is shown in **Figure 24-6.** The small diameter (minor axis) of the ellipse must always align with the axis of the

isometric line: Any line that is parallel to an axis in an isometric drawing.

nonisometric lines: Lines that are not parallel to the axes in an isometric drawing.

Figure 24-5.
Layout of the isometric axes.

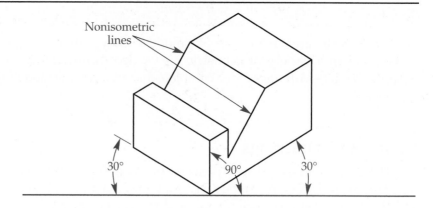

Nonisometric lines

30° 90° 30°

Figure 24-6.
Proper isometric circle (ellipse) orientation on isometric planes. The minor axis always aligns with the axis centerline.

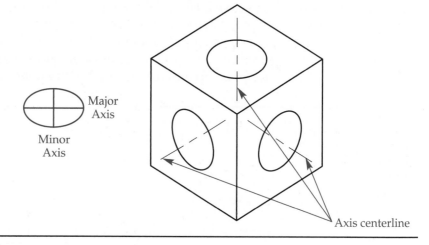

Major Axis

Minor Axis

Axis centerline

circular feature. Notice that the centerline axes of the holes in **Figure 24-6** are parallel to one of the isometric planes.

A basic rule to remember about isometric drawing is that lines that are parallel in an orthogonal view must also be parallel in the isometric view. AutoCAD's **ISOPLANE** tool makes that task, and the positioning of ellipses, easy.

PROFESSIONAL TIP

If you are ever in doubt about the proper orientation of an ellipse in an isometric drawing, remember that the minor axis of the ellipse must always be aligned with the centerline axis of the circular feature. This is shown clearly in **Figure 24-6**.

Settings for Isometric Drawing

Type
DSETTINGS
DS
SE

You can quickly set isometric variables in the **Snap and Grid** tab of the **Drafting Settings** dialog box. See **Figure 24-7**. This dialog box can also be accessed by right-clicking the **Snap Mode** or **Grid Display** status bar button and then selecting **Settings...** from the shortcut menu.

To activate the isometric snap grid, pick the **Isometric snap** radio button in the **Snap type** area. Notice that the **Snap X spacing** and **Grid X spacing** text boxes are now grayed out. Since X spacing relates to horizontal measurements, it is not used in **Isometric snap** mode. You can only set the Y spacing for grid and snap in isometric. Check the **Snap On (F9)** and **Grid On (F7)** check boxes if you want **Snap** and **Grid** modes to be activated. Pick the **OK** button to display the grid dots on the screen in an isometric orientation, as shown in **Figure 24-8**.

Figure 24-7.
The **Drafting Settings** dialog box allows you to specify the settings needed for isometric drawing.

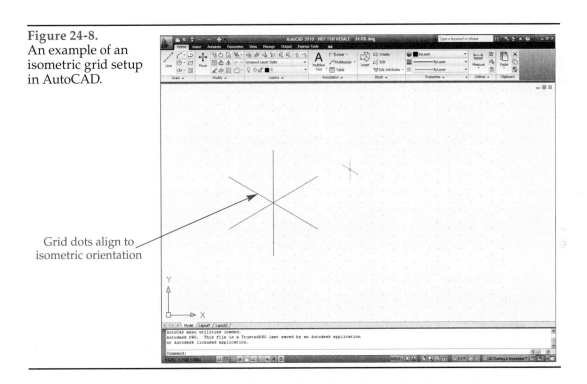

Pick to activate isometric snap grid

Figure 24-8.
An example of an isometric grid setup in AutoCAD.

Grid dots align to isometric orientation

Notice that the crosshairs appears angled. This aids you in drawing lines at the proper isometric angles. Try drawing a four-sided surface using the **LINE** tool. Draw the surface so it appears to be the left side of a box in an isometric layout. See **Figure 24-9**. To draw nonparallel surfaces, you can change the angle of the crosshairs to make your task easier, as discussed in the next section.

To turn off the **Isometric snap** mode, pick the **Rectangular snap** radio button in the **Snap type** area. The **Isometric snap** mode is turned off and you are returned to the drawing area when you pick the **OK** button.

NOTE

You can also set the **Isometric snap** mode by typing SNAP or SN, selecting the **Style** option, and then typing I to select **Isometric**.

Figure 24-9.
A four-sided object drawn with the **LINE** tool can be used as the left side of an isometric box.

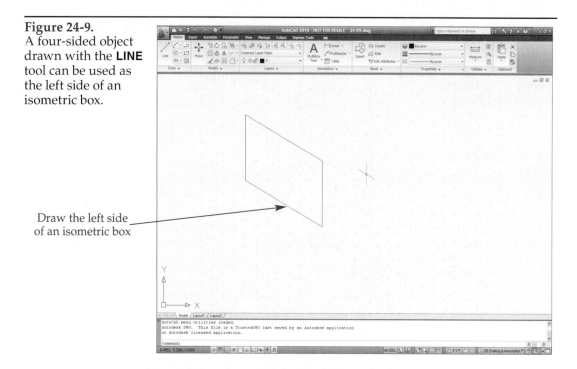

Draw the left side of an isometric box

Changing the Isometric Crosshairs Orientation

Drawing an isometric shape is possible without ever changing the angle of the crosshairs. However, the drawing process is easier and faster if the angle of the cross-hairs aligns with the isometric axes.

Whenever the isometric snap style is enabled, press the [F5] key or the [Ctrl]+[E] key combination to change the crosshairs immediately to the next isometric plane. AutoCAD refers to the isometric positions or planes as *isoplanes*. As you change among isoplanes, the current isoplane is displayed at the command line as a reference. The three crosshairs orientations and their angular values are shown in **Figure 24-10**.

isoplanes: The three isometric positions or planes.

Another method to toggle the crosshairs position is to use the **ISOPLANE** tool. Press [Enter] to toggle the crosshairs to the next position. The command line displays the new isoplane setting. You can toggle immediately to the next position by pressing [Enter] to repeat the **ISOPLANE** tool and pressing [Enter] again. To specify the plane of orientation, type the first letter of that position. The **ISOPLANE** tool can also be used transparently.

The crosshairs are always in one of the isoplane positions when **Isometric snap** mode is in effect. An exception occurs in display or editing tools when a multiple selection set method, such as a window, is used. In these cases, the crosshairs changes to the normal vertical and horizontal position. When the display or editing tool is closed, the crosshairs reverts to its former isoplane orientation.

Figure 24-10.
You can toggle among the three isometric crosshairs positions using the [F5] function key, the [Ctrl]+[E] key combination, or the **ISOPLANE** tool.

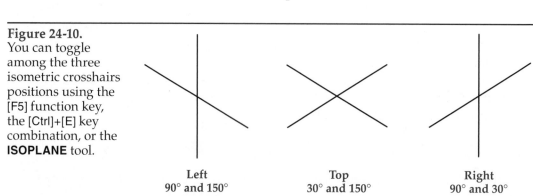

Left
90° and 150°

Top
30° and 150°

Right
90° and 30°

Exercise 24-1

Access the Student Web site (www.g-wlearning.com/CAD) and complete Exercise 24-1.

Isometric Ellipses

Placing an isometric ellipse on an object is easy with AutoCAD because of the **Isocircle** option of the **ELLIPSE** tool. An ellipse is positioned automatically on the current isoplane. To use the **ELLIPSE** tool, first make sure you are in **Isometric snap** mode. Access the **ELLIPSE** tool and select the **Isocircle** option. Then pick the center point and set the radius or diameter. Note that the **Isocircle** option only appears when you are in **Isometric snap** mode.

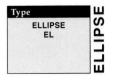

Always check the isoplane position before placing an ellipse (isocircle) on your drawing. You can dynamically view the three positions an ellipse can take. Access the **ELLIPSE** tool, enter the **Isocircle** option, pick a center point, and toggle the crosshairs orientation. See **Figure 24-11**. The ellipse rotates each time you toggle the crosshairs.

The isometric ellipse (isocircle) is a true ellipse. When an ellipse is selected, grips are displayed at the center and four quadrant points. See **Figure 24-12**. However, do not use grips to resize or otherwise adjust an isometric ellipse. As soon as you resize an isometric ellipse in this manner, its angular value is changed and it is no longer isometric. You can use the center grip to move the ellipse. Also, if you rotate an isometric ellipse while **Ortho** mode is on, it will not appear in a proper isometric plane. You *can* rotate an isometric ellipse from one isometric plane to another, but you must enter a value of 120°.

Figure 24-11.
The orientation of an isometric ellipse is determined by the current isometric plane.

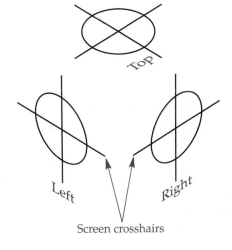

Figure 24-12.
An isometric ellipse has grips at its four quadrant points and center.

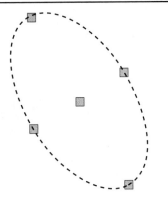

Exercise 24-2

Access the Student Web site (www.g-wlearning.com/CAD) and complete Exercise 24-2.

Constructing Isometric Arcs

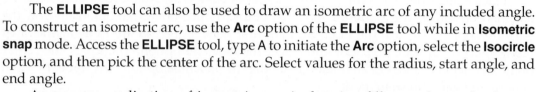

The **ELLIPSE** tool can also be used to draw an isometric arc of any included angle. To construct an isometric arc, use the **Arc** option of the **ELLIPSE** tool while in **Isometric snap** mode. Access the **ELLIPSE** tool, type A to initiate the **Arc** option, select the **Isocircle** option, and then pick the center of the arc. Select values for the radius, start angle, and end angle.

A common application of isometric arcs is drawing fillets and rounds. Once a round is created in isometric, the edge (corner) of the object sits back from its original, unfilleted position. See **Figure 24-13A.** You can draw the object first and then trim away the excess after locating the fillets, or you can draw the isometric arcs and then the connecting lines. Either way, the center point of the ellipse is a critical feature and should be located first. The left-hand arc in **Figure 24-13A** was drawn first and copied to the back position. Use **Ortho** mode to help quickly draw 90° arcs.

The next step is to move the original edge to its new position, which is tangent to the isometric arcs. You can do this by snapping the endpoint of the line to the quadrant point of the arc. See **Figure 24-13B.** Notice the grips on the line and on the arc. The endpoint of the line is snapped to the quadrant grip on the arc. The final step is to trim away the excess lines and arc segment. The completed feature is shown in **Figure 24-13C.**

Rounded edges, when viewed straight on, cannot be shown as complete-edge lines that extend to the ends of the object. Instead, a good technique to use is a broken line in the original location of the edge. This is clearly shown on the right-hand edge in **Figure 25-13C.**

Exercise 24-3

Access the Student Web site (www.g-wlearning.com/CAD) and complete Exercise 24-3.

Figure 24-13.
Fillets and rounds can be drawn with the **Arc** option of the **ELLIPSE** tool. A broken line is used to represent an edge that is viewed straight on.

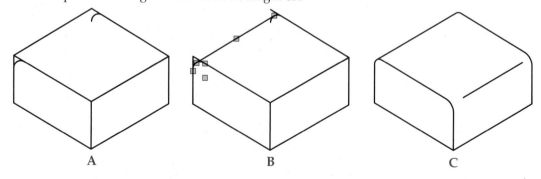

A   B   C

ELLIPSE
Ribbon
Home
> Draw
> Ellipse
Elliptical Arc
Type
ELLIPSE
EL

Creating Isometric Text Styles

Text placed in an isometric drawing should appear to be parallel to one of the isometric planes. It should align with the plane to which it applies. Text may be located on the object or positioned away from it as a note. Drafters and artists occasionally neglect this aspect of pictorial drawing, and it shows on the final product.

Properly placing text on an isometric drawing involves creating new text styles. **Figure 24-14** illustrates possible orientations of text on an isometric drawing. These examples were created using only two text styles. The text styles have an obliquing angle of either 30° or −30°. The labels in **Figure 24-14** refer to the following table. The angle in the figure indicates the rotation angle entered when using one of the text tools. For example, ISO-2 90 means that the ISO-2 style was used and the text was rotated 90°. This technique can be applied to any font.

Name	Font	Obliquing Angle
ISO-1	Romans	30°
ISO-2	Romans	−30°

Exercise 24-4

Access the Student Web site (www.g-wlearning.com/CAD) and complete Exercise 24-4.

Figure 24-14.
Isometric text applications. The text shown here indicates which style and angle were used.

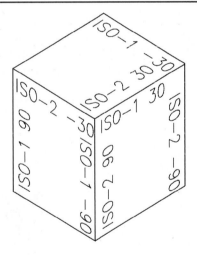

Isometric Dimensioning

An important aspect of dimensioning in isometric is to place dimension lines, text, and arrowheads on the proper plane. Remember these guidelines:

- Extension lines should always extend in the plane being dimensioned.
- The heel of the arrowhead should always be parallel to the extension line.
- Strokes of the text that would normally be vertical should always be parallel with the extension lines or dimension lines.

These techniques are shown on the dimensioned isometric part in **Figure 24-15**. AutoCAD does not automatically dimension isometric objects. You must first create isometric arrowheads and text styles. Then, manually draw the dimension lines and text as they should appear in each of the three isometric planes.

Figure 24-15.
A dimensioned
isometric part.
Note the text
and arrowhead
orientation in
relation to the
extension lines.

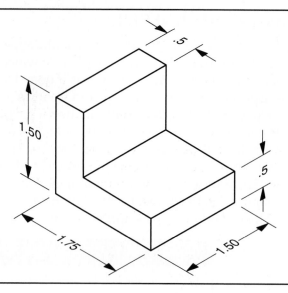

You have already learned how to create isometric text styles. You can set up appropriate text styles in an isometric template drawing if you draw isometrics often.

Isometric Arrowheads

You can draw isometric arrowheads and fill them in with a solid hatch pattern. A variable-width polyline cannot be used because the heel of the arrowhead will not be parallel to the extension lines. Examples of arrowheads for the three isometric planes are shown in **Figure 24-16.**

You do not need to draw every arrowhead individually. First, draw two isometric axes, as shown in **Figure 24-17A.** Then draw one arrowhead like the one shown in **Figure 24-17B.** Use the **MIRROR** tool to create additional arrowheads. As you create new arrowheads, move each one to its proper plane.

Figure 24-16.
Examples of
arrowheads in
each of the three
isometric planes.

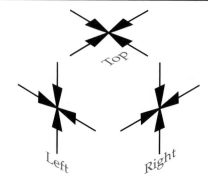

Figure 24-17.
Creating isometric
arrowheads.
A—Draw the two
isometric axes
for arrowhead
placement. B—Draw
the first arrowhead
on one of the axis
lines. Then mirror
the arrowhead to
create the others.

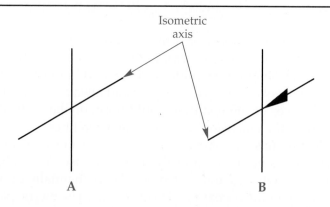

You can save each arrowhead as a block in your isometric template or prototype. Use block names that are easy to remember. Blocks are discussed in Chapters 25–28.

Oblique Dimensioning

AutoCAD has a way to dimension isometric and oblique lines semiautomatically. First, draw the dimensions using any of the linear dimensioning tools. The object in **Figure 24-18A** was dimensioned using the **DIMALIGNED** and **DIMLINEAR** tools. Then use the **Oblique** option of the **DIMEDIT** tool to rotate the extension lines. See **Figure 24-18B**.

To use the **Oblique** option, access the **DIMEDIT** tool and then type O for **Oblique**. When prompted, select the dimension and enter the obliquing angle. This technique creates suitable dimensions for an isometric drawing and is quicker than the method previously discussed. However, this method does not rotate the arrows to align the arrowhead heels with the extension lines. It also does not draw the dimension text aligned in the plane of the dimension. Therefore, this method does not produce technically correct dimensions.

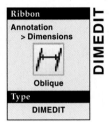

Figure 24-18.
Using the **Oblique** option of the **DIMEDIT** tool, you can create semiautomatic isometric dimensions by editing existing dimensions. A—Create the dimensions using the **DIMALIGNED** or **DIMLINEAR** tool. B—Adjust the positions of the dimensions using **DIMEDIT**. C—The obliquing angles used to achieve the results shown.

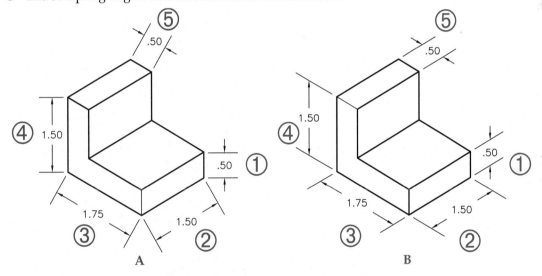

Dimension	Obliquing Angle
1	30°
2	−30°
3	30°
4	−30°
5	30°

C

Chapter Test

Answer the following questions. Write your answers on a separate sheet of paper or go to the Student Web site (www.g-wlearning.com/CAD) and complete the electronic chapter test.

1. Which is the simplest form of pictorial drawing?
2. How does isometric drawing differ from oblique drawing?
3. How do dimetric and trimetric drawings differ from isometric drawings?
4. Which is the most realistic form of pictorial drawing?
5. What must be set in the **Drafting Settings** dialog box to turn on **Isometric snap** mode and set a snap spacing of .2?
6. What function does the **ISOPLANE** tool perform?
7. Name the tool and option used to draw an isometric circle.
8. What factor determines the orientation of an isometric ellipse?
9. Where are grips located on a circle drawn in isometric?
10. Can grips be used to resize an isometric circle correctly? Explain your answer.
11. How are isometric arcs drawn?
12. Briefly explain how to create text that can be used on an isometric drawing.
13. How can you create isometric arrowheads?
14. What technique does AutoCAD provide for semiautomatically dimensioning isometric objects?
15. Why does the technique in Question 14 not produce technically correct dimensions?

Drawing Problems

▼ Basic

Create an isometric template drawing. Items that should be set in the template include grid spacing, snap spacing, ortho setting, and text size. Save the template as isoproto.dwt. *Use the template to construct the isometric drawings in Problems 1–10. Measure the drawings to obtain the dimensions. Save the drawing problems as* P24-*(problem number).*

1.
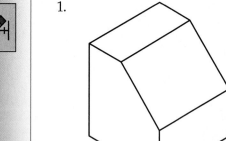

2.
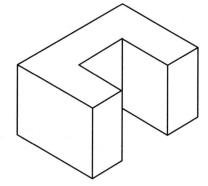

3.
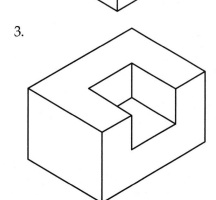

4.
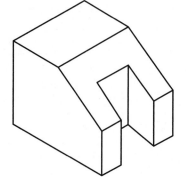

5.

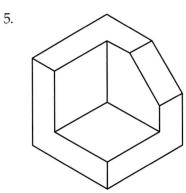

6.

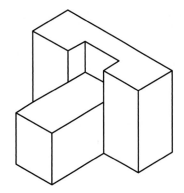

7.

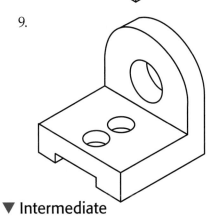

8.

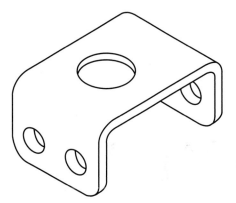

9.

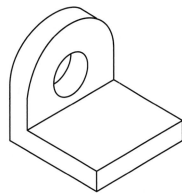

10.

▼ Intermediate

For Problems 11–14, create isometric drawings using the views shown. Measure the drawings to obtain the dimensions.

11.

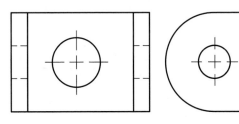

12.

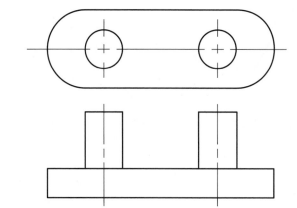

13.

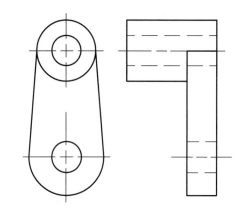

14.

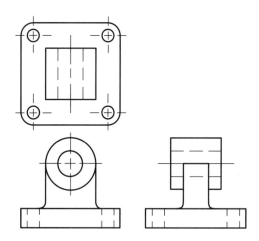

15. Construct a set of isometric arrowheads to use when dimensioning isometric drawings. Load your isometric template drawing. Create arrowheads for each of the three isometric planes. Name them with the first letter indicating the plane: T for top, L for left, and R for right. Also, number them clockwise from the top. See the example below for the right isometric plane. Do not include the labels in your drawing. Save the template again when you are finished.

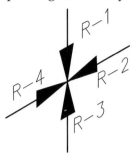

16. Create a set of isometric text styles like those shown in **Figure 24-14.** Load your template drawing and make a complete set in one font. Make additional sets in other fonts if you wish. Enter a text height of 0 so you can specify the height when placing the text. Save the template when finished.

17. Begin a new drawing using your isometric template. Select one of the following problems from this chapter and dimension it: Problem 5, 7, 8, or 9. When adding dimensions, be sure to use the proper arrowhead and text style for the plane on which you are working. Save the drawing as P24-17.

▼ Advanced

18. Create an isometric drawing of the switch plate shown below. Select a view that best displays the features of the object. Do not include dimensions. Save the drawing as P24-18.

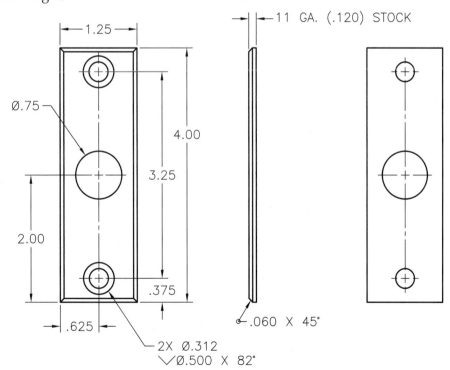

19. Create an isometric drawing of the retainer shown below. Select a view that best displays the features of the object. Do not include dimensions. Save the drawing as P24-19.

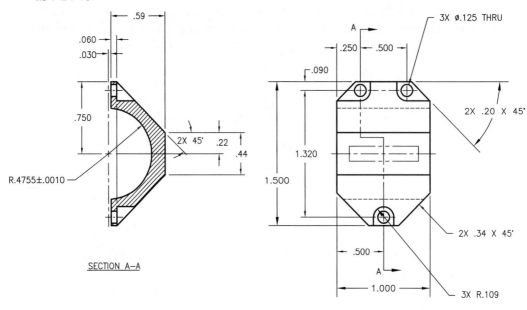

SECTION A—A

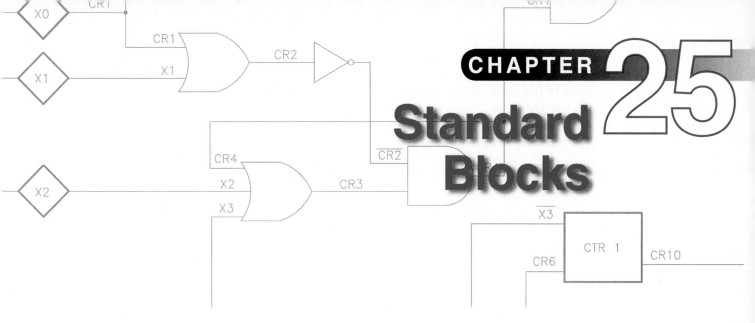

CHAPTER 25

Standard Blocks

Learning Objectives

After completing this chapter, you will be able to do the following:

✓ Create and save blocks.
✓ Insert blocks into a drawing.
✓ Edit a block and update the block in a drawing.
✓ Create blocks as drawing files.
✓ Construct and use a symbol library of blocks.
✓ Purge unused items from a drawing.

The ability to create and use *blocks* is one of the greatest benefits of drawing with AutoCAD. The **BLOCK** tool stores a block within a drawing as a *block definition*. The **WBLOCK** tool saves a *wblock* as a separate drawing file. You can insert blocks as often as needed and share blocks between drawings. You also have the option to scale, rotate, and adjust blocks to meet specific drawing requirements.

block: A symbol saved and stored in a drawing for future use.

block definition: Information about a block that is stored within the drawing file.

wblock: A block definition saved as a separate drawing file.

Constructing Blocks

A block can consist of any shape, group of objects, symbol, annotation, view, or drawing. Review the drawing to identify any item you plan to use more than once. Screws, punches, subassemblies, plumbing fixtures, and appliances are examples of items you may want to convert to blocks. Draw the item once and then save the objects as a block for multiple use.

Selecting a Layer

Before you begin drawing block components, you should identify the appropriate layer on which to create the objects. It is critical that you understand how layers and object properties apply when using blocks. The 0 layer is the preferred layer on which to draw block objects. If you originally create block objects on layer 0, the block assumes, or inherits, the properties of the layer you assign to the block. Draw the objects for all blocks on layer 0 and then assign the appropriate layer to each block when you insert the block. If you draw block objects on a layer other than layer 0, place all the objects on layer 0 before creating the block.

A second method is to create block objects using one or more layers other than layer 0. If you originally create block objects on a layer other than layer 0, the block belongs to the layer you assign to the block, but the objects retain the properties of the layers used to create the objects. The difference is only noticeable if you place the block on a layer other than the layer used to draw the block objects.

A third technique is to create block objects using the ByBlock color, lineweight, and linetype. If you originally create block objects using ByBlock properties, the block belongs to the layer you assign to the block, but the objects take on the color, lineweight, and linetype you assign to the block, regardless of the layer on which you place the block. Using the ByBlock setting is only noticeable if you assign absolute values to the block using the properties in the **Properties** panel of the **Home** ribbon tab or the **Properties** palette.

Another option is to create block objects using an absolute color, lineweight, and linetype. If you originally create block objects using absolute values, such as a Blue color, a 0.05mm lineweight, and a Continuous linetype, the block belongs to the layer you assign to the block, but the objects display the specified absolute values regardless of the properties assigned to the drawing or the layer on which you place the block.

CAUTION

You should typically draw block objects on layer 0, and then assign a specific layer to each block. Drawing block objects on a layer other than layer 0, or using ByBlock or absolute properties, can cause significant confusion. The result is often a situation in which a block belongs to a layer, but the block objects display properties of a different layer, or absolute values.

Drawing Block Components

insertion base point: The point on a block that defines where the block is positioned during insertion.

Draw a block as you would any other geometry. When you finish drawing the objects, determine the best location for the *insertion base point*. When you insert the block into a drawing, the insertion base point positions the block. **Figure 25-1** shows several examples of common blocks and a possible insertion base point for each.

PROFESSIONAL TIP

A single block allows you to create multiple features that are identical except for scale. In these cases, draw the base block to fit inside a one-unit square. This makes it easy to scale the block when you insert it into a drawing to create variations of the block.

Creating Blocks

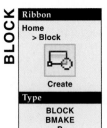

BLOCK

Ribbon
Home
> Block

Create

Type
BLOCK
BMAKE
B

Once you draw objects and identify an appropriate insertion base point, you are ready to save the objects as a block. Use the **BLOCK** tool and the corresponding **Block Definition** dialog box to create a block. See **Figure 25-2.**

Naming and Describing the Block

Enter a descriptive name for the block in the **Name:** text box. For example, name a vacuum pump PUMP or a 3′ × 6′-8″ door DOOR_3068. The block name cannot exceed 255 characters. It can include numbers, letters, and spaces, as well as the dollar sign ($), hyphen (-), and underscore (_). The drop-down list allows you to access an existing name to recreate a block, or to use as reference when naming a new block with a similar name.

Figure 25-1.
Common drafting symbols and their insertion points for placement on drawings. Colored dots indicate the insertion points for reference.

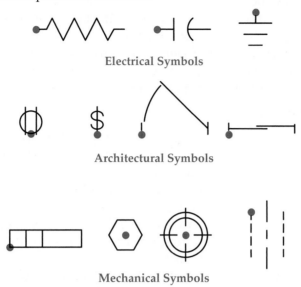

Electrical Symbols

Architectural Symbols

Mechanical Symbols

Figure 25-2.
Use the **Block Definition** dialog box to create a block.

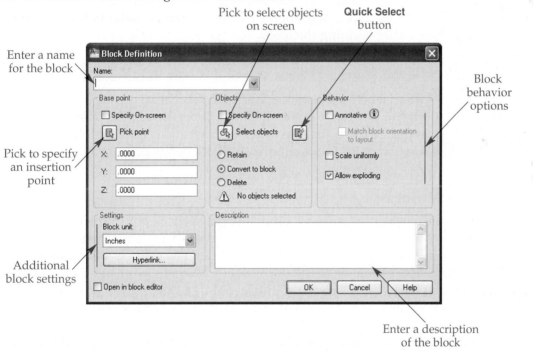

A block name is often descriptive enough to identify the block. However, you can enter a description of the block in the **Description:** text box to help identify the block. For example, the PUMP block might include the description This is a vacuum pump symbol, or the DOOR_3068 block might include the description This is 3' wide by 6'-8" tall interior single-swing door.

Defining the Block Insertion Base Point

The **Base point** area allows you to define the insertion base point. If you know the coordinates for the insertion base point, type values in the **X:**, **Y:**, and **Z:** text boxes. However, often the best way to specify the insertion base point is to use object snap to select a point on an object. Choose the **Pick point** button to return to the drawing and select an insertion base point. The **Block Definition** dialog box reappears after you select the insertion base point.

An alternative technique is to choose the **Specify On-screen** check box, which allows you to pick an insertion base point in the drawing after you pick the **OK** button to create the block and exit the **Block Definition** dialog box. This method can save time by allowing you to pick the insertion base point without using the **Pick point** button and re-entering the **Block Definition** dialog box.

Selecting Block Objects

The **Objects** area includes options for selecting objects for the block definition. Pick the **Select objects** button to return to the drawing and select the objects that will make up the block. Press [Enter] or the space bar or right-click to redisplay the **Block Definition** dialog box. The number of selected objects appears in the **Objects** area, and an image of the selection displays next to the **Name:** drop-down list. To create a selection set, use the **QuickSelect** button and **Quick Select** dialog box to define a filter.

An alternative method for selecting objects is to choose the **Specify On-screen** check box, which allows you to pick objects from the drawing after you pick the **OK** button to create the block and exit the **Block Definition** dialog box. This method can save time by allowing you to select objects without using the **Select objects** button and re-entering the **Block Definition** dialog box.

Pick the **Retain** radio button to keep the selected objects in the current drawing in their original, unblocked state. Select the **Convert to block** radio button to replace the selected objects with the block definition. Choose the **Delete** radio button to remove the selected objects after defining the block.

PROFESSIONAL TIP

If you select the **Delete** option and then decide to keep the original geometry in the drawing after defining the block, use the **OOPS** tool. This returns the original objects to the screen and keeps the block definition. Using the **UNDO** tool removes the block definition from the drawing.

Block Scale Settings

To make the block annotative, pick the **Annotative** check box in the **Behavior** area. AutoCAD scales annotative blocks according to the specified annotation scale, which eliminates the need to calculate the scale factor. When you select the **Annotative** check box, the **Match block orientation to layout** check box becomes available. Pick this check box to keep annotative blocks planar to the layout in a floating viewport, even if the drawing view is rotated, as it might be if you rotate the UCS. Selecting this option also prohibits you from using the **ROTATE** tool to rotate a block.

If you check the **Scale uniformly** check box in the **Behavior** area, you do not have the option of specifying different X and Y scale factors when you insert the block. Block scaling options are described later in this chapter.

Additional Block Definition Settings

If the **Allow exploding** check box in the **Behavior** area is checked, you have the option of exploding the block. If the box is not checked, you cannot explode the block, even after inserting it. Select a unit type from the **Block unit** drop-down list in the **Settings** area to specify the insertion units of the block. Pick the **Hyperlink...** button to access the **Insert Hyperlink** dialog box to insert a hyperlink in the block. If **Open in block editor** is checked, the new block immediately opens in the **Block Definition Editor** when you create the block and exit the **Block Definition** dialog box. The **Block Definition Editor** is described later in this chapter.

AutoCAD and Its Applications—Basics

PROFESSIONAL TIP

You can use blocks to create other blocks. Insert existing blocks into a view and then save all of the objects as a block. This is a process known as *nesting*. You must give the top-level block a name that is different from any nested block. Proper planning and knowledge of all existing blocks can speed up the drawing process and the creation of complex views.

nesting: Creating a block that includes other blocks.

Exercise 25-1

Access the Student Web site (www.g-wlearning.com/CAD) and complete Exercise 25-1.

Inserting Blocks

Once you create a block, you have several options for inserting it into a drawing. Remember to make the layer you want to assign to the block current *before* inserting the block. You should also determine the proper size and rotation angle for the block before insertion. The term *block reference* describes an inserted block. *Dependent symbols* are any named objects, such as blocks and layers. AutoCAD automatically updates dependent symbols in a drawing the next time you open the drawing.

Using the INSERT Tool

The **INSERT** tool is one of the most common methods for inserting blocks in a drawing. The **Insert** dialog box, shown in **Figure 25-3**, appears when you access the **INSERT** tool.

Selecting the Block to Insert

Pick the **Name:** drop-down list button to show the blocks defined in the current drawing and select the name of the block to insert. You can also type the name of the block in the **Name:** text box. Pick the **Browse...** button to display the **Select Drawing File** dialog box, used to locate and select a drawing or DXF file (wblock) for insertion as a block. This process is described later in this chapter.

Specifying the Block Insertion Point

The **Insertion point** area contains options for specifying where to insert the block. Select the **Specify On-screen** check box to specify a location in the drawing to insert the block when you pick the **OK** button. To insert the block using absolute coordinates, deselect the **Specify On-screen** check box and enter coordinates in the **X:**, **Y:**, and **Z:** text boxes.

block reference: A specific instance of a block inserted into a drawing.

dependent symbols: Named objects in a drawing that has been inserted or referenced into another drawing.

Ribbon
Home > Block
Insert
Type
INSERT DDINSERT I

INSERT

Figure 25-3.
The **Insert** dialog box allows you to select and prepare a block for insertion. Select the block to insert from the drop-down list or enter the block name in the **Name:** text box.

Pick to access defined blocks in the drawing

Activate to use X scale factor for Y and Z axes

Activate to explode block upon insertion

Pick to access the **Select Drawing File** dialog box

Preview

Enter a rotation angle

Block units

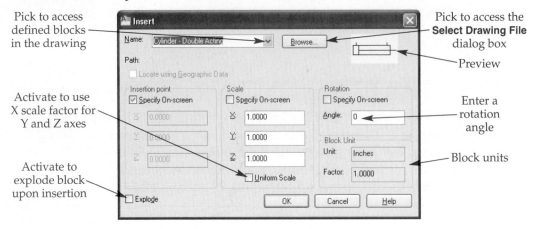

Scaling Blocks

The **Scale** area allows you to specify scale values for the block in relation to the X, Y, and Z axes. Deselect the **Specify On-screen** check box to enter scale values in the **X:, Y:,** and **Z:** text boxes. If you activate the **Uniform Scale** check box, you can specify a scale value for the X axis that also applies to the scale of the Y and Z axes. The **X** value is the only active axis value if you created the block with **Scale uniformly** checked in the **Block Definition** dialog box. Select the **Specify On-screen** check box to receive prompts for scaling the block during insertion.

It is possible to create a mirror image of a block by entering a negative scale factor values. For example, enter –1 for the X and Y scale factor to mirror the block to the opposite quadrant of the original orientation, but retain the original size. **Figure 25-4** shows different scale and mirroring techniques.

Blocks can be classified as real blocks, schematic blocks, or unit blocks, depending on how you scale the block during insertion. Examples of *real blocks* include a bolt, a bathtub, a pipe fitting, and the car shown in **Figure 25-5A.** Examples of *schematic blocks* include notes, detail bubbles, and section symbols. See **Figure 25-5B.** Schematic blocks typically include annotative blocks. When you insert an annotative schematic block, AutoCAD automatically determines the block scale based on the annotation scale. When you insert a non-annotative schematic block, you must specify the scale factor.

real block: A block originally drawn at a 1:1 scale and then inserted using 1 for both the X and Y scale factors.

schematic block: A block originally drawn at a 1:1 scale and then inserted using the drawing scale factor for both the X and Y scale values.

Figure 25-4.
Negative and positive scale factors have different effects when used to insert a block. Colored dots indicate the insertion points for reference.

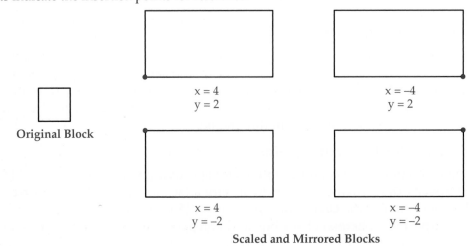

Original Block

x = 4
y = 2

x = –4
y = 2

x = 4
y = –2

x = –4
y = –2

Scaled and Mirrored Blocks

Figure 25-5.
A—Real blocks, such as this car, are drawn at full scale and inserted using a scale factor of 1 for both the X and Y axes. B—A schematic block is inserted using the scale factor of the drawing for the X and Y axes. C—A 2D unit block is often inserted at different scales for the X and Y axes.

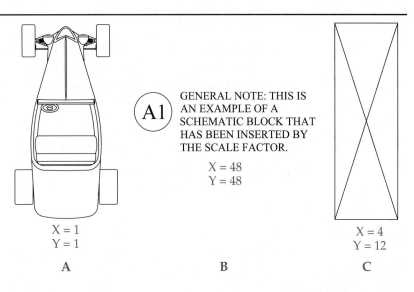

GENERAL NOTE: THIS IS AN EXAMPLE OF A SCHEMATIC BLOCK THAT HAS BEEN INSERTED BY THE SCALE FACTOR.

X = 48
Y = 48

X = 1
Y = 1

X = 4
Y = 12

A   B   C

NOTE

For most applications, you should insert annotative blocks at a scale of 1 in order for the annotation scale to apply correctly. Entering a scale other than 1 adjusts the scale of the block by multiplying the scale value by the annotative scale factor.

There are three different types of *unit blocks*. An example of a *1D unit block* is a 1-unit (1″, for example) blocked line object. A *2D unit block* is any blocked object that can fit inside a 1-unit × 1-unit (1″ × 1″, for example) square. A *3D unit block* is any blocked object that can fit inside a 1-unit × 1-unit × 1-unit (1″ × 1″ × 1″ for example) cube. To use a unit block, insert the block and determine the individual scale factors for each axis. For example, insert a 1″ 1D unit block line at a scale of 4 to create a 4″ line. When inserting a 2D unit block, assign different scale factors for the X and Y axes, such as 4 for the X axis and 12 for the Y axis, to create the 4″ × 12″ beam shown in **Figure 25-5C.** A 3D unit block allows you to adjust the scale of the X, Y, and Z axes.

unit block: A 1D, 2D, or 3D block drawn to fit in a 1-unit, 1-unit-square, or 1-unit-cubed area so that it can be scaled easily.

1D unit block: A 1-unit, one-dimensional object, such as a straight line segment, saved as a block.

2D unit block: A 2D object that fits into a 1-unit × 1-unit square, saved as a block.

3D unit block: A 3D object that fits into a 1-unit × 1-unit × 1-unit cube, saved as a block.

Rotating Blocks

The **Rotation** area allows you to insert the block at a specific angle. Deselect the **Specify On-screen** check box to enter a value in the **Angle:** text box. The default angle of 0° inserts the block as saved. Select the **Specify On-screen** check box to receive a prompt for rotating the block during insertion.

NOTE

You cannot rotate a block defined using the **Match block orientation to layout** option.

PROFESSIONAL TIP

You can rotate a block based on the current UCS. Be sure the proper UCS is active, and then insert the block using a rotation angle of 0°. If you decide to change the UCS later, any inserted blocks retain their original angle.

Additional Block Insertion Items

A block is saved as a single object, no matter how many objects the block includes. Select the **Explode** check box to explode the block into the original objects for editing purposes. If you explode the block on insertion, it assumes its original properties, such as its original layer, color, and linetype. If **Allow exploding** was unchecked when you defined the block, the **Explode** check box is inactive.

The **Block Unit** area displays read-only information about the selected block. The **Unit:** display box indicates the units for the block. The **Factor:** display box indicates the scale factor. The **Locate using Geographic Data** check box is active when the block and current drawing include *geographic data*. Pick the check box to position the block using geographic data.

geographic data: Information added to a drawing to describe specific locations and directions on Earth.

Working with Specify On-Screen Prompts

When you pick the **OK** button, prompts appear for any values defined as **Specify On-screen** in the **Insert** dialog box. If you specify the insertion point on-screen, the Specify insertion point or [Basepoint/Scale/X/Y/Z/Rotate/PScale/PX/PY/PZ/PRotate]: prompt appears. Enter or select a point to insert the block. The options allow you to specify a different base point; enter a value for the overall scale; enter independent scale factors for the X, Y, and Z axes; enter a rotation angle; and preview the scale of the X, Y, and Z axes or the rotation angle before entering actual values. If you use an available option, the new value overrides the related setting in the **Insert** dialog box.

If you specify the X scale factor on screen, the Enter X scale factor, specify opposite corner, or [Corner/XYZ] <1>: prompt appears. Pick a point or enter a value for the scale. You can also use the **Corner** option to scale the block. The Enter Y scale factor <use X scale factor>: prompt appears if you enter an X scale factor. Press [Enter] or the space bar, or right-click to accept the default scale value. Specify a value different from the X scale factor, or press [Enter] or the space bar, or right-click to accept the same scale specified for the X axis.

The X and Y scale factors allow you to stretch or compress the block to create modified versions of the block. See **Figure 25-6.** This is why it is a good idea to draw blocks to fit inside a one-unit square when appropriate. It makes the block easy to scale because you can enter the exact number of units for the X and Y dimensions. For example, if you want the block to be three units long and two units high, enter 3 when prompted to enter the X scale factor, and enter 2 when prompted to enter the Y scale factor.

The insertion base point specified when the block was created may not always be the best point when you actually insert the block. Instead of inserting and then moving the block, you can use the **Basepoint** option to specify a different base point before locating the block. Select the **Basepoint** option when prompted to specify the insertion point. The block temporarily appears on-screen, allowing you to choose an alternate insertion base point. The block reattaches to the crosshairs at the new point and a message appears indicating that the tool is resuming, allowing you to pick the insertion point in the drawing.

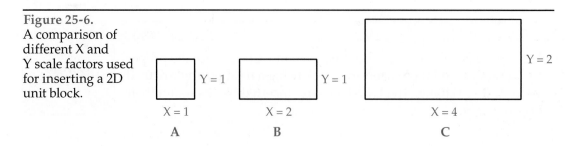

Figure 25-6.
A comparison of different X and Y scale factors used for inserting a 2D unit block.

Y = 1   Y = 1   Y = 2

X = 1   X = 2   X = 4

A   B   C

Exercise 25-2

Access the Student Web site (www.g-wlearning.com/CAD) and complete Exercise 25-2.

Inserting Multiple Arranged Copies of a Block

Type
MINSERT

MINSERT

The **MINSERT** tool combines the functions of the **INSERT** and **ARRAY** tools. **Figure 25-7** shows an example of an **MINSERT** tool application. To follow this example, set architectural units, draw a 4′ × 3′ rectangle, and save the rectangle as a block named DESK. Then access the **MINSERT** tool and enter DESK. Pick a point as the insertion point and then accept the X scale factor of 1, the Y scale factor of use X scale factor, and the rotation angle of 0. The arrangement is to be three rows and four columns. In order to make the horizontal spacing between desks 2′ and the vertical spacing 4′, you must consider the size of the desk when entering the distance between rows and columns. Enter 7′ (3′ desk depth + 4′ space between desks) at the Enter distance between rows of specify unit cell: prompt. Enter 6′ (4′ desk width × 2′ space between desks) at the Specify distance between columns: prompt.

The complete pattern takes on the characteristics of a block, except that you cannot explode the pattern. Therefore, you must use the **Properties** palette to modify the number of rows and columns, change the spacing between objects, or change other properties. If you rotate the initial block, all objects in the pattern rotate about their insertion points. If you rotate the patterned objects about the insertion point while using the **MINSERT** tool, all objects align on that point.

PROFESSIONAL TIP

As an alternative to the previous example, if you were working with different desk sizes, a 2D unit block may serve your purposes better than an exact size block. To create a 5′ × 3′-6″ (60″ × 42″) desk, for example, insert a one-unit-square block using either the **INSERT** or **MINSERT** tool, and enter 60 for the X scale factor and 42 for the Y scale factor.

Figure 25-7.
Creating an arrangement of desks using the **MINSERT** tool.

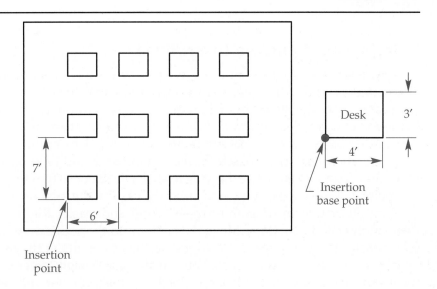

Exercise 25-3

Access the Student Web site (www.g-wlearning.com/CAD) and complete Exercise 25-3.

Inserting Entire Drawings

The **INSERT** tool also allows you to insert an entire drawing into the current drawing as a block. Access the **INSERT** tool and pick the **Browse...** button in the **Insert** dialog box. Use the **Select Drawing File** dialog box to select a drawing or DXF file to insert.

When you insert one drawing into another, the inserted drawing becomes a block reference and functions as a single object. The drawing is inserted on the current layer, but it does not inherit the color, linetype, or thickness properties of that layer. You can explode the inserted drawing back to its original objects if desired. Once exploded, the drawing objects revert to their original layers. An inserted drawing brings any existing block definitions and other drawing content, such as layers and dimension styles, into the current drawing.

By default, every drawing has an insertion base point of 0,0,0 when you insert the drawing into another drawing. To change the insertion base point of the drawing, access the **BASE** tool and select a new insertion base point. Save the drawing before inserting it into another drawing.

When inserting a drawing as a block, you have the option of using the existing drawing to create a block with a different name. For example, to define a block named BOLT from an existing drawing named Fastener.dwg, access the **INSERT** tool and use the **Browse...** button to select the Fastener.dwg file. Use the **Name:** text box to change the name from Fastener to BOLT, and pick the **OK** button. You can then insert the file into the drawing or press [Esc] to exit the tool. A BOLT block definition is now available to use as desired.

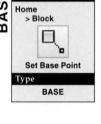

BASE

Ribbon
Home
> Block

Set Base Point
Type
BASE

Exercise 25-4

Access the Student Web site (www.g-wlearning.com/CAD) and complete Exercise 25-4.

Using DesignCenter to Insert Blocks

DesignCenter provides an effective way to insert blocks or entire drawings as blocks in the current drawing. To insert a block, locate the file containing the block to insert and select the **Blocks** branch in the tree view, or double-click on the **Blocks** icon in the content area. See **Figure 25-8.** Usually the quickest and most effective technique for transferring a block from **DesignCenter** into the active drawing is to use a drag-and-drop operation. Pick the block from the content area and hold down the pick button. Move the cursor into the active drawing to attach the block to the cursor at the block insertion point. Release the pick button to insert the block at the location of the cursor.

An alternative to drag-and-drop is copy and paste. Right-click on a block in **DesignCenter** and pick **Copy**. Move the cursor into the active drawing, right-click, and select **Paste**. The block appears attached to the crosshairs at the insertion base point. Specify a point to insert the block. You can also use **DesignCenter** in combination with the **Insert** dialog box. Right-click on a block in **DesignCenter** and select **Insert Block...** to access the **Insert** dialog box with the selected block active. This allows you to scale, rotate, or explode the block during insertion.

Figure 25-8.
Use the **DesignCenter** to insert blocks from files or drawings from folders. Several example blocks are available from the AutoCAD Sample folder, as shown.

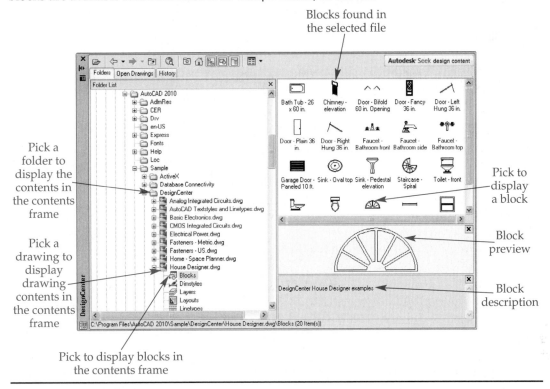

Blocks found in the selected file

Pick a folder to display the contents in the contents frame

Pick a drawing to display drawing contents in the contents frame

Pick to display blocks in the contents frame

Pick to display a block

Block preview

Block description

To insert a drawing or DXF file using **DesignCenter**, select the folder in the tree view that contains the file to display the contents in the content area. Drag and drop or copy and paste the file into the current drawing. You can also right-click a file icon in the content area and select **Insert as Block...**.

NOTE

Blocks are inserted from **DesignCenter** based on the type of block units you specified when you created the block. For example, if the original block was a 1 × 1 square and you specified the block units as feet when you created the block, then the block will insert as a 12″ × 12″ square.

Using Tool Palettes to Insert Blocks

The **Tool Palettes** palette, shown in **Figure 25-9**, provides another means of storing and inserting blocks. Blocks located in a tool palette are known as *block insertion tools.* Tool palettes can also store and activate many other types of drawing content and tools. For more information on AutoCAD customization and using tool palettes, refer to *AutoCAD and Its Applications—Advanced.*

block insertion tools: Blocks located on a tool palette.

To insert a block from the **Tool Palettes** palette, access the tool palette in which the block resides. Place the cursor over the block icon to display the name and description. Hold down the pick button on the block image. Move the cursor into the active drawing to attach the block to the cursor at the block insertion point. Release the pick button to insert the block at the location of the cursor.

An alternative to drag-and-drop is to pick once on the block image to attach the block to the crosshairs, and then pick a location for the block. This method offers an

Figure 25-9.
The **Tool Palettes** palette provides another means of inserting blocks.

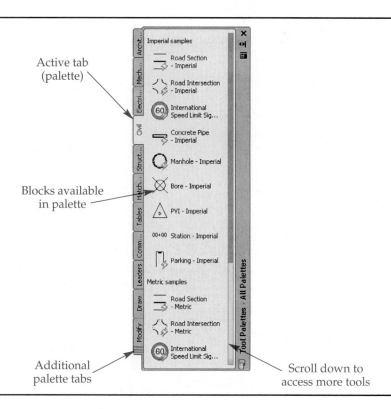

Active tab (palette)

Blocks available in palette

Additional palette tabs

Scroll down to access more tools

advantage over drag-and-drop by presenting options for adjusting the insertion base point, scale, and rotation. These options function the same as when you insert a block from the **Insert** dialog box.

Exercise 25-5

Access the Student Web site (www.g-wlearning.com/CAD) and complete Exercise 25-5.

Editing Blocks

The first form of block editing involves modifying a block that has already been inserted into a drawing using tools such as **MOVE**, **COPY**, **ROTATE**, or **MIRROR**. You can use grip editing by selecting the grip box that appears at the insertion base point of the block. You can also use the **Properties** palette and **Quick Properties** panel to make limited changes to inserted blocks. Remember, once you insert a block, it is treated as a single object.

The second type of block editing involves redefining the block by editing the block definition or changing the objects within the block. You can redefine a block using the **Block Editor** or by exploding and then recreating the block.

Changing Block Properties to ByLayer

If block component properties such as color and linetype were originally set to absolute values, and you want to change the properties to ByLayer, you can edit the block definition or use the **SETBYLAYER** tool to accomplish the same task without editing the block definition.

Access the **SETBYLAYER** tool and use the **Settings** option to display the **SetByLayer Settings** dialog box. Select the check boxes that correspond to the object properties to convert to ByLayer. Pick the **OK** button to exit the **SetByLayer Settings** dialog box.

Next, select the blocks with the properties to set to ByLayer and press [Enter] or the space bar, or right-click to display the Change ByBlock to ByLayer? prompt. Select the **Yes** option to change all object properties currently set to ByBlock to ByLayer. Pick the **No** option to change all object properties set to values other than ByBlock to ByLayer. The next prompt asks to include blocks in the conversion. If the selected object is a block, choose **Yes** to convert the properties of all references of the same block in the drawing to ByLayer. If you pick **No**, only the properties of the selected block are converted. All other references of the same block remain unchanged.

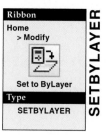

SETBYLAYER

Ribbon
Home
> Modify

Set to ByLayer
Type
SETBYLAYER

PROFESSIONAL TIP

To change the properties of several blocks, you can use the **Quick Select** tool to create a selection set of block reference objects.

Using the Block Editor

The **BEDIT** tool allows you to edit a block using the **Block Editor**. Access the **BEDIT** tool to display the **Edit Block Definition** dialog box, shown in **Figure 25-10**. To edit an existing block, select the name of the block from the list box. Pick the <Current Drawing> option to edit a block saved as the current drawing, such as a wblock. A preview and the description of the selected block appear. You can create a new block by typing a unique name in the **Block to create or edit** field. Pick the **OK** button to open the selected block in the **Block Editor**, as shown in **Figure 25-11**. If you typed a new block name, the drawing area is empty, allowing you to create a new block.

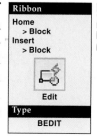

BEDIT

Ribbon
Home
> Block
Insert
> Block

Edit
Type
BEDIT

Modifying a Block

Use drawing and editing tools to modify or create the block. Specify the UCS origin, or 0,0,0 point, as the block insertion base point. The tools in the panels of the **Block Editor** ribbon tab are specific for modifying and creating block geometry. **Figure 25-12** describes some of the basic tools available in the **Block Editor** ribbon. The parametric tools allow you to constrain block geometry and form block tables. Many of the tools and options found on the **Block Editor** ribbon tab relate to dynamic blocks. This textbook explains dynamic blocks, block tables, and other block editing tools in later chapters.

Figure 25-10.
The **Edit Block Definition** dialog box.

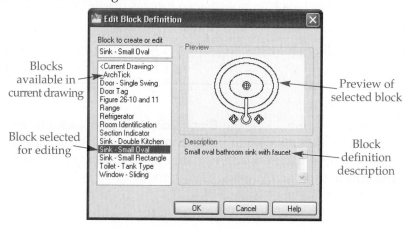

Figure 25-11.
The **Block Editor** ribbon tab and the **Block Authoring Palettes** palette are available in block editing mode. Only the block geometry appears in the **Block Editor**.

Block editing and construction tools available in the **Block Editor** ribbon tab

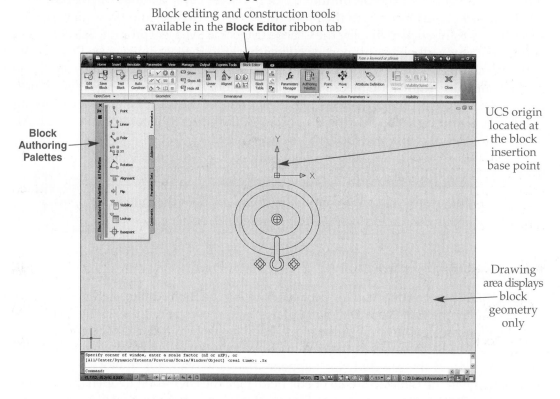

Block Authoring Palettes

UCS origin located at the block insertion base point

Drawing area displays block geometry only

Figure 25-12.
The **Block Editor** ribbon tab contains several tools and options specifically for editing and constructing blocks. This table describes the most basic functions.

Button	Description
	Saves changes to the block and updates the block definition.
	Opens the **Save Block As** dialog box, allowing you to save the block as a new block, using a different name.
	Opens the **Edit Block Definition** dialog box, which is the same dialog box displayed when you enter block editing mode. You can select a different block to edit or specify the name of a new block to create from scratch.
	Toggles the **Block Authoring Palettes** palette off and on.
	Closes the **Block Editor**.

When you finish editing, close the **Block Editor** to exit block editing mode and return to the drawing. If you have not saved changes, a dialog box appears asking if you want to save changes. Pick the appropriate option to save or discard changes, or pick the **Cancel** button to return to block editing mode.

Adding a Block Description

To change the description assigned to a block when it was originally created, open the block in the **Block Editor** and then display the **Properties** palette with no objects selected. Make changes to the description using the **Description** property in the **Block** category. Pick the **Save Block Definition** button and the **Close Block Editor** button to return to the drawing.

NOTE

Blocks can also be edited "in-place" using the **REFEDIT** tool. Chapter 32 describes in-place editing using the **REFEDIT** tool, as it applies to external references. You can use the same techniques to edit blocks.

Exercise 25-6

Access the Student Web site (www.g-wlearning.com/CAD) and complete Exercise 25-6.

Exploding and Redefining a Block

You have the option of exploding a block during insertion using the **Insert** dialog box. This is useful when you want to edit the individual objects of the block. You can also use the **EXPLODE** tool after inserting the block to break it into the original objects. Access the **EXPLODE** tool, select the objects to explode and press [Enter] or the space bar or right-click to complete the operation.

Ribbon
Home
> Modify
Explode
Type
EXPLODE
X

EXPLODE

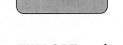

NOTE

You cannot explode a block that was created with **Allow exploding** unchecked in the **Block Definition** dialog box.

Follow this procedure to redefine an existing block using the **EXPLODE** and **BLOCK** tools:
1. Insert the block to redefine.
2. Make sure you know the exact location of the insertion point, because the point is lost during explosion.
3. Use the **EXPLODE** tool to explode the block.
4. Edit the components of the block as needed.
5. Recreate the block definition using the **BLOCK** tool.

6. Assign the block the same original name and, if appropriate, the same insertion point.
7. Select the objects to include in the block.
8. Pick the **OK** button in the **Block Definition** dialog box to save the block. When a message appears asking if you want to redefine the block, pick **Yes**.

A common mistake is to forget to use the **EXPLODE** tool before redefining the block. When you try to create the block again with the same name, an alert box indicates that the block references itself. This means you are trying to create a block that already exists. When you pick the **OK** button, the alert box disappears and the **Block Definition** dialog box redisplays. Press the **Cancel** button, explode the block, and try again to redefine the block.

> **NOTE**
>
> Once a block is modified, whether from changes made using the **BEDIT** tool or from redefinition using the **EXPLODE** and **BLOCK** tools, all instances of that block in the drawing update according to the changes.

Exercise 25-7

Access the Student Web site (www.g-wlearning.com/CAD) and complete Exercise 25-7.

Understanding the Circular Reference Error

circular reference error: An error that occurs when a block definition references itself.

When you try to redefine a block that already exists using the same name, a *circular reference error* occurs. AutoCAD informs you that the block references itself or that it has not been modified. The concept of a block referencing itself may be confusing unless you fully understand how AutoCAD works with blocks. A block can be composed of many objects, including other blocks. When you use the **BLOCK** tool to incorporate an existing block into a new block, AutoCAD makes a list of all the objects that compose the new block. This means AutoCAD refers to any existing block definitions added to the new block. A problem occurs if you select an instance, or reference, of the redefined block as a component object for the new definition. The new block refers to a block of the same name, or references itself. **Figure 25-13A** illustrates the process of correctly redefining a block named BOX to avoid a circular reference error. **Figure 25-13B** shows an incorrect redefinition resulting in a circular reference error.

Renaming Blocks

Use the **RENAME** tool to rename a block without editing the block definition. Access the **RENAME** tool to display the **Rename** dialog box shown in **Figure 25-14**. Select Blocks from the **Named Objects** list, and then pick the block to rename in the **Items** list. The current name appears in the **Old Name:** text box. Type the new block name in the **Rename To:** text box. Pick the **Rename To:** button to display the new name in the **Items** list. Pick the **OK** button to exit the **Rename** dialog box.

Figure 25-13.
A—The correct procedure for redefining a block. B—Redefining a block without first exploding the block creates an invalid circular reference.

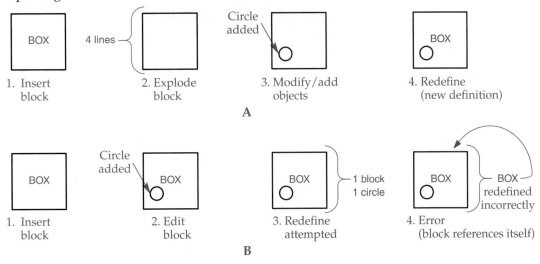

Figure 25-14.
The **Rename** dialog box allows you to change the name of blocks and other named objects.

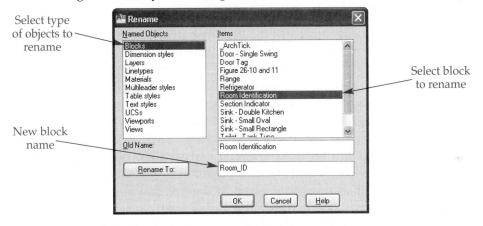

Updating Block Icons

A block icon forms when you build a block. The icon appears when you insert and edit blocks to help you recognize the block. Block icons require updating when an icon does not appear, as is often the case when you store a block in a drawing created with an older version of AutoCAD, or if the icon does not reflect changes made to the block. Use the **BLOCKICON** tool to create or update a block icon. Open the drawing in which the block is stored, access the **BLOCKICON** tool, enter the name of the block, and press [Enter].

Creating Blocks as Drawing Files

Blocks created with the **BLOCK** tool are stored in the drawing in which they are defined. A write block, created using the **WBLOCK** tool, saves the block as a separate drawing (DWG) file. You can also use the **WBLOCK** tool to create a global block from any object. It does not have to be a previously saved block. You can insert the resulting drawing file as a block into any drawing. Access the **WBLOCK** tool to display the **Write Block** dialog box shown in **Figure 25-15**.

Figure 25-15.
Using the **Write Block** dialog box to create a wblock from selected objects without first defining a block.

Pick to save selected objects as a wblock

Pick to select the insertion point

File location and name

Pick to select the objects defining the wblock

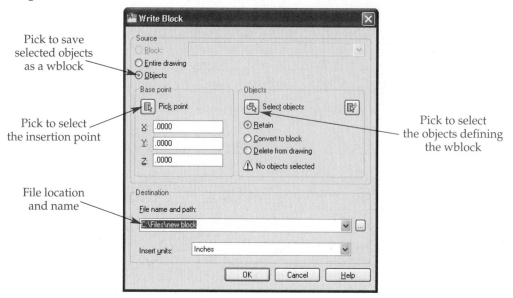

Creating a New Wblock

One method of creating a wblock is to create a drawing file from existing objects that you have not converted to a block. To use this technique, pick the **Objects** radio button in the **Source** area. The process of creating a wblock from existing non-block objects is similar to the process of creating a block using the **BLOCK** tool. Specify an insertion base point using options in the **Base point** area, and select the objects and the disposition of the objects using options in the **Objects** area. The **Base point** and **Objects** areas function the same as those found in the **Block Definition** dialog box.

In contrast to a block, a wblock is saved as a drawing file, *not* as a block in the current drawing. Enter a path and file name for the block in the **File name and path:** text box or pick the ellipsis (**...**) button next to the text box to display the **Browse for Drawing File** dialog box. Navigate to the folder where you want to save the file, confirm the name of the file in the **File name:** text box, and pick the **Save** button. The **Write Block** dialog box redisplays with the path and file name shown in the **File name and path:** text box. Finally, select the type of units that **DesignCenter** should use to insert the block in the **Insert units:** drop-down list. This is also located in the **Destination** area. Pick the **OK** button to finish. The objects are saved as a wblock in the specified folder. Now you can use the **INSERT** tool in any drawing to insert the block.

Saving an Existing Block As a Wblock

To create a wblock from an existing block, pick the **Block** radio button in the **Source** area. See **Figure 25-16.** Select the block to save as a wblock from the drop-down list. Use the options in the **Destination** area to locate the wblock, and pick the **OK** button to finish.

Storing a Drawing As a Wblock

To store an entire drawing as a wblock, pick the **Entire drawing** radio button in the **Source** area. Use the options in the **Destination** area to locate the wblock. In this case, the whole drawing is saved as if you were using the **SAVE** tool. However, all uninserted, or unused, blocks in the drawing are deleted. If the drawing contains any unused blocks, this method may reduce the size of a drawing considerably. Pick the **OK** button to finish.

Figure 25-16.
Using the **Write Block** dialog box to create a wblock from an existing block.

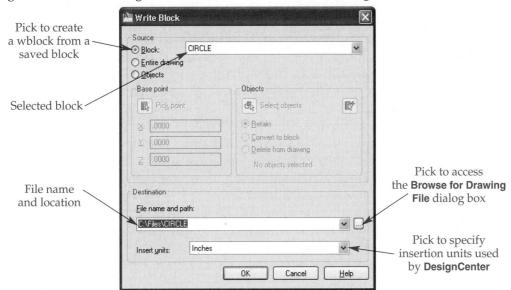

Pick to create a wblock from a saved block

Selected block

File name and location

Pick to access the **Browse for Drawing File** dialog box

Pick to specify insertion units used by **DesignCenter**

Exercise 25-8

Access the Student Web site (www.g-wlearning.com/CAD) and complete Exercise 25-8.

Revising an Inserted Drawing

If you insert a wblock into multiple drawings and then need to make changes to the wblock, use the **INSERT** tool to access the original drawing file with the **Select Drawing File** dialog box. Then activate the **Specify On-screen** check box in the **Insertion point** area and pick the **OK** button. When a message asks if you want to redefine the block, pick the **Yes** option. All of the wblock references automatically update. Press [Esc] to cancel the tool so that you do not insert a new block.

PROFESSIONAL TIP

If you work on projects in which inserted drawings require revisions, it is far more productive to use reference drawings instead of inserted drawing files. Chapter 32 explains reference drawings placed using the **XREF** tool. All referenced drawings automatically update when you open a drawing file that contains the externally referenced material.

Symbol Libraries

As you become proficient with AutoCAD, you may want to start constructing *symbol libraries*. Arranging a storage system for frequently used symbols significantly increases productivity. Establish how to store the symbols, as either blocks or drawing files, and identify a storage location and system.

symbol library: A collection of related blocks, shapes, views, symbols, or other content.

Creating a Symbol Library

The two basic options for creating a symbol library are to save all of the blocks in a single drawing or to save each block to a separate wblock file. Once you create a set of related block definitions, you can arrange the blocks in a symbol library. Identify each block with a name and insertion point location. Whether the blocks are being stored in a single drawing file or as individual files, follow these guidelines to create the symbol library:

- Assign one person to create the symbols for each specialty.
- Follow school or company standards for blocks and symbols.
- When saving multiple blocks in a drawing file, save one group of symbols per drawing file.
- When using wblocks, give the drawing files meaningful names and assign the files to separate folders on the hard drive.
- Provide all users with a hard copy of the symbol library showing each symbol, its insertion point, where it is located, and any other necessary information. See **Figure 25-17**.
- If a network is not in use, place the symbol library file(s) on each workstation in the classroom or office.
- Keep backup copies of all files in a secure place.
- When you revise symbols, update all files containing the edited symbols.
- Inform all users of any changes to saved symbols.

Storing Symbol Drawings

The local or network hard drive is one of the best places to store a symbol library. It is easy to access, quick, and more convenient to use than portable media. Removable media, such as a removable hard drive, USB flash drive, or a CD, are most appropriate

Figure 25-17.
A printed copy of electrical blocks stored in a symbol library. The "X" symbols indicate insertion points and are not part of the blocks.

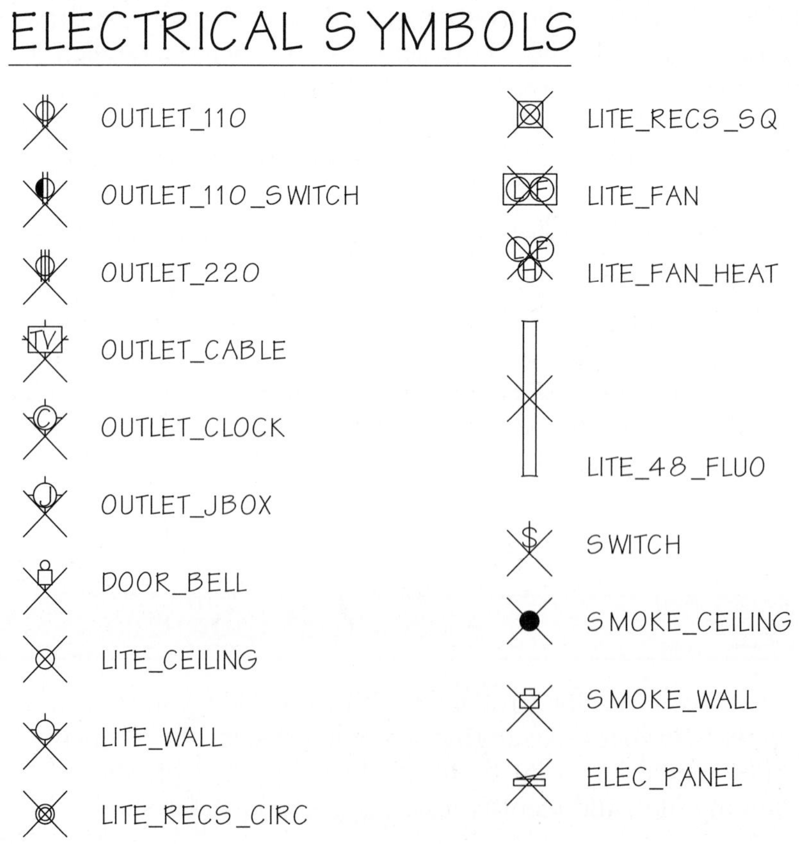

ELECTRICAL SYMBOLS

OUTLET_110	LITE_RECS_SQ
OUTLET_110_SWITCH	LITE_FAN
OUTLET_220	LITE_FAN_HEAT
OUTLET_CABLE	LITE_48_FLUO
OUTLET_CLOCK	
OUTLET_JBOX	SWITCH
DOOR_BELL	SMOKE_CEILING
LITE_CEILING	SMOKE_WALL
LITE_WALL	ELEC_PANEL
LITE_RECS_CIRC	

for backup purposes if a network drive with an automatic backup function is not available. In the absence of a network or modem, you can also use removable media to transport files from one workstation to another.

There are several methods of storing symbols on the hard drive. One option is to save symbols as wblocks in organized folders. You should store content outside of the AutoCAD folder to keep the system folder uncluttered and to allow you to differentiate your folders and files from AutoCAD system folders and files. A good method is to create a \Blocks folder for storing your blocks, as shown in Figure 25-18.

If you save multiple symbols within a single drawing, use **DesignCenter** or the **Tool Palettes** palette to insert the symbols as needed. When you use this system, it is often a good idea to use several drawing files to group similar symbols. For example, you may want to create different symbol libraries based on electronic, electrical, piping, mechanical, structural, architectural, landscaping, and mapping symbols. Limit the symbols in a drawing to a reasonable number so you can easily find the symbols. If there are too many blocks in a drawing, it may be difficult to locate the desired symbol.

You should arrange drawing files saved on the hard drive in a logical manner. All workstations in a non-networked classroom or office should have folders with the same names. Assign one person to update and copy symbol libraries to all workstations. Copy drawing files onto each workstation from a master CD. Keep the master and backup versions of the symbol libraries in separate locations.

Figure 25-18.
An efficient way to store blocks saved as drawing files is to set up a Blocks folder containing folders for each type of block on the hard drive.

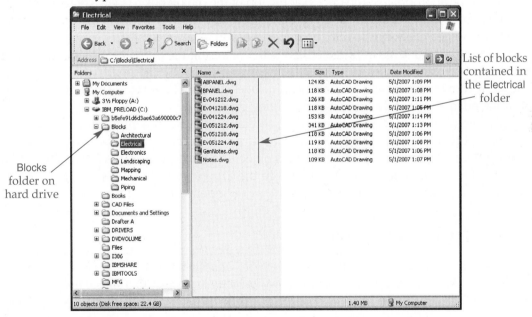

Blocks folder on hard drive

List of blocks contained in the Electrical folder

Purging Named Objects

A drawing often accumulates several *named objects* that are unused and may be unnecessary. Unused named objects increase the drawing file size and may make it difficult to locate and use items that are often required or referenced in the drawing. As a result, you may want to use the **PURGE** tool to delete or *purge* unused objects from the drawing. Access the **PURGE** tool to display the **Purge** dialog box, shown in Figure 25-19.

named objects: Blocks, dimension styles, layers, linetypes, materials, multileader styles, plot styles, shapes, table styles, text styles, and visual styles that have specific names.

purge: Delete unused named objects from a drawing file.

Type

PURGE
PU

Figure 25-19.
The **Purge** dialog box.

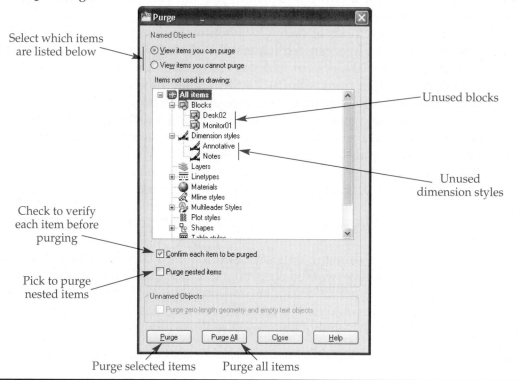

Select which items
are listed below →

→ Unused blocks

→ Unused
dimension styles

Check to verify
each item before
purging →

Pick to purge
nested items →

Purge selected items   Purge all items

Select the appropriate radio button at the top of the dialog box to view content that you can purge or to view content that you cannot purge. Before purging, select the **Confirm each item to be purged** check box to have an opportunity to review each item before it is deleted. Check **Purge nested items** to purge nested items. Selecting the **Purge zero-length geometry and empty text objects** check box is an effective way to erase all zero-length objects, such as a line or arc drawn as a dot and text that only includes spaces. These objects are often mistakes or unintended results of the drawing and editing processes.

To purge only some items, use the tree view to locate and highlight the items to purge, and then pick the **Purge** button. To purge all unused items, pick the **Purge All** button. Purging may cause other named objects to become unreferenced. Thus, you may need to purge more than once to purge the drawing of all unused named objects. Messages appear to guide you through the purge operation.

Chapter Test

Answer the following questions. Write your answers on a separate sheet of paper or go to the Student Web site (www.g-wlearning.com/CAD) and complete the electronic chapter test.

1. Why would you draw blocks on layer 0?
2. What properties do blocks drawn on a layer other than layer 0 assume when they are inserted?
3. What is the maximum number of characters allowed in a block name?
4. Define the term *nesting* in relation to blocks.
5. What is a block reference?
6. How can you access a listing of all blocks in the current drawing?
7. Describe the effect of entering negative scale factors when inserting a block.
8. What type of block is a one-unit line object?
9. How do you preset block insertion variables using the **Insert** dialog box?
10. Name a limitation of an array pattern created with the **MINSERT** tool.
11. What is the purpose of the **BASE** tool?

12. Briefly explain how to insert a block into a drawing from **DesignCenter**.
13. What tool allows you to change a block's layer without editing the block definition?
14. Identify the tool that allows you to break an inserted block into its individual objects for editing purposes.
15. Suppose you have found that a block was incorrectly drawn. Unfortunately, you have already inserted the block 30 times. How can you edit all of the blocks quickly?
16. What is the primary difference between blocks created with the **BLOCK** and **WBLOCK** tools?
17. Explain the advantage of storing a drawing as a wblock if you anticipate the need to insert the drawing into other drawings.
18. Define *symbol library*.
19. Explain two ways to remove all unused blocks from a drawing.
20. What is the purpose of the **PURGE** tool?

Drawing Problems

Start AutoCAD if it is not already started. Start a new drawing using an appropriate template of your choice. The template should include layers, text styles, dimension styles, and multileader styles appropriate for drawing the given objects. Add layers, text styles, dimension styles, and multileader styles as needed. Draw all objects using appropriate layers, text styles, dimension styles, multileader styles, justification, and format. Follow the specific instructions for each problem. Use your own judgment and approximate dimensions when necessary.

▼ Basic

1. Open P14-17 and save as P25-1. The P25-1 file should be active. The sketch for this drawing is shown below. Erase all copies of the symbols, leaving the original objects intact. These include the steel column symbols and the bay and column line tags. Then do the following:
   A. Make blocks of the steel column symbol and the tag symbols.
   B. Use the **MINSERT** tool or the **ARRAY** tool to place the symbols in the drawing.
   C. Dimension the drawing as shown in the sketch.
   D. Resave the drawing.

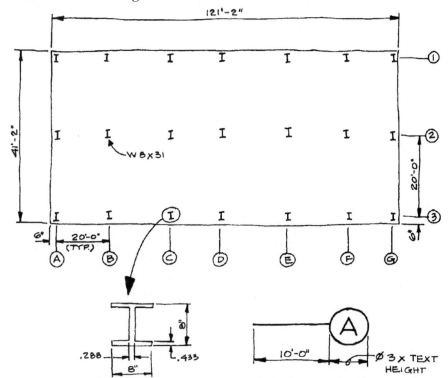

2. Open P14-18 and save as P25-2. The P25-2 file should be active. The sketch for this drawing is shown below. Erase all of the desk workstations except one. Then do the following:
   A. Create a block of the workstation.
   B. Insert the block into the drawing using the **MINSERT** tool.
   C. Dimension one of the workstations as shown in the sketch.
   D. Resave the drawing.

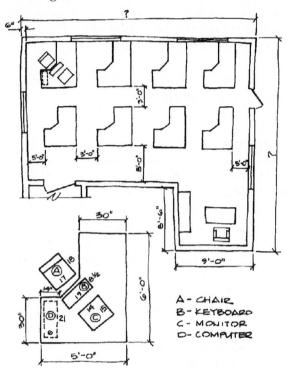

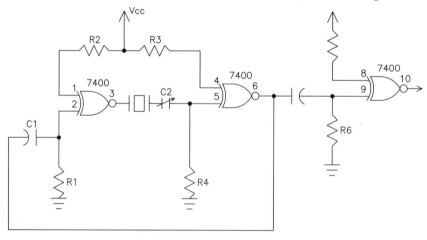

A – CHAIR
B – KEYBOARD
C – MONITOR
D – COMPUTER

3. Complete this problem after completing Problem 25-7. Open P25-7 and save as P25-3. The P25-3 file should be active. Modify the NAND gates to become XNOR gates, as shown below, by modifying the block definition. Save the drawing as P25-3.

AutoCAD and Its Applications—Basics

▼ Intermediate

Problems 4–7 represent a variety of diagrams created using symbols as blocks. Create each drawing as shown. (The drawings are not to scale.) Create the symbols first as blocks or wblocks and then save them in a symbol library using one of the methods described in this chapter.

4. Draw the integrated circuit schematic for a clock as shown. Save the drawing as P25-4.

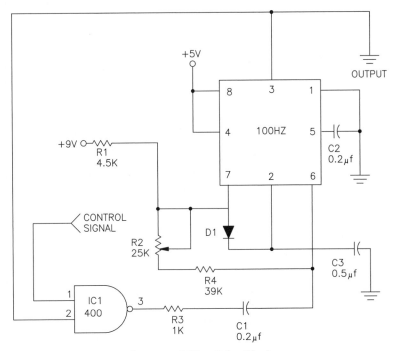

Integrated Circuit for Clock

5. Draw the piping flow diagram as shown. Save the drawing as P25-5.

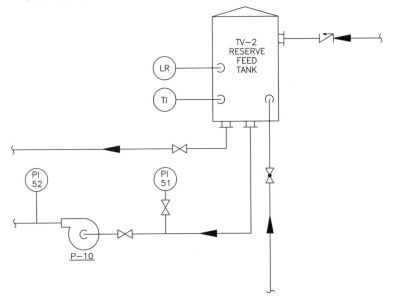

Piping Flow Diagram

6. Draw the logic diagram of a marking system as shown. Save the drawing as P25-6.

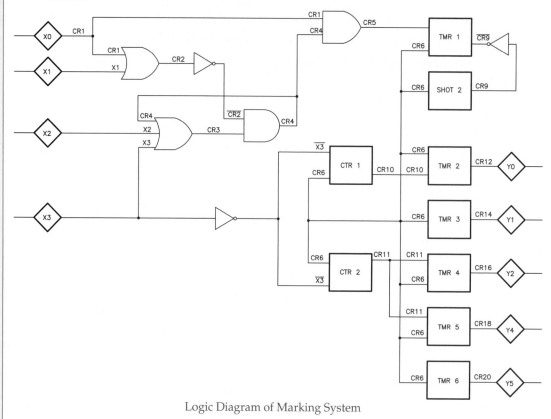

Logic Diagram of Marking System

7. Draw the digital logic circuit shown. Create each type of component in the circuit as a block. Save the drawing as P25-7.

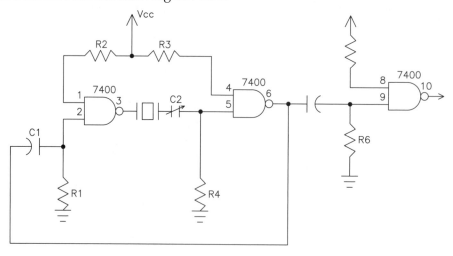

▼ Advanced

Problems 8–12 are presented as engineering sketches. They are schematic drawings created using symbols and are not drawn to scale. Create the symbols first as blocks or wblocks and then save them in a symbol library using one of the methods described in this chapter.

8. The rough sketch shown below is a logic diagram of a portion of the internal components of a computer. Save the drawing as P25-8.

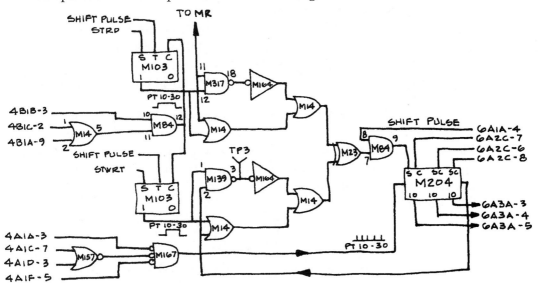

9. The rough sketch shown below is a piping flow diagram of a cooling water system. Look closely at this drawing before you begin. Draw the thick flow lines with polylines. Save the drawing as P25-9.

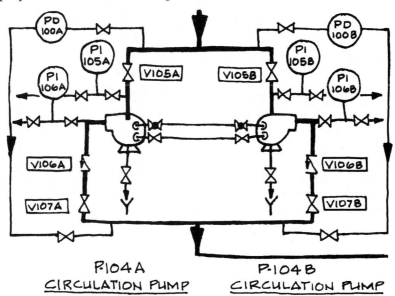

P-104A
CIRCULATION PUMP

P-104B
CIRCULATION PUMP

10. The rough sketch shown below is the general arrangement of a basement floor plan for a new building. The engineer has shown one example of each type of equipment. Use the following instructions to complete the drawing.

A. All text should be 1/8" high, except the text for the bay and column line tags, which should be 3/16" high. The diameter of the line balloons for the bay and column lines should be twice the diameter of the text height.

B. The column and bay steel symbols represent wide-flange structural shapes and should be 8" wide × 12" high.

C. The PUMP and CHILLER installations (except PUMP #4 and PUMP #5) should be drawn per the dimensions given for PUMP #1 and CHILLER #1. Use the dimensions shown for the other PUMP units.

D. TANK #2 and PUMP #5 (P-5) should be drawn per the dimensions given for TANK #1 and PUMP #4.

E. Tanks T-3, T-4, T-5, and T-6 are all the same size and are aligned 12' from column line A.

F. Plan this drawing carefully and create as many blocks as necessary to increase your productivity. Dimension the drawing as shown, and provide location dimensions for all equipment not shown in the engineer's sketch.

G. Save the drawing as P25-10.

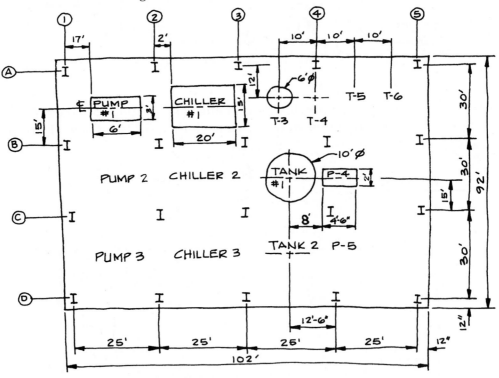

11. Open P25-10 and save as P25-11. The P25-11 file should be active. The engineer has provided you with a sketch of the necessary revisions to the drawing. It is up to you to alter the drawing as quickly and efficiently as possible. The dimensions shown on the sketch below *do not* need to be added to the drawing; they are provided for construction purposes only. Revise the drawing so all chillers and the four tanks reflect the changes. Save the drawing as P25-11.

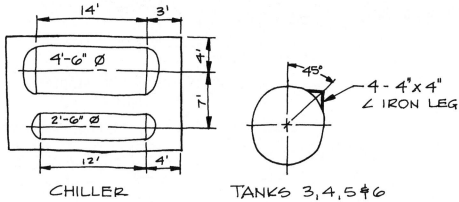

CHILLER   TANKS 3, 4, 5 & 6

12. The rough sketch of a piping flow diagram shown below is part of an industrial effluent treatment system. Eliminate as many bends in the flow lines as possible. Place arrowheads at all flow line intersections and bends. The flow lines should not run through any valves or equipment. Use polylines for the thick flow lines. Save the drawing as P25-12.

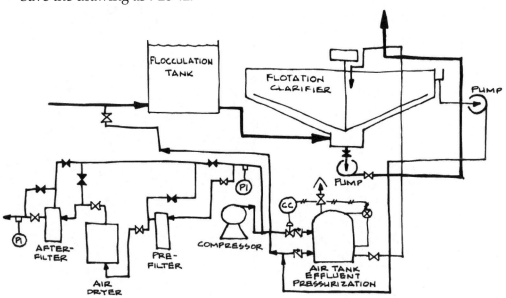

13. Create computer, plotter, and printer/copier blocks and then draw the network diagram. Save the drawing as P25-13.

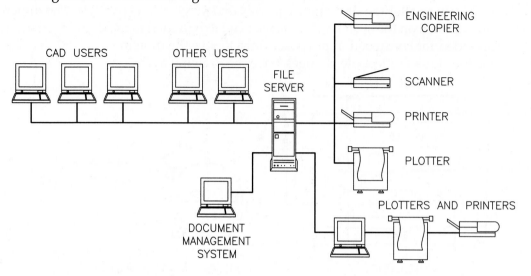

14. Draw the piping diagram shown, creating blocks for each type of fitting. Save the drawing as P25-14.

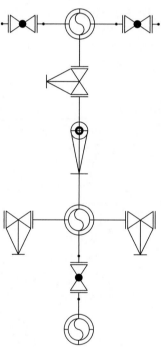

15. Create component blocks based on the dimensions shown. Then use the blocks to draw the schematic below. Save the drawing as P25-15.

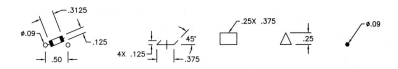

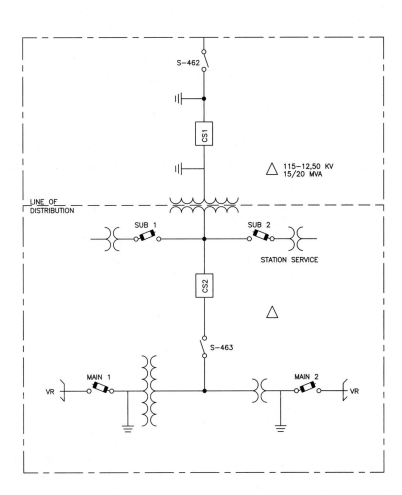

16. Create a symbol library for one of the drafting disciplines listed below and save it as a template or drawing file. Then, after checking with your instructor, draw a problem using the library. If you save the symbol library as a template, start the problem with the template. If you save it as a drawing file, start a new drawing and insert the symbol library into it. Specialty areas you might create symbols for include:

- Mechanical (machine features, fasteners, tolerance symbols)
- Architectural (doors, windows, fixtures)
- Structural (steel shapes, bolts, standard footings)
- Industrial piping (fittings, valves)
- Piping flow diagrams (tanks, valves, pumps)
- Electrical schematics (resistors, capacitors, switches)
- Electrical one-line (transformers, switches)
- Electronics (IC chips, test points, components)
- Logic diagrams (AND gates, NAND gates, buffers)
- Mapping, civil (survey markers, piping)
- Geometric tolerancing (feature control frames)

Save the drawing as P25-16 or choose an appropriate file name, such as ARCH-PRO or ELEC-PRO. Display the symbol library created in this problem and print a hard copy. Put the printed copy in your notebook as a reference.

Mode

☑ Invisible

☐ Constant

☑ Verify

☐ Preset

☑ Lock position

Attribute

Prompt: Which manufacturer?

Default:

CHAPTER 26

Block Attributes

Learning Objectives

After completing this chapter, you will be able to do the following:

✓ Define attributes.
✓ Create and insert blocks that contain attributes.
✓ Edit attribute values and definitions in existing blocks.
✓ Create title blocks, revision blocks, and parts lists with attributes.
✓ Display attribute values in fields.

Attributes significantly enhance blocks that require text or numerical information. For example, a door identification block contains a letter or number that links the door to a door schedule. Adding an attribute to the door identification symbol allows you to include any letter or number with the symbol, without adding block definitions. You can also *extract* attribute data to automate drawing applications, such as preparing schedules, parts lists, and bills of materials.

attributes:
Text-based data assigned to a specific object. Attributes turn a drawing into a graphical database.

extract: Gathering content from the drawing file database to display in the drawing or in an external document.

Defining Attributes

Attributes and geometry are often used together to create a block. See **Figure 26-1**. However, you can prepare stand-alone blocks that only include attributes. Create attributes along with other objects during the initial phase of block development. You can add as many attributes as needed to describe the symbol or product, such as the name, number, manufacturer, type, size, price, and weight of an item. To assign attributes, access the **ATTDEF** tool to display the **Attribute Definition** dialog box. See **Figure 26-2**.

Setting Attribute Modes

The **Mode** area allows you to set attribute modes. Symbols often require attributes to appear with the block. Select the **Invisible** check box to hide attributes, but still include attribute data in the drawing that you can reference and extract. The geranium symbol in **Figure 26-1** is an example of a block with attributes that you may want to hide. The other blocks show examples in which the attributes should appear. Blocks often include both visible and invisible attributes.

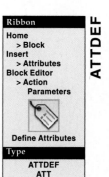

Ribbon

Home
> Block
Insert
> Attributes
Block Editor
> Action
Parameters

Define Attributes

Type
ATTDEF
ATT

ATTDEF

Figure 26-1.
Examples of blocks with defined attributes.

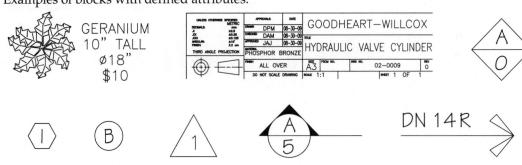

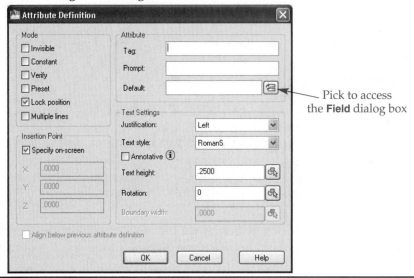

Figure 26-2.
Use the **Attribute Definition** dialog box to assign attributes to blocks.

Pick the **Constant** check box if the value of the attribute should always be the same. All insertions of the block display the same value for the attribute, without prompting for a new value when you insert the block. Deselect the **Constant** check box to use different attribute values for multiple insertions of the block. Pick the **Verify** check box to display a prompt that asks if attribute value is correct when you insert the block. Choose the **Preset** check box to have the attribute assume preset values during block insertion. This option disables the attribute prompt. Uncheck **Preset** to display the normal prompt.

Deselect the **Lock position** check box to have the ability to move the attribute independently of the block after insertion. In addition, you must deselect the **Lock position** check box to include the attribute with the action selection set when you assign an action to a dynamic block. If the box is checked, the attribute filters out when you assign the action to the dynamic block. Dynamic blocks are covered later in this textbook.

You can create single-line or multiple-line attributes. Pick the **Multiple lines** check box to activate options for creating a multiple-line attribute. Deselect the check box to create a single-line attribute.

Using the Attribute Area

The **Attribute** area provides text boxes for assigning a tag, prompt, and default value to the attribute. Attribute values can include up to 256 characters. If the first character in an entry is a space, start the string with a backslash (\). If the first character is a backslash, begin the entry with two backslashes (\\).

Use the **Tag** text box to enter the attribute name, or tag. For example, the tag for a size attribute for a valve block could be SIZE. You must enter a tag in order to create an attribute. The tag cannot include spaces. The attribute definition applies uppercase characters to the tag, even if you type lowercase characters in the text box.

Enter a statement in the **Prompt** text box that will display when you insert or edit the block. For example, if you specify SIZE as the attribute tag, you might specify What is the valve size? or Enter valve size: as the prompt. You have the option to leave the prompt blank. The **Prompt** text box is disabled when you select the **Constant** attribute mode.

The **Default** text box allows you to enter a default attribute value, or a description of an acceptable value for reference. For example, you might enter the most common size for the SIZE attribute, or a message regarding the type of information needed, such as 10 SPACES MAX or NUMBERS ONLY. If you deselect the **Multiple lines** attribute mode, enter the default value directly in the text box. When using the **Multiple lines** attribute mode, select the ellipsis (**...**) button to enter the drawing area and place multiline text. The **Text Editor** ribbon tab appears, along with the **Text Formatting** toolbar shown in **Figure 26-3**. Enter the default text, and then pick the **OK** button on the toolbar to return to the **Attribute Definition** dialog box. Use the **Insert field** button to include a field in the default value. You also have the option to leave the default value blank.

NOTE

The abbreviated **Text Formatting** toolbar shown in **Figure 26-3** appears by default. Set the **ATTIPE** system variable to 1 to display the complete **Text Formatting** toolbar. The **ATTIPE** system variable is set to 0 by default.

Adjusting Attribute Text Options

The **Text Options** area allows you to specify attribute text settings. Many of these options function like the text settings for single-line and multiline text. Use the **Justification** drop-down list to select a justification for the attribute text. The default option is Left. In single-line attributes, the text itself is justified. In the **Multiple lines** attribute mode, the text boundary is justified.

Use the **Text Style** drop-down list to select a text style for the attribute from the styles available in the current drawing. Pick the **Annotative** check box to make the attribute text height annotative. AutoCAD scales annotative attributes according to the specified annotation scale, which eliminates the need to calculate the scale factor.

Figure 26-3.
You can define multiple-line attributes directly on-screen. The abbreviated **Text Formatting** toolbar appears in addition to the **Text Editor** ribbon tab.

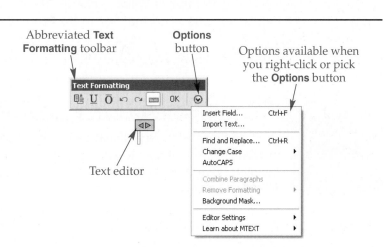

Specify the height of the attribute text in the **Height** text box, or pick the **Text Height** button next to the text box to pick two points in the drawing to set the text height. Identify the rotation angle for the attribute text in the **Rotation** text box, or pick the **Rotation** button next to the text box to pick two points in the drawing to set the text rotation. The **Boundary width** option is available only in the **Multiple lines** attribute mode. Enter a width for the multiple-line attribute boundary in the **Boundary width** text box, or pick the **Boundary width** button next to the text box to pick two points in the drawing to set a text boundary width.

Defining the Attribute Insertion Point

The **Insertion Point** area of the **Attribute Definition** dialog box provides options for defining how and where to position the attribute during insertion. Choose the **Specify On-screen** check box to pick an insertion point in the drawing after you pick the **OK** button to create the attribute and exit the **Attribute Definition** dialog box. This method can save time by allowing you to pick the insertion base point without using the **Pick point** button and then reentering the **Block Definition** dialog box. As an alternative, if you know the coordinates for the insertion point, deselect the **Specify On-screen** check box and type values in the **X:**, **Y:**, and **Z:** text boxes.

The **Align below previous attribute definition** check box becomes enabled if the drawing already contains at least one attribute. Check the box to place the new attribute directly below the most recently created attribute using the justification of that attribute. This is an effective technique for placing a group of different attributes in the same block. When this box is checked, the **Text Options** and **Insertion Point** areas are deactivated.

Placing the Attribute

After defining all elements of the attribute, pick **OK** to close the **Attribute Definition** dialog box. The attribute tag appears on-screen automatically if coordinates specify the insertion point, or if you are using the **Align below previous attribute definition** option. Otherwise, AutoCAD prompts you to select a location. If the attribute mode is set to **Invisible**, do not be concerned that the tag is visible; this is the only time the tag appears.

Editing Attribute Properties

The **Properties** palette provides expanded options for editing attributes. **Figure 26-4** shows the **Properties** palette with an attribute selected. You can change the color, linetype, or layer of the selected attribute in the **General** section. In the **Text** section, you can select **Tag**, **Prompt**, or **Value** to change the corresponding entries. If the value contains a field, it appears as normal text in the **Properties** palette. Modified field text automatically converts to text. The **Text** section also contains options to change the attribute text settings. Additional text and attribute options are available in the **Misc** section.

PROFESSIONAL TIP

Perhaps the most powerful feature of the **Properties** palette for editing attributes is the ability to change the original attribute modes. The **Invisible**, **Constant**, **Verify**, and **Preset** mode settings are available in the **Misc** category.

Figure 26-4.
Using the **Properties** palette to modify attributes.

Selected object to edit

Pick to change the attribute tag

Pick to change an attribute mode setting

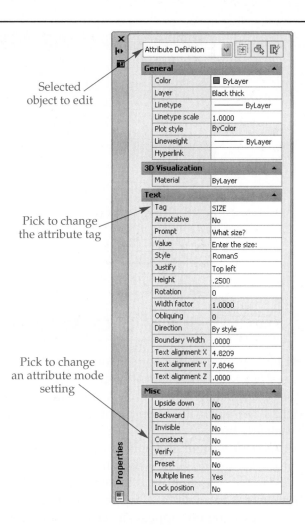

Creating Blocks with Attributes

Once you create attributes, use the **BLOCK** or **WBLOCK** tool to define a block with attributes. When creating the block, be sure to select all of the objects and attributes to include with the block. The order in which you select the attribute definitions is the order of prompts, or the order in which the attributes appear in the **Edit Attributes** dialog box. If you select the **Convert to Block** radio button in the **Block Definition** dialog box, the **Edit Attributes** dialog box appears when you create the block. See **Figure 26-5.** This dialog box allows you to adjust attribute values when you insert or edit the block.

PROFESSIONAL TIP

If you create attributes in the order in which you want to receive prompts and then use window or crossing selection to select the attributes, the attribute prompts are displayed in the *reverse* order of the desired prompting. To change the order, insert, explode, and then redefine the block, using window or crossing selection to pick the attributes. The attribute prompt order reverses again, placing the prompts in the desired order.

Figure 26-5.
The **Enter Attributes** dialog box allows you to enter attribute definitions when you insert or edit a block.

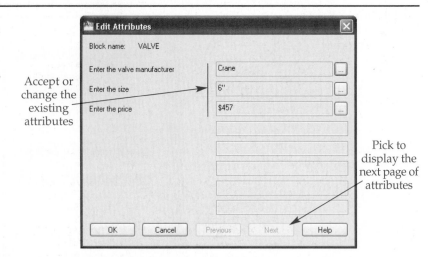

Accept or change the existing attributes

Pick to display the next page of attributes

Inserting Blocks with Attributes

Use the **INSERT** tool or another block insertion method, such as **DesignCenter** or a tool palette, to insert a block that contains attributes. The process of inserting a block with attributes is the same as inserting a block without attributes. The only difference is that after you define the block insertion point, scale, and rotation angle, prompts request values for each attribute.

By default, the **ATTDIA** system variable is set to 0, which displays single-line attribute prompts at the command line or dynamic input, and multiple-line attribute prompts using the AutoCAD text window. A better method of entering attribute values is to set the **ATTDIA** system variable to 1 before inserting blocks, to enable the **Edit Attributes** dialog box. The dialog box appears after you enter the insertion point, scale, and rotation angle, allowing you to answer each attribute prompt. Type single-line attribute values in the text boxes. To define multiple-line attributes, select the ellipsis (**...**) button next to the text boxes to enter values on-screen as multiline text. If a value includes a field, you can right-click on the field to edit it or convert it to text.

You can quickly move forward through the attributes and buttons in the **Edit Attributes** dialog box by pressing [Tab]. Press [Shift]+[Tab] to cycle through the attributes and buttons in reverse order. If the block includes more than eight attributes, pick the **Next** button at the bottom of the **Edit Attributes** dialog box to display the next page of attributes. When you finish entering values, pick the **OK** button to close the dialog box and create the block with all visible, defined attributes.

Exercise 26-1

Access the Student Web site (www.g-wlearning.com/CAD) and complete Exercise 26-1.

Attribute Prompt Suppression

Some drawings may use blocks with attributes that always retain their default values. In this case, there is no need to answer prompts for the attribute values when you insert the block. You can turn off the attribute prompts by setting the **ATTREQ** system variable to 0. After making this setting, try inserting the VALVE block created in Exercise 26-1. Notice that none of the attribute prompts appear. To display attribute prompts again, change the setting back to 1. The **ATTREQ** system variable setting is saved with the drawing.

Controlling Attribute Display

Attributes contain valuable drawing information. Some attributes only provide content to generate parts lists or bills of materials and to speed accounting. These types of attributes usually do not display on-screen or plot. Use the **ATTDISP** tool to control the display of attributes on-screen. The easiest way to activate an **ATTDISP** tool option is to pick the corresponding button from **Attributes** panel of the **Insert** ribbon tab.

Use the **Normal** (**Retain display**) option to display attributes exactly as created. This is the default setting. Use the **ON** (**Display all**) option to display *all* attributes. Apply the **OFF** (**Hide all**) option to suppress the display of all attributes, including visible attributes.

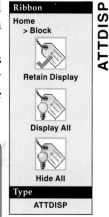

Ribbon
Home
> Block
Retain Display
Display All
Hide All
Type
ATTDISP

ATTDISP

Ribbon
Home
> Block
Insert
> Attributes
Edit Attributes (Single)
Type
EATTEDIT

EATTEDIT

Changing Attribute Values

Once you create a block with attributes, tools are available for editing attribute values and settings. One option is modify the attributes of a single block using the **EATTEDIT** tool. Access the **EATTEDIT** tool and pick the block containing the attributes you want to modify to display the **Enhanced Attribute Editor**. See **Figure 26-6.**

The **Attribute** tab, shown in **Figure 26-6,** displays all attributes assigned to the selected block. Pick the attribute to modify and enter a new value in the **Value:** text box. If the attribute is a multiple-line attribute, the ellipsis (**...**) button is available for selection, allowing you to modify the text on-screen. Pick the **Apply** button after adjusting the value.

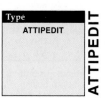

Type
ATTIPEDIT

ATTIPEDIT

Select the attribute to be modified

Value of the selected attribute

Pick to apply changes

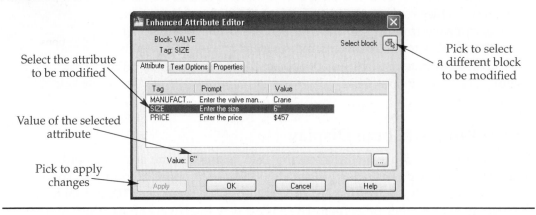

Pick to select a different block to be modified

To select a different block to modify, pick the **Select block** button in the dialog box. The dialog box hides to allow you to select a different block in the drawing. Then the dialog box reappears and displays the attributes for the selected block.

The **Text Options** tab, shown in **Figure 26-7A** allows you to modify the text properties of an attribute. The **Properties** tab, shown in **Figure 26-7B,** provides object property adjustments for an attribute. Each attribute in a block is a separate item. The settings you apply in the **Text Options** and **Properties** tabs affect the active attribute in the **Attribute** tab. Pick the **Apply** button to view changes made to attributes. Pick the **OK** button to close the dialog box.

Exercise 26-2

Access the Student Web site (www.g-wlearning.com/CAD) and complete Exercise 26-2.

Using the FIND Tool to Edit Attributes

One of the quickest ways to edit attributes is to use the **FIND** tool. With no tool active, right-click in the drawing area and select **Find...** to display the **Find and Replace** dialog box. You can search the entire drawing or a selected group of objects for an attribute.

Editing Attribute Values and Properties Globally

The **Enhanced Attribute Editor** allows you to edit attribute values by selecting blocks one at a time. You can use the **-ATTEDIT** tool to edit the attributes of several blocks. When you access the **-ATTEDIT** tool, a prompt asks if you want to edit attributes individually. Use the default Yes option to select specific blocks with attributes to edit. Use the No option to apply *global attribute editing.*

If you choose the Yes option, prompts appear to specify the block name, attribute tag, and attribute value. To edit attribute values selectively, respond to each prompt with the correct name or value, and then select one or more attributes. If you receive the message "0 found" after selecting attributes, you picked an incorrectly specified attribute. It is often quicker to press [Enter] at each of the three specification prompts and then pick the attribute to edit. Select an option and follow the prompts to edit the attribute(s).

-ATTEDIT

Ribbon
Home
> Block
Insert
> Attributes

Edit Attribute (Multiple)

Type
-ATTEDIT
-ATE

global attribute editing: Editing or changing all insertions, or instances, of the same block in a single operation.

Figure 26-7.
A—The **Text Options** tab provides options in addition to those set in the **Attribute Definition** dialog box. B—The **Properties** tab allows you to modify the properties of an attribute.

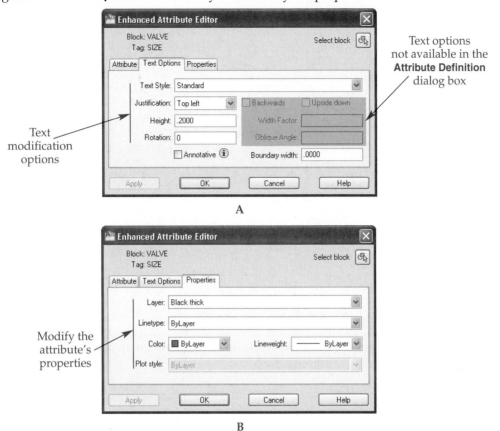

A

B

If you choose the No option, the Edit only attributes visible on screen? prompt appears. Select Yes to edit all visible attributes or No to edit all attributes, including those that are invisible. The same three prompts previously described for individual block editing now appear.

Figure 26-8A shows the VALVE block from Exercise 26-1 inserted three times with the manufacturer specified as CRANE. In this example, the manufacturer was supposed to be POWELL. To change the attribute for each insertion, enter the **-ATTEDIT** tool and specify global editing. Press [Enter] at each of the three specification prompts. When the Select attributes: prompt appears, pick CRANE on each of the VALVE blocks and press [Enter]. At the Enter string to change: prompt, enter CRANE, and at the Enter new string: prompt, enter POWELL. See the result in **Figure 26-8B.**

PROFESSIONAL TIP

Use care when assigning the **Constant** mode to attribute definitions. The **-ATTEDIT** tool displays 0 found if you attempt to edit a block attribute that has a **Constant** mode setting. Assign the **Constant** mode only to attributes you know will not change.

NOTE

You can also use the **-ATTEDIT** tool to edit individual attribute values and properties. However, it is more efficient to use the **Enhanced Attribute Editor** to change individual attributes.

Figure 26-8.
Using the global editing technique with the **-ATTEDIT** tool allows you to change the same attribute on several block insertions.

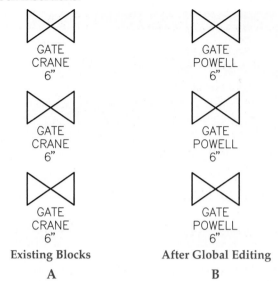

GATE CRANE 6"   GATE POWELL 6"

GATE CRANE 6"   GATE POWELL 6"

GATE CRANE 6"   GATE POWELL 6"

Existing Blocks   **After Global Editing**

A   B

Exercise 26-3

Access the Student Web site (www.g-wlearning.com/CAD) and complete Exercise 26-3.

Changing Attribute Definitions

Ribbon
Home
> Block
Insert
> Attributes

Manage Attributes
Type
-BATTMAN

Once you create a block with attributes, tools are available for modifying attribute definitions. One option is to modify attribute definitions using the **BATTMAN** tool, which displays the **Block Attribute Manager**. See **Figure 26-9**. To manage the attributes in a block, choose the block name from the **Block:** drop-down list or pick the **Select block** button to return to the drawing and pick a block.

The tag, prompt, default value, and modes for each attribute are listed by default. To select the attribute properties listed in the **Block Attribute Manager**, pick the **Settings...** button to open the **Block Attribute Settings** dialog box. See **Figure 26-10**. Check the

Figure 26-9.
Use the **Block Attribute Manager** to change attribute definitions, delete attributes, and change the order of attribute prompts.

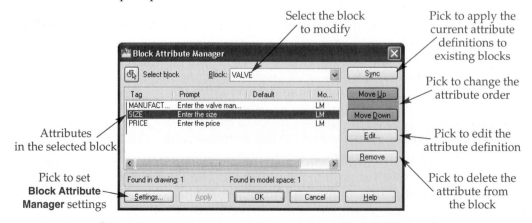

Select the block to modify

Pick to apply the current attribute definitions to existing blocks

Pick to change the attribute order

Pick to edit the attribute definition

Pick to delete the attribute from the block

Attributes in the selected block

Pick to set **Block Attribute Manager** settings

Figure 26-10.
The **Block Attribute Settings** dialog box controls the types of attributes displayed in the **Block Attribute Manager**.

Select the attribute properties to list in the **Block Attribute Manager**

Identifies duplicate tags

Updates existing blocks

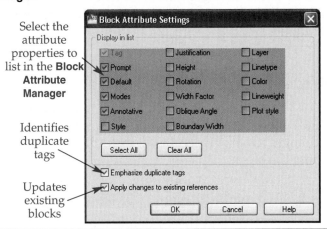

properties to list in the **Display in list** area. When you select the **Emphasize duplicate tags** check box, attributes with identical tags highlight in red. To apply the changes you make in the **Block Attribute Manager** to existing blocks, check **Apply changes to existing references**. Pick the **OK** button to return to the **Block Attribute Manager**.

The attribute list in the **Block Attribute Manager** reflects the order in which prompts appear when you insert a block. Use the **Move Up** and **Move Down** buttons to change the order of the selected attribute within the list, modifying the prompt order. To delete an attribute, pick the **Remove** button. To modify an attribute, select the attribute to edit and pick the **Edit...** button to display the **Edit Attribute** dialog box. See **Figure 26-11.** The **Attribute** tab allows you to modify the modes, tag, prompt, and default value. The **Text Options** and **Properties** tabs of the **Edit Attribute** dialog box are identical to the tabs found in the **Enhanced Attribute Editor**. If you check **Auto preview changes** at the bottom of the dialog box, changes to attributes display immediately in the drawing area.

After modifying the attribute definition in the **Edit Attribute** dialog box, pick the **OK** button to return to the **Block Attribute Manager**. Then pick the **OK** button to return to the drawing. When you modify attributes within a block, future insertions of the block reflect the changes. Existing blocks update only if you select the **Apply changes to existing references** check box in the **Settings** dialog box.

Figure 26-11.
Use the **Edit Attribute** dialog box to modify attribute definitions and properties.

Use these tabs to modify attribute properties

Select modes

Modify attribute definition

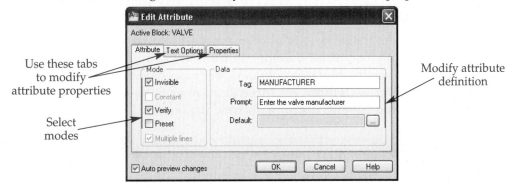

Redefining a Block and Its Attributes

To add attributes to, or revise the geometry of, a block, edit the block definition using the **BEDIT** or **REFEDIT** tools. The **REFEDIT** tool is covered in Chapter 32. Both tools allow you to make changes to a block definition, including attributes assigned to the block, without exploding the block.

Synchronizing Attributes

Redefining a block automatically updates the properties of all of the same blocks in the drawing, but not changes made to attributes. For example, if you add an object to a block, all existing blocks of the same name update to display the new object. However, if you add an attribute to a block, all existing blocks of the same name continue to display the original attributes, without the new attribute. Synchronize the blocks to update the attribute redefinition.

You can synchronize blocks in the **Block Attribute Manager** by picking the **Sync** button. This is convenient because of the ability to make changes to and remove attributes using the **Block Attribute Manager**. Use the **ATTSYNC** tool to synchronize attributes from outside the **Block Attribute Manager**. Access the **ATTSYNC** tool and use the default **Select** option to pick any of the blocks containing the attributes to synchronize. An alternative is to use the **Name** option to type the block name, or use the **?** option to list the names of all blocks in the drawing. Then choose the **Yes** option to synchronize attributes, or **No** to select a different block.

ATTSYNC

Ribbon
Home
> Block
Insert
> Attributes

Synchronize

Type
ATTSYNC

Automating Drafting Documentation

Attributes automate the process of placing symbols that require textual information. They are especially useful for automating common detailing or documentation tasks such as preparing title block information, revision block data, schedules, or a parts list or bill of materials. Filling out these items is usually one of the more time-consuming tasks associated with drafting documentation.

Creating Title Blocks

To create an automated title block, first use the correct layer(s), typically layer 0, to draw title block objects and add text that does not change, such as titles. Format the title block in accordance with industry or company standards. Include your company or school logo if appropriate. If you work in an industry that produces items for the federal government, also include the applicable *Federal Supply Code for Manufacturers (FSCM)*. **Figure 26-12** shows a title block drawn in accordance with the ASME Y14.1 *Decimal Inch Drawing Sheet Size and Format* standard.

Next, define attributes for each area of the title block. As you create attributes, determine the appropriate text height and justification for each definition. Common title block attributes include drawing title, drawing number, drafter, checker, dates, drawing scale, sheet size, material, finish, revision letter, and tolerance information. See **Figure 26-13**. Create approval attributes with a prompt such as ENTER INITIALS OR SEEK SIGNATURE, providing the flexibility to type initials or leave the cell blank for written initials. Apply the same practice to date attributes. Include any other information that may be specific to your organization or application. Assign default values to the attributes wherever possible. For example, if your organization consistently specifies the same overall tolerances for drawing dimensions, assign default values to the tolerance attributes.

> **Federal Supply Code for Manufacturers (FSCM):** A five-digit numerical code identifier applicable to any organization that produces items used by the federal government.

PROFESSIONAL TIP

The size of each area within the title block limits the number of characters displayed in a line of text. You may want to include a reminder about the maximum number of characters in the attribute prompt. For example, the prompt could read Enter drawing name (15 characters max). Each time you insert a block or drawing containing the attribute, the prompt displays the reminder.

Figure 26-12.
A title block must comply with applicable standards. This title block complies with the ASME Y14.1 standard, *Decimal Inch Drawing Sheet Size and Format*.

Figure 26-13.
Define attributes for each area of the title block. Attributes should define all information that might possibly change, including general tolerances.

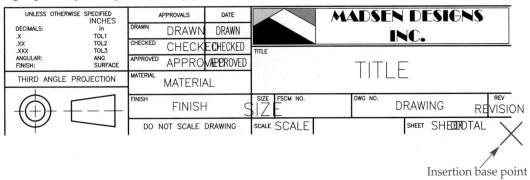

Insertion base point

After you define each attribute in the title block, you are ready to create the block. One option is to use the **BLOCK** tool to create a block of the title block within the current file. When specifying the insertion base point, pick a corner of the title block that is convenient to use each time you insert the block. The point indicated in **Figure 26-13** is an insertion base point for this particular title block. Use the **Delete** option in the **Block Definition** dialog box to remove the selected objects from the drawing. Another option is to use the **WBLOCK** tool to save the drawing as a file. Give the file a descriptive name, such as TITLE_B or FORMAT_B for a B-size title block. **Figure 26-14** shows the attribute block created in **Figure 26-13**, inserted and completely filled out using attributes.

NOTE

If you are creating a template, insert the block at the appropriate location and save the file as a drawing template. Edit the values in an existing title block using the **Enhanced Attribute Editor**.

Creating Revision Blocks

It is almost certain that a detail drawing will require revision at some time. Typical changes include design improvements and the correction of drafting errors. The first revision usually receives the revision letter *A*. If necessary, revision letters continue with *B* through *Y*, but the letters *I, O, Q, S, X,* and *Z* are not used because they might be confused with numbers.

Drawing layout formats include an area with columns specifically designated to record all drawing changes. This area, commonly called the *revision block*, is normally located at the upper-right corner of the drawing sheet. A column for *zones* is included only if applicable.

The **TABLE** tool is an excellent tool for preparing a revision block. An alternative is to use blocks and attributes to document revisions. The process is similar to creating a title block, but a revision block requires two separate blocks. The first block consists of only lines and text and forms the title and heading rows. See **Figure 26-15A**. The second block includes attributes and is inserted whenever a revision is required. See **Figure 26-15B**.

Format the revision block according to industry or company standards, and use the correct layer, typically layer 0. As you create attributes, determine the appropriate text height and justification for each definition. Define attributes for the zone (if necessary), revision letter, description, date, and approval. Assign the APPROVED attribute a prompt such as ENTER INITIALS OR SEEK SIGNATURE, providing the flexibility to type initials or leave the cell blank for written initials. Apply the same practice to the date attribute.

> **revision block:** A block that provides space for the revision letter, a description of the change, the date, and approvals.

> **zones:** A system of letters and numbers used on large drawings to help direct the print reader's attention to the correction location on the drawing.

Figure 26-14.
The title block after insertion of the attributes. Dates and approvals are added when the drawing is complete.

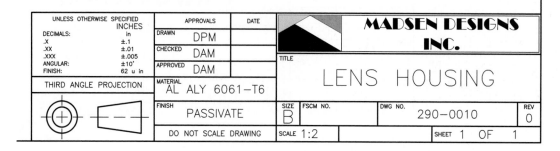

Figure 26-15.
You can create a revision block using two separate blocks. A—The first block forms the title and heading rows. B—The second block includes attributes and is added each time an engineering change is employed. The revision block shown complies with the ASME Y14.1 standard, *Decimal Inch Drawing Sheet Size and Format.*

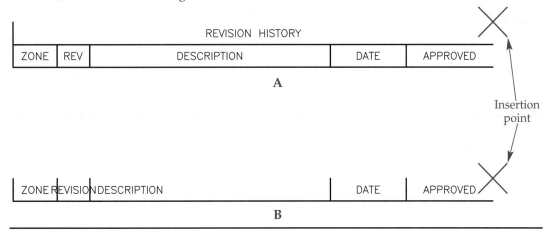

Use the **BLOCK** or **WBLOCK** tool to create the blocks. If you create wblocks, use descriptive file names such as REVBLK or REV. **Figure 26-16** shows an example of revision information added by inserting the two blocks created in **Figure 26-15** in the upper-left inside corner of the border.

Creating Parts Lists

Assembly drawings require a parts list, or bill of materials, that provides information about each component of the assembly or subassembly. This information includes the quantity, FSCM (when necessary), part number, description, and item number for each component. In some organizations, the parts list is a separate document, usually in an 8-1/2″ × 11″ format. At other companies, it is common practice to include the parts list on the face of the assembly drawing. A parts list on an assembly drawing usually appears directly above the title block, depending on industry and company standards.

The **TABLE** tool is an excellent tool for preparing a parts list. An alternative is to use blocks and attributes. The process is very similar to creating a revision block. The first block consists of only lines and text and forms the title (if used) and heading rows. See **Figure 26-17A**. The second block includes attributes and is inserted as many times as necessary to document each assembly component. See **Figure 26-17B**.

Format the parts list according to industry or company standards, and use the correct layer, typically layer 0. As you create attributes, select the appropriate text height and justification for each definition. Define attributes for the item number, quantity, FSCM (when necessary), part number, item description, and material specification.

Figure 26-16.
The completed revision block after inserting two blocks.

ZONE	REV	REVISION HISTORY		
		DESCRIPTION	DATE	APPROVED
C3	A	ADDED .125 CHAMFER	08−30−10	

Figure 26-17.
Creating a parts list using two separate blocks. A—The first block forms the title (if used) and heading rows. B—The second block includes attributes and is inserted as many times as necessary to define each assembly component. The revision block shown complies with the ASME Y14.1 standard, *Decimal Inch Drawing Sheet Size and Format*.

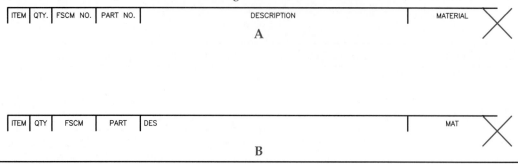

ITEM	QTY.	FSCM NO.	PART NO.	DESCRIPTION	MATERIAL

A

ITEM	QTY	FSCM	PART	DES	MAT

B

Use the **BLOCK** or **WBLOCK** tool to create the blocks. If you save wblocks, use descriptive file names, such as PL for parts list or BOM for bill of materials. **Figure 26-18** shows an example of the beginning of a parts list developed by inserting the blocks shown in **Figure 26-17**.

Figure 26-18.
The beginning of a parts list after inserting blocks and editing attribute values.

4	4		74–0080	SLEEVE	SAE 1020
3	12		85741	8–32UNC–2 X .50 HEX SOC CAP SCREW	SAE 4320
2	2		2569–01	RACK PAD	UHMW
1	1		52451	PLATE, MOUNTING	6061–T6 ALUM
ITEM	QTY.	FSCM NO.	PART NO.	DESCRIPTION	MATERIAL

UNLESS OTHERWISE SPECIFIED
INCHES
DECIMALS: in
.X   ±.1
.XX   ±.01
.XXX   ±.005
ANGULAR: ±10'
FINISH: 62 u in

THIRD ANGLE PROJECTION

APPROVALS   DATE
DRAWN   DPM
CHECKED DAM
APPROVED DAM
MATERIAL VARIES
FINISH   ALL OVER

MADSEN DESIGNS INC.

TITLE
VRF MULTIPLIER

SIZE C   FSCM NO.   DWG NO. 290010–A   REV 0

SCALE 1:2   DO NOT SCALE DRAWING   SHEET 25 OF 1

Using Fields to Reference Attributes

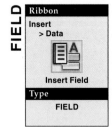
Use fields to display the value of an attribute in a location away from the block. To display an attribute value in a field, access the **Field** dialog box from within the **MTEXT** or **TEXT** tool, from the ribbon, or by typing FIELD. In the **Field** dialog box, pick **Objects** from the **Field category:** drop-down list, and pick **Object** in the **Field names:** list box. Then pick the **Select object** button to return to the drawing window and select the block containing the attribute.

When you select the block, the **Field** dialog box reappears with the available properties (attributes) listed. Pick the desired attribute tag to display the corresponding value in the **Preview:** box. Select the format and pick **OK** to insert the field in the text object.

Supplemental Material *Extracting Attribute Data*
For information about using attributes to create a table and exporting attribute data to an external file, go to the Student Web site (www.g-wlearning.com/CAD), select this chapter, and select **Extracting Attribute Data**.

Chapter Test

Answer the following questions. Write your answers on a separate sheet of paper or go to the Student Web site (www.g-wlearning.com/CAD) and complete the electronic chapter test.

1. What is an attribute?
2. Explain the purpose of the **ATTDEF** tool.
3. Define the function of the following attribute modes:
   A. **Invisible**
   B. **Constant**
   C. **Verify**
   D. **Preset**
4. What is the purpose of the **Default** text box in the **Attribute Definition** dialog box?
5. How can you edit attributes before they are included within a block?
6. How can you change an existing attribute from visible to invisible?
7. If you select attributes using the **Window** or **Crossing** selection method to define a block, in what order will attribute prompts appear?
8. What purpose does the **ATTREQ** system variable serve?
9. List the three options for attribute display.
10. Explain how to change the value of an inserted attribute.
11. What does *global attribute editing* mean?
12. After you save a block with attributes, what method can you use to change the order of prompts when you insert the block?
13. What three detailing or documentation tasks can be automated using attributes?
14. What section of an assembly drawing provides information about each component of the assembly or subassembly?
15. How can you display the value of an attribute in a location away from the associated block?

Drawing Problems

Start AutoCAD if it is not already started. Start a new drawing using an appropriate template of your choice. The template should include layers, text styles, dimension styles, and multileader styles appropriate for drawing the given objects. Add layers, text styles, dimension styles, and multileader styles as needed. Draw all objects using appropriate layers, text styles, dimension styles, multileader styles, justification, and format. Follow the specific instructions for each problem. Use your own judgment and approximate dimensions when necessary.

▼ Basic

1. Use a word processor to list each attribute mode. Provide a brief description of each.

2. Draw the structural steel wide flange shape shown below using the dimensions given. Do not dimension the drawing. Create attributes for the drawing using the information given. Make a block of the drawing and name it W12 X 40. Insert the block once to test the attributes. Save the drawing as P26-2.

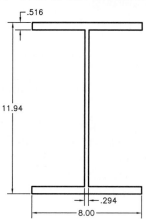

Attributes	Steel	W12 × 40	Visible
	Mfr.	Ryerson	Invisible
	Price	$.30/lb	Invisible
	Weight	40 lbs/ft	Invisible
	Length	10′	Invisible
	Code	03116WF	Invisible

▼ Intermediate

3. Open P26-2 and save it as P26-3. The P26-3 file should be active. Construct the floor plan shown using the dimensions given. Dimension the drawing. Insert the block W12 X 40 six times as shown. The chart below the drawing provides the required attribute data. Enter the appropriate information for the attributes as prompted. Note that the steel columns labeled 3 and 6 require slightly different attribute data. You can speed the drawing process by using **ARRAY** or **COPY**. Resave the drawing.

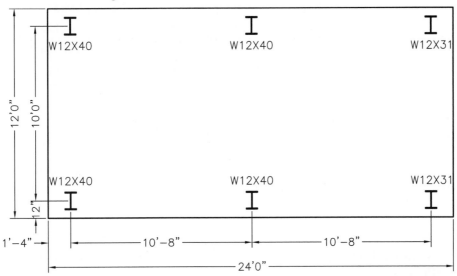

	Steel	Mfr.	Price	Weight	Length	Code
Blocks ①, ②, ④, & ⑤	W12 × 40	Ryerson	$.30/lb	40 lbs/ft	10'	03116WF
Blocks ③ & ⑥	W12 × 31	Ryerson	$.30/lb	31 lbs/ft	8.5'	03125WF

4. Open P26-2 and save it as P26-4. The P26-4 file should be active. Edit the W12 X 40 block in the newly saved drawing according to the following information. Resave the drawing.

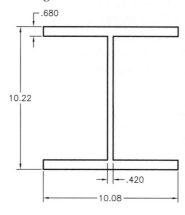

Attributes	Steel	W10 × 60	Visible
	Mfr.	Ryerson	Invisible
	Price	$.25/lb	Invisible
	Weight	60 lbs/ft	Invisible
	Length	10′	Invisible
	Code	02457WF	Invisible

▼ Advanced

5. Open P26-2 and save it as P26-4. The P26-4 file should be active. Create a tab-separated extraction file for the blocks in the drawing. Extract the following information for each block:
   - Block name
   - Steel
   - Manufacturer
   - Price
   - Weight
   - Length
   - Code

   Resave the drawing. Save the tab-separated extraction file as P26-4.

6. Open P26-2 and save it as P26-5. The P26-5 file should be active. Create a table from the block attribute data and insert it into the drawing. Resave the drawing.

7. Select a drawing from Chapter 25 and create a bill of materials for it using the **Data Extraction** wizard. Use the comma-separated format to display the file. Display the file in Windows Notepad. Save the drawing and the comma-separated extraction file as P26-6.

8. Create a drawing of the computer workstation layout in the classroom or office in which you are working. Provide attribute definitions for all of the items listed here.
   - Workstation ID number
   - Computer brand name
   - Model number
   - Processor chip
   - Amount of RAM
   - Hard disk capacity
   - Video graphics card brand and model
   - CD-ROM/DVD-ROM speed
   - Date purchased
   - Price
   - Vendor's phone number
   - Other data as you see fit

   Generate and extract a file for all of the computers in the drawing. Save the drawing and extracted file as P26-7.

Drawing Problems - Chapter 26

CHAPTER 27

Introduction to Dynamic Blocks

Learning Objectives

After completing this chapter, you will be able to do the following:

✓ Explain the function of dynamic blocks.
✓ Assign action parameters and actions to blocks.
✓ Modify parameters and actions.

A standard block typically represents a very specific item, such as a specific style of a 1" long bolt. In this example, if the same style of bolt is available in three other lengths, you must create three additional standard blocks. An alternative is to create a single *dynamic block* that adjusts according to each unique bolt length. Creating and using dynamic blocks can increase productivity and reduce the size of symbol libraries, making them more manageable.

Dynamic Block Fundamentals

A dynamic block is a parametric symbol that you can adjust to change the symbol size, shape, and even geometry, without drawing additional blocks, and without affecting other instances of the block reference. **Figure 27-1** shows an example of a dynamic block of a single-swing door symbol. In this example, the dynamic properties of the block allow you to create many different single-swing door symbols according to specific parameters, such as door size, wall thickness, swing location, swing angle representation, wall angle, and exterior or interior usage.

The process of constructing and using dynamic blocks is identical to the process for standard blocks, except for the addition of *action parameters* and (usually) *actions* that control block geometry. Action parameters are commonly known as *parameters* in the context of dynamic blocks. A dynamic block can contain multiple parameters, and a single parameter can include multiple actions. Geometric constraints and *constraint parameters* are available to use as an alternative or in addition to parameters and actions. Many different tools and options exist for constructing dynamic blocks, depending on the purpose of the block.

Figure 27-2A shows an example of a bolt symbol created as a dynamic block and selected for grip editing. The bolt shaft objects include a linear parameter with a stretch action, as indicated by the *parameter grips*. The length of the bolt increases when you stretch the right-hand linear parameter grip to the right. See **Figure 27-2B.**

dynamic block: An editable block that can be assigned parameters, actions, and/or geometric constraints and constraint parameters.

action parameter (parameter): A specification for block construction that controls block characteristics such as the positions, distances, and angles of dynamic block geometry.

action: A definition that controls how dynamic block parameters behave.

constraint parameters: Dimensional constraints available for block construction to control the size or location of block geometry numerically.

parameter grips: Special grips that allow you to change the parameters of a dynamic block.

Figure 27-1.
A—A dynamic block of a single swing door symbol. B—The dynamic block allows you to create many unique door symbols without creating new blocks or affecting other instances of the same block.

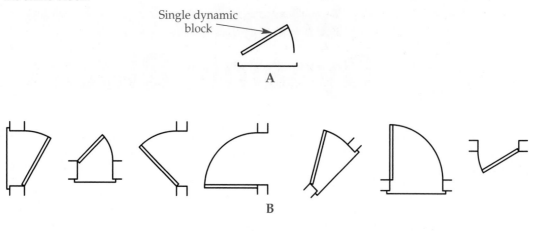

Single dynamic block

A

B

Figure 27-2.
A linear parameter with a stretch action assigned to the shaft objects in the block of a bolt. A—Selecting the block displays the linear grips. B—Selecting a linear grip and dragging it stretches the shaft of the bolt.

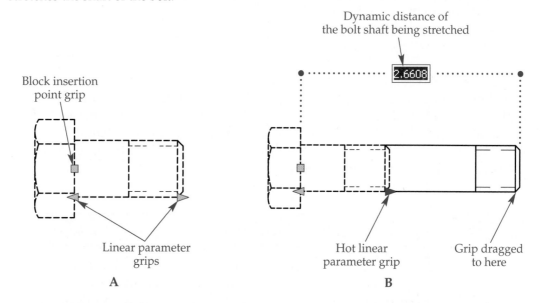

Dynamic distance of the bolt shaft being stretched

2.6608

Block insertion point grip

Linear parameter grips

A

Hot linear parameter grip

Grip dragged to here

B

NOTE

As you learn to create and use dynamic blocks, you will notice that many actions function as editing tools with which you are already familiar, allowing operations such as stretch, move, scale, array, and rotate.

PROFESSIONAL TIP

AutoCAD includes several files of dynamic block symbols. These samples are found in the Program Files/AutoCAD 2010/ Sample/ Dynamic Blocks folder. Dynamic blocks are also available from the **Tool Palettes** palette. Explore these sample symbols as you learn to create and use dynamic blocks.

Assigning Dynamic Properties

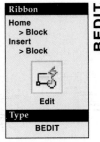

Ribbon

Home
> Block
Insert
> Block

Edit

Type

BEDIT

BEDIT

Edit a block in the **Block Editor** to assign dynamic properties. Access the **BEDIT** tool to display the **Edit Block Definition** dialog box shown in **Figure 27-3**. To create a dynamic block from scratch from within the **Block Editor,** type a name for the new block in the **Block to create or edit** field. To edit a block saved as the current drawing, such as a wblock, pick the <Current Drawing> option. To add dynamic properties to an existing block, select the block name from the list box. A preview and the description of the selected block appear. Pick the **OK** button to open the selection in the **Block Editor**. See **Figure 27-4**.

Figure 27-3.
The **Edit Block Definition** dialog box.

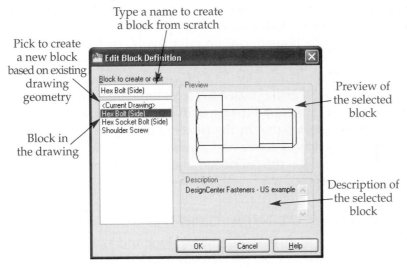

Figure 27-4.
In block editing mode, the **Block Editor** ribbon tab and the **Block Authoring Palettes** are available.

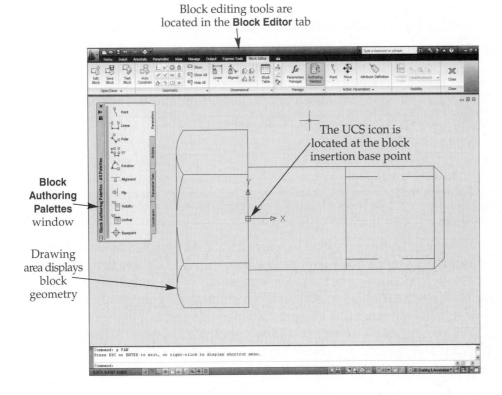

The **Block Editor** ribbon tab and **Block Authoring Palettes** window provide easy access to tools and options for assigning dynamic block properties and creating attributes. The **Block Authoring Palettes** contain parameter, action, and constraint tools. Though you can type BPARAMETER or BACTION to activate the **BPARAMETER** or **BACTION** tool and then select a parameter or action as an option, it is easier to use the **Block Editor** ribbon tab or the **Block Authoring Palettes**.

You can also assign actions to certain parameters, such as point parameters, by double-clicking on the parameter and selecting an action option. You can assign only specific actions to a given parameter. The process of assigning an action is slightly different, depending on the method used to access the action. If you type BACTION, you must first select the parameter and then specify the action type. If you pick the action from the **Action Parameters** panel in the **Block Editor** ribbon or the **Block Authoring Palettes**, the specific action is active and a prompt asks you to pick the parameter. Finally, if you double-click on the parameter, the parameter becomes selected, but you must choose the action type.

Saving a Block with Dynamic Properties

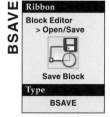

BSAVE

Ribbon
Block Editor > Open/Save
Save Block
Type
BSAVE

Once you add one or more parameters to a block and assign actions to the parameters, you are ready to save and use the dynamic block. Use the **BSAVE** tool to save the block, or use the **BSAVEAS** tool to save the block using a different name. Remember that saving changes to a block updates all blocks of the same name in the drawing. Use the **BCLOSE** tool to exit the **Block Editor** when you are finished.

BSAVEAS

Ribbon
Block Editor > Open/Save
Save Block As
Type
BSAVEAS

point parameter:
A parameter that defines an XY coordinate location in the drawing.

Ribbon
Block Editor > Action Parameters
Point
Type
BParameter > Point

Using Point Parameters

A *point parameter* creates a position property and can be assigned move and stretch actions. For example, assign a point parameter with a move action to a door tag that is part of a door block so you can move the tag independently of the door. Point parameters also provide multiple insertion point options. For example, add point parameters to the ends of a weld symbol reference line to create two insertion point options.

Figure 27-5 provides an example of adding a point parameter. Access the **Point** parameter option and specify a location for the parameter. The parameter location determines the base point from which dynamic actions occur. **Figure 27-5** shows

Exercise 27-2

Access the Student Web site (www.g-wlearning.com/CAD) and complete Exercise 27-2.

Using Linear Parameters

A *linear parameter* creates a distance property and can be assigned move, scale, stretch, and array actions. For example, assign a linear parameter with a stretch action to a bolt block so you can make the bolt shaft longer or shorter. Assign a second linear parameter and stretch action to the bolt head to control the bolt head diameter.

Figure 27-9 provides an example of adding a linear parameter. In this example, activate the **Linear** parameter option and use the **Label** function to name the linear parameter Shaft Length. Next, pick the start and endpoints of the linear parameter. The start and endpoints determine the locations from which dynamic actions occur. If you plan to assign a single action to the parameter, select the point associated with the action second. **Figure 27-9** shows picking the endpoint of the lower edge of the shaft, and then using polar tracking or the extension object snap to pick the point where the edge of the shaft would meet the end if extended. You must select points that are horizontal or vertical to each other to create a horizontal or vertical linear parameter.

Once you select the start and endpoints, pick a location for the parameter label. Next, enter the number of grips to associate with the parameter. The default **2** option creates grips at the start and endpoints, allowing you to use grip editing to carry out the action assigned to either point. Select the **1** option to assign a grip at the endpoint

linear parameter: A parameter that creates a measurement reference between two points.

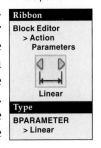

Ribbon

Block Editor > Action Parameters

Linear

Type

BPARAMETER > Linear

Figure 27-9.
Defining a linear parameter.

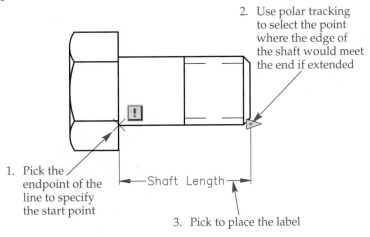

2. Use polar tracking to select the point where the edge of the shaft would meet the end if extended

1. Pick the endpoint of the line to specify the start point

Shaft Length

3. Pick to place the label

only, as shown in **Figure 27-9**. You will be able to grip-edit the block only if an action is associated with the endpoint. If you choose the **0** option, you can only use the **Properties** palette to adjust the block.

Name, **Label**, **Chain**, **Description**, **Base**, **Palette**, and **Value set** options are available before you specify points. Most of the options are also available from the **Properties** palette if you have already created the parameter. The **Base** option allows you to assign the start point or midpoint of the linear parameter as the action base point. The **Value set** option allows you to specify values for the action. Both options are described later in this chapter.

Assigning a Stretch Action

Ribbon

Block Editor > Action Parameters

Stretch

Type

BACTIONTOOL > Stretch

stretch action: An action used to change the size and shape of block objects with a stretch operation.

Figure 27-10 illustrates the process of adding a *stretch action* to the bolt symbol example. Access the **Stretch** action option and pick the Shaft Length parameter if it is not already selected. Then specify a parameter point to associate with the action. Move the crosshairs near the appropriate parameter point to display the red snap marker, and pick to select. An alternative is to choose the **sTart point** option to pick the start point of the linear parameter, or the **Second point** option to select the endpoint. If you plan to use grip editing to control the block, and added a single grip, pick the point with the grip.

Next, create a window to define the stretch frame. This is the same technique you apply when using the **STRETCH** tool. See **Figure 27-10A**. Pick the objects to stretch, including the associated parameter. You do not need to use a crossing window, because the previous operation defines the stretch. However, crossing selection is often quicker. See **Figure 27-10B**. Press [Enter] or the space bar or right-click to place the action icon. See **Figure 27-10C**. Test and save the block, and exit the **Block Editor**. The dynamic block is now ready to use.

Using a Stretch Action Dynamically

Figure 27-11 shows the bolt block reference selected for editing. The linear parameter grip displays as a light blue arrow at the far end of the bolt shaft. The insertion base point specified when the block was created appears as a standard unselected grip. Select the parameter grip and stretch the shaft to the new length. Use dynamic input to view the stretch dimension, and enter an exact length value in the distance field. You can also use the **Properties** palette to define the distance.

Figure 27-10.
Assigning a stretch action to a linear parameter. A—Specify the parameter and parameter grip, and create a window for the stretch frame. B—Use a crossing window to select the objects affected by the stretch action.

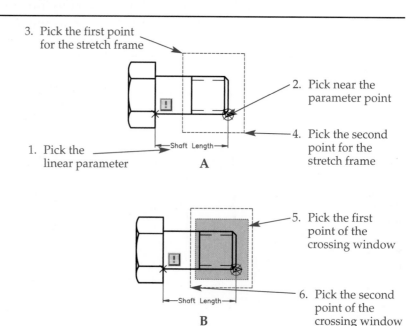

Figure 27-11.
Selecting the inserted block in the drawing displays the parameter grips.

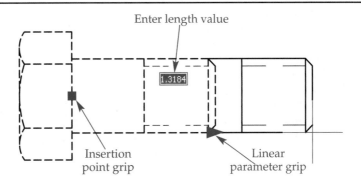

Enter length value

1.3184

Insertion point grip

Linear parameter grip

PROFESSIONAL TIP

The dynamic input distance field is a property of the linear parameter, allowing you to enter an exact distance. To get the best results when using a linear parameter, it is important that you locate the first and second points correctly.

Exercise 27-3

Access the Student Web site (www.g-wlearning.com/CAD) and complete Exercise 27-3.

Stretching Objects Symmetrically

The **Linear** parameter option includes a **Base** function that allows you to assign the start point or midpoint of the linear parameter as the action base point. Use the **Midpoint** setting to specify the midpoint as the action base point. This maintains symmetry when you adjust the block. You can set the **Base** preference before picking the first point or later using the **Properties** palette.

Figure 27-12 shows an example of a linear parameter with a stretch action assigned to the objects composing the bolt head. In this example, activate the **Linear** parameter option and use the **Base** option to choose the **Midpoint** setting. Next, use the **Label** option to change the label name to Head Diameter. Select the start and endpoints of the linear parameter to define the parameter and automatically calculate the midpoint. This example uses the upper-right and lower-left corners of the bolt head. After you select the start and endpoints, pick a location for the parameter label. Next, enter the number of grips to associate with the parameter. The **Figure 27-12** example uses the default **2** option to create grips at the start and endpoints.

Figure 27-12.
The base point of a linear parameter appears as an X. Use the **Midpoint** option to locate the base point halfway between the start and endpoints.

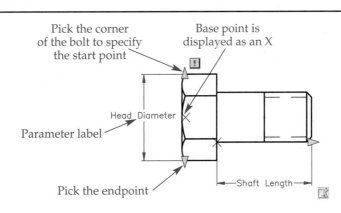

Pick the corner of the bolt to specify the start point

Base point is displayed as an X

Head Diameter

Parameter label

Pick the endpoint

Shaft Length

Figure 27-13.
Assigning a stretch action to one side of the bolt head. A—Create a crossing window around the top of the bolt head. B—Select the objects affected by the stretch action.

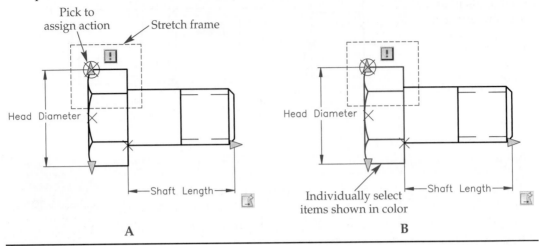

A

B

Figure 27-13 demonstrates the process of assigning a stretch action to one side of the bolt head. First, access the **Stretch** action option and pick the Head Diameter parameter if it is not already selected. Then pick the upper linear parameter point to associate with the action. Create a crossing window to define the stretch frame, as shown in **Figure 27-13A**. Then select the objects to stretch, including the associated parameter, as shown in **Figure 27-13B**. Press [Enter] or the space bar or right-click to place the action icon.

Repeat the previous sequence to assign a second stretch action to the opposite side of the bolt head. Test and save the block, and exit the **Block Editor**. The dynamic block is now ready to use. **Figure 27-14** illustrates using the lower grip point or dynamic input to stretch the bolt block reference. You can use either grip to stretch the bolt head. You can also use the **Properties** palette to define the distance.

Assigning a Scale Action

scale action: An action used to scale some of the objects within a block independently of the other objects.

Figure 27-15 shows a countertop and sink block. In this example, a *scale action* is assigned to a linear parameter to adjust the size of the sink while maintaining the dimensions of the countertop. Activate the **Linear** parameter option and use the **Base** option to choose the **Midpoint** setting. Next, use the **Label** option to change the label name to SINK LENGTH. Select the start and endpoints of the linear parameter to define the parameter and automatically calculate the midpoint. This example uses

Figure 27-14.
Dynamically stretching the bolt head. Notice that the head stretches symmetrically.

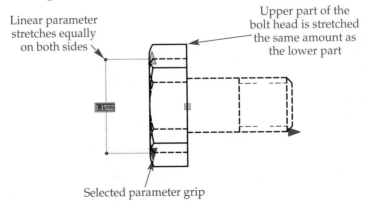

AutoCAD and Its Applications—Basics

Figure 27-15.
A linear parameter assigned to the sink objects in a block of a sink and countertop. You must locate the parameter base point and independent base type at the center of the sink to scale the sink symmetrically.

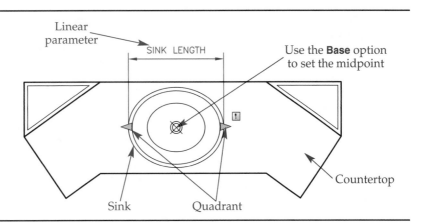

Linear parameter

SINK LENGTH

Use the **Base** option to set the midpoint

Sink

Quadrant

Countertop

two quadrants of the sink. After you select the start and endpoints, pick a location for the parameter label. Next, enter the number of grips to associate with the parameter. The **Figure 27-15** example uses the default **2** option to create grips at the start and endpoints.

Now assign a scale action to the parameter. First, access the **Scale** action option and pick the SINK LENGTH linear parameter if it is not already selected. Then select the objects to scale, including the associated parameter. Press [Enter] or the space bar or right-click to place the action.

When using a scale action, it is critical to scale objects relative to the correct base point. Access the **Properties** palette and display the properties of the scale action. The **Overrides** category includes options for adjusting the base point. The default **Base type** option is **Dependent**, which scales the objects relative to the base point of the associated parameter. Choose the **Independent** option to specify a different location. The **Base X** and **Base Y** values default to the parameter start point. Enter the coordinates relative to the block insertion base point, or use the pick button that appears when you select the **Base X** and **Base Y** values to choose points on-screen. For the sink example, it is important that the objects be scaled relative to the exact center of the sink to center the sink within the countertop as the scale changes. Test and save the block, and exit the **Block Editor**. The dynamic block is now ready to use.

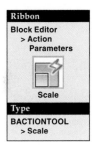

Ribbon

Block Editor
> Action
  Parameters

Scale

Type

BACTIONTOOL
> Scale

PROFESSIONAL TIP

In the previous example, the **Independent** option allows you to set the center of the sink as the base point for the scale action. This is necessary because the base point of the linear parameter is not the specified midpoint. The parameter uses a midpoint base to scale the parameter, and the parameter grips, from the parameter midpoint. The **Independent** option of the scale action controls the point from which the geometry, not the parameter, is scaled.

Using a Scale Action Dynamically

Figure 27-16 shows using the right grip or dynamic input to scale the sink in the countertop and sink block reference. You can use either grip to scale the sink. You can also use the **Properties** palette to define the distance.

Exercise 27-4

Access the Student Web site (www.g-wlearning.com/CAD) and complete Exercise 27-4.

Figure 27-16.
Scaling the sink dynamically. Notice that the countertop portion of the block remains the same.

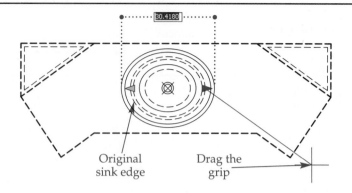

Using Polar Parameters

A *polar parameter* creates distance and angle properties and can be assigned move, scale, stretch, polar stretch, and array actions. For example, assign a polar parameter with a move action to the small circle in **Figure 27-17** to move the small circles a specified distance and angle without affecting the larger circle. To insert the polar parameter, access the **Polar** parameter option and specify the base point as the center of the large circle. Then pick the center of the small circle to specify the endpoint.

After you select the start and endpoints, pick a location for the parameter label. Next, enter the number of grips to associate with the parameter. Select the **1** option to assign a grip at the endpoint only, as shown in **Figure 27-17**. You will be able to grip-edit the block only if an action is associated with the endpoint.

NOTE

Name, **Label**, **Chain**, **Description**, **Palette**, and **Value set** options are available before you specify the parameter. Most of these options are also available from the **Properties** palette if you have already created the parameter.

Assigning a Move Action

Figure 27-18 illustrates the process of assigning a move action to the polar parameter. First, access the **Move** action option and pick the polar parameter if it is not already selected. Then select the parameter point in the center of the small circle to associate the point with the move action. Select the small circle and the associated parameter.

Figure 27-17.
Adding a polar parameter.

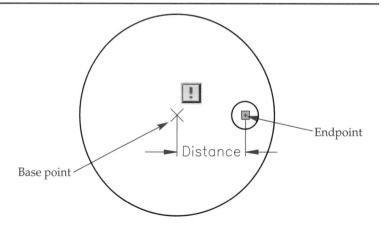

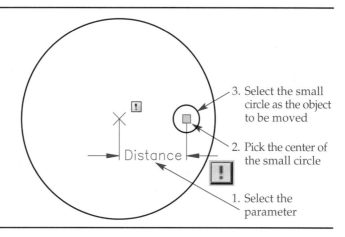

Figure 27-18.
Assigning a move action to a polar parameter.

3. Select the small circle as the object to be moved

2. Pick the center of the small circle

1. Select the parameter

Distance

Press [Enter] or the space bar or right-click to place the action. Test and save the block, and exit the **Block Editor**. The dynamic block is now ready to use.

Using a Move Action Dynamically

Figure 27-19 shows using the grip point or dynamic input to change the location and angle of the small circle from the base point in a block reference. For example, to move the small circle three inches away from the center of the large circle at 45°, type @3<45 and press [Enter], or enter 3 in the distance field and 45 in the angle field and press [Enter]. You can also use the **Properties** palette to define the distance.

> **NOTE**
>
> Move, stretch, and polar stretch actions include **Multiplier** and **Angle Offset** options available in the **Properties** palette. Enter a value in the **Multiplier** text box to multiply by the parameter value when adjusting the block. For example, if you assign a distance multiplier of 2 to a move action and move an object 4 units, the object actually moves 8 units. Enter an angle in the **Angle Offset** text box to change the parameter grip angle. For example, if you assign an offset angle of 45 to a move action and move an object 10°, the object actually moves 55°.

Figure 27-19.
Moving an object with a polar parameter displays the distance and the angle from the base point when dynamic input is on.

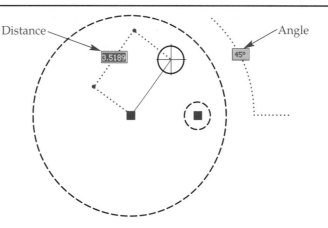

Distance

3.5189

Angle

45°

Exercise 27-5

Access the Student Web site (www.g-wlearning.com/CAD) and complete Exercise 27-5.

Using Rotation Parameters

Ribbon

Block Editor
> Action
Parameters

Rotation

Type

BPARAMETER
> Rotation

rotation parameter: A parameter that allows objects in a block to rotate independently of the block.

A *rotation parameter* creates an angle property to which you can assign a rotate action. For example, assign a rotation parameter with a rotate action to the needle in the speedometer block shown in **Figure 27-20** to rotate the needle around the circumference of the dial. To insert the rotation parameter, access the **Rotation** parameter option and pick the center of circular base of the needle as the rotation base point. Then pick a point, such as the needle endpoint shown, to specify the parameter radius. Set the default rotation angle from 0° east, or if rotation should originate from an angle other than 0°, use the **Base angle** option. **Figure 27-20** shows using the **Base angle** option to base the rotation at 0° and specify a default rotation angle of 200° to align the rotation with the 100 and 0 marks.

After you define the rotation parameter, pick a location for the parameter label. Next, enter the number of grips to associate with the parameter. The default **1** option creates a single grip at the parameter radius that allows you to use grip editing to carry out the rotate action.

NOTE

Name, **Label**, **Chain**, **Description**, **Palette**, and **Value set** options are available before you specify the parameter. Most of these options and the **Base angle** setting are also available from the **Properties** palette if you have already created the parameter.

Figure 27-20.
A rotation parameter with a rotate action allows you to rotate the needle in a speedometer block to indicate different speeds. The **Base angle** option allows you to set a base angle other than 0°.

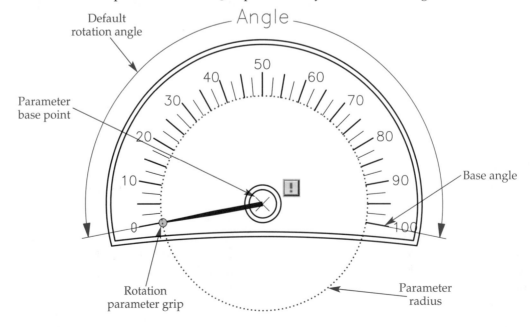

Assigning a Rotate Action

To assign a *rotate action* to the speedometer example, access the **Rotate Action** option and pick the rotation parameter if it is not already selected. Then select the objects that make up the needle and the rotation parameter. Press [Enter] or the space bar, or right-click to place the action. If necessary, access the **Properties** palette and adjust the **Base type** option. The default **Dependent** option sets the rotation point as the base point of the rotation parameter, which is appropriate for the speedometer example. Test and save the block, and exit the **Block Editor**. The dynamic block is now ready to use.

Ribbon
Block Editor
> Action
Parameters

Rotate

Type
BACTIONTOOL
> Rotate

rotate action: An action used to rotate individual objects within a block without affecting the other objects in the block.

Using a Rotate Action Dynamically

Figure 27-21 shows using the rotation parameter grip or dynamic input to rotate the needle inside a reference of the speedometer block. Selecting a specific speed using an endpoint object snap is most appropriate for this example. You can also use the **Properties** palette to define the distance.

Exercise 27-6

Access the Student Web site (www.g-wlearning.com/CAD) and complete Exercise 27-6.

Figure 27-21.
Dynamically rotating the needle in a speedometer block using a rotate action assigned to a rotation parameter.

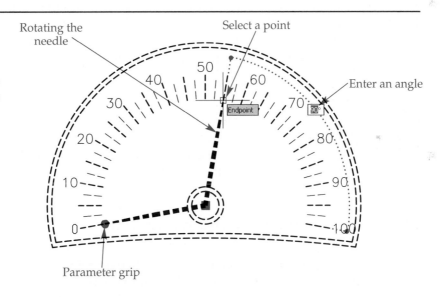

Using Alignment Parameters

An *alignment parameter* creates an alignment property. When you move a block with an alignment parameter near another object, the block rotates to align with the object based on the angle and alignment line defined in the block. This parameter saves time by eliminating the need to rotate a block or assign a rotation parameter. An alignment parameter affects the entire block, and therefore requires no action.

Figure 27-22 provides an example of adding an alignment parameter to the block of a gate valve symbol to align the gate valve with pipes. Access the **Alignment** parameter option and pick the point in the center of the valve to locate the parameter grip and define the first point of the alignment line. Next, specify the alignment direction, or use the **Type** option to specify the alignment type. Alignment type does not

alignment parameter: A parameter that aligns a block with another object in the drawing.

Ribbon
Block Editor
> Action
Parameters

Alignment

Type
BPARAMETER
> Alignment

Figure 27-22.
Adding an alignment parameter to a gate valve block.

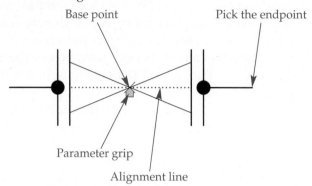

affect how the block aligns; it determines the direction of the alignment grip. Select the **Perpendicular** option to point the grip perpendicular to the alignment line, or choose the **Tangent** option to point the grip tangent to the alignment line. Set the **Tangent** option for the gate valve example.

After specifying the base point, and if necessary the alignment type, pick a second point to set the alignment direction. The angle between the first point and the second point defines the alignment line. The alignment line determines the default rotation angle. **Figure 27-22** shows selecting the endpoint of the value symbol. The alignment parameter grip is an arrow that points in the direction of alignment, perpendicular or tangent to the object to align. Test the block by drawing a line in the **Test Block Window** and attempting to align the block with the line. When you are finished, save the block and exit the **Block Editor**. The dynamic block is now ready to use.

> **NOTE**
>
> Use the **Name** option before you specify the parameter to rename the parameter. Alignment parameters do not include labels. You can also adjust the alignment **Type** from the **Properties** palette if you have already created the parameter.

Using an Alignment Parameter Dynamically

Figure 27-23 shows using the alignment parameter grip to align a reference of the gate valve with a pipeline. Select the block to display grips and then pick the parameter grip. Move the block near another object to align the block with the object. The rotation depends on the alignment path and type, and the angle of the other object.

Figure 27-23.
When you move the gate valve block near the angled line, the block aligns with the line.

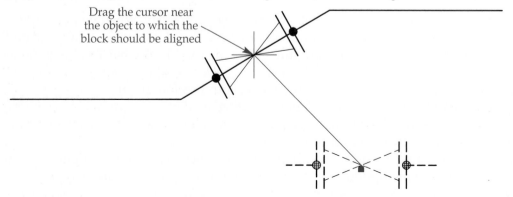

Exercise 27-7

Access the Student Web site (www.g-wlearning.com/CAD) and complete Exercise 27-7.

Using Flip Parameters

A *flip parameter* creates a flip property to which you can assign a flip action. For example, assign a flip parameter with a flip action to a door symbol to provide the option to place the door on either side of a wall. Another example is using a flip parameter to control the side of a reference line where a weld symbol displays for arrow side or other side applications.

Figure 27-24 provides an example of adding a flip parameter. Access the **Flip Parameter** option and pick the base point, followed by the endpoint of the reflection line. See **Figure 27-24A**. Pick a location for the parameter label, and then enter the number of grips to associate with the parameter. The default **1** option creates a single flip grip that allows you to use grip editing to carry out the flip action.

Flipping a block mirrors the block over the reflection line. However, for the door symbol, with the line in the current position, as shown in **Figure 27-24A**, an incorrect flip will result when you flip the block to the other side of a wall. To mirror the block properly, you must locate the reflection line to account for the wall thickness. To place the door on a 4″ wall, for example, use the **MOVE** tool to move the reflection line 2″ lower than the door. The label and parameter grip also move. In addition, you may want to move the parameter grip horizontally to the middle of the door opening. This may help in placing and flipping the block. See **Figure 27-24B**.

flip parameter:
A parameter that mirrors selected objects within a block.

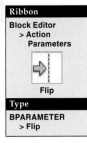

Ribbon
Block Editor > Action Parameters
Flip
Type
BPARAMETER > Flip

Figure 27-24.
A—Inserting a flip parameter. B—Moving the parameter so the block will correctly flip about the centerline of a wall.

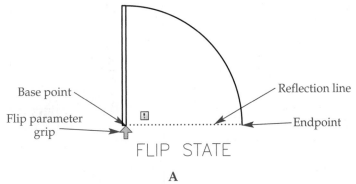
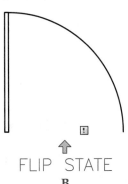

Name, **Label**, **Description**, and **Palette** options are available before you specify the parameter. Most of these options are also available from the **Properties** palette if you have already created the parameter.

A block reference with a flip parameter mirrors about the reflection line. You must place the reflection line in the correct location so the flip creates a symmetrical, or mirrored, copy. This typically requires the reflection line to be coincident with the block insertion point.

Assigning a Flip Action

Ribbon
Block Editor > Action Parameters
Flip
Type
BACTIONTOOL > Flip

flip action: An action used to flip the entire block.

To assign a *flip action* to the door example, access the **Flip** action option and pick the flip parameter if not already selected. Then select the objects that make up the door and the flip parameter. Press [Enter] or the space bar or right-click to place the action. Test and save the block, and exit the **Block Editor**. The dynamic block is now ready to use.

Using a Flip Action Dynamically

Figure 27-25A shows a reference of the door block, selected for editing. Pick the flip parameter grip to flip the block to the other side of the reflection line, as shown in **Figure 27-25B**. Unlike other parameters and actions that require stretching, moving, or rotating, a single pick initiates a flip action.

Add another flip parameter with a flip action to a door symbol to flip the door from side to side. In this way, one block takes the place of four blocks to accommodate different door positions.

Figure 27-25.
A—Select the block to display the flip parameter grip. B—Pick the flip parameter grip to flip the block about the reflection line. The entire block flips because all of the objects within the block are included in the selection set for the action.

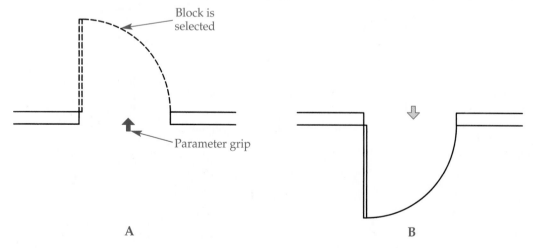

A   B

Exercise 27-8

Access the Student Web site (www.g-wlearning.com/CAD) and complete Exercise 27-8.

Using XY Parameters

An *XY parameter* creates horizontal and vertical distance properties and can be assigned move, scale, stretch, and array actions. The XY parameter can include up to four parameter grips—one at each corner of a rectangle defined by the parameter. You can use the XY parameter for a variety of applications, depending on the assigned actions.

Figure 27-26 provides an example of inserting an XY parameter. Access the **XY** parameter option and pick the base point. The base point is the origin of the X and Y distances. Next, pick a point to specify the XY point, which is the *corner* opposite the base point. Finally, enter the number of grips to associate with the parameter. The default **2** option creates grips at the start and endpoints, allowing you to use grip editing to carry out the action assigned to either point. Select the 4 option, as shown in **Figure 27-26**, to assign a grip at each XY corner to maximize flexibility, or choose a smaller number to limit dynamic options. If you choose the **0** option, you can only use the **Properties** palette to adjust the block.

> **XY parameter:**
> A parameter that specifies distance properties in the X and Y directions.

> **Ribbon**
> **Block Editor**
> **> Action**
> **Parameters**
>
>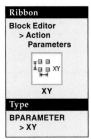
>
> **XY**
>
> **Type**
> **BPARAMETER**
> **> XY**

NOTE

Name, **Label**, **Chain**, **Description**, **Palette**, and **Value set** options are available before you specify the parameter. Most of these options and the **Base angle** setting are also available from the **Properties** palette if you have already created the parameter.

Assigning an Array Action

Figure 27-27 illustrates using an *array action* assigned to an XY parameter. This example shows dynamically arraying the block of an architectural glass block, allowing you to create an architectural feature of glass blocks without using a separate array operation. Access the **Array** action option and pick the XY parameter if it is not already selected. Then select the objects to include in the array, and press [Enter] or the space bar or right-click to accept the selection.

> **array action:**
> An action used to array objects within the block based on preset specifications.

> **Ribbon**
> **Block Editor**
> **> Action**
> **Parameters**
>
>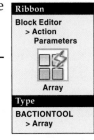
>
> **Array**
>
> **Type**
> **BACTIONTOOL**
> **> Array**

Figure 27-26.
Adding an XY parameter to a block of an architectural glass block. The XY parameter consists of X and Y distance properties and four grips.

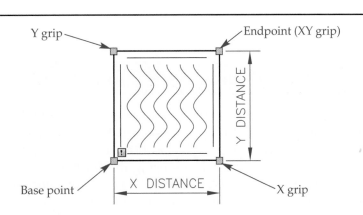

Figure 27-27.
Dynamically creating an array of architectural glass blocks using a block with an XY parameter and an array action. The pattern of rows and columns forms as you move the XY parameter. Notice the grout lines that form between the glass blocks because of proper action definition.

XY grip selected

Drag the XY grip

2.1057

Y distance

3.2128

X distance

At the Enter the distance between rows or specify unit cell: prompt, enter a value for the distance between rows or pick two points to set the row and column values. At the Enter the distance between columns: prompt, specify a value for the distance between columns. The second prompt does not appear if you select two points to define the row and column values. In the example of the glass block, be sure to allow for a grout joint when setting the row and column distance. Before assigning the action, you may want to draw a construction point offset from the block by the width of the grout joint. Then you can pick two points to define the row and column values. Be sure to erase the construction point before saving the block. Test and save the block, and exit the **Block Editor**. The dynamic block is now ready to use.

Using an Array Action Dynamically

Figure 27-27 illustrates using the upper-right grip or dynamic input to array a reference of the architectural glass block. You can use any available grip to apply the array, depending on where you want the array to occur. You can also use the **Properties** palette to define the array. Notice that proper action definition produces grout lines. The resulting array remains a single block.

Exercise 27-9

Access the Student Web site (www.g-wlearning.com/CAD) and complete Exercise 27-9.

Using Base Point Parameters

The **Block Editor** origin (0,0,0 point) determines the default location of the block insertion base point. Typically, you construct blocks in the **Block Editor** in reference to the origin, using the origin as the location of the insertion base point. The base point you choose when creating a block using the **BLOCK** tool attaches to the origin when you open the block in the **Block Editor**. You can add a *base point parameter* when it is necessary to override the base point of the default origin.

base point parameter: A parameter that defines an alternate base point for a block.

Access the **Basepoint** parameter option and pick a point to place the base point parameter. The parameter displays as a circle with crosshairs. After you save the block, the location of the base point parameter becomes the new base point for the block. You cannot assign actions to a base point parameter, but you can include a base point parameter in the selection set for actions.

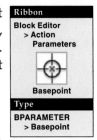

Ribbon

Block Editor
> Action
  Parameters

Basepoint

Type

BPARAMETER
> Basepoint

Using Parameter Value Sets

A *value set* helps to ensure that you select an appropriate value when editing a block, and can often increase the usefulness of a dynamic block. For example, if a window style is only available in widths of 36″, 42″, 48″, 54″, and 60″, add a value set to a linear parameter with a stretch action to limit selection to these sizes. See **Figure 27-28**. You can use a value set with linear, polar, XY, and rotation parameters.

> **value set:** A set of allowed values for a parameter.

To create a value set, select the **Value set** option available at the first prompt after you access a parameter option, and then pick a value set type. Choose the **List** option to create a list of possible sizes. Type all of the valid values for the parameter separated by commas. For the window block example, enter 36,42,48,54,60. Then press [Enter] or the space bar or right-click to return to the initial parameter prompt, and add the parameter as you normally would. After you insert the parameter, the valid values appear as tick marks.

Select the **Increment** option to specify an incremental value. Minimum and maximum values are also set to provide a limit for the increments. For the window block example, use the **Value set** option again to set 6″ width increments. This time choose the **Increment** option and type 6 for the distance increment, 36 for the minimum distance, and 60 for the maximum distance. The initial parameter prompt returns after you enter the maximum distance.

After you add a parameter with a value set, you must assign an action to the parameter. For the window block in **Figure 27-28**, assign a stretch action to the linear parameter. This allows the window to stretch to the valid widths specified in the value set. Test and save the block, and exit the **Block Editor**. The dynamic block is now ready to use.

NOTE

You can also use the options in the **Value Set** category of the **Properties** palette to specify value sets during block definition.

Figure 27-28.
When you adjust a block that includes a value set, tick marks appear at locations corresponding to the values in the value set. You can only stretch the block to one of the tick marks.

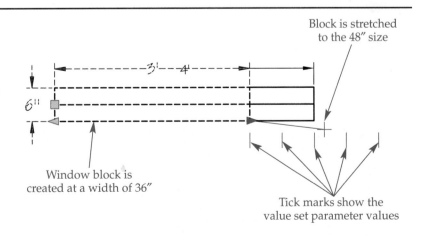

Block is stretched to the 48″ size

Window block is created at a width of 36″

Tick marks show the value set parameter values

Using a Value Set with a Parameter

Figure 27-28 shows using a linear parameter grip with a stretch action to specify the width of a window block reference. Tick marks appear, indicating the positions of valid values. As you stretch the grip, the modified block snaps to the nearest tick mark. When using dynamic input, you can also enter a value in the input field. If you type a value that is not included in the value set, the nearest valid value applies. You can also use the **Value Set** category of the **Properties** palette to select a value.

Exercise 27-10

Access the Student Web site (www.g-wlearning.com/CAD) and complete Exercise 27-10.

Using a Chain Action

chain action: An action that triggers another action when a parameter is modified.

A *chain action* limits the number of edits that you have to perform by allowing one action to trigger other actions. For example, **Figure 27-29** shows using a chain action to stretch the block of a table and chairs, and array the chairs along the table at the same time. Point, linear, polar, XY, and rotation parameters can be part of a chain action.

Creating a Chain Action

To create a chain action, select the **Chain** option available at the first prompt after you access a parameter option to display the Evaluate associated actions when parameter is edited by another action? [Yes/No]: prompt. The default **No** setting does not create a chain action. Select the **Yes** option to create a chain action.

Figure 27-29.
A—A block of a table with six chairs. B—Using a chain action with a linear parameter, you can array the chairs automatically when the table is stretched.

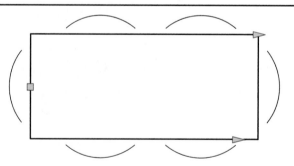

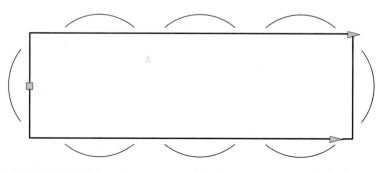

Figure 27-30 shows the default arrangement of the table and chairs block example. For this example, access the **Linear** parameter option and use the **Label** option to change the label name to CHAIR ARRAY. Next, choose the **Chain** option and select **Yes**. To complete the parameter, select the start and endpoints shown in **Figure 27-30A**, and assign a single grip to the endpoint. Then assign an array action to the parameter, selecting the chairs on the top and bottom of the table as the objects to array. At the

Figure 27-30.
A—Inserting a linear parameter to use with an array action for the chairs.
B—Inserting a linear parameter to stretch the table.
C—Assigning a stretch action to the linear parameter. When you specify the crossing window, be sure the CHAIR ARRAY parameter grip is included in the frame.

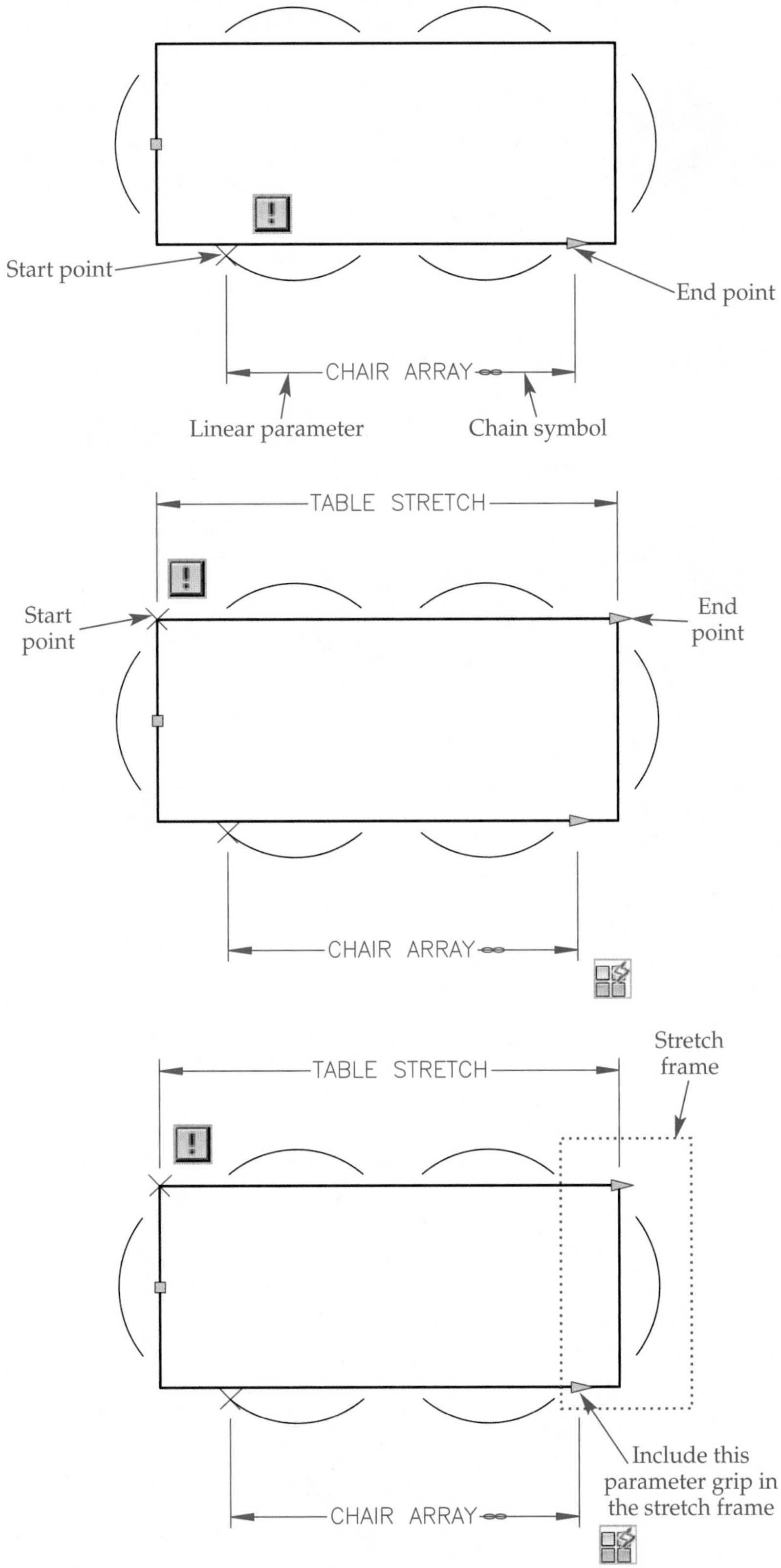

Enter the distance between rows or specify unit cell: prompt, use object snaps to snap to the endpoint of one of the chairs and then snap to the equivalent endpoint on the chair next to the first chair.

Add another single-grip linear parameter, labeled TABLE STRETCH, as shown in **Figure 27-30B.** Assign a stretch action to the parameter associated with the TABLE STRETCH parameter grip. Use a crossing window around the right end of the table and the CHAIR ARRAY parameter grip. See **Figure 27-30C.** Select the table, the chair at the right end of the table, and the CHAIR ARRAY parameter as the objects to stretch. Test and save the block, and exit the **Block Editor.** The dynamic block is now ready to use.

> **PROFESSIONAL TIP**
>
> The keys to successfully creating a chain action are to set the **Chain** option to **Yes** for the parameter that is affected automatically and to include the parameter in the object selection set when creating the action that drives the chain action.

Applying a Chain Action

Figure 27-31 shows using a linear parameter grip or dynamic input to stretch the table and chairs block reference. Stretching the table triggers the array action. You can also use the **Properties** palette to define the table length.

Exercise 27-11

Access the Student Web site (www.g-wlearning.com/CAD) and complete Exercise 27-11.

Figure 27-31.
As you stretch the parameter grip, the table stretches and the chairs are arrayed at the same time.

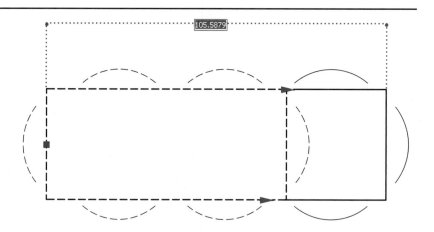

105.5879

Chapter Test

Answer the following questions. Write your answers on a separate sheet of paper or go to the Student Web site (www.g-wlearning.com/CAD) and complete the electronic chapter test.

1. Define *dynamic block*.
2. What is the function of a dynamic block?
3. How are standard blocks and dynamic blocks the same? How are they different?
4. Identify the property that forms when you create a point parameter and list the actions you can assign.
5. What is the purpose of a move action?
6. When do action bars appear?
7. What information can you get by hovering over or selecting an action bar?
8. What is the basic function of a linear parameter?
9. Explain the function of a stretch action.
10. Briefly describe how to use a stretch action symmetrically.
11. What is the basic function of a scale action?
12. Describe the polar parameter type and list the actions that can be assigned to it.
13. Identify the property that forms when you create a rotation parameter and list the action you can assign.
14. Describe what happens when you move a block with an alignment parameter near another object in the drawing. How does this save drawing time?
15. Give at least one practical example of using a flip parameter.
16. Briefly describe the properties an XY parameter creates and list the actions that can be assigned.
17. Give an example of using an array action assigned to an XY parameter.
18. When would you add a base point parameter?
19. Describe the basic use of a value set.
20. Explain the basic function of a chain action.

Drawing Problems

Start AutoCAD if it is not already started. Start a new drawing using an appropriate template of your choice. The template should include layers, text styles, dimension styles, and multileader styles appropriate for drawing the given objects. Add layers, text styles, dimension styles, and multileader styles as needed. Draw all objects using appropriate layers, text styles, dimension styles, multileader styles, justification, and format. Follow the specific instructions for each problem. Use your own judgment and approximate dimensions when necessary.

▼ Basic

1. Open P25-1 and save it as P27-1. The P27-1 file should be active. Erase all copies of the steel column symbols except for the one in the lower-left corner. Insert an XY parameter into the steel column block and associate an array action with the parameter. Use the proper values for the array action to dynamically array the block to match the drawing. Use the one dynamic block to create the rest of the steel columns in the drawing. Resave the drawing.

2. Create a block named WIRE ROLL as shown below. Do not include the dimensions. Insert a linear parameter on the entire length of the roll. Use a value set with the following values: 36″, 42″, 48″, and 54″. Assign a stretch action to the parameter and associate the action with either parameter grip. Create a crossing window that will allow the length of the roll to stretch. Select all of the objects on one end and the length lines as the objects to stretch. Insert the WIRE ROLL block four times into a drawing and stretch each block to use a different value set length. Save the drawing as P27-2.

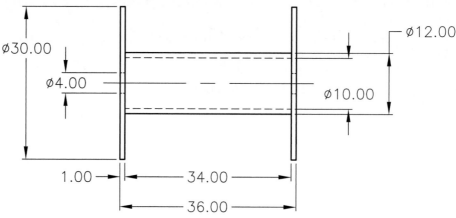

3. Create a block named 90D ELBOW as shown below on the left. Do not include the dimensions. Insert two flip parameters and two flip actions. One of the flip parameter/action combinations is to flip the elbow horizontally. The second flip parameter/action combination is to flip the elbow vertically. Use the dynamic block to create the drawing shown below on the right. Save the drawing as P27-3.

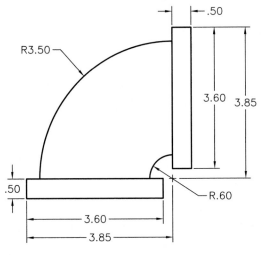

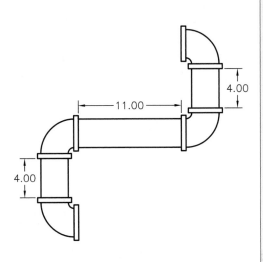

4. Create a block of the 48″ window shown below on the left. Do not include the dimensions. Insert an alignment parameter so the length of the window can align with a wall. Then draw the walls shown below on the right. Insert the window block as needed. Use the alignment parameter to align the window to the walls. Center the windows on wall segments unless dimensioned. Save the drawing as P27-4.

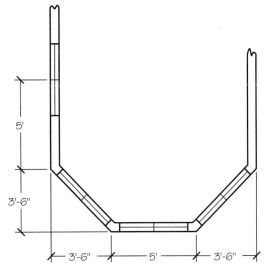

5. Create a block named **CONTROL VALVE** as shown below on the left. Include the label in the block. Insert a point parameter and assign a move action to it. Select the two lines of text as the objects to which the action applies. Insert the **CONTROL VALVE** block into the drawing three times. Use the point parameter to move the text to match the three positions shown below. Save the drawing as **P27-5**.

▼ Intermediate

6. Create a block named **FLANGE** as shown below. Do not include dimensions. Insert a rotation parameter specifying the center of the flange as the base point. Assign a rotate action to the parameter, selecting the six ∅.2 circles as the objects to which the action applies. Insert the **FLANGE** block into the drawing twice. Use the rotation parameter to create the two configurations shown below. Save the drawing as **P27-6**.

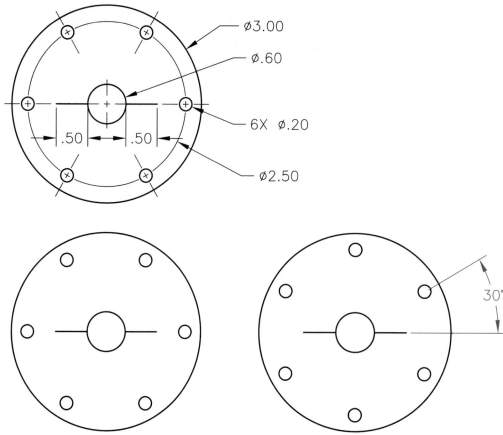

7. Open P27-6 and save it as P27-7. The P27-7 file should be active. Open the FLANGE block in the **Block Editor** and use the **Properties** palette to give the following settings to the rotation parameter:

A. **Angle label**—BOLT HOLES
B. **Angle description**—ROTATION OF BOLT HOLE PATTERN
C. **Ang type**—INCREMENT
D. **Ang increment**—30
E. Save the changes and exit the **Block Editor**. Save the drawing.

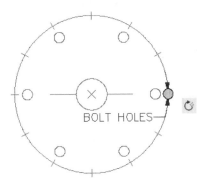

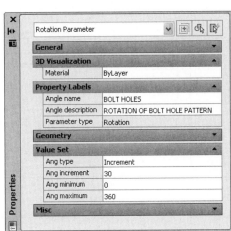

▼ Advanced

8. The drawing below shows a fan with an enlarged view of the motor. This fan can have one of three motors of different sizes. Create the fan as a dynamic block.
   A. Draw all the objects. Do not dimension the drawing or draw the enlarged view.
   B. Create a block named FAN consisting of the objects shown in the enlarged view.
   C. Open the block in the **Block Editor** and insert a linear parameter along the top of the motor (the 1.50" dimension). Use a value set with the following values: 1.5, 1.75, and 2.
   D. Assign a scale action to the linear parameter. Select all of the objects that make up the motor as the objects to which the action applies. Use an independent base point type and specify the base point as the lower-left corner of the motor (the implied intersection).
   E. Save the block and exit the **Block Editor**.
   F. Insert the block three times into the drawing. Use the linear parameter grip to scale the motor to the three different sizes, as shown below on the right.
   G. Save the drawing as P27-8.

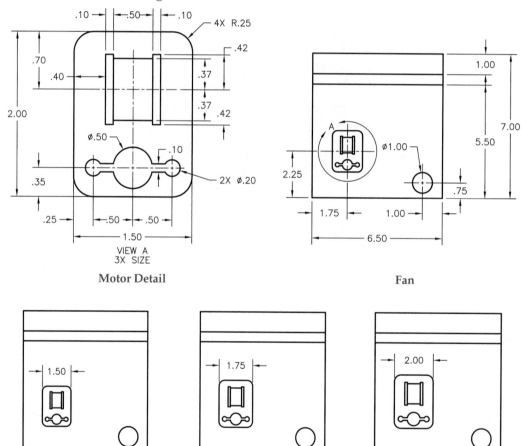

Motor Detail   Fan

ut Properties				
Middle Line	Start Line	End Line	Lookup	
0	.5000	.5000	0 Degrees	
10	.3700	.6300	10 Degrees	
20	.2400	.7600	20 Degrees	
<Unmatched>			Custom	
			Allow reverse lookup	

Additional Dynamic Block Tools

28

Learning Objectives

After completing this chapter, you will be able to do the following:

✓ Apply visibility and lookup parameters.
✓ Use parameter sets.
✓ Constrain block geometry.
✓ Use a block properties table.

This chapter describes adding visibility and lookup parameters to enhance the usefulness of blocks. You will also learn to apply geometric constraints and constraint parameters to blocks to use as an alternative or in addition to parameters and actions. Finally, this chapter explores the process of using a block properties table.

Using Visibility Parameters

A *visibility parameter* allows you to assign *visibility states* to objects within a block. Selecting a visibility state displays the only objects in the block that are associated with the visibility state. Visibility states expand on the ability to make a block into a symbol library by allowing you to hide or make visible specific objects and even completely different symbols. A block can include only one visibility parameter. Visibility parameters do not require an action.

Figure 28-1 provides a basic example of using a visibility parameter to create four different symbols from a single block. To create the block, you must draw all objects representing the different variations, as shown in **Figure 28-1A**. Then assign a visibility parameter and add visibility states that identify the objects that are visible with each variation. Insert the block and select a visibility state to display the associated objects. See **Figure 28-1B**.

To add a visibility parameter, access the **Visibility Parameter** option and pick a location for the parameter label. The parameter automatically includes a single grip. When you insert the block and select the grip, a shortcut menu appears listing visibility states. There is no prompt to select objects because the visibility parameter is associated with the entire block.

visibility parameter: A parameter that allows you to assign multiple different views to objects within a block.

visibility states: Views created by selecting block objects to display or hide.

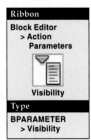

Ribbon

Block Editor
> Action
  Parameters

Visibility

Type

BPARAMETER
> Visibility

Figure 28-1.
A—All of the objects composing each unique valve are shown together. B—All four of these different valves are created from one block using a visibility parameter with different visibility states.

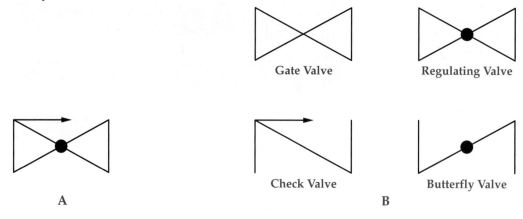

Gate Valve   Regulating Valve

Check Valve   Butterfly Valve

A   B

> **NOTE**
>
>
>
> **Name, Label, Description,** and **Palette** options are available before you specify the parameter. Most of the options are also available from the **Properties** palette if you have already created the parameter.

Creating Visibility States

BVSTATE

Ribbon

Block Editor
> Visibility

Visibility States

Type

BVSTATE

The tools in the **Visibility** panel of the **Block Editor** ribbon tab enable when you add a visibility parameter. See **Figure 28-2.** To create a visibility state, access the **BVSTATE** tool to display the **Visibility States** dialog box. See **Figure 28-3A.** Pick the **New...** button to open the **New Visibility State** dialog box shown in **Figure 28-3B.** Type the name of the new visibility state in the **Visibility state name:** text box. For the valve block example shown in **Figure 28-1,** an appropriate name could be GATE VALVE, REGULATING VALVE, CHECK VALVE, or BUTTERFLY VALVE, depending on which valve the visibility state represents.

Pick the **Hide all existing objects in new state** radio button to make all of the objects in the block invisible when you create the new visibility state. This allows you to display only the objects that should be visible for the visibility state. Pick the **Show all existing objects in new state** radio button to make all of the objects in the block visible when you create the new visibility state. This allows you to hide objects that should be invisible for the state. Select the **Leave visibility of existing objects unchanged in new state** radio button to display the objects that are currently visible when you create the new visibility state.

Figure 28-2.
The visibility tools found in the **Visibility** panel of the **Block Editor** ribbon tab.

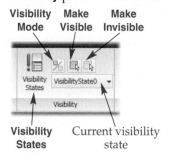

Visibility Make   Make
Mode   Visible Invisible

Visibility   Current visibility
States   state

Figure 28-3.
A—Manage
visibility states
using the **Visibility
States** dialog box.
B—Create new
visibility states
using the **New
Visibility State**
dialog box.

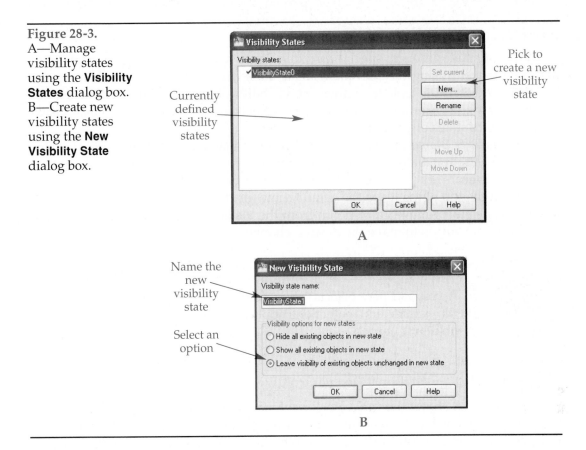

A

B

Pick the **OK** button to create the new visibility state. The new state adds to the list in the **Visibility States** dialog box and is current, as indicated by the check mark next to the name. Pick the **OK** button to return to block editing mode.

Next, use the **BVSHOW** and **BVHIDE** tools to display only the objects that should be visible in the current state. Pick the **Make Visible** button to select objects to make visible. Invisible objects temporarily display as semitransparent for selection. Pick the **Make Invisible** button to select objects to make invisible. For example, to make a visibility state to depict the gate valve shown in **Figure 28-4B** from the valve block shown in **Figure 28-4A**, use the **Make Invisible** tool to turn off the filled circle and the arrow. The changes are automatically saved to the visibility state. Use the **BVMODE** tool to toggle the visibility mode on and off. Turn on visibility mode to display invisible objects as semitransparent. Turn off visibility mode to display only visible objects.

Repeat the process to create additional visibility states for the block. The valve block example requires four visibility states. The **Current visibility state** drop-down list displays the current visibility state. Select a state from the drop-down list to make the state current. After you create all the visibility states, test and save the block, and exit the **Block Editor**. The dynamic block is now ready to use.

Ribbon
Block Editor
> Visibility

Make Visible

Type
BVSHOW

Ribbon
Block Editor
> Visibility

Make Invisible

Type
BVHIDE

Figure 28-4.
A—The VALVE block
with all objects
visible. B—The
VALVE block after
making the arrow
and filled circle
invisible to create
the GATE VALVE
visibility state.

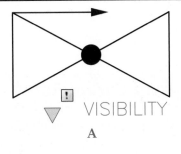

A

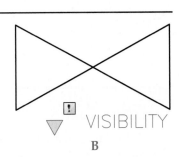

B

Modifying Visibility States

Visibility states require special consideration when modified. Set the state to be modified current using the **Current visibility state** drop-down list, and then use the **BVSHOW** and **BVHIDE** tools to change the visibility of objects as needed. When you add objects to the current visibility state, the objects are automatically set as invisible in all visibility states other than the current state.

The **Visibility States** dialog box allows you to rename and delete visibility states. You can also use the dialog box to arrange the order of visibility states in the shortcut menu that appears when you insert the block and pick the visibility parameter grip. The state at the top of the list is the default view for the block. Pick the visibility state to rename, delete, or move up or down from the **Visibility states:** list box. Then select the appropriate button to make the desired change.

PROFESSIONAL TIP

If you add new objects when modifying a state, be sure to update the parameters and actions applied to the block to include the new objects, if needed.

Using Visibility States Dynamically

Figure 28-5A shows the valve block reference selected for editing. Select the visibility grip to display a shortcut menu containing each visibility state. A check mark indicates the current visibility state. To switch to a different view of the block, select the name of the visibility state from the list. See **Figure 28-5B.** You can also use the **Properties** palette to select a visibility state.

Exercise 28-1

Access the Student Web site (www.g-wlearning.com/CAD) and complete Exercise 28-1.

Figure 28-5.
A—Pick the visibility parameter grip to display a shortcut menu with the available visibility states. The current state is checked.
B—Select a different visibility state from the shortcut menu to change the block appearance.

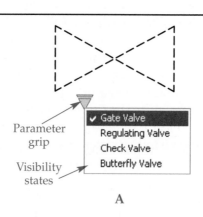

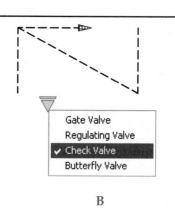

A   B

lookup parameter: A parameter that allows tabular properties to be used with existing parameter values.

lookup action: An action used to select a preset group of parameter values to carry out actions with stored values.

Using Lookup Parameters

A *lookup parameter* creates a lookup property to which you can assign a *lookup action.* For example, **Figure 28-6** shows three valve symbols created from a single block by adjusting the rotation parameter of the middle line. The lookup action allows the middle line rotation to control the length of the start and end lines.

                                          AutoCAD and Its Applications—Basics

Figure 28-6.
A lookup parameter allows you to create these three views of the same block. Notice how the geometry changes.

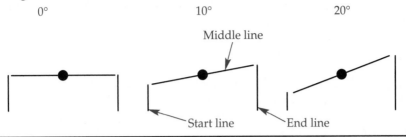

To create the valve block shown in **Figure 28-7,** first draw the geometry of the 0° symbol. Then add a linear parameter, labeled Start Line. Select the start point as the bottom of the start line, and the endpoint as the top of the start line. Assign a stretch action to the parameter that is associated with the top parameter grip. Draw the crossing window around the top of the start line and select the start line as the object to stretch.

Add another linear parameter, labeled End Line. Select the start point as the bottom of the end line and the endpoint as the top of the end line. Assign a stretch action to the parameter that is associated with the top parameter grip. Draw the crossing window around the top of the end line and select the end line as the object to stretch.

Next, add a rotation parameter, labeled Middle Line. Specify the base point as the center of the circle. Select the right endpoint of the middle line to set the radius, and specify the default rotation angle as 0. Assign a rotation action to the parameter. Pick the center of the circle as the rotation base point, and select the middle line as the object to rotate.

To add a lookup parameter, access the **Lookup Parameter** option and pick a location for the parameter label. Then enter the number of grips to associate with the parameter. The default **1** option creates a single lookup grip that allows you to use grip editing to carry out the lookup action. When you insert the block and select the grip, a shortcut menu appears listing rotation options. There is no prompt to select objects because a lookup parameter is associated with the entire block.

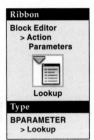

> **NOTE**
>
> **Name**, **Label**, **Description**, and **Palette** options are available before you specify the parameter. Most of the options are also available from the **Properties** palette if you have already created the parameter.

Figure 28-7.
The example block with linear parameters and stretch actions assigned to the start and end lines and a rotation parameter and rotate action assigned to the middle line.

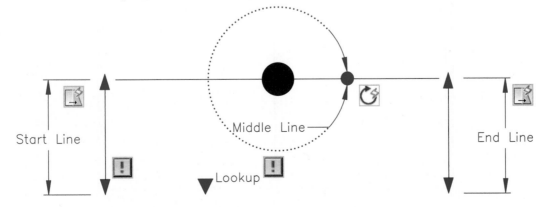

Ribbon

Block Editor
> Action
Parameters

Lookup

Type

BACTIONTOOL
> Lookup

lookup table: A table that groups the properties of parameters into custom-named lookup records.

Assigning a Lookup Action

To assign a lookup action, access the **Lookup Action** option and pick a lookup parameter if one is not already selected. The **Property Lookup Table** dialog box appears, allowing you to create a lookup table. See **Figure 28-8.**

Creating a Lookup Table

A *lookup table* groups the properties of parameters into custom-named lookup records. The **Action name:** display box indicates the name of the lookup action associated with the table. The table is initially blank. To add a parameter property, pick the **Add Properties...** button to open the **Add Parameter Properties** dialog box. See **Figure 28-9.**

All parameters in the block containing property values appear in the **Parameter properties:** list. Lookup, alignment, and base point parameters do not contain property values. Notice that the property name is the parameter label. The **Property type** area determines the type of property parameters shown in the list. By default, the **Add input properties** radio button is active, which displays available input property parameters. To display available lookup property parameters, select the **Add lookup properties** radio button.

To add parameter properties to the lookup table, select the properties in the **Parameter properties:** list and pick the **OK** button. A new column, named as the parameter property, forms for each parameter in the **Input Properties** area of the **Property Lookup Table** dialog box. See **Figure 28-10.** The **Input Properties** area allows you to specify a value for parameters added to the table. Type a value in each cell in the column. Add a custom name for each row, or record, in the **Lookup** column in the **Lookup Properties** area. This area displays the name that appears in the shortcut menu when you insert the block and select the lookup parameter grip.

For the valve block example, add the Middle Line, Start Line, and End Line parameter properties to the table. Then complete the lookup table as shown in **Figure 28-10.** Start with the Middle Line values. Press [Enter] after typing the value to add a new blank row and then type the remaining values in each cell. Use the [Enter], [Tab], arrow keys, or pick in a different cell to navigate through the table.

Figure 28-8.
The **Property Lookup Table** dialog box.

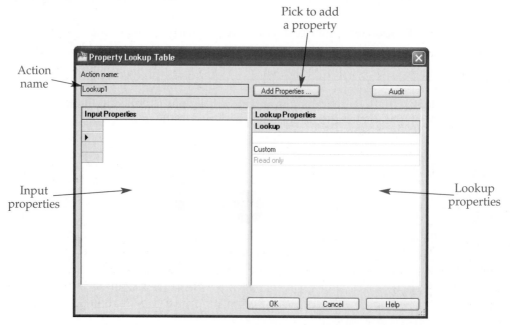

Figure 28-9.
Parameter properties are listed in the **Add Parameter Properties** dialog box.

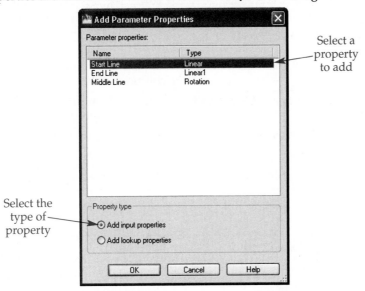

Select a property to add

Select the type of property

Figure 28-10.
A lookup table with multiple parameters and values added.

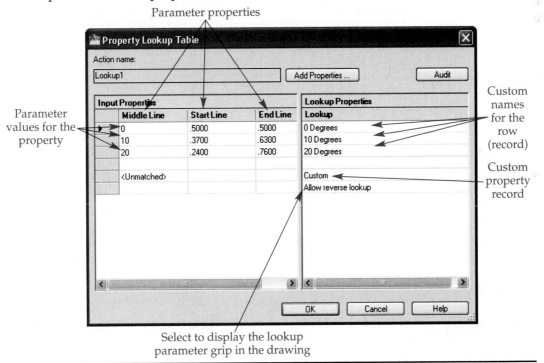

Parameter properties

Parameter values for the property

Select to display the lookup parameter grip in the drawing

Custom names for the row (record)

Custom property record

The row, or record, that contains the <Unmatched> value, named Custom in the **Lookup** column, applies when the current parameter values of the block do not match any of the records in the table. This allows you to adjust the block using parameter values other than those specified in the lookup table. You cannot add any values to the row, but you can change the name of **Custom**.

The Allow reverse lookup setting at the bottom of the **Lookup** column is available only if all of the names in the lookup table are unique. The option allows the lookup parameter grip to display when you select the block. Pick the grip to choose a specific lookup record. The Read only setting appears if you do not name a lookup property, or if two or more properties have the same name. Selecting **Read only** from the drop-down list disallows selecting a lookup record.

Figure 28-11.
A—Options available when you right-click on a column. B—Options available when you right-click on a row.

Menu Option	Function
Sort	Sorts the records (rows) in ascending or descending order. Pick again to reverse the sort order.
Maximize all headings	Adjusts all columns to the width of the column headings.
Maximize all data cells	Adjusts all columns to the width of the values in the cells.
Size columns equally	Makes all columns equal in width.
Delete property column	Deletes the column.
Clear contents	Deletes the cell values.

A

Menu Option	Function
Insert row	Inserts a new row above the selected row.
Delete row	Deletes the record (row).
Clear contents	Deletes the cell values.
Move up	Moves the row up by one row.
Move down	Moves the row down by one row.
Range syntax examples	Displays the online documentation examples of how to enter values into a lookup table.

B

You can right-click on a column heading to access a menu with options for adjusting columns, or right-click on a row to access a menu with options for adjusting rows. **Figure 28-11** briefly describes each option.

After you add all required properties to the table and assign values to each, pick the **Audit** button in the **Property Lookup Table** dialog box to check each record in the table to make sure they are all unique. If no errors are found, pick the **OK** button to return to the **Block Editor**. Test and save the block, and exit the **Block Editor**. The dynamic block is now ready to use.

NOTE

To redisplay the **Property Lookup Table** dialog box, right-click on a lookup action and pick **Display lookup table**.

Using a Lookup Action Dynamically

Figure 28-12 shows a valve block reference selected for editing. The figure shows the **Property Lookup Table** dialog box for reference only. Since Allow reverse lookup is set in the lookup table, the lookup parameter grip appears along with the other parameter grips. Pick the lookup parameter grip to display a shortcut menu containing each lookup record. The entries in this shortcut menu match the entries in the **Lookup** column of the **Property Lookup Table** dialog box. A check mark indicates the current record. To switch to a different view of the block, select the name of the record from the list.

AutoCAD and Its Applications—Basics

Figure 28-12.
The lookup parameter grip displays when you select the block. The list of available lookup records displays when you pick the lookup parameter grip. Notice the correlation between the available options and the lookup property names in the **Property Lookup Table** dialog box.

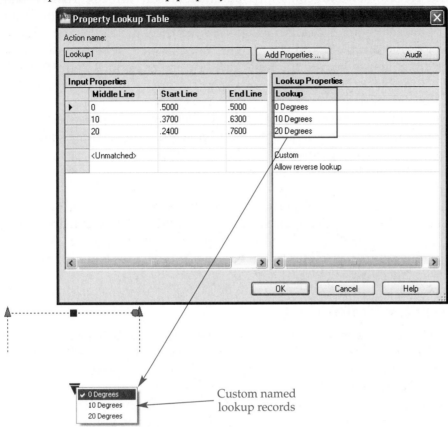

Custom named lookup records

You can change other parameters assigned to the block, such as the linear and rotation parameters of the example block, independently of the named records. When you change any of the parameters, the lookup parameter becomes Custom, because the current parameter values do not match one of the records in the lookup table.

Exercise 28-2

Access the Student Web site (www.g-wlearning.com/CAD) and complete Exercise 28-2.

Using Parameter Sets

The **Parameter Sets** tab of the **Block Authoring Palettes** window contains common parameter and actions grouped to enhance productivity. Follow the prompts to create the parameter and automatically associate an action with the parameter. The action forms without any selected objects, as is indicated by the yellow alert icon. If the parameter set contains an action that must include associated objects, as most do, double-click on the action icon and select objects. The prompts may differ depending on the type of action.

Constraining Block Geometry

Geometric constraints and constraint parameters can directly replace action parameters and actions. For example, the block of the cut framing member shown in **Figure 28-13A** uses geometric constraints to maintain geometric relationships and two linear constraint parameters to specify the member size. When you insert and select the block to edit, use the constraint parameter grips or options in the **Properties** palette to adjust the block. See **Figure 28-13B**. You can create exactly the same block using two linear parameters.

You may find that geometric constraints and constraint parameters are easier to use than action parameters and actions for certain tasks. However, for some blocks you will discover that action parameters and actions require less effort than adding geometric constraints and constraint parameters. You should decide which dynamic block tools and options are appropriate for the blocks you create.

A combination of dynamic properties is also effective. For example, parameters and actions such as alignment, array, and flip offer dynamic controls that are often not possible using geometric constraints and constraint parameters. **Figure 28-14** shows how adding an alignment parameter to the cut framing member block allows you to size and align instances of the block.

Using Geometric Constraints

The geometric constraint tools and options available in the **Block Editor** are identical to those you use to geometrically constrain a parametric drawing. The tools in the **Geometric** panel of the **Parametric** ribbon tab are duplicated in the **Geometric** panel of the **Block Editor** ribbon tab for convenience. Use the geometric constraints in the **Block Editor** as you would in the drawing environment, including the options for relaxing and deleting constraints. The same shortcut menu, **Constraint Settings** dialog box, and **Properties** palette functions apply. Review Chapter 22 for information on adding geometric constraints.

Figure 28-13.
A—A cut framing member block made dynamic using geometric constraints and constraint parameters. B—Using the default 2x4 block to create a 4x4 symbol.

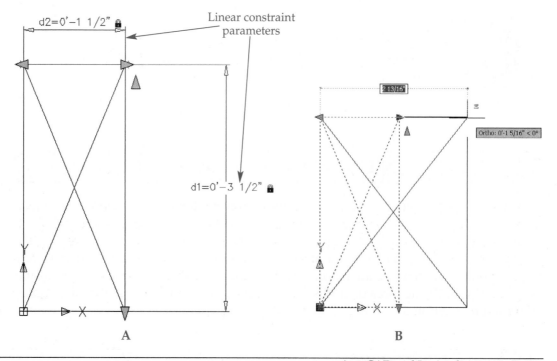

Figure 28-14.
An example of using geometric constraints and constraint parameters to adjust the cut framing member size. An alignment parameter aligns each member for unique applications.

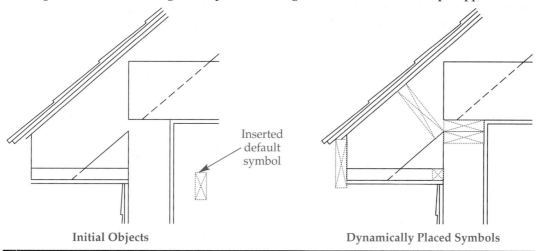

Inserted default symbol

Initial Objects

Dynamically Placed Symbols

You can assign constraints to block objects before you define the block or during block editing to create a dynamic block. See **Figure 28-15A.** Once you define and insert the block, only constraint parameters, action parameters, or actions influence geometric constraints. This allows you to use blocks as objects in parametric drawings. For example, you can insert and rotate the block, as shown in **Figure 28-15B,** even though the block definition includes a horizontal constraint. Use constraints in the drawing to locate blocks and establish geometric relationships between blocks and other objects. See **Figure 28-15C.**

NOTE

You can also use geometric constraints in the block environment to form geometric constructions in specific situations when standard AutoCAD tools are inefficient or ineffective.

Exercise 28-3

Access the Student Web site (www.g-wlearning.com/CAD) and complete Exercise 28-3.

Using Constraint Parameters

Constraint parameters replace dimensional constraints in the **Block Editor.** To help avoid confusion, remember that dimensional constraints constrain a parametric drawing, including block references, as shown in **Figure 28-15C.** Constraint parameters constrain the size and location of block components. By default, dimensional constraints are gray and constraint parameters are blue. You also have the option of converting dimensional constraints into constraint parameters.

You can often use constraint parameters instead of action parameters and actions. If you do not use action parameters, you must include constraint parameters to create a dynamic block. The constraint parameter tools and options available in the **Block Editor** function much like those you use to dimensionally constrain a parametric drawing. Review Chapter 22 for information on adding dimensional constraints.

> **constraint parameters:** Dimensional constraints available for block construction to control the size or location of block geometry numerically.

Figure 28-15.
A—A wide flange block made dynamic using geometric constraints, including a horizontal constraint, and constraint parameters. B—You can rotate the block, because constraints define the size and shape of block geometry during definition and when dynamically adjusting the block. C—Constrain blocks in a drawing as you would any other geometry.

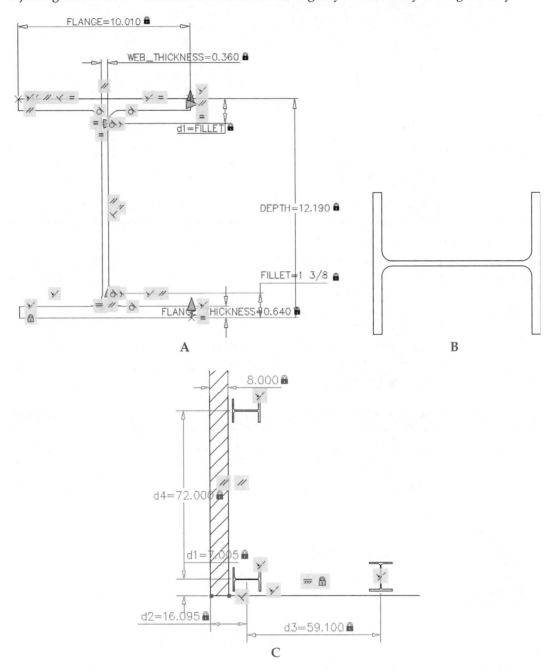

Using the BCPARAMETER Tool

The **BCPARAMETER** tool replaces the **DIMCONSTRAINT** tool in the **Block Editor**, and provides **Linear, Horizontal, Vertical, Aligned, Diameter,** and **Radius** options. The **Linear** option shows by default in the **Block Editor** ribbon tab. You can also use the **BCPARAMETER** tool to convert dimensional constraints to constraint parameters. Each constraint parameter is a separate **DIMCONSTRAINT** tool option. The quickest way to add or convert constraint parameters using this tool is to pick the appropriate button from the **Dimensional** panel of the **Block Editor** ribbon tab.

The process of adding constraint parameters is identical to adding dimensional constraints, except that constraint parameters can include grips. Constraint parameters are essentially a combination of dimensional constraints and action parameters. In the example shown in **Figure 28-16**, the constraint parameters given custom names are those that adjust for unique block references. As when creating a parametric drawing, the other constraint parameters are required to define the block and associate specific geometric relationships. Notice the expressions applied to these values.

To create a constraint parameter, follow the prompts to make the required selections, pick a location for the dimension line, and enter a value to form the constraint. When prompted, specify the number of grips. The radius constraint parameter allows you to add 0 or 1 grip. All other constraint parameters can include 0, 1, or 2 grips. If you plan to assign a single grip to a constraint parameter, select the point associated with the grip second. If you choose the **0** option, you can only use the **Properties** palette to adjust the block.

NOTE

If you attempt to over-constrain a block, a message appears indicating that adding the geometric constraint or constraint parameter is not allowed. You cannot create reference constraint parameters.

PROFESSIONAL TIP

As when adding dimensional constraints or action parameters, change the constraint parameter name to a custom, more descriptive name. Naming labels helps you organize parameters and recognize the parameter during editing. Custom parameters also appear in the **Custom** category of the **Properties** palette.

The **Convert** option of the **BCPARAMETER** tool allows you to convert a dimensional constraint to a constraint parameter. This allows you to prepare a dynamic block using existing dimensional constraints. Access the **Convert** option and pick the dimensional constraint to convert. The dimensional constraint becomes the corresponding constraint parameter and includes the default number of grips.

Ribbon
Block Editor > Dimensional
Convert
Type
BCPARAMETER > Convert

Figure 28-16.
Using constraint parameters to form a dynamic block of a spacer. A single grip is all that is required for these constraint parameters. Do not assign or rename constraint parameters that do not control geometry.

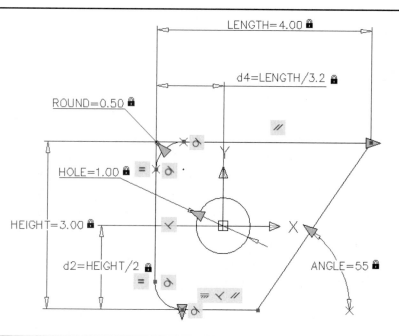

Controlling Constraint Parameters

Control and adjust constraint parameters using a combination of the same techniques you use to manage dimensional constraints and action parameters. Many of the options from shortcut menus, the **Constraint Settings** dialog box, and the **Properties** palette apply. Right-click with no objects selected to access options for displaying and hiding parametric constraints and for accessing the **Constraint Settings** dialog box. Select a constraint parameter and then right-click to display a shortcut menu with options for editing the constraint, changing the format of the name, and redefining the grips.

As with dimensional constraints and the action parameters, the **Properties** palette provides an effective way to control and enhance constraint parameters. You can also use the **Parameters Manager**. **Figure 28-17** shows a foundation detail block with linear constraint parameters. Notice the multiple options available in the **Properties** palette for adjusting the selected constraint parameter.

The options in the **Value set** category of the **Properties** palette allow you to assign value sets to a constraint parameter. Each constraint parameter in the **Figure 28-17** example uses an incremental value to help ensure that you select an appropriate value when adjusting a block reference. You can also create a list of possible sizes. The processes of creating a value set in the **Properties** palette and using value sets are identical for constraint parameters and action parameters.

Exercise 28-4

Access the Student Web site (www.g-wlearning.com/CAD) and complete Exercise 28-4.

Figure 28-17.
Adjust constraint parameters as you would dimensional constraints and action parameters. Use the **Properties** palette to add value sets.

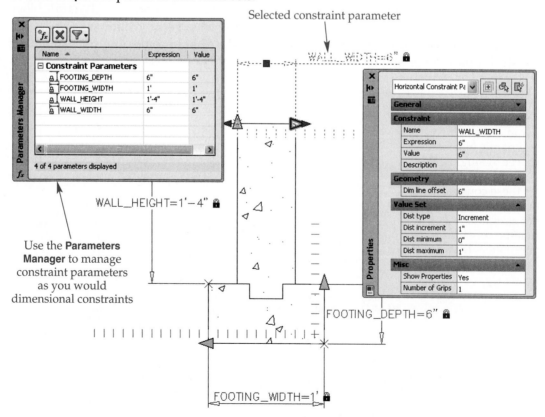

Additional Parametric Tools

The **Block Editor** offers additional options for adding constraints to blocks. Many of the tools, such as the **DELCONSTRAINT** tool, function the same in block editing mode as in drawing mode. However, the **Block Editor** does offer some unique parametric construction tools.

The **BCONSTRUCTION** tool allows you to create construction geometry to aid geometric construction and constraining. Construction geometry appears only in the block definition. See **Figure 28-18.** Access the **BCONSTRUCTION** tool and select the objects to convert to or revert from construction geometry. Press [Enter] or the space bar, or right-click and pick **Enter**. Next, choose the **Convert** option to convert non-construction objects to the construction format, or **Revert**, to return construction geometry to the standard format. You can also use the **Hide all** option to hide all existing construction geometry before selecting objects, or use the **Show all** option to display all construction geometry.

Use the **BCONSTATUSMODE** tool to toggle constraint status identification on and off. When you turn on constraint status mode, objects with no constraints appear white (black) by default, objects assigned some form of constraints are blue, and fully constrained geometry is magenta. If the block contains a constraint error, objects associated with the error are red. Using constraint status is helpful, especially if you want to constrain objects in a certain order or confirm that geometry has been fully constrained.

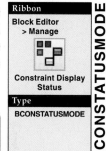

NOTE

Use the **BESETTINGS** tool to access the **Block Editor Settings** dialog box. There you can adjust parameter and parameter grip color and appearance, constraint status colors, and other **Block Editor** settings.

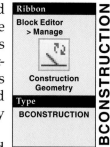

Using a Block Properties Table

A *block properties table* allows you to assign specific values to multiple block properties, and then select a unique group, or row, of properties to create block references. The concept is similar to using a lookup action parameter. A block properties table can include action parameters and/or constraint parameters. You can also add attributes to the table, which is often appropriate for naming each record, or row.

AutoCAD 2010 NEW

block properties table: A table of action parameters and/or constraint parameters that allows you to create multiple block properties and then select them to create block references.

Figure 28-18.
An example of a weld nut block that uses construction geometry to aid geometric construction and constraining.

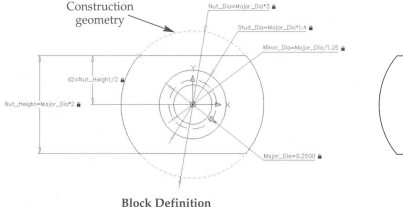
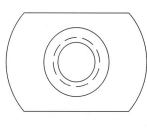

Figure 28-19.
A heavy hex nut block definition ready to use to create a block properties table.

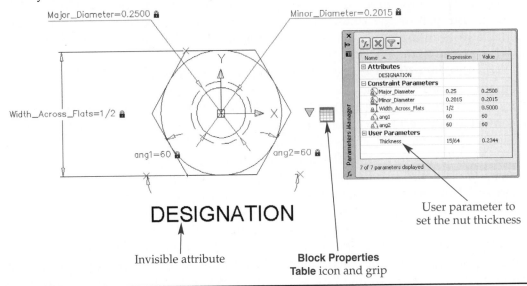

User parameter to
set the nut thickness

Invisible attribute

**Block Properties
Table** icon and grip

Figure 28-19 shows the block of the front view of a heavy hex nut in the **Block Editor**. The block includes an appropriate level of constraints and includes constraint parameters to direct dynamic changes. The block also includes an invisible and preset attribute for defining the designation of each different nut and, as shown in the **Parameters Manager**, a user-defined parameter for the nut thickness.

PROFESSIONAL TIP

It is critical that you assign the **Preset** mode to attributes that you include in a block properties table. This allows the attribute value to adjust to the selected block record. The **Preset** mode requires no default value, and you will not receive a prompt to adjust the value.

After you create parameters and attributes, access the **BTABLE** tool and select the parameter location. Next, enter the number of grips to associate with the parameter. The default **1** option creates a single grip that allows you to select a table record from the grip shortcut menu. If you choose the **0** option, you can only use the **Properties** palette to select a record. The **Palette** option, available before you specify the parameter location or from the **Properties** palette, determines whether the label displays in the **Properties** palette when you select the block reference. The **Block Properties Table** dialog box appears, allowing you to create a block properties table. See **Figure 28-20.**

Creating a Block Properties Table

A block properties table groups the properties of parameters into custom named records, or rows. To add parameter properties, pick the **Add Properties...** button to open the **Add Parameter Properties** dialog box. See **Figure 28-21.** All parameters in the block that contain property values appear in the **Parameter properties:** list. Lookup, alignment, and base point parameters do not contain property values. Notice that the property name is the parameter label.

To add parameter properties to the table, select the properties in the **Parameter properties:** list and pick the **OK** button. A column appears in the table for each parameter property. Type a value in each cell in the column. A new row forms automatically when you enter a value in a cell. See **Figure 28-22.** Press [Enter], [Tab], [Shift]+[Enter], arrow keys, or pick in a different cell to navigate through the table.

Figure 28-20.
The **Block Properties Table** dialog box.

Pick to add block properties

Pick to specify a user property

Block properties appear here

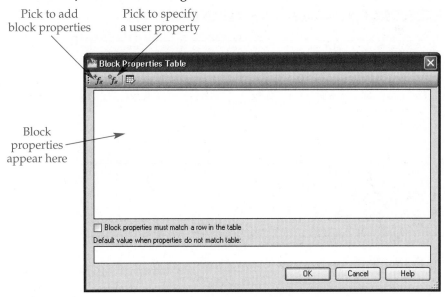

Figure 28-21.
Parameter properties are listed in the **Add Parameter Properties** dialog box.

Select the properties to include in the table

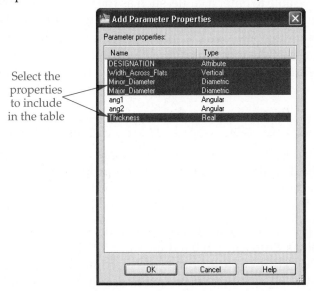

For the nut block example, complete the table as shown in **Figure 28-22**. The DESIGNATION column references the attribute property. The value you enter in this text box in each row specifies the row, or record, name. This value appears in the shortcut menu when you insert the block and select the block properties table parameter grip.

NOTE

Right-click on a column heading to access a menu with options for adjusting columns. Right-click on a row to access a menu with options for adjusting rows. The options function the same as those for adjusting lookup table columns and rows.

Figure 28-22.
A block properties table with multiple parameters and values added.

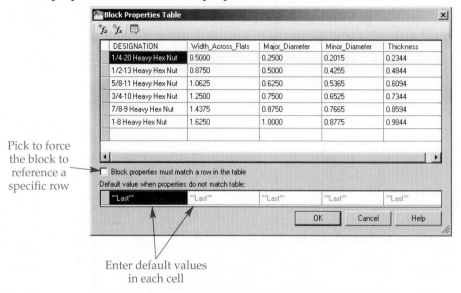

Pick to force the block to reference a specific row

Enter default values in each cell

By default, you can adjust the block using parameter values other than those specified in the table. You may be able to enter a value, such as the value of an attribute property, in a text box found in the **Default property when values do not match table** area of the **Block Properties Table** dialog box. Use the **Last** option to use the value assigned to the previous block reference when you specify a value not found in the table. Often it is appropriate to choose the **Block properties must match a row in the table** check box to force the selection of a specific record, matching all values in a row.

NOTE

It is critical that all block definition values match the values specified in the default block row in the block properties table, especially if you force the selection of a specific record.

After you add all required properties to the table and assign values to each, pick the **Audit** button in the **Block Properties Table** dialog box to check each record in the table to make sure they are all unique, and that there are no discrepancies between the block definition and the table values. If no errors are found, pick the **OK** button to return to the **Block Editor**. Test and save the block, and exit the **Block Editor**. The dynamic block is now ready to use.

NOTE

To redisplay the **Block Properties Table** dialog box, double-click on the parameter, or access the **BTABLE** tool.

Using a Block Properties Table Dynamically

Figure 28-23 shows the inserted nut block selected for editing. Since the block table parameter includes a grip, a grip appears that you can select to choose a specific block style. The entries in the grip shortcut menu match the rows in the **Block Properties Table** dialog box. A check mark indicates the current record. To switch to a different view of

Figure 28-23.
The lookup parameter grip displays when you select the block. The list of available lookup records displays when you pick the lookup parameter grip. Notice the correlation between the available options and the lookup property names in the **Property Lookup Table** dialog box.

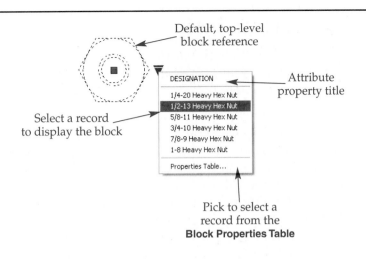

Default, top-level block reference

Attribute property title

Select a record to display the block

DESIGNATION

1/4-20 Heavy Hex Nut
1/2-13 Heavy Hex Nut
5/8-11 Heavy Hex Nut
3/4-10 Heavy Hex Nut
7/8-9 Heavy Hex Nut
1-8 Heavy Hex Nut

Properties Table...

Pick to select a record from the **Block Properties Table**

the block, select the name of the record from the list. You can also pick the **Properties Table...** option to display the **Block Properties Table** in drawing mode. Double-click a row to activate. In this example, no other grips were assigned to blocks. This makes the table and the **Properties** palette the only two methods of selecting a block reference.

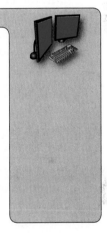

PROFESSIONAL TIP

The options for developing dynamic blocks and creating parametric drawings can become confusing. Keep the following concepts in mind as you proceed:
- Use constraints as an alternative or in addition to action parameters and actions.
- Constraints allow you to create a parametric drawing *or* a dynamic block.
- Assign constraints to create a dynamic block during block definition or while editing the block.
- Treat inserted blocks like any other object when preparing a parametric drawing.

Exercise 28-5

Access the Student Web site (www.g-wlearning.com/CAD) and complete Exercise 28-5.

Chapter Test

1. Define *visibility parameter*.
2. What are visibility states?
3. How do you display the shortcut menu that allows you to select from a block's existing visibility states?
4. When you display the shortcut menu when a block reference is selected for editing, what indicates the current visibility state?
5. Briefly describe a lookup parameter.
6. Explain the basic function of a lookup action.
7. Identify the basic function of a lookup table.
8. What is a parameter set?
9. What ribbon tab, in addition to the **Parametric** ribbon tab, contains geometric constraint tools?
10. When can you assign constraints to block objects to create a dynamic block?
11. What takes the place of dimensional constraints in the **Block Editor**?
12. What tool and option allow you to convert a dimensional constraint to a constraint parameter?
13. How can you create construction geometry to aid geometric construction in the **Block Editor**?
14. Explain how to toggle constraint status identification on and off in the **Block Editor**.
15. What does a block properties table allow you to do?

Drawing Problems

Start AutoCAD if it is not already started. Start a new drawing using an appropriate template of your choice. The template should include layers, text styles, dimension styles, and multileader styles appropriate for drawing the given objects. Add layers, text styles, dimension styles, and multileader styles as needed. Draw all objects using appropriate layers, text styles, dimension styles, multileader styles, justification, and format. Follow the specific instructions for each problem. Use your own judgment and approximate dimensions when necessary.

Note: Constraint parameters shown for reference are created using AutoCAD and may not comply with ASME standards.

▼ Basic

1. Use the **DesignCenter** tool to insert the HEAVY HEX NUT block you created in Exercise 28-5. Insert or copy the block to create six total symbols. Use the block table parameter to display each size nut as shown. Save the drawing as P22-1.

2. Open P22-2 and save as P28-2. The P28-2 file should be active. Create a block from the objects named FIXTURE, select all objects, and pick the center of the circle as the insertion base point. Open the block in the **Block Editor** and convert the dimensional constraints to constraint parameters. Insert the FIXTURE block into the drawing three times to create the 4x4, 6x6, and 4x5 symbols as shown. Resave the drawing.

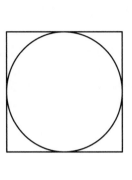
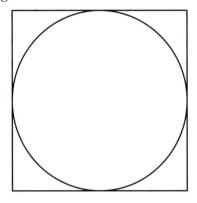
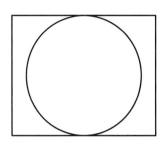

3. Open P22-5 and save as P28-3. The P28-3 file should be active. Create a block from the objects named PLATE, select all objects and pick the center of the plate as the insertion base point. Open the block in the **Block Editor** and convert the construction rectangle to construction geometry. Convert the dimensional constraints to constraint parameters. Insert the PLATE into the drawing and create the block shown. Do not add dimensions. Resave the drawing.

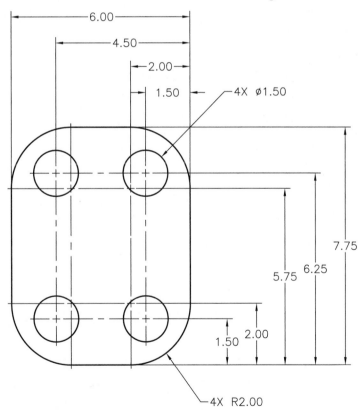

▼ Intermediate

4. Open P22-6 and save as P28-4. The P28-4 file should be active. Create a block named SELECTOR from the objects, select all objects, and pick the center of the center circle as the insertion base point. Open the block in the **Block Editor** and convert the construction lines to construction geometry. Convert the dimensional constraints to constraint parameters. Insert the SELECTOR block into the drawing three times and create three different symbols of your own design. Resave the drawing.

5. Create a single block that can be used to represent each of the three door blocks shown below. Name the block 30 INCH DOOR. Do not include labels. Create an appropriately named visibility state for each view: 90 OPEN, 60 OPEN, and 30 OPEN. Insert the 30 INCH DOOR block into the drawing three times. Set each block to a different visibility state. Save the drawing as P28-5.

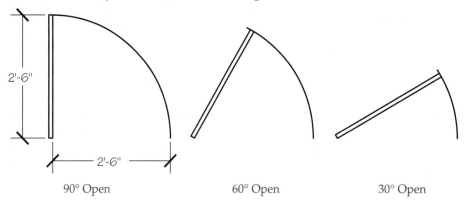

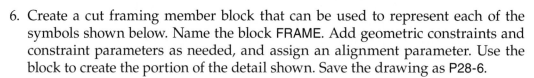

90° Open   60° Open   30° Open

6. Create a cut framing member block that can be used to represent each of the symbols shown below. Name the block FRAME. Add geometric constraints and constraint parameters as needed, and assign an alignment parameter. Use the block to create the portion of the detail shown. Save the drawing as P28-6.

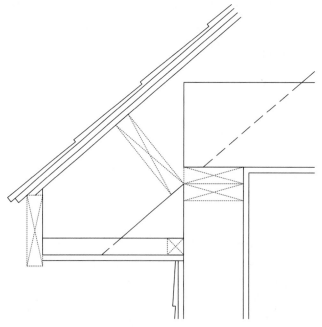

▼ Advanced

7. Create a foundation footing and wall block that can be used to represent unique construction requirements. Name the block Foundation. Add geometric constraints and constraint parameters as shown. Include 1″ increment value sets for each constraint parameter. The block should stretch symmetrically as shown. Save the drawing as P28-7.

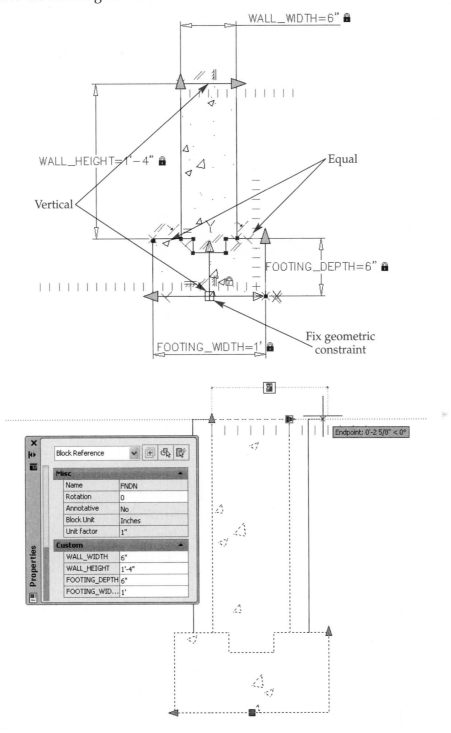

8. The bolt shown in the drawing below is available in four different lengths. As the length increases, the size of the bolt head increases for added strength. Create a dynamic block that will allow the length of the shaft and the size of the bolt head to be changed in a single operation.
   A. Draw the objects composing the bolt and create a block named BOLT. Do not include dimensions.
   B. Insert a linear parameter along the length of the shaft from the bottom of the bolt head to the end of the shaft. Label it SHAFT LENGTH.
   C. Assign a stretch action to the SHAFT LENGTH parameter. Associate the action with the parameter grip at the end of the shaft. Create a crossing window around the end of the shaft that includes the threads. Select the end of the shaft, threads, and edges of the shaft.
   D. Insert a linear parameter along the depth of the bolt head (the .3″ dimension). Label it HEAD THICKNESS.
   E. Assign a scale action to the HEAD THICKNESS parameter and select the objects that compose the bolt head. Use an independent base point type and specify the midpoint of the vertical line where the shaft meets the bolt head.
   F. Insert a lookup parameter and assign a lookup action to it.
   G. Add the SHAFT LENGTH and the HEAD THICKNESS parameters to the lookup table. Complete the table with the following properties:

Shaft Length	Head Thickness	Lookup
1	0.3	1″ Length
1.5	0.333	1.5″ Length
2	0.366	2″ Length
2.5	0.4	2.5″ Length

   H. Set the table to allow reverse lookup, save the block, and exit the **Block Editor**.
   I. Insert the block four times into the drawing. Specify a different lookup property for each block.
   J. Save the drawing as P28-8.

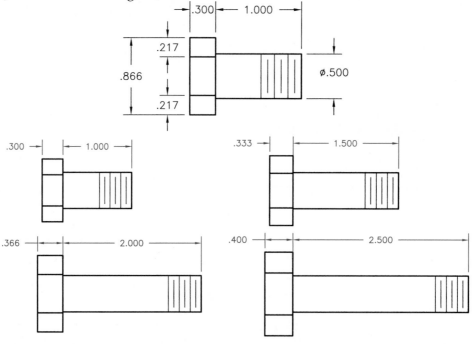

9. Repeat Problem 28-8, but this time use geometric constraints, constraint parameters, and a block properties table instead of action parameters. Save the drawing as P28-9.

Drawing Problems – Chapter 28

CHAPTER **29**

Layout Setup

Learning Objectives

After completing this chapter, you will be able to do the following:
- ✓ Describe the purpose for and proper use of layouts.
- ✓ Begin to prepare layouts for plotting.
- ✓ Manage layouts.
- ✓ Use the **Page Setup Manager** to define plot settings.
- ✓ Use plot styles and plot style tables.

You usually print or plot a drawing in model space to make a quick hard copy, often for check or reference purposes. Ordinarily, however, you create a drawing in model space and then lay out the drawing for plotting in paper space. Hard-copy plots are required for a variety of reasons. For example, it is typically easier for workers in a machine shop or a construction crew in the field to refer to a print than to use a computer to view the drawing file. Additionally, the process of preparing a drawing to plot uses many of the steps needed to publish or export a drawing.

Introduction to Layouts

The first step in making an AutoCAD drawing is to create a *model* in *model space*. See **Figure 29-1A.** Model space is usually active by default. You have been using model space throughout this textbook to create objects and dimensioned drawing views. Once you complete a model, use a *layout* in *paper space* to prepare the final drawing for plotting. See **Figure 29-1B.** A layout represents the sheet of paper used to lay out and plot a drawing. It may include the following items:
- Floating viewports
- Border
- Title block
- Revision block
- General notes
- Bill of materials, parts list, or schedules
- Page setup information

model: A 2D or 3D drawing composed of various objects, such as lines, circles, and text, usually created at full size.

model space: The environment in AutoCAD in which you create drawings and designs.

layout: A specific arrangement of views or drawings for plotting or printing on paper.

paper space: The environment in AutoCAD in which you create layouts.

Figure 29-1.
A—Create drawings and designs in model space. B—Use paper space to finalize and lay out drawings and designs on paper for plotting.

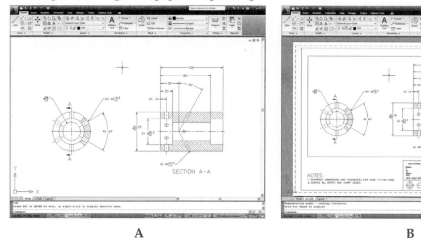

A   B

floating viewport:
A viewport added to a layout in paper space to display objects drawn in model space.

A major element of the layout system is the *floating viewport*. Consider a layout to be a virtual sheet of paper and a floating viewport as a hole cut into the paper to show objects drawn in model space. In **Figure 29-1B,** a single viewport exposes objects drawn in model space. You should usually draw the floating viewport on a layer that you can turn off or freeze so the viewport does not plot and is not displayed on-screen. Chapter 30 explains using floating viewports.

A single drawing can have multiple layouts, each representing a different paper space, or plot, definition. Each layout can include multiple floating viewports to provide additional or alternate drawing views, prepared at different scales if necessary. Layouts with floating viewports offer the ability to construct properly scaled drawings and use a single drawing file to prepare several unique final drawings and drawing views. For example, an architectural drawing file might include several details that are too large to place on a single sheet of paper. You can use multiple layouts, and if necessary differently scaled floating viewports, to prepare as many sheets as needed to plot all of the details found in the drawing.

Working with Layouts

Before preparing a layout for plotting, you should be familiar with tools and options for displaying and managing layouts. The layout and model tabs and model space and paper space tools in the status bar are available by default. These are the most effective tools for navigating to and from model space and layouts, and managing layouts. You can also return to model space from a layout by typing MODEL.

Using the Layout and Model Tabs

The model and layout tabs that appear directly below the drawing window by default are among the most useful options for accessing and preparing layouts. See **Figure 29-2A.** The model space tab is furthest to the left, followed by layout tabs arranged in the order created, from left to right. If the drawing includes so many layouts that tabs spread past the screen, use the forward and reverse buttons left of the tabs to access the appropriate layout or model space. Hover over an inactive tab to display a preview of the contents. Pick a layout tab to enter paper space with the selected layout current, or pick the **Model** tab to re-enter model space.

Figure 29-2.
A—Using the layout and model tabs to activate model space and paper space. B—Options available when you right-click on a tab.

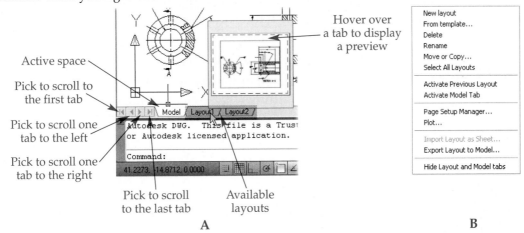

Active space

Pick to scroll to the first tab

Pick to scroll one tab to the left

Pick to scroll one tab to the right

Pick to scroll to the last tab

Available layouts

Hover over a tab to display a preview

New layout
From template...
Delete
Rename
Move or Copy...
Select All Layouts

Activate Previous Layout
Activate Model Tab

Page Setup Manager...
Plot...

Import Layout as Sheet...
Export Layout to Model...

Hide Layout and Model tabs

A   B

Right-click on a tab to access a shortcut menu of options to control layouts and an option to hide the layout and model tabs. See **Figure 29-2B.** Pick **Activate Model Tab** to enter model space. Select **Activate Previous Layout** to make the previously current layout current. Pick **Select All Layouts** to select all layouts in the drawing. This is a valuable option for selecting all layouts for editing purposes, such as deleting or publishing. In addition to these basic functions, the shortcut menu is the primary resource for adding layouts and moving, renaming, and deleting existing layouts.

Figure 29-3 shows an example of using the supplied acad.dwt drawing template and picking the **Layout1** tab to display the default **Layout1** layout. The layout uses default settings based on an 8.5″ × 11″ sheet of paper in a landscape (horizontal)

Figure 29-3.
Pick the **Layout1** tab to display **Layout1**, which is provided in the default acad.dwt template.

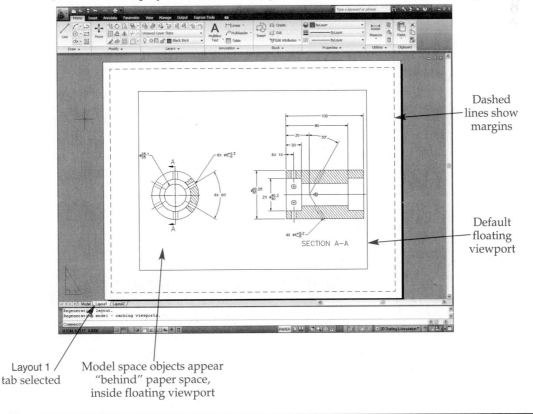

Dashed lines show margins

Default floating viewport

Layout 1 tab selected

Model space objects appear "behind" paper space, inside floating viewport orientation. The white rectangle you see on the gray background is a representation of the sheet. Dashed lines mark the sheet *margin*. A large, rectangular floating viewport reveals model space objects, in this example a dimensioned multiview drawing. The acad.dwt file includes an additional layout named **Layout2**.

NOTE

The **Layout elements** area of the **Display** tab in the **Options** dialog box includes several settings that affect the display and function of layouts. Use the default settings until you are comfortable working with layouts.

PROFESSIONAL TIP

Look at the user coordinate system (UCS) icon to confirm whether you are in model space or paper space. When you enter a layout, the UCS icon changes from two arrows to a triangle that indicates the X and Y coordinate directions.

Using the Model and Paper Buttons

The status bar provides other convenient tools for managing layouts. See **Figure 29-4**. The **Quick View Layouts** and **Quick View Drawings** tools are described later in this chapter. Pick the **MODEL** button to exit model space and enter paper space. If the file contains multiple layouts, the top-level layout displays unless you previously accessed a different layout. While a layout is active, pick the **PAPER** button to use the **MSPACE** tool, which activates a floating viewport. It does not return you to model space. Select the **PAPER** button to deactivate a floating viewport.

Type
MSPACE

Exercise 29-1

Access the Student Web site (www.g-wlearning.com/CAD) and complete Exercise 29-1.

Using the Quick View Layouts Tool

Type
QVLAYOUT

The **Quick View Layouts** tool is very similar to using the layout and model tabs. However, the tool provides additional options and a visual format for displaying and adjusting layouts in the current file. The quickest way to access the **Quick View Layouts** tool is to pick the **Quick View Layouts** button on the status bar.

Figure 29-4.
The status bar provides additional tools for activating model and paper space and managing layouts.

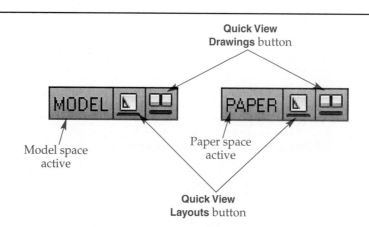

Quick View Drawings button

Model space active

Paper space active

Quick View Layouts button

Figure 29-5.
The **Quick View Layouts** tool offers an effective visual method for changing between model space and paper space and provides options for managing layouts.

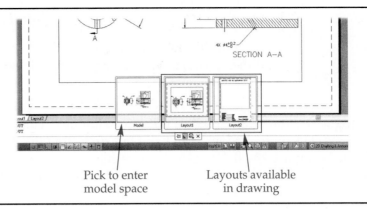

Pick to enter
model space

Layouts available
in drawing

When activated, the **Quick View Layouts** tool appears in the lower center of the AutoCAD window. See **Figure 29-5**. The **Model** thumbnail image is furthest to the left, followed by layout thumbnails in the order created, from left to right. If the drawing includes so many layouts that thumbnails spread past the screen, hover the cursor over the furthest right and left thumbnails to scroll though the options. Hover over a thumbnail to highlight the image and show additional options. See **Figure 29-6**. Pick a layout thumbnail to enter paper space with the selected layout current, or pick the **Model** thumbnail to re-enter model space.

> **NOTE**
>
> Icons represent model and layout thumbnails until you enter a layout for the first time (initialize the layout). The icon then changes to a thumbnail image of model space and eventually the layout.

The **Quick View Layouts** tool provides a small toolbar below the thumbnail images, as shown in **Figure 29-6**. By default, the **Quick View Layouts** tool disappears when you pick a thumbnail to switch layouts or enter model space. To keep the tool on-screen, pick the **Pin Quick View Layouts** button. Pick the **New Layout** button to create a new layout from scratch, as described later in this chapter. Pick the **Publish...** button to access the **Publish** dialog box, explained later in this textbook. Select the **Close** button to exit the **Quick View Layouts** tool.

Figure 29-6.
Hover over a thumbnail to display **Plot...** and **Publish...** buttons.

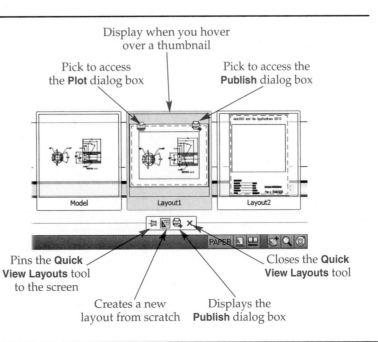

Display when you hover
over a thumbnail

Pick to access
the **Plot** dialog box

Pick to access the
Publish dialog box

Pins the **Quick View Layouts** tool to the screen

Closes the **Quick View Layouts** tool

Creates a new
layout from scratch

Displays the
Publish dialog box

Right-click on a tab to access the same shortcut menu of options available when you right-click on the model or a layout tab, excluding the **Hide Layout and Model tabs** option. See **Figure 29-2B.** You can also access some of the menu options from a shortcut menu displayed when you right-click directly on the **Quick View Layouts** button on the status bar. An option selected from this location applies to the current file and the current layout.

Using the Quick View Drawings Tool

The **Quick View Drawings** tool provides the same features for working with layouts as the **Quick View Layouts** tool, but it allows you to manage the layouts in all open drawings. Use this tool to increase productivity when you are working between existing drawings. The quickest way to access the **Quick View Drawings** tool is to pick the **Quick View Drawings** button on the status bar. Refer to Chapter 2 for information on basic **Quick View Drawings** tool features, such as using the tool to work with multiple open documents.

Access the **Quick View Drawings** tool and hover the cursor over the drawing file to control. The model and layout thumbnail images appear above the highlighted drawing. Move the cursor over the model or a layout thumbnail to enlarge the display. See **Figure 29-7.** Pick a layout thumbnail to switch to the highlighted file and enter paper space with the selected layout active, or pick the **Model** thumbnail to switch to the highlighted file in model space. Right-click on a thumbnail to access the same shortcut menu displayed when you right-click on a thumbnail using the **Quick View Layouts** tool. Right-clicking on the thumbnail makes the associated file current.

Figure 29-7.
Use the **Quick View Drawings** tool to manage layouts located in other open files. Hover over a file thumbnail to display model and layout thumbnails for the file.

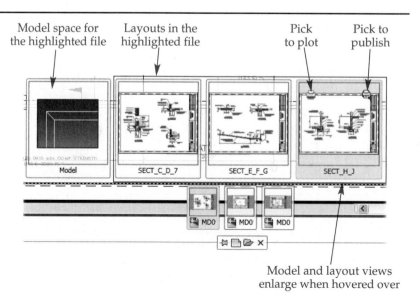

Model space for the highlighted file

Layouts in the highlighted file

Pick to plot

Pick to publish

Model and layout views enlarge when hovered over

Adding Layouts

To add a new layout to a drawing, create a new layout from scratch, use the **Create Layout** wizard, or reference an existing layout. Referencing an existing, preset layout is often the most effective approach. You can also insert a layout from a different DWG, DWT, or DXF file into the current file or create a copy of a layout from the current file.

Starting from Scratch

To create a new layout from scratch, right-click on the model or a layout tab, a **Quick View Layouts** or **Quick View Drawings** thumbnail image, or the **Quick View Layouts** button on the status bar, and pick **New Layout**. A new layout appears on the far right of the layout list. The settings applied to the new layout depend on the template used to create the original file. The name of the layout is set according to the names of other existing layouts. For example, when you add a new layout to a default drawing started from the acad.dwt template, a new layout named **Layout3** appears and includes an 8.5″ × 11″ sheet of paper, a landscape (horizontal) orientation, and a large floating viewport.

Using the Create Layout Wizard

Use the **Create Layout** wizard to build a layout from scratch using values and options you enter in the wizard. The pages of the wizard guide you through the process of developing the layout. They provide options for naming the new layout and selecting a printer, paper size, drawing units, paper orientation, title block, and viewport configuration.

Using a Template

To create a new layout from a layout stored in an existing DWG, DWT, or DXF file, right-click on the model or a layout tab, a **Quick View Layouts** or **Quick View Drawings** thumbnail image, or on the **Quick View Layouts** button on the status bar, and pick **From Template...**. The **Select Template From File** dialog box appears. See **Figure 29-8A**. The Template folder in the path set by the AutoCAD Drawing Template File Location is the default. Select the file containing the layout to add to the current drawing and pick the **Open** button. The **Insert Layout(s)** dialog box appears, listing all layouts in the selected file. See **Figure 29-8B**. Highlight the layout or layouts to copy and pick the **OK** button.

Using DesignCenter

DesignCenter provides an effective way to add existing layouts to the current drawing. To view the layouts found in a drawing, select the **Layouts** branch in the tree view or double-click on the **Layouts** icon in the content area. See **Figure 29-9**. Select the layout(s) to copy from the content area and then drag and drop, or use the **Add Layout(s)** or **Copy** and **Paste** options from the shortcut menu to insert the layouts in the current drawing.

Copying and Moving Layouts

To create a copy of a layout, right-click on a layout tab, a **Quick View Layouts** or **Quick View Drawings** thumbnail image, or make the layout to copy current and right-click on the **Quick View Layouts** button on the status bar. Then pick **Move or Copy...** to display the **Move or Copy** dialog box. See **Figure 29-10**. To create a copy, select the **Create a copy** check box and pick the layout that will appear to the right of the new layout, or pick (move to end) to place the copy right of all other layouts. The default name of the new layout is the name of the current or selected layout plus a number in parentheses.

Move a layout using the **Move or Copy...** dialog box without selecting the **Create a copy** check box. When you add and rename layouts, the layouts do not automatically rearrange into a predetermined order. Organize layouts in an appropriate order to reduce confusion and aid in the publishing process, as described in Chapter 33.

Figure 29-8.
Adding a layout using an existing layout stored in a different drawing, drawing template, or DXF file. A—Select the file containing the layout. B—Highlight the layout to add to the current drawing.

Template folder
opens by default

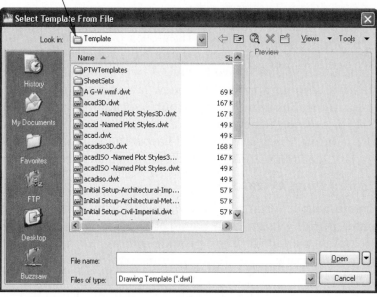

A

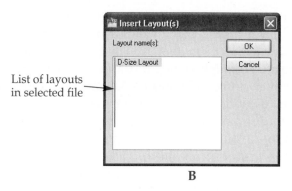

List of layouts
in selected file

B

Figure 29-9.
Using **DesignCenter** to share layouts between drawings.

Select
drawing

Pick to display
layouts defined
in the drawing

Copy layout to
current drawing

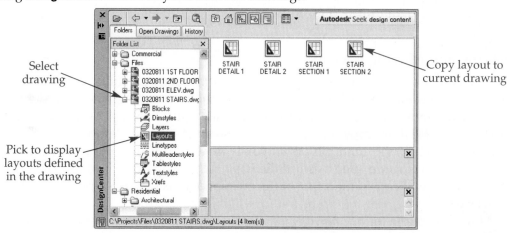

Figure 29-10.
The **Move or Copy** dialog box allows you to reorganize layouts and copy layouts within a drawing.

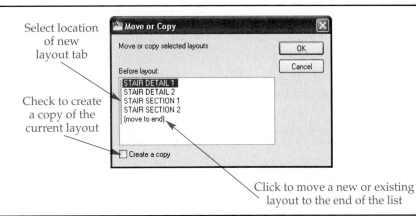

Select location of new layout tab

Check to create a copy of the current layout

Click to move a new or existing layout to the end of the list

Renaming Layouts

Layouts are easier to recognize and use when they have descriptive names. To rename a layout, right-click a layout tab or a **Quick View Layouts** or **Quick View Drawings** thumbnail image and pick **Rename**. You can also double-click slowly on the current name to activate it for editing. Once the layout name highlights, type a new name and press [Enter].

Deleting Layouts

To delete an unused layout from the drawing, right-click on the layout tab or the **Quick View Layouts** or **Quick View Drawings** thumbnail image and pick **Delete**. An alert message warns you that the layout will be deleted permanently. Pick the **OK** button to remove the layout.

Exporting a Layout to Model Space

The **EXPORTLAYOUT** tool allows you to save the layout display as a separate DWG file. This tool produces a "snapshot" of the current layout display that you can use for applications in which it is necessary to combine model space and paper space objects, such as when exporting a file as an image. (Model space and paper space do not export together as an image.)

Type
EXPORTLAYOUT

The quickest way to access the **EXPORTLAYOUT** tool is to right-click on a layout tab or a **Quick View Layouts** or **Quick View Drawings** thumbnail image and pick **Export Layout to Model…**. The **Export Layout to Model Space** dialog box appears and functions much like the **Save As** dialog box. Pick a location for the file, use the default file name or enter a different name, and pick the **SAVE** button. Everything shown in the layout, including objects drawn in model space, are converted to model space and are saved as a new file.

CAUTION

The **EXPORTLAYOUT** tool eliminates the relationship between model space and paper space. Export a layout only when it is necessary to export model space and paper space together as a single unit.

Exercise 29-2

Access the Student Web site (www.g-wlearning.com/CAD) and complete Exercise 29-2.

Initial Layout Setup

Preparing a layout for plotting involves creating and modifying floating viewports, adjusting plot settings, and adding layout content such as symbols, a border, and a title block. This chapter focuses on the process of preparing layouts for plotting using the **Page Setup Manager**. When you complete this initial phase, you will be better prepared to add content to layouts and create and manage floating viewports, as described in Chapter 30.

page setup: A saved collection of settings required to create a finished plot of a drawing.

A *page setup* establishes most of the settings that determine how a drawing plots. Plot settings include printer selection, paper size and orientation, plot area and offset, plot scale, and plot style. The **Page Setup Manager** and related **Page Setup** dialog box allow you to create and modify saved page setups that control how layouts appear on-screen and plot. This is where initial layout setup occurs. You then use the **Plot** dialog box to create the actual plot using the saved page setup. The **Page Setup** and **Plot** dialog boxes include most of the same settings.

PROFESSIONAL TIP

Layout setup usually involves several steps. A well-defined page setup decreases the amount of time required to prepare a drawing for plotting. Once a layout is set up, only a few steps are required to produce a plot. Add fully defined layouts to your drawing templates for convenient future use.

Working with Page Setups

PAGESETUP	
Ribbon	
Output > Plot	
Page Setup Manager	
Type	
PAGESETUP	

Access the **PAGESETUP** tool to create page setups using the **Page Setup Manager**. See **Figure 29-11**. The **Page setups** area of the **Page Setup Manager** contains a list box that lists available page setups, as well as buttons to add and modify page setups. The **Selected page setup details** area provides information about the highlighted page setup.

NOTE

You can also access the **Page Setup Manager** by right-clicking on the model or a layout tab, a **Quick View Layouts** or **Quick View Drawings** thumbnail image, or on the **Quick View Layouts** button on the status bar, and selecting **Page Setup Manager...**.

When you access the **Page Setup Manager** in model space, *Model* appears in the **Page Setups** list box. When you access the **Page Setup Manager** in paper space, the name of the current layout appears in the **Page Setups** list box. When preparing a layout for plotting, check to be sure that you are in paper space and that the appropriate layout is current. Each layout can have a unique page setup. Asterisks (*) before and after the layout name indicate the page setup assigned to the current layout. You have the option of creating or using other page setups instead of the page setup associated with the layout.

Create a new page setup to use different plot characteristics without overriding plot settings or spending time making page setup changes. For example, you can create two page setups to plot to two different printers or plotters. Pick the **New...** button to create a new page setup using the **New Page Setup** dialog box. See **Figure 29-12**. Type a name for the new page setup and choose an option from the **Start with:** list box. Pick

Figure 29-11.
Use the **Page Setup Manager** to modify existing page setups and to create and import page setups.

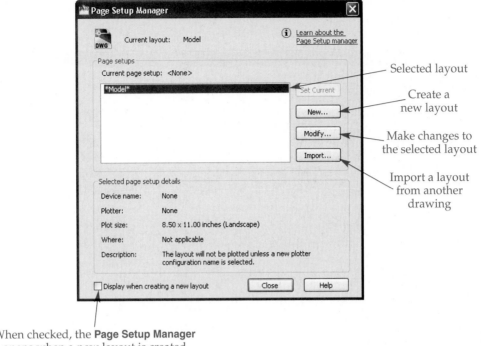

Selected layout

Create a
new layout

Make changes to
the selected layout

Import a layout
from another
drawing

When checked, the **Page Setup Manager**
opens when a new layout is created

Figure 29-12.
The **New Page Setup** dialog box appears when you create a new page setup. Selecting **None** in
the **Start with:** list box does not select a printer. **Default** selects the computer's default printer.

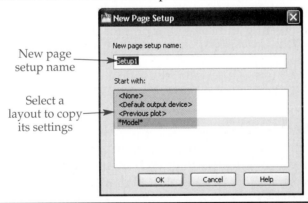

New page
setup name

Select a
layout to copy
its settings

the **OK** button to create the page setup and display the **Page Setup** dialog box. Pick the
Import... button to use existing page setups from a DWG, DWT, or DXF file.

To attach a different page setup to the current layout, select a page setup from the
list in the **Page Setup Manager** and pick the **Set Current** button, or right-click on a page
setup and choose **Set Current**. The layout will now plot according to the selected page
setup. When you make a different page setup current, the selected page setup overrides the layout page setup. The page setup name appears in parentheses next to the
layout name. To rename or delete an existing page setup, right-click on the page setup
and pick **Rename** or **Delete**.

Select the **Modify...** button in the **Page Setup Manager** to change the settings of an
existing page setup using the **Page Setup** dialog box. See **Figure 29-13.** The **Page Setup**
dialog box defines page setup characteristics. The settings control layout appearance
and plot function. Each area, as described in the following sections, controls a specific
plot setting.

Figure 29-13.
The **Page Setup** dialog box allows you to adjust the plot settings for the selected page setup. This is where the initial phase of layout setup begins.

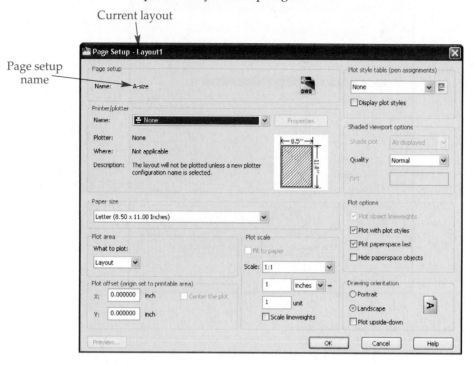

Current layout

Page setup name

Selecting a Plot Device

The **Printer/plotter** area of the **Page Setup** dialog box, shown in **Figure 29-14**, allows you to select the appropriate *plot device* and adjust the plot device configuration if necessary. The default None setting indicates that no plot device has been specified. If a plot device has been *configured*, select the plot device you want from the **Name:** drop-down list.

plot device: The printer, plotter, or alternative plotting system to which the drawing is sent.

configured: Installed and ready to use.

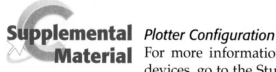

Supplemental Material

Plotter Configuration
For more information about managing and configuring plot devices, go to the Student Web site (www.g-wlearning.com/CAD), select this chapter, and select **Plotter Configuration**.

Figure 29-14.
The **Printer/plotter** area of the **Page Setup** dialog box.

Select a printer

After a printer is selected, additional changes can be made

Information about the selected printer

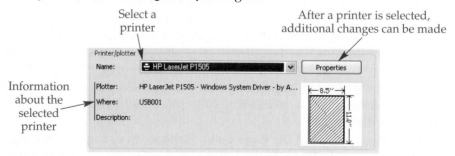

Figure 29-15.
The **Paper size** area allows you to define the sheet size applied to the layout, which corresponds to the sheet size on which you plan to plot.

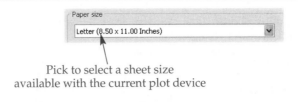

Paper size

Letter (8.50 x 11.00 Inches)

Pick to select a sheet size
available with the current plot device

Choosing a Sheet Size

The **Paper size** area of the **Page Setup** dialog box, shown in **Figure 29-15**, controls the *sheet size*. The sheet size determines the size of the virtual sheet of paper displayed in a layout, as well as the actual sheet size you plan to use when plotting. When determining sheet size, you should take into account the size of the drawing and additional space for dimensions, notes, border, clear space between the drawing and border, title block, revision block, zoning, and an area for general notes. Select the appropriate sheet size from the drop-down list in the **Paper size** area.

sheet size: The size of the paper you use to lay out and plot the final drawing.

Standard Sheet Sizes

The ASME Y14.1, *Decimal Inch Drawing Sheet Size and Format,* and ASME Y14.1M, *Metric Drawing Sheet Size and Format* documents specify the American Society of Mechanical Engineers (ASME) and American National Standards Institute (ANSI) standard sheet sizes and formats. ASME Y14.1 lists sheet size specifications in inches, as follows:

Size Designation	Size (in inches)	
	Vertical	**Horizontal**
A	8 1/2	11 (horizontal format)
	11	8 1/2 (vertical format)
B	11	17
C	17	22
D	22	34
E	34	44
F	28	40
Sizes G, H, J, and K are roll sizes.		

ASME Y14.1M lists sheet size specifications in metric units, as follows:

Size Designation	Size (in millimeters)	
	Vertical	**Horizontal**
A0	841	1189
A1	594	841
A2	420	594
A3	297	420
A4	210	297

Longer lengths are known as elongated and extra-elongated drawing sizes. These are available in multiples of the short side of the sheet size. **Figure 29-16** shows standard ASME/ANSI sheet sizes. To describe sheet size values verbally, generally state the vertical measurement and then the horizontal measurement. For example, describe a C-size sheet as 22 (horizontal) × 17 (vertical).

Figure 29-16.
A—Standard drawing sheet sizes (ASME Y14.1). B—Standard metric drawing sheet sizes
(ASME Y14.1M).

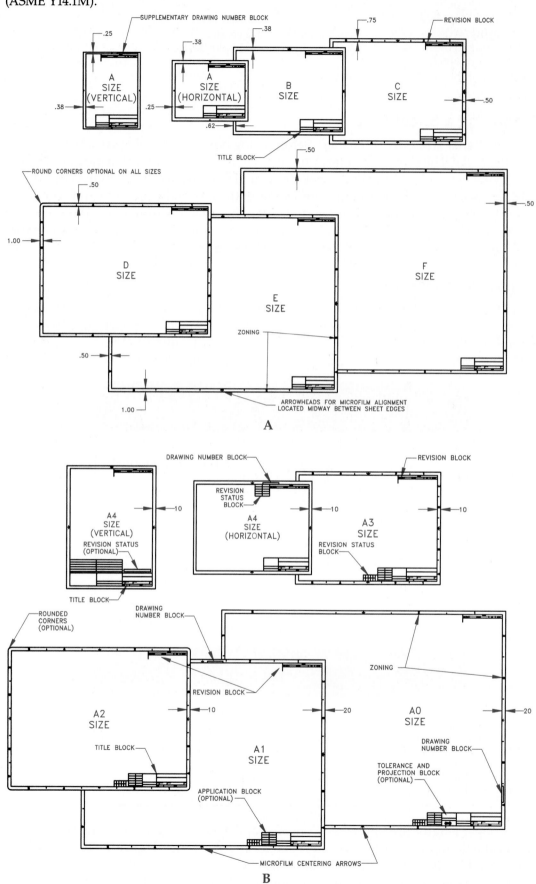

AutoCAD and Its Applications—Basics

Reference Material *Drawing Sheets*
For tables describing sheet characteristics, including sheet size, drawing scale, and drawing limits, go to the **Reference Material** section of the Student Web site (www.g-wlearning.com/CAD) and select **Drawing Sheets**.

Specifying the Drawing Orientation

The **Drawing orientation** area controls the plot rotation. See **Figure 29-17**. Landscape orientation is the most common engineering drawing orientation and is the default in most AutoCAD-supplied templates. Portrait orientation is the standard for most written documents printed on 8.5″ × 11″ paper.

Pick the **Plot upside-down** check box to produce variations of the standard landscape and portrait orientations. When you select an upside-down orientation, it may help to consider landscape format to be a rotation angle of 0° and portrait format to be a rotation angle of 90°. Therefore, an upside-down landscape format rotates the drawing 180°, and an upside-down portrait orientation rotates the drawing 270°. Use the preview image to help select the appropriate orientation.

PROFESSIONAL TIP

The way in which a sheet feeds into a printer or plotter can affect the sheet size and drawing orientation you select. Sheets of paper, especially large sheets, often feed into a plotter with the short side of the sheet entering first. This may require you to use a sheet size that orients the sheet in a portrait format, for example D-Size 22x34 instead of D-Size 34x22, while still using a landscape drawing orientation.

 **Exercise 29-3**

Access the Student Web site (www.g-wlearning.com/CAD) and complete Exercise 29-3.

Choosing the Plot Area

The **Plot area** section, shown in **Figure 29-18**, allows you to choose the portion of the drawing to plot. Select an option from the **What to plot:** drop-down list. The **Layout** option is available when you plot a layout. When you select this option, everything inside the margins of the layout plots. Plotting the layout using the **Layout** option is the default and is the most common setting for plotting a layout. Use other plot area

Figure 29-17.
The **Drawing orientation** area contains options for adjusting the drawing angle of rotation.

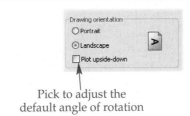

Pick to adjust the
default angle of rotation

Figure 29-18.
Use the **Plot area**
section to define
what portion of the
drawing to plot.

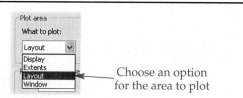

Choose an option
for the area to plot

options primarily for plotting in model space or when it is necessary to adjust the area plotted in paper space.

The **View** option is available when named views exist in the drawing, and if the model space or paper space environment associated with the current page setup includes a named view. When you pick the **View** option, an additional drop-down list appears in the **Plot area** section, allowing you to select a specific view to define as the plot area. Refer to Chapter 5 for information about other **Plot area** options.

Defining the Plot Offset

The **Plot offset** area, shown in **Figure 29-19,** controls how far the drawing is offset from the plot origin. You can specify the plot origin as the lower-left corner of the printable area or the lower-left corner of the sheet by selecting the appropriate radio button in the **Specify plot offset relative to** area in the **Plot and Publish** tab of the **Options** dialog box.

The **Plot offset (origin set to printable area)** title appears when you use the default **Printable area** option of the **Options** dialog box. The values you enter in the **X:** and **Y:** text boxes define the offset from the printable area. Use the default values of 0 to locate the plot origin at the lower-left corner of the printable area, which corresponds to the lower-left corner of the layout margin (dashed rectangle). To move the drawing away from the default printable area origin, enter positive or negative values in the text boxes. For example, to move the drawing one unit to the right and two units above the lower-left corner of the margin, enter 1 in the **X:** text box and 2 in the **Y:** text box.

Select the **Edge of paper** radio button in the **Options** dialog box to display the **Plot offset (origin set to layout border)** title. This causes the values you enter in the **X:** and **Y:** text boxes to define the offset from the edge of the sheet. Use the default values of 0 to locate the plot origin at the lower-left corner of the sheet. Change the values in the text boxes to move the drawing away from the sheet origin.

When you select any of the options from the **What to plot:** drop-down list in the **Plot area** section, except the **Layout** option, the **Center the plot** check box becomes enabled. Pick this check box to shift the plot origin automatically as needed to center the selected plot area in the printable area.

Figure 29-19.
The **Plot offset** area
controls how far
the drawing offsets
from the lower-
left corner of the
printable area or
layout border.

Enter offset from
lower-left corner

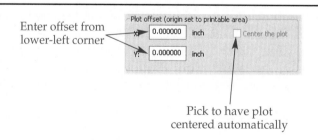

Pick to have plot
centered automatically

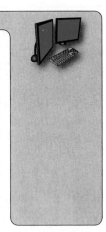

If your drawing does not center on the sheet when you plot using the **Layout** option, open the **Options** dialog box and select the **Plot and Publish** tab. Pick the **Edge of paper** radio button in the **Specify plot offset relative to** area. Then use values of 0 in the X and Y offset text boxes in the **Page Setup** dialog box to locate the plot origin at the lower-left corner of the sheet. Depending on the specific plot configuration, you may also need to pick the **Plot upside-down** check box in the **Drawing orientation** area of the **Page Setup** dialog box to locate the origin exactly at the lower-left corner of the sheet. Alternatively, resolve the issue by changing the plot offset to the X and Y values of the lower-left corner of the printable area.

Selecting a Plot Scale

You should always draw objects at their actual size, or full scale, in model space, regardless of the size of the objects. For example, if you draw a small machine part and the length of a line in the drawing is 2 mm, draw the line 2 mm long in model space. If you draw a building and the length of a line in the drawing is 80′, draw the line 80′ long in model space. For layout and printing purposes, you must then scale most drawings to fit properly on a sheet, according to a specific *drawing scale*.

drawing scale: The ratio between the actual size of objects in the drawing and the size at which the objects plot on a sheet of paper.

When you scale a drawing, you increase or decrease the *displayed* size of model space objects. A properly scaled floating viewport in a layout allows for this process, as described in Chapter 30. When setting up a layout for plotting, remember that the layout is also at full scale. One difference between model space and paper space is that objects in model space can be very large or very small, while objects in paper space always correspond to sheet size. In order for objects on the layout and in model space to appear correct when plotted, you must plot a layout at full scale, or 1:1.

The **Plot scale** area of the **Page Setup** dialog box allows you to specify the plot scale. See **Figure 29-20**. The **Scale:** drop-down list provides several predefined decimal and architectural scales, as well as a **Custom** option. For most applications, when setting up a layout for plotting, the plot scale is set to 1:1. This ensures that the layout and scaled floating viewports plot correctly. If you choose a scale other than 1:1, the layout does not plot to scale.

When preparing to plot in model space, you may choose to select a scale other than 1:1. For example, to plot an architectural floor plan, you might want to set the plot scale to 1/4″ = 1′-0″. If the desired scale is not available, enter values in the text boxes below the predefined scales drop-down list and select the correct unit of measure from the drop-down list. The **Custom** option is automatically displayed when you enter values. For example, 1 inch = 600 units is a custom scale entry used to plot at a scale of 1″ = 50′ (50′ x 12″ = 600). Refer to Chapter 9 for more information about drawing scale and scale factors.

Pick the **Fit to paper** check box to adjust the plot scale automatically to fit on the selected sheet. This is useful if you are not concerned about plotting to scale, such as when you are creating a "check copy" on a sheet that is too small to plot at the appropriate scale. Select the **Scale lineweights** check box to scale (increase or decrease the weight of) lines when the plot scale changes.

Figure 29-20.
Use the **Plot scale** area to adjust the scale at which the drawing plots. The plot scale is typically set to 1:1 to plot a layout even though the drawing scale may not be 1:1.

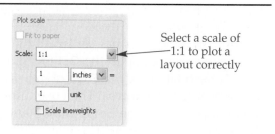

Select a scale of 1:1 to plot a layout correctly

PROFESSIONAL TIP

In the rare event that you need to plot an inch drawing on a metric sheet, use a custom scale of 1 inch = 25.4 units. Conversely, use a custom scale of 25.4 mm = 1 unit to plot a metric drawing on an inch sheet.

Exercise 29-4

Access the Student Web site (www.g-wlearning.com/CAD) and complete Exercise 29-4.

Using Plot Styles

Object properties control the appearance of objects on-screen. By default, what you see on-screen is what plots. For example, if you draw objects on a layer that uses a Red color, Continuous linetype, and 0.60 mm lineweight, the objects display and plot red, continuous, and thick, assuming you show lineweights on-screen and use a color plotter. If this situation is desirable, you are ready to continue with the page setup and plotting process.

plot styles: Properties, including color, linetype, lineweight, line end treatment, and fill style, that are applied to objects for plotting purposes only.

However, to define exactly how objects plot regardless of what displays on-screen, you must assign *plot styles* to objects. Use plot styles to maintain object properties in the drawing, but plot objects according to specific plotting properties. For example, you may want all of the objects in a drawing to plot as dark as possible, which requires that all objects use the color Black when plotted. In this example, plot styles allow you to plot all objects black without making them black on-screen. Plot styles also allow you to plot objects using shades of gray instead of color, or to plot objects lighter or darker than they display on-screen.

Plot Style Tables

plot style table: A configuration, saved as a separate file, that groups plot styles and provides complete control over plot style settings.

Plot styles are contained in *plot style tables*. You can choose to use either a *color-dependent plot style table* or a *named plot style table*. When you use a color-dependent plot style table, objects plot according to object color. Color-dependent plot style tables contain 256 preset plot styles—one for each AutoCAD Color Index (ACI) color. Each color-dependent plot style links to an index color. Plot style properties control how to treat objects of a certain color when they are plotted. For example, the plot style Color 1, which is Red, defines how all objects that are red on-screen plot. If you assign a Black plot color to plot style Color 1, all objects drawn using a Red color plot black, even though the objects are red on-screen.

color-dependent plot style table: A file that contains plot style settings used to assign plot values to object colors.

named plot style table: A file that contains plot style settings used to assign plot values to objects or layers.

When you use a named plot style table, drawing objects plot according to named plot style values, which you can assign to a layer or object. Any layer or object assigned a named plot style plots using the settings specified for that plot style. For example, create a layer named OBJECT that uses a Red color, and a plot style named BLACK that uses a Black color. Then assign the BLACK plot style to the OBJECT layer. Objects drawn on the OBJECT layer plot using the BLACK plot style and plot black in color, even though the objects are red on-screen.

Ideally, you should decide which plot style table is appropriate for your application before you begin drawing. The templates created in the Template Development feature of this textbook, for example, assume that drawings plot so that all objects appear dark, or black, with different object linetypes and lineweights. You can use a color-dependent plot style table or a named plot style table to create this effect. Default AutoCAD plot style behavior is set to use color-dependent plot style tables. For basic applications, it is usually best to use color-dependent plot style tables, because they are the default and do not require you to assign named plot styles to layers or objects.

Configuring Plot Style Table Type

You must choose a type of plot style table (color-dependent or named) to apply to new drawings *before* you start a new drawing file, unless you use a template that already is assigned a plot style table type. To configure the plot style type used by default when you create new drawings, pick the **Plot Style Table Settings...** button in the **Plot and Publish** tab of the **Options** dialog box. This opens the **Plot Style Table Settings** dialog box shown in **Figure 29-21.**

To use a named plot style table, pick the **Use named plot styles** radio button from the **Default plot style behavior for new drawings** area *before* you start a new drawing file. Select a specific plot style table to use as the default for new drawings from the **Default plot style table:** drop-down list in the **Current plot style table settings** area. When you select the **Use named plot styles** radio button, the **Default plot style for layer 0:** and **Default plot style for objects:** options become enabled.

NOTE

When you use a template to create a new drawing, the plot style settings defined in the template override the settings you specify in the **Plot Style Table Settings** dialog box. For example, if you configure the template to use named plot style tables, you can only select a named plot style table to apply to the plot, even if you select the **Use color dependent plot styles** radio button in the **Plot Style Table Settings** dialog box.

Figure 29-21.
Use the **Plot and Publish** tab of the **Options** dialog box to setup default plot style types and tables for new drawings.

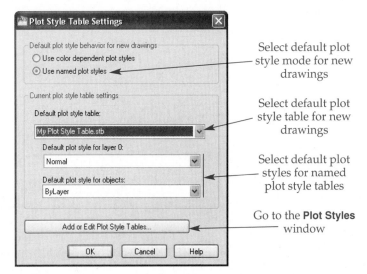

Pick the **Add or Edit Plot Style Tables...** button in the **Plot Style Table Settings** dialog box to open the **Plot Styles** window. See **Figure 29-22.** The **Plot Styles** window lists all available color-dependent and named plot style tables saved in the Plot Style Table Search Path, as defined in the expanded **Printer Support File Path** option in the **Files** tab of the **Options** dialog box. Color-dependent plot style table files (CTB) use the .ctb extension. Named plot style table files (STB) include an .stb extension. Double-click on **Add-A-Plot Style Table Wizard** to create a new plot style table, or double-click on an existing plot style table file to edit the file.

Supplemental Material

Creating and Editing Plot Style Tables

For detailed information about creating and editing plot style tables, go to the Student Web site (www.g-wlearning.com/CAD), select this chapter, and select **Creating and Editing Plot Style Tables**.

Selecting a Plot Style Table

Use the **Plot style table (pen assignments)** area of the **Page Setup** dialog box, shown in **Figure 29-23,** to activate and manage plot style tables. Select a plot style table to use from the drop-down list. The **None** option is default and can be used instead of selecting a specific color-dependent or named plot style table. Select **None** to plot exactly what appears on-screen, without using plot styles, assuming you use a color plotter to plot objects with color.

Only color-dependent or named plot style tables appear, depending on the type of plot style tables assigned to the current drawing. The most often used color-dependent plot style tables are:

- Monochrome.ctb—plots the drawing in monochrome (black and white).
- Grayscale.ctb—plots the drawing using shades of gray.
- Screening files—plot the drawing using faded, or screened, colors.

When using named plot style tables, it is common to select one of the following:

- Autodesk-Color.stb—provides access to named plot styles for plotting the drawing using solid and faded colors.

Figure 29-22.
The **Plot Styles** window lists all available plot style files and allows you to create and edit plot style tables.

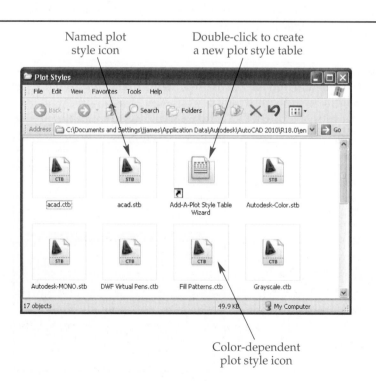

Named plot style icon

Double-click to create a new plot style table

Color-dependent plot style icon

Figure 29-23.
Use the **Plot
style table (pen
assignments)** area
to select, create, and
edit plot style tables.

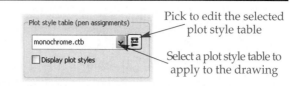

Pick to edit the selected
plot style table

Select a plot style table to
apply to the drawing

- Monochrome.stb—references a named plot style for plotting the drawing in monochrome.
- Autodesk-MONO.stb—provides access to named plot styles for plotting objects monochrome, in color, and in faded monochrome.

PROFESSIONAL TIP

You cannot select named plot style tables if you started the drawing with color-dependent plot style tables. Conversely, you cannot select color-dependent plot style tables if you started the drawing with named plot style tables. You can type CONVERTPSTYLES to access the **CONVERTPSTYLES** tool, which allows you to switch between table modes in a drawing. However, you should avoid converting plot style tables when possible. Instead, use the appropriate plot style type when beginning a new drawing.

Exercise 29-5

Access the Student Web site (www.g-wlearning.com/CAD) and complete Exercise 29-5.

Applying Plot Styles

After you select a plot style table from the drop-down list in the **Plot style table (pen assignments)** area of the **Page Setup** dialog box, the plot styles contained in the selected plot style table are ready to assign to objects in the drawing. When you select a color-dependent plot style table, plot styles are automatically applied to objects in the drawing according to the color of the objects. No additional steps are required to apply color-dependent plot styles to objects.

When you select a named plot style table, the named plot styles contained in the table are ready to assign to layers or individual objects. Any layer or object assigned a named plot style plots using the settings specified for that style. Named plot style tables contain as many plot styles as have been created. For example, the monochrome.stb plot style table contains the default Normal plot style and a style named Style 1. When you apply the Normal plot style to layers or objects, objects plot exactly as they appear on-screen. Style 1 assigns the color Black to all layers or objects that use Style 1 for plotting. In order to plot all objects in the drawing black, even if the objects display different colors on-screen, you must apply Style 1 to all layers.

Assign named plot styles to layers in the **Layer Properties Manager** palette. See **Figure 29-24.** Identify the layer to assign a named plot style and select the default plot style, such as **Normal,** from the **Plot Style** column. The **Select Plot Style** dialog box appears. See **Figure 29-25.** Select a named style to assign it to the highlighted layer. Any object drawn on a layer assigned to a named plot style plots using the settings in the named plot style.

Figure 29-24.
Assign named plot styles to layers using the **Layer Properties Manager**.

Pick plot style name for layer
to select a different plot style

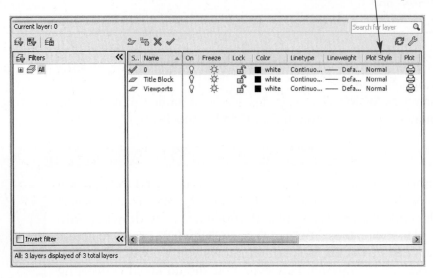

Figure 29-25.
Use the **Select Plot Style** dialog box to select a plot style to assign to a layer.

Pick plot style
for layer from list
of plot styles in
plot style table

Current plot
style table

Identifies
current
layout tab

Access **Plot Style
Table Editor**
dialog box

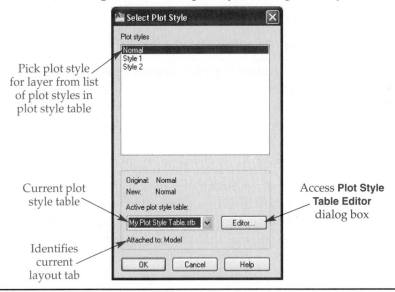

You can also assign named plot styles to individual objects. If you assign a plot style to an object drawn on a layer given a named plot style, the plot style assigned to the object overrides the plot style settings given to the layer. Assign plot styles to objects using the **Properties** palette or the **Plot Style Control** drop-down list in the **Properties** panel on the **Home** ribbon tab. See **Figure 29-26.**

NOTE

If the current drawing is set to use a named plot style table, the default plot style for all layers is Normal. The default plot style for all objects is ByLayer. Objects plotted with these settings keep their original properties.

Figure 29-26.
You can assign a plot style to individual objects using A— the **Properties** palette or B—the **Plot Style Control** drop-down list in the ribbon.

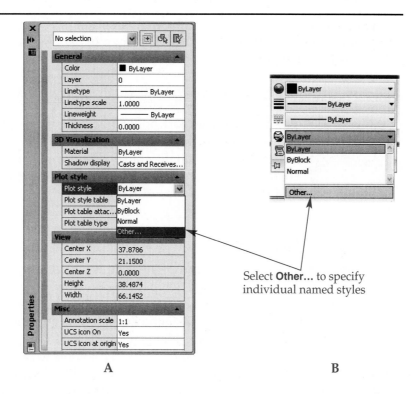

Select **Other...** to specify individual named styles

A   B

Exercise 29-6

Access the Student Web site (www.g-wlearning.com/CAD) and complete Exercise 29-6.

Viewing Plot Style Effects On-Screen

To display plot style effects on-screen, pick the **Display plot styles** check box in the **Plot style table** area in the **Page Setup** dialog box. Objects on-screen appear as they will plot. Typically, it is not appropriate to work with objects displayed as they will plot, especially if you print in monochrome. In this example, the color assigned to a layer appears black, which defeats the purpose of assigning unique colors to layers. A better practice is to use the preview feature of the **Page Setup** or **Plot** dialog box, as described later in this chapter, to preview the effects of plot styles before plotting.

Other Plotting Options

The **Plot options** area of the **Page Setup** dialog box contains additional plot options that affect how specific items plot. If you plan to plot objects using a plot style table, be sure to select the **Plot with plot styles** check box. The main purpose of this check box is to toggle the use of plot styles on and off. This allows you to create a plot quickly with or without using plot styles.

When you deselect the **Plot with plot styles** check box, the **Plot object lineweights** check box is enabled and selected by default. When this box is checked, all objects with a lineweight greater than 0 plot using the assigned lineweight. Deselect the check box to plot all objects using a 0, or thin, lineweight.

Pick the **Plot paperspace last** check box to plot paper space objects after model space objects. This option ensures that objects in paper space that overlap objects in model space plot on top of model space objects. The **Plot paperspace last** check box is disabled when you plot in model space, because model space does not include paper space objects. Select the **Hide paperspace objects** check box to remove hidden lines from 3D objects created in paper space. This option is only available when you plot from a layout tab and affects only objects drawn in paper space. It does not affect any 3D objects in a viewport.

NOTE

The **Shaded viewport options** area of the **Page Setup** dialog box provides settings that control viewport shading. These options set the type and quality of shading for plotting 3D models from a shaded or rendered viewport. 3D models are explained in *AutoCAD and Its Applications—Advanced*.

Completing Page Setup

After you have selected all appropriate page setup options, pick the **Preview** button in the lower-left corner of the **Page Setup** dialog box to preview the effects of the page setup. What you see on-screen is the exact plot appearance, assuming you use a color plotter to make color prints. The **Realtime Zoom** tool is activated automatically in the preview window. Additional view tools are available from the toolbar near the top of the window or from a shortcut menu. Use these tools to help confirm that the plot settings are correct. When you finish previewing the plot, press [Esc] or [Enter], or right-click and select **Exit** to return to the **Page Setup** dialog box. Pick the **OK** button to exit the **Page Setup** dialog box, and pick **Close** button to exit the **Page Setup Manager**.

Exercise 29-7

Access the Student Web site (www.g-wlearning.com/CAD) and complete Exercise 29-7.

Chapter Test

Answer the following questions. Write your answers on a separate sheet of paper or go to the Student Web site (www.g-wlearning.com/CAD) and complete the electronic chapter test.

1. What is a model?
2. Define *model space*.
3. How is a layout used?
4. What is paper space?
5. What is the purpose of a floating viewport?
6. What is the purpose of the dashed rectangle that appears on the default layout?
7. If you pick the **MODEL** button to enter paper space and the file contains multiple layouts, none of which has previously been accessed, which layout displays by default? Which layout displays if a different layout has already been opened?
8. Briefly describe the function of the **Quick View Layouts** tool.
9. What does the **Quick View Drawings** tool allow you to do?
10. What is a page setup?
11. Briefly describe the basic function of the **Page Setup Manager** and the related **Page Setup** dialog box.
12. Briefly describe the function of the **Plot** dialog box and explain when it is used in relation to the **Page Setup Manager** and **Page Setup** dialog box.

13. Explain the importance of a well-defined page setup.
14. How can you identify the page setup that is tied to the current layout?
15. What is a plot device?
16. Define *sheet size*.
17. What factors should you consider when you select a sheet size for a drawing?
18. Which ANSI standards specify sheet sizes?
19. What is the plot offset of a layout?
20. Define *drawing scale*.
21. What are plot styles?
22. Briefly explain the purpose of a plot style table.
23. How are plot styles assigned in a color-dependent plot style?
24. Briefly explain the function of a named plot style table.
25. How can you be certain that your page setup options will produce the desired plot?

Drawing Problems

Start AutoCAD if it is not already started. Follow the specific instructions for each problem.

▼ Basic

1. Use a word processor to list five items commonly found in a layout. Provide a brief description of each item.

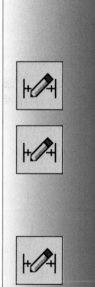

2. Start a new drawing using the acad.dwt template. Create a plot style table named Black35mm.cbt that will plot all colors in the AutoCAD drawing in black ink on the paper, with a lineweight of 0.35 mm. (*Hint:* To make the same change to a property of all the plot styles, select the first plot style in the list, in this case Color 1, then scroll to the end of the list, hold down the [Shift] key, and select the last plot style in the list, in this case Color 255). Save the drawing as P29-2.

3. Start a new drawing using the acad -Named Plot Styles.dwt template. Create a plot style table named BlackShades.stb. Create the following plot styles:
   • Black100% with color set to black, all other properties set to their default values.
   • Black50% with color set to black, screening set to 50, all other properties set to their default values.
   • Black25% with color set to black, screening set to 25, all other properties set to their default values.
   Save the drawing as P29-3.

▼ Intermediate

4. Open P8-14 and save as P29-4. The P29-4 file should be active. Delete the default **Layout2**. Create a new B-size sheet layout by following these steps:
   A. Rename the default **Layout1** to **B-SIZE**.
   B. Select the **B-SIZE** layout and access the **Page Setup Manager**.
   C. Modify the **B-SIZE** page setup according to the following settings:
      - **Printer/Plotter:** Select a printer or plotter that can plot a B-size sheet
      - **Paper size:** Select the appropriate B-size sheet (varies with printer or plotter)
      - **Plot area:** Layout
      - **Plot offset:** 0,0
      - **Plot scale:** 1:1 (1 inch = 1 unit)
      - **Plot style table:** monochrome.ctb
      - **Plot with plot styles**
      - **Plot paperspace last**
      - Do not check **Hide paperspace objects**
      - **Drawing orientation:** Select the appropriate orientation (varies with printer or plotter)

      Resave P29-4.

5. Open P8-15 and save as P29-5. The P29-5 file should be active. Delete the default **Layout2**. Create a new A2-size sheet layout according to the following steps:
   A. Rename the default **Layout1** to **A2-SIZE**.
   B. Select the **A2-SIZE** layout and access the **Page Setup Manager**.
   C. Modify the **A2-SIZE** page setup according to the following settings:
      - **Printer/Plotter:** Select a printer or plotter that can plot an A2-size sheet
      - **Paper size:** Select the appropriate A2-size sheet (varies with printer or plotter)
      - **Plot area:** Layout
      - **Plot offset:** 0,0
      - **Plot scale:** 1:1 (1 mm = 1 unit)
      - **Plot style table:** monochrome.ctb
      - **Plot with plot styles**
      - **Plot paperspace last**
      - Do not check **Hide paperspace objects**
      - **Drawing orientation:** Select the appropriate orientation (varies with printer or plotter)

      Resave P29-5.

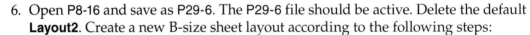

6. Open P8-16 and save as P29-6. The P29-6 file should be active. Delete the default **Layout2**. Create a new B-size sheet layout according to the following steps:
   A. Rename the default **Layout1** to **B-SIZE**.
   B. Select the **B-SIZE** layout and access the **Page Setup Manager**.
   C. Modify the **B-SIZE** page setup according to the following settings:
      - **Printer/Plotter:** Select a printer or plotter that can plot a B-size sheet
      - **Paper size:** Select the appropriate B-size sheet (varies with printer or plotter)
      - **Plot area:** Layout
      - **Plot offset:** 0,0
      - **Plot scale:** 1:1 (1 inch = 1 unit)
      - **Plot style table:** monochrome.ctb
      - **Plot with plot styles**
      - **Plot paperspace last**
      - Do not check **Hide paperspace objects**
      - **Drawing orientation:** Select the appropriate orientation (varies with printer or plotter)

      Resave P29-6.

▼ Advanced

7. Draw the wood beam details using the dimensions and notes provided. Establish the missing information using your own specifications, or determine the correct size of items not dimensioned. Prepare a single layout for plotting on a C-size sheet, using the monochrome.ctb plot style. Delete all other layouts. Save the drawing as P29-7.

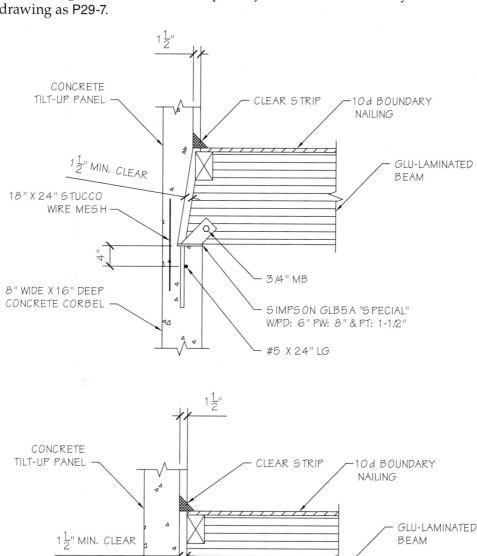

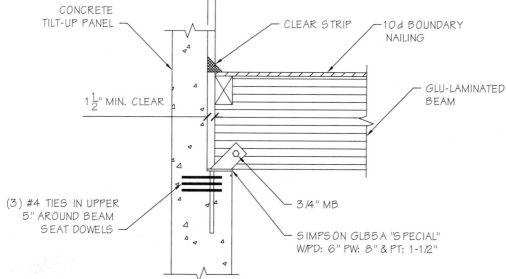

8. Use a word processor to write a report of approximately 250 words explaining the difference between model space and paper space and describing the importance of using layouts. Cite at least three examples from actual industry applications of using layouts to prepare a multisheet drawing. Use at least four drawings to illustrate your report.

9. Create the multiview drawing using the dimensions provided. Prepare a single layout for plotting on a C-size sheet, using the monochrom.ctb plot style. Delete all other layouts. Save the drawing as P29-9.

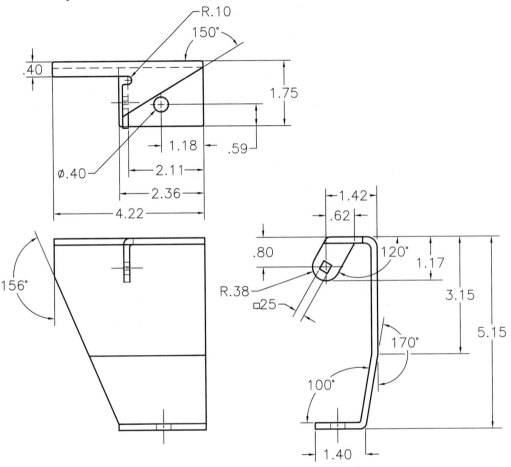

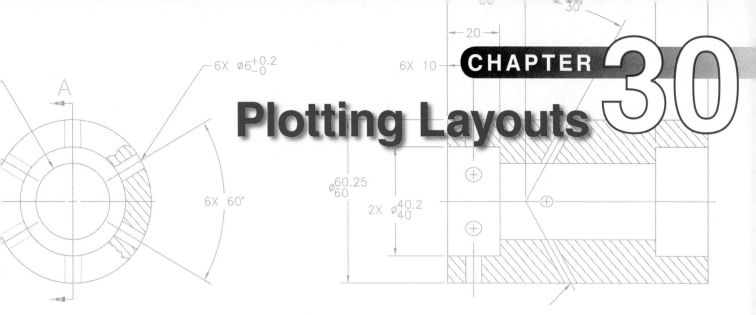

Plotting Layouts

Learning Objectives

After completing this chapter, you will be able to do the following:

✓ Add layout content.
✓ Use floating viewports.
✓ Create properly scaled final drawings.
✓ Preview and plot layouts.

This chapter explores the additional steps to complete layout setup. You will learn to add content to a layout and place and use floating viewports. You will also use the **PLOT** tool to preview the plot and send the layout to a printer or plotter.

Layout Content

Model space provides an environment to create drawing views and add dimensions and annotations directly to views. Layouts provide an effective method to display model space content and add items such as a border, title block, revision block, general notes, and a bill of materials, parts list, or schedules. Layouts provide flexibility for laying out, scaling, and preparing a final drawing. Consider the objects placed in model space to be *drawing* content and the objects you add to a layout to be *sheet* content. A complete drawing forms when you bring sheet and drawing content together. See **Figure 30-1.**

Drawing in Paper Space

A layout is a representation of a flat piece of paper. As a result, paper space is a 2D drawing environment. Most 2D drawing and editing tools and options described throughout this textbook function the same in paper space as in model space. Some tools, however, are specific to or are most often used in either model space or paper space. For example, the **Full Navigation Wheel** appears by default when you access the **SteeringWheel** view tool in model space. The **2D Navigation Wheel** appears in paper space and is specific to 2D drafting.

Figure 30-1.
Dimensioned drawing views created in model space, combined with a border, title block, revision block, and general notes created in paper space, form the final drawing.

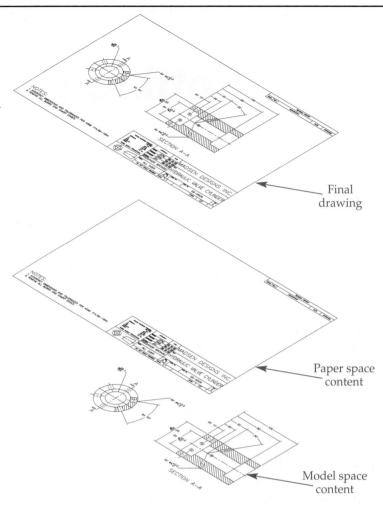

Final drawing

Paper space content

Model space content

NOTE

Although paper space is a 2D environment, you can display 3D models created in model space in paper space floating viewports.

As in model space, you typically draw geometry on a layout at full scale. One difference between model space and paper space is that objects in model space may be very large or very small, while paper space objects always correspond to the sheet size. Draw all layout content using the actual size you want the objects to appear on the plotted sheet. Add layout content such as general notes, view titles, and similar annotations as multiline or single-line text or as blocks with attributes. Place a bill of materials, parts list, schedule, or similar tabular information using the **TABLE** tool or blocks with attributes. You typically create most other items, such as a border, title block, and revision block as blocks, often with attributes, and then insert these items into the layout.

When you draw in paper space, use layers appropriate for the layout and the objects added to the layout. You may want to create a single layer named SHEET, for example, on which you draw all layout content. Another option is to use layers specific to layout items, such as a BORDER layer for the border and a TITLE or A-ANNO-TTBL layer for the title block. A layer named Viewport, VPORT, or A-ANNO-NPLT is typically assigned to floating viewports so the viewport boundary can be turned off, frozen, or set to "no plot" before plotting.

NOTE

The default floating viewport boundary assumes the layer that is current when you first access a layout. If you use the default viewport, it may be necessary to change the layer on which it is drawn.

The Layout Origin

Drawing and editing in paper space is most often done on the sheet, which is the white rectangle you see on the gray background. However, it is possible, and necessary in some applications, to create objects off the sheet. When drawing and editing on a layout, remember that the origin (0,0) is controlled by the X and Y values you enter in the **Plot offset** area of the **Page Setup** dialog box. For example, if you draw a line with a start point of (1,1), the line begins 1 unit to the left and 1 unit up from the plot origin. The default origin position is at the lower-left corner of the printable area. See **Figure 30-2.** Refer to Chapter 29 for more information on plot origin and the **Plot offset** area of the **Page Setup** dialog box.

Layout content often references the edge of the sheet. For example, you might position a border 1/2″ inside of the sheet edge. In this situation, it is usually best to define the layout origin as the lower-left corner of the sheet. The best option is to select the **Edge of paper** radio button in the **Specify plot offset relative to** area in the **Plot and Publish** tab of the **Options** dialog box. In the **Plot offset** area of the **Page Setup** dialog box, use values of 0.000 in the X and Y offset text boxes to locate the plot origin at the lower-left corner of the sheet. This is an excellent way to center the drawing on the sheet.

NOTE

Depending on the specific plot configuration, you may need to pick the **Plot upside-down** check box in the **Drawing orientation** area of the **Page Setup** dialog box in addition to using the **Edge of paper** offset option. Draw an object to see if (0,0) is actually located exactly at the lower-left corner of the sheet. If not, pick the **Plot upside-down** check box to solve the problem.

PROFESSIONAL TIP

Defining the plot offset relative to the edge of the paper has the benefit of maintaining the plot offset location from the sheet edge even when you use a different plotter or plot configuration, as long as you use the same sheet size.

A second option to relate content to the lower-left corner of the sheet is to select the **Printable area** radio button in the **Specify plot offset relative to** area in the **Plot and Publish** tab of the **Options** dialog box. Then identify the values of the lower-right corner of

Figure 30-2.
The default location of the plot origin is often not appropriate, because all point entry is in reference to the lower-left corner of the printable area. Change the location of the plot origin to define the lower-left corner of the sheet as the (0,0) point.

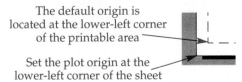

The default origin is located at the lower-left corner of the printable area

Set the plot origin at the lower-left corner of the sheet

the printable area. This information is available in the **Device and Document Settings** tab of the plotter's **Configuration Editor**. Next, in the **Plot offset** area of the **Page Setup** dialog box, change the X and Y plot offset values to the values of the lower-left corner of the printable area. The values you enter must be negative. The origin is offset from the printable area, which means if you use a different plotter or plot configuration, the printable area may change, causing the offset to shift.

Exercise 30-1

Access the Student Web site (www.g-wlearning.com/CAD) and complete Exercise 30-1.

Using Floating Viewports

The primary advantage of using floating viewports in a paper space layout is the ability to prepare scaled drawings without increasing or decreasing the actual size of drawing views or sheet content. You can also create multiple viewports on a single layout to show uniquely scaled or alternate drawing views. For example, a single sheet might contain a floor plan drawn at a 1/4″ = 1′-0″ scale, an eave detail drawn at a 3/4″ = 1′-0″ scale, and a foundation detail drawn at a 3/4″ = 1′-0″ scale. See **Figure 30-3.**

A floating viewport boundary is the portion of the viewport that you see. Everything inside the viewport is "showing through" from model space. Use tools such as **MOVE**, **ERASE**, **STRETCH**, and **COPY** to modify the viewport boundary in

Figure 30-3.
An example of a layout, ready to plot, that includes three floating viewports used to display drawing views at different scales. The layer on which the viewport is drawn is off or frozen for plotting.

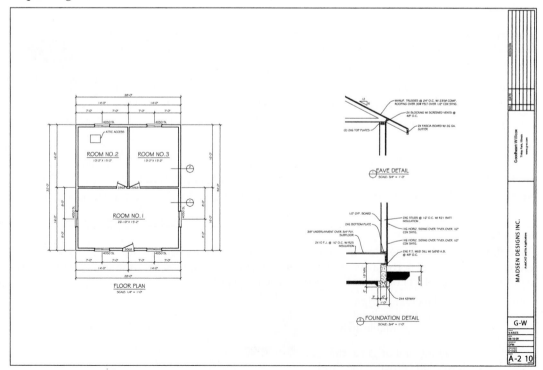

AutoCAD and Its Applications—Basics

paper space. Display tools such as **VIEW**, **PAN**, and **ZOOM** modify the display of model space objects in the floating viewport. Additional options are available for adjusting how objects appear, according to the layers on which you draw objects. This allows you to define how the drawing appears within the viewport. Be sure a layout tab is current as you work through the following sections describing floating viewports.

PROFESSIONAL TIP

When you first select a layout and enter paper space, a rectangular floating viewport appears around objects in model space. As part of layout setup, you may want to use the **ERASE** tool to erase the default viewport and draw a new viewport (or several viewports). Erasing the default viewport, or erasing everything in the layout, does not erase objects in model space.

Creating Floating Viewports

You can use a variety of techniques to create new floating viewports in paper space. The **Viewports** dialog box provides options for creating one to four new viewports according to a specific viewport configuration. The **MVIEW** tool is a text-based method for creating new viewports. The **MVIEW** tool provides the same options found in the **Viewports** dialog box, plus additional viewport definition options. You can also access many of the methods for creating new viewports directly from the **Viewports** panel on the **View** ribbon tab.

Using the Viewports Dialog Box

The **Viewports** dialog box, shown in **Figure 30-4**, looks and functions the same in paper space as in model space, except for a few differences. One difference is that the **Viewport spacing:** text box replaces the **Apply to:** drop-down list in the **New Viewports** tab in paper space. This text box is available only when you select a standard viewport configuration that places two or more viewports, as shown in **Figure 30-4**. Enter a value to define the space between multiple viewports.

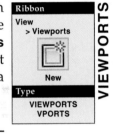

Ribbon
View
> Viewports
New
Type
VIEWPORTS
VPORTS

VIEWPORTS

Figure 30-4.
The **Viewports** dialog box with the **New Viewports** tab selected. The **Viewport Spacing:** setting is available when paper space is active.

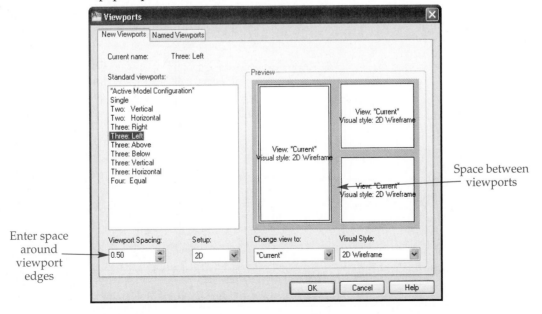

Enter space around viewport edges

Space between viewports

Another difference is that when you pick the **OK** button to create floating viewports, the viewports do not automatically appear, as in model space. Instead, you must specify a first and second corner to define the area occupied by the viewport configuration. See **Figure 30-5.** If you use the **Fit** option, AutoCAD fits the viewport(s) into the printable area without requiring you to pick points. Refer to Chapter 6 for more information about the **Viewports** dialog box.

NOTE

You can also create new floating viewports by selecting a specific configuration from the Application Menu. Select **View > Viewports** and then select the **1 Viewport**, **2 Viewports**, **3 Viewports**, or **4 Viewports** option and follow the prompts to create the appropriate configuration.

Exercise 30-2

Access the Student Web site (www.g-wlearning.com/CAD) and complete Exercise 30-2.

MVIEW

| Type |
| MVIEW |
| MV |

Using the MVIEW Tool

Once you access the **MVIEW** tool, you can create a single viewport by selecting opposite corners of the viewport, press [Enter] or the space bar, or right-click and choose **Enter** to activate the **Fit** option. The **Fit** option creates a viewport that fills the printable area. The **2**, **3**, and **4** options provide preset viewport configurations similar to those available from the **Viewports** dialog box. The **Fit** function is available when you create multiple viewports. The **Shadeplot** option includes the same options available from the **Visual Style** drop-down list in the **Viewports** dialog box, which is explained in *AutoCAD and Its Applications—Advanced*.

Figure 30-5.
Enter or select two points on the layout to specify the area filled by the viewport configuration. Notice the viewport spacing that forms as specified in the **Viewports** dialog box.

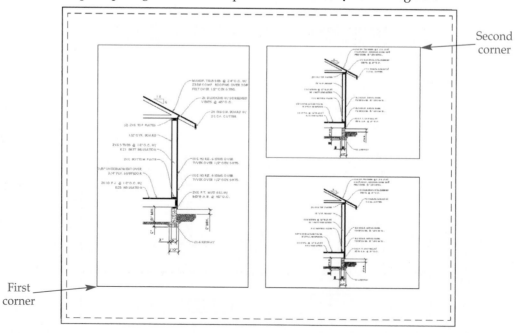

Choose the **Restore** option to convert a saved viewport configuration into individual floating viewports. This option is typically used to convert tiled viewports into floating viewports. For example, if model space displays two tiled viewports, use the **Restore** option to create two floating viewports. See **Figure 30-6.** After you access the **Restore** option, enter the viewport configuration name. In the previous example, you would select the **Active** option to use the two tiled viewports from model space. This is the active model space viewport configuration. Next, select opposite corners of the viewport, press [Enter] or the space bar, or right-click and choose **Enter** to activate the **Fit** option. The same model space tiled viewport configuration now displays in paper space as floating viewports.

Forming Polygonal Floating Viewports

The most common shape for a floating viewport is rectangular, which is suitable for many applications. As an alternative, use the **Polygonal** option of the **MVIEW** tool to form a polygonal floating viewport boundary. This option is easily accessed from the ribbon. Construct a polygonal viewport using the same techniques as when drawing a closed polyline object. The viewport can be any closed shape composed of lines and arcs. **Figure 30-7** shows an example of a polygonal floating viewport used to define the maximum drawing view area 1/2″ in from the border and title block.

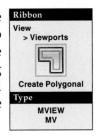

Ribbon

View
> Viewports

Create Polygonal

Type

MVIEW
MV

Figure 30-6.
A—A **Two: Horizontal** tiled viewport configuration in model space. B—The model space tiled viewports converted to floating viewports using the **Restore** option of the **MVIEW** tool.

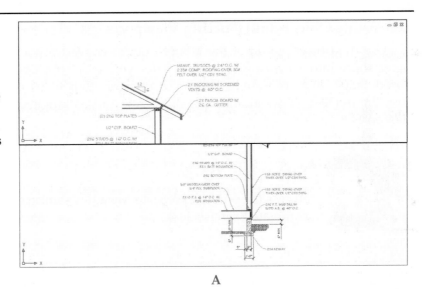

A

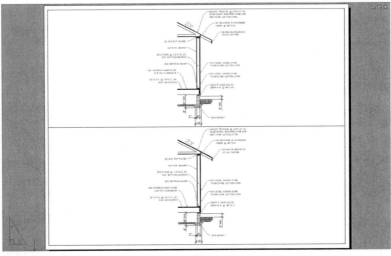

B

Figure 30-7.
An example of a polygonal floating viewport. The layer on which the viewport is drawn is turned off or frozen for plotting.

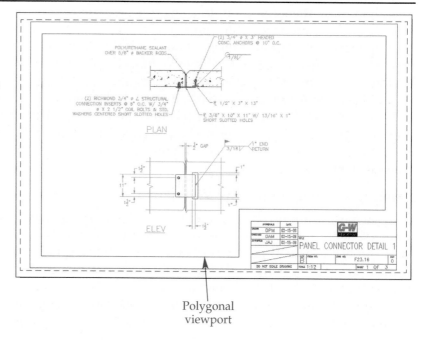

Polygonal viewport

Converting Objects into Floating Viewports

Use the **Object** option of the **MVIEW** tool to convert any closed object drawn in paper space into a floating viewport. After you select the **Object** option, select any closed shape, such as a circle, ellipse, or polygon, to convert the object into a viewport. **Figure 30-8** shows an example of a circle and rectangle converted into floating viewports.

Exercise 30-3

Access the Student Web site (www.g-wlearning.com/CAD) and complete Exercise 30-3.

Figure 30-8.
You can convert any closed object into a floating viewport. The layer on which these particular viewports are drawn remains on and thawed for plotting.

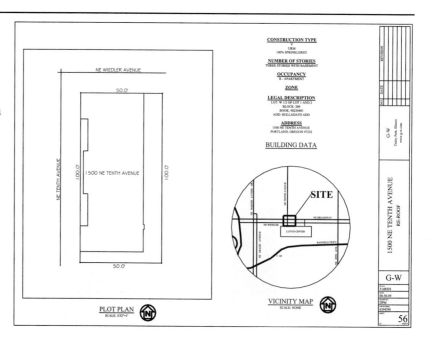

AutoCAD and Its Applications—Basics

Adjusting the Floating Viewport Boundary

For purposes of adjusting a floating viewport boundary, you should consider the boundary a closed object. For example, treat rectangular viewports like rectangles, polygonal viewports like closed polyline objects, circular viewports like circles, and elliptical viewports like ellipses. Use grips or tools such as **MOVE**, **ERASE**, **STRETCH**, and **COPY** as needed to modify the size, shape, and location of floating viewports. When you adjust a floating viewport, the "hole" cut through the sheet changes.

Exercise 30-4

Access the Student Web site (www.g-wlearning.com/CAD) and complete Exercise 30-4.

Clipping Viewports

The **VPCLIP** tool allows you to redefine the boundary of an existing viewport. To access the tool from a shortcut menu, select a viewport and then right-click and pick **Viewport Clip**. You can clip a floating viewport to an existing closed object that you draw before accessing the **VPCLIP** tool, or you can clip the viewport to a polygonal shape that you create while using the tool.

After you access the **VPCLIP** tool, select the viewport to clip. Then select an existing closed shape, such as a circle, ellipse, or polygon, to recreate the viewport in the shape of the selected object. See **Figure 30-9**. An alternative, once you select the existing closed shape, is to use the **Polygonal** option to redefine the viewport to a polygonal shape. This option functions the same as the **Polygonal** option of the **MVIEW** tool, except the existing viewport transforms into the new shape.

AutoCAD recognizes a clipped floating viewport as clipped. The **VPCLIP** tool offers a **Delete** option when you select a clipped viewport. Use the **Delete** option to remove the clipped definition and convert the shape into a rectangle sized to fit the extents of the original clipping object or polygonal shape.

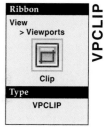

Ribbon
View
> Viewports

Clip

Type
VPCLIP

Figure 30-9.
An example of clipping a viewport to an existing rectangle. The original viewport is removed and the rectangle converts into a viewport.

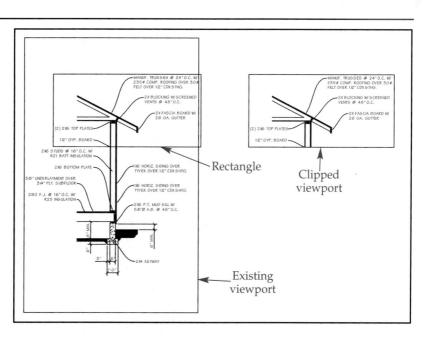

Rotating Model Space Content

The **VPROTATEASSOC** system variable setting determines what happens to model space content when you rotate a floating viewport. By default, the variable uses a value of 1. As a result, when you rotate a viewport, objects shown in model space rotate to align with the viewport. See **Figure 30-10A.** The orientation of objects in model space does not change. In order to maintain the original alignment of model space content in a floating viewport, as shown in **Figure 30-10B,** access the **VPROTATEASSOC** system variable *before* rotating the viewport, and enter a value of 0.

Figure 30-10.
A—The default **VPROTATEASSOC** setting of 1 rotates model space content on a floating viewport to align with viewport rotation. B—Change the value to 0 to maintain the original model space orientation when you rotate a viewport.

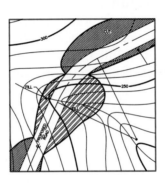
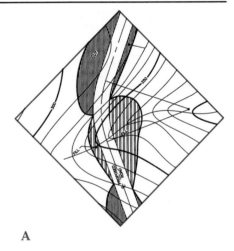

A

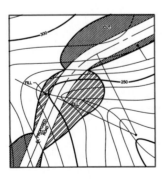
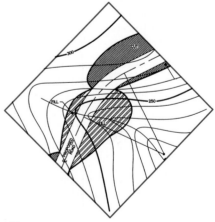

B

Other tools and options, such as **UCS** and **MVSETUP**, include options for rotating items shown through a viewport. The **VPROTATEASSOC** system variable automates the process. The entire display of model space rotates with the rotation of the viewport. Drawing characteristics, such as unidirectional dimensions, do not update according to the rotation.

Activating and Deactivating Floating Viewports

Activate a floating viewport to work with model space objects while in paper space. This allows you to adjust the display of the model space drawing shown in the viewport. Repeat the process of activating and adjusting a viewport for every floating viewport in the layout to achieve the final drawing.

To activate a floating viewport, double-click inside the viewport area, press the **PAPER** button on the status bar, or type MSPACE or MS. If the layout contains a single viewport, the viewport appears highlighted, indicating that it is current. On the layout, the UCS icon disappears, and the model space UCS icon displays in the corner of each layout viewport. You are now working directly in model space, through the paper space viewport. The active and highlighted viewport is the viewport you double-click on or the newest viewport depending on how you access the **MSPACE** tool. See **Figure 30-11**. To make a different viewport active, pick once inside the viewport.

After you adjust the display of all floating viewports, you must re-enter paper space to plot and continue working with the layout. To activate paper space, double-click outside the viewport area, press the **MODEL** button on the status bar, or type PSPACE or PS. The layout space UCS icon reappears, and the model space UCS icon disappears from the corners of the viewports.

Figure 30-11.
The active viewport appears highlighted and allows you to work in model space while AutoCAD displays paper space.

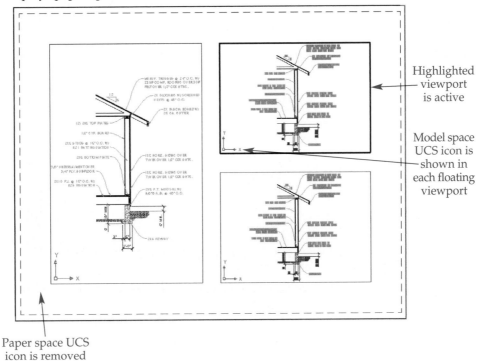

Highlighted viewport is active

Model space UCS icon is shown in each floating viewport

Paper space UCS icon is removed

Scaling a Floating Viewport

The scale you assign to a floating viewport is the same as the drawing scale. The quickest way to set viewport scale is to activate a viewport or pick a viewport boundary in paper space, *without* activating the viewport. Then select the appropriate scale from the **Viewport Scale** flyout on the status bar. See **Figure 30-12.** An alternative is to pick the viewport to scale in paper space and access the **Properties** palette. Then choose a viewport scale from the **Standard scale** drop-down list.

If a certain scale is not available from the **Viewport Scale** or **Standard scale** list, or if you want to change existing scales, pick the **Annotation Scale** flyout on the status bar and choose **Custom...** to access the **Edit Scale List** dialog box. From this dialog box, you can move the highlighted scale up or down in the list by picking the **Move Up** or **Move Down** button. To remove the highlighted scale from the list, pick the **Delete** button.

Select the **Edit...** button to open the **Edit Scale** dialog box. Here you can change the name of the scale and adjust the scale by entering the paper and drawing units. For example, a scale of 1/4″ = 1′-0″ uses a paper units value of .25 or 1 and a drawing units value of 12 or 48.

To create a new annotation scale, pick the **Add...** button to display the **Add Scale** dialog box, which functions the same as the **Edit Scale** dialog box. Pick the **Reset** button to restore the default annotation scale. When the annotation scale is set current, you are ready to type annotative text that automatically displays at the correct text height according to the drawing scale.

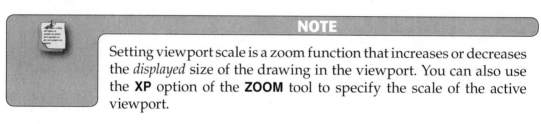

NOTE

Setting viewport scale is a zoom function that increases or decreases the *displayed* size of the drawing in the viewport. You can also use the **XP** option of the **ZOOM** tool to specify the scale of the active viewport.

Figure 30-12.
Using the **Viewport Scale** flyout button on the status bar to set the drawing scale.

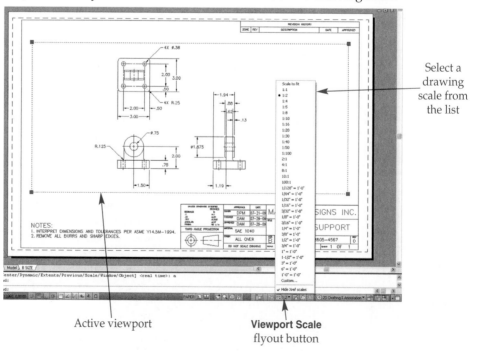

Select a drawing scale from the list

Active viewport

Viewport Scale
flyout button

If you use an option of the **ZOOM** tool other than a specific **XP** value to adjust the drawing inside an active floating viewport, the drawing loses the correct scale. Once the viewport scale is set, do not zoom in or out. Lock the viewport, as described later in this chapter, to help ensure that the drawing remains properly scaled.

Scaling Annotations

You should always draw objects at their actual size, or full scale, in model space, regardless of the size of the objects. However, this method requires special consideration for annotations, hatches, and similar items added to objects in model space. You can adjust the appearance of these items manually, but most often, it is best to use annotative objects to automate the process. Scaled viewports and annotative objects function together to scale drawings properly and increase multiview drawing flexibility. Chapter 31 explains annotative objects.

Controlling Linetype Scale

The **LTSCALE** system variable allows you to make a global change to the linetype scale to increase or decrease the lengths of the dashes and spaces found in some linetypes. You must often modify the **LTSCALE** value to make linetypes match standard drafting practices. However, depending on the size of objects in model space and the specified floating viewport scale, an **LTSCALE** value in model space may not be appropriate for paper space.

For example, an **LTSCALE** value of .5 is appropriate for a U.S. customary mechanical drawing plotted at full scale. In this example, apply an **LTSCALE** value of .5 to model space and paper space because both environments function at full scale. If you scale the drawing to 2:1, a linetype scale of .25 (scale factor of 1/2 × **LTSCALE** value of .5 = .25) is needed in model space and paper space in order for lines to appear correct in both environments. By default, AutoCAD calculates the appropriate linetype scale display in model space and paper space according to the **LTSCALE** setting.

The **CELTSCALE**, **PSLTSCALE**, and **MSLTSCALE** system variables control how the **LTSCALE** system variable applies, or does not apply, to linetypes in model space and paper space. The **CELTSCALE**, **PSLTSCALE**, and **MSLTSCALE** system variables are set to 1 by default, and should be set to 1 in order for the **LTSCALE** value to apply correctly in model space and paper space. All linetypes will then appear with the same lengths of dashes and dots regardless of the floating viewport scale, and no matter whether you are in paper space or model space.

Using the previous example, lines will appear correctly in model space and at a scale of 1:1 and 2:1 in paper space. However, when you scale a floating viewport or change the annotation scale in model space, you must remember to use the **REGEN** tool to regenerate the display. Otherwise, the linetype scale will not update according to the new scale. The **MSLTSCALE** system variable is associated with the selected annotation scale and is further described in Chapter 31.

Adjusting the View

When you first create a floating viewport, AutoCAD performs a **ZOOM Extents** to display everything in model space through the viewport. The **Scale to fit viewport scale** option accomplishes the same task. When you scale a viewport, AutoCAD adjusts the view from the center of the viewport. This is often the appropriate display. However, if you change the size or shape of the viewport, if a centered view is not appropriate, or if you want to display a specific portion of the drawing, you must adjust the view. Use the **PAN** tool in an active viewport to redefine the location of the view.

Viewport edges can "cut off" a scaled model space drawing. This may be acceptable to display a portion of a view. However, to display the entire view, you can increase the size of the viewport boundary or select a different scale to reduce the displayed size of the view to fit the viewport. If it is not appropriate to increase the size of the viewport or decrease the scale, use a larger sheet size. **Figure 30-13** shows a drawing with two viewports. One shows everything in model space, the other "cuts off" model space objects and displays objects at a higher zoom level to create a detail.

Locking and Unlocking Floating Viewports

Once you adjust the drawing in the viewport to reflect the proper scale and view, lock the viewport so the scale or view orientation does not accidentally change. This allows you to use display tools such as **ZOOM** and **PAN** to aid in working with objects in model space without changing the scale or position of the view.

The quickest way to lock or unlock a viewport is to select a viewport in paper space and right-click. From the **Display Locked** cascading submenu, select **Yes** to lock the viewport or select **No** to unlock the viewport. A second option is to select a viewport in paper space and access the **Properties** palette. From the **Display Locked** drop-down list, select **Yes** to lock the viewport or select **No** to unlock the viewport. You can also use the **Lock** option of the **MVIEW** tool. Follow the prompts to select the viewport(s) to lock or unlock.

Exercise 30-5

Access the Student Web site (www.g-wlearning.com/CAD) and complete Exercise 30-5.

Controlling Layer Display

Layers function the same in paper space as in model space. The **On, Freeze, Color, Linetype, Lineweight, Plot Style,** and **Plot** settings described throughout this textbook are *global layer settings*. **On, Freeze,** and **Plot** are global layer states. **Color, Linetype, Lineweight,** and **Plot Style** are global layer properties. Changing a global layer setting

global layer settings: Layer settings applied to both model space and paper space.

Figure 30-13.
This drawing shows examples of when it is appropriate to show all model space objects, and when it is necessary to display only a portion of model space. Use the **PAN** tool to adjust the position of model space objects in floating viewports.

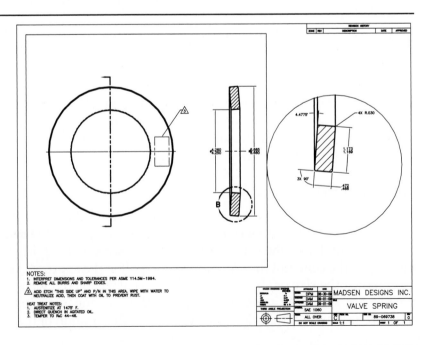

affects objects drawn in model space and paper space. For example, if you change the color of a layer in model space and lock the layer, all objects drawn on that layer in paper space also change color and become locked.

AutoCAD provides the option to freeze layers in a floating viewport and apply *layer property overrides.* These features expand the function of the layer system and improve your ability to reuse drawing content.

Use the **LAYER** tool and the corresponding **Layer Properties Manager** palette to control layer display in floating viewports. See **Figure 30-14.** This is the same palette used to manage layers throughout this textbook. The **NEW VP Freeze, VP Freeze, VP Color, VP Linetype, VP Lineweight,** and **VP Plot Style** columns control layer display options for floating viewports. Except for the **NEW VP Freeze** column, these columns appear only in layout mode. You probably need to use the scroll bar at the bottom of the palette to see the columns. The options can apply to layout content, such as the viewport boundary. However, layer settings typically apply to an active floating viewport. Be sure the floating viewport to which you want to apply layer control settings is active as you work through the following sections.

layer property overrides: Color, linetype, lineweight, and plot style properties applied to specific viewports in paper space.

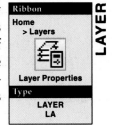

Ribbon
Home
> Layers

Layer Properties

Type
LAYER
LA

LAYER

NOTE

The **VPLAYER** tool is a text-based tool that also controls layer display in floating viewports. The **Layer Properties Manager** palette is faster and easier to use than the **VPLAYER** tool.

Freezing and Thawing

Freeze layers in the active viewport to create unique views using a single drawing. For example, **Figure 30-15A** shows the model space display of a floor plan with electrical plan content added directly to the floor plan using electrical plan layers. **Figure 30-15B** shows two layouts from the same drawing file. One layout creates a floor plan with no electrical information, and the other layout creates an electrical plan without specific floor plan content.

In this example, you draw many objects, such as doors, walls, and windows on layers that maintain the global **Thaw** setting. As a result, these objects appear in model space and both floating viewports. Freeze layers in specific viewports (**VP Freeze**) to create two different drawings. This example shows viewport layer freezing in two different viewports, each viewport in a different layout, but you can apply the same concept to multiple viewports in the same layout.

Figure 30-14.
Use the **Layer Properties Manager** to control the display of layers in floating viewports.

Controls freezing in new viewports   Controls freezing in the active viewports   Layer property overrides

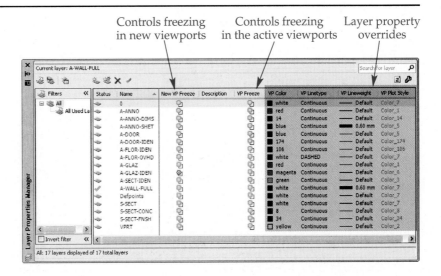

VP **VP**
Freeze **Thaw**

The **VP Freeze** column of the **Layer Properties Manager** palette controls freezing and thawing layers in the current viewport. Pick the **VP Thaw** icon or the **VP Freeze** icon to toggle freezing and thawing in the current viewport. Using the **VP Freeze** icon freezes layers only in the selected floating viewport, while the **Freeze** icon freezes layers globally in all floating viewports. You can freeze or thaw a layer in all layout viewports, including those created before picking the **VP Freeze** icon or **VP Thaw** icon, by right-clicking and picking **VP Freeze Layer in All Viewports** or **VP Thaw Layer in All Viewports**.

> **PROFESSIONAL TIP**
>
> The **VP Freeze** function is also available in the **Layer Control** drop-down list in the **Layers** panel on the **Home** ribbon tab. This provides a quick way to freeze and thaw layers in a viewport without accessing the **Layer Properties Manager** palette.

Figure 30-15.
A—An example of "overlapping" layers in model space. B—Layers frozen in separate layouts to create unique drawing views.

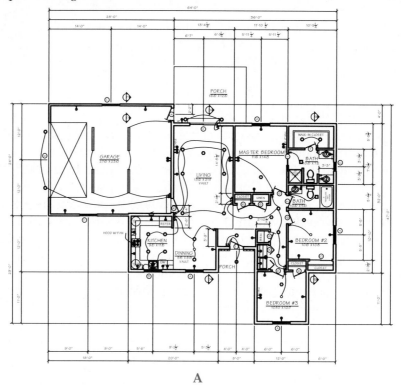

A

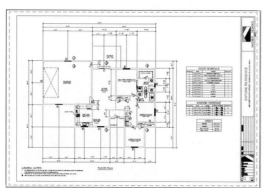

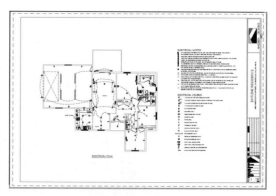

Floor Plan Layout
Created by freezing electrical layers in the floating viewport

Electrical Plan Layout
Created by freezing floor plan layers in the floating viewport

B

The **New VP Freeze** column of the **Layer Properties Manager** palette controls freezing and thawing of layers in *newly created* floating viewports. Pick the **VP Thaw** icon or the **VP Freeze** icon to toggle freezing and thawing in any new floating viewport. This feature has no effect on the active viewport. Use the **New VP Freeze** option to freeze specific layers in any new floating viewports.

New VP Freeze   New VP Thaw

NOTE

Right-click on a layer in the **Layer Properties Manager** palette and select **New Layer VP Frozen in All Viewports** to create a new layer preset with the **VP Freeze** and **New VP Freeze** icons selected.

Exercise 30-6

Access the Student Web site (www.g-wlearning.com/CAD) and complete Exercise 30-6.

Layer Property Overrides

Use layer property overrides to create unique views without changing individual object properties, creating separate drawing files, or readjusting global layer properties. For example, **Figure 30-16A** shows the model space display of a hopper and conveyer system with unique layers assigned to the hopper and conveyer. **Figure 30-16B** shows a layout with two floating viewports. The viewport on the left shows the hopper and

Figure 30-16.
A—A hopper and conveyer drawn in model space. B—Using property overrides to create a unique layout view.

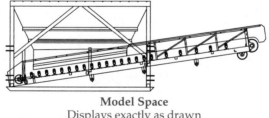

Model Space
Displays exactly as drawn

A

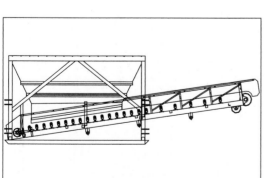

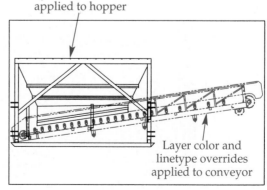

Left Viewport
Displays model space
objects exactly as drawn

Right Viewport
Layer property
overrides applied

B

conveyer with global layer settings applied, as in model space. The viewport on the right shows the hopper with a layer color override and the conveyer with a layer color and linetype override. In this example, layer property overrides create a view that clearly shows the two separate components. Phantom lines highlight the conveyer as the mechanism.

The **VP Color**, **VP Linetype**, **VP Lineweight**, and **VP Plot Style** columns in the **Layer Properties Manager** palette control the property overrides assigned to layers. The **VP Plot Style** column appears only when a named plot style is in use. Layer property overrides apply only to floating viewports in paper space. Layers that contain layer property overrides are not uniquely identified in model space.

The process of overriding a layer property is just like that for changing a global value. For example, to override the color assigned to a layer, pick the color swatch and choose a color from the **Select Color** dialog box. The difference is that layer property overrides apply only to specific layers in an active floating viewport. Object properties do not change from **Bylayer**, and the model space display does not change.

When viewed in paper space, the **Properties** palette, **Layer Properties Manager** palette, and **Layer Control** drop-down list on the ribbon indicate which layers include layer property overrides. See **Figure 30-17**. The **Layer Properties Manager** palette identifies layers that contain layer property overrides with a sheet and viewport icon in the status column. The layer names, global properties affected by the overrides, and the property overrides are highlighted. Use the **Viewport Overrides** filter to quickly display and manage only those layers that include layer property overrides. You can also save layer property overrides in a layer state.

The **Properties** palette identifies layers that contain layer property overrides with a highlighted layer name. Properties affected by the override also appear highlighted and are defined as **Bylayer (VP)**. Layers that include layer property overrides also appear highlighted in the **Layer Control** drop-down list on the ribbon.

Figure 30-17.
Layers with property overrides are highlighted in the **Properties Manager** palette, **Layer Properties Manager** palette, and **Layer Control** drop-down list on the ribbon.

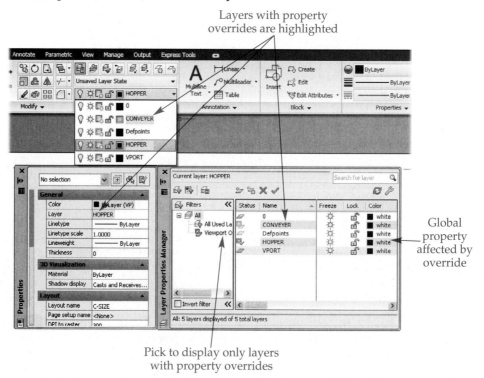

Layers with property overrides are highlighted

Global property affected by override

Pick to display only layers with property overrides

The **Viewport Overrides** icon appears in the status bar when you activate a floating viewport or assign layer property overrides to the active viewport.

If layer property overrides are no longer necessary, you must remove the overrides from the layer. Changing a property back to the original, or global, value does not remove the override. Right-click on a layer that contains layer property overrides in the **Layer Properties Manager** palette and pick **Remove Viewport Layer Overrides for** to access a cascading submenu of options for removing layer property overrides. Pick **Selected Layers** and then **In Current Viewport Only** or **In All Viewports** to remove layer property overrides from the current viewport or from all viewports that include overrides.

You can also use the **Layer** option of the **MVIEW** tool and the **Reset** option of the **VPLAYER** tool to remove layer property overrides.

Exercise 30-7

Access the Student Web site (www.g-wlearning.com/CAD) and complete Exercise 30-7.

Turning Off Floating Viewport Objects

By default, objects appear in floating viewports, allowing you to view model space through the viewports You can hide objects in the floating viewport without removing the viewport, which is convenient if, for example, you want to plot a certain view, but still have access to the viewport. One option to toggle the display of objects in the viewport on and off is to select a viewport in paper space and right-click. From the **Display Viewport Objects** cascading menu, select **No** to hide objects or select **Yes** to display objects.

Another option is to select a viewport in paper space and access the **Properties** palette. From the **On** drop-down list, select **Yes** to show objects or select **No** to hide objects. You can also use the **ON** and **OFF** options of the **MVIEW** tool. Follow the prompts to select the viewport(s) to display or hide objects.

Maximizing Floating Viewports

When you activate a floating viewport, you are working in model space from within the paper space display. The primary function of activating a floating viewport is to adjust the display of model space to prepare a final drawing. Typically, you should avoid working inside an active viewport to make changes to model space objects.

One alternative to activating a floating viewport is to maximize it by picking the **Maximize Viewport** button on the status bar or by selecting a viewport, right-clicking, and choosing **Maximize Viewport**. When you maximize a viewport, you fill the entire drawing window with the selected floating viewport. See **Figure 30-18**. This allows you to work more effectively than when the layout content covers much of the window. In addition, a maximized viewport displays objects exactly as they appear in the floating viewport, including frozen layers and layer overrides. Typically, you should maximize a floating viewport to use view tools such as **ZOOM** and **PAN** and make changes to objects in model space, while remaining in paper space.

Figure 30-18.
Maximize a floating viewport to work in a model-space-like environment, but with layout characteristics, such as layers frozen in the viewport and layer property overrides.

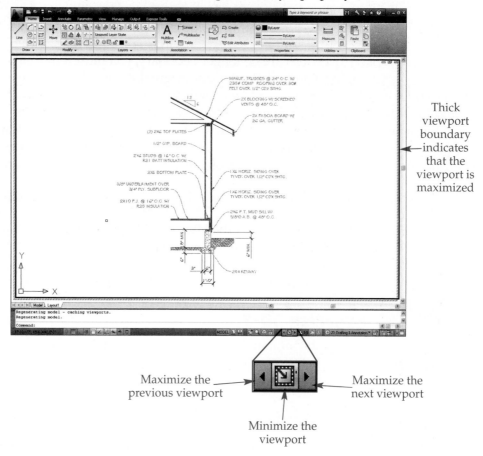

Thick viewport boundary indicates that the viewport is maximized

Maximize the previous viewport

Maximize the next viewport

Minimize the viewport

If the drawing includes multiple viewports, use the **Maximize Previous Viewport** and **Maximize Next Viewport** buttons on the status bar to change to other floating viewports in a maximized display. To redisplay the entire layout, pick the **Minimize Viewport** button on the status bar, right-click and choose **Minimize Viewport**, or type **VPMIN**.

NOTE

You can maximize a floating viewport even if the viewport is not active.

PROFESSIONAL TIP

If you do not want to see floating viewport boundaries on the plotted sheet, remember to freeze or turn off the layer assigned to the floating viewport before plotting.

Exercise 30-8

Access the Student Web site (www.g-wlearning.com/CAD) and complete Exercise 30-8.

Plotting

Quick Access

Plot

Ribbon

Output
> Plot
File
> Print

Plot

Type

PLOT
[CTRL]+[P]

After you prepare a layout for plotting, you are ready to plot. If you develop an appropriate page setup and layout, the process of creating the actual print should be almost automatic. Select the layout to plot and access the **PLOT** tool. The **Plot** dialog box appears with the name of the layout displayed on the title bar. See **Figure 30-19**.

> **NOTE**
>
> You can also access the **Plot** dialog box by selecting from the shortcut menu available from the model or layout tab, or by picking the **Plot** button in a **Model** or a layout thumbnail image in the **Quick View Layouts** or **Quick View Drawings** tool display.

The **Page Setup** and **Plot** dialog boxes are very similar, except the **Plot** dialog box provides additional options specific to creating a print. All the settings in the **Plot** dialog box correspond to those in the **Page Setup** dialog box. Pick the **>**, or **More Options**, button in the lower-right corner of the **Plot** dialog box to toggle the display of additional dialog box areas, as shown in **Figure 30-19**. Enter a number in the text box in the **Number of copies** area to specify how many copies of the layout to plot. The **Plot options** area provides other plot settings. Pick the **Plot in background** check box to continue working while the plot processes.

Most of the **Plot** dialog box settings are the same as those found in the **Page Setup** dialog box. Changing plot settings in the **Plot** dialog box is an effective way to override the page setup for a unique plotting requirement. This is a convenient way to make a plot using slightly modified plot settings without creating a new page setup. For

---

Figure 30-19.
Use the **Plot** dialog box to finalize the layout and send the drawing to the printer, plotter, or file.

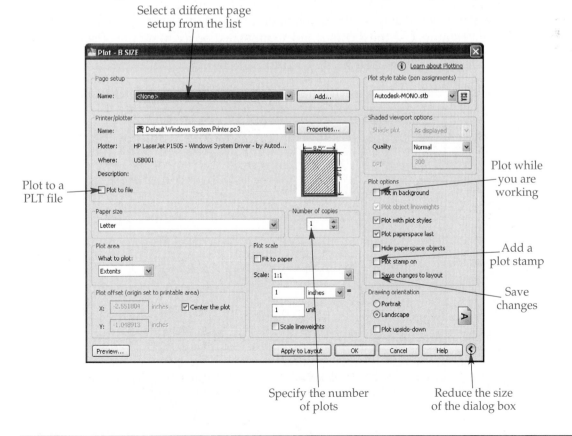

Select a different page setup from the list

Plot to a PLT file

Specify the number of plots

Plot while you are working

Add a plot stamp

Save changes

Reduce the size of the dialog box

---

example, you can make a "check print" by selecting a printer, using an A- or B-size sheet and scaling the plot to fit the paper. After printing, the settings return to those originally assigned in the page setup, allowing you to plot the final drawing using the appropriate printer, sheet size, and scale (1:1).

Adding a Plot Stamp

Pick the **Plot stamp on** check box in the **Plot options** area of the **Plot** dialog box to add a *plot stamp* to the plot. When you select the check box, the **Plot Stamp Settings...** button appears. Pick the button to display the **Plot Stamp** dialog box. See **Figure 30-20**.

Pick the check boxes located in the **Plot stamp fields** area to identify the information to include in the plot stamp. To create additional plot stamp items, pick the **Add/Edit** button in the **User defined fields** area, and use the **User Defined Fields** dialog box to add, edit, and delete custom fields. For example, add a field for the client name, project name, or contractor who uses the drawing. Select the fields from the drop-down lists in the **User defined fields** area.

The **Preview** area provides a preview of the location and orientation of the plot stamp. The preview does not show the actual plot stamp text. Plot stamp settings are saved in a plot stamp parameter (PSS) file. Pick the **Save As** button to save the current settings as a new PSS file, or pick the **Load** button to access and use an existing PSS file.

> **NOTE**
>
> The log file settings are independent of the plot stamp settings. You can produce a log file without creating a plot stamp or have a plot stamp without producing a log file.

Pick the **Advanced** button to display the **Advanced Options** dialog box shown in **Figure 30-21**. The **Location and offset** area includes options to define the position of the plot stamp. Use the **Location** drop-down list to select the corner where the plot stamp begins. To print the plot stamp upside-down, pick the **Stamp upside-down** check box. Pick **Horizontal** or **Vertical** from the **Orientation** drop-down list to specify the orientation of the plot stamp. Use the **X Offset** and **Y Offset** text boxes to set the offset distances

---

Figure 30-20.
Use the **Plot Stamp** dialog box to specify the information included in the plot stamp.

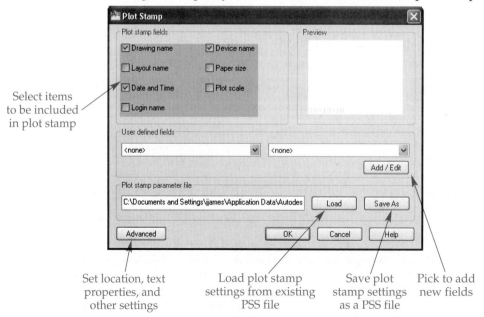

Select items to be included in plot stamp

Set location, text properties, and other settings

Load plot stamp settings from existing PSS file

Save plot stamp settings as a PSS file

Pick to add new fields

Figure 30-21.
The **Advanced Options** dialog box allows you to define the plot stamp location, orientation, text font and size, and units.

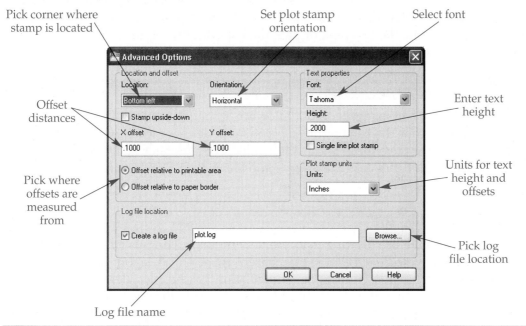

Pick corner where stamp is located

Set plot stamp orientation

Select font

Offset distances

Pick where offsets are measured from

Enter text height

Units for text height and offsets

Pick log file location

Log file name

for the plot stamp and pick whether the distances measure from the edge of the printable area or the paper border.

The **Text properties** area provides options for controlling plot stamp text characteristics. Use the **Font** drop-down list to select a font and the **Height** text box to specify the text height. Pick the **Single line plot stamp** check box to contain the plot stamp to a single line. If left unchecked, the plot stamp prints on two lines.

Use the **Units** drop-down list to select the units for the plot stamp offset and text height. The plot stamp units can be different from the drawing units. Select the **Log file location** check box to create a log file of plotted items. Specify the name of the log file in the text box. Pick the **Browse...** button to specify the location of the log file.

NOTE

You can also configure plot stamp settings by picking the **Plot Stamp Settings...** button on the **Plot and Publish** tab of the **Options** dialog box.

Saving Changes to the Layout

If you make changes in the **Plot** dialog box and want to save changes to the layout page setup for future plots, pick the **Save changes to layout** check box in the **Plot options** area. You can also save changes by picking the **Apply to Layout** button. If you do not select the **Save changes to layout** check box or pick the **Apply to Layout** button, changes made in the **Plot** dialog box are discarded, and the original page setup is used the next time you open the **Plot** dialog box.

Page Setup Options

The **Plot** dialog box provides an alternate means of creating a page setup. To apply this technique, access the **Plot** dialog box and make changes to plot settings, just as you would in the **Page Setup** dialog box. Then select the **Add...** button in the **Page setup**

---

area to display the **Add Page Setup** dialog box. Enter a name for the page setup in the **New page setup name:** text box. All current settings in the **Plot** dialog box are saved with the new page setup. Select a page setup from the **Name:** drop-down list to restore the settings in the **Plot** dialog box. Pick the **<Previous plot>** option to reference the setting used to create the last plot, or pick the **Import...** button to import a page setup from a DWG, DWT, or DXF file.

NOTE

When using the **Plot** dialog box to define settings for a page setup, name the page setup *after* you make changes to settings. If you name the page setup and want to make changes later, such as changes to a plot style, use the **Page Setup Manager** dialog box instead.

Previewing the Plot

The final step before plotting is to preview the plot. The plot preview shows you exactly what your plot *should* look like, based on plot and layout settings. Always preview the plot before sending the information to the plot device to check the drawing for errors, view the effects of plot settings, and eliminate unnecessary plots. To preview the plot, pick the **Preview** button in the lower-left corner of the **Plot** dialog box to enter preview mode. See **Figure 30-22**. What you see on-screen is exactly what will plot, assuming you use a color plotter to make color prints and load the correct sheet size in the plot device.

The **Realtime Zoom** tool is activated automatically. Additional view tools are available from the toolbar near the top of the window or from a shortcut menu. Use these tools to help confirm that the plot settings are correct. When you finish previewing the plot and are ready to plot, pick the **Plot** button on the toolbar, or right-click and select **Plot**. To exit the preview without plotting, and return to the **Plot** dialog box, pick the **Close** button on the toolbar, press [Esc] or [Enter], or right-click and select **Exit**. Pick the **OK** button to send the plot to the plot device and close the **Plot** dialog box.

Figure 30-22.
Previewing a plot is an excellent way to confirm that the plot will be correct before sending the information to the plot device.

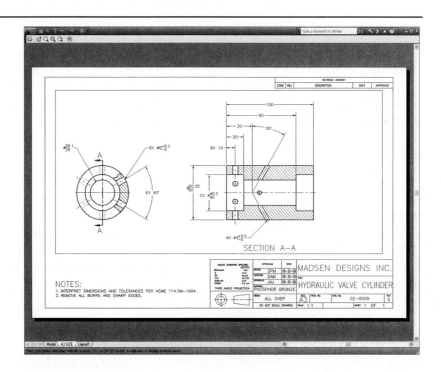

Exercise 30-9

Access the Student Web site (www.g-wlearning.com/CAD) and complete Exercise 30-9.

Plotting to a File

If a plot device is not available, but you are ready to plot, an alternative is to plot to a file. A plot file saves with a PLT extension. The file stores all the drawing geometry, plot styles, and plot settings assigned to the drawing. In offices or schools with only one printer or plotter, a *plot spooler* can be attached to the printer or plotter to plot a PLT file. This device usually allows you to take a PLT file from a storage disk and copy it to the plot spooler, which in turn plots the drawing.

plot spooler: A disk drive with memory that allows you to plot files.

To plot to a file, open the **Plot** dialog box, select the plot device from the **Name:** drop-down list, and check the **Plot to file** check box. The location in which the plot file is saved is set in the **Plot and Publish** tab of the **Options** dialog box. To specify the path, pick the ellipsis (...) button for the **Select default location for all plot-to-file operations** dialog box.

Additional Plotting Options
For information about several additional plot settings in the **Plot and Publish** tab of the **Options** dialog box, go to the Student Web site (www.g-wlearning.com/CAD), select this chapter, and select **Additional Plotting Options**.

Chapter 30

For detailed instructions on adding layouts to each drawing template, go to the Student Web site (www.g-wlearning.com/CAD), select this chapter, and select **Template Development**.

Chapter Test

Answer the following questions. Write your answers on a separate sheet of paper or go to the Student Web site (www.g-wlearning.com/CAD) and complete the electronic chapter test.

1. Name the two types of content that are brought together to create a complete drawing.
2. What tools can you use to modify the boundary of a floating viewport?
3. What **MVIEW** option can form a floating viewport outline using a polyline?
4. What **MVIEW** option can you use to convert any closed object drawn in paper space into a floating viewport?
5. How do you activate a floating viewport?
6. How can you tell that a viewport is activated in paper space?
7. How do you reactivate paper space after activating a floating viewport for editing?
8. How does the scale you assign to a floating viewport compare with the drawing scale?
9. To what value should the **CELTSCALE**, **PSLTSCALE**, and **MSLTSCALE** system variables be set so that the **LTSCALE** value applies correctly in model space and paper space?
10. Viewport edges may "cut off" the drawing when the viewport is correctly scaled. List three things you can do if you want to display the entire view.
11. Why should you lock a viewport after you have adjusted the drawing in the viewport to reflect the proper scale and view?
12. Give an example of why you would want to hide objects in the floating viewport without removing the viewport.
13. What is a plot stamp?
14. If you make changes to the page setup using the **Plot** dialog box, how can you save these changes to the page setup so that the changes apply to future plots?
15. Give at least two reasons why you should always preview a plot before sending the information to the plot device.

Drawing Problems

Note: Some of the following problems refer to drawings or templates created in previous chapters. If you have not yet created those drawings, you will need to do so before working these problems.

▼ Basic

1. Follow the instructions in the Template Development portion of the Student Web site to add and set up layouts for the Mechanical-Inch template file.

2. Follow the instructions in the Template Development portion of the Student Web site to add and set up layouts for the Mechanical-Metric template file.

3. Follow the instructions in the Template Development portion of the Student Web site to add and set up layouts for the Architectural-US template file.

4. Follow the instructions in the Template Development portion of the Student Web site to add and set up layouts for the Architectural-METRIC template file.

5. Follow the instructions in the Template Development portion of the Student Web site to add and set up layouts for the Civil-US template file.

6. Follow the instructions in the Template Development portion of the Student Web site to add and set up layouts for the Civil-METRIC template file.

7. Open P29-4 and save as P30-7. The P30-7 file should be active. Make the **B-SIZE** layout current. Create a new layer named **VPORT**. Delete the default floating viewport and create a single floating viewport .5" in from the edges of the sheet on the **VPORT** layer. Scale model space in the viewport to 1:1. Plot the layout, leaving the **VPORT** layer on and thawed. Resave the problem.

8. Open P29-5 and save as P30-8. The P30-8 file should be active. Activate the **A2-SIZE** layout. Create a new layer named **VPORT**. Delete the default floating viewport and create a single floating viewport 10 mm from the edges of the sheet on the **VPORT** layer. Scale model space in the viewport to 1:1. Plot the layout, leaving the **VPORT** layer on and thawed. Resave the problem.

9. Open P29-6 and save as P30-9. The P30-9 file should be active. Activate the **B-SIZE** layout. Create a new layer named **VPORT**. Delete the default floating viewport and create a single floating viewport .5" from the edges of the sheet on the **VPORT** layer. Scale model space in the viewport to 1:1. Plot the layout, leaving the **VPORT** layer on and thawed. Resave the problem.

▼ Intermediate

10. Open P29-7 and save as P30-10. The P30-10 file should be active. Create a floating viewport and scale model space in the viewport using an appropriate scale. Plot the layout, leaving the **VPORT** layer on and thawed. Resave the problem.

11. Open P8-1 and save as P30-11. The P30-11 file should be active. Delete the default **Layout2**. Create a new A-size sheet layout according to the following steps:
    A. Rename the default **Layout1** to **A-SIZE**.
    B. Select the **A-SIZE** layout and access the **Page Setup Manager**.
    C. Modify the **A-SIZE** page setup according to the following settings:
    - **Printer/Plotter:** Select a printer or plotter that can plot an A-size sheet
    - **Paper size:** Select the appropriate A-size sheet (varies with printer or plotter)
    - **Plot area:** Layout
    - **Plot offset:** 0,0
    - **Plot scale:** 1:1 (1 in. = 1 unit)
    - **Plot style table:** monochrome.ctb
    - **Plot with plot styles**
    - **Plot paper space last**
    - Do not check **Hide paper space objects**
    - **Drawing orientation:** Select the appropriate orientation (varies with printer or plotter)
    D. Create a new layer named **VPORT**.
    E. Delete the default floating viewport and create a single floating viewport .5" from the edges of the sheet on the **VPORT** layer.
    F. Scale model space in the viewport to 1:2. Plot the layout, leaving the **VPORT** layer on and thawed.
    Resave the problem.

12. Open P8-7 and save as P30-12. The P30-12 file should be active. Create layouts and floating viewports as needed to plot the drawing at an appropriate scale.

13. Open P30-13 from the Student Web site supplied with this textbook. Create a layout, plot style, and page setup so the layout can be plotted as follows: Using color-dependent plot styles, have the equipment (shown in color in the diagram) plot with a lineweight of 0.8 mm and 80% screening on an A-size sheet oriented horizontally. Plotted text height should be 1/8″. Plot in paper space at 1:1. Save the drawing as P30-13.

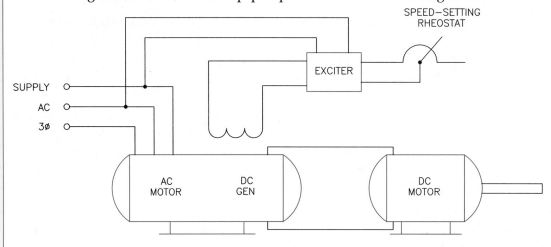

14. Open P30-14 from the Student Web site supplied with this textbook. Create four layouts with the names and displays as follows:
    - The **Entire Schematic** layout plots the entire schematic on a B-size sheet.
    - The **3 Wire Control** layout plots only the 3 Wire Control diagram on an A-size sheet, horizontally oriented.
    - The **Motor** layout plots the motor symbol and connections in the lower center of the schematic on an A-size sheet, oriented vertically.
    - The **Schematic** layout plots schematic without the 3 Wire Control and motor components on an A-size sheet, oriented horizontally.

Set up the layouts so they can plot with a text height of 1/8″. Plot in paper space at a scale of 1:1. Save the drawing as P30-14.

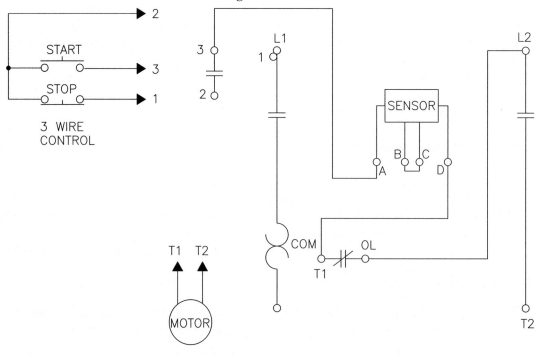

CHAPTER 31

Annotative Objects

Learning Objectives

After completing this chapter, you will be able to do the following:

✓ Explain the differences between manual and annotative object scaling.
✓ Specify objects as annotative.
✓ Create and use annotative objects in model space.
✓ Display annotative objects in scaled layout viewports.
✓ Adjust the scale of annotations according to a new drawing scale.
✓ Use annotative objects to help prepare multiview drawings.

You must scale *annotations* and related items, such as dimension objects and hatch patterns, so that information appears on-screen and plots correctly relative to scaled objects. AutoCAD provides annotative tools to automate the process of scaling *annotative objects*. Annotative tools also provide additional flexibility for working with layouts to create multiview drawings.

annotations: Letters, numbers, words, and notes used to describe information on a drawing.

annotative objects: AutoCAD objects that can be made to adapt automatically to the current drawing scale.

Introduction to Annotative Objects

As explained in previous chapters, you should always draw objects at their actual size, or full scale, in model space, regardless of the size of the objects. For example, if you draw a small machine part and the length of a line in the drawing is 2 mm, draw the line 2 mm long in model space. If you draw a building and the length of a line in the drawing is 80′, draw the line 80′ long in model space. These examples describe drawing objects that are too small or too large for layout and printing purposes. You must *scale* the objects to fit properly on a sheet, according to a specific drawing scale.

When you scale a drawing, you increase or decrease the *displayed* size of model space objects. A properly scaled floating viewport in a layout allows for this process. Scaling a drawing greatly affects the display of items added to objects in model space, such as annotations, because these items should be the same size on a plotted sheet, regardless of the displayed size, or scale, of the rest of the drawing. See **Figure 31-1.**

Traditionally, annotations, hatches, and other objects are scaled manually, which means you determine the scale factor of the drawing scale and then multiply the scale factor by the plotted size of the objects. In contrast, annotative objects are scaled

scale: The ratio between the actual size of drawing objects and the size at which objects plot on a sheet of paper. Also the process of enlarging or reducing objects to fit properly on a sheet of paper.

Figure 31-1.
The large drawing features in this example require scaling in order to fit on a standard size sheet. Annotations are scaled according to the plotted size of the drawing; otherwise, they would be too small to see.

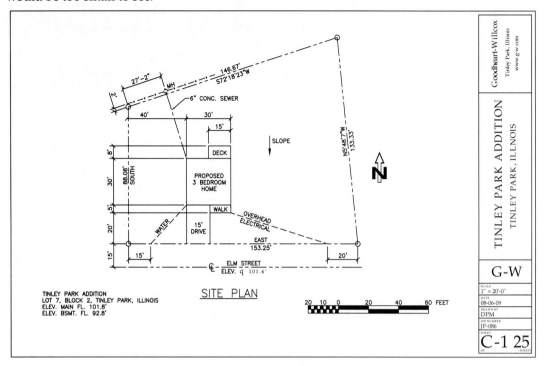

automatically according to the selected annotation scale, which is the same as the drawing scale. This eliminates the need for you to calculate the scale factor and manually adjust the size of objects according to the drawing scale.

PROFESSIONAL TIP

You should use annotative objects instead of traditional manual scaling even if you do not anticipate using a drawing scale other than 1:1.

Defining Annotative Objects

Annotative objects include single-line and multiline text, dimensions, leaders and multileaders, GD&T symbols created using the **TOLERANCE** tool, hatch patterns, blocks, and attributes. The method used to define objects as annotative varies depending on the object type. You can make objects annotative when you first draw them or convert non-annotative objects to annotative status.

Creating New Annotative Objects

Single-line and multiline text is annotative when drawn using an annotative text style. To make a text style annotative, pick the **Annotative** check box in the **Size** area of the **Text Style** dialog box. See **Figure 31-2.** A drawing may include a combination of annotative and non-annotative text, dimension, and multileader styles. An example of text that is typically *not* annotative is text added directly to a layout printed at a scale of 1:1.

Figure 31-2.
Single-line and multiline text objects are annotative if they are drawn using an annotative text style.

Pick to make the text style annotative

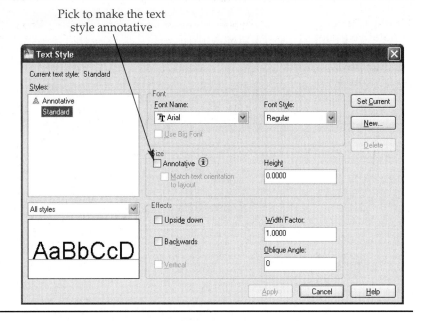

Dimensions, standard leaders, and GD&T symbols created using the **TOLERANCE** tool are annotative when drawn using an annotative dimension style. To make a dimension style annotative, pick the **Annotative** check box in the **Fit** tab of the **New** (or **Modify**) **Dimension Style** dialog box. See **Figure 31-3.**

Multileaders are annotative when drawn using an annotative multileader style. To make a multileader style annotative, pick the **Annotative** check box in the **Leader Structure** tab of the **Modify Multileader Style** dialog box. See **Figure 31-4.**

Figure 31-3.
Dimensions, leaders, and GD&T symbols created using the **TOLERANCE** tool are annotative if they are drawn using an annotative dimension style.

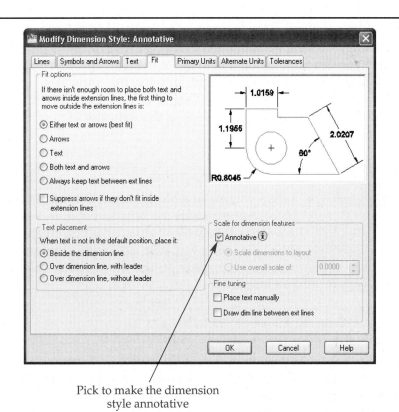

Pick to make the dimension style annotative

Figure 31-4.
Multileaders are
annotative if they
are drawn using
an annotative
multileader style.

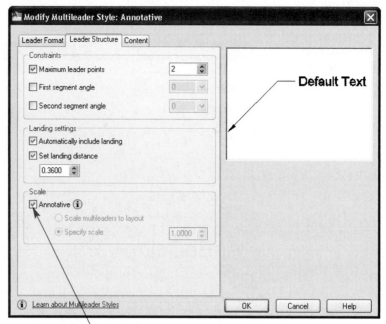

Pick to make the multileader
style annotative

 **NOTE**

When you create an annotative multileader using the block multi-leader type, the block automatically becomes annotative, even if the block is not set as annotative.

Hatch patterns are annotative when you set the hatch scale as annotative during hatching. Pick the **Annotative** check box in the **Options** area on the **Hatch** tab of the **Hatch and Gradient** dialog box to make the hatch pattern annotative. See **Figure 31-5.**

To make attribute text height and spacing annotative, pick the **Annotative** check box in the **Attribute Definition** dialog box. See **Figure 31-6A.** To make a block annotative, pick the **Annotative** check box in the **Behavior** area of the **Block Definition** dialog box. See **Figure 31-6B.**

 NOTE

When you make a block annotative, any attributes included in the block automatically become annotative, even if the attributes are not set as annotative. However, if you create a non-annotative block that contains annotative attributes, the annotative attribute scale changes according to the annotation scale, while the size of the block remains fixed.

Making Existing Objects Annotative

Specify objects as annotative when you first create them in model space when possible. However, you can assign annotative status to any objects originally drawn as non-annotative. The appropriate style controls the annotative status of single-line and multiline text, dimensions, standard leaders and multileaders, and GD&T symbols created using the **TOLERANCE** tool. Change the style assigned to the object to an annotative style to make these objects annotative. You must edit or recreate existing hatch patterns, blocks, and attributes in order to make the objects annotative.

Figure 31-5.
Set the hatch pattern scale to annotative when you create the hatch pattern.

Pick to make the hatch scale annotative

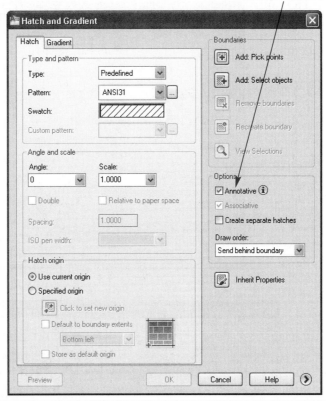

One method you can use to make existing objects annotative is to override the non-annotative status using the **Properties** palette. This technique is most effective to make a limited number of objects annotative. The location of the annotative properties in the **Properties** palette varies depending on the selected object. The **Annotative** and **Annotative scale** properties are common to all annotative objects. Select **Yes** from the **Annotative** drop-down list to make non-annotative objects annotative, or choose **No** to make annotative objects non-annotative.

CAUTION

Use caution when overriding an object to annotative status. Assign annotative objects an appropriate annotative style or status instead of overriding specific objects when possible.

NOTE

The **MATCHPROP** tool allows you to select the properties of annotative objects and apply those properties to existing objects, making the objects annotative.

Exercise 31-1

Access the Student Web site (www.g-wlearning.com/CAD) and complete Exercise 31-1.

Figure 31-6.
Set attributes (A) and blocks (B) as annotative during definition.

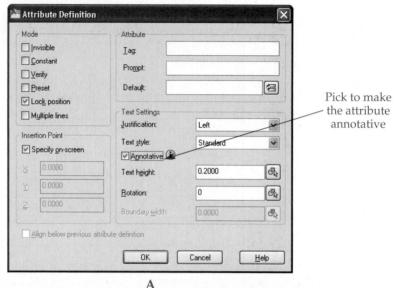

Pick to make the attribute annotative

A

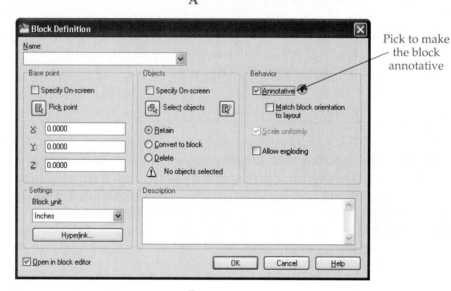

Pick to make the block annotative

B

Drawing Annotative Objects

Using annotative objects reduces the need to determine the drawing scale factor. However, you must still identify the appropriate drawing scale, which is the same as the *annotation scale*. Ideally, determine drawing scale during template development and incorporate the scale into the settings in your template files. If you do not apply drawing scale to settings in your templates, identify the scale before beginning a drawing, or at least before you begin placing annotations.

annotation scale: The scale AutoCAD uses to calculate the scale factor applied to annotative objects.

Setting Annotation Scale

You should set the annotation scale before you begin adding annotations so that annotations scale automatically. However, it may be necessary to adjust the annotation scale throughout the drawing process, especially if the drawing scale changes or when preparing multiple drawings with different scales on one sheet. Approach scaling annotations in model space by first selecting an annotation scale and then placing annotative objects. To draw differently scaled annotations, select the new annotation scale before placing the annotative objects.

When you add an annotative object, the **Select Annotation Scale** dialog box may appear. This is a very convenient way to set annotation scale while creating the object. The other primary means of specifying the annotation scale is to choose a scale from the **Annotation Scale** flyout on the status bar. See **Figure 31-7.** Remember that the annotation scale is typically the same as the drawing scale. You can also set the annotation scale in the **Properties** palette by selecting the annotation scale from the **Annotation Scale** option in the **Misc** category. This option is available when no objects are selected.

If a certain scale is not available, or if you want to change existing scales, pick the **Annotation Scale** flyout on the status bar and choose **Custom...** to access the **Edit Scale List** dialog box. The **Edit Scale List** dialog box is also available by picking the **Edit Scale List...** button in the **User Preferences** tab of the **Options** dialog box. The **Edit Scale List** dialog box is the same dialog box used to edit floating viewport scales, as explained in Chapter 30.

> **NOTE**
>
> Annotation scale sets the drawing scale in model space for controlling annotative objects. Viewport scale sets the drawing scale in a layout floating viewport to define the drawing scale. Both scales should be the same and should match the drawing scale.

Figure 31-7.
The status bar includes several annotation scale options. If you display the drawing status bar, the **Annotation Scale** button moves from the application status bar to the drawing status bar.

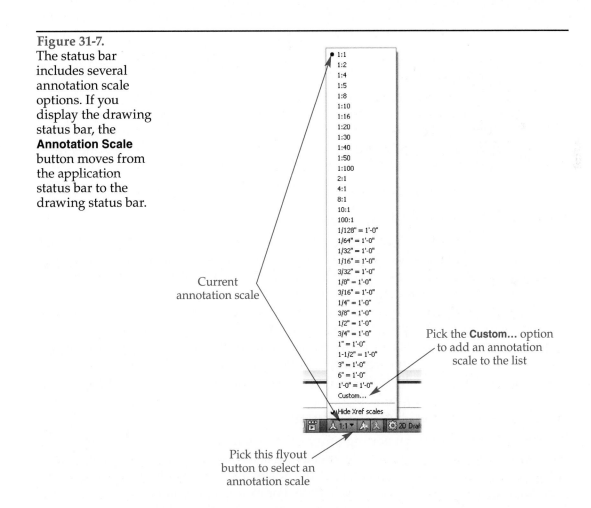

Current annotation scale

Pick the **Custom...** option to add an annotation scale to the list

Pick this flyout button to select an annotation scale

Controlling Model Space Linetype Scale

The **CELTSCALE**, **PSLTSCALE**, and **MSLTSCALE** system variables control how the **LTSCALE** system variable applies to linetypes in model space and paper space. Leave the **CELTSCALE**, **PSLTSCALE**, and **MSLTSCALE** system variables at their default setting of 1 to apply the **LTSCALE** value correctly according to the current annotation scale. However, when you change the annotation scale, you must remember to use the **REGEN** tool to regenerate the display. Otherwise, the linetype scale will not update according to the new scale.

> **PROFESSIONAL TIP**
>
> When you open a drawing in AutoCAD 2010 that was created in an AutoCAD version earlier than AutoCAD 2008, the **MSLTSCALE** system variable is set to 0. Change the value to 1 to take advantage of annotative linetype scaling.

Drawing Annotative Text

Draw annotative text using the same tools as non-annotative text. The difference is the value you enter for text height. To create annotative multiline text, select the **Annotative** button and enter the paper text height, such as 1/4″, in the **Size** text box. See Figure 31-8. The text scale, which includes spacing, width, and paragraph settings, automatically adjusts according to the current annotation scale. To create annotative single-line text, after you pick the start point, specify the paper height. The text scale automatically adjusts according to the current annotation scale.

> **NOTE**
>
> The **Properties** palette contains specific annotative text properties in addition to those displayed for all annotative objects. For example, use the **Paper text height** property to enter a paper text height. The **Model text height** property is a reference value that identifies the height of the text after the scale factor is automatically applied.

Drawing Annotative Dimensions

Draw annotative dimensions, leaders, GD&T symbols created using the **TOLERANCE** tool, and multileaders using the same tools as non-annotative dimensions. Once you activate an annotative dimension or multileader style and select the appropriate annotation scale, the process of placing correctly scaled dimensions is automatic.

However, you must still determine the correct dimension and text location and spacing from objects when you add dimensions and text to scaled drawings. This involves multiplying the scale factor by the plotted spacing. For example, if the first dimension line should be 3/4″ from an object when plotted, using a 1/4″ = 1′-0″ scale, the correct spacing in model space is 36″ from the object (a scale factor of 48 × 3/4″ = 36″).

Adding Annotative Hatch Patterns

The difference between adding annotative and non-annotative hatch patterns is the way in which the drawing scale affects the hatch scale. When you create annotative hatch patterns, the scale you enter in the **Scale:** text box produces the same results regardless of the specified annotation scale. For example, if you enter a value in the **Scale:** text box that is appropriate for an annotation scale of 1/4″ = 1′-0″, and then

Figure 31-8.
Creating annotative multiline text. Multiline text can contain annotative and non-annotative text.

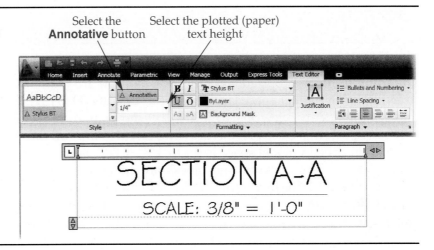

Select the **Annotative** button

Select the plotted (paper) text height

SECTION A-A
SCALE: 3/8" = 1'-0"

change the annotation scale to 1" = 1'-0", the hatch pattern scale does not change relative to the drawing display. It looks the same on the 1/4" = 1'-0" scaled drawing as on the 1" = 1'-0" scaled drawing.

In contrast, when you create non-annotative hatch patterns, if you enter a value in the **Scale:** text box that is appropriate for a drawing scaled to 1/4" = 1'-0" and then change the drawing scale to 1" = 1'-0", the displayed scale of the hatch pattern increases. It looks four times as large on the 1" = 1'-0"scaled drawing as on the 1/4" = 1'-0" scaled drawing.

Placing Annotative Blocks and Attributes

Annotative blocks, often classified as *schematic blocks*, are commonly used for annotation purposes. When you insert an annotative schematic block, AutoCAD determines the block scale based on the current annotation scale, eliminating the need for you to enter a scale factor. For most applications, insert annotative blocks at a scale of 1 to apply the annotation scale correctly. Entering a scale other than 1 adjusts the scale of the block by multiplying the block scale by the annotation scale.

schematic block:
A block originally drawn at a 1:1 scale.

PROFESSIONAL TIP

When you create unit and schematic blocks that contain text and attributes, you should usually not make the text and attributes annotative. The height you specify is set according to the full-scale size of the block, not necessarily the paper height. Any non-annotative text and attributes selected when you make a block annotative also automatically become annotative.

Exercise 31-2

Access the Student Web site (www.g-wlearning.com/CAD) and complete Exercise 31-2.

Displaying Annotative Objects in Layouts

Once you create drawing features and symbols and add annotative objects according to the appropriate annotation scale, you are ready to display and plot the drawing using a paper space layout. Refer to Chapter 30 to review the process of using and scaling floating viewports. **Figure 31-9** shows an example of a drawing scaled to 3/8″ = 1′-0″. In this example, the drawing features are full scale in model space. The annotation scale in model space is set to 3/8″ = 1′-0″, and annotative text, dimensions, multileaders, hatch patterns, and blocks are added. The annotative objects are automatically scaled according to the 3/8″ = 1′-0″ annotation scale.

In the **Figure 31-9** example, the viewport scale and the annotation scale are the same, which is typical when scaling annotative objects. If you select a different viewport scale from the **Viewport Scale** flyout button, the annotation scale automatically adjusts according to the viewport scale. However, if you adjust the viewport scale by zooming, for example, the annotation scale does not change. The viewport scale and the annotation scale must match in order for your drawing and annotative objects to be scaled correctly. You can pick the button to the right of the **Viewport Scale** flyout button, identified in **Figure 31-9,** to synchronize the viewport and annotation scales.

The **Properties** palette also allows you to control viewport and annotation scale. In order to use this method, you must be in paper space to access the viewport properties. Then choose a viewport scale from the **Standard scale** drop-down list. Adjust the annotation scale using the **Annotation scale** option. See **Figure 31-10.**

PROFESSIONAL TIP

Lock the viewport display to avoid zooming and disassociating the viewport scale from the annotation scale. Refer to Chapter 30 for more information on locking and unlocking floating viewports.

Figure 31-9.
Scale a drawing in a floating paper space viewport. Picking the **Viewport Scale** flyout button is one of the easiest ways to set the viewport scale. A button is also available to synchronize the viewport and annotation scale if they do not match.

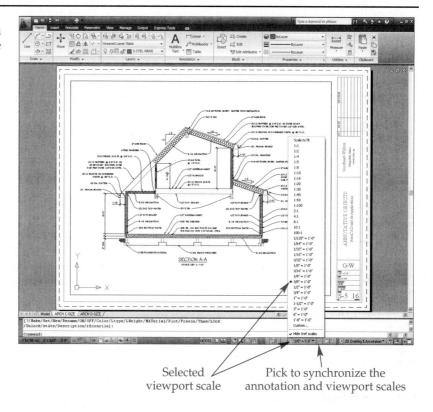

Selected viewport scale

Pick to synchronize the annotation and viewport scales

Figure 31-10.
The **Properties** palette also allows you to set the viewport and annotation scale.

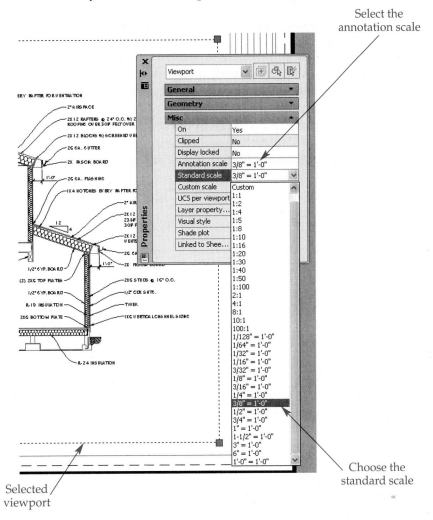

Select the annotation scale

Selected viewport

Choose the standard scale

Exercise 31-3

Access the Student Web site (www.g-wlearning.com/CAD) and complete Exercise 31-3.

Changing Drawing Scale

No matter how much you plan a drawing, drawing scale can change throughout the drawing process. Drawing scale may be reduced if it is necessary to use a smaller sheet. Drawing scale may increase if drawing features are redesigned and become larger, or if additional drawing detail is required.

Changing the drawing scale affects the size and position of annotations. A major advantage of using annotative objects is the ease with which the annotation scale adjusts to different drawing scales. When changing drawing scale, remember that the annotation scale is the same as the drawing scale.

To change annotation scale in model space, select a new annotation scale from the **Annotation Scale** flyout button. To change the annotation scale in an active viewport in a layout, adjust the viewport scale by selecting the drawing scale from the **Viewport**

Scale flyout button. Again, the viewport and annotation scale should be set to the same scale for most applications.

Using the ANNOUPDATE Tool

When you create single-line text using a non-annotative text style, and then change the style to annotative, text drawn using the style becomes annotative. However, the properties of the annotative text remain set according to the non-annotative text style. When you create annotative text using an annotative text style, and then change the style to non-annotative, text drawn in the style becomes non-annotative. However, the properties of the non-annotative text remain set according to the annotative style.

Use the **ANNOUPDATE** tool to update text properties to reflect the current properties of the text style in which the text is drawn. When prompted to select objects, pick the text to update to the current, modified text style. After making the selections, press [Enter] to exit the tool and update the text.

Introduction to Scale Representations

The previous content of this chapter assumes that you develop a drawing using a single annotation scale. In order for annotative object scale to change when the drawing scale changes, annotative objects must support the new scale. This involves assigning new annotation scales to annotative objects. If annotative objects do not support the new scale, the annotative object scale does not change, and can actually cause the objects to become invisible.

Figure 31-11A shows an example of a drawing prepared at a 3/8″ = 1′-0″ scale and placed on an architectural C-size sheet. The annotation scale in this example is set to 3/8″ = 1′-0″, to automatically scale annotative objects according to a 3/8″ = 1′-0″ drawing scale. In order to change the scale of the drawing to 1/2″ = 1′-0″ to display additional detail, you must ensure that the annotative objects support a 1/2″ = 1′-0″ scale.

Once you add the 1/2″ = 1′-0″ annotation scale to the annotative objects, you can change the annotation scale or the viewport scale to 1/2″ = 1′-0″ to correctly scale annotative objects. See **Figure 31-11B**. The annotative objects in this example support two annotation scales: 3/8″ = 1′-0″ and 1/2″ = 1′-0″. As a result, two *annotative object representations* are available.

annotative object representation: Display of an annotative object at an annotation scale that the object supports.

NOTE

Annotative objects display an icon when you hover the crosshairs over the objects. Objects that support a single annotation scale display the annotative icon shown in **Figure 31-12A**. Annotative objects that support more than one annotation scale display the annotative icon shown in **Figure 31-12B**. These icons appear by default according to selection preview settings in the **Selection** tab of the **Options** dialog box.

Understanding Annotation Visibility

Before changing the current annotation scale, you should understand how the annotation scale effects annotative object visibility. The annotative object scale does not change if annotative objects do not support an annotation scale. In addition, annotative objects disappear when an annotation scale that the objects do not support is current. For example, if annotative objects only support an annotation scale of 3/8″ = 1′-0″, and an annotation scale of 1/2″ = 1′-0″ is set current, the annotative object scale remains set at 3/8″ = 1′-0″, and the objects become invisible.

Figure 31-11.
A—A drawing created using an annotation scale of 3/8″ = 1′-0″ on an architectural C-size sheet. Annotative objects automatically appear at the correct scale. B—The same drawing shown in A, modified to an annotation scale of 1/2″ = 1′-0″ and placed on an architectural D-size sheet. An annotation scale of 1/2″ = 1′-0″ is added to all of the annotative objects, allowing the objects to adapt to the new scale automatically.

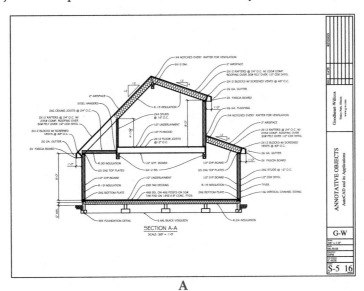

A

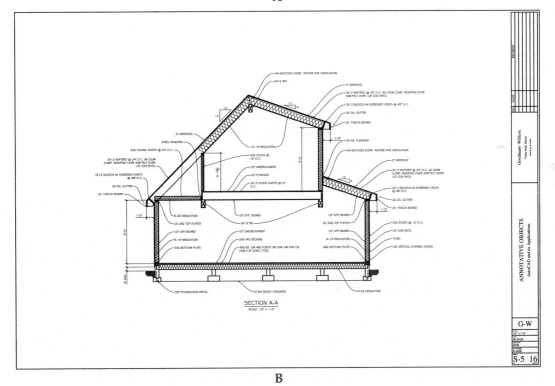

B

Figure 31-12.
Examples of annotative objects that support single and multiple annotation scales.

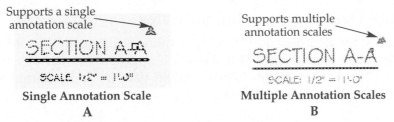

Single Annotation Scale	Multiple Annotation Scales
A	**B**

Figure 31-13.
A—The annotative objects in this example support only a 3/8″ = 1′-0″ annotation scale. However, with annotation visibility turned on, all annotative objects appear, even with the annotation scale set to 1/2″ = 1′-0″. B—The **Annotation Visibility** button on the status bar controls this feature. If you display the drawing status bar, the **Annotation Visibility** button moves from the application status bar to the drawing status bar.

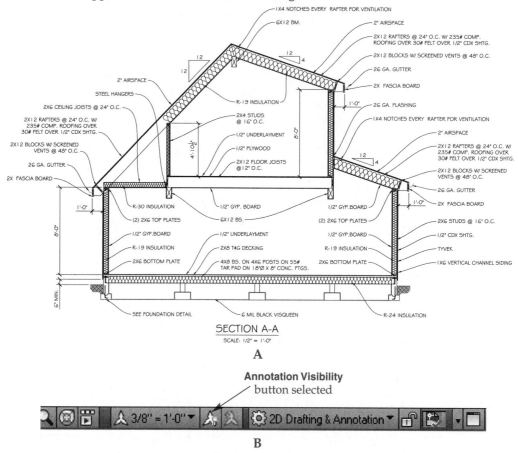

The easiest way to turn on and off annotative object visibility according to the current annotation scale is to pick the **Annotation Visibility** button on the status bar. See **Figure 31-13.** This is most effective when adding and deleting annotation scales to or from annotative objects. If you add multiple annotation scales to annotative objects, the annotative object representation appears based on the current scale.

Deselect the **Annotation Visibility** button to display only the annotative objects that support the current annotation scale. Any annotative objects unsupported by the current annotation scale become invisible. See **Figure 31-14.** This process is most effective when you want to annotate a drawing, or a portion of a drawing, using a different annotation scale without showing annotative object representations specific to a different annotation scale. Turning off the visibility of annotative objects that do not support the current annotation scale is also extremely effective for preparing multiview drawings because it eliminates the need to create separate layers for objects displayed at different scales. This practice is described later in this chapter.

Adding and Deleting Annotation Scales

One method for assigning additional annotation scales to annotative objects is to add the scales to selected objects. This method is appropriate whenever the drawing scale changes, but it is especially effective when you are adding annotation scales only to specific objects, such as when creating multiview drawings. Examples that

Figure 31-14.
Deselect the **Annotation Visibility** button to display only those annotative objects that support the current annotation scale. The annotative objects in this example do not appear because they support only a 3/8″ = 1′-0″ annotation scale, and the current annotation scale is 1/2″ = 1′-0″.

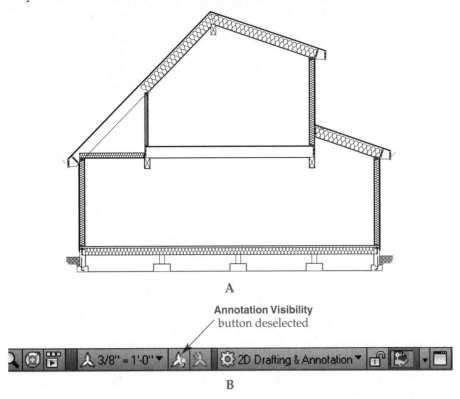

A

Annotation Visibility
button deselected

B

demonstrate this practice are described later in this chapter. You can add annotation scales to selected objects using annotation scaling tools or the **Properties** palette.

You can delete an annotation scale from annotative objects if the annotation scale is no longer in use, should not display in a specific view, or makes it difficult to work with annotative objects. When you delete an annotation scale from annotative objects, the scale no longer applies. You can delete annotation scales from selected objects using annotation scaling tools or the **Properties** palette.

Using the OBJECTSCALE Tool

The **OBJECTSCALE** tool provides one method of adding and deleting annotation scales supported by annotative objects. A quick way to access the **OBJECTSCALE** tool is to select an annotative object and then right-click and pick **Add/Delete Scale...** from the **Annotative Objects Scales** cascading submenu. If you activate the **OBJECTSCALE** tool by right-clicking on objects, the **Annotation Object Scale** dialog box appears, allowing you to add or remove annotation scales from the selected objects. See **Figure 31-15.** If you access the tool before selecting objects, all annotative objects display, even those objects that do not support the current annotation scale. Select the annotative objects to modify and press [Enter] to display the **Annotation Object Scale** dialog box.

The **Object Scale List** shows all of the annotation scales associated with the selected annotative object. A scale must appear in the list in order for the scale to apply to the annotative object. If you select a different annotation scale, and that scale does not display in the **Object Scale List**, annotative objects do not adapt to the new annotation scale, and you have the option of turning off the annotative objects' visibility. Using the previous example, 1/2″ = 1′-0″ must appear in the **Object Scale List** in order for the annotative objects to adapt to the new annotation scale of 1/2″ = 1′-0″.

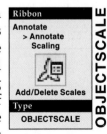

Ribbon
Annotate
> Annotate
Scaling

Add/Delete Scales
Type
OBJECTSCALE

OBJECTSCALE

Figure 31-15.
The **Annotation Object Scale** dialog box allows you to add and delete annotation scales to and from annotative objects.

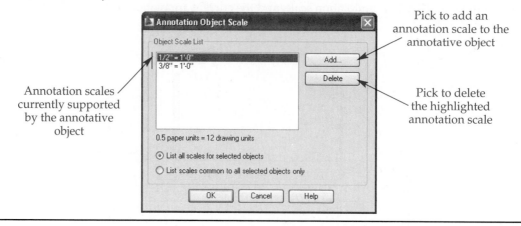

Annotation scales currently supported by the annotative object

Pick to add an annotation scale to the annotative object

Pick to delete the highlighted annotation scale

Pick the **Add...** button to add a scale to the **Object Scale List**. This opens the **Add Scales to Object** dialog box. Highlight scales in the **Scale List** and pick the **OK** button to add the scales to the **Object Scale List**. Once you add a scale to the **Object Scale List**, picking an annotation scale that corresponds to any of the listed scales automatically scales the selected annotative object. To remove a scale from the **Object Scale List**, highlight the scale to remove and pick the **Delete** button.

If you select multiple annotative objects, it may be helpful to display only the annotative scales that are common to the selected objects by picking the **List scales common to all selected objects only** radio button. To show all the annotation scales associated with any of the selected objects, even if some of the objects do not support the listed scales, pick the **List all scales for selected objects** radio button. Picking this option is helpful to delete a listed scale that applies only to certain objects.

> **NOTE**
>
> If a desired scale is not available in the **Add Scales to Object** dialog box, you must close the **Annotation Object Scale** dialog box and access the **Edit Scale List** dialog box to add a new scale to the list of available scales.

Using the Properties Palette

The **Properties** palette also allows you to add annotation scales to selected annotative objects. See **Figure 31-16.** The location of the annotative properties in the **Properties** palette varies depending on the selected object. The **Annotative scale** property displays the annotation scale currently applied to the selected annotative object and contains an ellipsis button (...) that opens the **Annotation Object Scale** dialog box when selected.

Automatically Adding Annotation Scales

Another technique for assigning additional annotation scales to annotative objects is to add a selected annotation scale automatically to all annotative objects in the drawing. This eliminates the need to add annotation scales to individual annotative objects and quickly produces newly scaled drawings.

The **ANNOAUTOSCALE** system variable controls the ability to add an annotation scale to all existing annotative objects. Enter 1, –1, 2, –2, 3, –3, 4, or –4, depending on the desired effect. **Figure 31-17A** describes each option. Once you enter the initial value, the easiest way to toggle this system variable on and off is to pick the button on the status bar shown in **Figure 31-17B.**

Figure 31-16.
The **Annotation scale** property in the **Properties** palette is another way to access the **Annotation Object Scale** dialog box.

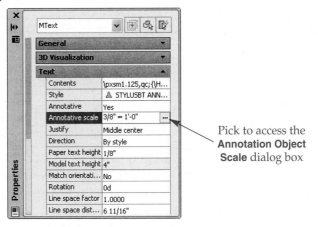

Pick to access the **Annotation Object Scale** dialog box

Figure 31-17.
A—**ANNOAUTOSCALE** system variable options. B—Once you enter the initial **ANNOAUTOSCALE** system variable, use the button on the status bar to toggle **ANNOAUTOSCALE** on and off. If you display the drawing status bar, the **ANNOAUTOSCALE** button moves from the application status bar to the drawing status bar.

Value	Mode	Description
1	On	Adds the selected annotation scale to annotative objects, not including those drawn on a layer that is turned off, frozen, locked, or frozen in a viewport.
−1	Off	1 behavior is used when **ANNOAUTOSCALE** is turned back on.
2	On	Adds the selected annotation scale to annotative objects, not including those drawn on a layer that is turned off, frozen, or frozen in a viewport.
−2	Off	2 behavior is used when **ANNOAUTOSCALE** is turned back on.
3	On	Adds the selected annotation scale to annotative objects, not including those drawn on a layer that is locked.
−3	Off	3 behavior is used when **ANNOAUTOSCALE** is turned back on.
4	On	Adds the selected annotation scale to all annotative objects regardless of the status of the layer on which the annotative object is drawn. 4 is the AutoCAD default setting when toggled on.
−4	Off	4 behavior is used when **ANNOAUTOSCALE** is turned back on. −4 is the AutoCAD default setting when toggled off.

A

Pick to toggle the
ANNOAUTOSCALE system
variable on or off

B

CAUTION

Use caution when adding annotation scales automatically. Due to the effectiveness and transparency of the tool, annotation scales are often added to annotative objects unintentionally. Though you can later delete scales, this causes additional work and confusion.

Exercise 31-4

Access the Student Web site (www.g-wlearning.com/CAD) and complete Exercise 31-4.

Preparing Multiview Drawings

Mechanical drawings and architectural construction drawings often contain sections and details drawn at different scales. Using annotative objects offers several advantages, especially when views in model space appear at different scales in layouts. Use scaled viewports to display multiple views using a single file. You can assign a different annotation scale to each drawing view that contains annotative objects, reducing the need to calculate multiple drawing scale factors, while maintaining the appropriate scale of previously drawn annotative objects. Additionally, by adjusting annotative scale representation visibility and position, you can prepare differently scaled multiview drawings, while eliminating the need to use separate, scale-specific layers and annotations.

Creating Differently Scaled Drawings

Figure 31-18A shows an example of two different drawing views, both drawn at full scale in model space. The full section in **Figure 31-18A** uses a 3/8″ = 1′-0″ scale. To prepare this view, set the annotation scale in model space to 3/8″ = 1′-0″, and then add annotative objects. The annotative objects automatically scale according to the 3/8″ = 1′-0″ annotation scale. The stair section in **Figure 31-18A** uses a 1/2″ = 1′-0″ scale. To prepare this view, change the annotation scale in model space from 3/8″ = 1′-0″ to 1/2″ = 1′-0″, and then add annotative objects. The annotative objects automatically scale according to the 1/2″ = 1′-0″ annotation scale. If you look closely, you can see the different scales applied to the drawing views.

With annotation visibility on, as shown in **Figure 31-18A,** you can see all annotative objects, and observe the effects of using different scales. With annotation visibility off, as shown in **Figure 31-18B,** only annotative objects that support the current annotation scale, which is 1/2″ = 1′-0″ in this example, appear.

The next step is to display and plot the drawing using multiple paper space viewports. **Figure 31-19** shows an architectural D-size sheet layout with two floating viewports. One viewport displays the full section at a viewport scale of 3/8″ = 1′-0″. The other viewport displays the stair section at a viewport scale of 1/2″ = 1′-0″. Notice how the annotative objects are the same size in both views.

Exercise 31-5

Access the Student Web site (www.g-wlearning.com/CAD) and complete Exercise 31-5.

Figure 31-18.
Two different drawing views drawn at full scale in model space. The full section uses an annotation scale of 3/8″ = 1′-0″, and the stair section uses an annotation scale of 1/2″ = 1′-0″. A—Annotation visibility is on. B—Annotation visibility is off with the current annotation scale set to 1/2″ = 1′-0″ (stair section view scale).

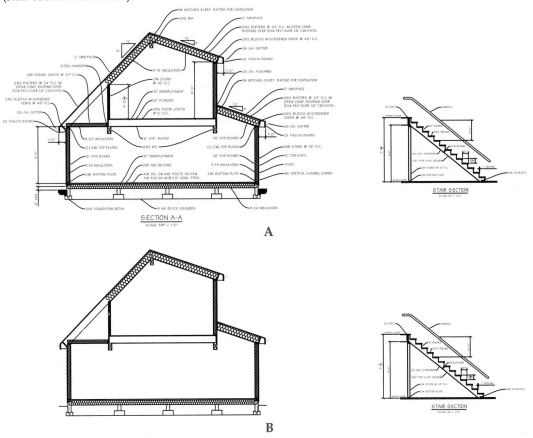

A

B

Figure 31-19.
Using viewports with different scales to create a multiview drawing. Notice that the annotative objects are the same size in both views.

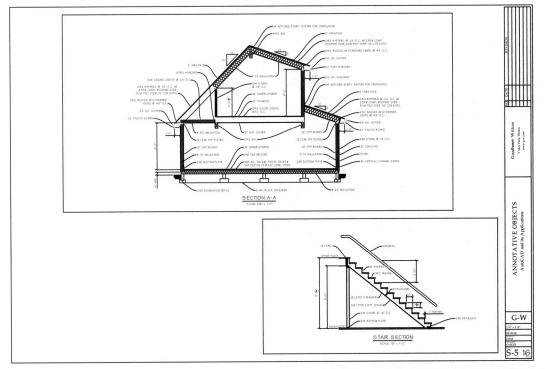

Reusing Annotative Objects

Often the same drawing features appear in different views at different scales. For example, you may want to plot a drawing on a large sheet using a large scale, and plot the same drawing on a smaller sheet using a smaller scale. Another example is preparing a view enlargement or detail.

Using annotative objects significantly improves the ability to reuse existing drawing features. You can use annotation visibility to hide annotative objects not supported by the current annotation scale. You can also adjust the position of scale representations according to the appropriate annotation scale. These options allow you to include differently scaled annotative objects on the same drawing sheet without creating copies of the objects and without using scale-specific layers.

Using Invisible Scale Representations

If annotative objects do not support an annotation scale, the annotative objects disappear when the annotation scale that the objects do not support is current. This is a valuable technique for displaying certain items at a specific scale. Pick the **Annotation Visibility** button on the status bar to turn on and off annotative object visibility.

The following example shows how adjusting the visibility of annotative objects that only support the current annotation scale allows you to create an additional view from existing drawing features. This example uses an annotation scale of 3/4″ = 1′-0″ to create a foundation detail. To begin constructing the foundation detail, add the 3/4″ = 1′-0″ annotation scale to the existing earth hatch pattern so it will appear on the full section and the foundation detail. See **Figure 31-20**. Next, with the current annotation scale set to 3/4″ = 1′-0″, add annotative objects to the foundation detail. See **Figure 31-21**. These objects only support the 3/4″ = 1′-0″ annotation scale, hiding the objects on the full section, which uses a 3/8″ = 1′-0″ scale.

Figure 31-20.
Reuse the earth hatch pattern by adding the 3/4″ = 1′-0″ foundation detail scale to the annotative hatch pattern.

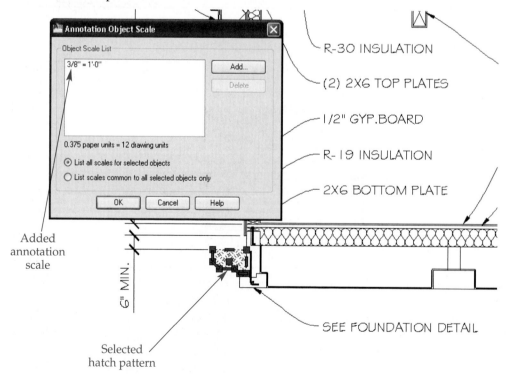

Figure 31-21.
Adding annotative text, dimensions, multileaders, and hatch patterns using a 3/4″ = 1′-0″
annotation scale.

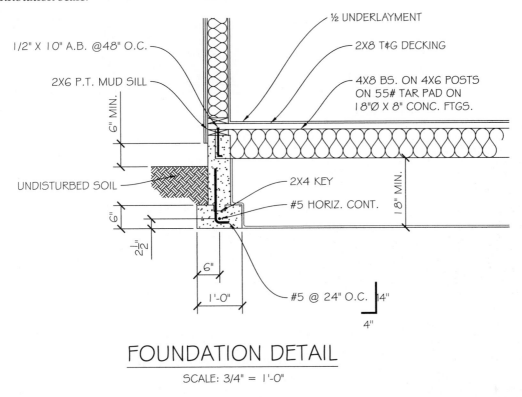

FOUNDATION DETAIL

SCALE: 3/4" = 1'-0"

PROFESSIONAL TIP

If objects already support an annotation scale, but you do not want
to display those annotations at the current scale, delete the annotation scale from the objects.

Adjusting Scale Representation Position

A major benefit of using annotative objects is the ability to reuse objects for differently scaled drawing views. The previous example of adding a 3/4″ = 1′-0″ annotation
scale to the earth hatch pattern highlights this concept. When you reuse annotative objects,
the location and spacing of annotative objects on one scale are often not appropriate for
another scale. You can reposition each scale representation to overcome this issue.

In the foundation detail example, some of the existing 3/8″ = 1′-0″ scaled full
section dimensions and multileaders are reused in the foundation detail. The first step
is to add a 3/4″ = 1′-0″ annotation scale to the objects. Next, with **Annotation Visibility**
turned off, as shown in **Figure 31-22,** you can see the resulting position of the selected
objects, which is initially the same location as the 3/8″ = 1′-0″ objects. The only difference is that now the 3/8″ = 1′-0″ objects also support a 3/4″ = 1′-0″ scale.

Adjust the position of annotation scale representations using grip editing methods.
When you select annotative objects that support more than one annotation scale, all
scale representations appear by default. See **Figure 31-23.** An annotative object is a
single object, but it can contain several scale representations. Grips attach to the scale
representation that corresponds to the current annotation scale. Using grips to edit
scale representations is similar to editing the object used to create the scale representation. The difference when editing a scale representation is that you are adjusting
a scaled "copy" of the object. **Figure 31-24** shows the effects of editing the position
of dimension and multileader scale representations on the foundation detail. The

Figure 31-22.
A—Reusing some of the existing 3/8″ = 1′-0″ scaled objects to create another drawing view. B—Adding a 3/4″ = 1′-0″ annotation scale to existing objects and setting the annotation scale to 3/4″ = 1′-0″.

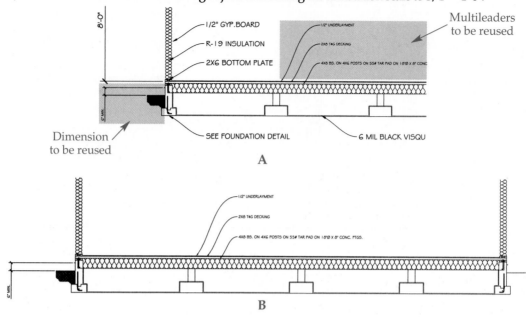

A

B

Figure 31-23.
Adjust the position of annotation scale representations using grip editing techniques. When you select annotative objects that support more than one annotation scale, all scale representations appear by default.

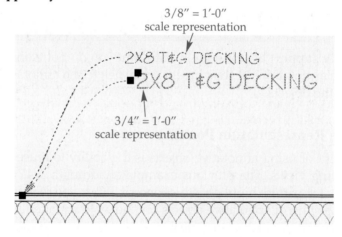

Figure 31-24.
Editing the position of scale representations is much like creating scaled copies of existing annotations.

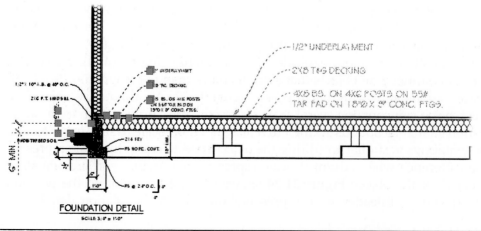

representations are selected to help demonstrate the effects of editing scale representation position. Notice that you can edit all elements of the scale representation to produce the desired annotations at the appropriate locations.

PROFESSIONAL TIP

Use the **DIMSPACE** and **MLEADERALIGN** tools to adjust dimension spacing and multileader alignment after the drawing scale changes.

The **SELECTIONANNODISPLAY** system variable controls the display of selected scale representations and is set to 1 by default. As a result, all scale representations display and appear dimmed when you pick an annotative object that supports multiple annotation scales. Refer again to **Figure 31-23.** The display can be confusing if the selected object supports several annotation scales. Set the **SELECTIONANNODISPLAY** system variable to 0 to display only the scale representation that corresponds to the current annotation scale.

NOTE

You can only edit scale representations individually using grip editing techniques. When you use modify tools to edit an annotative object, all scale representations are edited at once.

Resetting Scale Representation Position

The **ANNORESET** tool removes all unique scale representation positions, allowing you to change the position of all selected scale representations to the position of the scale representation that is set according to the current annotation scale. A quick way to access the **ANNORESET** tool is to select annotative objects, right-click and pick the option from the **Annotative Object Scale** cascading submenu. If you activate the **ANNORESET** tool by right-clicking on objects, the position of the selected objects resets. If you access the tool before selecting objects, pick the annotative objects. After making your selections, press [Enter] to exit the tool and reset the scale representation positions.

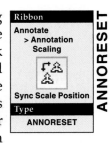

Ribbon
Annotate > Annotation Scaling
Sync Scale Position
Type
ANNORESET

Completing a Multiview Drawing

The last step in creating a multiview drawing is to display and plot the drawing using multiple paper space viewports. **Figure 31-25** shows an architectural D-size sheet layout with three floating viewports. One viewport displays the full section at a 3/8″ = 1′-0″ viewport scale. A second viewport displays the stair section at a 1/2″ = 1′-0″ viewport scale. A third viewport displays the foundation detail at a 3/4″ = 1′-0″ viewport scale.

PROFESSIONAL TIP

When you save drawings using annotative objects to earlier versions of AutoCAD that do not support annotative objects, scale representations may convert to non-annotative objects, but automatically become assigned to unique layers. To use this function, select the **Maintain visual fidelity for annotative objects** check box in the **Open and Save** tab of the **Options...** dialog box.

Figure 31-25.
A complete multiview drawing created using annotative objects.

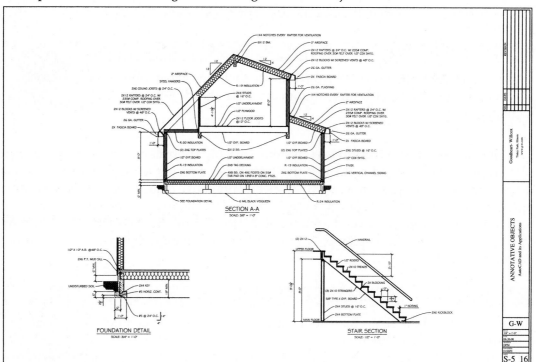

Exercise 31-6

Access the Student Web site (www.g-wlearning.com/CAD) and complete Exercise 31-6.

Chapter Test

Answer the following questions. Write your answers on a separate sheet of paper or go to the Student Web site (www.g-wlearning.com/CAD) and complete the electronic chapter test.

1. What are annotative objects?
2. Explain the practical differences between manual and annotative object scaling.
3. Identify at least four types of objects that can be made annotative.
4. How do you set the text scale, including spacing, width, and paragraph settings, to adjust automatically according to the current annotation scale?
5. Identify an important relationship between the viewport scale and the annotation scale.
6. Which **MSLTSCALE** system variable setting should you use so you do not have to calculate the drawing scale factor when entering an **LTSCALE** value?
7. Name the tool used to update text properties according to the current properties of the text style on which the text is drawn.
8. What is an annotative object representation?
9. Briefly describe the result of setting the **ANNOAUTOSCALE** system variable to a value of 4.
10. Briefly explain the effect of turning annotation visibility on and off.

Drawing Problems

Start AutoCAD if it is not already started. Start a new drawing using an appropriate template of your choice. The template should include layers, text styles, dimension styles, and multileader styles appropriate for drawing the given objects. Add layers, text styles, dimension styles, and multileader styles as needed. Draw all objects using appropriate layers, text styles, dimension styles, multileader styles, justification, and format. Follow the specific instructions for each problem. Use your own judgment and approximate dimensions when necessary.

▼ Basic

1. Open P23-9 and save as P31-1. The P31-1 file should be active. Convert all the non-annotative objects to annotative objects. Resave the drawing.

2. Open P25-10 and save as P31-2. The P31-2 file should be active. Convert all of the non-annotative objects to annotative objects. Resave the drawing.

▼ Intermediate

3. Create the section view and side view shown. Use annotative objects to prepare a full-scale drawing of the part. Change the annotation scale to 2:1 and adjust the scale representations as needed according to the new scale. Save the drawing as P31-3.

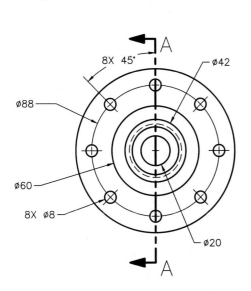

Name: Hub
Material: Cast Iron

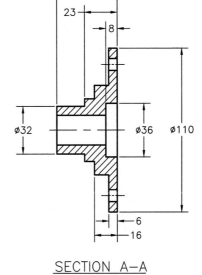

SECTION A–A

4. Create the section view and side views shown. Use annotative objects to prepare a full-scale drawing of the part. Change the annotation scale to 2:1 and adjust the scale representations as needed according to the new scale. Save the drawing as P31-4.

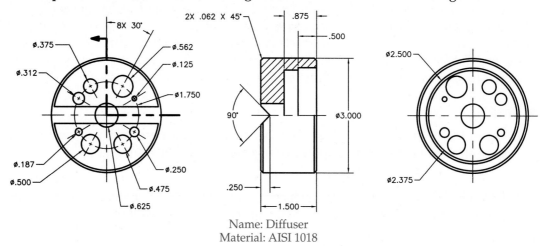

Name: Diffuser
Material: AISI 1018

5. Draw the fan shown at full scale in model space. Use annotative objects to prepare a full-scale view of the fan as shown and a view enlargement of the motor. You should not have to create a copy of the motor or develop scale specific layers. Save the drawing as P31-5.

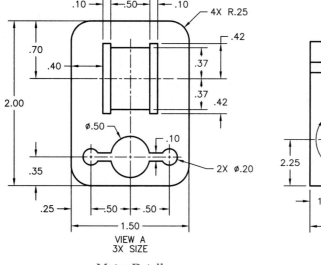

VIEW A
3X SIZE

Motor Detail

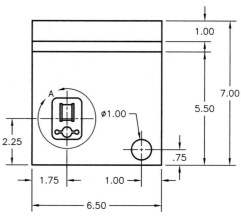

Fan

▼ Advanced

6. Draw the floor plan shown at full scale in model space. Use annotative objects to prepare a 1/4″ = 1′-0″ view. Change the annotation scale to 1/8″ = 1′-0″ and adjust the scale representations as needed according to the new scale. Save the drawing as P31-6.

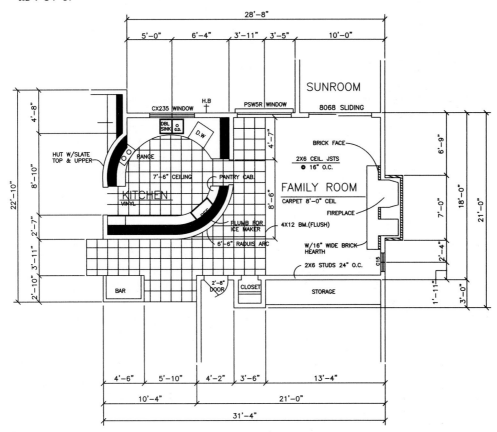

7. Draw the part shown at full scale in model space. Use annotative objects to prepare a full-scale view and a view enlargement of the part as shown. You should not have to create a copy of the part or develop scale specific layers. Save the drawing as P31-7.

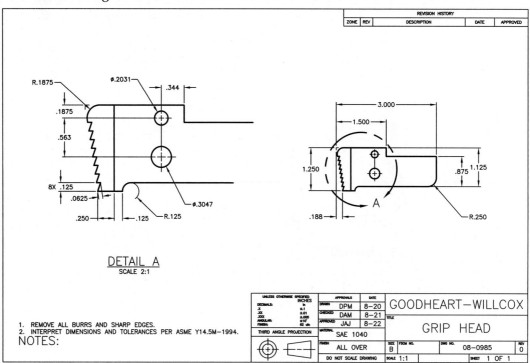

CHAPTER 32

External References

Learning Objectives

After completing this chapter, you will be able to do the following:

✓ Explain the function of external references.
✓ Attach an existing drawing to the current drawing.
✓ Use **DesignCenter** and tool palettes to attach external references.
✓ Bind external references and selected dependent objects to a drawing.
✓ Edit external references in the current drawing.

External references (xrefs) expand on the concept of reusing existing content, which is a major benefit of designing and drafting with AutoCAD. Xrefs are excellent for applications in which existing base drawings, complex symbols, images, and details are used often, are required to develop new drawings, or are shared by several users.

external reference (xref): A DWG, DWF, raster image, DNG, or PDF file incorporated into a drawing for reference only.

Introduction to Xrefs

You can reference drawing (DWG), design web format (DWF and DWFx), raster image, digital negative (DNG), and portable document format (PDF) files into the *host drawing*, also known as the *master drawing*. Using an xref is similar to inserting an entire drawing as a block. However, unlike a block, which is actually located in the file in which you insert it, xref file geometry is not added to the host drawing. File data appears on-screen for reference only. The result is usable information, but a much smaller host file size than would occur if you inserted a block or copied and pasted objects.

host (master) drawing: The drawing into which xrefs are incorporated.

In addition to reducing file size, one of the greatest benefits of using xrefs is that changes made to original xref files update in the host drawing to display the most recent reference content. AutoCAD reloads each xref whenever the host drawing loads. This allows an individual or a group of people to work on a multi-file project, with the assurance that any revisions to xrefs are displayed in any drawing in which the xrefs are used.

Types of Xref Files

Files that you can reference into a current drawing include existing DWG, DWF, DWFx, raster image, DNG, and PDF files. DWF and DWFx files are AutoCAD drawing files or other application files compressed for publication and viewing on the Web. DWF and DWFx file references are commonly used for sharing information from the Web or from an application other than AutoCAD.

Raster image and DNG file reference is appropriate whenever you need to add an image to a drawing, such as for a company logo in a title block. PDF file reference allows you to reuse PDF file content. Externally referencing an image or PDF file into a drawing is an excellent technique, because the large file sizes often associated with images and PDF files are not reproduced in the current drawing.

Xref Applications

DWG files are the most common externally referenced files and are the focus of this chapter. The term *xref* often applies specifically to referenced DWG files. In general, use external references to reuse existing drawing information to develop another drawing. There are countless applications for xref drawings in every drafting field. The following sections provide typical xref drawing applications. As you work with AutoCAD, you will discover a variety of uses for xref drawings.

Reference Existing Geometry

One of the most common applications for xref drawings is to reference existing geometry into the current drawing to use as a pattern or source of needed information. For example, a floor plan includes size and shape information required to prepare additional plans, elevations, sections, and details. **Figure 32-1** shows an example of referencing a floor plan file into a new drawing to use as an outline for creating a roof plan file. In this example, the floor plan xref attaches to the roof plan file to serve as an outline for creating the roof plan.

Figure 32-2 shows an example of the roof plan file created in **Figure 32-1** attached to an elevation file as an xref and then used to project an elevation. The roof plan xref includes a *nested* floor plan xref. In this example, the elevation file references the roof plan. The roof plan in turn references the floor plan.

nested xrefs: Xrefs contained within other xrefs.

Create a Multiview Drawing

Another xref application is to create commonly used drawings, such as sections and details, as separate drawing files and then attach each drawing as an xref to a host drawing. This method is similar to using blocks. Use floating viewports and layer viewport freezing to create a multiview layout. You can prepare a multiview drawing entirely from existing xref drawings or from a combination of objects created "in place" and attached xrefs.

Figure 32-1.
An example of using a floor plan xref drawing as a pattern, or outline, to draw a roof plan.

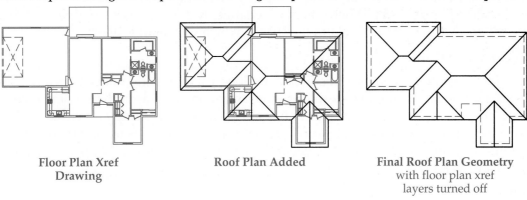

Floor Plan Xref
Drawing

Roof Plan Added

Final Roof Plan Geometry
with floor plan xref
layers turned off

Figure 32-2.
Using a roof plan xref drawing that contains a nested floor plan xref drawing as a pattern for projecting geometry needed to create an elevation. Projection lines, drawn using the **XLINE** tool, are for reference.

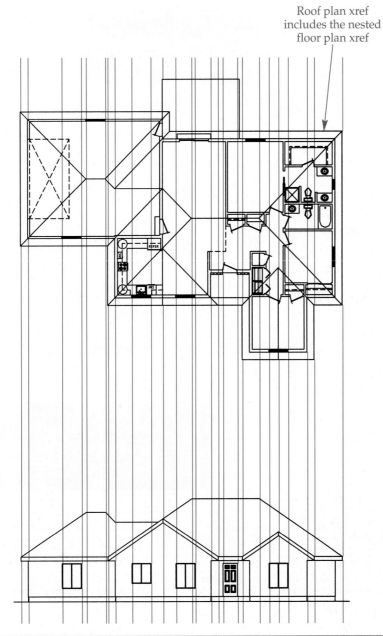

Roof plan xref
includes the nested
floor plan xref

Figure 32-3A shows an example of several stock details referenced into the model space environment of a new drawing. Floating viewports arrange the details in a layout as shown in **Figure 32-3B.** When you make changes to a detail in the original drawing file, all xrefs of the detail update in the files in which the detail is used.

Add Layout Content

Layout content, such as a title block or a long list of general notes, typically uses a standard format. If the format requires modification, such as adding a new note to a list of general notes, you can make changes to the xref drawing and easily update each host file that references the xref. This is the same concept as using xref drawings to build a multiview drawing. The general notes shown in **Figure 32-3B** are added to the layout as an xref.

Figure 32-3.
A—Xref frequently used drawing views into model space, reducing the size of the file and providing the ability to change instances of the view used in multiple drawings. B—Arrange referenced views in floating viewports like other model space objects.

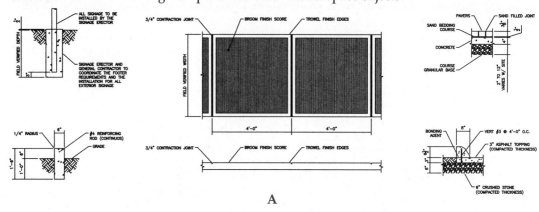

A

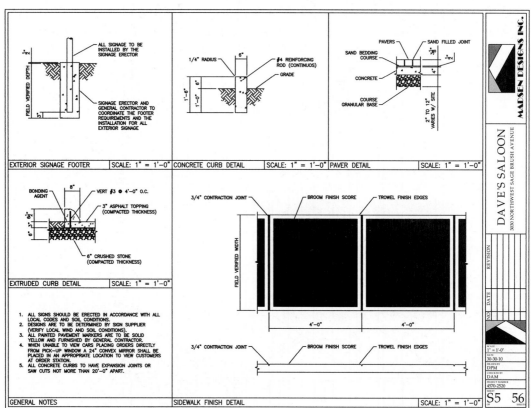

B

Arrange Sheet Views

You can use external references to arrange sheet views in layouts when you are working with sheet sets. Sheet sets are described in Chapter 33.

Preparing Xref and Host Drawings

Before you begin placing xref drawings, you should prepare the xref and host drawing files for xref insertion. When you place an xref drawing, everything you see in model space is inserted into the host file as a single item. Layout content is not xrefed. The default insertion base point for an xref file is the model space origin, or 0,0,0 point. This is the point attached to the crosshairs or located at the specified insertion point when you insert the xref into the host drawing. If it is critical that xref objects coincide with the 0,0,0 point for insertion, move all objects in model space as needed.

An alternative to moving objects to the origin is to use the **BASE** tool to change the insertion base point of the drawing. Access the **BASE** tool, and then select a new insertion base point. Save the drawing before using it as an xref.

If you use an appropriate template, little preparation should be necessary to prepare the host file to accept an xref. The host file should include a unique layer (named XREF or A-ANNO-REFR, for example) on which to place xrefs. As you will learn, layers in a referenced drawing file remain intact when you add the xref to a host drawing. Therefore, properties and states that you assign to the XREF layer have no effect on xref objects. The main purpose of the XREF layer is to contain the xref on a specific layer. Set the XREF layer current and proceed to place the xref drawing.

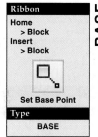

Placing External Reference Drawings

To place an xref, access the **ATTACH** tool to display the **Select Reference File** dialog box. The dialog box is set to display all file types by default. Pick **Drawing (*.dwg)** from the **Files of type:** drop-down list show and reference only drawings. Use the **Select Reference File** dialog box to locate the drawing file to add to the host file as an xref. Then pick the **Open** button to display the **Attach External Reference** dialog box. See **Figure 32-4.**

The **Attach External Reference** dialog box includes options for specifying how and where to place the selected file in the host drawing as an xref. If an external reference already exists in the current drawing, you can place another copy by choosing the file from the **Name:** drop-down. To place a different xref drawing, pick the **Browse...** button and select the new file in the **Select Reference File** dialog box.

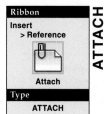

Figure 32-4.
The **Attach External Reference** dialog box allows you to specify how to place an xref in the host drawing. Pick the **Show Details** button to display additional file details, as shown.

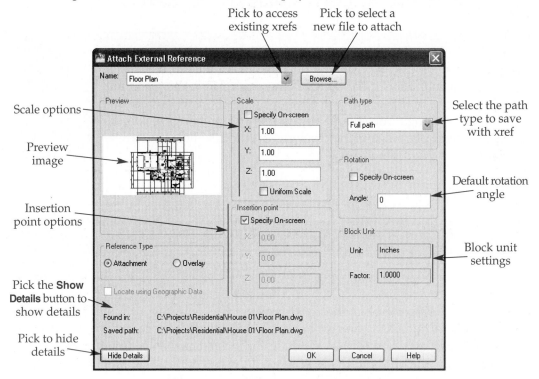

Pick to access existing xrefs

Pick to select a new file to attach

Scale options

Preview image

Insertion point options

Pick the **Show Details** button to show details

Pick to hide details

Select the path type to save with xref

Default rotation angle

Block unit settings

You can also place an xref using the **External References** palette shown in **Figure 32-5.** The **External References** palette is a complete external reference management tool. To place an xref drawing using the **External References** palette, pick the **Attach DWG** button from the **Attach** flyout, or right-click on the **File References** pane and select **Attach DWG...**. The **Select Reference File** dialog box appears, displaying only drawing files. Locate and select a file to add as an xref and pick the **Open** button to display the **Attach External Reference** dialog box.

> **NOTE**
>
> The **XATTACH** tool is identical to the **ATTACH** tool, but it initially displays only drawing files in the **Select Reference File** dialog box.

Attachment vs. Overlay

You can specify to insert an xref drawing as an *attachment* or an *overlay* by selecting the **Attachment** or **Overlay** radio button in the **Reference Type** area. Attach xrefs for most applications. Most often, an xref overlay allows you to share content with others in a design drafting team, typically while working in a networked environment. In this situation, you can overlay drawings without referencing nested xrefs.

Nesting occurs when an xref file contains, or references, another xref file. An attached xref with nested xrefs is the *parent xref*. When you attach an xref, the host drawing receives any nested xrefs that the xref contains. This does not happen when you overlay an xref. Furthermore, if you overlay an xref in a host drawing and then attach the host drawing to another drawing, the overlaid xref does not appear, even though the host drawing is attached.

For example, you should typically attach a floor plan xref to a host file to create a foundation plan. You can then attach the foundation plan xref to a host file to draw a

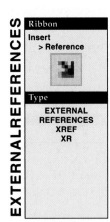

EXTERNALREFERENCES

Ribbon
Insert
> Reference

Type
EXTERNAL REFERENCES
XREF
XR

XATTACH

Type
XATTACH
XA

attachment: An xref linked with or referenced into the current drawing.

overlay: An xref displayed as an xref without being attached to the current drawing.

parent xref: An xref that contains one or more other xrefs.

Figure 32-5.
The **External References** palette provides access to all options for externally referenced files.

Pick to add an xref
drawing to the host file

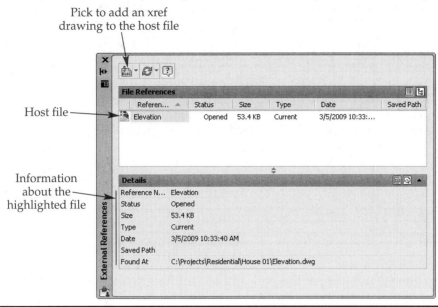

Host file

Information
about the
highlighted file

section. Attaching the foundation plan brings the foundation and floor plan geometry into the section file for reference. If a member of your design team uses your section, or is working on a drawing that already has the floor and/or foundation plan attached, she or he can overlay the section xref into a drawing without bringing in the floor plan and foundation plan.

> **NOTE**
>
> You can change an overlay to an attachment or an attachment to an overlay after insertion. Managing xrefs is described later in this chapter.

Selecting the Path Type

Use the **Path type** drop-down list in the **Path Type** area to set how AutoCAD stores the path to the xref file. The path locates the xref file when you open the host file. The path appears in the **Attach External Reference** dialog box when you pick the **Show Details** button, and later appears in the **External References** palette. See **Figure 32-6.**

The **Full path** option saves an *absolute path* and is active by default. When using this option, you must locate xref drawings in the same drive and folder specified in the saved path. You can move the host drawing to any location, but the xref drawings must remain in the saved path. This option is acceptable if it is unlikely that you will move or copy the host and xref drawings to another computer, drive, or folder.

absolute path: A path to a file defined by the location of the file on the computer system.

If you share your drawings with a client or eventually archive the drawings, the **Relative path** option is often more appropriate. This option saves a *relative path*, which you cannot use if the xref file is on a local or network drive other than the drive that stores the host file. If the host drawing and xref files are located in a single folder and subfolders, you can copy the folder to any location without losing the connection between files. For example, you can copy the folder from the C: drive of one computer to the D: drive of another computer, to a folder on a CD, or to an archive server. If you perform these types of transfers with the **Full path** option, you need to open the host drawing after copying and redefine the saved paths for all xref files.

relative path: A path to a file defined according to its location relative to the host drawing.

Figure 32-6.
You can reference a file using a full path, a relative path, or no path. The path type displays in the **Save Path** column in the **External References** palette.

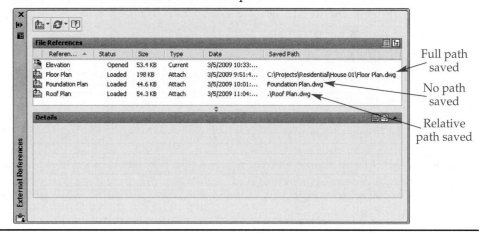

Full path saved

No path saved

Relative path saved

You can also choose not to save the path to the xref file by selecting the **No Path** option. If you choose this option, the xref file locates and loads only if you include the path to the file in one of the Support File Search Path locations or if the xref file is in the same folder as the host file. Specify the Support File Search Path locations in the **Files** tab of the **Options** dialog box.

> **NOTE**
>
> AutoCAD also searches for xref files in all paths of the current project name. These paths are listed under the Project Files Search Path in the **Files** tab of the **Options** dialog box. You can create a new project as follows:
> 1. Pick Project Files Search Path to highlight it, and then pick the **Add...** button.
> 2. Enter a project name if desired.
> 3. Pick the plus sign icon (+), and then pick the word Empty.
> 4. Pick the **Browse...** button and locate the folder that is to become part of the project search path. Then pick **OK**.
> 5. Complete the project search path definition by entering the **PROJECTNAME** system variable and specifying the same name used in the **Options** dialog box.

Additional Xref Placement Options

The remaining items in the **Attach External Reference** dialog box allow you to control or identify xref insertion location, scaling, rotation angle, and block unit settings. Use the text boxes in the **Insertion point** area to enter 2D or 3D coordinates for insertion of the xref if the **Specify On-screen** check box is deselected. Activate the **Specify On-screen** check box to specify the insertion location on-screen. The **Locate using Geographic Data** check box is active if the xref and host drawings include *geographic data*. Pick the check box to position the xref using geographic data.

geographic data: Information added to a drawing to describe specific locations and directions on Earth.

Use the **Scale** area to set xref scale factor. By default, the X, Y, and Z scale factors are set to 1. Enter different values in the corresponding text boxes or activate the **Specify On-screen** check box to display scaling prompts when you insert the xref. Select the **Uniform Scale** check box to apply the X scale factor to the Y and Z scale factors.

AutoCAD and Its Applications—Basics

The rotation angle for the inserted xref is 0 by default. Specify a different rotation angle in the **Angle:** text box, or select the **Specify On-screen** check box to display a rotation prompt for the rotation angle. The **Block Unit** area displays the unit and scale factor stored with the selected drawing file.

Inserting the Xref

After adjusting xref specifications in the **Attach External Reference** dialog box, pick the **OK** button to insert the xref into the host drawing. If you chose the **Specify On-screen** check box in the **Insertion point** area, the xref attaches to the crosshairs and a prompt asks for the insertion point. Specify an appropriate insertion point for the xref.

The options for attaching an xref are essentially the same as those used to insert a block. However, remember that xrefs are not added to the database of the host drawing file, as are inserted blocks. Therefore, using external references helps keep your drawing file size to a minimum.

Exercise 32-1

Access the Student Web site (www.g-wlearning.com/CAD) and complete Exercise 32-1.

Placing Xrefs with DesignCenter and Tool Palettes

To place an xref into the current drawing using **DesignCenter**, first use the **Tree View** area to locate the folder containing the drawing to attach. Then display the drawing files located in the selected folder in the **Content** area. Right-click on the drawing file in the **Content** area and select **Attach as Xref...**. Another method is to drag and drop the drawing into the current drawing area using the *right* mouse button. When you release the button, select the **Attach as Xref...** option. The **Attach External Reference** dialog box appears. Enter the appropriate values and pick the **OK** button to place the xref.

You must add an xref or drawing file to a tool palette in order to use the **Tool Palettes** palette to place the file as an xref. To add an xref to a tool palette, drag an existing xref from the current drawing or an xref from the **Content** area of **DesignCenter** into the **Tool Palettes** palette. Use drag and drop to attach the xref to the current drawing from the palette.

Xref files in tool palettes receive an external reference icon. A drawing file, not an xref, added to a tool palette from the current drawing or using **DesignCenter**, is designated as a block tool. To convert the block tool to an xref tool, right-click on the image in the **Tool Palettes** palette and select **Properties...** to display the **Tool Properties** dialog box. Then change the **Insert as** field status from Block to Xref using the **Insert as** drop-down list. The **Reference type** row controls whether the xref is inserted as an attachment or an overlay.

Working with Xref Objects

An xref is inserted as a single object. You can use editing tools such as **MOVE** and **COPY** to modify the xref as needed. However, there are some significant differences between xrefs and other objects. For example, if you erase an xref, the xref definition remains in the file, similar to an erased block. You must *detach* an xref to remove the xref from the file completely.

Xref drawings appear faded by default to help differentiate the xref from the host drawing. Xref fading is an on-screen display function only. Xrefs do not plot faded. The **XDWGFADECTL** system variable controls fading of xref drawings on-screen. The easiest way to adjust fading is to use the options in the expanded **Reference** panel of the **Insert** ribbon tab. See **Figure 32-7A**. Pick the **Xref Fading** button to activate or deactivate xref fading, and use the slider or text box to increase or decrease fading. The default value of 70% creates significant fading. See **Figure 32-7B**.

NOTE

In order to select an xref, you must pick an object displayed on-screen that is part of the xref.

Dependent Objects

dependent objects: Objects displayed in the host drawing, but defined in the xref drawing.

When you place an xref in a drawing, the host file receives all named objects in the xref file, such as layers and blocks, as *dependent objects*, even if the xref file does not use the objects. Dependent objects display in the host drawing for reference only. The xref drawing stores the actual object definitions.

Dependent objects are renamed when you attach an xref so that the xref file name precedes the actual object name, with a vertical bar symbol (|) separating the names. For example, a layer named A-DOOR within an xref drawing file named Floor Plan comes into the host drawing as Floor plan|A-DOOR. See **Figure 32-8**. The unique name distinguishes xref-dependent layers from layers that may have the same name in the host drawing. This also makes it easier to manage layers with several xrefs attached to the host drawing, because file names prefix the layers from each reference file. You cannot rename xref-dependent objects.

When you attach an xref, dependent objects such as layers are added to the host drawing only in order to support the display of the objects in the xref file. You cannot set xref layers current, and as a result, you cannot draw on xref layers. However, you can turn them on and off, thaw and freeze them, and lock or unlock them as needed. You can also change the colors and linetypes of xref layers.

Figure 32-7.
A—Use options in the ribbon to control xref fading. B—Default fading applied to xref objects.

Deselect to remove fading · Slide right to increase fading · Type a fade percentage

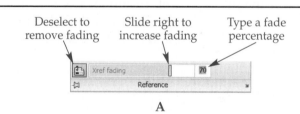

A

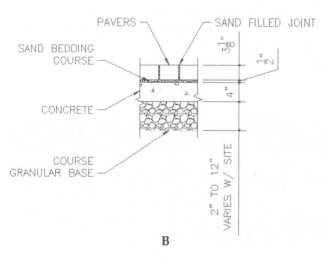

B

Figure 32-8.
The xref drawing name and a vertical bar symbol (|) precede xref-dependent layer names in the host drawing.

Filters available by default when you place an xref

Pick to display only xref layers

Xref-dependent layers

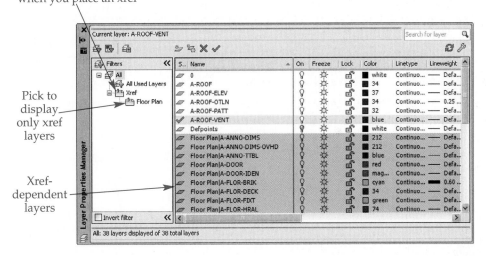

NOTE

Use the xref filter in the **Layer Properties Manager** palette to display and manage dependent layers. You can also save dependent layers in a layer state.

PROFESSIONAL TIP

When you attach a drawing as an xref, the reference file comes into the host drawing with the same layer colors and linetypes used in the original file. If you reference a drawing to check the relationship of objects between two drawings, it is a good idea to change the xref layer colors to make it easier to differentiate between the content of the host drawing and the xref drawing. Changing xref layer colors affects only the display in the current drawing and does not alter the original reference file.

Exercise 32-2

Access the Student Web site (www.g-wlearning.com/CAD) and complete Exercise 32-2.

Managing Xrefs

The **External References** palette is the primary tool for managing and accessing current information about xrefs found in a host drawing. The **External References** palette displays an upper **File References** pane and a lower **Details** pane. See **Figure 32-9.** You can display the **File References** pane in list view or tree view and with details or a preview.

Figure 32-9.
The **External References** palette allows you to view and manage referenced files. The **File References** pane appears in **List View** mode and the **Details** pane displays in **Details** mode.

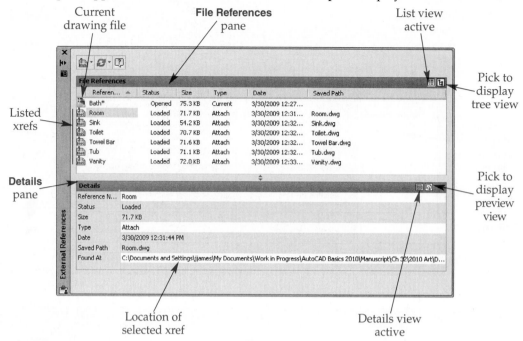

List View Display

The list view display shown in **Figure 32-9** is active by default. Pick the **List View** button or press the [F3] key to activate list view mode while in tree view mode. The labeled columns displayed in list view provide information about and management options for xrefs.

The **Reference Name** column displays the current drawing file name followed by the names of all existing xrefs in alphabetical or chronological order. The standard AutoCAD drawing file icon identifies the host drawing, and a sheet of paper with a paper clip icon identifies xref drawings. Each xref type displays a unique icon. The **Status** column describes the status of each xref, which can be:

- **Loaded.** The xref is attached to the drawing.
- **Unloaded.** The xref is attached but does not display or regenerate.
- **Unreferenced.** The xref has nested xrefs that are not found or are unresolved. An unreferenced xref does not display.
- **Not Found.** The xref file is not found in the specified search paths.
- **Unresolved.** The xref file is missing or cannot be found.
- **Orphaned.** The parent of the nested xref cannot be found.

The **Size** column lists the file size for each xref. The **Type** column indicates whether the xref is attached or referenced as an overlay. The **Date** column indicates the date the xref was last modified.

The **Saved Path** column lists the path name saved with the xref. If only a file name appears, the path is not saved. Prefixes describe the relative paths to xref files. In **Figure 32-6**, the characters .\ precede the Roof Plan reference file. The period (.) represents the folder containing the host drawing. From that folder, AutoCAD looks in the House 01 folder that contains the Roof Plan drawing. The Elevation reference file in **Figure 32-10** uses a similar specification. In this example, the same folder contains the Elevation xref and the host drawing. The characters ..\ precede the specification for the Wall xref. The double period instructs AutoCAD to move up one folder level from the current location. The double period repeats to move up multiple folder levels. For example, AutoCAD locates the Panel xref in **Figure 32-10** by moving up two folder levels from the folder of the host drawing and opening the Symbols folder.

Figure 32-10.
Relationship between the symbols in the **Saved Path** list and file locations within the folder structure.

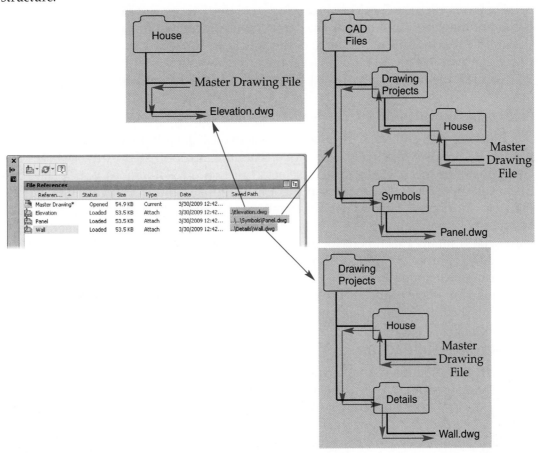

NOTE

The path saved to the xref is one of several locations AutoCAD searches when you open a host drawing and an xref requires loading. AutoCAD searches path locations to load xref files in the following order:

1. The path associated with the xref (full or relative path)
2. The current folder of the host drawing
3. The project paths specified in the Project Files Search Path
4. The support paths specified in the Support File Search Path
5. The Start in: folder path specified for the AutoCAD application shortcut, associated using the **Properties** option in the desktop icon shortcut menu.

PROFESSIONAL TIP

Adjust the column widths in list view mode as necessary to view complete information. To adjust the width of a column, move the cursor to the edge of the button at the top of the column until the cursor changes to a horizontal resizing cursor. Press and hold the left mouse button and drag the column to the desired width. If the columns extend beyond the width of the dialog box window, a horizontal scroll bar appears at the bottom of the list.

Tree View Display

To see a list of externally referenced files in the **File References** pane, and show nesting levels, pick the **Tree View** button or press the [F4] key. See **Figure 32-11.** Nesting levels are displayed in a format similar to the arrangement of folders. The status of the xref determines the appearance of the xref icon. An xref with an unloaded or not found status has a grayed-out icon. An upward arrow shown with the icon means the xref was reloaded, and a downward arrow means the xref was unloaded.

Viewing Details or a Preview

Details mode, shown in **Figure 32-9,** is active by default. Pick the **Details** button to display details while in **Preview** mode. The information listed in the **Details** pane corresponds to the host file or xref selected in the **File References** pane. The rows displayed in **Detail** mode are the same as the columns found in **List View** mode of the **File References** pane. However, in the **Details** pane, you can modify the reference name by entering a new name in the **Reference Name** text box. You can also adjust the reference type from an attachment to an overlay or from an overlay to an attachment by picking the appropriate option from the **Type** drop-down list. In addition, the **Details** pane contains a **Found At** row that you can use to update the location of an xref path. To display an image of the xref selected in the **File References** pane, pick the **Preview** button on the **Details** pane. Refer again to **Figure 32-11.**

Detaching, Reloading, and Unloading Xrefs

Each time you open a host drawing containing an attached xref, the xref loads and appears on-screen. This association remains permanent until you *detach* the xref. Erasing an xref does not remove the xref from the host drawing. To detach an xref, right-click on the reference name in the **File References** pane of the **External References** palette and pick **Detach.** All instances of the xref and all nested xrefs are removed from the current drawing, along with all referenced data.

In some situations, you may need to update, or *reload*, an xref file in the host drawing. For example, if you edit an xref while the host drawing is open, the updated version may be different from the version you see. To update the xref, right-click the

detach: Remove an xref from a host drawing.

reload: Update an xref in the host drawing.

Figure 32-11.
The **File References** pane in **Tree View** mode shows nested xref levels. The **Details** pane in **Preview** mode shows a thumbnail preview of the selected xref.

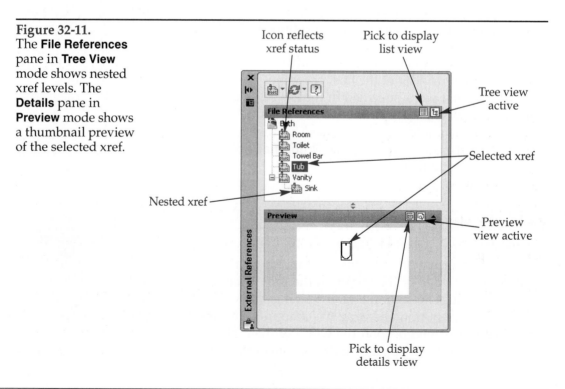

AutoCAD and Its Applications—Basics

reference name in the **File References** pane of the **External References** palette and pick **Reload**, or pick the **Reload All References** button from the flyout to reload all unloaded xrefs. Reloading xrefs forces AutoCAD to read and display the most recently saved version of each xref.

To *unload* an xref, right-click on the reference name in the **File References** pane of the **External References** palette and pick **Unload**. An unloaded xref does not display or regenerate, increasing performance. Reload the xref to redisplay it.

unload: Suppress the display of an xref without removing it from the host drawing.

NOTE

If AutoCAD cannot find an xref, an alert appears when you open the host drawing. Choose the appropriate option to ignore the problem or fix the problem using the **External References** palette.

Updating the Xref Path

A file path saved with an xref is displayed in the **Saved Path** column of the **File References** pane and the **Saved Path** row of the **Details** pane in the **External References** palette. If the **Saved Path** location does not include an xref file, when you open the host drawing, AutoCAD searches along the *library path*. A link to the xref forms if a file with a matching name is found. In such a case, the **Saved Path** location differs from where the file was actually found.

library path: The path AutoCAD searches by default to find an xref file, including the current folder and locations set in the **Options** dialog box.

Check for matching paths in the **External References** palette by comparing the path listed in the **Saved Path** column of the **File References** pane and **Saved Path** row of the **Details** pane with the listing in the **Found At** row of the **Details** pane. When you move an xref and the new location is not in the library path, the xref status is Not Found. To update or find the **Saved Path** location, select the path in the **Found At** edit box and pick the **Browse...** button to the right of the edit box to access the **Select new path** dialog box. Use this dialog box to locate the new folder and select the desired file. Then pick the **Open** button to update the path.

The Manage Xrefs Icon

By default, when you edit, save, and close an xref, and then open the host drawing, changes made to the xref automatically appear without any notification. When you make changes to an xref while the host drawing is open, a notification appears in the status bar tray. Changes indicate by the appearance of the **Manage Xrefs** icon, a balloon message, or both.

The **Tray Settings** dialog box controls notifications in the status bar tray for xref changes and other system updates. Select **Tray Settings...** from the status bar shortcut menu to access the **Tray Settings** dialog box. Select the **Display icons from services** check box to display the **Manage Xrefs** icon in the status bar tray when you attach an xref to the current drawing. If you modified an xref in the current file since opening the file, the **Manage Xrefs** icon appears with an exclamation sign. Pick the **Manage Xrefs** icon or right-click on the **Manage Xrefs** icon and select **External References...** to open the **External References** palette to reload the xref.

Select the **Display notifications from services** check box in the **Tray Settings** dialog box to display a balloon message notification with the name of the modified xref file. See **Figure 32-12A**. You can then pick the xref name in the balloon message to reload the file. The example in **Figure 32-12** shows adding a Towel Bar xref to the Room parent xref drawing. The xref is then reloaded in the host drawing named Bath. See **Figure 32-12B**. You can also reload xrefs by right-clicking on the **Manage Xrefs** icon and selecting **Reload DWG Xrefs**.

Figure 32-12.
The **Manage Xrefs** icon in the AutoCAD status bar tray provides a notification when you modify and save an xref file. A—A balloon message appears with an exclamation point over the icon. B—Reloading the xref file updates the current drawing and changes the appearance of the icon.

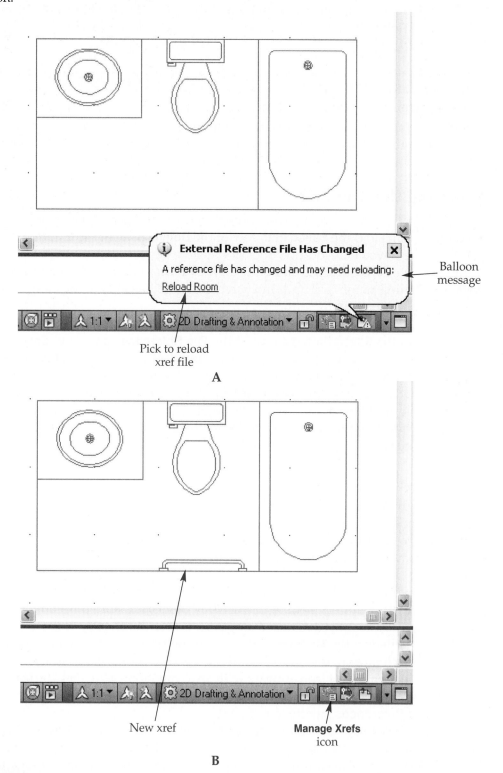

Exercise 32-3

Access the Student Web site (www.g-wlearning.com/CAD) and complete Exercise 32-3.

Clipping Xrefs

To display only a specific portion of an xref, AutoCAD allows you to create a boundary that displays an xref *subregion*. All geometry that falls outside the boundary is invisible, and objects that are partially within the subregion appear trimmed at the boundary. Although clipped objects appear trimmed, the xref file does not change. Clipping applies to a selected instance of an xref, not to the actual xref definition.

Use the **XCLIP** tool to create and modify clipping boundaries. A quick way to access the **XCLIP** tool is to select an object that is part of the xref file, then right-click and select **Clip Xref**. If you access the **XCLIP** tool while the xref is not selected, pick an object associated with the xref to clip. Then press [Enter] to accept the default **New boundary** option and select the clipping boundary.

When you select the **New boundary** option, a prompt asks you to specify the clipping boundary. Use the default **Rectangular** option to create a rectangular boundary. Then pick the corners of the rectangular boundary. **Figure 32-13** shows an example of using the **XCLIP** tool and a rectangular boundary. Note that the geometry outside the clipping boundary is no longer displayed after the clip is completed. The **New boundary** option includes additional options for specifying the clip boundary and area to clip, as briefly described in **Figure 32-14**.

> **subregion:** The displayed portion of a clipped xref.

Ribbon	XCLIP
Insert > Reference	
Clip	
Type	
XCLIP	
XC	

Figure 32-13.
A clipping boundary allows you to clip selected areas of an xref. A—Using the **Rectangular** boundary selection option. B—The clipped xref.

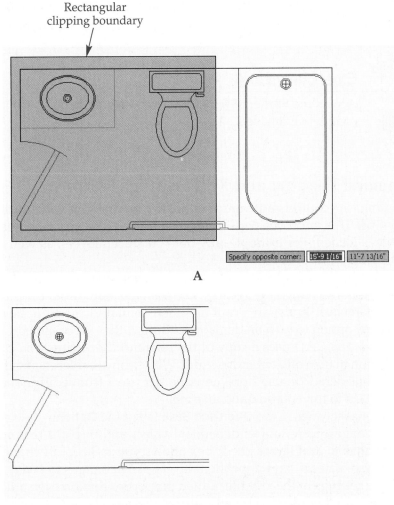

Rectangular clipping boundary

Specify opposite corner: 15'-9 1/16" 11'-7 13/16"

A

B

Figure 32-14.
Additional options available when using the **New boundary** function of the **XCLIP** tool.

Option	Description
Select Polyline	Select an existing polyline object as the clip boundary. If the polyline does not close, the start and endpoints of the boundary connect.
Polygonal	Draw an irregular polygon as a boundary.
Invert clip	Inverts the selection so that the portion of the xref that lies outside of the clipping boundary is clipped. Only the portion of the xref outside of the boundary is displayed.

You can edit a clipped xref as you would an unclipped xref. The clipping boundary moves with the xref. Note that nested xrefs are clipped according to the clipping boundary for the parent xref.

The clipping boundary, or frame, is invisible by default. Use the **XCLIPFRAME** system variable to toggle the display of the clipping boundary frame. Set the value of **XCLIPFRAME** to 1 to turn on the frame.

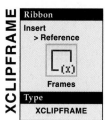

The other options of the **XCLIP** tool function after a clip boundary has been defined. The **ON** and **OFF** options turn the clipping feature on or off as needed. The **Clipdepth** option allows you to define front and back clipping planes to control the portion of a 3D drawing that displays. Clipping 3D models is described in *AutoCAD and Its Applications—Advanced*. Use the **Delete** option to remove an existing clipping boundary, returning the xref to its unclipped display. Use the **generate Polyline** option to create and display a polyline object at the clip boundary to frame the clipped portion.

Exercise 32-4

Access the Student Web site (www.g-wlearning.com/CAD) and complete Exercise 32-4.

Using Demand Loading and Xref Editing Controls

demand loading:
Loading only the part of an xref file necessary to regenerate the host drawing.

Demand loading controls how much of an xref loads when you attach the xref to the host drawing. This improves performance and saves disk space because only part of the xref file is loaded. For example, any data on frozen layers, as well as any data outside of clipping regions, does not load.

Demand loading occurs by default. Use the **Open and Save** tab of the **Options** dialog box to check or change the setting. The **Demand load Xrefs:** drop-down list in the **External References (Xrefs)** area contains each demand loading option. Select the **Enabled with copy** option to turn on demand loading. Other users can edit the original drawing because AutoCAD uses a copy of the referenced drawing. You can also pick the **Enabled** option to turn on demand loading. While you are referencing the drawing, the xref file is considered "in use," preventing other users from editing the file. Select the **Disabled** option to turn off demand loading.

Two additional settings in the **Open and Save** tab of the **Options** dialog box control the effects of changes made to xref-dependent layers and in-place reference editing. The **Retain changes to Xref layers** check box allows you to keep all changes made to the properties and states of xref-dependent layers. Any changes to layers take precedence over layer settings in the xref file. Edited properties remain even after the xref is reloaded. The **Allow other users to Refedit current drawing** check box controls whether the current drawing can be edited in place by others while it is open and when it is referenced by another file. Both check boxes are selected by default.

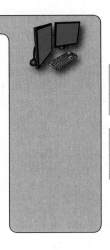

If you plan to use a drawing as an external reference, it is good practice to save the file with *spatial indexes* and *layer indexes*. These lists help improve performance when you reference drawings with frozen layers and clipping boundaries. Use the following procedure to create spatial and layer indexes:

1. Access the **Save Drawing As** dialog box.
2. Pick **Options...** from the **Tools** flyout button and select the **DWG Options** tab of the **Saveas Options** dialog box.
3. Select the type of index required from the **Index type:** drop-down list.
4. Pick the **OK** button and save the drawing.

spatial index: A list of objects ordered according to their locations in 3D space.

layer index: A list of objects ordered according to the layers on which they reside.

Binding an Xref

You can make an xref a permanent part of the host drawing as if you were inserting the file using the **INSERT** tool. This is called *binding* an xref. Binding is useful when you need to send the full drawing file to another location or user, such as a plotting service or client. To bind an xref using the **External References** palette, right-click on the reference name and pick **Bind....** This displays the **Bind Xrefs** dialog box, which contains **Bind** and **Insert** radio buttons.

binding: Converting an xref to a permanently inserted block in the host drawing.

Using the Insert and Bind Options

The **Insert** option converts the xref into a normal block, as if you had used the **INSERT** tool to place the file. In addition, the drawing is entered into the block definition table, and all named objects, such as layers, blocks, and styles, are incorporated into the host drawing as named in the xref. For example, if you bind an xref file named PLATE that contains a layer named OBJECT, the xref-dependent layer PLATE|OBJECT becomes the locally defined layer OBJECT. All other xref-dependent objects lose the xref name and assume the properties of the locally defined objects with the same name. The **Insert** binding option provides the best results for most purposes.

The **Bind** option also converts the xref into a normal block. However, the xref name remains with all dependent objects. Two dollar signs with a number in between replace the vertical line in each name. For example, an xref layer named Title|Notes becomes Title0Notes. The number inside the dollar signs automatically increments if a local object definition with the same name exists. For example, if Title0Notes already exists in the drawing, the layer is renamed to Title1Notes. In this manner, all xref-dependent object definitions that are bound receive unique names. Rename named objects using the **RENAME** tool.

Binding Specific Dependent Objects

Binding an xref allows you to make all dependent objects in an xref file a permanent part of the host drawing. Dependent objects include named items such as blocks, dimension styles, layers, linetypes, and text styles. Before binding, you cannot directly use any dependent objects from a referenced drawing in the host drawing. For example, you cannot make an xref layer or text style current in the host drawing.

In some cases, you may only need to incorporate one or more specific named objects, such as a layer or block, from an xref into the host drawing, instead of binding the entire xref. If you only need selected items, it can be counterproductive to bind an entire drawing. Instead, use the **XBIND** tool and corresponding **Xbind** dialog box, shown in **Figure 32-15,** to select specific named objects to bind.

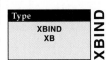

Type
XBIND
XB

XBIND

Figure 32-15.
Use the **Xbind** dialog box to bind xref-dependent objects individually to the host drawing.

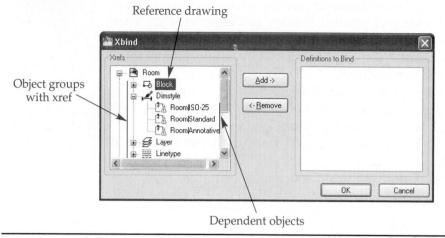

Reference drawing

Object groups with xref

Dependent objects

Xrefs are displayed using AutoCAD drawing file icons. Expand a group to select an individually named object from the corresponding group. To select an object for binding, highlight it and pick the **Add** button. The names of all objects selected and added display in the **Definitions to Bind** list. Pick the **OK** button to complete the operation. A message displayed on the command line indicates how many objects of each type are bound.

Individual objects bound using the **XBIND** tool are renamed in the same manner as objects bound using the **Bind** option in the **Bind Xrefs** dialog box. An automatic linetype bind performs so that a layer that includes a linetype not loaded in the host drawing can reference the required linetype definition. The linetype includes a new linetype name, such as xref1$0$hidden. In a similar manner, a previously undefined block may automatically bind to the host drawing because of binding nested blocks. Rename bound objects using the **RENAME** tool.

Exercise 32-5

Access the Student Web site (www.g-wlearning.com/CAD) and complete Exercise 32-5.

Editing Xref Drawings

reference editing:
Editing reference drawings from within the host file.

One option for editing an xref drawing is to use in-place, or *reference editing,* within the host drawing. You can save any changes made to the xref to the original drawing from within the host drawing. Alternatively, you can edit the xref in a separate drawing window as you would any other drawing file.

Reference Editing

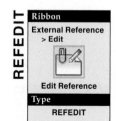

REFEDIT

Ribbon
External Reference > Edit

Edit Reference

Type
REFEDIT

The **REFEDIT** tool allows you to edit xref drawings in place. A quick way to initiate reference editing is to double-click on an xref, or select an xref and then right-click and select **Edit Xref In-place**. If you access the **REFEDIT** tool without first selecting an xref, you must then pick the xref to edit. The **Reference Edit** dialog box opens with the **Identify Reference** tab active. See **Figure 32-16.** The example shows the Room reference

Figure 32-16.
The **Reference Edit** dialog box lists the name of the selected reference drawing and displays an image preview.

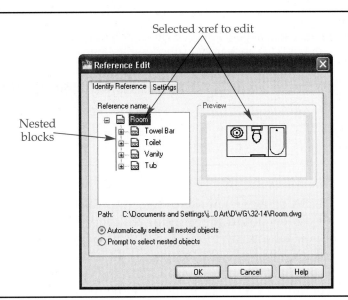

Selected xref to edit

Nested blocks

drawing selected for editing. Notice that nested blocks, like the Bath Tub 26 x 60 in. block found in the Tub reference, are listed under the parent xref.

The **Automatically select all nested objects** radio button in the **Path:** area is selected by default. Use this option to make all xref objects available for editing. To edit specific xref objects, pick the **Prompt to select nested objects** radio button. The Select nested objects: prompt displays after you pick the **OK** button, allowing you to pick objects that belong to the previously selected xref. Pick all geometry to edit and press [Enter]. The nested objects you select make up the *working set*. If multiple instances of the same xref appear, be sure to pick objects from the original xref you select.

Additional options for reference editing are available in the **Settings** tab of the **Reference Edit** dialog box. The **Create unique layer, style, and block names** option controls the naming of selected layers and *extracted* objects. Check the box to assign the prefix n, with n representing an incremental number, to object names. This is similar to the renaming method used when you bind an xref.

The **Display attribute definitions for editing** option is available if you select a block object in the **Identify Reference** tab of the **Reference Edit** dialog box. Check the box to edit any attribute definitions included in the reference. To prevent accidental changes to objects that do not belong to the working set, check the **Lock objects not in working set** option. This makes all objects outside of the working set unavailable for selection in reference editing mode.

If the selected xref file contains other references, the **Reference name:** area lists all nested xrefs and blocks in tree view. In the example given, Toilet, Tub, Vanity, and Towel Bar are nested xrefs in the Room xref. If you pick the drawing file icon next to Vanity in the tree view, for example, an image preview appears and the selected xref becomes highlighted in the drawing window.

When you finish adjusting settings, pick the **OK** button to begin editing the xref. The primary difference between the drawing and reference editing environments is the **Edit Reference** panel that appears in each ribbon tab. See **Figure 32-17.** Use the tools in the **Edit Reference** panel to add objects to the working set, remove objects from the working set, and save or discard changes to the original xref file.

Any object drawn during the in-place edit is automatically added to the working set. Use the **Add to Working Set** button to add existing objects to the working set. When you add an object to the working set, it is extracted, or removed, from the host drawing. The **Remove from Working Set** button allows you to remove selected objects from the working set. Removing a previously extracted object adds the object back to the host drawing.

working set:
Nested objects selected for editing during a **REFEDIT** operation.

extracted:
Temporarily removed from the drawing for editing purposes.

Figure 32-17.
The **Edit Reference** panel appears in each ribbon tab during reference editing. Most other drawing tools and options appear and function the same in the drawing and reference editing environments.

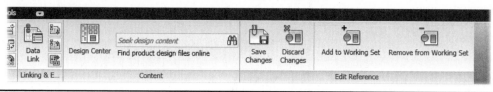

Figure 32-18A shows reference-editing the Vanity xref nested in the Room xref. Notice that all objects not in the working set fade, while the objects in the working set appear in normal display mode. Once you define the working set, use drawing and editing tools to alter the xref. When the edit is complete and you want to save the changes, pick the **Save Changes** button. Pick the **OK** button when AutoCAD asks if you want to continue with the save and redefine the xref. All instances of the xref are updated. **Figure 32-18B** shows the xref after editing to redesign the sink and add a faucet. To exit reference editing without saving changes, pick the **Discard Changes** button.

Figure 32-18.
Reference editing. A—Objects in the drawing that are not a part of the working set are grayed out during the reference-editing session. B—All instances of the xref immediately update after reference editing.

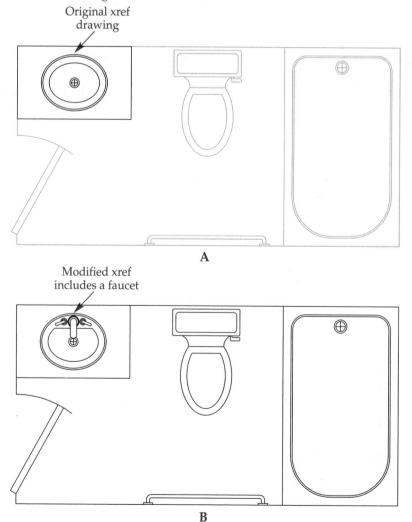

In-place reference editing is best suited for minor revisions. Conduct major xref revisions in the original drawing file.

All reference edits made using reference editing are saved back to the original drawing file and affect any host drawing that references the file. For this reason, it is critically important that you edit external references only with the permission of your instructor or supervisor.

Opening an Xref File

The **XOPEN** tool allows you to open an xref in a separate drawing window from within the host drawing. A quick way to access the **XOPEN** tool is to select an xref, and then right-click and select **Open Xref**. This is essentially the same procedure as using the **OPEN** tool, but faster. If you access the **XOPEN** tool without first selecting an xref, you must then pick the xref to open.

The xref drawing file opens in a separate drawing window. After you make changes and save the xref file, you must reload the xref file in the host file. Use the **External References** palette or the **Manage Xrefs** icon in the status bar to reload the modified xref file. This ensures that the host file is up-to-date.

You can also open an xref in the **External References** palette by right-clicking the xref name and selecting **Open**.

Select an xref to display the **External References** ribbon tab. This is a convenient location to access tools for reference editing, opening, and clipping an xref. An option is also available for accessing the **External References** palette.

Exercise 32-6

Access the Student Web site (www.g-wlearning.com/CAD) and complete Exercise 32-6.

Chapter Test

Answer the following questions. Write your answers on a separate sheet of paper or go to the Student Web site (www.g-wlearning.com/CAD) and complete the electronic chapter test.

1. What five types of files can you reference into an AutoCAD drawing?
2. What effect does the use of referenced drawings have on drawing file size?
3. What is a nested xref?
4. List at least three common applications for xrefs.
5. On what layer should you insert xrefs into a drawing?
6. Which tool allows you to attach an xref drawing to the current file?
7. What is the difference between an overlaid xref and an attached xref?
8. What is the difference between an absolute path and a relative path?
9. Describe the process of placing an xref using **DesignCenter**.
10. What must you do before you can use a tool palette to place an xref?
11. If you attach an xref file named FPLAN to the current drawing, and FPLAN contains a layer called ELECTRICAL, what name will appear for this layer in the **Layer Properties Manager**?
12. What is the purpose of the **Detach** option in the **External References** palette?
13. When do xrefs update in the host drawing?
14. What could you do to suppress an xref temporarily without detaching it from the master drawing?
15. Which tool allows you to display only a specific portion of an externally referenced drawing?
16. What are spatial and layer indexes, and what function do they perform?
17. Why would you want to bind a dependent object to a master drawing?
18. What does the layer name WALL0NOTES mean?
19. What tool allows you to edit external references in place?
20. What tool allows you to open a parent xref drawing into a new AutoCAD drawing window by selecting the xref in the master drawing?

882

AutoCAD and Its Applications—Basics

Drawing Problems

Start AutoCAD if it is not already started. Start a new drawing using an appropriate template of your choice. The template should include layers, text styles, dimension styles, and multileader styles appropriate for drawing the given objects. Add layers, text styles, dimension styles, and multileader styles as needed. Draw all objects using appropriate layers, text styles, dimension styles, multileader styles, justification, and format. Follow the specific instructions for each problem. Use your own judgment and approximate dimensions when necessary.

▼ Basic

1. Attach a dimensioned problem from Chapter 18 into a new drawing as an xref. Save the drawing as P32-1.

2. Attach a dimensioned problem from Chapter 19 into a new drawing as an xref. Save the drawing as P32-2.

3. Attach a dimensioned problem from Chapter 20 into a new drawing as an xref. Save the drawing as P32-3.

▼ Intermediate

4. Attach the EX30-9.dwg file used in Exercise 30-9 into a new drawing as an xref. Copy the xref three times. Use the **XCLIP** tool to create a clipping boundary on each view. Apply an inverted rectangular clip to the original xref, a polyline boundary on the first copy, and a polygonal boundary on the second copy. Save the drawing as P32-4.

5. Attach the EX30-9.dwg file used in Exercise 30-9 into a new drawing as an xref. Bind the xref to the new drawing. Rename the layers to the names assigned to the original EX30-9 (xref) file. Explode the block created by binding the xref. Save the drawing as P32-5.

6. Create the multi-detail drawing shown according to the following information:
   - Use the MECHANICAL-INCH.dwt drawing template file available on the Student Web site.
   - Set drawing units to fractional.
   - Xref the following files into model space: Detail-Item 1.dwg, Detail-Item 2.dwg, Detail-Item 3.dwg, Detail-Item 4.dwg, Detail-Item 5.dwg, and Detail-Item 6.dwg. These files are available on the Student Web site.
   - Use six floating viewports on the **C-SIZE** layout to arrange and scale the details. Use a 1:2 scale.
   - Plot the drawing.

   Save the drawing as P32-6.

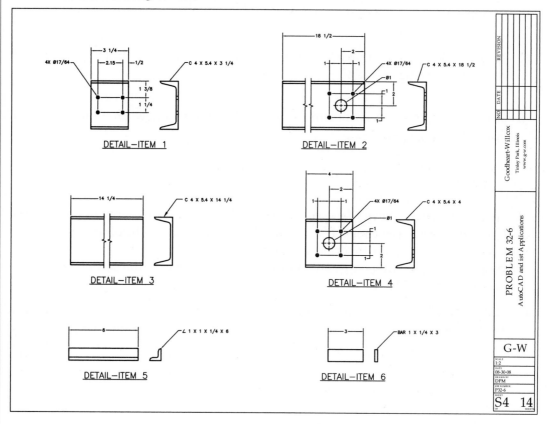

▼ Advanced

7. Design and draw a basic residential floor plan using an appropriate template. Save the file as P32-7FLOOR. Xref the P32-7FLOOR file into a new file as an attachment. Use the xref to help draw a roof plan. Save the roof plan file as P32-7ROOF.

8. Xref the P32-7ROOF file into a new file as an attachment. Use the xref to help draw front and rear elevations. Save the elevation file as P32-8.

9. Use a word processor to write a report of approximately 250 words explaining the purpose of external references. Include a brief description of the types of files that you can reference. Cite at least three examples from actual industry applications of using external references to help prepare drawings. Use at least four sketches to illustrate your report.

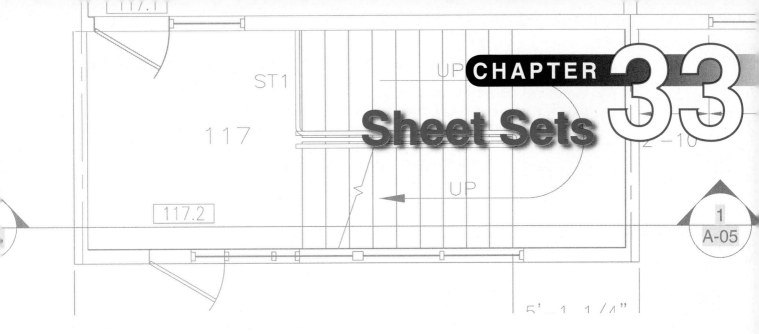

CHAPTER 33

Sheet Sets

Learning Objectives

After completing this chapter, you will be able to do the following:

✓ Identify and describe the functions of the **Sheet Set Manager**.
✓ Create sheet sets and subsets.
✓ Add sheets and sheet views to a sheet set.
✓ Plot or publish a set of sheets.
✓ Insert callout blocks and view labels into sheet views.
✓ Set up custom properties for a sheet set.
✓ Create a sheet list table.
✓ Archive a set of electronic files for a sheet set.

A design project typically requires a set of drawings and documents that completely specify design requirements. Often a number of sources, such as clients and vendors, share project information. Effectively organizing and distributing drawings during the course of a project is critical to delivering accurate drawings in an orderly and timely manner. *Sheet sets* help simplify the management of a project that contains multiple drawings and views. This chapter describes how to use sheet sets to structure different drawing layouts into groups of files for reviewing, plotting, and publishing.

> **sheet set:** A collection of drawing sheets for a project.

Introduction to Sheet Sets

AutoCAD *sheets* created in drawing file layouts can have additional project-specific properties. All sheets in a sheet set can use a single template. The template can include a title block and other layout content with attributes containing *fields*. Fields allow values such as project name and sheet number to change during the course of the project. The field modifies in the template, automatically applying changes to all sheets within the set. When you insert a new sheet into a sheet set, the sheet numbers and all sheet references update. This automation saves time and improves drawing set accuracy.

> **sheet:** A printed drawing or electronic layout produced for a project.

> **fields:** Special text objects that display values that update automatically.

Once a sheet set is complete, you can print, *publish*, and archive the entire set in a single operation. This is very efficient. For example, it is far easier to plot a sheet set containing twenty sheets than to open and plot twenty separate drawings.

> **publish:** Create electronic files for distribution or plotting.

SHEETSET

Quick Access

Sheet Set Manager

Ribbon

View
> Palettes

Sheet Set Manager

Type

**SHEETSET
SSM**

The Sheet Set Manager

The **Sheet Set Manager** palette, shown in **Figure 33-1,** includes three tabs that allow you to create, organize, and access sheet sets. Use the **Sheet Set Control** drop-down list at the top of the **Sheet Set Manager** to open and create sheet sets. The buttons next to the drop-down list control and manage the items listed in the **Sheet Set Manager.** The buttons vary depending on the current tab. Like other palettes, you can resize, dock, and set the **Sheet Set Manager** to auto-hide. The **Sheet Set Manager** also makes use of extensive tooltips that describe and allow you to preview items in the palette, as well as shortcut menus for accessing tools and options.

Figure 33-1.
The **Sheet Set Manager** palette contains the **Sheet List, Sheet Views,** and **Model Views** tabs. The **Sheet Set Control** drop-down list contains options for creating and opening a sheet set.

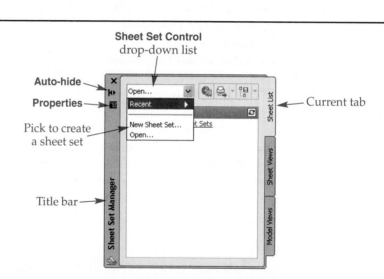

Creating Sheet Sets

Create sheet sets from within an open file using the **Create Sheet Set** wizard. See **Figure 33-2.** Access the wizard from within the **Sheet Set Manager** by picking **New Sheet Set...** from the **Sheet Set Control** drop-down list. You can create sheet sets from an example sheet set or by collecting existing drawing files.

Figure 33-2.
Select the **An example sheet set** radio button on the **Begin** page to use an AutoCAD sheet set or another existing sheet set as a model. Pick the **Existing drawings** radio button to create a new sheet set from an existing drawing project.

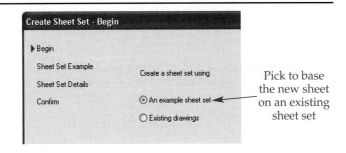

Using an Example Sheet Set

An example sheet set uses an existing sheet set as a model, similar to a template, for developing a new sheet set. You can modify the example sheet as needed to fit the needs of the new sheet set. AutoCAD provides several example sheet sets based on different drafting disciplines. You are not limited to the examples provided—you can use any existing sheet set as an example sheet set. To create a new sheet set from an example sheet set, access the **Create Sheet Set** wizard, and at the **Begin** page, pick the **An example sheet set** radio button.

Sheet Set Example Page

Pick the **Next** button to display the **Sheet Set Example** page. See **Figure 33-3.** Sheet set information is saved in a sheet set data file (DST). When you create a sheet set from an example sheet set, you start from an existing DST file. The **Select a sheet set to use as an example** radio button is active by default, and a list box displays all DST files in the default Template folder. Pick a sheet set from the list box, or select the **Browse to another sheet set to use as an example** radio button and then select the ellipsis (...) button and use the **Browse for sheet set** dialog box to locate a DST file in another folder. After selecting the DST file for the example sheet set, pick the **Next** button to display the **Sheet Set Details** page. See **Figure 33-4.**

Sheet Set Details Page

The **Sheet Set Details** page allows you to modify the existing sheet set data and create settings for the new project. Enter the name, or title, of the sheet set in the **Name of new sheet set** text box. The name is typically the project number or a short description of the project. Type a description for the sheet set in the **Description (optional)**

Figure 33-3.
Use the **Sheet Set Example** page to select an example sheet set.

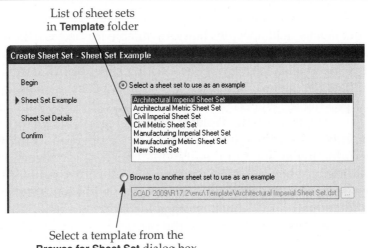

Figure 33-4.
Enter a name, description, and file path location for the new sheet set on the **Sheet Set Details** page.

Enter a name

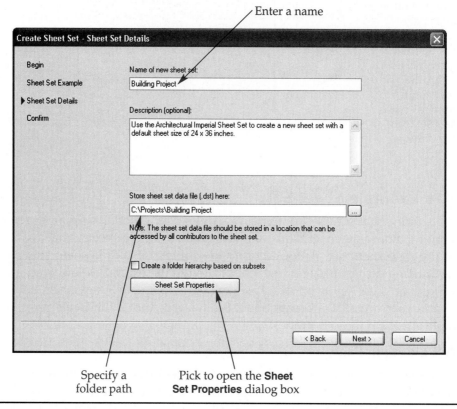

Specify a
folder path

Pick to open the **Sheet
Set Properties** dialog box

area if desired. The **Store sheet set data file (.dst) here** text box determines where the sheet set file is saved. Pick the ellipsis (**...**) button and select a folder in the **Browse for sheet set folder** dialog box to redefine the default sheet set file location. Pick the **Create a folder hierarchy based on subsets** check box to organize a sheet set so that folders group into *subsets*. Layouts in each folder form under each subset.

subsets: Groups of layouts based on folder hierarchy.

Pick the **Sheet Set Properties** button to open the **Sheet Set Properties** dialog box. See **Figure 33-5**. The **Sheet Set** category includes basic sheet set properties. The **Name** field contains the name entered in the **Name of new sheet set** text box of the **Sheet Set Details** page. The **Sheet set data file** field shows the location of the DST file, specified in the **Store sheet set data file (.dst) here** text box of the **Sheet Set Details** page. The **Description** field contains the description entered in the **Description** area of the **Sheet Set Details** page.

The **Model view** field specifies the folder(s) containing the drawing files used for the sheet set. The **Label block for views** field specifies the block used to label views. The **Callout blocks** field specifies blocks available for use as callout blocks. The **Page setup overrides file** field specifies the location of a drawing template (DWT) file containing a page setup used to override the existing sheet layout settings. You can change all of the **Sheet Set** category settings, except the **Sheet set data file** option, using the text box or by picking the ellipsis (**...**) button to navigate to a specific location.

The properties in the **Project Control** category allow you to store and update information based on the current project. The properties in the **Sheet Creation** category determine the location of the drawing files for new sheets and the template used to create them. When you add a new sheet to a sheet set, AutoCAD creates a new drawing file based on the template and layout specified in the **Sheet creation template** setting. The folder path in the **Sheet storage location** field determines where the new file is saved. It is important to specify the correct location so you know where the files are saved.

Figure 33-5.
The **Sheet Set Properties** dialog box stores the main properties of a sheet set.

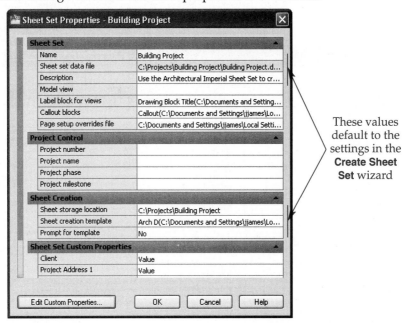

These values
default to the
settings in the
**Create Sheet
Set** wizard

When selecting the **Sheet creation template** value, you must specify a template file and a layout. To modify this setting, pick in the text box and then pick the ellipsis (**...**) button to display the **Select Layout as Sheet Template** dialog box. See **Figure 33-6.** All layouts in the selected template appear in the list box. Select the layout and then pick the **OK** button.

If the value in the **Prompt for template** property is set to the default **No**, AutoCAD automatically uses the template layout specified in the **Sheet creation template** field when creating a new sheet. Select **Yes** to choose a different layout when creating a new sheet. Information specific to the project is set up in the **Sheet Set Custom Properties** section, as described later in this chapter. Once all the values in the **Sheet Set Properties** dialog box are set, pick the **OK** button to return to the **Sheet Set Details** page. Pick the **Next** button to continue creating the new sheet set.

Figure 33-6.
Selecting an existing layout as a template for new sheets in a sheet set.

Pick to select a
different template file

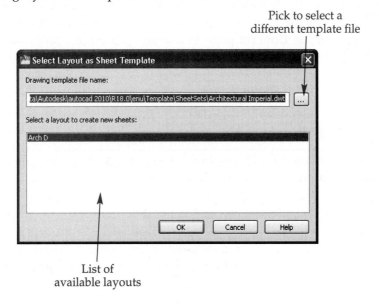

List of
available layouts

Figure 33-7.
Use the **Confirm** page to preview settings before creating the sheet set.

Subsets in new sheet set

Scroll down to preview more information

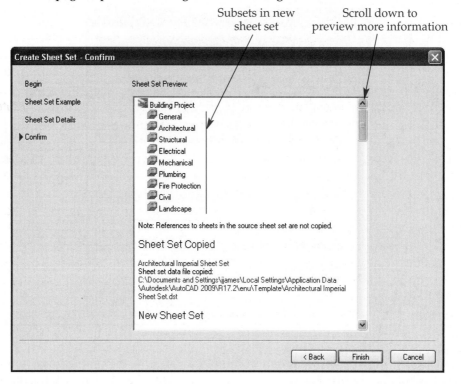

Confirm Page

The **Sheet Set Preview** area of the **Confirm** page displays all information associated with the sheet set. See **Figure 33-7.** The example shown includes a sheet set named Building Project. This sheet set contains a number of subsets related to the project, such as General and Architectural. After creating the sheet set, you can add sheets to each subset. Use the **Back** button to return to a previous page to make changes. After reviewing the information on the **Confirm** page, pick the **Finish** button to create the sheet set.

The sheet set data file is saved to the specified location. You can then open the sheet set in the **Sheet Set Manager**. Since a sheet set is not associated with a particular drawing file, you can open any sheet set, regardless of the active drawing file.

PROFESSIONAL TIP

You can copy and paste the information in the **Sheet Set Preview** area of the **Confirm** page to a word processing program for saving and printing.

Exercise 33-1

Access the Student Web site (www.g-wlearning.com/CAD) and complete Exercise 33-1.

Collecting Existing Drawing Files

You can also import existing layouts from drawing files to create sheets. Each layout in each drawing becomes a separate sheet. When you create a sheet set in this manner, organize all the files used in the project in a structured hierarchy of folders. You should place one layout in each drawing file to simplify access to different layout tabs. In addition, prepare a sheet creation template and specify the template in the **Sheet Set Properties** dialog box to ensure that all sheets use the same layout settings.

To create a new sheet set from an existing drawing project, access the **Create Sheet Set** wizard. At the **Begin** page, select the **Existing drawings** radio button and pick the **Next** button to display the **Sheet Set Details** page. Specify a name and description for the sheet set and the location to save the data file. Pick the **Sheet Set Properties** button to specify additional sheet set properties. When you are finished, pick the **Next** button to display the **Choose Layouts** page shown in **Figure 33-8**.

The **Choose Layouts** page allows you to specify the drawings and layouts added to the sheet set. Pick the **Browse...** button to display the **Browse for Folder** dialog box, and select the folder containing the drawing files with the desired layouts. The selected folder, the drawing files it contains, and all layouts within those drawings are displayed in the list box. When you first open a folder in the **Choose Layouts** page, all drawing files with layouts in the selected folder appear for selection.

The items you check are added to the new sheet set. Uncheck a layout box to remove the layout from the new sheet set. Uncheck a drawing file to uncheck all related layouts. Uncheck a folder to uncheck all related drawing files and layouts. In **Figure 33-8**, only the Furniture.dwg and associated layouts are not included in the sheet set. Use the **Browse for Folder** dialog box as needed to add more folders to the **Choose Layouts** page.

Figure 33-8.
Use the **Choose Layouts** page to import existing layouts to a new sheet set.

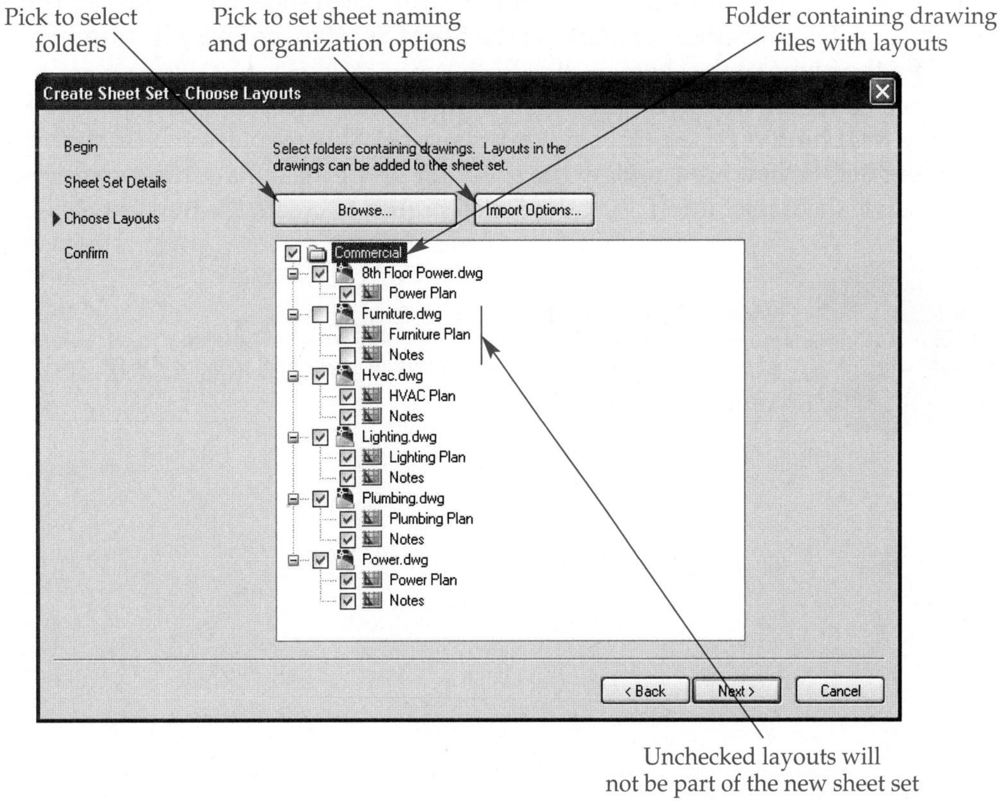

Pick to select folders

Pick to set sheet naming and organization options

Folder containing drawing files with layouts

Unchecked layouts will not be part of the new sheet set

Figure 33-9.
The **Import Options** dialog box allows you to specify naming conventions for sheets and folder structuring options.

Check to include drawing file name with layout name for new sheets

Check to create subsets from folders

Check to omit top folder name from subset structure

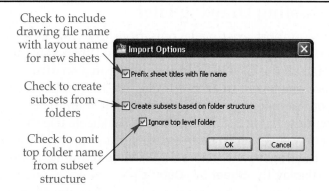

To adjust sheet naming and subset options, pick the **Import Options...** button to display the **Import Options** dialog box. See **Figure 33-9.** When you create a sheet set using existing layouts, the name for a new sheet can be the same as the layout name, or it can be the drawing file name combined with the layout name. If you select the **Prefix sheet titles with file name** check box, the name of the layouts that become sheets includes the drawing file name and the name of the layout. For example, the sheet name Electrical Plan – First Floor Electrical forms from a layout named First Floor Electrical imported from the drawing file Electrical Plan.dwg. To have sheets take on only the layout name, deselect the **Prefix sheet titles with file name** check box.

Select the **Create subsets based on folder structure** check box to organize a sheet set so that folders are grouped into subsets. Layouts in each folder form under each subset. The **Ignore top level folder** option determines whether the folder name at the top level creates a subset. **Figure 33-10A** shows the **Choose Layouts** page with layouts imported from the Residential folder for the Residential Project sheet set. This sheet set uses the **Create subsets based on folder structure** option. The **Sheet Set Manager** in **Figure 33-10B** shows the result of the configuration. Creating subsets for sheet sets helps organize sheets. As shown in **Figure 33-10B**, each sheet has a number preceding its name. By default, a sheet displays in the **Sheet Set Manager** with its number, a dash, and then the name of the sheet.

Once you select all folders and layouts for the new sheet set and adjust all settings, pick the **Next** button on the **Choose Layouts** page. This displays the **Confirm** page. In the **Sheet Set Preview** area, review the sheet set properties. If a setting requires modification, use the **Back** button. Pick the **Finish** button to create the new sheet set.

Exercise 33-2

Access the Student Web site (www.g-wlearning.com/CAD) and complete Exercise 33-2.

Figure 33-10.
Creating a sheet set named Residential Project with subsets. A—Layouts imported from the Architectural and Structural subfolders in the Residential folder. Subfolders are designated as subsets for the new sheet set. B—Subsets shown in the **Sheet Set Manager**.

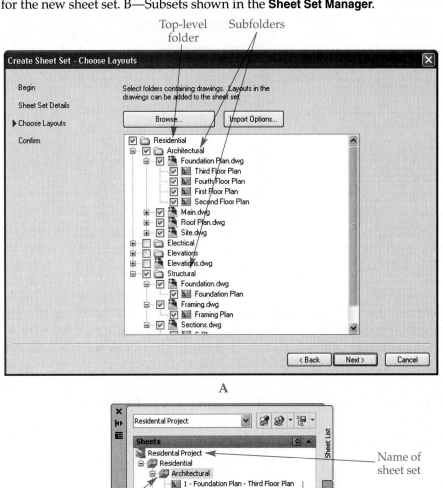

A

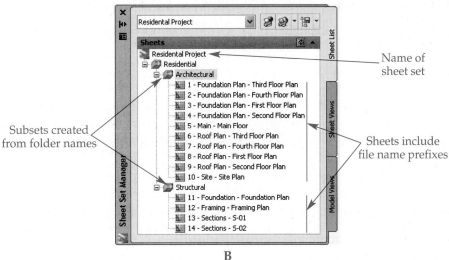

B

Working with Sheet Sets

Open an existing sheet set from within the **Sheet Set Manager** by picking a sheet set from the **Sheet Set Control** drop-down list. See **Figure 33-11.** The top portion of the drop-down list displays sheet sets opened in the current AutoCAD session. The area clears when you close AutoCAD. Select **Recent** to display a list of the most recently opened sheet sets. Select **Open...** to display the **Open Sheet Set** dialog box. Then navigate to a sheet set data (DST) file and open the file in the **Sheet Set Manager**.

Use the **Sheet List** tab to manage sheets, the **Sheet Views** tab to manage sheet views, and the **Model Views** tab to manage drawing files with layouts. Almost all of the

Figure 33-11.
The **Sheet Set Control** drop-down list displays currently open sheet sets. Pick **Open...** to browse for a sheet set that is not in the list.

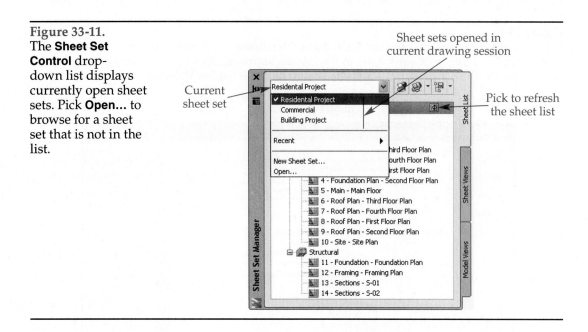

Current sheet set

Sheet sets opened in current drawing session

Pick to refresh the sheet list

options for working with sheet sets are available from shortcut menus. Right-click on a sheet set to access the **Close Sheet Set** option to remove the sheet set from the **Sheet Set Manager**. Select the **Resave All Sheets** option to update the drawing files associated with the current sheet set. You must close all files to update the list. Additional shortcut menu options are described when applicable throughout this chapter.

> **NOTE**
>
> To update changes to the sheet list manually, pick the **Refresh Sheet Status** button. Refer again to **Figure 33-11**.

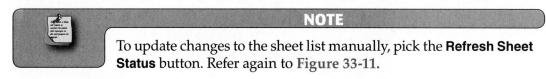

Working with Subsets

Subsets are similar to subfolders under a top-level folder in Windows Explorer. Subsets help manage the contents of the sheet set. For example, create a subset named Architectural to store all architectural sheets for a project, a subset named Electrical to store all electrical sheets, and a subset named Plumbing to store all plumbing sheets.

Creating a New Subset

To create a new subset, right-click on the sheet set name or an existing subset in the **Sheet Set Manager** and select **New Subset...** to open the **Subset Properties** dialog box. See **Figure 33-12**. Type the name of the new subset in the **Subset name** text box. For example, if a subset will contain all of the electrical sheets in a sheet set, name the subset Electrical. When you add a new sheet to the subset using a template, the sheet is saved as a drawing file. To create a new folder for the subset, select **Yes** from the **Create Folder Hierarchy** drop-down list. The folder structure mimics the subset structure.

The **Publish Sheets in Subset** option determines whether sheet sets in the subset are published. The **Do Not Publish Sheets** setting does not publish sheets in the subset. An icon identifies subsets set not to publish. The **New Sheet Location** setting determines the path to which new sheets are saved. The default value is the location specified when you created the sheet set.

Figure 33-12.
The **Subset Properties** dialog box allows you to define a new subset.

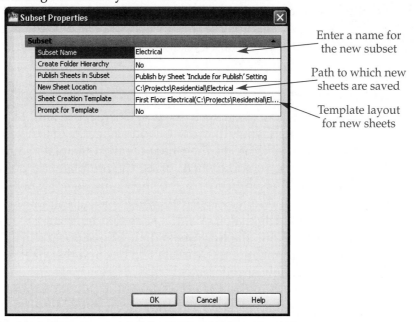

Enter a name for the new subset

Path to which new sheets are saved

Template layout for new sheets

Each subset can also have its own template and layout for new sheets, as specified in the **Sheet Creation Template** setting. For example, if electrical sheets use their own title block and notes, assign a template sheet with the appropriate settings. The procedure for specifying the template and layout for a subset is identical to the procedure used to select the sheet set properties. Use the **Prompt for Template** drop-down list to indicate if a prompt should ask for a sheet template instead of using the specified sheet creation template.

Modifying a Subset

To make changes to an existing subset, right-click on the subset and select **Properties...** to redisplay the **Subset Properties** dialog box. The **Rename Subset...** shortcut menu option also opens the **Subset Properties** dialog box. To delete a subset, right-click on the subset and select **Remove Subset**. This option is unavailable if the subset contains sheets.

Exercise 33-3

Access the Student Web site (www.g-wlearning.com/CAD) and complete Exercise 33-3.

Working with Sheets

One of the most useful features of the **Sheet Set Manager** is the ability to open a sheet quickly for review or modification. To open a sheet, double-click on the sheet or right-click on the sheet and select **Open**. The drawing file that contains the referenced layout tab opens and displays the layout corresponding to the sheet.

Adding a Sheet Using a Template

You can add a new sheet to a sheet set using the template layout sheet or by importing an existing layout. In order to add a sheet using the template layout sheet, you must specify a template in the **Sheet creation template** setting of the **Sheet Set** properties dialog box. To add a sheet using the template, right-click on the sheet set name or the subset that will contain the sheet, and select **New Sheet…**. If there is no template layout specified, an alert directs you to pick a template layout using the **Select Layout as Sheet Template** dialog box. If a specified template layout exists, or after you select the template layout, the **New Sheet** dialog box appears. See **Figure 33-13**.

Type the sheet number in the **Number** text box and the sheet name in the **Sheet title** text box. Creating a new sheet creates a new drawing file and the sheet title becomes the name of the layout in the drawing file. Enter the file name in the **File name** text box. The file name is the sheet number and title by default. The **Folder path** display box shows where the drawing file will be saved, as specified in the **Subset Properties** or **Sheet Set Properties** dialog box.

Adding an Existing Layout As a Sheet

You can add an existing drawing layout to a sheet set using the **Sheet Set Manager** or directly from an open drawing. To add an existing layout to a sheet set using the **Sheet Set Manager**, right-click on the sheet set name or the subset that will contain the sheet, and select **Import Layout as Sheet…**. This displays the **Import Layouts as Sheets** dialog box. See **Figure 33-14**. Pick the **Browse for Drawings** button to select a drawing file. The layouts from the drawing file appear in the list box. The **Status** field indicates whether the layout can be imported into the sheet set. You cannot import a layout that is already part of a sheet set. By default, all layouts in the drawing are checked for import. Uncheck a box to exclude a layout from import. Select the **Prefix sheet titles with file name** check box to include the name of the file in the sheet title. To import the sheets, pick the **Import Checked** button.

Use a layout tab or the **Quick View Layouts** or **Quick View Drawings** tool to add an existing layout to a sheet set directly from an open drawing. Right-click on the layout tab or thumbnail image to import and select **Import Layout as Sheet…**. The **Import Layouts as Sheets** dialog box appears with the selected layout automatically listed. You must save the drawing and set up the layout to make the **Import Layout as Sheet…** shortcut menu available.

Figure 33-13.
Use the **New Sheet** dialog box to define a new sheet from a template.

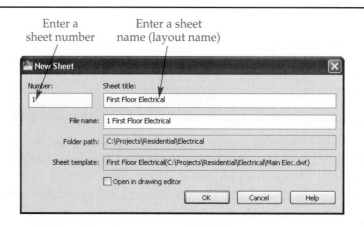

Figure 33-14.
Add existing layouts as sheets to a sheet set from the **Import Layouts as Sheets** dialog box.

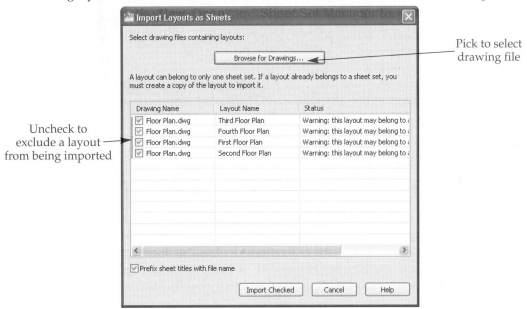

Pick to select drawing file

Uncheck to exclude a layout from being imported

Exercise 33-4

Access the Student Web site (www.g-wlearning.com/CAD) and complete Exercise 33-4.

Modifying Sheet Properties

To modify sheet properties such as the name, number, and description, right-click on the sheet name in the **Sheet Set Manager** to display a shortcut menu. Select **Rename & Renumber...** to change the sheet name and number using the **Rename & Renumber Sheet** dialog box, which is similar to the **New Sheet** dialog box. If the sheet is one of several in a subset, pick the **Next** and **Previous** buttons to access different sheets in the subset.

The **Sheet Properties** dialog box, shown in **Figure 33-15,** also allows you to change the sheet name and number, along with the description and the publish option. To open the **Sheet Properties** dialog box, right-click on the sheet name and select **Properties...**. Type a description of the sheet in the **Description** text box. The **Include for publish** option determines whether the sheet is published or plotted with the sheet set. The default value is **Yes**.

The **Expected layout** and **Found layout** text boxes display the file path where the sheet was originally saved and the file path where the sheet was found. If the paths are different, update the **Expected layout** setting by picking the ellipsis (...) button and using the **Import layout as sheet** dialog box. To delete a sheet from a sheet set, select the **Remove Sheet** option. This does not delete the drawing file; it only removes the sheet from the sheet set.

NOTE

If you modify the location of a drawing file and the drawing file has layouts associated with a sheet set, the association is broken. You must re-import the layouts into the sheet set or update the specified path to the drawing file in the **Sheet Properties** dialog box.

Figure 33-15.
The **Sheet Properties** dialog box allows you to modify sheet properties.

Determines whether the sheet is published or included in plot

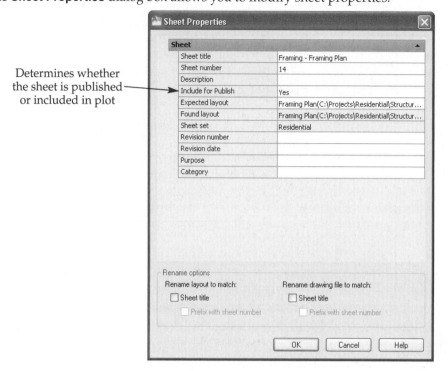

Publishing a Sheet Set

You can publish sheet sets by creating drawing web format (DWF) files, DWFx files supported by Windows Vista, or portable document format (PDF) files. DWF and DWFx files are compressed, vector-based files viewable with the Autodesk Design Review software. You can also publish a sheet set by sending the sheet set to a plotter. For more information about outputting DWF, DWFx, and PDF files, refer to *AutoCAD and Its Applications—Advanced.*

Using the Publish Shortcut Menu

A **Publish** shortcut menu is available in the **Sheet Set Manager** for publishing or plotting an entire sheet set. To access publish options, pick the **Publish** flyout on the **Sheet Set Manager** toolbar, as shown in **Figure 33-16,** or select the **Publish** cascading submenu option available when you right-click in the **Sheets** list.

> **NOTE**
>
> You can select a sheet set, subset, or individual sheets for publishing. Select the appropriate items by pressing [Shift] and [Ctrl] in the **Sheet Set Manager.**

Pick the appropriate **Publish to DWF, Publish to DWFx**, or **Publish to PDF** option to create a DWF, DWFx, or PDF file from the sheet set or the selected sheets. A dialog box appears, allowing you to specify a name and location for the file. The file creates multiple pages with each sheet on a separate page. Use the **Publish to Plotter** option to plot the sheet set or selected sheets to the default plotter or printer using the plot settings from each layout. Select an override from the **Publish using Page Setup Override** cascading submenu to force the sheet to use the selected page setup settings

Figure 33-16.
The **Publish** flyout provides options for preparing a sheet set for publishing or plotting.

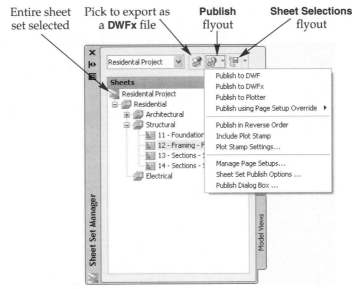

instead of the plot settings saved with the layout. This option is unavailable if you have not specified a page setup override for the sheet set or subset. Use the **Include for Publish** option to identify whether the selected items will publish. Pick the **Edit Subset and Sheet Publish Settings...** option to display the **Publish Sheets** dialog box. Check specific sheets to include for publishing.

The **Publish in Reverse Order** option publishes sheets in the opposite order from the order displayed in the **Sheet Set Manager**. When using certain printers, publishing a sheet set in reverse order is helpful so that the last sheet is on the bottom of the stack, at the end of the entire set. Select the **Include Plot Stamp** option to place the plot stamp information assigned to the layout on the sheet during plotting. Select the **Plot Stamp Settings** option to open the **Plot Stamp** dialog box to specify the plot stamp settings. The **Manage Page Setups** option opens the **Page Setup Manager**, allowing you to create a new page setup or modify an existing one. The **Sheet Set Publish Options** selection displays the **Sheet Set Publish Options** dialog box, which displays the available settings for creating a DWF, DWFx, or PDF file. The **Publish Dialog Box** option opens the **Publish** dialog box, which lists the sheets in the current sheet set or the sheet selection.

Creating Sheet Selection Sets

During the course of a project, you may need to publish the same set of sheets often. Save sheets as a selection set to accommodate this need. To save a sheet selection set, select the sheets or subsets to include in the set. Then pick the **Sheet Selections** flyout on the **Sheet Set Manager** toolbar and select **Create...** to access the **New Sheet Selection** dialog box. Enter a name for the selection set and pick the **OK** button. The new selection set appears when you pick the **Sheet Selections** button. **Figure 33-17** shows three different sheet selection sets suitable for common publishing or plotting requirements. When you select a selection set, the sheets become highlighted in the **Sheet Set Manager**.

To rename or delete a sheet selection set, pick **Manage...** from the **Sheet Selections** flyout to display the **Sheet Selections** dialog box. Select the sheet selection set and then pick the **Rename** or **Delete** button.

Figure 33-17.
You can create sheet selection sets from selected sheets or subsets in a sheet set. Access selection sets from the **Sheet Selections** flyout.

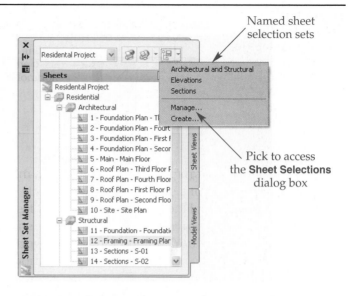

Named sheet selection sets

Pick to access the **Sheet Selections** dialog box

Using Sheet Views

sheet view: A referenced portion of a drawing set, such as an elevation, a section, or a detail.

Sheets can contain *sheet views* that you can use to provide links between sheets and drawing content in a sheet set. Sheet views provide automated labeling and referencing using blocks with attributes containing fields. Sheet view field values update automatically to reflect changes in sheet numbering and organization. Use the **Sheet Views** tab of the **Sheet Set Manager** to group views by category and open views for viewing and editing. Special tools in the **Sheet Set Manager** allow you to identify views with numbers, labels, and callout blocks.

Adding a View Category

View categories organize views in the **Sheet Views** tab, and function similar to the subsets created in the **Sheet List** tab. To create a new view category, make the **Sheet Views** tab current and select the **View by category** button. See **Figure 33-18.** Pick the **New View Category** button or right-click on the sheet set name and select **New View Category...** to open the **View Category** dialog box. See **Figure 33-19.** Enter a name for the category in the **Category name** text box. For example, specify a view category named Elevations and add four elevation views to the new category.

The **View Category** dialog box lists all available callout blocks for the current view category. Check the box next to the callout block to make the block available for all views added to the category. If a block is not in the list, use the **Add Blocks...** button to select the block from a drawing file. After you select the necessary callout blocks, pick the **OK** button to create the new category. Callout blocks are described later in this chapter.

Modifying a View Category

To modify view category properties, right-click on the category name in the **Sheet Set Manager** and select **Rename...** or **Properties...** to reopen the **View Category** dialog box. Change the category name and add different callout blocks to the category as needed. To delete a category, right-click on the category name and select **Remove Category**. The **Remove Category** option is unavailable if there are views in the category. You must remove all of the views before deleting the category.

Figure 33-18.
Create view categories in the **Sheet Views** tab of the **Sheet Set Manager**.

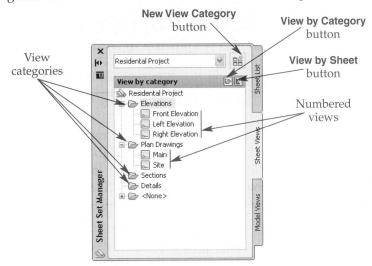

Figure 33-19.
The **View Category** dialog box allows you to name the category and select callout blocks for use with views.

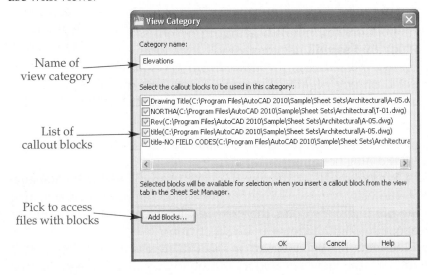

Creating Sheet Views in an Existing Sheet

The **Sheet Set Manager** allows you to add new views to sheets and organize views within sheet sets. Use the following procedure to add a view to an existing sheet set:

1. Open the desired sheet set and add a category for the view if a suitable category does not exist.
2. A sheet must be a part of the sheet set to include a view. If the sheet is not included with the sheet set, add the sheet now.
3. Open the drawing file and set the layout tab in which you will create the new view current.
4. Use display tools to orient the view as needed and then use the **VIEW** tool to access the **View Manager**.
5. In the **View Manager**, pick the **New...** button to open the **New View** dialog box.
6. Select the category to associate with the view from the **View category** drop-down list. See **Figure 33-20**.
7. Specify the remaining view settings and pick the **OK** button to save the view.

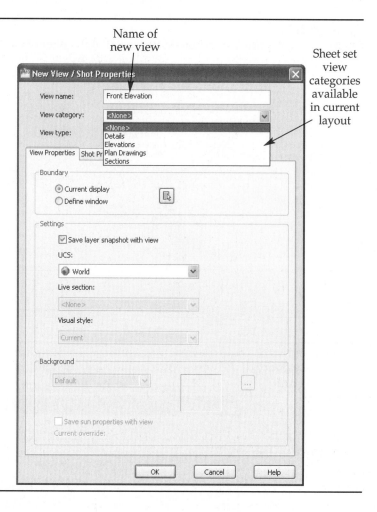

Figure 33-20.
The **View Category** drop-down list displays the available view categories for the view definition.

Name of new view

Sheet set view categories available in current layout

The newly saved view now appears in the **Sheet Set Manager** under the view category selected in the **New View** dialog box.

Once you add a view to a sheet set, display the view from the **Sheet Set Manager** by double-clicking on the view name or by right-clicking on the name and selecting **Display**. If the drawing file is already open, the view is set current. If the drawing file is not open, the file opens so that the view can be set current.

You can display the view list by category or sheet using the appropriate button shown in **Figure 33-18**. Pick the **View by category** button to display all categories, and access views by expanding the category and the sheet. Pick the **View by sheet** button to display the sheet name. Expand the sheet name to display saved views within the sheet.

Creating Sheet Views from Resource Drawings

resource drawings: Drawing files that contain model space views referenced for use as sheet views.

You can also create sheet views from *resource drawings* listed in the **Model Views** tab. The sheet view can be the entire model space drawing or a model space view. When you insert a model space view or drawing into a sheet, the resource drawing becomes an external reference of the sheet drawing.

Folders containing reference drawings appear in the **Model Views** tab, as shown in **Figure 33-21**. To add a new folder, double-click on the Add New Location entry or pick the **Add New Location** button and select a folder. You cannot select specific drawing files—you must select the folder containing the drawing. You can only insert drawings listed in the **Model Views** tab into a sheet to create a new sheet view. If a required drawing does not appear, you must add the folder containing the drawing to the resource drawing list.

Figure 33-21.
You can create sheet views by inserting model space views and drawings from the **Model Views** tab. To see a thumbnail and details, hover the cursor over the drawing or view.

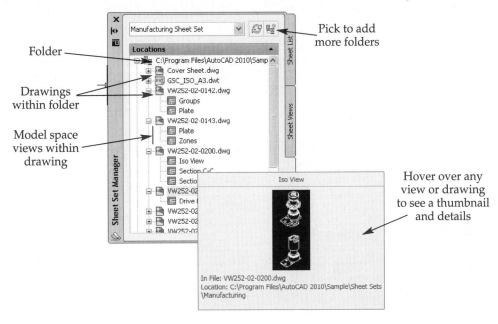

The folder and all of the drawing files found in the folder now appear in the **Locations** list area. The model space views saved in the drawing are listed under the drawing file. The options available for a drawing file are located in the drawing file shortcut menu. To display the menu, right-click on a drawing file.

Pick the **Open** option to open the drawing file and set model space current. Double-clicking on the drawing file also opens the file. Select the **Open read-only** option to open the drawing file as read-only so that file changes are not possible. The **Place on Sheet** option inserts the file into the current sheet as a sheet view. AutoCAD prompts you to specify an insertion point and creates a viewport automatically in the sheet. The **See Model Space Views** option expands the list of model space views in the drawing. This is the same as picking the + sign next to the drawing file. Select the **eTransmit** option to open the **Create Transmittal** dialog box to package the selected file and its associated files together.

If you have not saved a model space view in the drawing, it is listed under the drawing file name. You can insert the model space view as a sheet view in a sheet. To do so, right-click on the model space view name, select **Place on Sheet**, and pick an insertion point in the sheet.

When you insert a model space view or drawing into a sheet, AutoCAD creates a viewport and an xref to the selected drawing. AutoCAD assigns a scale for the viewport, or you can right-click before selecting the insertion point and select the scale for the sheet view. The scale is stored as the **ViewportScale** property of the **SheetView** field and is often displayed in the view label block.

When you create sheet views from resource drawings, an entry is added to the **Sheet Views** tab. If you insert a model space view, the view name is added to the **Sheet Views** tab. If you insert a drawing, the drawing name is added to the **Sheet Views** tab. To delete a location from a sheet set, right-click on the location and select **Remove Location**.

Naming and Numbering Sheet Views

Most projects contain several components, elevations, sections, or details. These items typically receive a number or other value associated with the drawing set for reference. For example, a set of architectural drawings may include a foundation plan and a sheet with foundation details. On the foundation detail sheet, a unique number identifies each detail. The foundation plan includes references to these numbers.

To change the name or number of a sheet view, right-click on the sheet view name in the **Sheet Views** tab and select **Rename & Renumber...** to display the **Rename & Renumber View** dialog box. See **Figure 33-22.** Enter a number for the view in the **Number** text box, and modify the view name in the **View title** text box. Use the **Next** and **Previous** buttons to access different sheets in the subset to move to different views in the view category. Pick the **OK** button when you are finished. The view number displays in front of the view name in the **Sheet Set Manager.**

Exercise 33-5

Access the Student Web site (www.g-wlearning.com/CAD) and complete Exercise 33-5.

Working with Sheet View Blocks

The components, elevations, sections, details, and other drawings shown in sheet views are often located on a specific sheet and referenced on different sheets. When using sheet views, you can insert blocks to identify the sheet view name, number, and scale on the sheet with the sheet view and the sheet that refers to the sheet views. Typically, callout blocks and view label blocks are required for these applications. Using sheet view blocks can help automate the process of adding drawing titles, labels, and similar annotations.

Using Callout Blocks

callout block: A block inserted to indicate a reference to another sheet.

A *callout block* refers to a sheet view. For example, a section line drawn through a building receives a callout block at the end of the section line. The callout block indicates the sheet or the location of the section view and provides information about the viewing direction. You can also use a callout block on a foundation plan, for example, to identify an area addressed by a detail drawing. The callout block is typically located on a different sheet from the sheet view it references.

Several styles of callout blocks are available from AutoCAD to use for different types of sheet views. See **Figure 33-23.** The upper value in a callout block is typically the sheet view number, and the lower value is the drawing on which the sheet view appears. In the default callout blocks, the upper value is an attribute containing the **ViewNumber** property of the **SheetView** field. See **Figure 33-24.** This lists the sheet view

Figure 33-22.
Use the **Rename & Renumber View** dialog box to renumber a view.

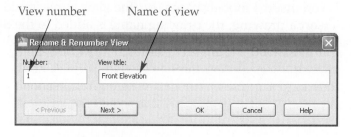

View number   Name of view

Figure 33-23.
Callout blocks provide reference information for views and sheets. A—Elements of an elevation symbol. B—Examples of commonly used callout blocks.

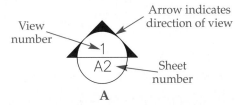

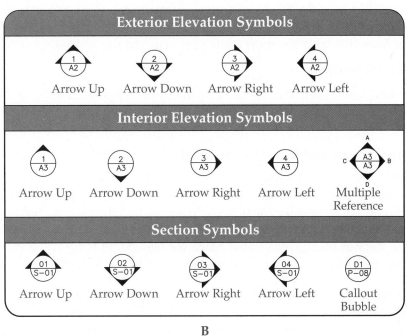

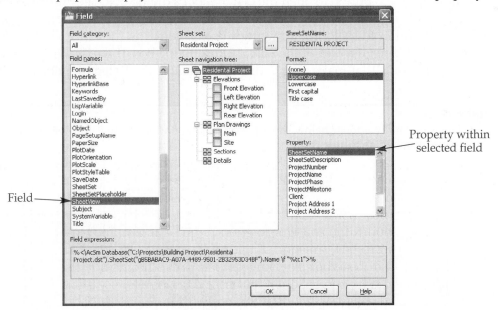

Figure 33-24.
The **ViewNumber** property displays the sheet view number. Callout blocks use this field property.

number specified for the sheet view. The lower value is the **SheetNumber** property of the **SheetSet** value. This lists the sheet number of the sheet containing the sheet view.

Sheet view blocks use fields to update attribute values if changes occur to the sheet set. For example, if you add a new sheet in the middle of a sheet set, all later sheets require renumbering. The sheet view block values update automatically as the sheet numbers change.

Using View Label Blocks

view label block: A block that contains view information such as the view name, number, and scale.

View label blocks often appear below the sheet view, as shown in **Figure 33-25**. Like callout blocks, view label blocks include attributes with fields that update to reflect changes to the sheet set or sheet views. View label blocks typically include the **ViewNumber**, **ViewTitle**, and **Viewport Scale** properties of the **SheetView** fields.

Using Block Hyperlinks

hyperlinks: Links in a document connected to related information in other documents or on the Internet.

Callout and view label blocks can also include hyperlink fields, or *hyperlinks*. You can pick the hyperlink on a callout block to access the referenced detail, section, or elevation. This greatly simplifies the process of accessing sheet views.

Associating Callout and View Label Blocks

Before you can insert a callout or view label block from the **Sheet Set Manager**, the block must be available to the sheet set containing the view. Use the **Sheet Set Properties** dialog box to specify the blocks. To access this dialog box, right-click on the sheet set name in the **Sheet Set Manager** and select **Properties…**. Specify the available blocks in the **Callout blocks** and **Label block for views** text boxes. The name of each block appears, followed by the path to the drawing file where the block is saved.

> **PROFESSIONAL TIP**
>
> A sheet set or view category can have multiple callout blocks available, but only one view label block.

To add a callout block to a sheet set, pick in the **Callout blocks** text box and then pick the ellipsis (**…**) button to open the **List of Blocks** dialog box. See **Figure 33-26.** Pick the **Add…** button to display the **Select Block** dialog box. In this dialog box, pick the ellipsis (**…**) button to select the drawing file that contains the block. Then select the block from the block list area of the **Select Block** dialog box. If the drawing file consists of only the objects that make up the drawing file, use the **Select the drawing file as a**

Figure 33-25.
View labels normally appear below the view on a sheet and indicate information such as the view name, number, and scale.

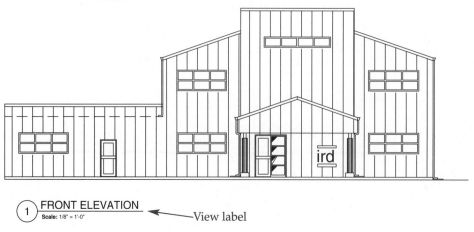

Figure 33-26.
All callout blocks available to a sheet set appear in the **List of Blocks** dialog box.

Pick to access **Select Block** dialog box

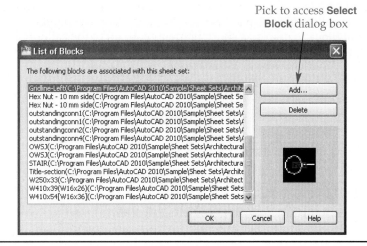

List of Blocks

The following blocks are associated with this sheet set:

Gridline-Left(C:\Program Files\AutoCAD 2010\Sample\Sheet Sets\Archite
Hex Nut - 10 mm side(C:\Program Files\AutoCAD 2010\Sample\Sheet Se
Hex Nut - 10 mm side(C:\Program Files\AutoCAD 2010\Sample\Sheet Se
outstandingconn1(C:\Program Files\AutoCAD 2010\Sample\Sheet Sets\/
outstandingconn1(C:\Program Files\AutoCAD 2010\Sample\Sheet Sets\/
outstandingconn2(C:\Program Files\AutoCAD 2010\Sample\Sheet Sets\/
outstandingconn4(C:\Program Files\AutoCAD 2010\Sample\Sheet Sets\/
OWSJ(C:\Program Files\AutoCAD 2010\Sample\Sheet Sets\Architectural
OWSJ(C:\Program Files\AutoCAD 2010\Sample\Sheet Sets\Architectural
STAIR(C:\Program Files\AutoCAD 2010\Sample\Sheet Sets\Architectura
Title-section(C:\Program Files\AutoCAD 2010\Sample\Sheet Sets\Archite
W250x33(C:\Program Files\AutoCAD 2010\Sample\Sheet Sets\Architect
W410x39(W16x26)(C:\Program Files\AutoCAD 2010\Sample\Sheet Sets
W410x54[W16x36](C:\Program Files\AutoCAD 2010\Sample\Sheet Sets

Add...
Delete

OK   Cancel   Help

block option. To delete a block from the block list, select it in the **List of Blocks** dialog box and pick the **Delete** button.

Specifying a view title block is similar to specifying a callout block. However, only one view title block can be specified for the sheet set, so the **List of Blocks** dialog box does not display. You can assign each view category its own callout blocks. This way, only the blocks needed for the views in a category are available. For example, a category named Section may only need a section callout bubble, while a category named Elevation may need ten different types of elevation symbols. To modify the callout blocks available for a view category, right-click on the category name and select **Properties...** to open the **View Category** dialog box.

By default, the callout blocks and view title block assigned to a sheet set appear in the block list area. To make a block available to the view category, check the box next to the block. Refer again to **Figure 33-19.** This makes the block available to all of the views in the view category. To add new blocks to the view category, pick the **Add Blocks...** button to access the **Select Block** dialog box.

Inserting Callout and View Label Blocks

To insert a callout block, open the sheet on which the reference is to appear. In the **Sheet Views** tab of the **Sheet Set Manager**, right-click on the sheet view name and select the block from the **Place Callout Block** cascading submenu. See **Figure 33-27A.** Then specify an insertion point for the block. Follow the prompts to scale and rotate the block as needed. When you insert the block, AutoCAD gives the block the same sheet view number and sheet number as the reference view and sheet. See **Figure 33-27B.** AutoCAD automatically renumbers the block if the reference information changes.

The process of inserting a view label block is similar to that for inserting a callout block. In the **Sheet Set Manager**, right-click on the sheet view name and select **Place View Label Block**. Follow the prompts to specify an insertion point, scale, and rotation angle. After you insert the block, the label appears with the view name and number. If the view name or number changes in the sheet set, AutoCAD automatically updates the information.

Exercise 33-6

Access the Student Web site (www.g-wlearning.com/CAD) and complete Exercise 33-6.

Figure 33-27.
Placing callout blocks in a view. A—Right-click on the reference view name and select **Place Callout Block** to display a shortcut menu with all of the callout blocks available. B—Callout blocks placed in the 1-Main Floor Plan view in the A-01 sheet to reference the section view named 1-Section in the A-05 sheet.

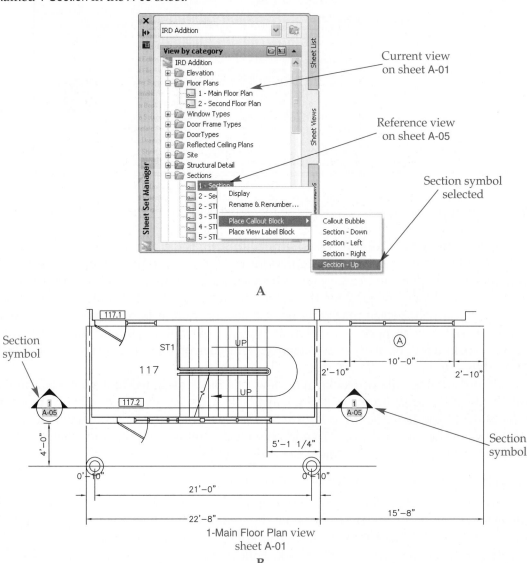

A

B

1-Main Floor Plan view
sheet A-01

Sheet Set Fields

Fields are valuable features for sheet sets. As the project develops, you can set field text on sheets to display up-to-date information when changes occur. For example, create title block items such as the number of sheets, drawing number, project number, and the date the sheet was plotted as fields to automate the process of making changes to the values. To create a field, access the **Field** dialog box from within the **MTEXT** or **TEXT** tool, from the ribbon, or by typing FIELD.

Specifying Fields for Sheet Sets

AutoCAD provides specific field types for use with sheet sets. In the **Field** dialog box, pick **SheetSet** from the **Field category:** drop-down list to display a list of predefined field types in the **Field names:** list box. See **Figure 33-28.** Use these fields to display

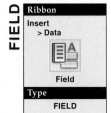

FIELD

Ribbon
Insert
> Data

Field

Type
FIELD

Figure 33-28.
Select **SheetSet** in the **Field category:** list in the **Field** dialog box to display the many fields related to sheet sets.

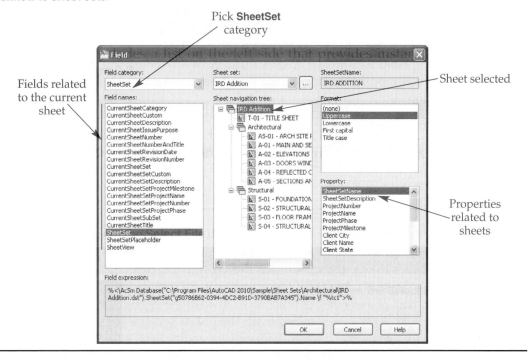

Pick **SheetSet** category

Fields related to the current sheet

Sheet selected

Properties related to sheets

values defined in the sheet, sheet view, or sheet set, such as the sheet title, number, and description. Some of the fields also have several properties. Select one of the field types or properties to display the related value in the **Field** dialog box. For example, select the **CurrentSheetNumber** field to insert a field that displays the sheet number of the current sheet. If the sheet renumbers at a later date, the field changes to display the most current information.

Select the **SheetSet** field to display the **Sheet navigation tree**, which provides options for inserting several values. If you select the sheet set at the top of the tree, a set of properties related to the entire set appears in the **Property** list box. These properties include settings that can apply to all sheets in the set, such as project information and client information. **SheetSet** field properties display the same values on all sheets.

Pick a sheet in the **Sheet navigation tree** to display properties related to the individual sheet, including **SheetTitle**, **SheetNumber**, **Drawn By**, and **Checked By** values. When you include these fields in the sheet set template title block, each sheet can display a unique value. If you add a new sheet to a sheet set, the fields automatically update.

Select the **SheetSetPlaceholder** field to insert a field that contains a *placeholder*. Choose a placeholder from the **Placeholder type:** list box to assign a temporary value to the associated field, such as SheetNumber. Placeholders insert temporary field values in user-defined callout blocks and view labels. When defined with attributes in a callout block, placeholders update to display the correct values when you insert the block onto a sheet from the **Sheet Set Manager**.

Select the **SheetView** field to display the **Sheet navigation tree** with a view list for the sheet set. Sheet set properties appear when you pick the sheet set name in the **Sheet navigation tree**. These properties are identical to those displayed with the **SheetSet** field. Sheet view properties display when you pick a sheet view name in the **Sheet navigation tree**. These properties are specific to a sheet view and include **ViewTitle**, **ViewNumber**, and **ViewScale**. Sheet view fields are common in callout and view label blocks.

placeholder: A temporary value for a field.

Using Custom Properties

Select the **CurrentSheetCustom** or **CurrentSheetSetCustom** field to insert a field that links to a custom property defined for a sheet or sheet set. Information about the sheet set or a specific sheet is stored electronically with fields and custom properties and is viewable from the **Sheet Set Manager**. You can also insert the data into the drawing using the **Field** tool, which creates a link between the text data and the custom field data. When you modify the data in the **Sheet Set Manager**, the linked data updates in the drawing files.

Adding a Custom Property Field

Use the **Sheet Set Properties** dialog box to manage custom properties. To add a custom property field to a sheet set, right-click on the sheet set name in the **Sheet Set Manager** and select **Properties....** In the **Sheet Set Properties** dialog box, pick the **Edit Custom Properties...** button to open the **Custom Properties** dialog box. See **Figure 33-29.**

To add a custom property field to the sheet set, pick the **Add...** button to display the **Add Custom Property** dialog box. See **Figure 33-30.** Enter a name for the custom property in the **Name** field. Examples of a custom property include Job Number, Client Name, Checked By, and Date. If the data for the custom property is usually the same value, enter it in the **Default value** field. For example, if the custom property is Checked by, and DAM checks most of the sheets in this project, then enter DAM as the default value.

If the custom property is associated with the entire project, select **Sheet Set** from the **Owner** area. To assign the custom property to each sheet, select **Sheet** from the **Owner** area. When you select **Sheet**, the custom property is available in the **Sheet Properties** dialog box and the data attaches to each sheet. Pick the **OK** button to add the custom property to the sheet set.

Entering Custom Property Data

To modify or enter information into a custom property field for a sheet set, open the **Sheet Set Properties** dialog box and modify the value. If a sheet includes custom properties, individual sheets display the custom fields. To modify or enter information into a sheet custom property field, right-click on the sheet and select **Properties....** Custom properties are listed under the **Sheet Custom Properties** heading of the **Sheet Properties** dialog box. See **Figure 33-31.**

Deleting a Custom Property

To delete a custom property field that you no longer need, right-click on the sheet set and select **Properties...** to open the **Sheet Set Properties** dialog box. Then pick the **Edit Custom Properties...** button. In the **Custom Properties** dialog box, select the custom property to remove and pick the **Delete** button.

Figure 33-29.
Attach information to a sheet set in the **Custom Properties** dialog box.

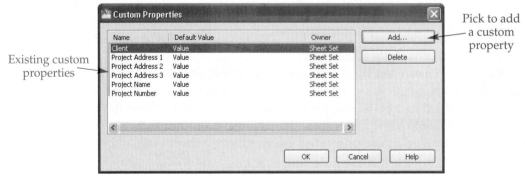

Existing custom properties

Pick to add a custom property

Figure 33-30.
Enter the information for the custom property in the **Add Custom Property** dialog box.

Custom property name → Default value → Pick custom property type →

Figure 33-31.
After you add custom properties to individual sheets, the properties are available in the **Sheet Properties** dialog box.

Custom properties →

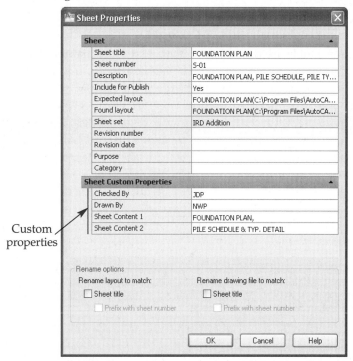

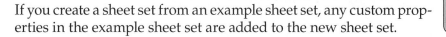

NOTE

If you create a sheet set from an example sheet set, any custom properties in the example sheet set are added to the new sheet set.

Exercise 33-7

Access the Student Web site (www.g-wlearning.com/CAD) and complete Exercise 33-7.

One of the first pages of a sheet set typically includes a *sheet list*. A sheet list displays all of the pages in the sheet set and indicates the type of information found on each sheet. You can create a sheet list as a table object using information from sheet properties. The information in the table links directly to sheet properties, allowing the sheet list table to update automatically if sheet information in the **Sheet Set Manager** changes.

Inserting a Sheet List Table

To insert a sheet list table, access the **Sheet Set Manager** and open the sheet that will receive the table. Once you open a sheet, right-click on the sheet set name in the **Sheet Set Manager** and select **Insert Sheet List Table...** to open the **Sheet List Table** dialog box. See **Figure 33-32**.

Figure 33-32.
Properties for the sheet list table are set up in the **Sheet List Table** dialog box.

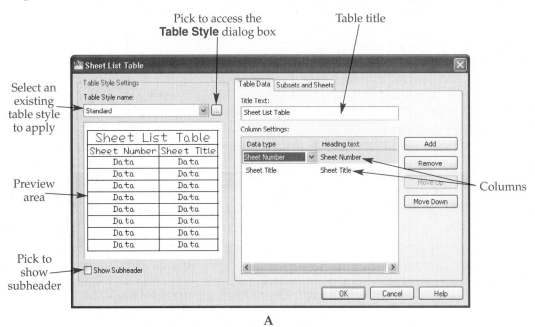

A

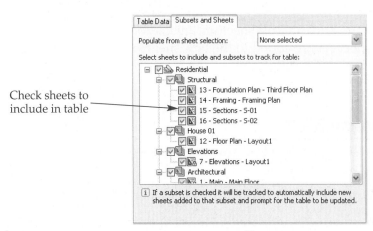

B

Select a preset table style for the sheet list from the **Table Style name** drop-down list, or pick the ellipsis (**...**) button to access the **Table Style** dialog box. A representation of the table is displayed in the preview area. The **Show Subheader** check box determines whether the table includes a subheader row. **Figure 33-33** shows a table that includes Architectural and Structural subheaders created by referencing subset names.

Specifying the Title and Columns

Use the **Table Data** tab, shown in **Figure 33-32A,** to specify the table title, columns, and column organization. Enter the title for the sheet list in the **Title Text** text box. A sheet list table can include various types of information from the drawing file and the sheet set. The default table includes the Sheet Number and Sheet Title fields. To specify the data type, pick the name in the **Data type** column to activate a drop-down list and then select a property from the list. Select the heading text in the **Heading text** column to enter a new value for the column heading.

The data types that are available in the drop-down list come from sheet set properties and drawing properties. To add a different data type to the list, you need to add a custom property to the sheet set. Use the **Add, Remove, Move Up,** and **Move Down** buttons as necessary to add, remove, and organize columns. A new column initially displays under the last column in the list. The column at the top of the list appears on the far right side of the table as the first column in the table.

Specifying Rows

The **Subsets and Sheets** tab, shown in **Figure 33-32B,** allows you to specify table rows by selecting check boxes corresponding to sheets to include in the table. Each row is a sheet reference. If you right-click on a sheet set to access the **Sheet List Table** dialog box, all subsets and sheets in the sheet set are automatically checked for addition to the table. Right-clicking on a subset or sheet initially limits the sheets in the table, although you can add and remove sheets from the table using the appropriate check boxes.

Checked sheets are the only items that appear in the table. You do not need to check subsets to display subheaders. However, you must check sheet sets and subsets to include sheet sets and subsets during updates. If you make changes to the sheet set, a prompt appears to update only the subsets you check in the **Subsets and Sheets** tab. If a sheet selection is available from the **Populate from sheet selection** drop-down list, choosing the selection checks only the sheets associated with the saved selection.

After you specify the table content, pick the **OK** button and select an insertion point to insert the table. **Figure 33-33** shows a sheet list table that uses the sheet number and sheet description fields.

Figure 33-33.
A sheet list table displays information about selected sheets in the sheet set.

SHEET INDEX	
Sheet Number	Sheet Description
T–01	SHEET INDEX, VICINITY MAP, BUILDING CODE ANALYSIS
Architectural	
AS–01	ARCHITECTURAL SITE PLAN, NOTES
A–01	MAIN FLOOR PLAN, SECOND FLOOR PLAN, WALL TYPE NOTES
A–02	EXTERIOR ELEVATIONS
A–03	DOOR & FRAME SCHEDULE, ROOM FINISH SCHEDULE, DOOR, DOOR FRAME & WINDOW TYPES
A–04	MAIN & SECOND FLOOR REFLECTED CEILING PLANS
A–05	STAIR SECTIONS AND DETAILS
Structural	
S–01	FOUNDATION PLAN, PILE SCHEDULE, PILE TYPICAL DETAIL
S–02	STRUCTURAL SECTIONS AND DETAILS
S–03	FLOOR FRAMING PLAN AND SECTIONS
S–04	STRUCTURAL SECTIONS

Editing a Sheet List Table

The information in the sheet list table links directly to the data source field. For example, if you modify the sheet numbers in the **Sheet Set Manager**, the sheet list table can update to reflect those changes. To update, select the sheet list table in the drawing, right-click, and select **Update Sheet List Table**. The option is also available from the **Sheet List Table** cascading submenu that appears when you select inside a table header or data cell and then right-click.

To modify the properties for the table, select inside a table header or data cell and then right-click. Choose the **Edit Sheet List Table Settings…** option from the **Sheet List Table** cascading submenu to reopen the **Sheet List Table** dialog box. After making the changes, pick the **OK** button to update the sheet list table.

Using Sheet List Table Hyperlinks

If you include the **Sheet Number** or **Sheet Title** columns in the sheet list table, hyperlinks are assigned to the data automatically. To open a sheet using a hyperlink, hover the crosshairs over a sheet number or sheet title until the hyperlink icon and tooltip appear. The tooltip displays the message CTRL + click to follow link. Hold down the [Ctrl] key and pick the hyperlink to open the selected sheet. This is another way to open a sheet quickly.

Exercise 33-8

Access the Student Web site (www.g-wlearning.com/CAD) and complete Exercise 33-8.

Archiving a Sheet Set

archiving:
Gathering and storing all of the electronic drawing files related to a project.

At different periods throughout a project, you may want to *archive* the drawing set. For example, when you present a set of drawings in a project to the client for the first time, the client may want to make some changes. It may be wise to archive the files for future reference before making modifications. Archive copies of all drawing and related files to a single location. Related files include external references, font files, plot style table files, and template files.

Figure 33-34.
A—The **Archive a Sheet Set** dialog box allows you to specify files to archive and archive settings. B—Use the **Files Tree** tab to archive documents that relate to a project along with the AutoCAD files.

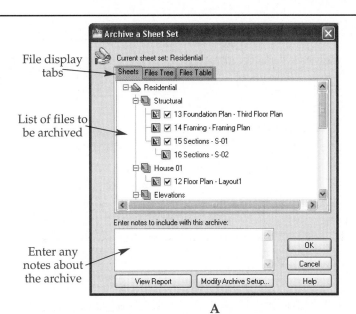

File display tabs

List of files to be archived

Enter any notes about the archive

A

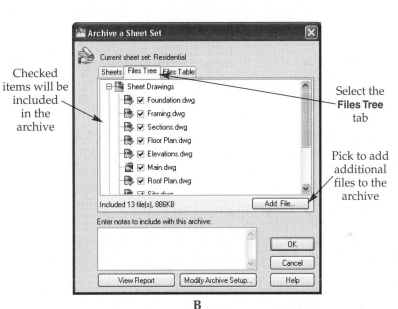

Checked items will be included in the archive

Select the **Files Tree** tab

Pick to add additional files to the archive

B

To archive a sheet set, right-click on the sheet set name in the **Sheet Set Manager** and select **Archive...**, or type ARCHIVE. The **Archive a Sheet Set** dialog box opens. See **Figure 33-34.** The **Sheets** tab, shown in **Figure 33-34A,** displays all of the subsets and sheets in the sheet set. Check the sheets to archive. Drawing files and related files are listed in the **Files Tree** tab, shown in **Figure 33-34B.** Pick the + sign next to a file to display related files.

You can include a file that is not part of the sheet set in the archive by picking the **Add a File** button on the **Files Tree** tab. This opens the **Add File to Archive** dialog box. You can include any type of file with the archive—the archive is not limited to AutoCAD files. To include a file, type the file in the **Enter notes to include with this archive** text box. The **View Report** button lists all of the files included in the archive. Pick the **Save As...** button to save the information to a text file.

The **Modify Archive Setup** dialog box, shown in **Figure 33-35** allows you to specify the location to which the archive is saved, the type of archive, and additional settings. To open the dialog box, pick the **Modify Archive Setup...** button. Use the **Archive package type** drop-down list to specify the archive format. Choose the **Folder (set of files)** option to copy all archived files into a single folder. Select the **Self-extracting executable (*.exe)**

Figure 33-35.
Specify archive file settings using the **Modify Archive Setup** dialog box.

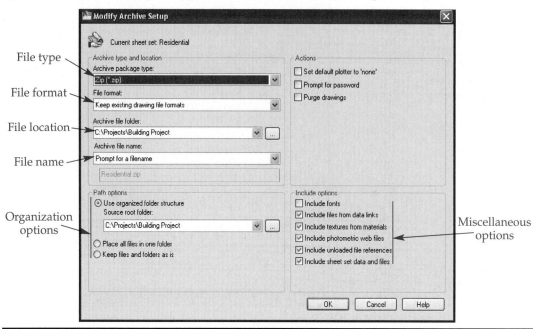

File type

File format

File location

File name

Organization options

Miscellaneous options

option to compress all files into a self-extracting *zip file*. Self-extracting zip files have an EXE extension. You can extract the files later by double-clicking on the file. Choose the **Zip (*.zip)** option to compress all files into a normal zip file. You must use a program that works with zip files to extract the files.

Select an earlier version of AutoCAD in the **File Format** drop-down list to convert the archived files to the selected version. The **Archive file folder** drop-down list defines where the archive is saved. Select a location from the list or pick the **Browse...** button to choose a different location.

The **Archive file name** drop-down list provides options for naming the archive. Choose the **Prompt for a filename** option to display the **Specify Zip File** dialog box so that you can specify a name for the archive package. Pick the **Overwrite if necessary** option to overwrite the file name if a file with the same name already exists. Select the **Increment file name if necessary** option to create a new file with an incremental number added to the file name if a file with the same name already exists. You can save multiple versions of the archive package using this option.

Use the **Archive Options** area to specify additional settings for the archive package. The first option determines how the folder structure is saved. If you select the **Use organized folder structure** radio button, the archive file duplicates the folder structure for the files, and the **Source root folder** setting determines the root folder for files that use relative paths, such as xrefs. To archive all files into a single folder, pick the **Place all files in one folder** radio button. Select the **Keep files and folders as is** radio button to use the same folder structure for all the files in the sheet set. Pick the **Include fonts** check box to include all fonts used in the drawings in the archive. The **Set default plotter to 'none'** check box disassociates the plotter name from the drawing files. This is useful if you send files to someone using a different plotter. Select the **Prompt for password** check box to set a password for the archive. The password is then required to open the archive package. Pick the **Include sheet set data and files** check box to include the sheet set data file with the archive package.

Chapter Test

Answer the following questions. Write your answers on a separate sheet of paper or go to the Student Web site (www.g-wlearning.com/CAD) and complete the electronic chapter test.

1. What is a sheet set?
2. What does the term *sheet* refer to in relation to a sheet set and a drawing file?
3. Briefly explain two methods for creating a new sheet set.
4. What file extension applies to sheet sets?
5. What are subsets in relation to a sheet set?
6. What is the purpose of the **Create subsets based on folder structure** option in the **Import Options** dialog box?
7. Explain how to create a new subset in a sheet set and specify a template file and layout for creating new sheets in the subset.
8. List two ways to open a sheet from the **Sheet Set Manager**.
9. What is the purpose of the **Import Layouts as Sheets** dialog box?
10. How do you modify a sheet name or number?
11. Briefly explain how to publish a sheet set to a DWF, DWFx ,or PDF file. How are the sheets organized in the resulting file?
12. How do you create a sheet selection set?
13. What are sheet views and how can they be referenced to each other within a sheet set?
14. What tab in the **Sheet Set Manager** allows you to manage sheet views?
15. Briefly explain how to create a view category for a sheet set and associate callout blocks to the category.
16. Explain how to add a drawing file to a sheet set to place views in the drawing on a sheet.
17. What information do the upper and lower values in a callout block typically provide?
18. Explain why AutoCAD callout blocks and view labels automatically update when you make changes to the related sheet set.
19. Explain how to insert a callout block into a drawing.
20. List three examples of fields that are often used in a sheet set.
21. What is the purpose of custom sheet set properties?
22. How do you add a custom property to a sheet set?
23. What is a sheet list?
24. Explain how to add a column heading to a sheet list table.
25. How can you update a sheet list table to reflect changes made in the **Sheet Set Manager**?
26. What happens to edits made manually to a sheet list table when the **Update Sheet List Table** tool is used?
27. What is the purpose of archiving a sheet set?
28. What is a zip file?
29. List the three packaging types available for archiving a sheet set.
30. How can you password-protect an archive?

Drawing Problems

▼ Basic

1. Create a new sheet set using the **Create Sheet Set** wizard and an example sheet set. Use the Civil Imperial Sheet Set example sheet set. Name the new sheet set Civil Sheet Set. Finish creating the sheet set.

2. Create a new sheet set using the **Create Sheet Set** wizard and an example sheet set. Use the New Sheet Set example sheet set. Name the new sheet set My Sheet Set. Finish creating the sheet set.

▼ Intermediate

3. Create a new sheet set using the **Create Sheet Set** wizard and the **Existing drawings** option. Name the new sheet set Schematic Drawings. On the **Choose Layouts** page, pick the **Browse...** button and browse to the folder where the P30-14.dwg file from Chapter 30 is saved. Import all of the layouts from the file into the new sheet set. Continue creating the sheet set as follows:

   A. In the **Sheet Set Properties** dialog box, assign the layout named ISO A1 Layout from the Tutorial-mMfg.dwt template file in the AutoCAD 2010 Template folder as the sheet creation template.

   B. Open a new drawing file using the template of your choice and create a block for a view label. Save the drawing file and then assign the block to the sheet set using the **Label block for views** setting in the **Sheet Set Properties** dialog box.

   C. Create a new view category and name it Schematics.

   D. Open the 3 Wire Control layout, create a new view, and add it to the Schematics view category. Double-click on the new view name in the **Sheet Views** tab and insert the view label block you previously created. Renumber the view and save the drawing.

   E. Add a custom property to the sheet set named Checked by and set the **Owner** type to **Sheet**. Add another custom property named Client and set the **Owner** type to **Sheet Set**.

4. Create a new sheet set using the **Create Sheet Set** wizard and an example sheet set. Use the Architectural Imperial Sheet Set example sheet set. Name the new sheet set Floor Plan Drawings. Finish creating the sheet set. Under the Architectural subset, create a new sheet named Floor Plan. Number the sheet A1. In the **Model Views** tab, add a new location by browsing to the folder where the P18-16.dwg file from Chapter 18 is saved. Open the P18-16.dwg file and continue as follows:

   A. Create three model space views named Kitchen, Living Room, and Dining Room. Orient each display as needed to describe the area of the floor plan. Save and close the drawing.

   B. Open the A1-Floor Plan sheet. Create a new layer named Viewport and set it current.

   C. In the **Model Views** tab, expand the listing under the P18-16.dwg file. Right-click on each view name and select **Place on Sheet**. Insert each view into the layout. Delete the default view labels inserted with the views. Double-click inside each viewport and set the viewport scale as desired.

   D. In the **Sheet Views** tab, renumber the views. Insert a new view label block under each view.

   E. Save and close the drawing.

5. Open the Floor Plan Drawings sheet set created in Problem 33-4. Create an archive of the sheet set using the self-extracting zip executable (EXE) file format.

▼ Advanced

6. Use a word processor to write a report of approximately 250 words explaining the purpose of sheet sets. Cite at least three examples from actual industry applications of using sheet sets to help manage a set of drawings. Use at least two sketches to illustrate your report.

7. Plan a new shopping center for your area. Determine how many stores will be included. If possible, obtain a copy of a survey for vacant land in your area suitable for building the shopping center. Determine the components of a complete set of plans for the shopping center, including a site plan, floor plans, foundation plans, roof plans, elevations, and any needed sections. Establish the components for a new sheet set to organize the drawings and layouts.

8. Plan a new residence with approximately 3500–4000 square feet, four bedrooms, three baths, a den/office, kitchen, dining room, nook, family room, and three-car garage. Create a complete set of plans for the residence, including a site plan, floor plans, foundation plans, roof plans, elevations, and any needed sections. Prepare layouts for each drawing. Then create a new sheet set to organize the drawings and layouts. Archive the final sheet set.

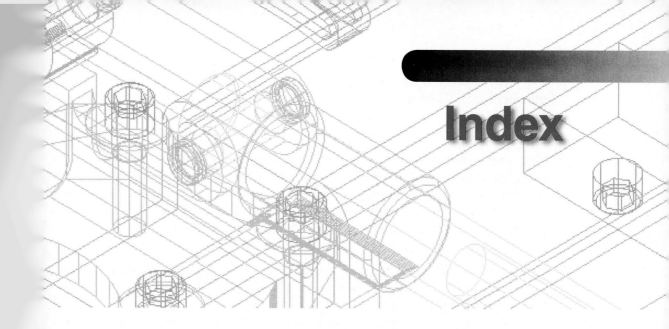

Index

B

background mask, 262
Background Mask dialog box, 262
backup files, 57
BACTION tool, 722
balloons, 576
base dimension, 568
base point, 339, 381–382
base point parameters, 740–741
baseline dimensioning, 485
BASE tool, 676, 863
basic dimensions, 462, 543
basic object tools, 105–125
    using, 105–106
basic tolerance method, 543
BATTMAN tool, 708
BCLOSE tool, 722
BCONSTATUSMODE tool, 765
BCONSTRUCTION tool, 765
BCPARAMETER tool, 762–763
BEDIT tool, 679, 710, 721
BESETTINGS tool, 765
big fonts, 252
bilateral tolerance, 537
binding, 877–878
Bind Xrefs dialog box, 877
Block Attribute Manager, 708–710
Block Attribute Settings dialog box,
    708–709
Block Authoring Palettes, 722
    Parameter Sets tab, 759
block definition, 67
Block Definition dialog box, 669–671, 834
Block Editor, 679–681, 721–722
    geometric constraints tools, 760
    insertion base point, 740
Block Editor ribbon tab, 679–681
    Dimensional panel, 762
    Geometric panel, 760
    Manage Parameters panel, 727
    Visibility panel, 752
block geometry, constraining, 760–765
block icons, updating, 683
BLOCKICON tool, 683
block insertion tools, 677
block properties table, 765–769
    creating, 766–768
    using dynamically, 768–769
Block Properties Table dialog box, 766–768
block reference, 671
blocks, 225, 667
    adding description, 681

additional block definition settings,
    670
additional block insertion items, 674
annotative, 834, 839
attaching to leaders, 511
changing properties to ByLayer,
    678–679
circular reference error, 682
constructing, 667–671
creating, 668–671
creating as drawing files, 683–685
creating with attributes, 703
defining insertion base point, 669–670
drawing components, 668
editing, 678–683
exploding and redefining, 681–682
inserting, 671–678
inserting entire drawings, 676
inserting in table cells, 315
inserting multiple arranged copies, 675
inserting with attributes, 704–705
modifying with **Block Editor**, 679–681
naming and describing, 668–669
nesting, 671
placing at specified increments, 225
real, 672
redefining, 710
renaming, 682–683
rotating, 673
saving with dynamic properties, 722
scale settings, 670
scaling, 672–673
schematic, 672
selecting block objects, 670
selecting block to insert, 671
selecting layers, 667–668
specifying block insertion point, 671
using **DesignCenter** to insert, 676–677
using **Tool Palettes** to insert, 677–678
working with **Specify On-screen**, 674
BLOCK tool, 668, 681, 703, 712, 713, 714
boundaries, adding and removing, 637
boundary, 620
Boundary Creation dialog box, 359
boundary edges, 337
boundary errors, 631
boundary hatching speed, 629
boundary set, 359
BOUNDARY tool, 359–360
BPARAMETER tool, 722
break, 480–481
breaking objects, 332–333
BREAK tool, 332–333
broken-out sections, 619

U